ISBN 978-0-265-93470-8
PIBN 10912113

PROCEEDINGS

OF THE

CALIFORNIA

ACADEMY OF SCIENCES.

SECOND SERIES.

VOLUME V.

1895.

San Francisco, 1896.

Ranzania makua

PROCEEDINGS

OF THE

CALIFORNIA ACADEMY OF SCIENCES.

SECOND SERIES.

VOLUME V.

1895.

San Francisco, 1896.

Committee of Publication:

E. W. HARKNESS. GULIAN P. RIXFORD. CARLOS TROYER.

EDITOR: FRANK H. VASLIT.

TABLE OF CONTENTS.

LIST OF PLATES.

PROCEEDINGS

OF THE

CALIFORNIA ACADEMY

OF

SCIENCES.

ON THE VARIOUS STAGES OF DEVELOPMENT OF SPERMATOBIUM, WITH NOTES ON OTHER PARASITIC SPOROZOA.

BY GUSTAV EISEN.

[With Plate i.]

Spermatobium nov. gen.

The hosts of this parasitic sporozoan are two oligochæta, *Eclipidrilus frigidus* and *Phœnicodrilus taste*, both found on the Pacific Coast of North America, the former in the Californian Sierra Nevada at high altitudes, the latter in the mountains of the Cape Region of Baja California, Mexico, at an altitude of about 4,000 feet. In both hosts the Spermatobium is confined to the sperm-sacs, where in the young stages it occupies the interior of a sperm cell, but in later stages lives free in the sperm-sac outside the sperm cells.

Although in the following I have described all the various stages as belonging to one and the same form, it is evident that we here have to deal with two distinct species, of similar structure, but differing very markedly in size of the adult, but principally in the size of the cytospheres. That this difference is not dependent upon the host in which they live is evident from the fact that the Spermatobium inhabiting the smallest host is the largest and possesses the largest cytospheres, while *vice versa* the larger host houses the smaller species with the smaller cytospheres. While in one and the same individual host the size of the adult Spermatobia may differ some, I have never observed the cytospheres of the respective specimens to differ; this of course makes the difference in size of the cytospheres in the two species to be of great importance. I believe, therefore, I am justified in establishing two species, especially as I may thereby call attention to similar differences in other sporozoa.

Spermatobium Freundi n. sp. Figs. 1 to 4, 6 to 9, 11 to 18, 20 to 32, 34 to 37, 39 to 41, 43 to 45.

Host, *Phœnicodrilus taste*, an oligochæt from the Sierra El Taste in the Cape Region of Baja California, Mexico.

Adult and sporogonium (pansporoblast) about one-half the diameter of those of the following species. Cytospheres about half the diameter of those of the following species.

Spermatobium eclipidrili n. sp. Figs. 5, 10, 19, 38, 38, 42.

Host, *Eclipidrilus? frigidus*, a oligochæt from the Sierra Nevada, Alpine Meadow, about 11,000 feet altitude, on middle fork of King's River, Fresno county, Cal. Sperm-sac of the host.

Adult and sporogonium (pansporoblast) about twice or more the diameter of the former species. Cytospheres about twice the diameter of those of the former species.

In the following I have not separated the two species, as the various stages of development are evidently the same in both. I have considered them together, and placed the figures illustrating my remarks in consecutive order, regardless of the species to which they refer.

Habitat. In my late paper on Eclipidrilus I have referred to the occurrence of this annelid in two separate localities at different elevations. One is a cluster of springs on the south slope of the middle fork of King's River in California, at an altitude of about 11,000 feet. The water in these springs is very cool—icy, in fact—very clear, transparent, without apparent trace of suspended sediment. The bottom is very sandy, here and there covered with water mosses. The hosts of Spermatobium live in the sand or fine sediment among the roots of the moss, etc.

In another locality, the Three Spring Meadow on the north fork of King's River, this protozoa was not found, though Eclipidrilus is common there, too. The water in these springs is less pure with more sediment, and the Eclipidrili were found crawling in decayed wood, etc. The altitude was only about 7,000 to 8,000 feet. I refer thus in detail to these localities, because other protozoa *Hæmagregarina nasuta* were found to infest in countless numbers the same hosts, from the higher altitude and the purer and cooler water, while in those from the lower locality and the less pure water they were totally absent—that is, the *Hæmagregarina nasuta* as well as *Spermatobium eclipidrili*.

Phœnicodrilus taste occurs in the mountains of the Cape Region of Baja California, at an altitude of 4,000 feet, and lower down to the coast, about fifty to sixty miles north of Cape San Lucas.

METHOD OF INVESTIGATION.

The Eclipidrili were in rather poor state of preservation, having been hardened and kept in alcohol. The Phœnicodrili had been carefully fixed in a solution of corrosive sublimate and afterwards hardened, some in absolute alcohol, others in formaline. But notwithstanding the different conditions of the hosts, the protozoa did not show any prominent difference in structure, the caryoplasm only being less distinctly stained in the alcoholic specimens. I am therefore satisfied that the structure here represented and figured is the real one. This I believe with the more confidence as Beddard has shown that the protozoa—Gregarina of Perichæta—do not generally change their form and structure, even when attacked by powerful reagents. Much of the success in observing protozoa depends upon the staining, especially so in this form, which was not greatly sensitive to the common stains of hæmatoxylon, methyl green, safranine, etc. After having tried a dozen or more stains at my disposal, I found the following method to be superior to any other, and to give by far the finest nuclear images:

1. Staining of the hosts in toto in very weak Delafield's hæmatoxylon or in Ehrlich's ammonia hæmatoxylon.

2. Hardening and sectioning in paraffine.

3. The slide fixing consisted simply of distilled water or of formaline and gelatin (½ per cent.) in water. This fixing is used as follows:

1. Cover the space of the cover-glass on the slide with several drops of the fixing, so that the sections will float.

2. Warm gently over a plate until the paraffine becomes slightly transparent, but not so long that it begins to melt. If it melts, the sections shrink at once and spoil, but if just heated sufficiently they stretch out, even if ruffled by the knife.

3. Let the fixing harden in the air during at least four hours, or better, during the night. Sections heated this way never loosen, and are always straight. They should

never be melted, but the paraffine dissolved in pure turpentine or xylol.

When the latter is at last removed by alcohol the slides are stained by a saturated solution of orange G. in 33 per cent. alcohol. The stain should be left on only a few seconds, then immediately washed off in 95 per cent. alcohol. If too darkly stained wash gradually off with weaker alcohol until the proper tint has been found. It is better not to have the shade too bright yellow. Pure water will wash out all of the stain.

If it is found that nuclei of the hosts are not sufficiently brightly or darkly stained by the hæmatoxylon, the slide may be again stained by a weak solution of Ehrlich's ammonia hæmatoxylon, under the microscope. Clear with oil of bergamot, mount in gum-thus in xylol. Such sections give exceedingly good images. The nuclei of the hosts were stained by the hæmatoxylon, while the nuclei of the protozoa were stained by the orange and well differentiated. The nucleoli were nearly always stained deep yellow, while the other part of the nucleus was lighter; chromosomes and microsomes in the nuclei were always stained very dark brown. The cell plasma of the protozoa were stained lightly by the hæmatoxylon. A prolonged staining with only hæmatoxylon would stain the protozoa nuclei, but such prepared sections did never show the elements and structure of these nuclei.

I will also here call the attention to the very great advantages of gum-thus in xylol as a mounting medium, it giving images far superior to those by canada balsam or damar.

IMMATURE INTRACELLULAR STAGES. Fig. 1*a*, *b*, *c*.

The youngest stages of Spermatobium which I have been able to observe are seen in the spermatogonium, or the mother sperm cell after it has dropped from the testis into the sperm-sac. This stage is intracellular, the parasites having entered the cell and are feeding on its contents. In size the young Spermatobium is there not any larger than the fully developed spore (fig. *1a, b, c*), though I do not thereby wish to say that it is the spore which has been transformed into an immature Spermatobium, as on this point I am yet in doubt. This stage is but slightly larger than the nucleus of the spermatogonium of the host (fig. *1a*).

The protozoan may at this stage be described as a small cell with nucleus and nucleolus, the cell being of various forms, round or oblong, the nucleus always round and the nucleolus globular and well defined. Generally, however, the protozoan is pointed oviform, as in the fig. *1a*. In the following I will always refer to the two or three principal parts of the protozoan as cell or cytosome, nucleus or caryosome and nucleolus, the definitions of which are the general accepted ones. The cytotheca is thin and structureless, frequently wavy and ruffled.

THE CYTOPLASM.

The cytoplasm consists of at least two distinct parts, which, however, are not always localized, and which can in no way be designated as ectoplasm and entoplasm. At times the cytoplasm appears uniform, slightly grainy; at other times, or generally, there is a coarser granulation in the pointed end of the cytosome, as in fig. *1a*. In this figure the nucleus of the spermatogonium is seen to the right in the upper corner. In the lower corner of the protozoan the cytoplasm is seen to be coarser, though it is not always darker. The fact that at this stage the protozoan is intracellular makes observation difficult and less exact. I think, however, that these two differentiations of the cytoplasm correspond to those found in the free and adult form, and soon to be described.

As the intracellular protozoan grows, it gradually fills out the larger part of the host's spermatogonium. The nucleus of the latter, however, remains intact to the last, and I am doubtful if it is really at any time consumed by the parasite. Fig. *1b* shows this stage of the protozoan. The remains of the spermatogonial cytosome is seen outside of the edges of the protozoan. On the figure they are slightly colored yellow; on the slide, however, they were stained light blue by the hæmatoxylon, only the nucleolus of the spermatobium having taken the yellow stain. As will be seen from this figure, the nucleus and nucleolus of the spermatobium have obtained their full size, which, however, is variable in different individuals, in this instance being unusually large. The nucleus of the spermatogonium shows yet plainly its chromosomes, which apparently have not been in the least affected by the parasite.

FREE IMMATURE STAGE. Figs. 2, 3.

In this stage the protozoan is seen free from the spermatogonial host, living an intercellular life in the fluid surrounding the various developmental stages of the spermatozoå in the sperm-sac of the host. This mode of existence is kept on until after the formation of the spores or sporocysts. In fig. *2a* is represented a young spermatobium lately having left the spermatogonium of the host. Its nucleolus is large and has taken the stain deeply. The boundary of the caryosome is at this stage often even forming a globe; but this is not always the case, nor is it typical of this stage, as both previously in the intracellular form (fig. *1b*), as later (fig. *2b*), may the caryotheca be wavy and irregular in outline. In fig. *2a* the cytoplasm is seen to be differentiated, there appearing several pellucid vacuoles at the apex. In fig. *2b* the cytoplasm forms a network, consisting entirely of a grainy matter with a few more deeply staining microsomes. If this stage is previous to or later than the one represented in fig. 2*a*, I am not able to say.

After these stages are passed the spermatobium assumes a broader shape, and at the same time the cytoplasm is seen to be differentiated. Clear, rounded, pellucid sacs or vacuoles of different size begin to form in the center of the cell, while at the surface below the cytotheca are seen accumulating very small, extremely regular, globular bodies. The former I term for the present simply vacuoles, the latter cytospheres. The vacuoles are first seen in the vicinity of the nucleus and opposite to it (figs. 3*a* and 38*b*). Their contents do not stain, but in some I could detect a faint shading in their center. They are of different sizes; the largest appear nearest the center, the smaller further away, or mixed in with the other in an irregular manner (fig. 3*a*). The cytospheres are at first few and gradually increase in number. Correspondingly the vacuoles decrease in number and size, and at last the protozoan cytoplasm contains nothing but cytospheres of the same size and shape, the vacuoles having disappeared entirely. If there exists any connection between the vacuoles and the cytospheres I cannot say.

ADULT STAGE. Figs. 4, 5.

The stage when all the vacuoles have disappeared and the whole space outside the nucleus consists of cytospheres, or at least is apparently filled with cytospheres, may be termed the mature stage of this protozoan. While the protozoan varies in size and shape in the same host, the cytospheres appear to be always of the same size respectively in each species, at least from the beginning of the adult stage to the forming of the spores. In *Spermatobium eclipidrili* these cytospheres are several times larger than in *Spermatobium Freundi*.

In the first stages when the first cytospheres are formed their bodies are somewhat smaller, but they quickly increase in size; at least, the more numerous the larger they are up to a certain point. The first cytospheres are seen only near the surface, below the cytotheca, irregularly distributed in groups, like land and oceans on a map. What I consider the adult individual is solid, so to say, with cytospheres, the vacuoles then having entirely disappeared, and even the interior containing cytospheres.

CYTOSPHERES.

The cytospheres are small, perfectly globular, glassy, pellucid bodies, which do not stain with orange G., and only very faintly with hæmotoxylon, though the latter can hardly be called staining, but may rather be termed soiling. In fact, they remain pellucid to the last, with the exception of a central spot of darker color, the latter, however, not being due to any stain. Soon after the appearance of the cytospheres this spot is seen to be very small, exceedingly well defined and very dark, merely appearing as a single point (figs. 10*b*, 14), sometimes surrounded by a white zone.

At what I suppose a later stage in the cytosphere, this central dot or cytosphero-center enlarges and appears as a small circular disc (fig. 16), also well defined, the boundary being much the darkest. At a later stage yet, the cytosphero-center becomes diffuse, and gradually occupies a large part of the cytosphere. This is the character of the cytospheres at the end of the formation of the sporogonia and sporoblasts (figs. 18, 20). After the pseudo navicella spore is formed, the cytospheres diminish in size, and finally are seen to possess only about one-half the diameter of the original cytosphere (fig. 26). I believe this diminished size is caused by a division of the cytosphere into four parts, as I have observed a number of cytospheres in a spore showing the appearance of division (fig. 33*a*), the dark center remaining outside of the four new microcytospheres. If the division goes still further I am unable to state. It is not unlikely that it does, as no such microcytospheres are seen in the young protozoan. The cytospheres probably correspond with the amyloid granules of Bütschli, and appear to be present in most sporozoa. Besides the cytospheres and the vacuoles the cytosome contains, especially in the earlier stages, a diffuse darker staining plasma (fig. 2*a*, etc.), which, however, mostly disappears from view as the cytospheres accumulate and increase in number. But little of this plasma is seen in the fully developed form, and only rarely is any found in the sporulation stage (fig. 10*a*), and then generally around the nucleus.

NUCLEUS.

A nucleus is nearly always present and well developed, though the chromatin bodies are not well definable. The nucleus shows some very decided phases of development and differentiation proceeding along two different lines, accordingly as its division is caused by simple budding and subsequent contraction, or by caryokinesis. The former phase is found in the early stages of sporulation, the latter again in the last stages of this process. In the resting nucleus we find especially prominent a single nucleolus. In the adult stage the nucleus is furnished with a distinct caryotheca.

The nucleus is not always present in a fully developed form and in some instances apparently absent. I think this is due not to the total absence of the nucleus but rather to the fact that it has disseminated itself all through the elements of the cytosome or rather scattered its fragment between the cytospheres, as I will endeavor to prove directly. The nucleus when fully developed does always possess a nucleolus, of distinct form and appearance, which I will describe later on. In its earliest stage the nucleus is clear, globular, surrounded by a circular caryotheca of considerable thickness. As, however, I have frequently found the outline rosette-like (figs. 41*c*; 1*b*), I believe that it possesses an amœboid movement, already in the very early stage, when the protozoan is yet confined to the spermogonium of the host (fig. 1*b*). In most of the intercellular Spermatobia the caryotheca was wavy, slightly folded, showing signs of having altered its shape (fig. 41*c* and *d*). Some, however, possessed the regular circular outline. The caryotheca always stains readily but not deeply with the orange G. Of the contents of the caryosome—disregarding the nucleolus for the present—I could sometimes distinguish two different substances: one protoplasmatic, by far the most abundant, and also a darker staining, more regularly grained part, probably the chromosomes. How far these respective substances in the resting nucleus correspond with the chromosomes and microcaryosomes, etc., of higher nuclei I am unable to say, as they are not well differentiated until in the latter stages of sporulation, where chromosomes and filaments may be distinctly recognized. We may distinguish several distinct stages of nuclear development, each one of which presents some characteristics of importance:

1. *Resting macronucleus*, with perfect caryotheca, diffuse caryoplasma, single large nucleolus with several intranucleolar bodies.

2. *Amœboid nucleus*, or the first stage of sporulation in which the former resting macronucleus divides itself in numerous micronuclei by an apparently amœboid budding or diffusion of the caryoplasm in among the cytospheres. The amitotic stage.

3. *Micronuclei, or contracting stage*, in which the diffused caryoplasm is concentrated at regular intervals, forming a large number of micronuclei.

4. *Caryokinetic stage*, in which the micronuclei further divide by caryokinesis.

In the early stages of the macronucleus the caryoplasm is small in quantity, generally arranged along the inner wall of the caryotheca (fig. 3*b*), staining in places more darkly, probably as small chromosomes.

In the more developed protozoan the caryoplasm fills a large part of the caryosome, is of a streaky, ramified nature, the ramifications evidently proceeding from a center near or around the nucleolus (figs. 7, 41). In unstained specimens the nucleus appears always as a light round spot, with the darker nucleolus in the center. It is first only at a more advanced stage that it takes the stain.

In the fully developed Spermatobium the caryoplasma stains freely yellow, but not as deeply as the nucleolus. Even a prolonged exposure to hæmatoxylon fails to stain it in a distinct way, and it is entirely due to its affinity to orange G. that it becomes well defined. The most interesting and striking character of the caryoplasm is its growth, amœboid extension, budding and division, by which it extends itself far outside of the caryotheca— in fact, diffuses all through or rather between the cytospheres. This diffusion of the caryoplasma is undoubtedly connected with the formation of spores and sporoblasts, and appears to begin as soon as the Spermatobium is fully developed. Of the very great importance of the caryoplasm in the early stages of the formation of the sporagonia I believe I have made several demonstrative observations. In several instances I have observed how after division one part of the sporogonium remains in an undeveloped state while the other part develops sporoblasts and shuttle spore. In every such instance I found the arrested or undeveloped sporogonium deficient in or entirely without nucleus, only possessing cytospheres. I believe that it is want of sufficient nuclear matter which has caused the development of sporoblasts to cease. In fig. 45 I have figured such a Spermatobium in which a part has been arrested in its development while the other has already produced spores. How this deficiency in caryoplasm originated in this instance I cannot say, but it may depend on two distinct causes: either the caryoplasm moved the larger part of its bulk to the part which later on developed, this as I believe causing a division of the original sporogonium into two smaller sporoblast, or the caryoplasm may from some cause or other have been destroyed in one sporoblast while not in the other. That the two sporoblasts originally belonged to a single pansporoblast I judge from the remains of the original cytotheca, which is yet seen surrounding the two sporoblasts.

NUCLEOLUS.

But before I describe this diffusion and subdivision of the nucleus proper, it will be in order to consider the form, structure and nature of the nucleolus. I believe it safe to say that the nucleolus is always present, even if not always under the same form and of the same size. I have never seen a single fully developed Spermatobium which did not possess a nucleolus of some size, small or large, and when the caryoplasm diffuses the nucleolus remains, though sometimes in greatly diminished form, until the very last, when its final division or disintegration takes place; it apparently does not move with the caryoplasm.

In its perfect form, even in the intracellular stage of the Spermatobium, the nucleolus consists of one single, globular body, varying in size from one-third to three-fourths the diameter of the nucleus. It is nearly always very regular in outline, strictly globular, and in its early stages shows a homogeneous consistency, its plasma staining intensely and evenly with orange G., but not with hæmatoxylon nor with methyl green, and only faintly with safranine (figs. 1*b*; 4*b*; 7*b*; 41*d*, etc.). In the free Spermatobium the nucleolus always contains one or more, up to six, minute round intranucleolar bodies of different sizes and of an intensely light-refracting nature.

These intranucleolar bodies are often, but not always, surrounded by a transparent sphere, like a vacuole (fig. 4*b*). In very young Spermatobia they are absent or few in number; in adult specimens again they are more numerous, and I have counted six or eight—some larger, some smaller (fig. 41*d*.) In a young Spermatobium the nucleolus appears homogeneous, but in older specimens, especially those which are in the stage of division by sporulation (fig. 10*b*), I have nearly always found the nucleolus to contain a number of round, lighter-appearing globules, which certainly do not appear as if they were vacuoles, but rather as differentiated nucleolar matter.

I have also but rarely seen a vacuole. At other times again (fig. 23) the nucleolus seemed to be composed of a few nearly round globules with irregular outlines and a darker center.

BUDDING OF THE NUCLEUS.

The diffusion of the macronucleus and the formation of micronuclei in different parts of the Spermatobium is the most interesting fact connected with this protozoa. I may state at once that I have in no instance in this stage of the nucleus observed regular caryokinesis. The division appears to take place only by diffusion or budding. The process of division of the macronucleus in Spermatobium is effected by at least five phases:

1. Budding proper of the caryoplasm, by means of fine, thread-like elongations from the nucleus proper.

2. Formation of caryoplasmic nodes at certain at first irregular, later on at regular intervals.

3. Growth of these nodules by attraction and contraction of the outlying caryoplasmic threads.

4. Final division of these secondary or micronuclei by caryokinesis and their moving apart, forming the final nuclei preparatory to the formations of the pseudonavicella spores.

5. To this may be added the division or disintegration of the nucleolus, which takes place later on, and which does not appear to be of importance in the formation of the micronuclei. In some instances the nucleolus remains intact for some time after the division and redistribution of the caryoplasm.

The first indication of a division of the macronucleus is seen in the unequal distribution of the caryoplasm within the caryotheca. The latter at the same time assumes an irregular outline and soon disappears entirely (fig. 7*a* and *b*). But even before the caryotheca has vanished, the caryoplasm has penetrated its walls and accumulated outside of, but adjacent to the latter. From these agglomerations caryoplasmic filaments are seen extending irregularly in all directions (figs. 7 and 8). These caryoplasmic filaments when properly stained may be found extending all through the cytospheres, winding their way between them. At certain intervals there appear thicker nodes on the filaments and from these nodes other filaments radiate in various directions. At last a stage in radiation has arrived when nodes are found at fairly regular intervals throughout the cytosome (fig. 9*a* and *b*). At this stage there frequently or nearly always appear one or more darker staining bodies in the nodes, possibly developing chromosomes, preparatory to final caryokinesis.

These nodes appear to absorb the outlying caryoplasmic filaments by which they were at first connected with the mother or macronucleus. During all this radiation of the caryoplasm, the original caryosome appears to be growing in size, and it is able to furnish plasma for forty to sixty micronuclei before diminishing in size, while frequently it becomes much larger than in its early resting stage. Finally, however, the original caryosome generally entirely disappears, though this may not always be the case, as sometimes (as, for instance, in the case figured at 10*a*) a part of the caryosome as well as some cytoplasm remains after the sporoblasts have already been formed. In this figure to the upper left of the nucleus is seen a bluish mass, consisting of caryoplasm and unused cytospheres. The nucleus, colored yellow, has evidently contracted its caryoplasmic filaments, as none could be seen either around the remains of the macronucleus or around the micronuclei. The nucleolus again has been broken up into one larger and three smaller nucleoli. The larger one of these (figured separately 10*b*) is seen to contain a number of semi-transparent globules of nearly equal size. After the stage in which the caryoplasmic filaments have been contracted (or disappeared), the small newly-formed micronuclei begin to divide. Previous to this division, however, four important points are to be noticed:

1. The micronuclei are all of the same size, or almost of the same size.

2. They are scattered at almost equal distances all through the cytosome of the sporogonium.

3. The cytospheres become grouped around each nucleus in such a way as to form separate little balls or

sporoblasts with inclosing membrane, the outline of which is more or less distinct, accordingly as the spheres are packed closer or looser together.

4. The division of the micronuclei is not effected by the budding process, but by division in equal parts by caryokinesis.

DIVISION BY CARYOKINESIS.

After the cytospheres have been attracted into sporoblasts, each one surrounded by a thin membrane, the micronucleus begins again to divide. But now the division is not effected by budding but by a distinct caryokinesis. Small chromosomes may be seen scattered about at first irregularly; later they congregate at the equator, and finally caryokinesis takes place. I have, however, not been able to observe either asters or centrosomes, the highly refractive cytospheres so far obscuring observation.

Fig. 12 represents such a sporoblast with a single micronucleus. In Fig. 13 the nucleus has divided into two, which have moved to opposite poles, and these secondary micronuclei have again divided. The two upper ones are yet connected by a caryoplasmic filament.

DISINTEGRATION OF THE NUCLEOLUS.

The first indication of a disintegration of the nucleolus is a blurred outline, caused by small irregular drops, staining exactly as the nucleolus, appearing on the outer circular surface of the nucleolus (fig. 8*b*). Smaller, more or less irregular globules are seen in the caryoplasm near by, and in more advanced specimens nucleolar fragments are seen in the various or in some of the caryoplasmic nodes. Sometimes one of these new nucleoli are larger than others, staining either darker or lighter than the surrounding caryoplasm. Around such caryoplasmic nodes the cytospheres are seen to arrange themselves regularly.

In the majority of the new cytoplasmic micronuclei I have not been able to see the nucleoli, but I think it is safe to say that they are frequently present. As to the relationship of the nucleoli found in the micronuclei with that of the macronuclei I cannot say anything with certainty from observation. I believe, however, that the nucleolus of the macronucleus dissolves sooner or later, and that the nuclei found in the micronuclei are really new elements not directly derived from the macronucleus. Generally, however, the nucleoli remain conspicuously, though of diminished size, long after the macronuclei have all diffused themselves through the cytoplasm. In the newly formed spores more or less numerous highly refractive bodies are seen, greatly resembling the internucleolar bodies and probably identical with them.

Since the above was written and presented for publication, I have received the paper by Dr. L. Rhumbler on "Die Enstehung und Bedeutung der Binnenkörper." and I am pleased to say that I find in his explanation of the structure and action of the nucleoli a satisfactory solution of the morphological importance and nature of these interesting bodies. I have observed in Spermatobium all the three stages he refers to, the liquid, the viscicous and the solid stage of the nucleolar contents, the above described highly refractive intranucleolar bodies belonging to the latter. The nucleoli of *Truncatulina lobatula*, as delineated by him (fig. 30, Taf. xviii), is almost exactly identical with some of the nucleoli observed by me. Judging from my own observations, the more solid parts of the nucleoli could form directly from the liquid part, or at least independently of the viscous part. Dr. Rhumbler's theory that the nucleoli are not organic structures but only accumulations of organic secretions of different states of liquidness, appears to explain the nucleolar structure of Spermatobium.

SPORULATION.

The sporulation is undoubtedly the chief object of the adult Spermatobium, and it depends chiefly on this process for the maintenance of the species. In the various phases of sporulation we have, in fact, the larger part of the life history of this protozoa. Sporulation, or the forming of spores, comprises again various stages of development. These are:

A. Preparatory stages and amitosis.

1. Diffusion or budding of macronucleus.
2. Formation of numerous micronuclei.

B. Formation of spores.

3. Attraction by the micronuclei of cytospheres, forming sporoblasts.
4. Divisions of micronuclei by caryokinesis.
5. Transformation of each sporoblast into a shuttle spore.
6. To this may be probably added another stage, the formation of sickle germs in the shuttle spore. This stage I have not observed, and its existence can only be inferred from what takes place in other protozoa.

When the adult Spermatobium has begun the process of sporulation it may be more properly called a sporogonium or rather macrosporogonium, as at a later stage this macrosporogonium divides into numerous microsporogonia. The smaller agglomerations of cytospheres and micronuclei may again be termed sporoblasts.

Strictly speaking, the sporulation begins with formation of sporoblasts. After the micronuclei have contracted their plasma filaments and attained their proper distribution as regards each other (figs. 10 to 13), the cytospheres appear to group themselves in certain num bers around each nucleus. The number of such cytospheres varies, the new sporoblasts being similarly of various sizes, not, however, differing greatly. some being nearly twice as large as others. It would appear as if the micronuclei possessed two distinct properties, one of which is to repel each other, which would cause them to be regularly distributed; and one to attract the cytospheres, which would explain the comparatively even distribution of the latter and their collecting to form sporoblasts. The sporoblasts become surrounded with a thin membrane, which becomes thicker just before the forming of the lunate and shuttle spore. As to the process of forming these spores I am unable to give any satisfactory account. It appears to me, however, as if in the final, smallest and ultimate sporoblast, we find several micronuclei scattered about among the cytospheres which at this stage begin to further divide. After the sporoblasts have begun to form—that is, after the cytospheres have begun to arrange themselves into agglomerate balls (sporoblasts)—the whole individual, now a sporogonium, increases in size, and finally divides itself into two or more smaller or microsporogonia. The ultimate size of these sporogonia varies. It is probable that all the microsporogonia are found at the same time from the macrosporogonium and not by successive divisions. Thus the individual Spermatobium transforms itself into a macrosporogonium, which later again divides into a number of microsporogonia, each one containing a number of sporoblasts, consisting each one of cytospheres, cytospherotheca and micronuclei. Each sporoblast converts itself into a lunate or shuttle spore. In each microsporogonium we may find from forty to sixty sporoblasts or shuttle spores, but generally very many less (figs. 24 to 31).

Frequently we find a restkörper consisting of a number of unused cytospheres scattered about, but principally situated in the center of the sporoblast. I think that the name sporophore is an unsuitable one for this body, because, as has frequently been remarked, the restkörper consists simply of the unused cytospheres, which perhaps for want of nuclei have not consolidated into sporoblasts.

THE SHUTTLE SPORES AND PSEUDONAVICELLÆ.

The ultimate transformation of the sporoblasts are into shuttle spores and later the pseudonavicellæ. At the earliest stage these bodies are, however, lunate, crescent-shaped, concave, considerably varying in form, but not in size. Each one contains a yellow staining crescent, possibly the accumulation of nuclear matter, always situated close to the convex side (figs. 24 and 25).

I have never been able to clearly make out the structure of the spores at this stage, and, while assuming that they really are only earlier stages of the shuttle spores, I have yet some hesitation as to this being the case; the reason for this is that I have never distinctly seen in them the cytospheres, which, however, are always very distinct in the shuttle and pseudonavicellæ spores.

One-half or more of the lunate spore is occupied by a transparent lunate sac, while the other half or less consists of a granulated crescent.

The real shuttle spores are more regular, but even they show a concave side, but their contents can always be clearly dissolved into cytospheres and micronuclear bodies of varying number (figs. 32, 33, etc.) This is their shape while yet enclosed in the sporogonium. Their cytothecas are there thin, hardly visible, but always well defined. Such shuttle spores are also found free among the spermatozoa of the host, and show these often irregular outlines, probably the indication of amœboid movements (fig. 35). In this free stage the shuttle spores nearly always contain a distinct nucleus situated generally at one of the poles, but sometimes also in the middle of the convex side.

While I have found these shuttle spores in great numbers, I have observed only very few real pseudonavicellæ forms. These pseudonavicellæ spores are of the same size as the former, but differ considerably in their structure.

Their cytotheca is very thick, projecting into a knob at each pole (fig. 27). The cytotheca contains a single row of very minute, dark, entirely opaque globular bodies, which in the knob-like projections at the poles are more thickly accumulated.

Their contents consist of cytospheres of much smaller size probably due to a further division of those found in the shuttle form of spore, in the manner as indicated in fig. 33*b*. A large nucleus appears present in at least some of the spores, but the scarcity of these spores has prevented me from studying its development and form. So far I have never observed the sickle germs found in Monocystis, as well as in a large number of other sporozoa.

In this description I have assumed that what I have here called pseudonavicellæ spore is a direct modification of the shuttle spore. I must, however, add that I have never found one of these pseudonavicellæ in the sporogonium, but always scattered loose among the shuttle spores.

DIVISION OF THE ADULT.

A division of the adult form, as well as of the sporogonium, is frequently observed. In figures 38 to 42 I have endeavored to figure a series of such divisions. Fig. 38 represents an adult with large cytospheres, and not yet in the sporogonium stage. The original nucleus has partly diffused and formed two secondary nuclei, in none of which, however, the nucleolus has assumed its globular form. A thin division of the cytotheca is discernible extending from *a* to *b*, and from *c* towards the center.

In fig. 39 we find a partly formed sporogonium, in which the division is more advanced. A number of nuclei have formed, and the original nucleolus, or what is left of it, is seen in a stage of division surrounded by a transparent zone, probably the remains of the macronucleus.

Fig. 40 represents an adult form, in which the division is more perfect; one-half of the figure is drawn from a focus set on the micronuclei and vacuoles, the left half again was focused on the cytotheca showing the accumulation of cytospheres.

Fig. 41*a*, *b*, *c*, represent one and the same individual in division, focussed at different depths. In 41*a* the focus is on the vacuoles, in 41*b* on the surface cytospheres and in 41*c* on the nuclei. Fig. 41*d* is a nucleus drawn on a larger scale. In fig. 42 is seen a Spermatobium in division, in which the left half consists of an undeveloped spermatogonium, while the right part consists of two separating spermatogonia, in one of which is seen remains of a larger nucleus.

My conclusions about the division of the adu lt may be summed up as follows:

1. The object of division is not the propagation of the species, but rather a convenient subdivision of the large forms.

2. New macronuclei are sometimes formed in the new individual, but not always.

3. Micronuclei are always formed previous to segregation of the new individual.

4. In one or more of the separating individuals, the process of sporulation—forming of a sporogonium—may be more advanced than in any of the other parts, while all are yet connected together.

5. The formation of a new perfect nucleus as in fig. 41, probably depends upon the accidental accumulation of sufficient nuclear matter in one place.

LARGER CYSTS.

In figs. 43 and 44 I have figured two very large and unusual forms. One of these is a very young form, irregular in outline, with very small cytospheres and no nuclei. The other is undoubtedly a sporogonium stage, with a very thick cytotheca, if we here have to deal with a formation of a different kind of cyst or with abnormal forms of the common cyst is undecided.

I found only few of these forms and only in the Eclipidrilus host.

AFFINITIES.

The characters of Spermatobium appears to be intermediate between Klossia and Monocystis, and I think demonstates that the gregarines cannot properly be systematically divided accordingly as their habitat as intracellular and cœlomic.

In Spermatobium the young individual inhabits the spermatogonium or mother cell, just as Monocystis, and the adult dwells free in the fluid surrounding the sperm cells. As in Monocystis, Spermatobium develops shuttle and pseudonavicella spores, the resemblance between the spores in the two genera being very great.

But the formation of the sporogonium, the sporoblasts and the spores resemble much more that of Klossia and Monocystis. While in Monocystis the cyst contains a few, generally two, sporogonia of unequal and irregular size, in Spermatobium are found numerous microsporogonia of nearly equal and very regular size as in Klossia. The adult Spermatobium differs from Monocystis in its general form. The pointed apices and epimerit are not seen in Spermatobium, which is more regular, oval, globular or slightly lunate with rarely projecting apex. The ciliated covering, consisting of abnormally developed spermatogonia of the host, found in Monocystis are not seen in Spermatobium.

Finally, the form and structure of the nucleus and nucleolus in Spermatobium agrees more with those of Klossia than with those of Monocystis.

The life history of Spermatobium is shortly as follows:

RÉSUMÉ.

The young Spermatobium is intracellular parasitic in the spermatogonium of the hosts Eclipidrilus and Phœnicodrilus, but it leaves these cells before the spermatoblasts have begun to grow. The free form is shuttle-like, later ovoid and finally globular, with extremely prominent nucleus and nucleolus. In the cytoplasm of the Spermatobium the cytospheres gradually develop at the expense of or from the other element.

The macronucleus, at first globular, becomes later irregular, and finally diffuses itself all through and between the cytospheres, forming at first nodes which later change into new secondary nuclei equidistant from each other. The cytospheres group themselves around these micronuclei, which latter again divide by caryokinesis, thus forming at first microsporogonia, then sporoblasts. The sporoblasts develop first into shuttle spores. Pseudonavicella spores are also found.

A division of the adult takes place sometimes, probably caused by the accumulation of a too large quantity of nuclear matter in one spot outside of the original nucleus. When the nuclear matter is evenly distributed, the same power of attraction which caused the adult to divide causes the sporoblasts to form, which latter is accomplished by the even grouping of the cytospheres around the secondary nuclei. A thin membrane is formed around the sporoblasts, after which the nucleus of the latter begins again to divide by mitosis.

This budding or amitosis of the nucleus has previously been observed in a large variety of cells, and the various views of respective investigators of this subject have been set forth by Dr. Richard Zander (4) with great clearness. The type of division found in the early sporogonial stage of Spermatobium must, with some allowance, be considered as related to Arnold's "fragmentation" type, though want of access to his paper (3) has prevented me from making a closer comparison.

Ziegler (5) holds that amitosis only takes place in meganuclei, and that these soon perish after the process is over. This is exactly the case with the macronucleus of Spermatobium. Fritz Schaudinn (6) again has described amitosis in the nucleus of various foraminifera, but this process, as observed by him, differs from the amitosis of Spermatobium through the presence of achromatic filaments which divide the caryosomic substance in various parts. Here then the division takes place inside the nuclear membrane, while in Spermatobium the amitosis is entirely extra nuclear or outside of the original nuclear membrane.

The amitotic division of the nucleus can thus take place in at least four different ways:

Segmentation. The nucleus divides itself in equal parts in the equatorial plane.

Fragmentation. The nucleus is beaded off in various

equal or unequal parts, not separated by regular division fields.

Radiation or Budding. The nucleus branches out, forms nodes, which latter by retraction of filaments become independent nuclei—Spermatobium.

Sporulation. The nuclear plasma become by the aid of achromatic filaments divided into numerous equal parts, which, through the bursting of the caryotheca, are set free and form independent nuclei—Foraminifera.

Hæmagregarina nasuta n. sp. Figs. 50 to 64.

This form infests the walls of the blood-vessels and surrounding mesenterium of *Eclipidrilus frigidus* in enormous numbers. The cysts lie so closely that they frequently touch each other, and totally obscure the structure of the tissue of the host, to such an extent that there appears to be more of the parasites than of the tissue. I have only found the parasite in specimens of Eclipidrilus from the locality on the middle fork of King's River, California, at an altitude of about 11,000 feet, while the Eclipidrili found at the lower altitude and in impure water were entirely free both from Hæmagregarina and Spermatobium.

Although the host from the former locality were infested at the rate of thousands, none contained protozoa of different stages of development. I could only observe the fully developed form, all cysts and all spores being absent, the more to be regretted as related forms are only imperfectly known. The relationship of *Hæmagregarina nasuta* must therefore remain in doubt, and my reasons for classing it with Hæmagregarina depend alone upon the appearance of the adult form as well as upon its habitat. As is well known, Hæmagregarinas are principally known through Danilewsky's description of

H. Stepanowi from the blood of Emys, the fresh-water turtle of Europe, and it may be considered fairly certain that all forms related to Hæmagregarina are true blood parasites, principally inhabiting the blood-corpuscles of the vertebrates, turtles, birds and lizards while adult. The spores again are found in the bone marrow of the turtle, the kidneys, spleen and bone marrow of the lizard. This habitat of all the species of this group makes the presence of a Hæmagregarina in the blood of an oligochæt all the more interesting. The want of large blood corpuscles in the blood of oligochæta has made it necessary for our present species to select another habitat, if indeed it is not the original one. Here it is the lining of the blood-vessels and the surrounding mesenterial tissues which are infested, especially so the lining of the blood lacunes in the alimentary canal; in no instance did I find any of these protozoas in the blood itself.

The youngest form was straight, slightly sigmoid, with no well defined nucleus, while the more advanced individuals were folded together like the blade and handle of a pocket-knife, as far as I can judge from drawings very similar to *Hæmagregarina Stepanowi*. The anterior end, however, differs from that of this species by having a slight prolongation, which in fully developed individuals was sharply pointed, but in less developed ones only appearing as a serrated surface of the thicker apex. The most advanced specimens possessed a circular nucleus near the thicker apex, while less advanced individuals showed an oblong, less well-defined nucleus nearer the middle of the body bend. Each individual was surrounded by a dry, thin cyst, considerably distant from its body, causing it to lie in a large pellucid vacuol.

EXPLANATION OF THE FIGURES.

PLATE I.

Spermatobium Freundi figures: 1 to 4, 6 to 9, 11 to 18, 20 to 32, 34 to 37, 39 to 41, 43 to 45.

Spermatobium Eclipidrili figures: 5, 10, 19, 35, 38, 42.

All the figures are drawn from parafine sections stained with Orange G. and Ehrlich Hæmotoxylon. All drawings were made under Zeiss Obj. 1-12 hom. imm. Oc. 2. Some of the detail figures were drawn on a larger scale, but not under a higher power.

Fig. 1A. A spermatogonium of the host containing a parasite in very early stage of development.

n. h. nucleus of host.

p. protozoa parasite.

Fig. 1B. Another spermatogonium from the same. The Spermatobium is more advanced in development, having nearly occupied the whole cell, the nucleus of which is yet intact.

n. h. nucleus of the host.

c. h. cytotheca of the host.

n. sp. nucleus of Spermatobium.

v. c. vacuole in the cytoplasm of the parasite.

Fig. 2A. A free Spermatobium. The nucleus is globular, the nucleolus is also globular and well defined. At the apex of the cytotheca are seen the commencement of vacuoles *c. s. b.*

Fig. 2B. Another nearly adult Spermatobium.

Fig. 3A. An adult Spermatobium, with commencing vacuoles.

Fig. 3B. Nucleus of the same.

Fig. 4. Another Spermatobium of more regular form.

Fig 4B. The nucleolus of the latter drawn on a larger scale in order to show the intranucleolar bodies.

n. nucleus.

no. nucleolus.

no. b. intranucleolar bodies.

Fig. 5. An adult Spermatobium from Eclipidrilus.

cys. cytospheres.

n. nucleus.

no. nucleolus.

Fig. 6. Surface view of a Spermatobium Freundi, showing the cytospheres and their relative size compared to those of Spermatobium eclipidrili.

Fig. 7A. An adult Spermatobium, with nucleus in a stage of budding.

Fig. 7B. The nucleus and nucleolus of the latter.

Fig. 8A. An adult Spermatobium, with a nucleus yet more advanced in budding, nodes and micronuclei already having formed in places.

Fig. 8B. The nucleus and nucleolus of the same drawn on a larger scale

no. nucleolus in a state of disintegration diffusing nucleolar matter.

n. m. nucleolar matter.

no. intranucleolar bodies.

nd. nodes or forming micronuclei, by contraction of the nuclear filaments, or by accumulation of the caryoplasm of two different kinds.

Fig. 9A. Another Spermatobium in which the caryoplasm has been partly, but yet irregularly distributed.

n. n. new larger nucleus with distinct nucleolus.

n. d. new nucleus in a state of mitosis.

Fig. 9B. A part of the former drawn on a larger scale.

n. remains of macronucleus.

n. d. macronuclear node division.

nc. macronucleus around which the cytospheres are grouping themselves. Smaller nodes are seen scattered about some of which are yet connected by caryosomic filaments.

Fig. 9C. A smaller specimen with diffusing nucleus around the vacuoles.

Fig. 9D. Spermatobium Freundi with large budding nucleus and large nucleolus. Vacuoles and cytospheres not drawn.

Fig. 9E. Part of a resting nucleus with two extra nuclear bodies of unknown nature, possibly nucleolar ejected matter.

cyt. cytoplasm.

nob. nucleolus.

en. ejected nucleoli.

Fig. 10. Spermatobium eclipidrili, transforming into an encysted sporogonium.

n. remains of macronucleus.

n. no. nucleoli.

sp. bl. sporoblasts.

r. k. "restkörper," consisting of unused cytoplasm.

n. n. micronuclei.

Fig. 10B. One of the new nucleoli and a cytosphere drawn on a larger scale.

no. nucleolus with intranucleolar transparent globules.

cys. cytospheres.

These intranucleolar globules are entirely distinct from the intranucleolar bodies elsewhere referred to.

Fig. 11. A sporogonium of Spermatobium Freundi. The micronuclei are resting.

Fig. 12. A sporoblast of the former drawn on a larger scale.

n. micronucleus.

cyt. cyototheca.

cys. cytospheres.

Fig. 13. A sporoblast with nuclei in mitosis.

Fig. 13B & C. micronuclei with chromosomes at end of mitosis.

Fig. 14. Another sporoblast with single nucleus.

Fig. 15. A sporoblast in which the centers of the cytospheres are more developed, micronuclei not eliminated.

Fig. 16A. A sporogonium, with developed sporoblasts, a stage further advanced than the one figured in 10A.

Fig. 16B. One of the sporoblasts drawn on a larger scale, showing three micronuclei.

Fig. 16C. A nucleus and a cytosphere of the former.

n. nucleus.

cy. cytosphere.

Fig. 17. Two sporoblasts with nuclei in a state of division. A large sporoblast has just divided itself in two almost equal parts.

Fig. 18A. A smaller, probably final sporoblast with nuclei in mitosis. The nuclei are yet connected by caryoplasm.

Fig. 18B. One of the cytospheres of the former.

Fig. 19. A sporogonium of Spermatobium eclipidrili. The nuclei are well scattered out, but remains of nucleoli are seen in two places. The cytospheres have not yet collected into sporoblasts.

Fig. 20A. A similar sporogonium of Spermatobium Freundi at the same stage of development, Figs. 19 and 20, are drawn under the same magnification and show the relative size of the sporogonia and cytospheres in the two species of Spermatobium.

Fig. 20B. A group of cytospheres.

Fig. 20C. A cytosphere drawn on a larger scale, both from the sporogonium figured in fig. 20.

Fig. 21A. A sporogonium in which some of the micronuclei are unusually small, by an error of engraver not shown.

Fig. 21B. A sporoblast of the same with five micronuclei.

Fig. 22. A sporoblast with dividing nuclei. These sporoblasts would have further divided.

Fig. 23. A remaining nucleolus, showing a spherical granulation, surrounded by cytospheres. The macronucleus has entirely disappeared.

Fig. 24. A sporogonium with partly developed lunate spores.

Fig. 25. Some of the spores drawn on a larger scale.

Fig. 26. A shuttle spore with nuclear or nucleolar bodies.

Fig. 27. A fully developed pseudonavicella spore with beaded margin, nucleus and cytospheres.

Fig. 28. An empty spore of unusual form.

Fig. 29. A sporogonium with spores in various stages of development. A central "restkörper" of unused cytospheres.

Fig. 29B. A sporogonium with shuttle spores and "restkörper."

Fig. 30. Another sporogonium with shuttle spores. The restkörper is scattered and divided.

Fig. 31. A sporogonium with multinucleated shuttle spores.

Fig. 32. Some of the spores more highly magnified. Zeiss 1-12, oc. 4.

Fig. 33. Lunate spore from Spermatobium eclipidrili.

Fig. 34. A shuttle spore from Spermatobium Freundi. The two smaller figures, A and B, represent the cytospheres, respectively from figs. 33 and 34, showing their relative size.

Fig. 35. A shuttle spore from Spermatobium Freundi, showing amœboid movement.

Fig. 36. A sporogonium just bursting, with fully nucleated shuttle spores.

Fig. 37. Two of the spores drawn on a larger scale.

Fig. 38. A Spermatobium eclipidrili, in division, with three nuclei. Between *a b c* the division lines are seen.

Fig. 39. A Spermatobium Freundi in division. The nuclei are already formed, the remaining nucleoli in division.

Fig. 40. A Spermatobium in division. Micronuclei are formed and nucleoli dispersed.

Fig. 41A, B, C. A divided Spermatobium with perfect nuclei, but not fully developed cytospheres. In 41A the focus is set on the vacuoles.

" B. Similarly focused on the surface cytospheres.

" C. Focused on the nuclei.

" D. A nucleus more magnified.

Fig. 42. A Spermatobium eclipidrili in division. In the smaller sporogonia the sporoblasts are already formed, while in the one to the left the cytospheres are yet diffused. The micronuclei in position.

Fig 43 A Spermatobium of unusual size and structure.

Fig. 44. A Spermatobium of similar shape enclosed in a thick cyst. Possibly the two last are in a stage of development for producing "resting cysts."

Fig. 45. A sporogonium of Spermatobium Freundi in which part of the sporogonium has developed ripe spores while the other part has remained undeveloped, probably from want of nuclear matter.

Fig. 46 to 59. Hæmagregarina nasuta. In 50 we see the youngest form observed, the body not yet having folded itself. In 62 and 63 the anterior edge is serrated or slightly lobed, while in 64, the most highly developed form observed by me, the anterior projection is very prominent. The cyst surrounding the protozoa consists probably of the remains of a cell wall of the host as the nucleus (fig. 58) would indicate. This Hæmagregarina stains best and principally with hæmatoxylon, it fails to take distinctly the orange stain. All figures drawn under Zeiss hom. im. 1-12, Oc. 3 and 4.

LITERATURE SPECIALLY REFERRED TO.

ARNOLD, J. Weitere Mitteilungen über Kern und Zellteilungen, etc. Archiv f. mikr. Anatomie. 31 Bd., 1888. (3)

BÜTSCHLI, O. Protozoa, Bronn's Klassen und Ordnungen, etc. 1882.

PFEIFFER, P. Die Protozoen als Krankheitserreger. Jena. 1891.

RHUMBLER, L. Ueber Entstehung u. Bedeutung der in den Kernen vieler Protozoen, Binnenkörper, etc. Zeitschrift f. Wissenschaftliche Zoologie 56 Bd., 1893, p. 328.

SCHAUDINN, FRITZ. Die Fortpflanzung der Foraminiferen und eine neue Art der Kernvermehrung. Biolog. Centralblatt. 14 Bd., 1894, p. 161.

ZANDER, RICHARD. Ueber den gegenwartigen Stand der Lehre von der Zellteilung. Biologisches Centralblatt. Bd. 12, p. 281. (4)

ZIEGLER, H. E. Die biologische Bedeutung der amitotischen Kernteilung, etc. Biolog. Centralblatt, xi Bd., Nos. 12–13. (5)

ZIEGLER, H. E. and O. VON RATH. Die amitotische Kernteilung bei den Arthropoden. Biolog. Centralbl., xi Bd., No. 24, p. 744. (6)

CATALOGUE OF MARINE SHELLS, COLLECTED CHIEFLY ON THE EASTERN SHORE OF LOWER CALIFORNIA FOR THE CALIFORNIA ACADEMY OF SCIENCES DURING 1891–2.

The Gulf of California is the richest field for molluscan collections outside of the tropics along the whole west American coast, principally for the reason that, being nearly landlocked and opening only southward, it is almost as tropical as the more southern waters toward the Equator, and perhaps even warmer than some regions where currents from the north have free circulation. The contrast is thus strongly shown between the gulf and the western coast of the peninsula, in the small proportion of tropical species found on the latter and their more limited range northward.

The length of the gulf is about 760 miles northward of the latitude of Cape St. Lucas (22° 52′), and of this only forty-four miles are south of the Tropic of Cancer, while the width averages about fifty miles. The influx of the Colorado and other smaller rivers serves to keep the water from becoming too salt for molluscan life, and, though evaporation must be enormous, it seems thus balanced, while the usual differences in the species found in brackish waters are observed to only a limited extent compared with gulfs of less depth, like the Gulf of Mexico. Still, there are many species identical in both gulfs, and many analogies with the species found in the Mediterranean and Red seas, which are the most similar waters of the eastern continent.

Some of the most important collections from the gulf previously made are mentioned in Carpenter's "Mollusks of Western North America," Smithsonian Edition, 1872, and will give some idea of the number of species found there. The first well recorded were collected in 1825, during Captain Beechey's voyage of discovery in the "Blossom," and published in London, 1839, in which twenty species from the gulf are mentioned, mostly new ones. These were chiefly shore shells and showy species. In 1836–1842 Captain Belcher, with the British ship "Sulphur," surveyed that coast, and the ship's surgeon, Richard B. Hinds, made some more thorough collections, partly by dredging, obtaining about thirty species, most of them published in the "Zoology of the Sulphur," 1836 to 1842.

The next important collections were made during the Mexican war by the American, Col. E. Jewett, traveling at his own expense. He touched at Mazatlan, and obtained six species, supposed to be from there, but probably many more were mixed with the shells collected by Major Rich, U. S. A., numbering 108 species, and 102 obtained by Lieutenant Green, U. S. N., in the gulf. These were catalogued by Dr. A. A. Gould, and about thirty supposed new species described by him as [new] "Mexican and Californian Shells," with figures, in the Boston Journal of Nat. Hist., vol. vi, 1853.

The most extensive collection ever made in the gulf was by Fred. Reigen at Mazatlan, which place, being only about twenty-five miles north of the latitude of Cape St. Lucas and close to the Tropic of Cancer, shows most perfectly the influence of a tropical climate on the mollusca. A special work on this collection of about 708 species, and also on all others then known from Mazatlan, was published by P. P. Carpenter, as the "Mazatlan Catalogue," 1855–7. The collection was the result of three years' work, and contained a few species that may have been imported on ships.

The Xantus collection, made at Cape St. Lucas, has been before mentioned, in the first article on land shells collected by Mr. W. E. Bryant, in these Proceedings, 1891, p. 100. In eighteen months Xantus obtained 361 marine species, some of them probably from the Socorro Islands and the coast of Mexico, and about sixty of them were described by Carpenter as new. He states that "Pacific [Polynesian] shells may have been given to Xantus by sailors; they were not distinguished from his own series in opening the packages." A larger proportion of Panama species were found than at Mazatlan by Reigen.

The next marine collections known from the gulf and also west of the peninsula are those mentioned in a paper by R. E. C. Stearns on "The Shells of the Tres Marias Islands and other localities along the shores of Lower California and the Gulf of California" (from the Proceedings of the U. S. National Museum, vol. xvii, pp. 134–204, 1894). The islands named are over 100 miles southeast of the gulf, and therefore have no relation to the present subject, except that many of the species reach the gulf (about fifty-eight out of eighty-nine). Out of 294 in the catalogue, about 200 occur in the gulf, and several others on the west coast. It is not, therefore, as complete a list of gulf shells as we might expect from collections made by the U. S. Fish Commission steamer "Albatross," with its facilities for dredging and collecting otherwise. The greater part of the species were obtained by the late Mr. W. J. Fisher, who was better fitted out for collecting than any other private collector, but only credited with about 130 species from the gulf. The Academy's museum is indebted to Mr. Fisher for many North Pacific shells, and perhaps some from the gulf, but the latter were left by him in such a confused condition that they can rarely be identified as his. Besides the two collections mentioned above, Mr. Stearns includes those contributed by ten other persons, who collected small numbers from both coasts of the peninsula and the main Mexican shores.

It thus appears that there are known about 700 to 800 species of mollusca from near the entrance of the gulf, and even there very little thorough work in collecting has been done and most of the shells obtained have been dead ones more or less imperfect.

From small collections hitherto made in the northern end of the gulf, quoted by Carpenter or Stearns, it appears that the species found there are more largely of the temperate fauna, many of them being identical with those from the same latitude on the west coast of the peninsula. This seems to indicate that the dividing ridge, now 3,000 feet or more in altitude, was crossed by one or more channels within geologically recent times.

The parties collecting for the Academy in 1891-2 were not well prepared for obtaining marine mollusca, being engaged chiefly in collecting vertebrate animals, insects and plants, on the peninsula and nearest islands, though also preserving such land shells as they met with, when not too busy otherwise. Mr. W. E. Bryant, being often on the seashore in pursuit of vertebrate animals, spent some time in collecting the shells along the beaches, living or dead, and when La Paz was reached, continued their pursuit onto Espiritu Santo and San José islands lying nearly in a line northward from that place, and each about fifteen to twenty miles long. The latter, crossed midway by the 25th parallel of latitude, proved to be the most productive of species of any point visited north of Cape St. Lucas, nearly 150 miles to the south. Besides a large number of beach shells perfect enough for identification, many were obtained in excellent condition through the aid of native divers, who not only dive for pearls, but will find other shells for money and collect many of the larger kinds for food, eating nearly every large mollusc that is tender and well flavored. The collections were thus made chiefly at San José del Cabo, twelve miles east of Cape St. Lucas, at La Paz and at San José Island. Mr. Brandegee and Dr. Eisen also contributed several species from the same places. These places are near enough to the locality of the Xantus collection to be considered as belonging to the same local fauna, yet several species occurred that are new to the region. As almost every collection of shells, however small and imperfect, adds some new facts to our knowledge of geographical distribution, a list of these is thought to be worthy of publication. Some of the doubtful forms were sent to Dr. Dall of the U. S. National Museum for comparison with authentic specimens and are given as identified by him. Duplicates of many of the shells (and also of many others, native and foreign) are ready for exchange, in return for species not contained in the Academy's museum, or not in good condition.

CLASS GASTROPODA—UNIVALVES, ETC.

1. Acmæa dalliana Pilsbry. 2, very near *A. scabra*.
2. Acmæa fascicularis Menke. 3, young only.
3. Acmæa pediculus Philippi. Large, beach-worn.
4. Alaba supralirata Carpenter. Also Mazatlan and Cape St. Lucas.
5. Anachis tæniata Philippi. Rare, San José del Cabo.
6. Anachis coronata? Sowerby. 5, larger than usual.
7. Aplysia ———? Several young. San José del Cabo.
8. Astralium olivaceum Wood. 15, San José del Cabo.

9. Astralium unguis Wood. 2, young. Also Cape St. Lucas.
10. Bulla adamsi Menke. 28. Many approach *nebulosa*.
11. Cassis coarctatus Gray. 5, good examples; common south.
12. Cassis sulcosa Bruguiere, var. *abbreviata* Lamk. 3, good.
13. Cassis tenuis Gray. 2, fresh examples. Perhaps imported.
14. Cerithium adustum Kiener. 35, common; living.
15. Cerithium gemmatum Hinds. 100, common; living.
16. Cerithium incisum Sowerby. 200, common; living.
17. Cerithium interruptum Menke. 2, San José del Cabo.
18. Cerithium ocellatum Bruguiere. 150, mostly La Paz.
19. Chlorostoma coronulatum C. B. Adams. A pint of beach shells.
20. Columbella cribraria Lamarck. Common on algæ.
21. Columbella fuscata Sowerby. San José del Cabo. Common.
22. Conus brunneus Wood. 30, beach shells; many perfect.
23. Conus princeps Linné. 32, many fine specimens.
24. Conus puncticulatus Hwass. 6, beach.
25. Conus purpurascens Broderip. 14, many very perfect.
26. Coralliophila nux Reeve. 1, San José del Cabo.
27. Crepidula aculeata Gmelin. 6.
28. Crepidula onyx Sowerby. Many small specimens.

29. Crucibulum scutellatum Gray. A pint or more.
30. Crucibulum tubiferum Lesson. A pint or more.
31. Cypræa albuginosa Mawe. 10, also San José del Cabo; good.
32. Cypræa arabicula Lamarck. 8, some good.
33. Cypræa pustulata Swainson. San José del Cabo.
34. Cypræa sowerbii Gray. 30, mostly beach-worn.
35. Drillia aterrima Sowerby. 2, beach shells.
36. Drillia maura Kiener. 1, west coast of Lower California.
37. Erato maugeriæ Gray. 1, also Cape St. Lucas. West Indies.
38. Engina reevei Tryon. 12, beach shells.
39. Euplerura muriciformis Broderip. 3, young; living.
40. Euryta aciculata Lamarck. San José del Cabo.
41. Fasciolaria princeps Lamarck. 10, eaten by natives.
42. Fissurella rugosa Sowerby. 1, known before from the gulf.
43. Fissurella virescens Sowerby. 7, known before from the gulf.
44. Fissuridea inæqualis Sowerby. A common gulf species.
45. Fusus cinereus Reeve (not of Say). 10, fresh specimens.
46. Fusus dupetithouarsi Kiener. 10, fresh; eaten by natives.
47. Gadinia reticulata Sowerby. Many, San José del Cabo.
48. Haliotis fulgens Philippi. 1, San José del Cabo; about half grown.
49. Harpa crenata Swainson. 2, one fresh, one beach-worn.

50. Latirus gracilis Reeve. 2, beach-worn examples.
51. Lepidopleurus (probably *L. pectinulatus* Carp.) 1.
52. Littorina aspera Philippi. Abundant; several varieties.
53. Malea ringens Swainson. 20, eaten by natives.
54. Melampus olivaceus Carpenter. 2, beach shells.
55. Melongena (Solenosteira) modificata Reeve. 6. This much discussed shell seems to belong chiefly to this region.
56. Mitra maura Swainson. 1, a wide-spread species.
57. Mitra tristis Sowerby. 1, a finely colored example.
58. Modulus cerodes A. Adams. Half-pint of good specimens.
59. Muricidea dubia Sowerby. 8, chiefly beach shells.
60. Murex bicolor Valenciennes. 5, perhaps *brassica lamarck*.
61. Murex plicatus Sowerby. 4, in good condition.
62. Murex radix Gmelin. 4. This and *bicolor* are eaten.
63. Nassa tegula Reeve. 2, more common northward.
64. Nassa versicolor C. B. Adams. 70, also Mazatlan (Reigen).
65. Natica bifasciata Recluz. 45, common beach shells.
66. Natica glauca Humboldt. 2, also Mazatlan (Reigen).
67. Natica marochiensis Gmelin. 1, at La Paz only.
68. Natica uber Valenciennes. 13, common beach shells.
69. Nerita bernhardi Gmelin. 25, common at low water.
70. Nerita scabricosta Lamarck. 50, common at low water.

71. Neritina picta Broderip. 25, San José del Cabo only.
72. Neritina californica Reeve. Same place, in creek.
73. Nuttalina scabra Reeve. 1, a dry beach specimen.
74. Oliva araneosa Lamarck. 15, mostly beach-worn.
75. Oliva porphyria Linné. 15, fresh specimens.
76. Olivella dama Duclos. 150, common at low water.
77. Olivella myriadina Duclos. 45, called "rice shell."
78. Oniscidia tuberculosa Reeve. 2, good specimens.
79. Opalia crenatoides Carpenter. 1, a rare species.
80. Pleurotoma nobilis Hinds. 1, a fresh specimen.
81. Pleurotoma olivacea Sowerby. Several, beach-worn.
82. Potamides montagnei Orbigny. 2, rare in the gulf.
83. Purpura biseriata Blainville. 8, living at low water.
84. Purpura patula Linné. 60, living at low water.
85. Purpura triserialis Blainville. 4, living at low water.
86. Pyrula decussata Wood. 5, eaten by the natives.
87. Rissoina stricta Menke. On algæ, San José del Cabo.
88. Scalaria hexagona Sowerby. 2, five-angled examples.
89. Siphonaria peltoides Carpenter. 1, beach specimen.
90. Siphonaria lecanium Philippi. 1, rare in gulf?
91. Sistrum carbonarium Reeve. 1, rare; identified by Dall.

92. Solarium granulatum Lamarck. 2, beach shells.

93. Strombina angulosa Sowerby. 20, San José del Cabo.

94. Strombina maculosa Sowerby. 50, mostly from San José del Cabo.

95. Strombus galeatus Swainson. 10, eaten by natives.

96. Strombus gracilior Sowerby. Several beach shells.

97. Strombus granulatus Lamarck. 100, beach shells.

98. Terebra lingualis Hinds. 1, also Cape St. Lucas (Xantus).

99. Terebra variegata Gray. 3, beach specimens only.

100. Tornatina culcitella Gould. 1, more common in California.

101. Tritonium gibbosum Broderip. 5, young specimens.

102. Tritonium vestitum Hinds. 1, rare in the gulf.

103. Trivia pacifica Gray. 1, rare in the gulf.

104. Trivia sanguinea Gray. 12, common at low tide.

105. Trivia solandri Sowerby. 100, common at low tide.

106. Turbo fluctuosus Wood. 20, common at low tide on algæ.

107. Turritella sanguinea Reeve. 2, a variety of next?

108. Turritella tigrina Kiener. 12, *T. goniostoma* Valenc.?

109. Vermetus centiquadrus Valenciennes. 15, very variable.

110. Voluta cumingi Broderip. 3, also var. *pederseni* Verrill, 1.

111. Volvarina varia Sowerby. 1, *Marginella varia* (Tryon).

CLASS PELECYPODA—BIVALVES.

112. Anomia lampe Gray. Valves common; none entire.

113. Arca grandis Broderip. 10 valves; 3 young entire.

114. Arca multicostata Sowerby. 4 valves; 3 young entire.

115. Arca pacifica Sowerby. Several valves and 3 large fresh examples.

116. Arca tuberculosa Sowerby. Several middle sized examples.

117. Avicula peruviana Reeve. Common and very large.

118. Axinæa gigantea Reeve. 12 valves; eaten by natives.

119. Barbatia fusca Bruguiere? A few, perhaps *B. solida* Sby.

120. Barbatia gradata Broderip. A few beach-worn valves.

121. Callista aurantia Hanley. 30, many fresh; eaten by natives.

122. Callista chionæa Menke. Common, eatable; *C. squalida* Sby. (in part).

123. Callista pollicaris Carpenter. One specimen, determined by Dall.

124. Cardita crassa Lamarck. Several valves, large and small.

125. Cardita flammea Michelin. Several valves, large and small.

126. Carditamera affinis Broderip. Valves common and 1 entire.

127. Cardium consors Broderip. 1 valve, San José del Cabo.

128. Cardium procerum Sowerby. Valves only.

129. Cardium senticosum Sowerby. Valves only.

130. Chama echinata Broderip. Valves, and entire young.

131. Chama exogyra Conrad. A few seen of this form.

132. Chama frondosa Broderip. 4, the most common form.

133. Clidiophora punctata Conrad. Two flat valves, beach.

134. Chione pulicaria Sowerby. 30, probably a variety of next.

135. Chione succincta Valenciennes. 20, common, eatable.

136. Chione undatella Sowerby. 30, common variety? of the last.

137. Codakia tigrina Linné. Many fresh ones, eatable.

138. Crassatella gibbosa Sowerby. A few, large and small.

139. Diplodonta semiaspera Philippi. 1 valve, San José del Cabo.

140. Donax californicus Conrad. 2 valves.

141. Dosinia ponderosa Gray. A few small ones and valves.

142. Hemicardium biangulatum Sowerby. Small ones; rare.

143. Heterodonax bimaculatus. 50 valves, scarcely two alike.

144. Kellia suborbicularis Montagu. 5 specimens.

145. Labiosa undulata Conrad. 2 valves.

146. Lima squamosa Lamarck. Some entire, valves common.

147. LIOCARDIUM APICINUM Carpenter. Many valves, 30 entire.

148. LIOCARDIUM ELATUM Sowerby. A few valves.

149. LITHOPHAGUS PLUMULA Hanley. Burrowing in spondylus, etc.

150. LUCINA EBURNEA Reeve. 30 valves, San José del Cabo.

151. LUCINA EXCAVATA Carpenter. Valves, very near *L. nuttalli.*

152. LUCINA PECTINATA Carpenter. 4 valves. This and last also found at Mazatlan.

153. LUCINA UNDATA Carpenter. 3, common in the gulf.

154. MACOMA OCHRACEA Carpenter. A few valves. (Dall identified.)

155. MACTRA DOLABRIFORMIS Conrad. A few valves. (Dall identified.)

156. MARGARITIPHORA FIMBRIATA Dunker. 30, pearl shells; animal, eatable.

157. MODIOLA CAPAX Gould. 10, entire; many valves.

158. MYTILUS MULTIFORMIS Carpenter. 2, San José del Cabo only.

159. MYTILUS PALLIOPUNCTATUS Dunker. 100, San José del Cabo only.

160. OPALIA FUNICULATA Carpenter. 2, one from Mazatlan.

161. OSTREA AMARA Carpenter. 4, identified by Dall.

162. OSTREA IRIDESCENS Gray. 5 valves, identified by Dall.

163. PAPYRIDEA ASPERSA Sowerby. Probably same as *P. bullata.*

164. PECTEN SUBNODOSUS Sowerby. 24, common, collected for food.

165. PECTEN VENTRICOSUS Sowerby. 8, common, collected for food.

166. PECTUNCULUS MULTICOSTATUS Sowerby. Small but perfect examples.

167. PERIPLOMA PLANIUSCULA Sowerby. Valves, same as *P. argentaria* Con.

168. PERNA CHEMNITZIANA Orbigny. Valves, and some perfect examples.

169. PINNA LANCEOLATA Sowerby. Several specimens.

170. PINNA MAURA Sowerby. A few specimens.

171. PLACUNANOMIA CUMINGI Broderip. Only 1 valve.

172. PLACUNANOMIA MACROCHISMA Deshayes. Common.

173. SANGUINOLARIA MINIATA Gould. 2, entire shells.

174. SANGUINOLARIA NUTTALLI Conrad. A few valves.

175. SEMELE BICOLOR C. B. Adams. 1 entire, 2 valves. (Dall identified.)

176. SEMELE FLAVESCENS Gould. 10 valves.

177. SEMELE VENUSTA A. Adams. 2 entire shells.

178. SPONDYLUS CALCIFER Carpenter. Living, very massive.

179. SPONDYLUS LIMBATUS Sowerby. 7 living. All spondyli are used as food.

180. SPONDYLUS PRINCEPS Broderip. Living.

181. STRIGILLA CARNARIA Linné. Valves, common on beach.

182. STRIGILLA LENTICULA Philippi. San José del Cabo.

183. TELLINA INTERRUPTA Wood. Several whole shells and valves.

184. TELLINA PURA Gould. 6 valves.

185. TELLINA PURPUREA Broderip. 1 valve, San José del Cabo.

186. TEREBRA LUCTUOSA Hinds. 1, San Jose del Cabo.

187. TIVELA RADIATA Sowerby. Several good specimens.

188. TIVELA ELEGANS Verrill. Several valves.

189. THRACIA CURTA Conrad. 1 specimen (Dall ident.)

190. VENUS SUBIMBRICATA Sowerby. Valves (S. G. Anomalocardia).

191. VOLA DENTATA Sowerby. Valves; common, eatable.

NOTE.—In the late "Catalogue of Shells in the Museum of the Academy," etc., including the geographical distribution of species known from Sitka to Cape St. Lucas, many are given as from "West Coast, lat. 26°," and southward. These were collected by the late Dr. Gabb at San Juanico on the west coast of the peninsula, but the locality having been confused with one on the east side, in the gulf, at about the same latitude, it was uncertain which coast they were from. In his late paper Dr. Stearns credits them all to the west coast. (See "Zoe" for April, 1892.)

NOTES ON A SPECIMEN OF ALEPISAURUS ÆSCULAPIUS BEAN, FROM THE COAST OF SAN LUIS OBISPO COUNTY, CALIFORNIA.

BY FLORA HARTLEY.

[With Plate ii.]

The valuable specimen here described and figured was found on the coast of San Luis Obispo county, Cal., by W. P. Stevens, on September 19, 1894, having been thrown on the beach by the waves. It was presented by Mr. Stevens to the California Academy of Sciences. The specimen agrees in most particulars with Dr. Bean's original description of *Alepidosaurus æsculapius* (Proc. U. S. Nat. Mus. 1882, p. 661). The following descriptive notes contain the principal points wherein our specimen differs from the type, and present some additional details:

Head in body 5¼ times; width of head in length of head 4; height of head in length of head a little over 2; eye in head 6; interorbital area wider than diameter of eye; eye in snout 2½. Top of head with two prominent, sharp, bony ridges, that run from the tip of the snout to behind the eye. On the parietals striæ radiate from this ridge toward the median line and outward toward the eye. Just behind the orbit, and parallel with it, is a series of six small blunt spines.

Opercle with strong striæ radiating from the upper anterior angle; subopercle nearly as large as opercle and strongly striate, the striæ radiating from the anterior lower articulation; a thin membrane connects the opercle and subopercle; mandible with strong longitudinal striations and many mucous pores.

Outer row of teeth in upper jaw short and very sharp, increasing in length anteriorly. No teeth in the extreme front of upper jaw, but two small decurved teeth on each side of tip of jaw. Palatine teeth arranged in three groups.

In front are two very large compressed teeth on each side, 1½ inches long, behind which is a naked space 1¼ inches long; this is followed by three teeth ⅝ inches long, and finally eight short knife-like teeth. Lower jaw with symphysis bare, a bifid tooth-like spine on each side of it. Behind this come two long thorn-like teeth, the first ⅓ in. long, the second ½ in., followed by 15 short knife-like teeth. The teeth are all very sharp and directed backward and so long that the mouth cannot be completely closed. Upper pharyngeal teeth long and slender; upper pharyngeals formed by the coalesced parts of the 2, 3 and 4 pharyngobranchials.

Gill-rakers long, sharp and toothlike, bifid or trifid; on the anterior part of the lower gill-arches they appear as teeth where the arches join the hyoid. Upper lobe of caudal evidently not prolonged into a filament; middle rays longer than the height of the tail.

D 39, P 15, A 17, V 9, B 7. Length of specimen, 52¼ inches.

The type of this description is in the museum of the California Academy of Sciences. It is in good condition, only the tips of the dorsal rays being broken. The accompanying figure representing it was drawn by Miss Anna L. Brown. This specimen is the only one thus far preserved, excepting the original types from Alaska.

About two years ago, however, a specimen in fine condition was taken at Eureka, Humboldt county, Cal., by Mr. Augustus J. Wiley and Mr. J. B. Brown, of Eureka. Mr. Wiley was unable to preserve this specimen, but took a series of good photographs, which were presented by him to the Leland Stanford Jr. University. From one of these photographs a drawing was made by Miss Anna L. Brown. In view of the great interest attached to this rare and singular inhabitant of the deep seas, a copy of this drawing has been given.

DESCRIPTION OF A NEW JACK-RABBIT FROM SAN PEDRO MARTIR MOUNTAIN, LOWER CALIFORNIA.

BY JOHN M. STOWELL.

[With Plate iii.]

Lepus martirensis sp. nov.

Type, ad. ♂ No. 748, Museum Leland Stanford Junior University; La Grulla, San Pedro Martir Mountains, Lower California, June 30, 1893. Collectors, John M. Stowell and Samuel C. Lunt.

Total length 603; tail vertebræ 95; hind foot 126; ear, from crown, 184.

Cranial Measurements.

Collection number.	Basilar length.	Total length.	Greatest breadth.	Distance between orbits.	Nasal bones, length.	Nasal bones, width behind.	Nasal bones, width before.	Upper incisors, from front to molars.	Upper incisors, from front to hinder margin of palate.	Upper incisors, height.	Upper incisors, width between external edges.	Upper molars, length taken together.	Upper molars, distance between.	Lower jaw, length.	Lower jaw, height.
748	79.5	101	42.	29	41.	20	14.5	31.5	40.2	10.7	9.	17.5	13.	?	41.
749	79.5	99	42.5	29	37.	21	13.	31.5	41.	11.	8.5	17.5	12.5	72	42.
750	76.	95	42.	26	38.7	19	12.	31.7	41.	11.	9.	17.	11.5	71	43.5

About the size of *Lepus californicus*, but with much larger ears and darker coloration.

Color above, steel gray, strongly mixed with black, without the rufous tinge of *L. californicus*. Sides lighter, with traces of rufous. Chin and throat yellowish white. Under side of neck same color as sides, a black tip and a subterminal white zone on the longer hairs giving a decided gray tinge. Breast and inner side of legs salmon color, not cinnamon as in *L. californicus*. Belly whitish, but strongly washed with light salmon. A white patch between hind legs. The tail darker on lower side than in *L. californicus*.

The ears are sparsely haired and the hair is very short, in this respect much like *L. alleni*. The fringe on the anterior edge is dark gray, short and fine in texture. The tips of ears are black, the black extending down the posterior edge of the convex side for nearly one-half its length.

This species is based upon three specimens, all fully adult, collected by Samuel C. Lunt and John M. Stowell during June and July, 1893, in the San Pedro Martir Mountains, Lower California, at an altitude of about 7,000 or 8,000 feet.

Four other specimens were seen by the party but not secured, all in the vicinity of La Grulla, a large timber-inclosed meadow-tract, watered by mountain streams. Tracks were seen in several other places upon the mountains, but we did not start the animals from their hiding places and concluded that they must secrete themselves in the crevices among the huge rocks which are heaped up so numerously.

The San Pedro Martir Mountains form a range about 70 miles in length, their eastern slope passing into the deserts surrounding the Gulf of California, their western face so abrupt and precipitous as to admit of ascent at two points only: one from Agua Caliente, near the extreme northern end of the range, and the other sixty miles to the south, where the mountains are a barren waste inhabited only by mountain sheep.

Between Agua Caliente and Cape Colnett one passes over a series of mesas and low ridges, where jack-rabbits occur. Two of these were seen by us, though not secured, at San Telmo, and were easily recognized as different from those upon the mountains. The difference is well known also to the natives, who recognized our specimens at once, distinguishing them from the lowland form by their large ears and dark coloration.

A SUPPLEMENT TO THE BIBLIOGRAPHY OF THE PALÆOZOIC CRUSTACEA.*

BY ANTHONY W. VOGDES.

In offering this supplement to the literature of this special subject the author has brought the catalogue up to date, as far as lies within his knowledge, and corrected some errors and omissions of the first edition.

A few may ask, what is the advantage of such a catalogue; but let them take up any special study, and the first thing that is wanted is a list of books, to know how and what to read.

Such compilations are dry and laborious, but like all things that lead to, or add to knowledge, we have to dig through a mass of details.

The palæontologist has to work with species and all such lists save him many a weary day of research through many pamphlets and books; then again the want of just such a catalogue as is herewith presented to the Academy leads to the making of new species, from which we pray to be delivered. These hastily made children retard the progress of knowledge, and sooner or later will have to be reclassified under some older name, delaying the progress of an advancing science, taking up the time and labor of the student, to say nothing of his temper.

Agassiz (L.) The Trilobites.
In Canadian Nat. Geol., vol. 6, 1872, pp. 358-361.

Ammon (Ludw. von). Devonische Versteinerungen von Lagoinha in Mato Grosso (Brasilien).
In Zeitsch. Ges. für Erdkunde, Berlin, vol. 28, 1893; No. 5, p. 352.
Phacops brasiliensis Clarke. *Harpes* sp.

---

*Occasional Papers of the California Academy of Sciences, No. IV, 1893.

Andersson (J. G.) Note on the Occurrence of the Paradoxides ölandicus zone in Nerike.

In Bull. Geol. Inst., Upsala, vol. 1, 1892, No. 1.

——— Ueber Blöcke aus dem jüngerex Untersilur auf der Insel Oland vorkommend.

In Kongl. Svenska Vetenskaps-Akad., Förhandlingar, No. 8, 1893, p. 521.

Trinucleus seticornis His.

——— Ueber das Alter der *Isochilina canaliculata* Fauna.

In Ofversigt af Kongl. Vetenskaps-Akad., Förhandlingar, 1893, No. 2, p. 125.

Isochilina canaliculata Kr. *Primitia distans* Kr. *P. plana* var. *tuberculata* Kr. *P. plicata* Kr. *Entomis (Primitia?) flabellifera* Kr. *E. quadrispina* Kr.? *E. obliqua* Kr. *Primitia auricularis* Kr. *Entomis (Bursulella?) quadrispina* Kr. *Bollia minor* Kr. *B. major* Kr. *Tetradella harpa* Kr. var. *T. carinata* Kr. *T. rostrata* Kr. *T. erratica* Kr. var. *Ctenobolbina ciliata* Emmons? *Beyrichia radians* Kr. *B. (Ulrichia?) bidens* Kr.

Armstrong (J.), **Young** (J.) and **Robertson** (D.) Catalogue of Western Scottish Fossils, with introduction on the Geology and Palæontology of the District by John Young. Glasgow, 1876.

Aurivillius (C. W. S.) Ueber einige obersilurische Cirripeden aus Gotland.

In Bihang Svenska Akad., vol. 18, 1892-1893, Part iv, No. 3, pp. 1, 22, plate.

Pollicipes siynatus n. sp. *P. validus* n. sp. *Scalpellum fragile* n. sp. *S. sulcatum* n. sp. *S. varium* n. sp. *S. procerum* n. sp. *S. cylindricum* n. sp. *S. strobiloides* n. sp. *S. granulatum* n. sp. *S. distinctum* Hoek. *S. septentrionale* C. W. Auriv. *Turrilepas*.

Baily (W. H.) Explanation Sheet No. 135 Geological Survey of Ireland, 1860.

Cucullella angulata n. sp., now referred to *Aptychopsis angulata*.

Barrande (Joachim). Silurische Fauna aus der Umgeburg von Hof in Bayern.

In N. Jahrb. Min. Geol. Jahrgang 1868, p. 641, pls. 6-7.

For list of species see French edition, Bibliography Palæozoic Crustacea, p 14.

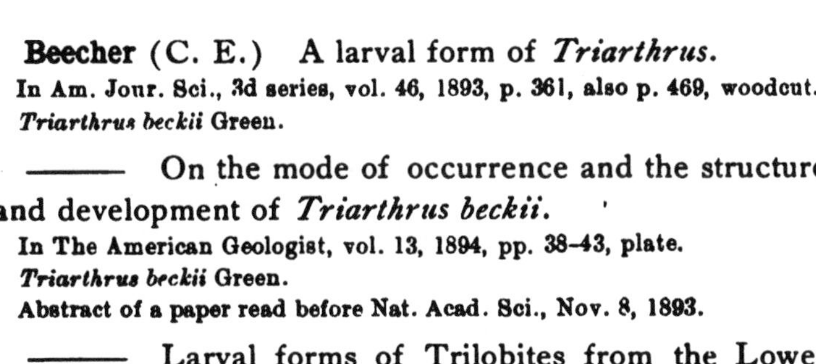

Beecher (C. E.) A larval form of *Triarthrus*.

In Am. Jour. Sci., 3d series, vol. 46, 1893, p. 361, also p. 469, woodcut.

Triarthrus beckii Green.

——— On the mode of occurrence and the structure and development of *Triarthrus beckii*.

In The American Geologist, vol. 13, 1894, pp. 38–43, plate.

Triarthrus beckii Green.

Abstract of a paper read before Nat. Acad. Sci., Nov. 8, 1893.

——— Larval forms of Trilobites from the Lower Helderberg Group.

In Am. Jour. Sci., 3d series, vol. 46, 1893, p. 142, plate.

The author illustrated the larval forms of *Acidaspis tuberculatus* Conrad and *Phaëthonides* sp.?

——— The appendages of the pygidium of *Triarthrus*.

In Am. Jour. Sci., 3d series, vol. 47, 1894, pp. 298-300, plate vii.

——— On the mode of occurrence and the structure and development of *Triarthrus beckii*.

In the American Geol., vol. 13, 1894, pp. 38–43, plate iii.

——— Further observations on the ventral structure of *Triarthrus*.

In the American Geol., vol. 15, 1895, pp. 91-100, plates iv-v.

——— Structure and appendages of *Trinucleus*.

In Am. Jour. Sci., 3d series, vol. 49, 1895, pp. 307-311, plate iii.

Bennie (James). On the prevalence of Eurypterid remains in the Carboniferous shales of Scotland.

In Proc. Royal Phy. Soc. Edinburgh, vol. 9, 1885–88, p. 499.

Not descriptive.

Benshauser (L.) Ueber Hypostomen von *Homalonotus*.

In Jahrb. Königl. preuss. Geol. Landesanst. für 1891, pp. 154-166, 1892.

Bergeron (Jules). Étude paléontologique et stratigraphique des terrains anciens de la Montagne Noire.

In Bull. Soc. Géol. France, 3d series, vol. 15, 1887, No. 5, p. 376.

Harpes Escoti n. sp. *Cheirurus Lenoiri* n. sp. *Phacops Munieri* n. sp. *P. Rouvillei* n. sp.

——— Etude géologique du massif ancien situé au sud du Plateau central.

Paris 1889, pp. 1-361, plates i-ix.

Conocoryphe coronata Barr. *C. Rouayrouxi* M. C. et Berg. *C. Heberti* M. C. et Berg. *C. Levyi* M. C. et Berg. *Conocoryphe* sp.? *Paradoxides rugulosus* Corda. *Agnostus Sallesi* M. C. et Berg. *Asaphelina Barroisi* M. C. et Berg. *Megalaspis Filacovi* M. C. et Berg. *Calymene Filacovi* M. C. et Berg. *Agnostus Ferralsensis* M. C. et Berg. *Phacops Potieri* Bayle. *Harpes Escoti* J. Berg. *Cheirurus Lenoiri* J. Berg. *Phacops Munieri* J. Berg. *P. Rouvillei* J. Berg. *Bronteus Gonzeni* J. Berg.

Fig. 2*b*, plate v, represents the pygidium of a *Lichas* n. sp. and not a *Cheirurus*. The author mentions the following undescribed species: *Asaphus Fourneti* Vern., *A. Graffi* Vern., *Illænus Lebescontei* Trom.

——— Crustacés.

In L'Annuaire Géologique Universel Tome, viii, 1891.

A review of Matthews', Clarke's and Delgado's papers on Trilobites.

——— La Faune dite "Primodiale"; est-elle la plus ancienne?

In Revue Gen. des Sci. Dec., 1891-1892, pp. 1-24.

Figures *Paradoxides bohemicus* Bœck. *Calymene senaria* Conrad. *Sao hirsuta* Barr. *Olenellus gilberti* Wal. *Agnostus rex* Barr. *Paradoxides spinosus* Bœck. *Conocoryphe coronata* Barr.

——— Notes Paléontologique 1. Crustacés.

In Bull. Géol. Soc. France, 3d series, vol. 21, 1893, pp. 333-347, plates vii, viii.

The author gives a generic description of the genus *Asaphelina*, *A. miqueli* n. sp. *Anthracopeltis crepini* Boulay from the Coal Measures is referred to the genus *Prestwichia crepini*.

——— Notes Paléontologiques 2. Crustacés. Description de quelques Trilobites de L'ordovicien D'Ecalgrain (Manche), 1 plate.

In Bull. Soc. Géol. de Normandie, vol. 15, p. 42, 1894.

Calymene aff. *Tristani*. *C. Lennieri* n. sp. *Trinucleus Grenieri* n. sp. *Dalmanites* sp.

Bernard (H. M.) The systematic position of the Trilobites.

In Quart. Jour. Geol. Soc. London, vol. 50, 1894, p. 411.

The author remarks that "the trilobites may thus be briefly described as fixed specialized stages in the evolution of the crustacea from an Annelidan ancestor which bent its mouth round ventrally so as to use its parapodia as jaws."

Beyrich (Ernst). Ueber eine Kohlenkalk-Fauna von Timor.

In Abhandl. d. k. Acad. Wissensch. zu Berlin, 1864, vol. 36, 1865.
Phillipsia parvula n. sp.

Bigot (A.) Note sur les *Homalonotus* des grès siluriens de Normandie.

In Bull. Soc. Géol. France, 3d series, vol. 16, 1888, p. 419, plates v–vii.
Homalonotus bonissenti Morière. *H. deslongchampsi* Trom. *H. brongniarti* Delong. *H. serrata* Trom. *H. vicaryi* Salt. *H. besnevillensis* n. sp. *H. morieri* n. sp. *H. incertus* n. sp. *H. viellardi* Trom. *Plæsiacomia brevicaudata* Delong.

Bolsche (W.) Ueber *Prestwichia rotundata* H. Woodward aus der Stein Kohlen formation der Piesberges bie Osnabrück.

In Jahr d. Naturn. Vereins zu Osnabrück.

Bolton (H.) On the occurrence of a Trilobite in the Skiddaw slates of the Isle of Man.

In Geol. Magazine, Decade 3, vol. 10, 1893, p. 29.
The author describes an imperfect specimen of the genus *Asaphus* or *Æglina*.

——— Catalogue of the types and figured specimens in the Geological Department of the Manchester Museum, Manchester, 1893, 35 pp.

Leaia Leidyi var. *Williamsoniana* Jones. *Carbonia Rœderiana* J. & K. *C. fabulina* J. & K. *C. Bairdoides* J. & K. *C. Salteriana* J. & K. *C. pungens* J. & K. *C. secans* J. & K. *Ceratiocaris minuta* J. & W. *Pygocephalus Cooperi* Huxley fig. *Hymenocaris vermicauda* Salter. *Cyclus Scotti* sp. nov. This article is by Henry Woodward. *Arionellus longicephalus* Hicks. *Niobe Menapiensis* Hicks. *N. Homfrayi* Salt. *Neseuretus Ramseyensis* Hicks. *N. quadratus* Hicks. *Neseuretus* sp. *N.* sp. *Phillipsia gemmulifera* Phillips. *Griffithides acanthiceps* H. W. *Eurypterus punctatus* Salt. *Pterygotus Lundensis* Salt. *P. Banksii* Salt. *Stylonurus megalops* Salt.

Brongniart (Charles). Note sur un nouveau genre d'Entomostracé fossile provenant du terrain Carbonifère des environs de St. Etienne.

In Annales Sci. Géol., vol. 7, 1876, plate vi.
Palæocypris edwardsii.

Burmeister (H.) Ueber *Gampsonychus fimbriatus* Jord.

In Abh. d. Naturforsch. Gesellschaft zu Halle, 1854, vol. 2, pp. 191–200, plate.

The author changes Jordon's generic name of *Gampsonyx* to *Gampsonychus*, the former having been used by Swainson for a genus of *Falconidæ*.

Clarke (J. M.) On the structure of the carapace in the Devonian Crustacean *Rhinocaris* and the relation of the genus to *Mesothyra* and the Phyllocaridæ.

In American Naturalist, Sept., 1893, p. 793.

The author places the genera *Mesothyra* and *Rhinocaris* under the Rhinocaridæ, remarking "that the two fossils are very closely related, and it will not do to separate them by more than a generic difference." He illustrates the carapace structure in *Rhinocaris*, *R. columbina*, *Mesothyra oceani*, *Hymenocaris vermicauda* and *Protocaris marshi*.

——— List of the original and illustrated specimens in the Palæontological Collections, Part 1. Crustacea.

In 11th Annual Report of the N. Y. State Geologist for the Year 1891. Albany, 1892.

The author gives a systematic classification of the fossil Crustacea.

See also 45th Report N. Y. State Museum for 1891.

——— On *Cordania*, a proposed new genus of Trilobites.

In 11th Annual Report N. York State Geologist for the Year 1891, p. 124.

The author proposes the new genus of *Cordania* with *Phæthonides cyclurus* Hall as the type.

——— Report of the Assistant Palæontologist.

In 12th Annual Report of the N. Y. State Geologist for the Year 1892. Albany, 1893.

The author describes a new species of *Dalmanites Dolphi*, p. 49, fig. 1.

——— The Lower Silurian Trilobites of Minnesota.

In vol. 3, part 2, of the Final Report of the Geol. and Nat. Hist. Sur. Minnesota, Sept. 27, 1894. Published in advance of the Report.

Calymene callicephala Gr. *Isotelus gigas* Dekay. *I. maximus* Locke. *I. canalis* Whitf. *I. [illegible]* Whitf. *Ptychopyge ulrichi* n. sp. *Gerasaphes* n. subgen. *G. ulrichana* n. sp. *Nileus vigilans* M. & W. *Illænus americanus* Billings. *Illænus* cf. *I. indeterminatus* Wal. *Thaleops ovata* Con. *Bumastus trentonensis* Emm. *B. orbicaudatus* Billings. *Bathyurus extans* Hall. *B. spiniger* Hall. *B. schucherti* n. sp. *Bronteus lunatus* Billings.

Dalmanites achates Billings. *Pterygometopus intermedius* Wal. *P. eboraceus* n. sp. *P. schmidti* n. sp. *P. callicephalus* Hall. *Ceraurus pleurexanthemus* Gr. *Pseudosphærexochus trentonensis* n. sp. *Cyrtometopus scofieldi*. *Encrinurus vannulus* n. sp. *E. raricostatus* Wal. *E. cristatus* n. sp. *Cybele winchelli* n. sp. *Odontopleura parvula* Wal. *Arges wesenbergensis* Schmidt var. *paulianus* n. var. *Platymetopus cucullus* M. & W. *P. robbinsi* Ulrich. *Proetus parviusculus* Hall *Harpina minnesotensis* n. sp. *Cyphaspis? galenensis* n. sp. *Proetus stonemani* Vogd.

Claypole (E. W.) Palæontological notes from Indianapolis.

In The American Geologist, vol. 6, 1890, p. 255.

Eurysoma n. g. *E. newlini* n. sp.

The term *Eurysoma* being preoccupied, the author changes the generic term to *Carcinosoma* in the same volume, p. 400.

——— A new species of *Carcinosoma*.

In the Am. Geologist, vol. 13, 1894, p. 77, plate iv.

Carcinosoma ingens n. sp.

Cole (G. A. J.) The story of Olenellus.

In Natural Science, London, vol. 1, July, 1892, pp. 340–346.

The author figures after Walcott. *Olenellus Thompsoni* Hall. *O. (Mesonacis) vermontana* Hall. *O. (Holmia) Kjerulfi* Linnarsson. This figure is taken from Holm. There are no detailed descriptions in this paper.

Cox (James C.) Note on the Moore Park Borings.

In Proc. Linn. Soc. N. S. Wales for 1880 (1881), vol. 5, part 3, p. 273.

Estheria coglani n. sp.

Delgado (J. F. N.) Fauna Silurica de Portugal. Descripcao de uma forma nova de Trilobite *Lichas (Uralichas) ribeiroi*.

In Comm. dos Trabalhos Geol. de Portugal, Lisboa, 1892, 16 pp. and 6 plates.

In French and Portuguese.

Edwards (H. Milne). Des nouvelles rescherches de M. Walcott relatives a la structure des Trilobites suivi de quelques considérations sur l'interprétation des facts ainsi constatés.

In Ann. Sci. Naturelle, series 6, vol. 12, 1881, No. 3, 31 pages, plates x, xi, xii.

Etheridge (R.) The Geology of the northern part of the English Lake District, by J, Clifton Ward, with an appendix of new species of fossils by R. Etheridge.

In Mem. Geol. Survey England and Wales, London, 1876, 132 pp., 12 plates.

Niobe doveri n. sp. *Æglina* sp.? *Asaphus* sp.? *Cybele ovata* n. sp.

Etheridge (R. Junr.) A monograph of the Carboniferous and Permo-Carboniferous Invertebrata of New South Wales, Part 2, Echinodermata, Annelida and Crustacea.

In Mem. Geol. Sur. N. S. Wales Palæontology, No. 5, Sydney, 1893, 131 pages, plates xii-xxii.

Carbonia australis n. sp. *Entomis Jonesi* Dekon. *Bairdia affinis* Morris. *B. curtus* McCoy. *Phillipsia dubia* Eth. jr. *P. Woodwardi* Eth. jr. *P. grandis* Eth. jr. *Phillipsia* sp.? *Griffithides Sweeti* Eth. jr. *Griffithides* sp.?

——— The invertebrate fauna of the Hawkesbury-Wianamatta series (Beds above the productive Coal Measures) of New South Wales.

In Mem. Geol. Survey N. S. Wales Palæontology, No. 1, Sydney, 1888, 21 pages, 2 plates.

Estheria coglani Cox.

——— *Hymenocaris salteri* McCoy.

In Records Geol. Survey N. S. Wales, 1892, vol. 3, pt. 1, pp. 5-8, plate iv.

The author states that this species is neither a *Hymenocaris* nor a *Caryocaris*, but is in all probability referable to another genus of Salter's, *Lingulocaris*. He gives the new name of *Lingulocaris McCoyi*.

——— Further additions to the Lower Silurian fauna of Central Australia.

In Ann. Rep. Government Geologist South Australia for the Year 1894, Adelaide, 1894, pp. 23-26, plate iii.

Asaphus (Megalaspis) howchini n. sp. Hypostome and glabella of an *Asaphus*.

——— On *Leaia mitchelli* from the Upper Coal Measures of the Newcastle District.

In Proc. Linn. Soc. N. S. Wales, 1892, vii (2), pt. 2, pp. 307-310, woodcut.

Etheridge (R. Jr.) and **Jacks** (R. L.) The Geology and Palæontology of Queensland and New Guinea.

Pp. xxx and 768, 68 plates and Geol. map of Queensland, London, 1892. *Beyrichia varicosa* Jones. *Phillipsia dubia* Ether. *P. woodwardi* Ether. Jr. *Phillipsia* sp.? *Griffithides seminiferus* Phill.

Etheridge (R. Jr.) and **Mitchell** (John) The Silurian Trilobites of New South Wales, with reference to those of other parts of Australia.

In Proc. Linn. Soc. N. S. Wales, vol. 8, 1893, p. 169, plates vi–vii, pt. 2. The genera *Proetus* and *Cyphaspis*. *Cyphaspis browningensis* Mitch. *C. yassensis* n. sp. *C. horani* n. sp. *C. rotunda* n. sp.

Feistmantel (O.) Ueber den Nürschener Gasschiefer dessen geologische Stellung und organische Einschlüsse.

In Zeitsch. deutsch. geol. Gesellschaft, vol. 25, 1873, p. 593, pl. xviii. *Gampsonychus fimbriatus* Jordon.

Fraipont (Julien). Euryptérides nouveau du Dévonien Supèrieur de Belgique.

In Ann. Soc. Geol. de Belgique, vol. 17, 1889, p. 53, plate. *Eurypterus Lohesti* Dewalque. *E.? dewalquei* n. sp.

Geinitz (H. B.) Ueber *Arthopleura armata* Jordon in der Steinkohlen Formation von Zwickau.

In N. Jahrb. Min. Geol. Jahrgang 1866, p. 144, pl. iii.

——— Dyas oder die zechsteinformation und das Rothliegende, 1861.

The author refigures *Hemitrochiscus paradoxus* Schaur, and referred it to the Decapoda. He also describes *Palæocrangon (Prosoponiscus) problematicus* Schaur, and refers it to the Isopoda. He adopted *Prosoponiscus* Kirkby in preference to *Palæocrangon* Salter, remarking that *Palæosphæroma* would have been a more appropriate name than either of the foregoing.

Gurich (G.) Ueber eine cambrische Fauna von Sandomir in Russisch-Polen.

In Neues Jahrb. Mineralogie, 1892, Bd. 1, p. 69.

Agnostus fallax Linn. *A. gibbus* Linn. *Agnostus* sp. *Liostracus Linnarssoni* Brögg. *Paradoxides* cf. *tessini* Brongn.

——— Ueber eine Cambrische Trilobiten Fauna bei Sandomir.

In J. Ber. Schles Gesellsch., No. 69, 1891, p. 55.

Hector (James). On a new Trilobite (*Homalonotus expansus*).

In Trans. New Zealand Inst., vol. 9, 1876, p. 602, plate xxvii.

Hicks (H.) The Fauna of the Olenellus zone in Wales.

In Geol. Magazine Decade 3, vol. 9, 1892, p. 21.

The author refers *Leperditia? cambrensis*, described in Quart. Journal, 1871, to the head of an *Olenellus*.

Huxley (T. H.) Description of a new Crustacean (*Pygocephalus cooperi* Huxley) from the Coal Measures.

In Quart. Jour. Geol. Soc., vol. 13, 1857, p. 363, plate xiii.

——— On a stalk-eyed Crustacean from the Carboniferous Strata near Paisley.

In Quart. Jour. Geol. Soc., vol. 18, 1862, p. 420.

Pygocephalus cooperi Huxley.

Jones (T. Rupert) Notes on the Palæozoic Bivalved Entomostraca, No. xxix.

In Annals Mag. Nat. Hist., 6th series, vol. 6, 1890, pp. 317–324, plate xi.

Entomis serratostriata Sandb. *E. Richteri* n. sp. *E. gyrata* Richter. *E. variostriata* Clarke.

——— Notes on the Palæozoic Bivalved Entomostraca, No. xxx.

In Annals Mag. Nat. Hist., 6th series, vol. 9, 1892, pp. 302–308, plate xvi.

Leperditia Okeni (Münster) and var. *inornata* McCoy. *Bythocypris bilobata* Münster. *B.? cuneola* J. & K. var. *Bairdia curta* McCoy. *B. subelongata* J. & K. *B. brevis* J. & K. *B. amputata* Kirkby. *B. ampla* Reuss *B. grandis* J. & K. *B. Hisingeri?* Munster var. *Mongoliensis*.

——— Notes on the Palæozoic Bivalved Entomostraca, No. xxxi. Some Devonian species.

In Annals Mag. Nat. Hist., 6th series, vol. 15, 1895, pp. 59–67, plate vii.

Aparchites reticulatus n. sp. *Primitia mundula* var. *sacculus* n. sp. *P. nitida* Roemer. *P. laevigata* n. sp. *Entomis serratostriata* Sandberger. *Barychilina? semen* n. sp. *Beyrichia strictisulcata* n. sp. *Bollia varians* n. sp. *Drepanella serotina* n. sp. *Strepula? annulata* n. sp.

——— Fossil Phyllopoda of the Palæozoic Rocks. Ninth Report of the committee, consisting of Prof.

T. Wiltshire, Dr. H. Woodward and Prof. T. Rupert Jones (secretary).

In Brit. Assoc. Adv. Sci., 1892, 62d Meeting, pp. 298–300.

The genera here treated of are: *Hymenocaris*, *Lingulocaris*, *Saccocaris*, *Caryocaris*, *Aptychopsis*, *Peltocaris*, *Pinnocaris* and *Discinocaris*.

——— Fossil Phyllopoda of the Palæozoic Rocks. Tenth Report of the Committee, consisting of Prof. T. Wiltshire, Dr. H. Woodward and Prof. T. Rupert Jones (secreta.y).

In Brit. Assoc. Adv. Sci., 1893, 63d Meeting, pp. 455–470, plate i.

Phyllocarida. *Estheria*. *E. striata* Munster var. *Muensteriana* n. var. *E. Reinachi* n. sp. *E. Geinitzii* n. sp. *E.* var. *Grebeana* n. var. *Anomalocaris*. *Caryocaris Salteri*. *Aptychopsis anatina* Salt. and *Peltocaris Marrii* n. sp. *Hymenocaris vermicauda* Salt. *Lingularocaris siliquiformis* Jones.

——— Fossil Phyllopoda of the Palæozoic Rocks. Eleventh Report of the Committee, consisting of Prof. T. Wiltshire (chairman), Dr. H. Woodward and Prof. T. Rupert Jones (secretary). Drawn up by Prof. T. Rupert Jones.

In Brist. Assoc. Adv. Sci., 64th Meeting, August, 1894, pp. 271-272.

Elymocaris Hindii n. sp. *Discinocaris* and *Aptychopsis*, *Macrocaris Gorbyi*. *Estheria* sp.? *E. Dawsoni* from Nova Scotia.

Jones (T. Rupert) and **Woodward** (Henry). A monograph of the British Palæozoic Phyllopoda (*Phyllocarida* Packard), part 2. Some bivalved and univalved species, pp. 73–124, plates xiii–xxvii.

Palæontological Soc. London, 1892.

Hymenocaris vermicauda Salt. *H.? lata* Salt. *Lingulocaris lingulæcomes* Salt. *L. siliquiformis* Jones. *L. salteriana* J. & W. *Lingulocaris* sp. *Saccocaris major* Salt. *S. minor* J. & W. *Caryocaris wrightii* Salt. *C. marrii* Hicks. *C.? salteri* McCoy. *Aptychopsis prima* Barr. *A. barrandeana* n. sp.; also var. *brevior*. *A. anatina* Salt. (corrected by authors, not *A. cordiformis* n. sp.) *A. lata* n. sp. *A. glabra* H. W. *A. wilsoni* H. W. *A. lapworthi* H. W. *A. ovata* n. sp. *A. salteri* H. W. *A. subquadrata* n. sp. *A. angulata* Baily. *A. oblata* n. sp. *Peltocaris aptychoides* Salt. *P. marrii* n. sp. (corrected by author, not *P. anatina* Salt.) *P. patula* n. sp. *P. carruthersii* n. sp. *P.? harknessii* Salt. *Pinnocaris lapworthi* R. Eth., jr. *Discinocaris browniana* H. W. *D. ovalis* n. sp. *D. undulata* n. sp. *D. gigas* H. W. *Ceratiocaris? Caryocaris?*

——— On some Palæozoic Phyllopodous and other fossils.

In Geol. Magazine Decade 3, vol. 10, 1893, p. 198, plate

Peltocaris Salteriana n. sp. *Dipterocaris Etheridge* J. & W. Fragments of a *Phyllocarid? Aptychopsis Williamsii* n. sp. *Ceratiocaris insperata* Salt.

——— On some Scandinavian Phyllocaridæ.

In Geol. Mag., new series, decade 3, vol. 5. London, 1888, pp. 145-150, plate vi.

Ceratiocaris Scharyi Barr. *C. pectinata* J. & W. *Phasganocaris pugio* (Barr.) var. *serrata* J. & W. *Ceratiocaris Angelini* J. & W.

Kiesow (J.) Die Cœlosphæridiengesteine und Backsteinkalke des westpreussischen Diluviums, ihre Versteinerungen und ihr geologisches Alter.

In Schriften d. Naturf Gesellsch. zu Danzig N. F. viii, vol. 3, 1893, 30 pp., 2 plates.

Entomis sigma Kr. *Beyrichia* sp. *B. marchica* var. *lata* Kr. *Lichas angusta* Beyr. *Cheirurus (Nieszkowskia) cephaloceras* Nieszkowski aff. *Phacops (Pterygometopus) lævigata* Schm. *Chasmops odeni* Eichw. *C. marginata* Schm. *C. macroura* Sjögren. *Chasmops* sp. *Asaphus (Isotelus)* sp. *Illænus roemeri* Volb. *I. gigas* Holm. *I. angustifrons* Holm.

——— Beitrag zur Kenntniss der in westpreussischen Silurgeschieben gefundenen Ostracoden.

In Jahrb. K preuss. geol. Landesanstalt 1889, p. 80, Berlin, 2 plates.

Leperditia phaseolus His. *L. phaseolus* His. var. *subpentagona* n. var. *L. gregaria* n. sp. *L. gregaria* var. *arcticoidea* n. var. *L. gregaria* var. *ardua* n. var. *L. baltica* His. *L. eichwaldi* Schm. *Leperditia* sp. *L. conspersa* n. sp. *Beyrichia gedanensis* Kiesow. *B. gedanensis* var. *pustulosa* Hall. *B. kockii* Boll. *B. borussica* n. sp. *B. (Klædenia) wilckensiana* var. *plicata* Jones.

Kirkby (J. W.) On some Permian fossils from Durham.

In Quart. Jour. Geol., vol. 13, 1857, p. 213, plate vii.

The author refers to Schlotherim's *Trilobites problematicus* under the new name of *Prosoponiscus problematicus* instead of adopting for it Schauroth's name of *Palæocrangon*.

——— On the fossil Crustacean found in the Magnesian Limestone of Durham by J. W. Kirkby, and on a new species of Amphiod by C. Spence Bate.

In Quart. Jour. Geol. Soc., vol. 15, 1858, p. 137, plate vi.

Prosoponiscus problematicus Schloth.

Koch (C.) Ueber das Vorkommen von *Homalonotus* Arten in den rheinischen Unterdevon.

In Verhandl. Naturhist. Vereins d. Preuss. Rheinlande und Westfalens. Vierte Folge, 7 Jahrgang, Bonn 1880, p. 132.

Homalonotus armata Burm. *H. rhenanus* Kock. *H. crassicauda* Sandb. *H. scabrosus* Kock. *H. obtusus* Sandb. *H. subarmatus* n. sp. *H. aculeatus* n. sp. *H. ornatus* n. sp. *H. multicostatus* n. sp. *H. mutabilis* n. sp. *H. römeri* DeKon. *H. planus* Sandb.

Kratow (P.) Geologische Forschungen am Westlichen Ural-Abhange in den Gebieten von Tscherdyn und Ssolikamsk.

In Mém. du Comité Géol. St. Petersburgh, vol. 6, part 2, 1888.

In addition to those mentioned on p. 129 Bibliography Palæozoic Crustacea the author describes *Estheria subconcentrica, Estheriella trapezoidalis, E. oblonge, Estheria* sp.?

Küntgen (Carl). Die Trilobiten des K. G. H. naturhistorischen Museums.

In Publications de l'Institut Royal Grand-Ducal de Luxembourg, Tome xvi, 1877, p. 127.

Proetus cuvieri Steininger. *Phacops latifrons* Bronn. *P. brevicauda* Sandb. *Dalmania caudata* Brünn. *Homalonotus platynotus* Dalm. *H. delphinocephalus* Murch. *H. laticauda. H. knightii* Kœnig. *H. obtusus* Sandb. *H. crassicauda* Sandb. *Homalonotus* sp.? *Calymene blumenbachii* Brong.

Laurie (M.) Some Eurypterid Remains from the Upper Silurian Rocks of the Pentland Hills.

In Trans. Royal Soc. Edinburgh, vol. 37, pt. 1, No. 10, pp. 151-161, plates i-iii.

Stylonurus ornatus n. sp. *S. macrophthalmus* n. sp. *Eurypterus scorpioides* H. W. *E. conicus* n. sp. *E. cyclophthalmus* n. sp. *Drepanopterus* n. g. *D. pentlandicus* n. sp.

——— The Anatomy and Relations of the Eurypteridæ.

In Trans. Royal Soc. Edinburgh, vol. 37, pt. 2, No. 24, 1893, pp. 509-528, plates i-ii.

——— Recent Additions to our Knowledge of the Eurypteridæ.

In Natural Science, vol. 3, No. 18, 1893, pp. 124-127.

Lima (W. De). Note sur un nouveau Eurypterus du Rothliegendes de Bussaco (Portugal).

In Comm. da Commissao dos Trabalhos Geol., vol. 2, fasc. 2, 1892, pp. 153–157, plate.

Eurypterus douvillei n. sp.

This Permian species is a small one, characterized by its broad cephalon and thorax and slender abdomen.

Lindström (G.) List of the Fossil Faunas of Sweden. II, Upper Silurian. Stockholm, 1888, 29 pp.

Locke (John). On the fossil *Cryptolithus tesselatus*.

In Proc. Phila. Acad., vol. 1, 1842, pp. 196–197.

Further observations on the same, with woodcut, in the Proceedings, 1843, p. 236.

Matthew (G. F.) Illustrations of the Fauna of the St. John Group, No. 7.

In Trans. Royal Soc. Canada, section 4, 1892, pp. 95–109, plate vii.

Parabolinella posthuma n. sp. *Parabolinella* ?sp. *Cyclognathus rotundifrons* Matt. *Euloma* sp.?

Matthew (W. D.) On Antennæ and other Appendages of *Triarthrus beckii* Green.

In Trans. N. Y. Acad. Sci., vol. 12, 1893, p. 137, plate; also Am. Jour. Sci., 3d series, vol. 46, 1893, p. 121, plate.

——— Sur le développement des premiers Trilobites.

In Annals Soc. Roy. Malacologique Belgique, vol. 23, 1888.

Ptychoparia linnarssoni Brögg.

Traduction faite sur le manuscrit anglais, par H. Foeir.

——— Illustrations of the Fauna of the St. John Group, No. 8.

In Trans. Royal Soc. Canada, vol. 11, section 4, 1893, pp. 85–129, plates vi–vii.

Beyrichona tinea Matt. *Primitia aurora* n. sp. *Lepiditta sigillata* n. sp. *L. auriculata* n. sp. *Hipponicharion cavatum* n. sp. *H. minus* n. sp. *Protolenus elegans* W. D. Matthew. *P. paradoxoides* Matt. *Ellipsocephalus galeatus* Matt. *E. articephalus* Matt. *E. grandis* n. sp. *Leptoplas-*[illegible] n. sp. *Sphærophthalmus alatus* Boock var. *Canadensis* n. var. [illegible]*dica* n. sp. *Conocephalites* sp.? *Agnostus trisectus* Salt.

Maurer (F.) Mittheilungen über Fauna and Gliederung des rechtsrheinischen Unterdevon.

In N. Jahr. Min. Geol., Jahrgang 1890, ii Band.

Proetus orbitatus Barr. *P. strengi* Maurer. *P. koeneni* Maurer. *Proetus* conf. *P. myops* Barr. *P. glaber* Maurer. *Proetus* conf. *P. neglectus* Barr. *P. crassirhachis*, A. Römer, *A. catillus* Maurer. *Homalonotus armatus* Burm. *H. ornatus* Kock. *Phacops fecundus* conf. var. *major* Barr. *P. latifrons* Bronn.

McCoy (F.) Ueber die Naturgeschichte von Victoria in alter und neuer Zeit.

In Die Colonie Victoria in Australieu; ihr Fortschritt, ihre Hilfsquellen und ihr physikalischer Charakter. Melbourne, 1861, p. 165.

Hymenocaris salteri n. sp.

This species has remained for over thirty years without a description. The name has appeared in print at intervals, first in McCoy's Ancient and Recent Nat. Hist. Victoria (Victoria Intercol. Exhib. Essays, 1861, p. 162), and also in Ann. Mag. Nat. Hist., 1867, vol. 20, p. 201, and in Smyth's Ann. Rep. 2d Geol. Sur., Vict., 1874, p. 33. Salter, in the Quart. Jour. Geol. Soc., 1863, vol. 19, p. 139, in a note refers the species to *Caryocaris salteri*. The species is now referred by Etheridge, Records Geol. Sur. N. S. Wales, vol. 3, 1892, pt. 1, p. 5, to *Lingulocaris McCoyi* sp. nov.

Miller (S. A.) Palæontology.

In 18th Annual Report Department of Geology and Natural Resources of Indiana. Indianapolis, 1894, pp. 257-356, 12 plates.

Mesothyra gurleyi n. sp. *Macrocaris* n. gen. *M. gorbyi* n. sp.

Miller (S. A.) and **Gurley** (Wm. F. E.) Bulletin No. 3 of the Illinois State Museum of Natural History. Description of some new species of Invertebrates from the Palæozoic rocks of Illinois and adjacent states. Springfield, 1894. 81 pp., 8 plates.

Illænus danielsi n. sp. *Lichas hanoverensis* n. sp. *L. byrnesanus* n. sp. *Ceraurus milleranus* n. sp.

Moriere (J.) Note sur un Homalonotus du grès de May.

In Bull. Soc. Linn., 3d series, vol. 8, 1884, p. 383, 2 plates.

Homalonotus deslongchampsi Trom.

——— Note sur quelques Trilobites de l'étage du grês de May.

In Bull. Soc. Linn. Normandie, vol. 9, 1884-85, 3d series, pp. 74-85, plates i-ii.

Homalonotus bonissenti n. sp. *H. serratus* Trom. *H. brongniarti* Deslong. *H. fugitivus* Trom. *Asaphus carabeufi.*

Novák (O.) On the occurrence of a new form of *Discinocaris* in the Graptolitic beds of the Colonie Haidinger in Bohemia.

In Geol. Magazine, Decade 3, vol. 9, 1892, p. 148.

Discinocaris dustiana n. sp.

D'Orbigny (A.)

In Bull. Soc. Géol. France, vol. 14, 1842-1843, p. 563.

This paper contains a discussion by D'Orbigny, Michelin and Huot regarding the habits of Trilobites in general. No genera are mentioned.

Peach (B. N.) On a new Eurypterid from the Upper Coal Measures of Radstock, Somersetshire.

In Proc. Royal Phy. Soc. Edinburgh, vol. 9, 1885-88, p. 438, plate xx.

Glyptoscorpius Kidstoni n. sp. *Eurypterus remipes* Dekay, ornamentation of *Illænus*, after Salter.

——— Additions to the fauna of the Olenellus zone of the Northwest Highlands.

In Quart. Jour. Geol. Soc. London, vol. 50, 1894, pp. 661-676, plates xxix-xxxii.

Olenellus lapworthi, also var. *elongatus*. *O. reticulatus*, *Olenelloides* n. subgen. *O. armatus*, *Olenellus intemedius*, *O. gibberti* Wal. *Mesonacis (O.) asaphoides* Emn. *Holmia (O.) kjerulfi* Linrs. *O. gigas.*

Postlethwaite (J.) Trilobites of the Skiddaw slates.

In Trans. Cumb. and Wertm. Assoc., 1884-85, No. 10.

Republished in Proceeding of the Geol. Assoc., vol. 9, 1886, No. 7, with plates illustrating the species.

Reed (F. R. Cowper) Woodwardian Museum notes on a new species of Cyclus.

In Geol. Magazine, Decade 3, vol. 10, 1893, p. 64.

Cyclus sp. *C. harknessi.*

——— Woodward Museum notes. New Trilobites from the Bala beds of Co. Waterford.

In Geol. Magazine, Decade 4, vol. 2, No. 368, p. 49, Feb., 1895.

Cybele tramorensis n. sp. *Trinucleus hibernicus* n. sp.

Remelé (A.) and **Dames** (W.) Rechtigstellung einer auf die Phacopiden species *Homalops altumii* Remelé, bezüglichen Augabe.

In Zeitschrift Deutsch. Geol. Gesellsch., vol. 40, 1888, p. 586.

Reuss (A. E.) Ueber Entomostraceen und Foraminiferen im Zechstein der Wetterau.

In Jahres. Ber. d. Wetterauer Gesellsch. für 1851-53, p. 59, plate i.

1, *Bairdia gracilis* McCoy. 2, *B. geinitziana* Jones. 3, *B. Kingi* n. sp. 4, *B. plebeia* n. sp. 5, *B. mucronata* n. sp. 6, *B. ampla* n. sp. 7, *B. frumentum* n. sp. 8? *Cytherella nuciformis* Jones. 9, *Cythere bituberculata* n. sp. 10, *C. rœssleri* n. sp.

Richter (R.) Beiträge zur Palæontologie Thüringens Waldes. Dresden und Leipzig.

The author describes from the Devonian of Saalfeld a badly preserved crustacean under the name of *Gitocrangon granulatus* in addition to those mentioned of p. 184 of the Bibliography Palæozoic Crustacea.

Roemer (Ferd.) Geologie von Oberschlesien. Breslau, 1870. Atlas of 50 plates, maps and sections.

The author illustrates the following species: *Homalonotus crassicauda* Sandb. *Cyphaspis* sp. *Phacops latifrons* Bronn. *Phillipsia latispinosa* Sandb. *P. marginlifera* F. Roemer. *P. mucronata* n. sp.

Sandberger (F. v.) Ueber die Entwickelung der unteren Abtheilung des devonischen Systems in Nassau, verglichen mit jener in anderen Ländern. Nebst einem paläontologischen Anhang.

In Jahrb. Nassauischen Vereins für Naturkunde Jahrg. xlii, 1889, pp. 1-108, plates i-v.

Phacops Ferdinandi Kayser, pl. iii, fig. 4.

The author also enumerates the fossil crustacea of the Devonian System in Nassau.

Schauroth (Baron von) Ein Beitrag zur Paläontologie des deutschen Zechsteingebirges.

In Zeitsch. deutschen Geol. Gesellschaft, vol. 6, 1854, pp. 539-577.

Palæocrangon n. g. *Hemitrochiscus paradoxus*.

Schenk (A.) Die geolische Entwickelung Sudafrikas.

In Petermann's Mitth., 34 Bd., 1888, p. 224.

Proetus ricardi. *Encrinurus cristagalli*. *Homalonotus herschelii*. *Phacops Africanus*. *P. kafir*.

Not descriptive.

Schmidt (Friedrich) Ueber eine neuentdeckte untercambrische Fauna.

In Mém. Acad. Imp. Sci. St. Petersburgh, vol. 36, No. 2, 1888, p. 27, plates i-ii.

Olenellus Mickwitzi n. sp.

——— Revision der ostbaltischen silurischen Trilobiten. Abtheilung iv. Calymeniden, Proetiden, Bronteiden, Harpediden, Trinucleiden, Remopleuriden und Agnostiden.

In Mém. Acad. Imp. des Sci., St. Petersb., 7th series, vol. 42, No. 5, pp. 1-93, plates i-vi, 1894.

Calymene tuberculata Brünn. *C. intermedia* Lindstr. *C. frontosa* Lindstr. *C. conspicua* n. sp. *C. ohhesaarensis* n. sp. *C. senaria* Conr. var. *Stacyi*. Subgen. *Pharostoma* Conr. *P. pediloba* F. Röm. *P. Nieszkowskii* n. sp. *P. denticulata* Eichw. Subgen. *Ptychometopus* Schm. *P. Solborthi* n. sp. *Bronteus laticauda* Wahlb. *B. estonicus* n. sp. *B. Marklini* Ang.? *Proetus concinnus* Dalm. var. *osiliensis*. *P. verrucosus* Lindstr. aff. *S. conspersus* Ang. *P. planedorsatus* n. sp. *P.* cf. *distans* Lindstr. *P. ramisulcatus* Nieszk. *P. Kertelensis* n. sp. *P. Wesenbergensis* n. sp. *Cyphaspis elegantula* Lov. *C. planifrons* Eichw. *Menocephalus minutus* Nieszk. *Harpides Plautini* n. sp. *Harpes Spasskii* Eichw. *H. Wegelini* Ang. *Trinucleus seticornis* His. *Ampyx nasutus* Dalm. *A. Volborthi* n. sp. *A. Linnarsoni* n. sp. *A. costatus* S. et B. *A. rostratus* Sars. *Remopleurides nanus* Leuchtb. *R.* var. *elongata* Schm. *R. emarginatus* Törnq. *Agnostus glaber* Ang. var. *ingrica*.

Smith (J.) English Upper Silurian Ostracoda.

In Trans. Nat. Hist. Soc. Glasgow, vol. 3, new series, part 2, 1889-90. Glasgow, 1892, p. 134.

The author gives a catalogue of Scottish Silurian Entomostraca, with brief descriptions of the following genera: *Bolia*, *Klædenia*, *Strepula*, *Placentura* and *Octonaria*.

Smyth (R. B.) Report of Progress Geological Survey of Victoria, 1874.

Hymenocaris salteri McCoy.

Tilesius (A. von) Sendschreiben an meinen collegen Herrn Staatsralt und Ritter von Severguine in St. Petersburgh über die Natur der Trilobiten.

In A. von Tilesius' Naturhistorische Abhandlungen Erlautermegen besonders die Petrefaktenkunde betreffend Cassel, 1826, pp. 27-37, 38-46, Taf. iv-v.

The author describes the Trilobites as relations to the Chitonidæ.

Toll (Edward von). Die paläozoischen Versteinerungen der Neusibirischen Insel Kotelny.

In Mém. Acad. Imp. Sci. St. Petersburgh, vol. 37, No. 3, 1889.

Phacops quadrilineata Ang. *Monorakos Schmidti*, n. sp. *Proetus* sp.? *Bronteus Andersoni* Eth. jr. et Nich. *Leperditia Kotelnyensis* n. sp. *L. arctica* Jones. *L. Czeskii* n. sp. *L. Sannikowi* n. sp. *L. Keyserlingi* Schm. *Leperditia* sp.?

Toula (F.) Eine Kohlenkalk-Fauna von der Barents-Inseln.

In Sitzungsberichte der Akad. der Wissenschaften, Wien, vol. 71, 1875, Abth. 1, p. 527, 6 plates.

Phillipsia Grünewaldti Möller.

Tromelin (G.) Étude de la faune du grès Silurien dans le Calvados.

In Bull. Soc. Linn. Normandie, 3d series, vol. 1, 1876–77.

Homalonotus deslongchampsi n. sp. *H. brongniarti* Delong. *H. serratus* n. sp. *H. vicaryi* Salt. *Plæsiacomia (Homalonotus) brevicaudata* Delong. *Illænus docens* Trom. *I. viducassianus* Trom. *Dalmanites incerta* Delong.

——— Under the heading of Phyllopodes belonging to the Grès des May (Calvados) the author cites two species of *Ribeiria.*

In Bull. Soc. Linn. Normandie, series 3, vol. 1, 1887, pp. 35, 74.

The author states that Mr. Salter has named but not described two species of *Ribeiria* from Mr. Vicary's collection, *R. conformis* Salt and *R. magnifica* Salt.

Tromelin (G. De) et **Lebesconte** (P.) Essai d'un catalogue raisonné des fossiles Siluriens des départments de Maine-et-Loire, de la Loire Inférieure et du Morbihan, (Anjou et Bretagne méridionale), avec des observations sur les terrains paléozoiques de l'ouest de la France.

(Extrait du compte-rendu de la 4th ,Session Nantes, 1875) de l'Assoc. française l'adv. Sc., pp. 601–661.

——— Présentation de fossiles paléoiques du départment d'Ille-et-Vilaine et Note additionelle sur la Fauna Silurien de l'ouest de la France.

In Assoc. franc. l'adv. Sc., 1875, pp. 683–687.

——— Observations sur les terrains primaires du Nord du Department d'Ille-et-Vilaine et de quelques autres parties du massif breton.

In Bull. Géol. Soc. France. 3d series. vol. 4. 1875-76. pp. 583-623.

Dalmanitis rouaulti n. sp. *D. incertus*. *D. minus* Salt. *D. phillipsi* Barr. *Calymene bayani* T. & L. *Homalonotus brongniarti* Desl. *H. vicaryi* Salt. *Trinucleus goldfussi*. *Homalonotus gahardensis* n. sp.

There is a note to this article on p. 612, in which *Bronteus thysanopeltis* Caill. (non Barr.) is mentioned.

Tromelin (Gaston Le Goarant de). Letter sur le terrain Silurien de la Sarthe.

In Bull. Agriculture Sci. at Arts de la Sarthe. vol. 22, 1874, pp. 582-590.

Cerateocaris Cenomanensis n. sp. *C. Bohemica* Barr. *C. inaequalis* Barr.

Ulrich (Arnold). Beiträge zur Geologie und Paläontologie von Südamerica unter Mitwirkung von Fachgenossen, herausgegeben von Dr. Gustav Steinmann. I Palæozoische Versteinerungen aus Bolivien von A. Ulrich.

In Neues Jahrb. fur Min. viii. Beilage Bd., 1892, Hefte 1-2; also published separately. Stuttgart. 1892, vol. 1. p. 116, 5 plates.

Acaste devonica n. sp. *Cryphaeus convexus* n. sp. *C. giganteus* n. sp. *Cryphaeus* sp. *Cyphaspis* sp. *Dalmanites Clarkei* n. sp. *Phacops Dagincourti* n. sp. *Phacops* sp.

Ulrich (E. O.) The Lower Silurian Ostracoda of Minnesota, vol. 3. Final Report.

In Geol. and Nat. Hist. Sur. Minnesota, July 24, 1894. Author's extra. 3 plates.

Leperditia Rouault. *L. fabulites* Conrad. *Leperditella* n. gen. *L. canalis* n. sp. *L. persimilis* n. sp. *L. macra* n. sp. *L. germana* Ul. *L.? dorsicornis* Ul. *Schmidtella* Ulrich. *S. crassimarginata* Ul. *S. affinis* n. sp. *S. umbonata* n. sp. *S. incompta* n. sp. *S. brevis* n. sp. *S. subrotunda* n. sp. *Aparchites* Jones. *A. ellipticus* n. sp. *A. granilabiatus* Ul. *A. millepunctatus* Ul. *A. fimbriatus* Ul. *A. arrectus* n. sp. *A. chatfieldensis* n. sp. *A. minutissimus* Hall var. *trentonensis* n. var. *Primitiella* n. gen. *P. constricta* n. sp. *P. limbata* n. sp. *P. simulans* n. sp. *P. fillmorensis* n. sp. *P. unicornis* Ul. *Primitia* Jones & Holl. *P. minutissima* n. sp. *P. uphami* n. sp. *P. mamata* n. sp. *P. sancti-pauli* n. sp. *P. micula* n. sp. *P. celata* n. sp. *P. duplicata* n. sp. *P. tumidula* n. sp. *P. gibbera* n. sp. *Halliella* Ulrich. *H. labiosa* n. sp. *Beyrichia* McCoy. *B. initialis* n. sp. *Eurychilina* Ulrich. *E. reticulata* Ulrich var. *incurva* n. var. *E. subradiata* Ul. *E. ventrosa* n. sp. *E.? subæquata* n. sp. *E.? symmetrica* n. sp. *Dicranella* n. gen. *D. bicornis* n. sp. *D. spinosa* n. sp. *D. marginata* n. sp. *D.? simplex* n. sp. *Jonesella* Ulrich. *J. obscura* n. sp. *Bollia* Jones & Holl. *B. subæquata* n. sp. *B. Unguloidea* n. sp. *Drepanella* Ulrich. *D. bilateralis* n. sp. *D. bigeneris* n. sp. *Dilobella* n. gen. *D. typa* n. sp. *Ctenobolbina* Ulrich. *C. fulcrata* n. sp. *C. crassa* Ul. *Ceratopsis* n. gen. *C. chambersi* Miller; also var. *robusta* n. var. *Tetradella* Ulrich. *T. quadrilirata* H. & W., and varieties. *T. lunatifera* Ul. *Moorea* Jones & Kirkby. *M. angularis* n. sp. *M. punctata* n. sp. *M.? perplexa* n. sp. *Macronotella* n. gen. *M. scofieldi* n. sp. *Cytherella* Jones & Bosquet. *C.? subrotunda* n. sp. *C.? rugosa* Jones and var. *arcta* n. var. *Bythocypris* Brady. *B. cylindrica* Hall. *B.? curta* n. sp. *B. granti* n. sp. *B.? robusta* n. sp. *Krausella* n. gen. *K. inæqualis* n. sp. *K. arcuata* n. sp.

——— Two new Lower Silurian species of *Lichas* (subgenus) *Hoplolichas*.

In the Am. Geologist, vol. 10, 1892, p. 270.

Lichas (Hoplolichas) robbinsi n. sp. *L. (Hoplolichas) bicornis* n. sp.

——— New Lower Silurian ostracoda. No. 1.

In the Am. Geologist, vol. 10, 1892, p. 263, plate ix.

Leperditia tumida n. sp. *L. mundula* n. sp. *L. æquilatera* n. sp. *L. inflata* n. sp. *L. germana* n. sp. *L. sulcata* n. sp. and *ventricornis* n. var. *L. (?Primitia) dorsicornis* n. sp. *L. granilabiata* n. sp. *L. millepunctata* n. sp. *L. fimbriata* n. sp. *Schmidtella* n. g. *S. crassimarginata* n. sp.

Vogdes (A. W.) Notes on Palæozoic Crustacea, No. 4. On a new Trilobite from Arkansas Lower Coal Measures.

In Proc. Cal. Acad. Sci., 2d series, vol. 4, 1895, pp. 589–591.

Griffithides ornata n. sp.

Waagen (W.) Salt Range Fossils.

In Mem. Geol. Sur. India, series 13, vol. 4, part 2. Calcutta, 1891.

Olenus indicus n. sp. *Conocephalites Warthi* n. sp.

Walcott (Chas. D.) Note on some appendages of the Trilobites.

In Proc. Biological Soc. Washington, vol. 9, 1894, pp. 89–97, plate.

Triarthrus becki Green. *Calymene senaria* Conrad.

Weitenweber (W. R.) Systemisch Verzeichniss der Böhmen Trilobiten.

In Lotos, vol. 7, 1857.

——— Einige historische Bemerkungen über die Silurische Fauna Böhmens, insbesondere über die Trilobiten.

In Sitzungsberichte der K. bohm. Ges. Wissenschaften. Prag 1881, pp. 12–14.

The author enumerates some new species and specimens of the collection of the Pralat Zeidler at Strabow. He gives only the names without descriptions or figures, as follows:

Paradoxides Sackeri, Proetus myops, P. eremita, P. astyanax, P. frontalis, P. Lyelli, Dalmanites McCoyi, D. Fletcheri, Cyphaspis Halli, Lichas simplex, Trinucleus ultimus, Asaphus alienus, Illænus transfuga, Acidaspis Laportei, Amphion senilis, Cromus transiens, Bronteus infaustus, B. furcifer.

Whidborne (G. F.) On some Devonian Crustaceans.

In Report Brit. Assoc., 1888, p. 681.

Entomis peregrina n. sp. *Acidaspis robertsii* n. sp. *A. hughesii* n. sp. *Proetus batillus* n. sp. *P. subfrontalis* n. sp. *P. audax*. *Cyphaspis ocellatus* n. sp. *Lichas devonianus* n. sp. *Bactropus decoratus* n. sp. *Cheirurus pengellii* n. sp. *Bronteus granulatus* Goldf.

——— On some Devonian Crustacea.

In Geol. Magazine, Decade 3, vol. 6, 1889, p. 28. Abstract of a paper read at the British Association. The author gives a brief description of the following new species from Wolborough and Lummaton:

Phacops batracheus, Proetus batillus, P. subfrontalis, P. audax, Cyphaspis ocellata, Lichas devonianus, Acidaspis robertsii, A. hughesii, Bronteus delicatus, B. pardalios, Entomis peregrina, Bactropus decoratus, Cheirurus pengellii, Dechenella setosa. Phacops granulatus Phil. *Proetus champernowni. Entomis peregrina* Whid.

——— A monograph of the Devonian Fauna of the south of England.

In Palæontographical Society London, vol. 1, pp. 1–344, plates i–xxx; vol. 2, parts 1 to 3, pp. 1–160, plates i–xvii, 1889–1893.

Phacops batracheus Whid. *P. latifrons* Bronn. *Cheirurus pengellii* Whid. *C. sternbergii?* Bœck. *Acidaspis robertsii* Whid. *A. pilata* Whid. *Lichas devonianus* Whid. *Cyphaspis ocellata* Whid. *Proetus batillus* Whid. *P. subfrontalis* Whid. *P. champernowni* n. sp. *P. audax* Whid. *Dechenella setosa* Whid. *Harpes macrocephalus* Goldf. *Bronteus delicatus* Whid. *B. tigrinus* n. sp. *B. pardalios* Whid. *B. alutaceus* Goldf. *B. flabellifer* Goldf. *B. granulatus* Goldf. *Bactropus decoratus* Whid. *Tropidocaris?* sp. *Cypridina?* sp., 3 species. *Cypridinella cæca* n. sp. *Cypridella* sp. *Polycope simplex* J. & K. *P. devonica* Jones, also var. *major*, *obliqua* and *concinna*. *P. hughesiæ* n. sp. *Entomis peregrina* Whid. *Cyprosina whidbornei* Jones.

Whiteaves (J. F.) Description of Four New Species of Fossils from the Silurian Rocks of the Southeastern Portion of the District of Saskatchewa.

In Canadian Record Sci., April, 1891.

Acidaspis perarmata n. sp.

Wood (Henry). Catalogue of the Fossils in the Students' Stratigraphical Series. Cambridge, 1893, 23 pp.

——— Additions to the Type Fossils in the Woodwardian Museum.

In Geol. Magazine, Decade 3, vol. 10, 1893, p. 111.

Woods (H.) and **Hughes** (T. McKenny). Catalogue of the Type Fossils in the Woodwardian Museum. Cambridge, 1891, 180 pp.

Woodward (H.) Note on a new British Species of *Cyclus* from the Coal Measures of Racup, Lancashire.

In Geol. Magazine, Decade 3, vol. 10, 1893, p. 28.

Cyclus Scotti n. sp.

——— Some points in the Life History of the Crustacea in early Palæozoic times.

In address delivered at meeting Geol. Soc. London, Feb. 15, 1895. Proceeding Geol. Soc., May, 1895, pp. lxx-lxxxviii.

Young (J.) Note on a Series of Trilobites of Caradoc age from the Silurian Strata of the Girvan Valley.

In Proc. Nat. Hist. Soc. Glasgow, vol. 2, 1876, pt. 2, p. 179.

Illænus bowmanni Salt. *I. thomsoni* Salt. *I. barriensis* Murch. *Cybele verrucosa* Dalm. *Proetus latifrons* McCoy. *Zethus rugosus* Portl. *Cheirurus clavifrons* Dalm. *Odontopleura ovata?* *Staurocephalus unicus* Wyv. Thomson.

The author notes the identification of the above species in the Gray collection.

——— On new forms of Crustacea from the Silurian rocks at Girvan.

In Proc. Nat. Hist. Soc. Glasgow, vol. 1, 1868, p. 169.

Cheirurus trispinosus n. sp. *Solenocaris* n. g. *S. solenoides* n. sp.

——— Note on a new species of Crustacean belonging to the genus *Solenocaris* from the Silurian Strata near Girvan, and on fragments probably the appendages of a Trilobite or Limulid Crustacean.

In Proc. Nat. Hist. Soc. Glasgow, vol. 2, 1875, p. 66.

——— Notes on the group of Carboniferous Ostracoda found in the Strata of Western Scotland.

In Trans. Glasgow Geol. Soc., vol. 9, pt. 2, 1890-91, 1891-93, p. 301.

A REVIEW OF THE HERPETOLOGY OF LOWER CALIFORNIA. PART I—REPTILES.

BY JOHN VAN DENBURGH,

Curator of the Department of Herpetology.

[With Plates iv–xiv.]

The peninsula of Lower California lies so far from the usual routes of travel that few collections of its animals have found their way into museums. Its reptiles have been known chiefly from the specimens secured by Botta, Xantus, and Belding. Within the past few years the California Academy of Sciences has sent several collectors to the peninsula, and among the specimens brought back each time have been a few reptiles. In this way the collection has been formed upon which this paper is primarily based.

A few remarks on the zoögeographical position of Lower California may not be out of place.

The Sonoran Subprovince, as defined by Dr. Allen, but excluding Lower California, is inhabited by the following forty genera of reptiles:*

Phyllodactylus,	Cnemidophorus,	Contia,
Dipsosaurus,	Verticaria,	Gyalopum,
Crotaphytus,	Eumeces,	Hypsiglena,
Callisaurus,	Rena,	Phyllorhynchus,
Holbrookia,	Leptotyphlops,	Salvadora,
Uma,	Lichanura,	Bascanion,
Sauromalus,	Charina,	Pituophis,
Uta,	Chilomeniscus,	Arizona,
Sceloporus,	Tantilla,	Thamnophis,
Phrynosoma,	Chionactis,	Natrix,
Heloderma,	Rhinochilus,	Trimorphodon,
Gerrhonotus,	Lampropeltis,	Elaps,
Anniella,	Diadophis,	Crotalus.
Xantusia,		

Twenty-nine of these range over a greater or less part

* The turtles are not considered in this discussion.

of the Campestrian Subprovince on the north, or of the tropical Central American Region on the south. They need not, therefore, be considered in the present connection. Eleven genera remain which are confined to the Sonoran Subprovince, and may be considered characteristic of that area. These genera are:

Dipsosaurus,	Sauromalus,	Rena,
Callisaurus,	Heloderma,	Lichanura,
Holbrookia,	Xantusia,	Phyllorhynchus.
Uma,	Verticaria,	

In the so-called Cape Region of Lower California, twenty-eight genera of reptiles occur, namely:

Phyllodactylus,	Cnemidophorus,	Hypsiglena,
Ctenosaura,	Verticaria,	Phyllorhynchus,
Dipsosaurus,	Eumeces,	Salvadora,
Crotaphytus,	Euchirotes,	Bascanion,
Callisaurus,	Rena,	Pituophis,
Uta,	Lichanura,	Thamnophis,
Sceloporus,	Chilomeniscus,	Natrix,
Phrynosoma,	Tantilla,	Trimorphodon,
Gerrhonotus,	Lampropeltis,	Crotalus.
Xantusia,		

Only two of these have not been obtained elsewhere in the Sonoran Subprovince, while, with the exception of Holbrookia, Uma, Sauromalus, and Heloderma, all the characteristic Sonoran genera are represented. The two Cape genera which have not been found in any other part of the Sonoran Subprovince are Euchirotes, a two-footed amphisbænian which has been secured only in southern Lower California, and Ctenosaura, a genus widely distributed in tropical America and here represented by a single species.

Considering now the species of these areas, it is found that seventy-eight have been obtained in the Sonoran Subprovince.* Twenty-one of these are of partly or

*Some of the eastern and southern species have not been included for lack of precise data.

chiefly Campestrian or tropical distribution, leaving fifty-seven species which may be considered distinctively Sonoran.

Thirty-eight species have thus far been found in the southern part of Lower California,* as follows:

Phyllodactylus tuberculosus,	Lichanura trivirgata,
Phyllodactylus unctus,	Chilomeniscus stramineus,
Ctenosaura hemilopha,	Chilomeniscus fasciatus,
Dipsosaurus dorsalis,	Tantilla planiceps,
Crotaphytus copeii,	Lampropeltis conjuncta,
Callisaurus draconoides,	Lampropeltis nitida,
Uta thalassina,	Hypsiglena ochrorhyncha,
Uta stansburiana,	Phyllorhynchus decurtatus,
Uta nigricauda,	Salvadora grahamiæ,
Sceloporus zosteromus,	Bascanion flagellum frenatum,
Sceloporus licki,	Bascanion aurigulum,
Phrynosoma coronatum,	Pituophis vertebralis,
Gerrhonotus multicarinatus,	Thamnophis cyrtopsis collaris,
Xantusia gilberti,	Natrix valida,
Cnemidophorus maximus,	Natrix celæno,
Verticaria hyperythra,	Trimorphodon lyrophanes,
Eumeces lagunensis,	Crotalus atrox,
Euchirotes biporus,	Crotalus enyo,
Rena humilis,	Crotalus mitchellii.

None of these have been found upon the tropical Mexican mainland. *Uta nigricauda*, *Sceloporus zosteromus*, *Phrynosoma coronatum*, and *Phyllorhynchus decurtatus*, range considerably north of the confines of the "Cape Region." Twenty-two of the thirty-eight species are of very limited distribution, having been found only in the extreme southern part of the peninsula. The remaining twelve forms, mentioned below, extend their range over a greater or less part of the Sonoran Subprovince, and are among those characteristic of that area.†

* Several species, as *Xantusia vigilis*, *Charina bottæ*, and *Lampropeltis californiæ*, have often been credited to Lower California without evidence of their occurring there.

† Except *Uta stansburiana*, which is also Campestrian.

Phyllodactylus tuberculosus,
Dipsosaurus dorsalis,
Uta stansburiana,
Rena humilis,
Hypsiglena ochrorhyncha,
Salvadora grahamiæ,
Bascanion flagellum frenatum,
Thamnophis cyrtopsis collaris,
Natrix valida,
Trimorphodon lyrophanes,
Crotalus atrox,
Crotalus mitchellii.

Formulating these data we have the following tables:

GENERA.

Total number in the Sonoran Subprovince	40
Confined to the Sonoran Subprovince	11
Total number in the "Cape Region"	29
Restricted to the "Cape Region"	1
Common to the "Cape Region" and Tropical America	1
Common to the "Cape Region" and the Sonoran Subprovince	26

SPECIES.

Total number in the Sonoran Subprovince	78
Confined to the Sonoran Subprovince	57
Total number in the "Cape Region"	38
Restricted to the "Cape Region"	22
Common to the "Cape Region" and Tropical America	0
Common to the "Cape Region" and the Sonoran Subprovince	12

From these it appears that the affinities of the reptiles which inhabit southern Lower California are almost entirely with those of the Sonoran Subprovince, of which the "Cape Region," therefore, forms a part. It is also shown that a strongly characterized center of reptilian distribution is located in the terminal part of the peninsula, entitling it to rank as one of the minor constituent life areas or faunæ of the Sonoran Subprovince. For this area Dr. Allen has already proposed the name "Saint Lucas Fauna."

It is unfortunate that so little material has been collected on that part of the peninsula which is just north of La Paz, for on this account the northern limit of this San Lucas Fauna cannot, at present, be determined with accuracy. A sufficient number of specimens has been obtained, however, to show that this fauna is apparently restricted to the rather mountainous area south of La Paz.

There is in this area, as has already been shown, a slight infusion of tropical forms, represented among reptiles by the genera Ctenosaura and Euchirotes, but probably best illustrated by the plants growing near the coast lagoons. These forms, however, are doubtless no more numerous or characteristic than those forms, of tropical origin, which will be found to intrude upon the entire southern border of the Sonoran Subprovince.

The northern part of Lower California is much more closely related to the rest of the Sonoran Subprovince than to the San Lucas Fauna. This is well shown by the presence of such forms as *Callisaurus ventralis*, *Crotaphytus wislizenii*, *Phrynosoma solare*, *Rhinochilus lecontei*, and *Cnemidophori* of the *tessellatus* group. While the known ranges of several species may be considerably enlarged in the future, the northwestern part of Lower California and the coastal slopes of San Diego (and Los Angeles?) County, California, seem to be so well characterized as to merit recognition as a distinct faunal area of the lowest rank. Its distinctive features are the presence of certain peculiar species, the absence of others occurring near by, and its forming the limit of distribution of species whose chief habitat is either north or south.*

Pending further evidence, this area may be known as the San Diegan Fanna.† Among the reptiles peculiar to it may be mentioned the following:‡

*In this connection I have had the use of a large collection of the reptiles of San Diego County, made by Messrs. Hyatt and Stoddard for the Leland Stanford Junior University, as well as the specimens belonging to the Academy.

†When this was written the author was not aware that this area had been previously recognized and mapped from study of other branches of animal life.

‡Birds apparently belonging to the same list are *Pipilo fuscus senicula*, *Harporhynchus cinereus mearnsi*, and *Helodytes brunneicapillus bryanti*.

Uta mearnsi (?),
Sceloporus orcutti,
Phrynosoma blainvillii,
Xantusia henshawi,
Cnemidophorus stejnegeri,
Verticaria hyperythra beldingi (?),
Lichanura roseofusca,
Lichanura orcutti.

Several forms of the neighboring areas have not been taken here, namely:

Uta repens,
Uta microscutata (?),
Cnemidophorus rubidus,
Cnemidophorus tigris,
Xantusia vigilis,
Sauromalus ater,
Sceloporus magister,
Sceloporus occidentalis,
Phrynosoma frontale,
Crotalus cerastes.

Northern species which have not been collected south of this area are:

Sceloporus biseriatus,
Sceloporus graciosus,
Gerrhonotus scincicauda,
Anniella pulchra,
Eumeces skiltonianus (?),
Lampropeltis boylii,
Bascanion laterale,
Crotalus lucifer.

A few southern forms are also limited by it, as:

Sceloporus zosteromus,
Crotalus atrox.

The reptiles of the islands which naturally belong to Lower California may be divided into two groups, as follows:

(a) Species which are purely insular; as, *Sauromalus hispidus, Uta palmeri, Phrynosoma cerroense, Cnemidophorus martyris, C. labialis*, and *Verticaria sericea.*

(b) Species which occur also on the northern part of the peninsula; as, *Uta microscutata, Uta stansburiana, Uta nigricauda, Sceloporus zosteromus, Verticaria hyperythra beldingi, Crotalus atrox, Crotalus mitchellii*, and *Callisaurus ventralis.*

No species characteristic of the San Lucas Fauna has been collected on any of the islands.*

It has been thought best to redescribe many of the species which have been known only from the very brief and often inadequate original characterizations. The descrip-

* With the possible exception of *Crotaphytus copeii.*

tions are all based upon alcoholic specimens. The colors have been determined by reference to Ridgway's "Nomenclature of Colors." Measurements are given in millimeters, unless otherwise stated. Only references to a species as it occurs in Lower California are included in the synonymies, except that the original description is cited in all cases. Whenever a citation has not been verified by actual reference to the original article, it has been given in quotation marks. When the article contains no original information about the species as it occurs in Lower California, the citation has been put in parenthesis. Most of the localities mentioned may be found on the map of Lower California, published in the second volume of the second series of these Proceedings.

I am indebted to Dr. Leonhard Stejneger for the reidentification of many of the specimens listed by Dr. Yarrow, and for the loan of specimens of *Sceloporus consobrinus*.

CARETTA IMBRICATA (L.)

Testudo imbricata.
(1766, Linn., Syst. Nat., 1, p. 350.)
Chelonia imbricata.
1887, Cope, Bull. U. S. Nat. Mus., No. 32, p. 24.

The Academy's collection contains a single carapace (No. 2249) of this turtle. It was obtained at San José del Cabo, by the Expedition of 1893. Mr. Bryant tells me that he has often seen them in the waters near the shore.

CHELONIA AGASSIZII Dum. & Boc.

Chelonia agassizii.
(1870, Duméril et Bocourt, Miss. Sci. au Mex., Reptiles, 1e livr., p. 26, pl. vi.)
1887, Cope, Bull. U. S. Nat. Mus., No. 32, p. 24.
Chelonia virgata.
1883, True, Bull. U. S. Nat. Mus., No. 24, p. 28.

The green turtle has been taken at Cape San Lucas. It doubtless occurs in many places along the coast of the peninsula.

CHRYSEMYS NEBULOSA, new species. Plates iv, v and vi.

Pseudemys ornata.

1883, True, Bull. U. S. Nat. Mus., No. 24, p. 33.

Diagnosis.—Allied to *C. ornata* (Gray), but without black centers in the costal ocelli, which are much more irregular and indistinct than in that form. The markings on the head, neck, and limbs, are much coarser, and the longitudinal lines less numerous. There are four yellow rays on the upper surface of the arm, instead of eleven.

Type.—Cal. Acad. Sci. No. 2244, "Mainland abreast of San José Island," Lower California,* W. E. Bryant.

Description of the Type.—The neck is clove brown with several pale longitudinal lines on each side. The highest one of these ends on the temple in a large oval spot of the same color. The lowest and largest is continued forward across the middle of the lower eyelid, giving off, at the lower edge of the inferior maxillary bone, a branch which, continuing forward, crosses to the upper jaw, runs past the anterior edge of the orbit, turns forward at a right angle, and terminates at the nostril. The five similar lines on the nape are continued forward over the top of the head, and, besides being more or less undulating, give rise to several short transverse branches. There are six longitudinal yellow rays on the forearm; one on each edge and two on each surface. Greenish yellow lines traverse the backs of the five fingers and four perfect toes. The vertebrals sometimes show black spots. All the marginals are ornamented with black ocelli. The plastron is marked with large longitudinal

* Mr Bryant informs me that the exact locality is Los Dolores, L. C., and that No 2245 was also taken there.

seal brown blotches, not at all like the double lines on this region in *C. ornata.*

Length of carapace 80 mm. Its greatest width 63 mm.

A carapace (No. 2246) 283 mm. in length is much less distinctly marked than the type, but has a rather indistinct black-centered ocellus on each of the last pair of costal scutes. Another (No. 2247) 273 mm. long shows no trace of these ocelli, nor are they visible in the other alcoholic specimen (No. 2245), the carapace of which measures 194 mm.

List of Specimens of Chrysemys nebulosa.

Cal. Acad. Sci. No.	Locality.	Date.	Collector.
2244	Lower California,* abreast of San José Island.		W. E. Bryant.
2245	No data.*		
2246	" †		
2247	San José del Cabo, L. C.		Gustav Eisen.

Phyllodactylus tuberculosus Wieg.

Phyllodactylus tuberculosus.

"(1835, Wiegmann, Acta. Acad. Cæs. Leop. Carol., xvii, 1, p. 241, pl. xviii, fig. 2.)"

(1887, Cope, Bull. U. S. Nat. Mus., No, 32, p. 28.)

Phyllodactylus xanti.

1863, Cope, Proc. Ac. Nat. Sci. Phila., p. 102.

(1866, Cope, Proc. Ac. Nat. Sci. Phila., p. 312.)

(1875, Cope, Bull. U. S. Nat. Mus., No. 1, pp. 50, 93.)

1883, Yarrow, Bull. U. S. Nat. Mus., No. 24, p. 73.

(1884, S. Garman, Bull. Essex Inst., xvi, 1, p. 12.)

(1887, Belding, West Am. Scientist, iii, 24, p. 98.)

The writer has not seen this species, which has been recorded from Cape San Lucas and La Paz.

* From Los Dolores, by W. E. Bryant, fide Bryant, from memory.

† From Agua Caliente, by W. E. Bryant, fide Bryant, from memory.

PHYLLODACTYLUS UNCTUS (Cope).

Diplodactylus unctus.

1863, Cope, Proc. Ac. Nat. Sci. Phila., p. 102.
(1866, Cope, Proc. Ac. Nat. Sci. Phila., p. 312.)
(1875, Cope, Bull. U. S. Nat. Mus., No. 1, pp. 50, 93.)
1877, Streets, Bull. U. S. Nat. Mus., No. 7, p. 35.
1883, Yarrow, Bull. U. S. Nat. Mus., No. 24, p. 73.
(1884, S. Garman, Bull. Essex Inst., xvi, 1, p. 12.)
(1887, Belding, West Am. Scientist, iii, 24, p. 98.)

Phyllodactylus unctus.

(1873, Bocourt, Miss. Sci. au Mex., Reptiles, 2e livr., p. 43.)
(1885, Boulenger, Cat. Lizards Brit. Mus., I, p. 94.)
(1887, Cope, Bull. U. S. Nat. Mus., No. 32, p. 28.)
(1890, Townsend, Proc. U. S. Nat. Mus., xiii, p. 144.

Description of No. 886.—The head is much longer than broad. The rounded snout is longer than the distance between the eye and the ear opening. The lips are very prominent. The ear opening is a narrow slit, about the length of the pupil, and has a slight denticulation posteriorly. The scales on the eyelids form a rather conspicuous comb. The slender digits are covered below with a series of transverse lamellæ, terminated by two large plates which are somewhat rounded and wider distally than proximally. The nostril is pierced between the rostral, first labial, and three nasals, the upper of which is in contact with its fellow of the opposite side. There are seven upper and six lower labials; the last of each, under the pupil, is very small. The two plates behind the large pentagonal mental are followed by several about the size of the dorsals, which are in turn gradually replaced by the small flat gulars. The back and limbs are covered with smooth, flat, rounded, equal sized scales, without tubercles or granules. The muzzle has convex plates, smaller than the dorsals, but larger than those on the occiput, which are also convex. The lower surfaces are covered with smooth flat scales, larger than those on the back. The conical tail is slightly flattened at its base, has large plates below, and is covered elsewhere with smooth flat scales which are somewhat larger than those on the back.

Variation.—There is great variation in the ground color of the head and back. In some specimens it is pale gray or creamy white, while in others the prevailing tint is a dark seal brown. There are, however, some fairly constant markings, brighter in young than in old individuals, but apparently subject like the ground color, though to a less extent, to modification in accordance with the amount of light, or perhaps in obedience to the will of the animal. These markings are of a deeper seal brown than the ground color of the darkest individuals. A line originates on the second labial plate, and, passing through the eye and the upper end of the ear opening, runs for some distance along the neck. The upper surface of the head is blotched and spotted, as are also the limbs. The tail has about nine cross-bars on its upper surface. All the lower surfaces are creamy white, slightly tinged with brown in the darkest specimens. The scales are everywhere minutely punctulated with dark brown.

	mm.	mm.	mm.	mm.
Length to anus	45	46	35	42
Tail	48*	40*	31*	—
Hind limb	18	18	15	17
Fore limb	15	15	11	14
Head to ear	12	12	10	10½
Snout to orbit	5	5½	4	4½
Diameter of orbit	3½	3	3	3

Phyllodactylus unctus has been previously recorded from Cape San Lucas (the type locality), by Mr. Xantus; from Triunfo, by Dr. Streets; and from La Paz, by Messrs. Belding and Townsend.

* Reproduced.

List of Specimens of Phyllodactylus unctus.

Cal. Acad. Sci. No.	Locality.	Date.	Collector.
857	San José del Cabo, L. C.	Oct., 1893.	Gustav Eisen.
885 to 893	San José del Cabo, L. C.	Jan. 25, 1893	Gustav Eisen.
1182	Miraflores, L. C.	Sept., 1894.	Eisen and Vaslit.
1663 to 1669	San José del Cabo, L. C.	"	"
2204 to 2207	Lower California.		

CTENOSAURA HEMILOPHA Cope.

Iguana acanthura.

"1835, Blainville, Nouv. Ann. Mus., iv, p. 288, pl. xxiv. fig. 1."

Cyclura acanthura.

1837, Duméril et Bibron, Erpétologie Générale, iv, p. 222 (part).

(1883, Yarrow, Bull. U. S. Nat. Mus., No. 24, p. 71.)

(1887, Belding, West Am. Scientist, iii, 24, p. 98.)

Ctenosaura.

1895, Baird, Proc. Ac. Nat. Sci. Phila., p. 300.

Cyclura (Ctenosaura) hemilopha.

1863, Cope, Proc. Ac. Nat. Sci. Phila., p. 105.

(1875, Cope, Bull. U. S. Nat. Mus., No. 1, pp. 50, 93.)

1883, Yarrow, Bull. U. S. Nat. Mus., No. 24, p. 71.

(1884, S. Garman, Bull. Essex Inst., xvi, 1, p. 19.)

(1887, Belding, West Am. Scientist, iii, 24, p. 98.)

Ctenosaura acanthura.

1874, Bocourt, Miss. Sci. au Mex., Reptiles, p. 138.

Cyclura teres.

(1883, Yarrow, Bull. U. S. Nat. Mus., No. 24, p. 71.)

Ctenosaura hemilopha.

(1866, Cope, Proc. Ac. Nat. Sci. Phila., p. 312.)

1885, Boulenger, Cat. Lizards Brit. Mus., ii, p. 197.

(1886, Cope, Proc. Am. Philos. Soc., xxiii, p. 266.)

(1887, Cope, Bull. U. S. Nat. Mus., No. 32, p. 33.)

Description of No. 463.—The body is considerably

compressed. The tail is conical except at its base, where it is almost square in section. The limbs and head are large, the latter sharply triangular and with flattened top and almost vertical sides. The large nostril is in a round plate, whose posterior edge is nearer to the orbit than to the end of the snout. The rostral and symphysial plates are very broad and low. There are ten labials. There is a very large plate below the eye, and a series of large superciliaries. The entire top and sides of the head are covered with small irregularly hexagonal plates, which are convex, except on the snout and lores. The very large ear opening is almost vertical and without denticulation. Several series of large sublabial plates pass gradually into the gulars. The dorsal crest begins some distance behind the shielded part of the head, is composed of high spines on the nape, and gradually diminishes in height posteriorly. It is continued on the middle third of the vertebral line as a series of enlarged flat plates, but is not traceable on the posterior third. The back and sides are covered with small, smooth, subquadrate scales, which pass gradually into the larger ventrals. The gular regions are covered with smooth scales which become gradually larger posteriorly. The smallest gulars are larger than the dorsals, the largest are smaller than the ventrals. The scales on the limbs are all smooth. The tail bears whorls of spinose scales. The first three of these whorls are separated from one another by three series of smaller smooth scales; the fourth, fifth and sixth spiny whorls are each preceded by two series of smooth scales, and the more distal whorls by single series which gradually become spinose.

The top and sides of the head are dull pea green. The back, sides, and hind limbs are pale straw color, heavily washed with pale olive, and spotted and reticulated with seal brown and black. There are five black blotches on the vertebral line, separated by areas paler than the general tint. The first of these black markings is very small; the second is broader than long; the third and fourth are very large and faintly continuous with the blackish brown of the ventral surfaces; the fifth is almost confined to the enlarged medial scales. There are two longitudinal black blotches on the side of the neck, and two corresponding lines on the temple. The chin, gular region, chest, and forelimbs, are blackish brown. The tail has a ground color of straw yellow clouded with olive, but is dull pea green on the spines, and barred with seal brown terminally.

Snout to vent 224 mm. Snout to ear 53 mm. Hind limb 129 mm. Snout to edge of fold 76 mm. Fore limb 84 mm. Highest dorsal spines 9 mm.

Variation.—The youngest individuals (58 to 76 mm. from snout to vent) are bright terre-verte green above, except on the tail which has broad rings of dark olive separated by narrow ones of broccoli brown. There are very faint indications of dark vertebral bars. The lower parts are yellowish white tinged with green. As the animals increase in size, the green gradually disappears and the dark markings increase in size and number until the adult coloration is assumed. The number of femoral pores ranges from four to eight. The dorsal crest seems to be higher in the males than in the females, but is never continued on the posterior part of the back.

This species was collected by Botta in "California." Xantus secured it at at Cape San Lucas, the type locality. Mr. Belding found it at La Paz.

List of Specimens of Ctenosaura hemilopha.

Cal. Acad. Sci. No.	Locality.	Date.	Collector.
463 to 466	San José del Cabo, L. C.	Mar., 1892.	Gustav Eisen.
703	Miraflores, L. C.	Oct., 1890.	W. E. Bryant.
704	"	"	"
709	San José del Cabo, L. C.	Oct. 10, 1890	"
718	"	Sept. 17, 1890	"
744	Agua Caliente, L. C.	Oct., 1890.	"
830 to 833	San José del Cabo, L. C.	Sept., 1893.	Gustav Eisen.
850 to 855		Oct., 1893.	"
858		Nov., 1893.	"
859		"	"
871		"	"
872	"	"	"
985	Pescadero, L. C.	Sept., 1893.	"
991 to 995	Miraflores, L. C.	"	"
998 to 1095		Sept., 1894.	Eisen and Vaslit.
1389	Sierra San Lazaro, L. C.	"	"
1620 to 1628	San José del Cabo, L. C.	"	"
2240	"	Mar., 1892.	W. E. Bryant.

DIPSOSAURUS DORSALIS (B. & G.)

Crotaphytus dorsalis.

(1852, Baird and Girard, Proc. Ac. Nat. Sci. Phila., p. 126.)

Dipsosaurus dorsalis.

1859, Baird, Proc. Ac. Nat. Sci. Phila., p. 299.
(1866, Cope, Proc. Ac. Nat. Sci. Phila., p. 312.)
(1875, Cope, Bull. U. S. Nat. Mus., No. 1, p. 48.)
(1880, Lockington, Am. Nat., xiv, p. 295.)
1883, Yarrow, Bull. U. S. Nat. Mus., No. 24, p. 54.
(1885, Boulenger, Cat. Lizards Brit. Mus., ii, p. 201.)
(1887, Cope, Bull. U. S. Nat. Mus., No. 32, p. 34.)
(1887, Belding, West Am. Scientist, iii, 24, p. 97.)
1890, Townsend, Proc. U. S. Nat. Mus., p. 144.
(1893, Stejneger, N. A. Fauna, No. 7, p. 164.)

Specimens of this lizard from the "Cape Region" show a tendency to have but one row of scales between the rostral and nasal plates, while those from northern Lower California and California more frequently have two rows. The following table, based upon two hundred and thirty-one specimens shows this quite plainly:*

Number of scale rows between rostral and nasal	2-2	2-1	1-1
Northern specimens	19	9	13
Specimens from the "Cape Region"	14	6	170

This appears to be the only difference, and is not constant enough to warrant recognition by name.

The species has been recorded from Cape San Lucas (*Xantus*), La Paz (*Belding*), and San Luis Gonzales Bay (*Townsend*), in Lower California.

* Dr Leonhard Stejneger kindly furnished me notes on forty of these specimens, which are in the U. S. National Museum.

List of Specimens of Dipsosaurus dorsalis.

Cal. Acad. Sci. No.	Locality.	Date.	Collector.
467	San José del Cabo, L. C.	Mar., 1892.	Gustav Eisen.
468	"	"	"
613	"	Sept. 18, 1890	W. E. Bryant.
640	Magdalena Island, L. C.	Mar., 1889.	"
644	Comondu to San Quintin, L.C.	April, 1889.	"
710 to 713	San José del Cabo, L. C.	Sept. 1, 1890	"
716		Sept. 9, 1890	"
717		Sept. 1, 1890	"
827 to 829		Sept., 1893.	Gustav Eisen.
875	"	Aug., 1893.	"
1180	Miraflores, L. C.	Sept., 1894.	Eisen and Vaslit.
1181	"	"	"
1847 to 2011	San José del Cabo, L. C.	"	"
67	Lower California.	——	W. J. Fisher.

CROTAPHYTUS COPEII Yarrow?

Crotaphytus copeii.
1882, Yarrow, Proc. U. S. Nat. Mus., p. 441.
(1883, Yarrow, Bull. U. S. Nat. Mus., No. 24, p. 53.)
(1890, Stejneger, N. A. Fauna, No. 3, p. 105.)
Crotaphytus copii.
(1884, S. Garman, Bull. Essex Inst., xvi, 1, p. 16.)
Crotaphytus copei.
(1887, Cope, Bull. U. S. Nat. Mus., No. 32, p. 45.)
(1887, Belding, West Am. Scientist, iii, 24, p. 97.)

Description of No. 638.—The head is large and considerably depressed. The large nostril is much nearer to the end of the snout than to the orbit. The ear opening is very large, oblique, and with an anterior denticulation of small scales. The scales on the upper surface of the head are very small; largest on the middle third of its length, smallest on the supraocular region. There are twelve superior and twelve inferior labials to below the middle of the eye. All the labials are rectangular, of about equal width, and longer than wide. There are several series of enlarged sublabials, which pass gradually into the granular gulars. The strong gular fold is covered centrally with larger, pointed, imbricate, scales, but has very small granular ones at its edge. The back is covered with small, smooth, weakly pointed, subgranular scales. There is a strong fold along each side of the body between the limbs, and several irregular folds on the neck. The chest and belly are covered with smooth flat scales. The weakly keeled scales on the tail are smaller than the ventrals. There are twenty femoral pores on each side. Male, with enlarged postanals.

The general ground color is pale hair brown, changing to broccoli brown centrally, finely dotted with white and cream-buff, and with small spots of very dark sepia which increase slightly in size medially and posteriorly. There are two faint pale clay-colored cross-bars on the back above the hind limbs, and several similar ones on the basal portion of the tail. Between these bars are pairs of rather large dark sepia spots, each with a small cream-colored center. The tail has brown rings separated by narrower pale cream-colored ones. The chest and belly are whitish, more or less flecked with slate. The throat has longitudinal olive-gray bands and blotches on a pale cream-colored ground.

Snout to vent 119 mm. Tail 240 mm. Snout to edge of fold 40 mm. Snout to anterior edge of ear 28 mm. Fore limb 46 mm. Hind limb 89 mm.

This species has been known from a single specimen, secured at La Paz, by Mr. L. Belding.

List of Specimens of Crotaphytus copeii ?

Cal. Acad. Sci. No.	Locality.	Date.	Collector.
637	Magdalena Island.	Mar., 1889.	W. E. Bryant.
638	"	"	"

Callisaurus draconoides Blain.

Callisaurus draconoides.

" 1835, Blainville, Nouv. Ann. Mus., iv, p. 286, pl. xxiv, fig. 2."
1837, Duméril et Bibron, Erpétologie Générale, iv, p. 326.
(1845, Gray, Cat. Lizards Brit. Mus., p. 227.)
1874, Bocourt, Miss. Sci. au Mex., Reptiles, 3e livr., p. 158, pl. xvii bis., figs. 10–10b.
(1893, Stejneger, N. A. Fauna, No. 7, p. 171.)

Callisaurus ventralis.

1859, Baird, Proc. Ac. Nat. Sci. Phila., p. 299.

Callisaurus dracontoides.

(1866, Cope, Proc. Ac. Nat. Sci. Phila., p. 312.)
(1887, Cope, Bull. U. S. Nat. Mus., No. 32, p. 38.)

Callisaurus dracontoides dracontoides.

(1875, Cope, Bull. U. S. Nat. Mus., No. 1, pp. 47, 93.)
1883, Yarrow, Bull. U. S. Nat. Mus., No. 24, p. 50.
(1887, Belding, West Am. Scientist, iii, 24, p. 97.)

The head is broad and low. The snout is rounded when viewed from above, but sharply pointed when seen in profile. The nostrils are large and superior. There is a strongly marked canthus rostralis. The very large interparietal plate is broader than long. There are two or three series of enlarged supraoculars. The other head plates are small and irregular, largest on the frontal and prefontal regions, everywhere very flat and smooth. There is a very long suborbital. The eyelids are heavily fringed. The labials are low, but long, imbricate, and projecting laterally. The infralabials are bordered below by from one to three series of large sublabials. The gulars are granular, smooth, and, except on the central part of the region, longer than wide. The gular fold is covered with imbricate scales, largest at its edge. The back and sides are covered with small flattened granules, which pass gradually into the much-larger smooth ventrals. A strong fold extends along each side between the limbs. The tail is of medium length, and considerably depressed at its base. The limbs are very long and slender. The number of femoral pores varies from twelve to eighteen. The males have enlarged postanal plates.

Measurements of the largest specimens in mm.

Sex	♂	♂	♂	♂	♀	♀
Snout to vent	67	68	65	67	60	55
Tail	80	—	—	—	73	62
Head to ear	14	14	14	14	13	12
Width of head	11	12	12	11	11	10
Head to posterior edge of interparietal	12	12	13	13	11	11
Hind limb	62	65	61	62	52	50
Fore limb	36	37	33	35	31	29
Base of 5th to end of 4th toe	27	28	28	27	22	22

There is so much variation in color that no exact description can be given. The males are grayish above, tinted with primrose yellow and ochraceous buff on sides, and thickly spotted with pale yellow or white. Two series of brown blotches on the back are united on the upper surface of the tail to form undulate brown crossbands. The lower surface of the tail is white, crossed by from six to eight black bars which correspond in position to, and are often united with, the brown bands of its upper surface. The chin and throat are marked with numerous oblique dusky lines. The latter has a large half-concealed patch of red. A large blue or green area along each side is crossed by two almost vertical black blotches, behind which is a small round black spot. The limbs are crossed by bands of dusky. The females have the larger markings on the upper surfaces more distinct than the males, lack the lateral blue blotch and posterior black spot of the males, and have a large bright cadmium orange spot behind the axilla.

This species was first described from a specimen collected by Botta in "California." It was afterwards found by Mr. Xantus at Cape San Lucas, and by Mr. Belding at La Paz.

List of specimens of Callisaurus draconoides.

Cal. Acad. Sci. No.	Locality.	Date.	Collector.
605	San José del Cabo, L. C.	Sept. 3, 1890	W. E. Bryant.
729	"	Sept. 16, 1890	"
730		"	"
732 to 735		Sept. 1, 1890	"
739		Sept. 2, 1890	"
740		"	"
754		"	"
894		Sept., 1893	Gustav Eisen.
1400 to 1403	Sierra San Lazaro, L. C.	Sept., 1894	Eisen and Vaslit.
1750 to 1846	San José del Cabo, L. C.	"	"
2212 to 2237		1892	W. E. Bryant.
2195		Mar., 1892	"

Callisaurus ventralis (Hallow.)

Homalosaurus ventralis.
(1852, Hallowell, Proc. Ac. Nat. Sci. Phila., p. 179.)
Callisaurus dracontoides gabbii.
(1875, Cope, Bull. U. S. Nat. Mus., No. 1, p. 47.)
(1883, Yarrow, Bull. U. S. Nat. Mus., No. 24, p. 189.)

Callisaurus dracontoides.
1880, Lockington, Am. Nat., p. 295.
(? 1889, Cope, Proc. U. S. Nat. Mus., p. 147.)
1890, Townsend, Proc. U. S. Nat. Mus., p. 144.
Callisaurus ventralis.
(1893, Stejneger, N. A. Fauna, No. 7, p. 171.)

This more northern species is quite distinct from *C. draconoides* of the "Cape Region," and may be readily distinguished from it by the following characters:

C. ventralis.	*C. draconoides.*
Large.	Small.
Snout short and rounded.	Snout longer and less rounded.
Supralabials prominent and very convex in lateral outline.	Supralabials much less prominent and convex.
Males with two large oblique black blotches on each side.	Males with two smaller almost vertical black blotches, followed by a small black spot.

No intergradation of the two forms has yet been shown, but two young females from San Ignacio, and one from Santa Margarita Island, are more nearly like *C. draconoides* than are any of the other specimens of *C. ventralis*, suggesting, but not showing, an instability of character farther to the south.

The following measurements of *C. ventralis* are given for comparison with those of *C. draconoides:*

Sex	♂	♂	♂	♂	♀	♀
Snout to vent	86	88	81	82	72	74
Tail	117	—	—	107	102	98
Head to ear	16	16	15	16	15	14
Width of head	14	13	13	14	13	13
Head to posterior edge of interparietal	15	15	14	14	13	13
Hind limb	79	—	75	76	70	65
Fore limb	49	46	40	45	42	41
Base of fifth to end of fourth toe	35	—	32	33	31	30

This species has been recorded from San Luis Gonzales Bay, and Angel de la Guardia Island. It is common on the deserts of the southwestern United States, and of Sonora.

List of specimens of Callisaurus ventralis.

Cal. Acad. Sci. No.	Locality.	Date.	Collector.
624 to 627	El Llano de Santano, L. C.	April, 1889.	W. E. Bryant.
628	San Ignacio, L. C.	"	"
629	"	"	"
675	Santa Margarita Island.	Mar. 5, 1889	"

SAUROMALUS HISPIDUS Stejn.

Sauromalus ater.

1877, Streets, Bull. U. S. Nat. Mus., No. 7, p. 36.
?1883, Yarrow, Bull. U. S. Nat. Mus., No. 24, p. 51.
(1887, Cope, Bull. U. S. Nat. Mus., No. 32, p. 35.)
?1887, Belding, West Am. Scientist, iii, 24, pp. 96, 97.
(1890, Townsend, Proc. U. S. Nat. Mus., p. 144.)

Sauromalus hispidus.

1891, Stejneger, Proc. U. S. Nat. Mus., p. 409.
(1893, Stejneger, N. A. Fauna, No. 7, p. 174.)

This species has been found only on Angel de la Guardia Island, Gulf of California. Mr. Belding secured a Sauromalus on Espiritu Santo Island, but it is not known to what species it belongs.

UTA THALASSINA Cope.

Uta thalassina.

1863, Cope, Proc. Ac. Nat. Sci. Phila., p. 104.
(1864, Cope, Proc. Ac. Nat. Sci. Phila., p. 177.)
(1866, Cope, Proc. Ac. Nat. Sci. Phila., p. 312.)
(1875, Cope, Bull. U. S. Nat. Mus., No. 1, pp. 48, 93.)
1883, Yarrow, Bull. U. S. Nat. Mus., No. 24, p. 54.
(1884, S. Garman, Bull. Essex Inst., xvi, 1, p. 16.)
(1887, Cope, Bull. U. S. Nat. Mus., No. 32, p. 35.)
1887, Belding, West Am. Scientist, iii, 24, pp. 96, 98.
1894, Stejneger, Proc. U. S. Nat. Mus., pp. 589, 591

Petrosaurus thalassinus.

1885, Boulenger, Cat. Lizards Brit. Mus., ii, p. 205.

Description of No. 1472.—The head is flattened, swollen at temples, and with rounded snout. The nostrils are large, superior, and a little nearer to the end of the snout than to the orbit. The large ear opening has a very weak anterior denticulation. The head scales are smooth, and slightly convex anteriorly. The frontal is transversely divided. The largest supraoculars are separated from the frontals, frontoparietals, and parietals, by two series of small plates. The interparietal is very large. There are six superior and seven inferior labials to below the middle of the orbit. Several series of enlarged sublabials pass gradually into the granular gulars which are slightly largest centrally. The first of the two strong gular folds ends in a large pouch at each side. The second is continued as a flap in front of each arm, and is covered with smooth flat plates the largest of which, at its edge, are somewhat larger than the scales on the chest. The back and sides are covered with small smooth round granules, much larger medially than laterally. The tail, conical except where depressed at its base, is covered with whorls of weakly keeled scales, which are a little smaller than those of the belly. The scales of the chest and belly are smooth. Those on the limbs have distinct keels. Femoral pores 17–19.

Most of the large specimens have lost their original tails, and are now provided with regrowths. The number of femoral pores varies from fifteen to twenty-one.

There is considerable variation in the intensity, and some in the distribution, of color in the large series at hand, but the general pattern is the same in all the specimens. Very young specimens are fully as brightly colored as older ones, and females as brightly as males. The largest specimens, however, appear somewhat duller than others, especially on the posterior part of the back.

One of the brightest individuals, which has been in alcohol little more than a month, may be described thus: On the anterior half of the back are three transverse bands of intense black, bordered posteriorly by others of olivaceous yellow. The first of these bars connects the shoulders. The second is the shortest and narrowest. Near its anterior edge are two round yellow spots, about half the size of the tympanum. The third is the largest and best defined. It is bordered in front by a narrow band of plumbeous, which separates it from another of olivaceous yellow. The remaining space between these black bands is finely dotted and reticulated with black, sepia, and azure. The posterior half of the back is similarly banded, but the colors are here so dull as to appear as if viewed through a thick and discolored epidermis. In front of each shoulder is an azure spot about the size of the tympanum. Half-way between the upper edges of these spots and the tympana are smaller spots of the same color, and others may be seen on the dorsal median line of the neck. The chin and gular regions, except a large central patch of greenish olive (pale turquoise blue in some specimens), are Indian yellow, which color is continued over the sides, and faintly over the back of the neck, just in front of the first black dorsal band. The eyelids and a small area surrounding the pineal "eye" are also yellow. The hind limbs are pale sepia, with indications of seven faint yellowish crossbars. The upper surface of the tail is bluish, greenish, and brownish, crossed by twenty-one broad dark olive or greenish olive bars. The lower surfaces of the tail, limbs, abdomen, and chest, are creamy white, tinged on the chest with olive-green and indian yellow. (In very young individuals there are three transverse greenish bars on a yellow ground.) In the pouches at each end of the middle gular fold are patches of flame scarlet, but close examination shows that this color is due to the presence of multitudes of minute parasites.

Three specimens (smallest, medium, and largest) measure in millimeters as follows:

Total length	162	378	(tail reproduced.)
Snout to vent	52	130	154
Hind limb	38	90	95
Fore limb	26	63	67
Head to ear	14	32	36
Width of head	11	26	30

This beautiful species was originally described from specimens collected by Mr. Xantus at Cape San Lucas. Mr. Belding found it at Playitas, San Lazaro, and in the Victoria Mountains.

List of specimens of Uta thalassina.

Cal. Acad. Sci. No.	Locality.	Date.	Collector.
741	San José del Cabo, L. C.	Sept. 1, 1890	W. E. Bryant.
748	San Bartolome, L. C.	Oct., 1890.	"
751	"	"	"
974	Corral de Piedras, Sierra El Taste, L. C.	Sept., 1893	Gustav Eisen.
975	"	"	"
976	"		
1191 to 1236	Miraflores, L. C.	Sept., 1894	Eisen and Vaslit.
1452 to 1532	Sierra San Lazaro, L. C.		

Uta repens, new species. Plates vii. and viii figs. A–E.

Diagnosis.—Allied to *U. thalassina*, but with hind limb much shorter, snout shorter and more truncate, and four transverse black dorsal bars in place of the anterior three of that species.

Type.—Cal. Acad. Sci. No. 633, Comondu, Lower California, W. E. Bryant, April, 1889.

Description of the Type.—The head is broad, short, and depressed. The snout is short and truncate. The nostrils are large, superior, and much nearer to the end of the snout than to the orbit. The ear opening is large, and has an anterior denticulation of three pointed scales. The head scales are smooth, and slightly convex anteriorly. The rostral is very broad and low, with a median superior projection. The frontal is transversely divided. The largest supraoculars are separated from the frontals, frontoparietals, and parietals, by two series of small plates. The interparietal is very large. There are five superior and seven inferior labials to below the middle of the eye. There are several series of enlarged sublabials. The gular region is covered with small granules which are slightly largest centrally. There is a weak anterior gular fold followed by a strong posterior fold. The latter is covered with small subgranular plates, the largest of which, on its edge, are about equal in size to the first scales on the chest. The back and sides are covered with round granules, which are larger medially than laterally. The tail is somewhat depressed and expanded at its base, and is covered with whorls of small weakly keeled scales. The scales on the anterior surfaces of the limbs are large and weakly keeled. The ventral plates are larger than the caudals.

The color above is dull grayish olive, with four distinct anterior, and three fainter posterior, transverse black bands. The tail is similarly barred with dusky. The throat is brownish marked with blackish slate centrally. The chest and abdomen are white clouded with slate.

Measurements in mm. of	*Uta repens.*	*Uta thalassina.*	
Snout to vent	103	103	95
Snout to fold	34	35	33
Snout to orbit	7	9	8
Snout to ear	22	24	23
Snout to back of interparietal	18	20	19
Width of head	18	18	15
Fore limb	45	42	41
Hind limb	62	69	63
Base of fifth to end of fourth toe	22	27	25

This species is represented by a single specimen. Its general aspect is very much like that of *U. thalassina*, not at all like *U. mearnsi*.

UTA STANSBURIANA B. and G.

Uta stansburiana.

(1853, Baird & Girard, Stansbury's Report, p. 345, pl. v, figs. 1-6.)
1859, Baird, Proc. Ac. Nat. Sci. Phila., p. 299.
(1864, Cope, Proc. Ac. Nat. Sci. Phila., p. 177.)
(1866, Cope, Proc. Ac. Nat. Sci. Phila., p. 312.)
(1875, Cope, Bull. U. S. Nat. Mus., No. 1, p. 48.)
1877, Streets, Bull. U. S. Nat. Mus., No. 7, p. 37.
1880, Lockington, Am. Nat., p. 295.
(1885, Boulenger, Cat. Lizards Brit. Mus., ii, p. 211.)
(1887, Cope, Bull. U. S. Nat. Mus., No. 32, p. 35.)
(1887, Belding, West Am. Scientist, iii, 24, p. 98.)
(1889, Cope, Proc. U. S. Nat. Mus., 1889, p. 147.)

Uta elegans.

1882, Yarrow, Proc. U. S. Nat. Mus., p. 442.
(1883, Yarrow, Bull. U. S. Nat. Mus., No. 24, p. 55.)
(1885, Boulenger, Cat. Lizards Brit. Mus., ii, p. 211.)
(1887, Belding, West Am. Scientist, iii, 24, p. 98.)
1890, Townsend, Proc. U. S. Nat. Mus., p. 144.

Uta schotti.

[1883, Yarrow, Bull. U. S. Nat. Mus., No. 24, p. 55 (part.)]
(1887, Belding, West Am. Scientist, iii, 24, p. 98.)

An examination of large series has failed to reveal any

character, which will separate Lower Californian specimens of this form from specimens collected in California and Arizona. Dr. Yarrow gives no character which will separate them, and, in his Check List, refers specimens from the Cape Region to *U. stansburiana* and *U. elegans* indifferently.

Two specimens from Espiritu Santo Island do not seem to differ from those collected on the peninsula.

Uta stansburiana was first described from the valley of the Great Salt Lake, Utah. *Uta elegans* was established upon specimens collected by Mr. Belding at La Paz. Dr. Streets found the species on Cerros Island; and Mr. Townsend at San Luis Gonzales Bay, San Bartolome Bay, and on Carmen Island in the Gulf of California. Mr. Xantus found it at Cape San Lucas.

List of specimens of Uta stansburiana.

Cal. Acad. Sci. No.	Locality.	Date.	Collector.
438	Espiritu Santo Island.	April, 1892	W. E. Bryant.
439	"	"	"
487 to 491	San José del Cabo, L. C.	Mar.10, 1892	
581	Guadalupe to Colnett, L. C.	Apr.28, 1893	A. W. Anthony.
582	Valladares, L. C.	May 29, 1893	"
585	San Tomas to Guadalupe, L.C.	Apr.27, 1893	
588	"	"	
594	Guadalupe to Colnett, L. C.	Apr.28, 1893	
599	"	"	"
632	Comondu to San Quintin, L.C.	April, 1889	W. E. Bryant.
643	"	"	"
1700 to 1721	San José del Cabo, L. C.	Sept., 1894	Eisen and Vaslit.

UTA PALMERI Stejn.

Uta palmeri.

1890, Stejneger, N. A. Fauna, No. 3, p. 106.

Under this name, Dr. Stejneger has described a Uta from San Pietro Martir Island, Gulf of California. The species belongs to the *U. stansburiana* group.

UTA MICROSCUTATA Van D. Plates viii, figs. F, G, and ix.

Uta microscutata.

1894, Van Denburgh, Proc. Cal. Acad. Sci., Ser. ii, Vol. iv, Part 1, p. 298.

Two specimens from San José Island, in the Gulf of California, have been compared with the type from San Pedro Martir Mountain, Lower California. They differ only in having the bluish white dots on the ventral surfaces more numerous, and the femoral pores eleven and twelve respectively. One is somewhat larger than the type, being 119 mm. in total length. A third specimen, labeled Comondu to San Quintin, is also identical with the type.

List of specimens of Uta microscutata.

Cal. Acad. Sci. No.	Locality.	Date.	Collector.
433	San José Island, L. C.	April, 1892.	W. E. Bryant.
434	"	"	"
631	Comondu to San Quintin, L.C.	"	"

UTA NIGRICAUDA Cope.

Uta ornata.

1859, Baird, Proc. Ac. Nat. Sci. Phila., p. 299.

[1883, Yarrow, Bull. U. S. Nat. Mus., No. 24, p. 57 (part).]

(1887, Belding, West Am. Scientist, iii, 24, p. 98.)

Uta nigricauda.

1864, Cope, Proc. Ac. Nat. Sci. Phila., p. 176.

(1866, Cope, Proc. Ac. Nat. Sci., Phila., p. 312.)

(1875, Cope, Bull. U. S. Nat. Mus., No. 1, pp. 48, 93.)

1883, Yarrow, Bull. U. S. Nat. Mus., No. 24, p. 55.)
(1884, S. Garman, Bull. Essex Inst., xvi, 1, p. 16.)
1885, Boulenger, Cat. Lizards Brit. Mus., ii, p. 212.
(1887, Cope, Bull. U. S. Nat. Mus., No. 32, p. 35.)
(1887, Belding, West Am. Scientist, iii, 24, p. 98.)

There is a very great amount of variation in the size, shape, and number of the head plates, even the frontal being sometimes divided transversely. The largest dorsal scales are along the median line. They are replaced, sometimes gradually and sometimes abruptly, by granules on the sides. Seventeen to twenty-four of the largest dorsals are equal to the length of the head to the posterior edge of the interparietal (occipital) plate. The ventral scales are larger than the dorsals, and perfectly smooth. The caudals are the largest of all, and are very strongly keeled and mucronate. The number of femoral pores ranges from nine to thirteen. The color of the throats of the males varies from canary yellow to deep Chinese orange.

List of specimens of Uta nigricauda.

Cal. Acad. Sci. No.	Locality.	Date.	Collector.
621	Magdalena Island, L. C.	1888	W. E. Bryant.
622	"	"	"
673		Mar. 12, 1889	"
676 to 679		Mar. 11, 1889	"
724	San José del Cabo, L. C.	Sept. 6, 1890	"
737	"	Sept. 2, 1890	"
758		Sept. 3, 1890	"
856	"	Oct., 1893	Gustav Eisen.
986	Miraflores, L. C.	"	"
987	"	"	"
1227 to 1358		Sept., 1894	Eisen and Vaslit.
1390 to 1399	Sierra San Lazaro, L. C.	"	"
2012 to 2108	San José del Cabo, L. C.	"	"
2191		Mar., 1892	W. E. Bryant.
2211		"	"
2282 to 2373		"	"

Sceloporus zosteromus Cope.

Sceloporus zosteromus.

1863, Cope, Proc. Ac. Nat. Sci. Phila., p. 105.
(1866, Cope, Proc. Ac. Nat. Sci. Phila., p. 312.)
(1885, Cope, Proc. Am. Philos. Soc., xxii, pp. 395, 399.)
(1885, Boulenger, Cat. Lizards Brit. Mus., ii, p. 225.)
(1887, Cope, Bull. U. S. Nat. Mus., No. 32, p. 37.)
(1889, Cope, Proc. U. S. Nat. Mus., p. 147.)
1893, Stejneger, N. A. Fauna, No. 7, p. 178, pl. i, fig. 3.

Sceloporus clarkii zosteromus.

(1875, Cope, Bull. U. S. Nat. Mus., No. 1, pp. 49, 93.)
(1880, Lockington, Am. Nat. p. 295.)

(1883, Yarrow, Bull. U. S. Nat. Mus., No. 24, p. 64.)
(1884, S. Garman, Bull. Essex Inst., xvi, 1, p. 17.)
(1887, Belding, West Am. Scientist, iii, 24, p. 98.)

Sceloporus rufidorsum.

1882, Yarrow, Proc. U. S. Nat. Mus., p. 442.
(1883, Yarrow, Bull. U. S. Nat. Mus., No. 24, p. 64.)
(1887, Belding, West Am. Scientist, iii, 24, pp. 96, 98.)
(1890, Townsend, Proc. U. S. Nat. Mus., p. 144.)

Sceloporus consobrinus.

[1883, Yarrow, Bull. U. S. Nat. Mus., No. 24, p. 61 (part).]
(1887, Belding, West Am. Scientist, iii, 24, p. 98.)

Sceloporus clarki clarki.

[1883, Yarrow, Bull. U. S. Nat. Mus., No. 24, p. 63 (part).]
(1887, Belding, West Am. Scientist, iii, 24, p. 99.)

This very distinct species belongs to the *S. magister* group. Specimens from the northern part of the peninsula and from several of the neighboring islands seem to be like those from the "Cape Region" in all respects. The following table may be useful for comparison with other species:

Cal. Acad. Sci. No.	Sex.	No. of scales on back.	Scales equal shielded part of head.	Femoral pores.	Measurements in mm.					Locality.
					Snout to vent.	Tail.	Hind limb.	Fore limb.	Shielded part of head.	
630	♂	29	6	19	106	143	73	48	20	San Pablo.
666	♂	28	6	22	106	148	78	49	21	Magdalena Island.
665	♂	30	5	19	104	141	78	51	20	"
664	♀	29	5	18	95	121	65	44	18	Sta. Margarita Island.
663	♀	29	5	18	93	126	70	46	18	"
437	♂	30	5	19	100	135	71	48	19	San José Island.
1691	♂	30	5	20	93	—	66	42	18	San José del Cabo.
1690	♂	29	5	19	83	—	62	40	17	"
1688	♂	29	5	20	99	—	66	43	18	
1686	♀	30	6	18	74	107	57	38	15	

Sceloporus zosteromus is not confined to the "Cape Region" of Lower California. It has been recorded from Cape San Lucas (type locality), La Paz, Cerros Island, and San Quintin Bay (the type locality of S. *rufidorsum*). The present collection contains specimens from San José Island, San Pablo, Santa Margarita Island, Miraflores, Magdalena Island, and San José del Cabo.

List of specimens of Sceloporus zosteromus.

Cal. Acad. Sci. No.	Locality.	Date.	Collector.
436	San José Island, L. C.	April, 1892	W. E. Bryant.
437	"	May, 1892	"
539	San José del Cabo, L. C.	April, 1892	Gustav Eisen.
604	"	Sept. 3, 1890	W. E. Bryant.
608	"	"	"
609	"	"	"
630	San Pablo, L. C.	April, 1889	"
641	Magdalena Island ?	Mar., 1889	"
663	Santa Margarita Island.	Mar. 1, 1889	"
664	"	"	
665	Magdalena Island, L. C.	Mar. 11, 1889	"
666	"	"	"
667	"	"	"
668		Feb. 25, 1889	"
669		"	"
670		"	"
671		Mar. 11, 1889	"
672	"	"	"
707	San José del Cabo, L. C.	Sept. 13, 1890	"
708	"	"	"
714	"	Sept. 3, 1890	"
736		Sept. 5, 1890	"
738		Sept. 2, 1890	"
844 to 848		Sept., 1893	Gustav Eisen.
1146 to 1151	Miraflores, L. C.	Sept., 1894	Eisen and Vaslit.
1678 to 1699	San José del Cabo, L. C.	"	"
2192	"	Mar., 1892	W. E. Bryant.

SCELOPORUS LICKI new species. Plate x.

Diagnosis.—Allied to *S. consobrinus*, but much larger, with more strongly mucronate scales, with larger scales on the back, with much larger scales on the posterior surface of thigh, and never with two blue patches on throat.

Type.—Cal. Acad. Sci. No. 1436, Sierra San Lazaro, Lower California, Eisen and Vaslit, Sept., 1894.

Description of the Type.—The head is considerably depressed, with rounded snout. There are two scales on the canthus rostralis. The nostrils are large, almost superior, and nearer to the end of the snout than to the orbit. The ear opening is very large, almost vertical, and with a strong anterior denticulation of six pointed scales. The head shields are smooth and somewhat convex. The supraoculars are very broad. The superciliaries are very long, narrow, and strongly imbricate. There are two series of small, and one of large, sublabial plates, bordered below by the large, imbricate, bicuspid gulars. There is a strong fold on each side of the neck. The dorsal scales are slightly smaller than the caudals, strongly keeled, very strongly mucronate, and with serrate edges. The lateral scales are similar to, but smaller than, the dorsals, arranged in oblique series, and graduating into the dorsals and ventrals. The ventrals are much smaller than the dorsals, smooth, and bi- or tricuspid. The caudals are very strongly keeled and mucronate. The posterior surface of the thigh is covered with large, pointed, keeled scales. There are fifteen femoral pores. Male, with enlarged postanal plates. There are thirty-three dorsal scales between the interparietal plate and the base of the tail.

The back and sides are olive brown, many of the scales having central markings of deep blue or green. A narrow line of verdigris green runs along each side from the eye to the base of the tail. Below this, a narrower similarly colored line runs from the ear to a point a short distance above and behind the axilla. A patch in front of the shoulder, the central part of the belly, and the anterior and lower surfaces of the thigh, are black, which color gradually fades into the cyanine blue of the sides of the belly. The throat is olive gray with greenish white lines which converge to a point midway between the neck pouches. The tail is brown suffused with campanula blue and beryl green towards its base.

Snout to vent 74 mm. Fore limb 37 mm. Tail 105 mm. Shielded part of head, 15 mm. Hind limb 54 mm. Base of 5th to end of 4th toe 22 mm.

Variation.—There is very little variation in color, either individual, sexual, or in accordance with age. One male from Miraflores has a single large blue patch on the throat, through which the ordinarily whitish lines show as lines of paler blue.

The following table will serve to show the variation in structural characters:

Cal. Acad. Sci. No.	Sex.	Scales on back.	Scales equal shielded part of head.	No. of scales in ear denticulation.	Femoral pores.	Measurements in mm.					
						Snout to vent.	Tail.	Hind limb.	Fore limb.	Shielded part of head.	Base of 5th to end of 4th toe.
1436	♂	33	6	6	15	74	105	54	37	15	22
1419	♀	33	7	5	14	65	93	46	29	13	19
1409	♀	38	7	5	14	70	—	47	32	14	19
1410	♂	32	6	5	14	73	—	53	36	15	22
1425	♂	34	7	7	15	67	105	52	36	13	22
1411	♀	34	8	5	16	65	97	44	31	13	18
1435	♂	34	8	5	14	63	102	48	33	13	20
1422	♂	36	7	5	14	74	117	50	35	14	20
1412	♂	36	6	5	16	83	112	55	39	16	22
1426	♂	37	8	5	15	59	—	44	31	13	19
1439	♂	34	9	5	16	40	52	30	19	9	14
1433	♂	34	10	5	15	32	—	24	17	8	11
1157	♂	34	7	5	15	62	103	47	32	14	19
1152	♀	36	7	5	15	70	—	45	29	13	18
1155	♂	36	6	5	15	72	—	53	35	14	22
1160	♂	36	7	5	16	76	—	55	35	15	22
1159	♀	35	7	6	16	57	—	43	29	12	17

The next table shows the same characters of *Sceloporus consobrinus:*[*]

U. S. Nat. Mus. No.	Sex.	Scales on back.	Scales equal the shielded part of head.	Scales in ear denticulation.	Femoral pores.	Measurements in mm.						Locality.
						Snout to vent.	Tail.	Hind limb.	Fore limb.	Shielded part of head.	Base of 5th to end of 4th toe.	
15697	♂	39	8	5	18	62	—	48	31	13	20	Prescott, Ariz.
15696	♀	39	9	4	15	58	86	44	28	12	18	"
15695	♀	41	10	5	18	55	—	39	25	12	16	"
17235	♀	41	7	5	16	67	98	45	28	13	19	Tucson, Ariz.
17234	♂	38	8	5	17	54	91	43	28	12	18	"
16958	♂	39	8	5	17	65	95	45	27	13	18	"
16959	♂	—	8	5	16	—	103	47	28	13	20	"
16960	♀	38	9	5	14	46	—	32	21	10	14	"
2895	♂	—	9	5	14	48	62	30	20	10	12	Nebraska.
2895	♀	40	8	4	14	63	80	37	26	13	15	"
2895	♂	42	9	4	15	45	—	30	19	10	13	"
8163	♀	47	9	5	13	—	—	42	27	12	17	Utah.
8163	♂	43	9	4	13	61	—	43	28	12	18	"
8491	♀	45	10	5	18	63	—	41	27	13	17	New Mexico.
8491	♂	46	10	5	16	49	—	34	23	10	14	"

Comparison.—This species may be easily distinguished from *S. consobrinus* by its larger scales on the back of the thigh; from *S. biseriatus* by its larger scales on the border of the ear and the back of the thigh; from *S. orcutti* by its smaller and much rougher dorsals; and from *S. magister* and *S. zosteromus* by its smaller and more sharply mucronate scales. It differs from all these in coloration.

This species is named in honor of Mr. James Lick, who has done so much to foster Science in California.

[*] I am greatly indebted to Dr. Leonhard Stejneger for the opportunity to examine these specimens.

List of specimens of Sceloporus licki.

Cal. Acad. Sci. No.	Locality.	Date.	Collector.
731	San José del Cabo, L. C.	Sept. 16, 1890	W. E. Bryant.
979	Corral de Piedras, Sierra El Taste, L. C.	Sept., 1893	Gustav Eisen.
1162 to 1168	Miraflores, L. C.	Sept., 1894	Eisen and Vaslit.
1409 to 1413	Sierra San Lazaro, L. C.	"	
1415 to 1437		"	
1439		"	

Sceloporus biseriatus Hallow.

Sceloporus biseriatus.
(1854, Hallowell, Proc. Ac. Nat. Sci. Phila., p. 93.)

Specimens of this species from northern Lower California do not seem to differ from Californian ones.

List of specimens of Sceloporus biseriatus.

Cal. Acad. Sci. No.	Locality.	Date.	Collector.
583	San Pedro Martir Mt., L. C.	May 27, 1893	A. W. Anthony.
589	Valladares, L. C.	May 29, 1893	"
593	San Pedro Martir Mt., L. C.		
595	"	May, 1893	"

Sceloporus graciosus B. & G.

Sceloporus graciosus.
(1852, Baird and Girard, Proc. Ac. Nat. Sci. Phila., p. 69.)

Mr. Anthony has collected a number of lizards of this species on San Pedro Martir Mountain in the northern part of the peninsula.

List of specimens of Sceloporus graciosus.

Cal. Acad. Sci. No.	Locality.	Date.	Collector.
584	San Pedro Martir Mt., L. C.	May, 1893.	A. W. Anthony.
586	"	May 12, 1893	"
587		May, 1893.	"
590		May 11, 1893	"
591			
592			

PHRYNOSOMA SOLARE Gray.

Phrynosoma solaris.

(1845, Gray, Cat. Lizards Brit. Mus., p. 229.)

(1894, Van Denburgh, Proc. Cal. Acad. Sci., Ser. 2, Vol. iv, Pt. 1, p. 456.)

Phrynosoma regale.

1880, Lockington, Am. Nat., p. 295.

The Academy has one specimen (No. 90) from Las Animas Bay, Lower California, the locality from which Mr. Lockington has recorded the species.

PHRYNOSOMA CORONATUM Blainv.

Phrynosoma coronatum.

" 1835, Blainville, Nouv. Ann. Mus., iv, p. 284, pl. xxv, figs. 1-1c."

1837, Duméril et Bibron, Erpétologie Générale, iv, p. 318.

1870, Bocourt, Miss. Sci. au Mex., Reptiles, 1e livr., pl. xii, fig. 10.

1874, Bocourt, Miss. Sci. au Mex., Reptiles, 4e livr., p. 239 (part).

(1866, Cope, Proc. Ac. Nat. Sci. Phila., p. 312.)

(1875, Cope, Bull. U. S. Nat. Mus., No. 1, pp. 50, 93.)

[1883, Yarrow, Bull. U. S. Nat. Mus., No. 24, p. 70 (part).]

(1887, Cope, Bull. U. S. Nat. Mus., No. 32, p. 39).

1893, Stejneger, N. A. Fauna, No. 7, p. 187.

1894, Van Denburgh, Proc. Cal. Acad. Sci., Ser. 2, Vol. iv, Pt. 1, p. 296.

Phrynosoma ———

1859, Baird, Proc. Ac. Nat. Sci. Phila., p. 299.

Phrynosoma cornutum.
1883, Yarrow, Bull. U. S. Nat. Mus., No. 24, pp. 66, 67, (part).
(1887, Belding, West Am. Scientist, iii, 24, p. 98.)
1893, Stejneger, N. A. Fauna, No. 7, pl. ii, figs. 1-1c.
Phrynosoma asio.
(1883, Yarrow, Bull. U. S. Nat. Mus., No. 24, p. 67.)
[1885, Boulenger, Cat. Lizards, Brit. Mus., ii, p. 244 (part).]
(1887, Belding, West Am. Scientist, iii, 24, p. 98.)
Phrynosoma hernandezi.
[1883, Yarrow, Bull. U. S. Nat. Mus., No. 24, p. 68 (part).]

The nostrils are pierced in the lines joining the superciliary ridges with the end of the snout. There are several longitudinal series of large pointed gular scales, the exterior of which are continued back upon the gular folds. There is a series of five very large pointed sublabial plates. The head spines are very large. They are four temporals, one occipital and one postorbital, on each side. and one large interoccipital. Occasionally small spines are developed between the temporals. Below the rictus is a broad spine usually without any, but sometimes with a very small. spine behind it. There is a row of four or five spinose scales in front of the occipital spines. The other head scales. with few exceptions, are flat and rugose. usually with irregular ridges radiating from near the center of each scale. There are two groups of spines on each side of the neck, the lower larger. The tail is bordered with a single row of lateral spines. and bears a group of smaller ones behind the insertion of the thigh. There are two series of pepiphero-abdominal spines: the lower shorter than the upper, and formed of smaller spines. The scales on the chest are sometimes faintly keeled. Those on the abdomen and basal part of tail are smooth: on the terminal part of tail, keeled. The tympanum is naked. There are from sixteen to twenty-two femoro-preanal pores. The males have enlarged postanal plates. The tails of the females are shorter than the distance from the axilla to the front of the thigh, but those of the males are considerably longer than this distance. The young of both sexes have short tails. The color above is brownish, yellowish or grayish, darker laterally. There is a large brown patch on each side of the neck and a series of three more or less distinct brown bars on each side of the back. These bars are light bordered posteriorly. The tail is transversely banded with brown. The belly is often dotted or blotched with black or brown. All these markings are more distinct in the young. The larger dorsal tubercles are often tipped with orange-rufous, and those on each side of the median line have seal brown or black keels. The occipital spines are ribbed with very dark brown. The temporals are yellow tinged with rufous. In very young individuals the scales of the vertex are grayish or yellowish white, with a few minute brown or black spots. These spots, which are on the raised portions of the scales, become more numerous as the animals increase in size, until the whole crown appears black or dark brown crossed by irregular lines formed by the yellow posterior edges of the scales.

Phrynosoma coronatum was first described from a specimen collected by Botta in "California." It has since been recorded from Cape San Lucas, and La Paz. The specimens enumerated below show that it ranges far north of the limits of the "Cape Region."

List of specimens of Phrynosoma coronatum.

Cal. Acad. Sci. No.	Locality.	Date.	Collector.
645 to 649	Comondu to San Quintin, L. C.	April, 1889	W. E. Bryant.
659	Poso Grande, L. C.	Mar. 18, 1889	"
660	"	Mar. 20, 1889	"
719	San José del Cabo, L. C.	Sept. 7, 1890	"
720	"	"	"
721		"	"
722		Sept. 6, 1890	"
723		"	"
862		Oct., 1893	Gustav Eisen.
863		"	"
864		"	
901		"	
902		"	
903	"	Jan. 25, 1893	
904	San Francisquito, Sierra Laguna, L. C.	Mar. 28, 1892	
905	San José del Cabo, L. C.	Nov., 1893	
906	"	"	
907			
908		..	
909 to 924		Sept., 1893	
925		April, 1892	
926	"	"	"
1169	Miraflores, L. C.	Sept., 1894	Eisen and Vaslit.
1630 to 1662	San José del Cabo, L. C.	"	

PHRYNOSOMA BLAINVILLII Gray.

Phrynosoma blainvillii.

(1839, Gray, Zoölogy Beechey's Voyage, p. 96, pl. xxix, fig. 1.)

1894, Van Denburgh, Proc. Cal. Acad. Sci., Ser. 2. Vol. iv, pt. 1, p. 296.

Phrynosoma coronatum.

1880, Lockington, Am. Nat., p. 295.

This species has been found as far south as San Tomas, in the northern part of the peninsula.

A key to the North American species of the *Ph. coronatum* group * is given.

I.—A long spine just behind the broad subrictal. Head plates of adults yellow, sparsely dotted with brown.

A.—Head plates convex and almost smooth. *Ph. blainvillii* Gray.

B.—Head plates flat and rugose. *Ph. frontale* Van D.

II.—No spine, or a very small one, behind the broad subrictal. Head plates, of adults, chiefly black or dark brown.

Ph. coronatum Blainv.

The present collection contains a single specimen of this species. It (No. 579) was secured by Mr. A. W. Anthony, at Valladares, L. C.

PHRYNOSOMA CERROENSE Stejn.

Phrynosoma cerroense.

1893, Stejneger, N. A. Fauna, No. 7, p. 187.

This species, from Cerros Island, is characterized by having the lower row of peripheral spines "only indicated by a few scattered small spines."

GERRHONOTUS MULTICARINATUS Blainv.

Gerrhonotus multicarinatus.

" 1835, Blainville, Nouv. Ann. Mus., iv, p. 289, pl. xxv, fig. 2."

1839, Duméril et Bibron, Erpétologie Générale, v, p. 404.

1866, Cope, Proc. Ac. Nat. Sci. Phila., p. 312.

[1875, Cope, Bull. U. S. Nat. Mus., No. 1, p. 46 (part).]

1878, Bocourt, Miss. Sci. au Mex., Reptiles, 5e livr., p. 357, pl. xxi e, fig. 5–5a.

1883, Yarrow, Bull. U. S. Nat. Mus., No. 24, p. 47 (part).

[1887, Cope, Bull. U. S. Nat. Mus., No. 32, p. 41 (part).]

(1887, Belding, West Am. Scientist, iii, 24, p. 97).

1893, Stejneger, N. A. Fauna, No. 7, p. 195.

Without larger series of the other species than are at hand, the status of the *Gerrhonoti* from the "Cape Region" of Lower California cannot be satisfactorily determined.

* Not including the insular *Ph. cerroense*.

It seems probable, however, that they are distinct from the more northern *G. scincicauda*, and are referable to the name *G. multicarinatus* Blainv.

The type of *G. multicarinatus* is one of the specimens collected by Botta in "California." Mr. Belding found the species at La Paz.

List of specimens of Gerrhonotus multicarinatus.

Cal. Acad. Sci. No.	Locality.	Date.	Collector.
536	San José del Cabo, L. C.		Gustav Eisen.
869	"	Oct., 1893	"
874		Aug., 1893	"
884	"	July, 1893	"
980	{ Corral de Piedras, Sierra El Taste, L. C. }	Sept., 1893	"
981	"	"	"
982	"	"	"
983	Sierra El Taste, L. C.	"	"
984	"	"	"
1387	Sierra San Lazaro, L. C.	Sept., 1894	Eisen and Vaslit.
1388	"	"	"
1619	San José del Cabo, L. C.	"	"
2196	"	Mar., 1892	W. E. Bryant.
2197	"	"	"
2203	Miraflores, L. C.	Oct., 1893	Gustav Eisen.
2251 to 2255	Sierra Laguna, L. C.	"	"

GERRHONOTUS SCINCICAUDA (Skilt.)

Tropidolepis scincicauda.

(1849, Skilton, Am. Journ. Sci. Arts., Ser. 2, Vol. vii, pp. 202, 312 plate, figs. 1-3.)

There are two specimens of this species in the collection made by Mr. Anthony in northern Lower California.

List of specimens of Gerrhonotus scincicauda.

Cal. Acad. Sci. No.	Locality.	Date.	Collector.
580	San Pedro Martir Mt., L. C.	May 4, 1893	A. W. Anthony.
596	Valladares?, L. C.	May 29, 1893	"

XANTUSIA GILBERTI new species. Plate xi.

Diagnosis.—Similar to *X. vigilis*, but with smaller eye, two plates occupying the position of the frontal of that species, and the interfrontonasal larger and continued posteriorly to completely separate the frontonasals.

Type.—Cal. Acad. Sci. No. 401, San Francisquito, Sierra Laguna, Lower California, Gustav Eisen, March 28, 1892.

Description of the Type.—The eye is very small, without lids, and with vertical pupil. The nostrils are pierced at the junction of the rostral, internasal, first superior labial, and first loreal plates. There are three loreals, increasing in size posteriorly. There are two internasals. The two frontonasals are separated by the interfrontonasal, which is in contact, also, with the two frontal plates. The other head plates are two frontoparietals, two parietals, two large occipitals, and one interparietal. The eye is surrounded by a ring of small scales, of which the superciliaries are largest. This ring is separated from the third loreal by two small scales. There are eight superior and eight inferior labials. The anterior border of the ear is slightly denticulate. The ventral plates are arranged in thirty-two transverse and ten or twelve longitudinal series. The caudal scales are smooth, convex, and in whorls of about equal length. The back and sides are covered with smooth convex granules of about uniform size. There are eight and nine femoral pores. The gular regions are covered with smooth, flattened, subhexagonal granules which are slightly larger than those on the back and sides.

The color above is dark brownish clay, dotted with black on single granules. A pale yellowish line, two granules wide, runs posteriorly from each occipital plate, but is soon lost on the back to reappear over the thigh.

Length to vent (about) 39 mm. Tail (about) 38 mm. Hind limb 14 mm. Fore limb 10 mm. Diameter of eye 1+ mm. Shielded part of head 8½ mm. Head to posterior edge of ear 8½ mm. Head to anterior gular fold 7¼ mm. Head to posterior edge of anterior fold 11¼ mm. Head to posterior edge of posterior fold 12¾ mm.

The single specimen of *X. gilberti* has been compared with one hundred and forty-four of *X. vigilis* without any approach to its distinctive characters having been found. It is of great interest, for it extends the known range of the genus *Xantusia* several hundred miles to the southward, introducing it for the first time into Mexican territory, and affording another link between the "Cape Region" and the Sonoran Subprovince.

It gives me great pleasure to name this interesting lizard in honor of Dr. Charles H. Gilbert, to whom my interest in herpetology is entirely due.

CNEMIDOPHORUS MAXIMUS Cope.

Cnemidophorus maximus.

1863, Cope, Proc. Ac. Nat. Sci. Phila, p. 104.
(1866, Cope, Proc. Ac. Nat. Sci. Phila., p. 312.)
(1875, Cope, Bull. U. S. Nat. Mus., No. 1, pp. 45, 93.)
(1880, Lockington, Am. Nat., xiv, 4, p. 295.)
1883, Yarrow, Bull. U. S. Nat. Mus., No. 24, pp. 42, 188.
(1884, S. Garman, Bull. Essex Inst., xvi, 1, p. 13.)
(1885, Boulenger, Cat. Lizards Brit. Mus., ii, p. 369.)
(1887, Cope, Bull. U. S. Nat. Mus., No. 32, p. 45.)
(1887, Belding, West Am. Scientist, iii, 24, p. 97.)
1892, Cope, Trans. Am. Philos. Soc., xvii, 1, p. 32.

Description of No. 835.—The nostrils are pierced in the large anterior nasal plates, which are in contact on top of the snout. The posterior nasal forms sutures with the anterior nasal, first second and third labials, loreal, prefrontal, and frontonasal plates. The loreal is in contact with the third and fourth labials, first subocular, preocular, first superciliary, first supraocular, prefrontal, and anterior nasal. There are four supraoculars; the first, long and narrow, the fourth rather small. The second, third, and fourth supraoculars are separated from the superciliaries, and the third and fourth from the frontoparietal and parietal, by small convex granules. There are two transverse rows of small plates behind the parietals and interparietal. There are five superior and six inferior labials to below the middle of the eye. The sublabials are very large and are separated from the infralabials by a series of small granules and plates. The anterior gulars are large centrally, become gradually smaller laterally, and are abruptly separated from the medium sized posterior gulars. The central scales of the collar are quite large, those on its edge, smaller. The back is covered with small uniform granules. There are eight longitudinal and thirty-seven transverse rows of ventral plates, and four series of large preanals. There are seven rows of brachials, three of antebrachials, seven of femorals, and four of tibials, but no postantebrachials. The tail is covered with whorls of obliquely keeled scales. There are twenty-three and twenty-four femoral pores.

The color above is grayish sepia fading to olive gray laterally, with three longitudinal dark chestnut bands on each side, which (bands) are twice as wide as the intervals between them, and are so invaded by spots of the ground color, as to resemble series of confluent brown maculations. The limbs are reticulated with coarse chestnut lines. The upper surface of the head is olive, palest on the snout. The gular region and the sides of

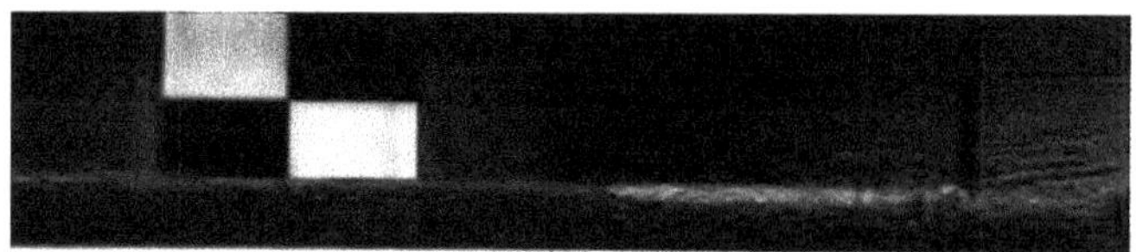

the head are blotched with walnut brown. Many of the ventral plates have black basal markings. The tail is tawny olive, tinged and spotted with dark chestnut.

Snout to vent 131 mm. Tail (regrown) 302 mm. Hind limb 87 mm. Fore limb 48 mm. Snout to ear 30 mm. Greatest width of head 19 mm. Snout to edge of collar 41 mm.

Young.—Young individuals have five bluish white longitudinal lines on a black ground which is more or less broken by spots of the same color as the lines. Their tails and hind limbs are suffused with bright flesh color. In one there are six instead of five pale lines.

Variation.—The femoral pores vary in number from twenty to twenty-six. The scales of the collar are sometimes largest at its edge. The general ground color is at times quite gray, and the dark markings often more or less obsolete, particularly on the anterior part of the body. The number of plates on the limbs is very variable.

List of specimens of Cnemidophorus maximus.

Cal. Acad. Sci. No.	Locality.	Date.	Collector.
108	Lower California.		W. J. Fisher.
705	San José del Cabo, L. C.	Sept. 1, 1890	W. E. Bryant.
706	"	"	"
715		Sept. 9, 1890	"
742		Sept., 1890	"
834 to 843		Sept., 1893	Gustav Eisen.
1096 to 1119	Miraflores, L. C.	Sept., 1894	Eisen and Vaslit.
1171 to 1179		"	"
1438	Sierra San Lazaro, L. C.	"	"
1440 to 1451		"	"
1670 to 1677	San José del Cabo, L. C.	"	"
2194		March, 1892	W. E. Bryant.
2198		"	Gustav Eisen.
2193		"	W. E. Bryant.
2241	"	"	"

CNEMIDOPHORUS MARTYRIS Stejn.

Cnemidophorus martyris.

1891, Stejneger, Proc. U. S. Nat. Mus., p. 407.

(1892, Cope, Trans. Am. Philos. Soc., xvii, 1, p. 36.)

This species has been described by Dr. Stejneger from two specimens taken on San Pedro Martir Island, Gulf of California.

It is most closely allied to *C. melanostethus* Cope, but has the blackish suffusion extended over the entire lower surface.

CNEMIDOPHORUS MULTISCUTATUS (Cope).

Cnemidophorus tessellatus melanostethus.
[?1883, Yarrow, Bull. U. S. Nat. Mus., No. 24, p. 45 (part).]
?1887, Belding, West Am. Sci., iii, 24, p. 99.
Cnemidophorus tessellatus tigris.
(1889, Cope, Proc. U. S. Nat. Mus., p. 147.)
Cnemidophorus tessellatus multiscutatus.
1892, Cope, Trans. Am. Philos. Soc., xvii, 1, p. 38.

Prof. Cope has proposed this name for specimens from Cerros Island, characterized by the large number of brachial (7 to 8 rows) and femoral (8 to 9 rows) plates.

CNEMIDOPHORUS STEJNEGERI Van D.

Cnemidophorus tessellatus tessellatus.
?1880, Lockington, Am. Nat., xiv, 4, p. 295.
(?1889, Cope, Proc. U. S. Nat. Mus., p. 147.)
[?1892, Cope, Trans. Am. Philos. Soc., xvii, 1, p. 34 (part).]
Cnemidophorus stejnegeri.
1894, Van Denburgh, Proc. Cal. Acad. Sci., Ser. 2, Vol. iv, Pt. 1, p. 300.

There are two specimens of this species in the present collection. One (No. 642) was obtained by Mr. Bryant and labeled Comondu to San Quintin, L. C. The other (No. 597) was secured by Mr. Anthony and is labeled Lower California.

This form has already been recorded from San Telmo, the "foothills of San Pedro Martir Mountain," and "between San Rafael and Ensenada," in Lower California.

CNEMIDOPHORUS RUBIDUS (Cope).

Cnemidophorus tessellatus rubidus.
1892, Cope, Trans. Am. Philos. Soc., xvii, 1, p. 27, pl. xii, f.

Description of No. 661.—The nostrils are anterior to the nasal suture. There are three parietals, two fronto-parietals, and four supraoculars. The postnasal is in contact with the first, second, and third labials. The loreal is very large, longer than high. There are six superior and five inferior labial plates to below the middle of the orbit. The infralabials are separated from the large sublabials by a series of granules. The gular scales are rather large centrally, and the posterior part of the region is not very distinct from the anterior. The scales on the collar are of medium size, largest centrally, small at edge, and in about seven transverse rows. There are no large postantebrachials. The caudal scales are large and provided with prominent diagonal keels. There are twenty femoral pores.

The color above is brownish olive, paler on the sides, overlaid with tawny olive posteriorly, and crossed by narrow transverse black bands. The posterior six of these bands extend entirely across the back, but the others are interrupted, forming a dorsal series of black spots, with corresponding vertical bars upon the sides. The black markings on the neck are reduced to six longitudinal series of more or less obsolete spots. The posterior limbs are faintly reticulated with black, and illuminated with numerous white spots above and posteriorly. The inferior surfaces of the limbs are deep flesh color, with a slightly purplish tinge. This color appears, also, on the gular region, about the ears, and on the lower surface of the tail. There are no large or distinct markings on the gular region.

Snout to vent 100 mm. Snout to ear 23 mm. Fore limb 36 mm. Width of head 17 mm. Hind limb 69 mm. Snout to edge of color 33 mm.

This name was established upon seven specimens from Santa Margarita Island. The present collection contains one from Magdalena Island, and two fine adults from Comondu on the peninsula, thus greatly increasing the known range of the species. The Magdalena specimen is much smaller than those from Comondu, and differs from them in the presence of small, well defined black blotches on the gular region. Professor Cope is not folfowed in his use of a trinomial because no intergradation of this with other forms has been shown.

List of specimens of Cnemidophorus rubidus.

Cal. Acad. Sci. No.	Locality.	Date.	Collector.
661	Comoudu, L. C.	Mar. 22, 1889	W. E. Bryant.
662	"	"	"
674	Magdalena Island, L. C.	Mar., 1889	"

CNEMIDOPHORUS LABIALIS Stejn.

Cnemidophorus labialis.
1889, Stejneger, Proc. U. S. Nat. Mus., p. 643.
(1892, Cope, Trans. Am. Philos. Soc., xvii, 1, p. 51.)

This species has been found only on Cerros Island.

VERTICARIA HYPERYTHRA Cope.

Cnemidophorus hyperythrus.
1863, Cope, Proc. Ac. Nat. Sci. Phila., p. 103.
(1866, Cope, Proc. Ac. Nat. Sci. Phila., p. 312.)
(1885, Boulenger, Cat. Lizards Brit. Mus., ii, p. 371.)
Verticaria hyperythra.
" 1869, Cope, Proc. Am. Philos. Soc., xi, p. 158."
(1875, Cope, Bull. U. S. Nat. Mus., No. 1, pp. 46, 93.)
1883, Yarrow, Bull. U. S. Nat. Mus., No. 24, p. 45.
(1887, Cope, Bull. U. S. Nat. Mus., No. 32, 45.)
(1887, Belding, West Am. Scientist, iii, 24, p. 97.)
1894, Stejneger, Proc. U. S. Nat. Mus., p. 17.
Cnemidophorus hyperethra.
(1884, S. Garman, Bull. Essex Inst., xvi, 1, p. 13.)

Description of No. 1567.—The nostrils are in the large anterior nasal plates which meet on top of the snout. The posterior nasal forms sutures with the anterior nasal, first and second labials, loreal, prefrontal, and frontonasal plates. The loreal is in contact with the second, third and fourth labials, first subocular, preocular, first superciliary, first supraocular, prefrontal, and posterior nasal, plates. There are four supraoculars; the first is in contact with the first superciliary, loreal, prefrontal, frontal and second supraocular; the second touches the frontal; the third forms sutures with the frontal and frontoparietal; the fourth is separated from the parietal by a series of granules. The frontoparietal is more than half as large as the frontal. There is a transverse row of small occipital plates. The sublabials are separated from the infralabials by granules. There are five superior and five inferior labials to below the middle of the eye. The ear opening is not denticulated. The anterior gulars are quite large, and abruptly separated from the small posterior granules. The scales on the collar are very large, largest on its edge. The ventral plates are in thirty transverse, and eight longitudinal rows. The back and sides are covered with small equal sized granules. The tail is somewhat flattened at its base, and is covered with whorls of diagonally keeled scales. The lower caudals are smooth. The hind limb is as long as the distance between the anus and the front of the collar. There are fourteen and fifteen femoral pores.

The plates on the head are pale olive. There are two narrow longitudinal wood brown lines on the back, separated by an area of sepia. The sides are dark olive with two bluish white longitudinal lines, The upper, of these lateral lines, arises on the superciliary plates and is continued for some distance on the tail. The lower originates on the posterior nasal plate, and ends on the thigh. A light stripe on the back of the thigh is continued along the tail. The first and half of the second longitudinal rows of ventral plates are grayish pale blue. The entire lower surface, except of the hind limbs, is reddish orange-crome.

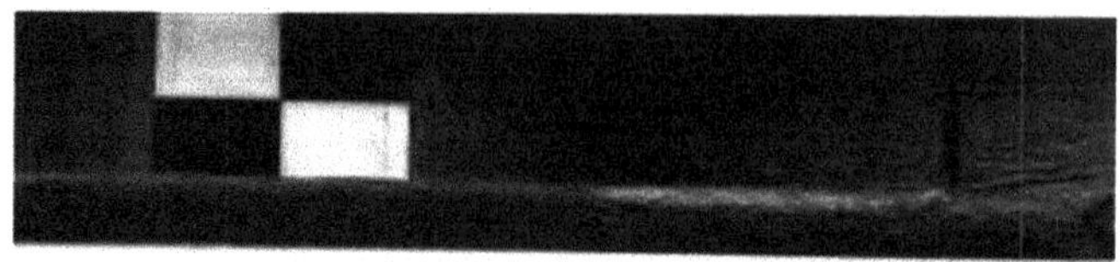

Snout to vent 61 mm. Snout to ear 14 mm. Hind limb 42 mm. Anus to gular fold 41 mm. Fore limb 19 mm. Anus to anterior gulars 47 mm. Width of head 9 mm.

The types of this species were collected at Cape San Lucas, by Mr. John Xantus. Mr. Belding secured others at La Paz.

List of specimens of Verticaria hyperythra.

Cal. Acad. Sci. No.	Locality.	Date.	Collector.
451	San José del Cabo, L. C.	April, 1892	Gustav Eisen.
452	"	"	"
534		Sept. 20, 1890	W. E. Bryant.
606		Sept. 3, 1890	"
607		"	
611		Sept. 1, 1890	
725		Sept. 6, 1890	
726		Sept. 1, 1890	
727		"	"
728		Sept. 16, 1890	
743		Sept., 1890	"
879		Sept., 1893	Gustav Eisen.
1120 to 1145	Miraflores, L. C.	Sept., 1894	Eisen and Vaslit.
1567	San José del Cabo, L. C.	"	"
1568	"		
1722 to 1749		"	"

VERTICARIA HYPERYTHRA BELDINGI (Stejn.)

Verticaria beldingi.
1894, Stejneger, Proc. U. S. Nat. Mus., p. 17.

Ninety-eight Verticarias from the "Cape Region" of Lower California, and thirty-eight (including one of the specimens upon which *V. beldingi* was established) from northern Lower California and San Diego County, California, have been examined with a view to determining the status of this form. The character originally depended upon for the distinction of *V. beldingi* from *V. hyperythra*, viz., the small size of the scales on the collar in *V. beldingi*, was found to be valueless, since many of the northern specimens have these scales as large as in individuals collected near Cape San Lucas, and since much individual variation exists in both. There appears to be not even an average difference, in this respect, between the northern and southern forms. The difference in the extent to which granules intrude between the supraoculars and the large medial head plates, seems, however, to present a good average distinction between the two forms, as is shown in the following table:

	Number of Specimens of	
	hyperythra.	*beldingi.*
Second supraocular separated from median head scales	3	20
Second supraocular partly separated	5	15
Third supraocular separated	8	0
Third supraocular partly separated	82	3
Total number examined	98	38

As this difference is merely an average one, it becomes necessary to regard *V. beldingi* as a subspecies of *V. hyperythra*. A trinomial is therefore used.

The type of this form came from Cerros Island.

List of specimens of Verticaria hyperythra beldingi.

Cal. Acad. Sci. No.	Locality.	Date.	Collector.
639	Magdalena Island, L. C.	Mar., 1889	W. E. Bryant.
680	"	Mar. 11, 1889	"
681		"	"

VERTICARIA SERICEA, new species. Plate xii.

Diagnosis.—Hind limb relatively much longer, and scales on collar, especially on its edge, much smaller, than in *V. hyperythra* and *V. hyperythra beldingi.* A single median dorsal line, as light distinct and well defined as the lateral ones, instead of two faint brownish lines as in *V. hyperythra* and *V. hyperythra beldingi.*

Type.—Cal. Acad. Sci. No. 435, San José Island, Gulf of California, Walter E. Bryant, April, 1892.

Description of the Type.—The nostrils are in the large anterior nasal plates, which meet on top of the snout. The posterior nasal forms sutures with the anterior nasal, first and second labials, loreal, prefrontal, and frontonasal plates. The loreal is in contact with the second third and fourth labials, first subocular, preocular, first superciliary, first supraocular, prefrontal, and posterior nasal. There are three supraoculars, the first is in contact with the first and second superciliaries, loreal, prefrontal, frontal, and second supraocular; the second is in contact with the frontal: the third is separated from the frontal and the frontoparietal by a series of granules. The interparietal is very narrow. There is a series of occipital plates. There are five superior and six inferior labials to below the middle of the eye. The ear opening is not denticulated. The sublabials are separated from the infralabials by granules. The anterior gulars are rather large, and abruptly separated from the small posterior gulars. The scales on the collar are very small, largest centrally, smaller on edge. The ventral plates are arranged in eight longitudinal and thirty transverse rows. The back is covered with small equal-sized granules. The conical tail is provided with scales arranged in whorls. The upper caudals have strong diagonal keels, but the lower are smooth. There are sixteen femoral pores. The hind limb is longer than the distance between the anus and the line of separation of the anterior and posterior gulars.

The back is clove brown, dotted with gray on single granules posteriorly, with a median bluish white line which bifurcates on the neck about a fourth of an inch behind the occipital plates. There are two similar lines on each side; the first originating on the superciliaries and with a faint continuation on the tail; the second starting at the nostril and ending on the thigh. The ground color of the sides is much paler than in *V. hyperythra*, being pale sepia. The general tint of the tail is hair brown above, pale blue below. The ventral and sublabial plates, the chin, gular region, and collar, are all pale blue.

Length to anus 54 mm. Hind limb 44 mm. Fore limb 22 mm. Head to ear 13 mm. Anus to gular fold 36 mm. Anus to anterior gulars 42 mm. Width of head 8 mm.

The single specimen of *Verticaria sericea* has been compared with ninety-eight of *Verticaria hyperythra* and thirty-eight of *Verticaria hyperythra beldingi*, without any approach to its distinctive characters having been found.

EUMECES LAGUNENSIS, new species. Plate xiii.

Eumeces skiltonianus.

?1883, Yarrow, Bull. U. S. Nat. Mus., No. 24, p. 41.

(?1887, Belding, West Am. Scientist, iii, 24, p. 99.)

Diagnosis.—Similar to *E. skiltonianus*, but tail salmon color instead of blue, and with interparietal smaller than either frontoparietal instead of larger.

Type.—Cal. Acad. Sci. No. 400, San Francisquito, Sierra Laguna, Gustav Eisen, March 28, 1892.

Description of the Type.—The nasal is small, in contact with the internasal, postnasal, first labial, and rostral plates. The postnasal touches the nasal, internasal, anterior loreal, and the first and second labials. The anterior loreal forms sutures with the postnasal, internasal, frontonasal, prefrontal, second loreal, and second and third labials. The three anterior of the four supraoculars are in contact with the frontal. The interparietal is smaller than either of the frontoparietals. The parietals are in contact posteriorly. The last of the seven labials is largest. There are two azygos postmentals. The limbs overlap when pressed against the body. There are twenty-four longitudinal rows of scales. The dorsal scales are larger than the laterals and ventrals. There is a median series of transversely enlarged subcaudals, on each side of which the other caudals become gradually smaller dorsally.

The ground color above and on the sides is dark olive. There are two bluish gray lines on each side. The upper of these lines originates on the internasal plate, crosses the anterior loreal, prefrontal, supraocular, and parietal plates, and runs along the dorsal scales (second and third rows from the median line) to the tail. The lower traverses the labial plates, crosses the ear opening, and runs along the side of the neck and body to the hind limb, forming the lower boundary of the olive ground color. The lower labials, chin, throat, chest, preanal region, the lower surfaces of the limbs, and the proximal half of the tail, are dull pinkish buff. The belly and a faint bar across the throat, are bluish gray. The tail is salmon or bright flesh color, marked, except on its terminal fourth, with three narrow poorly defined lines of slaty heliotrope, in continuation of the olive ground color of the back.

Snout to vent 52 mm. Tail (about) 95 mm. Hind limb 18 mm. Fore limb 14 mm. Head to posterior edge of ear 10 mm.

List of specimens of Eumeces lagunensis.

Cal. Acad. Sci. No.	Locality.	Date.	Collector.
400	San Francisquito, Sierra Laguna, L. C.	Mar. 28, 1892	Gustav Eisen.
402	"	"	"

EUCHIROTES BIPORUS Cope.

Chirotes canaliculatus.
1877, Streets, Bull. U. S. Nat. Mus., No. 7, p. 37.
(1883, Yarrow, Bull. U. S. Nat. Mus., No. 24, p. 38.)
1887, Cope, Bull. U. S. Nat. Mus., No. 32, p. 47.
(1887, Belding, West Am. Scientist, iii, 24, p. 97.)

Chirotes sp.?
(1880, Lockington, Am. Nat., p. 295.)

Euchirotes biporus.
1894, Cope, Am. Nat., p. 437, figs. 5–5e.

The snout is short, rounded and very convex. The limbs are very broad and short, with five perfect clawed digits. The larger head plates are a rostral, three labials, a nasal, an ocular, a preocular, two suboculars, one supraocular, a very large prefrontal, and a pair of frontals. There are also two small plates between the third labial and the suboculars. The anus is preceded by several rows of granules, in front of which is a transverse series of six large plates. There is a single preanal pore in a large plate in front of the external preanal plate of each side.

Total length 199 mm. Limb 8 mm. Tail 18 mm. Head 7 mm.

The Academy has a single specimen (No. 128) from La Paz. The type came from Cape San Lucas.

RENA HUMILIS B. and G.

Rena humilis.

(1853, Baird and Girard, Cat. N. A. Reptiles, i, Serpents, p. 143.)
(1887, Cope, Bull. U. S. Nat. Mus., No. 32, p. 64.)
(1891, Stejneger, Proc. U. S. Nat. Mus., 1891, p. 501.)
(1892, Cope, Proc. U. S. Nat. Mus., 1891, p. 590.)

Stenostoma humile.

1861, Cope, Proc. Ac. Nat. Sci. Phila., p. 305.
1882, Yarrow, Bull. U. S. Nat. Mus., No. 24, p. 142.
(1887, Belding, West Am. Scientist, iii, No. 24, p. 98.)

Glauconia humilis.

(1893, Boulenger, Cat. Snakes Brit. Mus., i, p. 70.)

The twenty-three Lower Californian specimens of this curious little reptile in the Academy's collection are all from the "Cape Region." They show that the species lives both in the mountains, and at the level of the sea.

There is very little variation in color. The lower parts are creamy white, the upper (five to seven rows of scales) Prout's brown. The smallest individual is 91 mm. long, while the largest measures 305 mm., of which the tail forms 12 mm.

The type locality is Valliecitas (Colorado Desert), California. The species has been taken in Lower California, at Cape San Lucas, by Mr. Xantus, and at La Paz, by Mr. Belding.

List of specimens of Rena humilis.

Cal. Acad. Sci. No.	Locality.	Date.	Collector.
447 to 450	San José del Cabo, L. C.		Gustav Eisen.
469		March, 1892	"
817 to 821		Sept., 1893	"
823 to 826		"	"
880		May, 1893	
881		"	"
882		"	"
883		July, 1893	"
1547		Sept., 1894	Eisen and Vaslit.
1629		"	"
822	"	Sept., 1893	Gustav Eisen.
2200	San Francisquito, Sierra Laguna, L. C.	Mar. 28, 1892	"
2201	"	"	"

LICHANURA TRIVIRGATA Cope.

Lichanura trivirgata.

1861, Cope, Proc. Ac. Nat. Sci. Phila., p. 304.
" 1865, Jan, Iconogr. génér. Oph., 2e livr., pp. 69, 70."
(1875, Cope, Bull. U. S. Nat. Mus., No. 1, pp. 43, 93.)
(1882, Bocourt, Miss. Sci. au Mex., Reptiles, 8e livr., p. 514.)
1883, Yarrow, Bull. U. S. Nat. Mus., No. 24, p. 142.
(1887, Belding, West. Am. Scientist, iii, No. 24, p. 98.)
(1887, Cope, Bull. U. S. Nat. Mus., No. 32, p. 65.)
1889, Stejneger, Proc. U. S. Nat. Mus., p. 98, fig. 3.
1891, Stejneger, Proc. U. S. Nat. Mus., pp. 512, 514, 515.
(1892, Cope, Proc. U. S. Nat. Mus., 1891, p. 591.)
[1893, Boulenger, Cat. Snakes, Brit. Mus., i, p. 129 (part).]

Charina trivirgata.

(1883, S. Garman, Mem. Mus. Comp. Zool., viii, 3, pp. 8, 131.)
(1884, S. Garman, Bull. Essex Inst., xvi, 1, p. 22.)

The Academy's collectors have failed to find this snake, which seems to be a very distinct species.

The types were collected, by Mr. Xantus, in swamps among the mountains near Cape San Lucas. Mr. Belding obtained an individual near La Paz.

LICHANURA ROSEOFUSCA Cope.

Lichanura roseofusca.

1868, Cope, Proc. Ac. Nat. Sci. Phila., p. 2.
(1875, Cope, Bull. U. S. Nat. Mus., No. 1, p. 43.)
(1887, Cope, Bull. U. S. Nat. Mus., No. 32, p. 65.)
(1889, Stejneger, Proc. U. S. Nat. Mus., pp. 94, 97, 98.)
1891, Stejneger, Proc. U. S. Nat. Mus., pp. 512–515.)
(1891, Cope, Proc. U. S. Nat. Mus., p. 591.)

Lichanura myriolepis.

1868, Cope, Proc. Ac. Nat. Sci. Phila., p. 2.
(1875, Cope, Bull. U. S. Nat. Mus., No. 1, p. 43.)
1887, Cope, Bull. U. S. Nat. Mus., No. 32, p. 65.)
(1889, Stejneger, Proc. U. S. Nat. Mus., pp. 94, 97, 98.)
1891, Stejneger, Proc. U. S. Nat. Mus., pp. 512–515.

The type of this species was collected in "northern Lower California," by Wm. M. Gabb.

CHILOMENISCUS STRAMINEUS Cope.

Chilomeniscus stramineus.

1860, Cope, Proc. Ac. Nat. Sci. Phila., p. 339.
(1861, Cope, Proc. Ac. Nat. Sci. Phila., p. 302.)
(1875, Cope, Bull. U. S. Nat. Mus., No. 1, pp. 35, 92.
1883, Yarrow, Bull. U. S. Nat. Mus., No. 24, pp. 13, 86.
(1887, Cope, Bull. U. S., Nat. Mus., No. 32, p. 81.)
(1887, Belding, West Am. Scientist, iii, No. 24, p. 98.)
(1892, Cope, Proc. U. S. Nat. Mus., 1891, p. 594.)
[1894, Boulenger, Cat. Snakes, Brit. Mus., ii, p. 573 (part).]

Carphophis straminea.

(1883, S. Garman, Mem. Mus. Compr. Zoöl. Cambr., viii, 3, pp. 166, 99.)
(1884, S. Garman, Bull. Essex Inst., xvi, 1, p. 32.)

The specimens enumerated below agree with Prof. Cope's original description, except in the number of temporal plates. These are 1–1, as stated by Cope in his Critical Review of the Characters and Variations of the Snakes of North America.

The ground color of the upper surfaces of the adult specimens varies from brownish drab to bright yellowish cinnamon. It is cream buff in a younger individual. The dark dots near the tips of the scales are present in all the specimens, but do not appear upon the first row of scales. They are rarely present upon the scales of the second row, but constantly upon those of the third. The first, second, and half of the third rows of scales are yellowish white or straw color, as are also the gastrosteges.

This beautiful little snake was first described from specimens collected at Cape San Lucas by Mr. Xantus. It was afterwards found by Mr. Belding, at La Paz.

List of specimens of Chilomeniscus stramineus.

Cal. Acad. Sci. No.	Locality.	Date.	Collector.
453	San José del Cabo, L. C.	March, 1892	Gustav Eisen.
814	"	Sept., 1893	"
815		"	"
816		"	"
877	"	May, 1893	"
990	Miraflores, L. C.	Oct., 1893	"
1170	"	Sept., 1894	Eisen and Vaslit.
2199		Oct., 1893	Gustav Eisen.

Chilomeniscus fasciatus (Cope).

Chilomeniscus cinctus.

1883, Yarrow, Bull. U. S. Nat. Mus., No. 24, p. 86.

(1887, Belding, West. Am. Scientist, iii, 24, p. 98.)

Chilomeniscus stramineus fasciatus.

1892, Cope, Proc. U. S. Nat. Mus., 1891, p. 595.

Chilomeniscus stramineus.

[1894, Boulenger, Cat. Snakes Brit. Mus., ii, p. 273 (part).]

This species is known only from two specimens collected by Mr. Belding at La Paz, in 1882.

As no intergradation with other forms has been shown, Professor Cope is not followed in the use of a trinomial.

TANTILLA PLANICEPS (Blainv.)

Coluber planiceps.
" 1835, Blainville, Nouv. Ann. Mus., iv, p. 294, pl. 27, figs. 3–3b.
(1853, Baird and Girard, Cat. N. A. Reptiles, i, Serpents, p. 154.)
Homalocranion planiceps.
1854, Duméril et Bibron, Erpétologie Générale, vii, p. 857.
" 1863, Jan, Elenco sist. degli Ofidi, p. 40."
" 1866, Jan and Sordelli, Iconogr. génér. des Ophid., 15e livr., pl. ii, 2."
1883, Bocourt, Miss. Sci. au Mex., p. 581, pl. xxxvi, fig. 7–7d.
Tantilla planiceps.
(1875, Cope, Bull. U. S. Nat. Mus., No. 1, p. 35.)
" 1875, Cope, Journ. Ac. Nat. Sci. Phila., p. 143."
(1883, Yarrow, Bull. U. S. Nat. Mus., No. 24. pp. 13, 190.)
(1883, S. Garman, Mem. Mus. Compr. Zoöl. Cambr., viii, 3, pp. 89, 163.)
(1884, S. Garman, Bull. Essex Inst., xvi, 1, p. 31.)
(1887, Cope, Bull. U. S. Nat. Mus., No. 32, p. 84.)
(1892, Cope, Proc. U. S. Nat. Mus., 1891, p. 598.)

It is interesting to be able to assign a definite habitat to this long lost species, which has been known from a single specimen collected by Botta, in "California," early in the second quarter of our century.

There seems to be no doubt that the specimens before me are referable to this name, although they have, with one exception, two postocular plates.

The head is very flat, and the snout considerably prolonged beyond the lower jaw. The rostral is somewhat recurved on top of the snout. Behind it are two small internasals, followed by two prefrontals of about twice the size of the internasals. The large frontal presents six edges, but is, in the main, triangular. The parietals are very large, and much broader anteriorly than posteriorly. The nostril is pierced between two nasal plates, which are united above, but distinct below the nostril. One pre- and two postoculars on each side. The parietal is separated from the labials by two longitudinally placed temporals. There are seven superior labials (the third and fourth entering the orbit) and six infralabials (the first pair in contact on the middle line). The anal plate is divided.

There is little variation in color. A specimen from San José del Cabo may be described thus: The top of the head, the temporal regions, and the first five transverse rows of scales on the neck are brown, changing gradually from hair brown, on the snout, to deep clove brown posteriorly. On the sixth and seventh rows of scales of the neck is a whitish collar about as wide as the length of one scale. The rest of the upper surface is bright broccoli brown, slightly vinaceous on the tail. The posterior three-fourths of the ventral surface are tinged with coral red, brightest immediately in front of the anus. The anterior fourth of the ventral surface is pale grayish clay color, but may have been red in life, as this color has entirely disappeared from all parts of the belly in other specimens.

The specimens mentioned in the following table are all from San José del Cabo except the first, which was secured in the Sierra de la Laguna.

Postoculars.	Scale rows.	Urosteges.	Gastrosteges	Length of tail in mm.	Total length in mm.
2-2	15	57	139	57	224
2-2	15	57	138	34	155
2-2	15	49	139	23	123
2-2	15	55	139	67	260
2-2	15	58	140	64	251
1-1	15	55	139	63	251

No. 446 has the frontal plate partially united with the prefrontals. The first pair of infralabials are separated below in No. 2208.

List of specimens of Tantilla planiceps.

Cal. Acad. Sci. No.	Locality.	Date.	Collector.
440 to 445	San José del Cabo, L. C.	1892	W. E. Bryant.
446	"		Gustav Eisen.
537	Sierra Laguna, L. C.	Mar. 27, 1892	"
996	San José del Cabo, L. C.	May, 1893	
2208	Lower California.		

RHINOCHILUS LECONTEI B. & G.

Rhinochilus lecontei.

(1853, Baird and Girard, Cat. N. A. Reptiles, i, Serpents, p. 120.)
1880, Lockington, Am. Nat., p. 295.

Mr. Lockington has recorded this species as having been collected by Mr. W. J. Fisher, "at or to the south of Magdalena Bay," Lower California.

The type came from San Diego, Cal.

LAMPROPELTIS CONJUNCTA (Cope).

Lampropeltis boylii.

1860, Cope, Proc. Ac. Nat. Sci. Phila., p. 255.

Lampropeltis boylii var. *conjuncta.*

1861, Cope, Proc. Ac. Nat. Sci. Phila., p. 301.

Ophibolus getulus conjunctus.

(1875, Cope, Bull. U. S. Nat. Mus., No. 1, pp. 37, 92.)
1878, Yarrow and Henshaw, U. S. G. G. Surv. W. 100th Mer., Appendix N N, p. 212.
(1887, Cope, Bull. U. S. Nat. Mus., No. 32, p. 78.)

Ophibolus getulus boyli.

1883, Yarrow, Bull. U. S. Nat. Mus., No. 24, p. 92 (part).
(1887, Belding, West Am. Scientist, iii, 24, p. 98.)

Young specimens of this form are not distinguishable from those of *L. boylii.* However, all the larger individuals from Lower California differ from those collected in California, in having the scales of the white rings marked basally with black or dark brown. This black edging seems to appear first upon those scales which are nearest the median dorsal line, and to extend to the lateral ones and over more and more of the surface of each scale, as the animal increases in size.

The type was taken by John Xantus near Cape San Lucas. Mr. Belding found the species at La Paz.

List of specimens of Lampropeltis conjuncta.

Cal. Acad. Sci. No.	Locality.	Date.	Collector.
618	San José del Cabo, L. C.		W. E. Bryant.
801 to 809		Sept., 1893	Gustav Eisen.
865		Oct., 1893	"
1560		Sept., 1894	Eisen and Vaslit.
1561		"	"
1562		"	"

LAMPROPELTIS NITIDA new species. Plate xiv.

Diagnosis.—Allied to *L. californiæ*, but with the gastrosteges, urosteges, and upper surfaces of head and snout, entirely brownish black.

Type.—Cal. Acad. Sci. No. 800, San José del Cabo, Lower California, Gustav Eisen, September, 1893.

Description of the Type.—The head is slightly distinct, considerably depressed, its plates normal; one loreal; one pre- and two postoculars; scales in twenty-three rows, smooth, with two apical pits; postgeneials very small; anal entire; seven superior labials, the third and fourth entering the orbit; two hundred and twenty-seven gastrosteges; fifty-six pairs of urosteges.

The back and sides are blackish brown; the former, with a rather indistinct longitudinal line composed of cinnamon colored spots upon the centers of the scales of the median series, and upon the inner edges of those forming the first row on each side of this series; the latter, with a few scales of the first and second rows dotted, centrally, with cinnamon or yellowish white. A band of cinnamon crosses the nape. The gulars, geneials, and inferior labials, are blackish brown with paler centers. The plates on the top and sides of the head are brownish black, with faintly indicated dots of raw umber upon the loreal, pre- and postocular plates, and near the posterior edges of the supraoculars and parietals. There are six cinnamon colored blotches on the upper surface of the tail. The gastrosteges and urosteges are entirely brownish black, with the exception of the first ten gastrosteges, which show faint cinnamon colored dots.

Total length 965 mm. Tail 125 mm.

A small specimen (290 mm.) has, on the sides, rather numerous cinnamon colored blotches or enlargements of a similarly colored longitudinal line. This line is of about the width of one row of scales, and occupies the tips of the gastrosteges and the lower half of each scale of the first series.

List of specimens of Lampropeltis nitida.

Cal. Acad. Sci. No.	Locality.	Date.	Collector.
800	San José del Cabo, L. C.	Sept., 1893	Gustav Eisen.
1533	" "	Sept., 1894	Eisen and Vaslit.

HYPSIGLENA OCHRORHYNCHA Cope.

Hypsiglena ochrorhynchus.

1860, Cope, Proc. Ac. Nat. Sci. Phila., p. 246.

(1894, Boulenger, Cat. Snakes Brit. Mus., ii, p. 209 [part].)

Hypsiglena ochrorhyncha.

(1875, Cope, Bull. U. S. Nat. Mus., No. 1, pp. 38, 92.)
(1880, Lockington, Am. Nat., xiv, p. 295.)
1883, Yarrow, Bull. U. S. Nat. Mus., No. 24, pp. 15, 97.
(1883, Garman, Mem. Mus. Compr. Zoöl. Cambr., viii, 3, pp. 80, 161.)
(1884, Garman, Bull. Essex Inst., xvi, 1, p. 30.)
(1887, Cope, Bull. U. S. Nat. Mus., No. 32, p. 78.)
(1887, Belding. West Am. Scientist, iii, 24, p. 98.)
(1892, Cope, Proc. U. S. Nat. Mus., 1891, p. 617.)

The present specimens all agree in having "pseudo-preoculars," flat heads, and dark postocular stripes covering less than half of the sixth supralabial plates—the characters said to distinguish *H. ochrorhyncha* from *H. texana* Stejneger, and *H. chlorophæa* Cope.

The total length of the largest specimen is 525 mm.

This species was originally described from specimens collected by Mr. Xantus at Cape San Lucas. Mr. Belding obtained others at La Paz, in 1882. It is not confined to Lower California.

List of specimens of Hypsiglena ochrorhyncha.

Cal. Acad. Sci. No.	Locality.	Date.	Collector.
757	San José del Cabo, L. C.	Oct. 7, 1890	W. E. Bryant.
811	"	Sept., 1893	Gustav Eisen.
868	"	Oct., 1893	"
1407	Sierra San Lazaro, L. C.	Sept., 1894	Eisen and Vaslit.
1408	"	"	"
1548	San José del Cabo, L. C.	"	"
2202	San Francisquito, Sierra Laguna, L. C.	Mar. 28, 1892	Gustav Eisen.

Phyllorhynchus decurtatus (Cope).

Phimothyra decurtata.

1868, Cope, Proc. Ac. Nat. Sci. Phila., p. 310.
(1875, Cope, Bull. U. S. Nat. Mus., No. 1, pp. 38, 92.)
1883, Yarrow, Bull. U. S. Nat. Mus., No. 24, pp. 99, 191.
(1887, Belding, West Am. Scientist, iii, 24, p. 98.)

Salvadora decurtata.
(1883, Garman, Mem. Mus. Compr. Zoöl. Camb., viii, 3, pp. 39, 145.)
(1884, Garman, Bull. Essex Inst., xvi, p. 25.)
(1887, Cope, Bull. U. S. Nat. Mus., No. 32, p. 72.)
(1888, Bocourt, Miss. Sci. au Mex., Reptiles, 11e livr., p. 663.)
Phyllorhynchus decurtatus.
1890, Stejneger, Proc. U. S. Nat. Mus., p. 154.
(1892, Cope, Proc. U. S. Nat. Mus., 1891, p. 618.)
Lytorhynchus decurtatus.
(1893, Boulenger, Cat. Snakes Brit. Mus., i, p. 417.)

This peculiar snake has not been found by any of the California Academy's expeditions to Lower California.

The type was collected, by Mr. Wm. M. Gabb, in the "upper part of Lower California." Mr. L. Belding obtained a second specimen at La Paz.

SALVADORA GRAHAMIÆ B. and G.

Salvadora grahamiæ.
(1853, Baird and Girard, Cat. N. A. Reptiles, 1, Serpents, p. 104.)
(1887, Cope, Bull. U. S. Nat. Mus., No. 32, p. 72.)
(1892, Cope, Proc. U. S. Nat. Mus., 1891, p. 619.)
Phimothyra grahamiæ.
1861, Cope, Proc. Ac. Nat. Sci. Phila., p. 300.
(1875, Yarrow, U. S. G. G. Surv. W. 100th Mer., v, p. 538.)
(1875, Coues, U. S. G. G. Surv. W. 100th Mer., v, p. 620.)
(1875, Cope, Bull. U. S. Nat. Mus., No. 1, p. 38.)
Phimothyra grahami.
1883, Yarrow, Bull. U. S. Nat. Mus., No. 24, pp. 15, 98.
(1887, Belding, West. Am. Scientist, iii, 24, p. 98.)

Each of the specimens in the Academy's collection has nine upper labials, the number originally stated by Prof. Baird.* Two have a single small subocular plate on each side, as in the type of *S. grahamiæ hexalepis* and in one of the specimens referred to that name by Dr. Stejneger in his Annotated List of the Reptiles and Batrachians Collected by the Death Valley Expedition.

* Cope gives eight as the number in his key to the species of this genus (Proc. U. S. Nat. Mus., 1891, p. 619).

Another individual has this plate present on one side only, thus agreeing with the specimen from St. Thomas, Nevada, recorded by Dr. Stejneger (l. c., p. 206), who also mentions and figures specimens in which a second subocular is present. In view of these facts, it appears that this character is not constant enough to warrant the retention of the name *hexalepis*.

The largest individual measures 940 mm. in total length.

This species was first described from a specimen collected by Col. J. D. Graham, in "Sonora, Mex." Subsequently, Mr. Xantus found it at Cape San Lucas, and Mr. Belding, at La Paz.

List of specimens of Salvadora grahamiæ.

Cal. Acad. Sci. No.	Locality.	Date.	Collector.
652	Comondu, L. C.	Mar. 3, 1889	W. E. Bryant.
760	Agua Caliente, L. C.	Oct., 1890.	"
761	"	"	"
762	San José del Cabo, L. C.	Oct. 7, 1890	"
812	"	Sept., 1893	Gustav Eisen.
813		"	"
876	"	Aug., 1893	"
1406	Sierra San Lazaro, L. C.	Sept., 1894	Eisen and Vaslit.
1546	San José del Cabo, L. C.	"	"

BASCANION FLAGELLUM FRENATUM Stejn.

Bascanium flagelliforme testaceum.
1875, Cope, Bull. U. S. Nat. Mus., No. 1, p. 40.
1883, Yarrow, Bull. U. S. Nat. Mus., No. 24, p. 112.
(1887, Belding, West Am. Scientist, iii, No. 24, p. 89.)
(1887, Cope, Bull. U. S. Nat. Mus., No. 32, p. 71.)

Bascanium flagelliforme.
1891, Cope, Proc. U. S. Nat. Mus., p. 626 (part).

Bascanion flagellum frenatum.
1893, Stejneger, N. A. Fauna, No. 7, p. 208.

The specimens enumerated below have been compared

with a number from Southern California,* and found not appreciably different, except that the Lower Californian snakes seem to have the black markings at the bases of the dorsal and lateral scales often larger and more numerous. Even this character is, however, rather inconstant, and there seems, therefore, to be no reason for making the separation tentatively suggested by Dr. Stejneger.†

The type came from Mountain Spring, Colorado Desert, San Diego Co., California. Mr. Xantus found the species at Cape San Lucas. The three specimens collected by Mr. Belding at La Paz, and recorded by Dr. Yarrow under the name *B. flagelliforme testaceum*, doubtless belong here.

List of specimens of Bascanion flagellum frenatum.

Cal. Acad. Sci. No.	Locality.	Date.	Collector.
483	San José del Cabo, L. C.	April 4, 1892	Gustav Eisen.
484	"	April, 1892	W. E. Bryant.
535		Sept. 30, 1890	"
602		Oct. 6, 1890	"
610	"	Sept. 4, 1890	"
658	Lower California.	1889	"
793 to 799	San José del Cabo, L. C.	Sept., 1893	Gustav Eisen.
849		"	"
866		Oct., 1893	"
1571 to 1577		Sept., 1894	Eisen and Vaslit.
1618	"	"	"
1183	Miraflores, L. C.	"	"
1184	"	"	"

* One of these specimens, from Yosemite Valley, Cal., has a single anal plate.

† N. A. Fauna, No. 7, 1893, p. 208.

BASCANION AURIGULUM Cope.

Drymobius aurigulus.

1861, Cope, Proc. Ac. Nat. Sci. Phila., p. 301.

Bascanium aurigulum.

(1875, Cope, Bull. U. S. Nat. Mus., No. 1, pp. 40, 92.)

(1883, Yarrow, Bull. U. S. Nat. Mus., No. 24, pp. 113, 191.)

(1887, Cope, Bull. U. S. Nat. Mus., No. 32, p. 71.)

(1887, Belding, West. Am. Scientist, iii, 24, p. 99.)

Coluber flagelliformis var. *aurigulus.*

(1883, S. Garman, Mem. Mus. Compr. Zoöl. Cambr., viii, 3, pp. 44, 148.)

(1884, S. Garman, Bull. Essex Inst., xvi, 1, p. 26.)

Bascanium laterale auriqulum.

(1892, Cope, Proc. U. S., Nat. Mus., 1891, p. 629.)

The apparent rarity of this species may be due to ignorance of its habits. The various expeditions sent by the Academy to Lower California have secured one specimen. It agrees perfectly with Cope's original description of the only other known representative of the species, secured by Mr. Xantus at Cape San Lucas.

Although this species is undoubtedly closely related to *B. laterale*, no intergradation has been shown, and there seems to be, therefore, no reason for using a trinomial appellation in this connection.

The present specimen (Cal. Acad. Sci. No. 870) is 1.045 m. in length, of which the tail forms 348 mm. It was collected by Gustav Eisen, at San José del Cabo, in November, 1893.

PITUOPHIS CATENIFER DESERTICOLA Stejn.

Pityophis sayi bellona.

1877, Streets, Bull. U. S. Nat. Mus., No. 7, p. 40.

(1883, Yarrow, Bull. U. S. Nat. Mus., No. 24, p. 106.)

Pituophis catenifer deserticola.

1893, Stejneger, N. A. Fauna, No. 7, p. 206.

Dr. Streets has recorded a specimen of this snake from San Martin Island, off the Pacific coast of Lower California. It is a locality where *P. catenifer* might rather be expected to occur.

PITUOPHIS VERTEBRALIS (Blainv.)

Coluber vertebralis.

1835, Blainville, Nouv. Ann. Mus., iv, p. 293, pl. 27, fig. 2-2b.

(1853, Baird and Girard, Cat. N. A. Reptiles, 1, Serpents, p. 152.)

Pituophis vertebralis.

1854, Duméril et Bibron, Erpétologie Générale, vii, p. 238.

1888, Bocourt, Miss. Sci. au Mex., Reptiles, 11e livr., p. 672, pl. xlvii, figs. 1-1d.

Pityophis hæmatois.

1860, Cope, Proc. Ac. Nat. Sci. Phila., p. 342.

Pityophis melanoleucus vertebralis.

"1863, Jan, Elenco, sist. degli Ofidi, p. 59."

" , Jan, Iconogr. génér. Oph., 22e livr., pl. 1, fig. 3."

Pityophis vertebralis.

(1875, Cope, Bull. U. S. Nat. Mus., No. 1, pp. 39, 92.)

1883, Yarrow, Bull. U. S. Nat. Mus., No. 24, p. 107.

(1884, Garman, Bull. Essex Inst. xvi, p. 27.)

(1887, Cope, Bull. U. S. Nat. Mus., No. 32, p. 72.)

(1887, Belding, West Am. Scientist, iii, 24, p. 98.)

(1892, Cope, Proc. U. S. Nat. Mus., 1891, p. 642.)

Pityophis catenifer.

(1883, Garman, Mem. Mus. Compr. Zoöl. Cambr., viii, 3, pp. 52, 150 [part].)

Professor Cope's description (Proc. Ac. Phila., 1860, p. 342) gives a good idea of this species. Many of the scales of the red dorsal blotches have blackish centers, varying in extent in different specimens. The anterior urosteges are frequently undivided. The following table shows the variability of the scale characters. The specimens mentioned in it are all from San José del Cabo.

Scale rows.	Smooth scale rows.	Prefrontals.	Preoculars.	Postoculars.	Supralabials.	Infralabials.	Gastrosteges.	Urosteges
35	12	4	2-2	3-3	8-9	13-13	239	64
35	8-10	4	2-2	3-3	9-9	12-12	243	67
33	13	4	2-2	3-3	9-9	12-12	234	65
34	11	4	2-2	3-3	9-9	13-13	251	60
35	10	4	2-2	3-3	9-10	13-13	245	61
35	10-14	4	2-2	3-3	9-9	12-12	233	63
35	11	2	1-1	3-3	10-9	12-12	243	62

Pituophis vertebralis was originally described from a specimen which Botta collected in "California." Mr. Xantus found it at Cape San Lucas, and Mr. Belding at La Paz.

List of specimens of Pituophis vertebralis.

Cal. Acad. Sci. No.	Locality.	Date.	Collector.
221	San José del Cabo, L. C.	Mar., 1892	W. E. Bryant.
485	"	"	"
752		Sept. 20, 1890	"
790		Sept., 1893	Gustav Eisen.
791		"	"
792		"	"
867	"	Oct., 1893	"
1186	Miraflores, L. C.	Sept., 1894	Eisen and Vaslit.
1563 to 1566	San José del Cabo, L. C.	"	"
1569		"	"
1570	"	"	"

THAMNOPHIS CYRTOPSIS COLLARIS (Jan).

Thamnophis cyrtopsis var. *cyclides*.
1861, Cope, Proc. Ac. Nat. Sci. Phila., p. 299.
Eutænia cyrtopsis.
(1875, Cope, Bull. U. S. Nat. Mus., No. 1, p. 41.)
(1883, Yarrow, Bull. U. S. Nat. Mus., No. 24, p. 121.)
(1887, Cope, Bull. U. S. Nat. Mus., No. 32, p. 73.)
(1887, Belding, West. Am. Scientist, iii, 24, p. 99.)
Eutænia cyrtopsis collaris.
(1892, Cope, Proc. U. S. Nat. Mus., 1891, p. 657.)
Tropidonotus ordinatus var. *eques*.
(1893, Boulenger, Cat. Snakes Brit. Mus., i, p. 209.)

Since Mr. Xantus secured a single garter snake at Cape San Lucas, no representative of this genus has been recorded from Lower California. It seems hardly probable that the Lower Californian form is identical with that of the Mexican mainland.

NATRIX VALIDA (Kenn.)

Regina valida.

1860, Kennicott, Proc. Ac. Nat. Sci. Phila., p. 334.
(1887, Cope, Bull. U. S. Nat. Mus., No. 32, p. 74.)

Tropidonotus tephropleura.

1860, Cope, Proc. Ac. Nat. Sci. Phila., p. 341.

Tropidonotus validus validus.

(1875, Cope, Bull. U. S. Nat. Mus., No. 1, p. 42.)
(1883, Yarrow, Bull. U. S. Nat. Mus., No. 24, p. 132.)
(1887, Belding, West Am. Scientist, iii, 24, p. 99.)

Tropidonotus validus tephropleura.

1883, Yarrow, Bull. U. S. Nat. Mus., No. 24, p. 133.
(1887, Belding, West Am. Scientist, iii, 24, p. 98.)

Tropidonotus leberis validus.

[1883, Garman, Mem. Mus. Compr. Zoöl. Cambr., viii, 3, pp. 28, 143 (part).]

Natrix valida valida.

1892, Cope, Proc. U. S. Nat. Mus., 1891, p. 670.

Tropidonotus validus.

1893, Boulenger, Cat. Snakes, Brit. Mus., i, p. 237 (part).
(1894, Günther, Biologia C.-Am., p. 134.)

The following table, based upon fifteen specimens from San José del Cabo, will serve to show the variation in the scale characters of this species.

Urosteges.	Gastrosteges	Scale rows.	Rows of smooth scales.	Upper labials.	Lower labials.	Preoculars.	Postoculars.	Temporals.	
								Right.	Left.
72	145	19	0	8	10	2-2	3-3	1-2	1-2
73	143	19	0	8	10	?-2	3-3	1-2	1-2
74	146	19	0	8	10	1-1	3-3	1-2	1-2
75	144	19	0	8	10	1-1	3-3	1-3	1-2
—	142	19	0	8	10	1-2	3-3	1-2	1-2
77	142	19	0	7-8	10	1-1	3-3	1-2	1-2
78	144	19	0	8	10	1-1	3-2	1-2	1-3
79	144	19	1 (part)	8	10	2-2	3-3	1-2	1-2
80	142	19	1 (part)	8	10	1-1	3-3	1-3	1-2
81	145	19	0	8	10	2-1	3-3	1-2	1-2
82	139	19	0	8	10	2-2	3-3	1-3	1-2
—	141	19	0	8	10	1-1	3-3	1-2	1-2
73	140	19	0	7-8	10	3-1	3-3	1-2	1-2
78	146	19	0	8	10	1-1	3-2	1-2	1-3
81	147	19	0	8	10	1-1	3-3	1-2	1-2

Many specimens contained fish which Dr. Chas. H. Gilbert has identified with *Mugil brasiliensis*.

Natrix valida was first described by Robert Kennicott, in 1860, from a specimen collected in Durango, Mexico, by Lieut. Couch. Professor Cope described others contemporaneously from Cape San Lucas, under the name *Tropidonotus tephropleura*. Mr. Belding found the species at La Paz. It appears to be the most abundant snake of the country immediately surrounding San José del Cabo.

List of specimens of Natrix valida.

Cal. Acad. Sci. No.	Locality.	Date.	Collector.
454 to 462	San José del Cabo, L. C.	Mar., 1892	Gustav Eisen.
486		"	W. E. Bryant.
538		Mar. 16, 1892	"
603		Oct. 6, 1890	"
540		Mar., 1892	Gustav Eisen.
614		Sept. 6, 1890	W. E. Bryant.
615		Sept. 28, 1890	"
616		"	"
617		"	"
755		Oct., 1890	"
756		"	"
783 to 788		Sept., 1893	Gustav Eisen.
873		Nov., 1893	"
1580 to 1617		Sept., 1894	Eisen and Vaslit.

NATRIX CELÆNO (Cope).

Tropidonotus celæno.

1860, Cope, Proc. Ac. Nat. Sci. Phila., p. 341.

1861. Cope, Proc. Ac. Nat. Sci. Phila., p. 298.

Tropidonotus validus celæno.

(1875, Cope, Bull. U. S. Nat. Mus., No. 1, pp. 42, 93.)

(1883, Yarrow, Bull. U. S. Nat. Mus., No. 24, p. 133.)

(1887, Belding, West Am. Scientist, iii, 24, p. 99.)

Regina valida celæno.

(1887, Cope, Bull. U. S. Nat. Mus., No. 32, p. 74.)

Tropidonotus leberis validus.

[1883, Garman, Mem. Mus. Compr. Zoöl. Cambr., viii, 3, p. 143, (part).]

Natrix valida celæno.

1892, Cope, Proc. U. S. Nat. Mus., 1891, p. 670.

Tropidonotus validus.

[1893, Boulenger, Cat. Snakes Brit. Mus., i, p. 237 (part).]

There seem to be no structural differences between this and the preceding species. In fact, it is not improbable that *N. celæno* will ultimately be found to have been established upon melanistic individuals of *N. valida*. Until this has been shown to be the case, however, they must be regarded as distinct species, for they live in the same localities without any apparent tendency towards intergradation, so far as the specimens before me reveal. The number of preocular plates is either one or two, but the postoculars seem to be constantly three. The largest specimen is 890 mm. in total length, the tail being 208 mm. It has the internasal plate of each side united with the corresponding prefontal, but a groove extends halfway across between them. Some of the specimens contained small fish, *Mugil brasiliensis*.

List of specimens of Natrix celæno.

Cal. Acad. Sci. No.	Locality.	Date.	Collector.
612	San José del Cabo, L. C.	Sept. 18, 1890	W. E. Bryant.
789	"	Sept., 1893	Gustav Eisen.
1185	Miraflores, L. C.	Sept., 1894	Eisen and Vaslit.
1359	"	"	"
1578	San José del Cabo, L. C.	"	
1579	"	"	

Trimorphodon lyrophanes Cope.

Lycodon lyrophanes.

1860, Cope, Proc. Ac. Nat. Sci. Phila., p. 343.

Trimorphodon lyrophanes.

1861, Cope, Proc, Ac. Nat. Sci. Phila., p. 297.
(1875, Cope, Bull. U. S. Nat. Mus., No. 1, p. 38.)
1883, Yarrow, Bull. U. S. Nat. Mus., No. 24, pp. 15, 98.
1886, Cope, Proc. Am. Philos. Soc., p. 286.
(1887, Cope, Bull. U. S. Nat. Mus., p. 68.)
(1892, Cope, Proc. U. S. Nat. Mus., 1891, p. 679.)

The specimens in the collection of the California Academy of Sciences agree perfectly with Prof. Cope's original description, except in the number of scale rows and loreal plates. One of the seven has twenty-one rows of scales, two have twenty-two, and the remaining four have twenty-three rows. Each of four of these specimens has a small plate in the notch between the second and third supralabials, below the posterior loreal, and in front of the two inferior preoculars. The largest individual (No. 482) is 990 mm. in total length, 155 mm. representing the tail. It is a female, and contains eggs measuring about 9x21 mm.

The types of this species were secured by Mr. Xantus at Cape San Lucas. Mr. Belding found other specimens at La Paz, in 1882.

List of specimens of Trimorphodon lyrophanes.

Cal. Acad. Sci. No.	Locality.	Date.	Collector.
482	San José del Cabo, L. C.	Mar. 16, 1892	W. E. Bryant.
810	"	Sept., 1893	Gustav Eisen.
878		May, 1893	"
1405	Sierra San Lazaro, L. C.	Sept., 1894	Eisen and Vaslit.
1549	San José del Cabo, L. C.	"	"
2209	Lower California.		
2210	"		

CROTALIS ATROX B. and G.

Crotalus atrox.
(1853, Baird and Girard, Cat. N. A. Reptiles, i, Serpents, p. 5.)
Caudisona atrox sonoraensis.
1861, Cope, Proc. Ac. Nat. Sci. Phila., p. 292.
Crotalus adamanteus atrox.
(1875, Cope, Bull. U. S. Nat. Mus., No. 1, p. 33.)
1877, Streets, Bull. U. S. Nat. Mus., No. 7, p. 40.
1883, Yarrow, Bull. U. S. Nat. Mus., p. No. 24, p. 75.
(1889, Cope, Proc. U. S. Nat. Mus., p. 147.)
(1891, Cope, Proc. U. S. Nat. Mus., p. 690.)

Without specimens from other localities for comparison, nothing can be said about the status of Lower Californian snakes of the *Crotalus adamanteus* group. Prof. Cope is followed, therefore, in the use of the name *atrox*, although it seems scarcely probable that Lower Californian and Texan specimens are identical.

In the collection of 1894 are seven specimens about 170 mm. long. These, Dr. Eisen informs me, were taken from a large female caught at San José del Cabo.

This rattlesnake has been reported from Cape San Lucas, La Paz, and Los Coronados Islands, in Lower California.

List of specimens of Crotalus atrox.

Cal. Acad. Sci. No.	Locality.	Date.	Collector.
653	Lower California.	1889	
775 to 782	San José del Cabo, L. C.	Sept., 1893	Gustav Eisen.
860		Oct., 1893	"
861	"	"	"
973	Sierra El Taste, L. C.	Sept., 1893	"
1542 to 1545	San José del Cabo, L. C.	Sept., 1894	Eisen and Vaslit.
1550 to 1559		"	"

CROTALUS LUCIFER B. & G.

Crotalus lucifer.

(1852, Baird and Girard, Proc. Ac. Nat. Sci. Phila., p. 177.)
1880, Lockington, Am. Nat., p. 295.

Mr. Lockington has recorded this species from the northern part of the peninsula.

CROTALUS EXSUL Garman.

Crotalus exsul.

1883, S. Garman, Mem. Mus. Compr. Zoöl. Cambr., viii, 3, pp. 114, 174.
(1884, S. Garman, Bull. Essex Inst., xvi, 1, p. 35.)

Under this name, Garman has described a small rattlesnake from Cedros (= Cerros) Island, Lower California. It seems to be very closely related to *C. atrox*.

CROTALUS ENYO Cope.

Caudisona enyo.

1861, Cope, Proc. Ac. Nat. Sci. Phila., p. 293.
(1866, Cope, Proc. Ac. Nat. Sci. Phila., p. 309.)

Crotalus enyo.

(1875, Yarrow from Cope, U. S. G. G. Surv. W. 100th Mer., v, p. 534.)
(1875, Cope, Bull. U. S. Nat. Mus., No. 1, pp. 33, 92.)
(1883, Yarrow, Bull. U. S. Nat. Mus., No. 24, p. 74.)
(1887, Cope, Bull. U. S. Nat. Mus., No. 32, p. 90.)
(1887, Belding, West Am. Scientist, iii, 24, p. 98.)
(1892, Cope, Proc. U. S. Nat. Mus., 1891, p. 693.)

Crotalus oregonus enyo.

(1883, S. Garman, Mem. Mus. Compr. Zoöl. Cambr., viii, 3, p. 174.)
(1884, S. Garman, Bull. Essex Inst., xvi, 1, p. 35.)

This rare rattlesnake is apparently of small dimensions, the largest individual measuring only 810 mm. to the base of the rattle.

The scale rows vary in number from twenty-five to twenty-six; the gastrosteges, from one hundred and sixty-one to one hundred and seventy-two; the urosteges, from twenty to twenty-eight. The first and several (1 to 5) of the posterior urosteges are divided. The scales of the first row only, and not all of these, are smooth. There is considerable variation in the number, shape, and relative size of the head scales of different specimens.

Mr. Xantus collected the type at Cape St. Lucas. Mr. Belding found a single specimen at La Paz.

List of specimens of Crotalus enyo.

Cal. Acad. Sci. No.	Locality.	Date.	Collector.
749	San José del Cabo, L. C.	Sept. 27, 1890	W. E. Bryant.
750	"	"	"
772		Sept., 1893	Gustav Eisen.
773		"	"
774		"	"
1534		Sept., 1894	Eisen and Vaslit.

Crotalus mitchellii Cope.

Caudisona mitchellii.

1861, Cope, Proc. Ac. Nat. Sci. Phila., p. 293.

(1866, Cope, Proc. Ac. Nat. Sci. Phila., p. 310.)

Crotalus mitchellii.

(1875, Yarrow from Cope, U. S. G. G. Surv. W. 100th Mer., v, p. 535.)

(1875, Cope, Bull. U. S. Nat. Mus., No. 1, pp. 33, 92.)

1883, Yarrow, Bull. U. S. Nat. Mus., No. 24, pp. 73, 189.

(1887, Cope, Bull. U. S. Nat. Mus., No. 32, p. 90.)

(1887, Belding, West Am. Scientist, iii, 24, p. 98.)

(1892, Cope, Proc. U. S. Nat. Mus., 1891, p. 694.)

1894, Van Denburgh, Proc. Cal. Ac. Sci., Ser. 2, iv, p. 450.

Crotalus pyrrhus.

1877, Streets, Bull. U. S. Nat. Mus., No. 7, p. 39.

(1883, Yarrow, Bull. U. S. Nat. Mus., No. 24, p. 73.)

(1890, Townsend, Proc. U. S. Nat. Mus., xiii, p. 144.)

(1891, Stejneger, West Am. Scientist, vii, April, p. 165.)

(1892, Cope, Proc. U. S. Nat. Mus , 1891, p. 694.)

Crotalus oregonus mitchellii.

(1883, S. Garman, Mem. Mus. Compr. Zoöl. Cambr., viii, 3, p. 173.)

(1884, S. Garman, Bull. Essex Inst., xvi, 1, p. 35.)

Since the identity of *C. pyrrhus* with *C. mitchellii* was shown in these Proceedings (1894, p. 450), the Academy has obtained twelve additional specimens of this snake. These show as much individual variation as regards the scales on the head as the ten specimens already recorded. One is deep pinkish buff in general coloration, while the others are of the more usual grayish tint. Among them are three very young specimens which Mr. Vaslit informs me were taken from a large female secured at San José del Cabo, thus showing that this species is ovoviviparous. The largest specimen is somewhat longer than any previously recorded from Lower California, being forty-one inches in length.

This species, which seems to be most closely allied to *Crotalus cerastes* of our southwestern deserts, was originally described from a specimen obtained by Mr. John

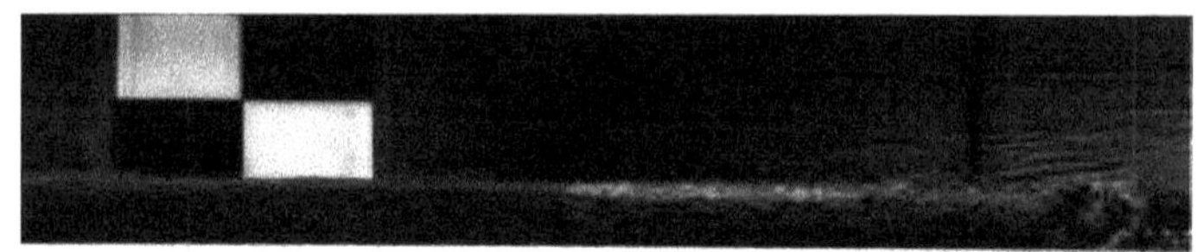

Xantus at Cape San Lucas. Mr. L. Belding secured another at La Paz, and Dr. Streets and Mr. Townsend each collected one on Angel de la Guardia Island. It has been found in California and Arizona,

List of specimens of Crotalus mitchellii.

Cal. Acad. Sci. No.	Locality.	Date.	Collector.
623	Santa Margarita Island, L. C.	Feb., 1889	W. E. Bryant.
654	Las Huavitas, L. C.	1889	"
764	Sierra El Taste, L. C.	Sept., 1893	Gustav Eisen.
765	San José del Cabo, L. C.	Oct., 1893	"
766	Lower California.		W. E. Bryant.
767 to 770	San José del Cabo, L. C.	Sept., 1893	Gustav Eisen.
771		Oct., 1889	"
1187 to 1190	Miraflores, L. C.	Sept., 1894	Eisen and Vaslit.
1404	Sierra San Lazaro, L. C.	"	"
1535 to 1541	San José del Cabo, L. C.	"	"

EXPLANATION OF PLATES.

PLATE IV.

Chrysemys nebulosa, new species.

Type. (No. 2244, Lower California, abreast of San José Island.)
(One and seven-tenths times natural size.)

PLATE V.

Chrysemys nebulosa, new species.

Type. (No. 2244, Lower California, abreast of San José Island.)
(One and seven-tenths times natural size.)

PLATE VI.

Chrysemys nebulosa, new species.

Type. (No. 2244, Lower California, abreast of San José Island.)

a. Head from side. (Twice natural size.)
b. Head from above. (Twice natural size.)
c. Fore limb from above. (Two and one-fourth times natural size.)
d. Hind limb from below. (Two and one-fourth times natural size.)
e. Tail from above. (Twice natural size.)

PLATE VII.

Uta repens, new species.

Type. (No. 633, Comondu, Lower California.)
General view. (About one and three-tenths times natural size.)

PLATE VIII.

Uta repens, new species.

Type. (No. 633, Comondu, Lower California.)

a. Head from side. (Two and three-fourths times natural size.)
b. Head from below. (Two and three-tenths times natural size.)
c. Fore limb. (One and eight-tenths times natural size.)
d. Scales of arm. (Three times natural size.)
e. Hind limb. (One and four-tenths times natural size.)

Uta microscutata Van D.

Type. (San Pedro Martir Mt., Lower California.)

f. Hind limb. (Three and four-tenths times natural size.)
g. Fore limb. (Three and four-tenths times natural size.)

PLATE IX.

Uta microscutata Van D.

Type. (San Pedro Martir Mt., Lower California.)

a. General view. (One and four-tenths times natural size.)
b. Head from side. (Four and nine-tenths times natural size.)
c. Head from above. (Four and nine-tenths times natural size.)
d. Head from below. (Four times natural size.)
e. Section of back. (Five and six-tenths times natural size.)

PLATE X.

Sceloporus licki, new species.

Type. (No. 1436, Sierra San Lazaro, Lower California.)

a. General view. (About natural size.)

b. Head from above. (Two and one-fourth times natural size.)

c. Head from side. (Two and one-fourth times natural size.)

d. Head from below. (Two and one-fourth times natural size.)

e. Fore limb. (One and one-half times natural size.)

f. Hind limb. (One and one-half times natural size.)

g. Scale from back of thigh. (Four and seven-tenths times natural size.)

h. Dorsal scale. (Four and seven-tenths times natural size.)

PLATE XI.

Xantusia gilberti, new species.

Type. (No. 401, San Francisquito, Sierra Laguna, Lower California.)

a. General view. (Two and one-half times natural size.)

b. Head from above. (Four times natural size.)

c. Head from side. (Six times natural size.)

d. Head from below. (Six and one-tenth times natural size.)

e. Fore limb. (Six times natural size.)

f. Hind limb. (Six times natural size.)

PLATE XII.

Verticaria sericea, new species.

Type. (No. 435, San José Island, Gulf of California.)

a. General view. (About twice natural size.)

b. Head from side. (Three times natural size.)

c. Head from below. (Three times natural size.)

d. Fore limb. (One and two-tenths times natural size.)

e. Hind limb. (One and two-tenths times natural size.)

PLATE XIII.

Eumeces lagunensis, new species.

Type. (No. 400, San Francisquito, Sierra Laguna, Lower California.)

a. General view. (About one and one-half times natural size.)

b. Head from above. (Three and one-half times natural size.)

c. Head from side. (Three and one-half times natural size.)

d. Head from below. (Three and two-thirds times natural size.)

e. Hind limb, etc. (Three times natural size.)

f. Fore limb. (Three times natural size.)

PLATE XIV.

Lampropeltis nitida, new species.

Type. (No. 800, San José del Cabo, Lower California.)

General view. (Seven-tenths natural size.)

Head from side. (One and one-third times natural size.)

Head from below. (One and one-fourth times natural size.)

Tail from below. (Seven-tenths natural size.)

ON LAND AND FRESH WATER SHELLS OF LOWER CALIFORNIA. NO. 5.

BY J. G. COOPER.

Dr. Eisen, accompanied by Mr. F. H. Vaslit, made a short journey to the Sierra San Lazaro, twenty-five miles north of Cape St. Lucas, in the early summer of 1894, collecting for the California Academy of Sciences. At that point they obtained a few land shells, among which is one new species, besides additional specimens of others. They then crossed the gulf to Mazatlan, and spent some months in the region south of there, as far as Tepic, the collections made there, up to 3000 or 4000 feet, serving as material for another article following this.

Bulimulus artemisia W. G. Binney.

One specimen is a fourth larger than any before obtained, but is bleached and shows no characters to distinguish it otherwise; 29 specimens brought.

Bulimulus cooperi Dall (*B. pilula* Crosse and Fisher, not of W. G. Binney).

One specimen, which I before referred to in article 2 (Proc. Cal. Acad., 2d series, vol. iii, p. 210), thus: "One specimen has faint traces of two bands on the body, but is otherwise bleached." After seeing the figures in C. & F.'s work, which are represented with two narrow bands on the body-whorl, I supposed that to be the normal condition of the fresh shell. But as Binney's type (which was from nearer the west coast) does not have bands, and also differs in form, while the bands do not quite disappear from bleaching, it seems that C. & F. were wrong in identifying their shells with *B. pilula* and a new name is needed. No more specimens are known to have been found lately, and, according to Dr. Dall, none of the specimens collected by the Academy's expeditions agree exactly with the types, although some show no more variation than we see in varieties of *B. xantusi*.

Bulimulus decipiens J. G. Cooper, n. sp.

I propose this name for a new form, of which three specimens were brought from San Lazaro Mt. They were living but apparently quite young, too immature to figure. The largest is nearly of the size and form of *B. pilula* as figured, but still more globular, being shorter and wider, with three whorls. It is more Heliciform, much resembling *H. californiensis* Young, and like that has a single vittiform band around periphery, which becomes hidden in the suture of two upper whorls. The band is, however, paler than the brownish epidermis (faded in alcohol?). To prove their affinity to the Bulimuli of the region, they show the vertical riblets on 1½ apical whorls, and a more sunken nucleus than in the Helix. No single-banded Helix is known for 200 miles north of the locality of this species.

Bulimulus inscendens W. G. Binney.

About 45 specimens were brought from San Lazaro, representing all its varieties, while the other large species were quite rare there compared with other localities.

Genus Plicolumna J. G. Cooper.

In article 2, p. 215, vol. iii, 2d series, I referred to Columna, the species before called *Rhodea* var. *ramentosa* by me in article 1, p. 102, vol. iii, April, 1891. The original Columna (Perry), is a quite different shell, but other authors have included with it shells with plications on the body whorl, complete spire, etc., as in my species. Dr. Dall, having succeeded in finding enough of the soft parts, in specimens I sent, to examine microscopically, writes as follows:

"I am now able to state that your *ramentosa* is a Bulimulus! The jaw is like that of Thysanophora exactly, and also like that of *Bul. artemisia*, while the teeth differ very little." He does not, however, notice the differences in the shell from that of the latter, and I therefore propose the above name, having before mentioned the resemblance in the nuclear whorls and epidermis to those of *B. artemisia* and suggested the affinity of the two forms on p. 138.*

Genus Pseudosubulina.

In describing *Melaniella eiseniana* in article 3, vol. iii, p. 339, of these Proceedings in 1893, I adopted the genus with a (?), remarking on the absence of jaw, but did not suspect it to have lingual teeth of the carnivorous type, These have been found in a specimen dissected by Dr. Dall, who refers this and *M. tastensis* to the above genus, and shows an apparent affinity in this respect to Oleacinidæ. Several genera resembling this species and *P. tastensis* in form of the shells have been referred to the same family, especially Megaspira and Balea. I would have put *P. eiseniana* in the latter genus if not misled by its external likeness to Stenogyra. But the fact of carnivorous dentition does not separate the families entirely in habits, as some species are known to eat vegetable food as a rule, not being able to obtain animal food suitable to their needs very often. They should rather be called omnivorous, and indeed there are probably few, if any, species even of the phyllovorous genera exclusively vegetivorous.

---

*The subgenus name *Peronæus* cannot be used for either of these species, the name being preoccupied. The same is true of *Leptobyrsus*, it being used in the feminine form in entomology two years before C. & F. used it (Scudder).

Orthalicus princeps Broderip.

But two of this genus were brought, which come nearest to this subspecies of the old species *undatus* Bruguiere, which they closely resemble. They are from Tepic, but some form of it inhabits all of Mexico and tropical America, the West Indies and Florida.

Conulus fulvus Müller.

One specimen is exactly like the large west coast form of this species found in California.

Hyalinia indentata Say.

Fourteen fresh shells from the Tepic Valley are larger than those from Lower California and darker than usual, otherwise similar to eastern shells.

Patula horni Gabb.

The five specimens brought from Tepic do not differ from the Arizona type, or from the one found on the peninsula. They differ from any allied shell figured or described as from Eastern Mexico.

Polygyra hindsi Pfeiffer.

Over fifty specimens were brought from Mazatlan, Tepic, etc., mostly fresh and full grown. All seem easily distinguishable from the next in size, being mostly about half as large, but there is a great difference in size among these small shells, varying in about the same proportions as the varieties of *H. monodon* called var. *leai* and var. *fraterna* as figured by Binney.

Polygyra ventrosula Pfeiffer.

Twenty-two specimens from the same region are about twice the size of the preceding. Both forms show specimens having a broad pale band on top of the body-whorl, while a narrow dark one runs along the suture and a wider one near the periphery, which in some becomes a reddish brown color covering the whole under surface of the shell. Generally these colors are faint or faded out, and they do not seem to have been described before.

Subulina octona Chemnitz.

Twenty-two from San Blas seem to be of this species, which is said to range over most of tropical America and adjacent islands.

Subulina lirifera? Morelet.

Seventy or more of a scalariform species from Tepic approach nearest to this species, which, however, was described as from Guatemala only. It is probably contained in the collections obtained from Tepic by the authors of Biol. Centrali–Americana, and I therefore defer its identification until the results of their study of this genus are printed.

SUCCINEA CALIFORNICA Crosse and Fischer.

This form was founded on specimens obtained near where Orcutt obtained those identified as *S. oregonensis* Lea, in Lower California, lat. 31°. Those found by others near Cape St. Lucas seemed to me to be nearer the *S. rusticana* Gould, perhaps a var. of last. Some of the latter had undulations more or less strongly marked, but they were not constant, and I did not consider them specific characters. It seems, however, that there is a *S. undulata* Say, from Mexico, probably from Acapulco. Those brought from Tepic are without undulations, and seem to me identical with *S. oregonensis*, while I cannot see any difference in the figures of *S. californica* C. & F. If the same, Say's name is prior to Lea's, but seems founded on an abnormal character, and not tenable.

TEBENNOPHORUS SALLEI C. & F.

Two specimens brought from Tepic are considered by Dr. Dall as probably this species.

VAGINULUS MORELETI C. & F.

One found at Tepic, and is said to be found also in Central America. Although somewhat contracted in alcohol, it is very similar to the figure of the living animal given by the authors, measuring about 1½ inches long and 0.6 wide.

LIMNÆA COLUMELLA Say.

Five specimens from Tepic cannot be distinguished as species from the typical form as figured by W. G. Binney in Land and Fresh Water Shells, part ii, p. 33, except that the largest is only about half the size of the northern shells. I believe it has not before been reported from Mexico. According to Binney, the *Succinea wilsoni* Lea, of Georgia, and *S. pellucida* Lea, are forms of this Limnea, but to avoid the chance of a mistake, I examined the animal taken from alcohol and found the jaw of Limnea.

PHYSA MEXICANA Philippi.

Two specimens two-thirds the size of Say's type of *P. heterostropha* figured by Binney in Fresh Water Shells, p. 84, seem to me to confirm identity of the species. They are less similar to Haldeman's *P. osculans* from Mexico.

PLANORBIS LIEBMANNI Dunker.

Over 70 specimens from Mazatlan do not differ in form or size from two Vera Cruz shells, whence the typical form was described. They are much smaller than Gould's *P. gracilentus* of the Colorado Desert, showing the same difference as in the figures given by Binney in Fresh Water Shells, p. 108. Binney says that the figure is enlarged, but Gould gives it as half an inch wide or larger than the figure, while the *P. liebmanni* is little over one-quarter inch. It comes nearer *P. havanensis*, which Pfeiffer says (l. c., p. 107) was found in Texas also, by Roemer. On p. 108, Roemer is quoted for only *P. liebmanni* from Texas, showing some confusion of the species.

CALYCULINA PARTUMEIA Say, var. TRUNCATA Linsley.

One specimen from Tepic seems to agree better with this form than with any other, but is even flatter in proportion and smaller than usual, showing that those characters are not confined to the most northern shells, while the opposite extremes as found in *lenticula* are not exclusively western. It measures 0.27 inch long, 0.23 high, and 0.13 in diameter.

The only *calyculate* species described by Prime from Mexico is *Sphærium subtransversum* from Tabasco, and differs in very small size, more elongation and other characters. The size is given by Prime as length 0.30 inch, height 0.20, diameter 0.10.

ON HETEROMORPHIC ORGANS OF SEQUOIA SEMPERVIRENS ENDL.

BY ALICE EASTWOOD,

Curator of the Herbarium.

[With Plates xv-xviii.]

The following investigations were instigated by the discovery of a branch of redwood with foliage so unlike the ordinary form of *Sequoia sempervirens* that at first it seemed probable that a new variety of Sequoia had been discovered. The trees from which the branch must have fallen grow at the head of Sequoia Cañon in Marin County, on the southern side of Mt. Tamalpais, and differ from the trees in the lower part of the cañon in a more open straggling habit, a weather-beaten appearance, and a preponderance of branches densely covered with short, stout, closely appressed leaves.

A careful examination was made of one of the trees, and branches were obtained from both the upper and lower parts. Plate xvi, fig. 1, shows a piece from an upper branch with the peculiar foliage; fig. 2, a piece from a lower branch. It will at once be evident that there are two quite different kinds of leaves on the same tree, the lower being the ordinary redwood foliage with broad distichous leaves, while the upper more nearly resembles that of *Sequoia gigantea* Decaisne. Two trees that had been overthrown in a storm in the lower part of the cañon showed the same characteristics.

Dr. Kellogg had noticed the scale-like leaves of *Sequoia sempervirens* in "Forest Trees of California," published by the State Mining Bureau in 1882, and so had Dr. Newberry in "Pacific Railroad Reports," part iii, p. 58; but neither had thoroughly investigated the matter, nor did they set forth the facts exactly.

In order to be sure that this was a universal and not a local characteristic, it was necessary to examine many trees in different localities. Besides the trees in Sequoia Cañon, I examined trees in Santa Cruz County, at Wright's and at Boulder Creek, and in Sonoma County at Duncan's Mills. At Boulder Creek the woodchoppers were at work destroying the forest, and I had an opportunity to examine many trees from the topmost to the lowest branches. Fine specimens were sent from Mendocino, Humboldt and Sonoma counties, in response to a request for branches from different trees and different parts of the same tree, also information concerning the size and location of the trees from which the specimens were obtained. Without any exception, the large trees—two, three or more feet in diameter—possessed heteromorphic foliage. Many small trees, a foot or so in diameter, were seen that had only the broad distichous leaves. There are always scale like leaves on young upper shoots; but on young trees they afterwards generally expand into the broad leaves. (Plate xv, fig. 2.)

From all these observations I conclude that all large trees of *Sequoia sempervirens* have the upper foliage quite different from the lower, with intermediate forms. This is not true of *Sequoia gigantea*, so far as my observation goes.

Among the forty or more fossil species of Sequoia described from the Northern Hemisphere it is interesting to note that the foliage of several species has been found to be heteromorphic. Of *Sequoia biformis*, Lesquereux, the author, says: "This species apparently bears two kinds of leaves even upon the same specimens, either long 2 cm. and very narrow linear, less than 1 mm. wide; or shorter and broader, decreasing gradually from the base to the point, linear-lanceolate nearly 1½ mm. wide and only 8 to 10 mm. long; the middle nerve is deeply marked upon both kinds of leaves." (Bull. No. 5, 2d series, Hayden's Geol. Survey of Terr., page 366.)

In Heer's "Flora Fossilis Arctica" illustrations are given of many fossil species. *Sequoia Langsdorfii* Heer, the species apparently most widely distributed in past ages, is similar to *S. sempervirens*, and the fossil specimens show dimorphic foliage. (Flora Fossilis Arctica, vol. ii, plate xliv, vol. iv, plates xiii–xiv.) The resemblance to *S. sempervirens* is quite evident. *S. Reichenbachi* Heer, vol. iii, plates xv–xxxvi, shows three forms of leaves on the same branch.

These go to prove that the genus Sequoia possesses a tendency towards heteromorphism, which the environment probably develops.

Along the coast *Sequoia sempervirens* does not grow above the altitude to which the moist sea air generally rises—the fog-line, as it is often called. This is more apparent near the southern limit of the species, which is between Pt. Gordo and San Simeon Bay. The luxuriance of the lower foliage may be due to the greater humidity of the lower strata of the atmosphere, while the stunted upper leaves indicate less nutrition. This view is strengthened by the appearance of the upland trees which begin to outgrow the distichous foliage when much younger than the valley trees, and by the consideration that the tall trees that rise to a height of two or three hundred feet must reach an elevation which is often above the fog, even when growing at a comparatively low altitude.

Nowhere is it more plainly shown than in California that amount of moisture is the most important factor in forest distribution. Many of the trees that abound near the coast disappear in the dry hills of the southern Coast

Range and are not found in the hot valley of the San Joaquin, but reappear in the Sierra Nevada mountains at an altitude of three or four thousand feet above the sea level, where they again come into almost the same conditions of humidity and temperature as at the coast.

The amount of light probably also exerts a great influence. The lower branches need more leaf surface because of the shade from the surrounding arboreal vegetation, while the upper boughs that rise above all the other trees have for themselves alone all the light and sunshine and so are not compelled to spread themselves out. Their energies are bent to the reproduction of the species, and with less nutrition economy is necessary.

As the parts of the flower are but transformed leaves, the same diversity might be expected. It is even greater. The carpellary scales of the pistillate aments take many forms. Plate xviii, figs. 2 and 3, show pistillate aments from different trees, figs. 4, 5, 6 carpellary scales. It will be noticed that in the same ament there is no uniformity, though the scales are longer and narrower in some than in others. The scales of the staminate aments range from the broad form shown in plate xviii, fig. 8, to one much narrower and more pointed. The number of pollen sacs is not invariable, three or four being the rule. The cones are round or oblong, and vary in size irrespective of the height of the tree. The seeds when ripe show many forms, figs. 9 and 11 being of one type, figs. 10 and 12 of another.

The proliferous cone shown in plate xvi, fig. 3, is interesting, as it illustrates the nature of the ament, a subject which has been discussed by many botanists. Such cones are not uncommon. Engelman says of a similar cone of *Sequoia gigantea:* "It seems to prove not only that the fruit scale in this species (and consequently in the whole tribe) is homologous with that of Abietineæ in so far as it consists of leaves of an axillary shoot, yet that these leaves are not a single pair, but, as A. Braun has long ago suggested in regard to Cupressineæ, that there is a number of leaves laterally co-ordinate and connate bearing a number of ovules on their back." (Bot. Gaz., vol. vii, pp. 104–105.)

Prof. Eichler regards the scales of the female ament in all coniferæ as representing nothing but simple leaves. (Bot. Gaz., vol. vii, p. 39. Review of Prof. Eichler's article by Prof. Geo. L. Goodale.) From the manner in which the scales are arranged on the axis of the cone, as well as the position of the resin ducts, as shown in plate xvii, sections 14–15, *S. sempervirens*, and 16–17 *S. gigantea*, this view seems the more reasonable.

Sequoia and Taxodium are the North American representatives of the tribe Taxodineæ, which is intermediate between Cupressineæ and Abietineæ. Several species of Juniperus and Cupressus of the Cupressineæ have heteromorphic foliage, also some species of Taxodium; but in none is the difference so marked as in *Sequoia sempervirens*, which approaches the Cupressineæ in its upper leaves and the Abietineæ in the lower. It may be that the hetermorphic character of its foliage indicates its relationship to the two tribes and that environment may have no influence. However, while it is interesting to theorize concerning the cause of the heteromorphism and the theories may be suggestive, too little is yet known to form a safe basis for conclusions.

EXPLANATION OF PLATES.

The figures in plates xv and xvi are the natural size. Plates xvii and xviii are magnified.

Plate XV.

Fig. 1. Fruiting twig from a low branch of a small tree in Mill Valley, Marin County. The tree had grown from one of the suckers of a tree that had been felled many years ago. It shows the well-known broad distichous foliage.

Fig. 2. Flowering branch from the upper part of a small tree in the same locality. The young upper leaves are scale-like, but afterwards broaden out.

Plate XVI.

Fig. 1. Fruiting branch from the upper part of a medium sized tree growing at the head of Sequoia Cañon. This is the characteristic upper foliage of the large trees—short, stout, scale-like leaves, similar to those of *S. gigantea*.

Fig. 2. Lower branch of the same tree with distichous foliage.

Fig. 3. Small twig with proliferous cone.

Plate XVII.

Fig. 1. Broad leaf from the twig shown in plate xv, fig. 1. The canal down the center which is near the surface makes the central ridge seem depressed. Figs. 5 and 10 are cross-sections of similar leaves. The position and number of the resin ducts are seen to be variable in the different cross-sections of leaves of *S. sempervirens*, figs. 5, 10, 6, 7, 8, 9. In these sections the epidermis is represented by the parallel vertical dotted lines, the cross-sections of the resin duct by the openings, and the wood cells by the horizontal parallel lines.

Fig. 2. Young appressed leaf of branch shown in plate xv, fig. 2. The resin duct is nearer the surface at the upper and lower part of the leaf. The small irregularly placed dots represent the stomata. Fig. 6 cross-section of similar leaf.

Fig. 3. Leaf from the branch shown in plate xvi, fig. 2. Figs. 8 and 9 sections of same.

Fig. 4. Leaf from branch shown in Plate xvi, fig. 1. Fig. 7 cross-section of similar leaf. This is the form that resembles *S. gigantea*, as can be seen by comparing cross-sections. Figs. 11 and 12 sections of leaves of *S. gigantea*. Fig. 13 cross-section of the bract of a staminate ament.

Figs. 14 and 15. Sections of pistillate scales of *S. sempervirens*. It is not usual to find more than one resin duct in these scales. Figs. 16 and 17, sections of pistillate scales of *S. gigantea*.

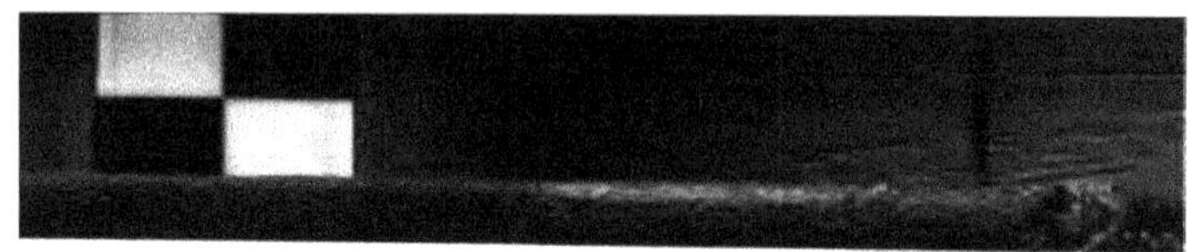

PLATE XVIII.

Fig. 1. Staminate ament with bracts at base and the staminate scales attached to the axis.

Figs. 2 and 3. Pistillate aments to show variation in the scales.

Figs. 4 and 6. Pistillate scales, back view.

Fig. 5. Same, showing the naked ovules; the number is variable. The micropyle which receives the pollen can be seen at the top of the naked ovules. The long point on these scales becomes the inconspicuous bristle of the ripe cone.

Fig. 7. A bract of a pistillate ament.

Fig. 8. Scale of staminate ament, showing the stem that unites the scale to the axis, the pollen sacs, one of which is ripe and discharging pollen. The number is variable, but generally three.

Fig. 8c shows some pollen grains magnified.

Figs. 9, 10, 11, 12 are seeds and show two types of the variable seeds.

CALIFORNIA WATER BIRDS. No. I.—MONTEREY AND VICINITY FROM THE MIDDLE OF JUNE TO THE END OF AUGUST.

BY LEVERETT M. LOOMIS,

Curator of the Department of Ornithology.

[With Plate xix.]

A glance at a map of California will reveal that Monterey Bay is not a sheltered roadstead, but merely an abrupt inward bend in the coast-line. In the vicinity of the town of Monterey there is quite a heavy surf at all times, except in coves protected by little promontories that intercept and break the swell from the ocean. To the northward of the town, in the direction of Pt. Santa Cruz, the beach is sandy with high dunes immediately back. In the opposite direction, toward Pacific Grove, Pt. Pinos, and southward, the coast-line is rock-bound. Sunken rocks and miniature islets are numerous a few miles south of Pt. Pinos. These rocky islets are favorite resorts of seals, sea-lions, and water birds. The larger ones whose surfaces are above the reach of the surf are occupied as rookeries by sea birds during the breeding season.

While there is no rain during the summer months, cold fogs are frequent, and so dense are they sometimes, that they almost amount to drizzling rain. There were but few sunny days, owing to the "high fogs" and "low fogs," during the two months and a half of my stay in 1894. During a fortnight in June the sun was seen only on one day. These fogs hide the coast mountains, and consequently must have a marked influence upon the migratory birds that follow the coast-line in their journey southward. The locality is a very favorable one for observing the early southward movements, for there are no storms as upon the Atlantic seaboard. Disturbances far out on the ocean, however, produce quite heavy seas at times, necessitating some skill in the management of a row-boat when there is a strong breeze. The tides are not an obstacle. There is no wind during the forenoon until after nine o'clock, when a breeze usually springs up. The fishermen take advantage of this wind, and at midday the fleet stands in to the land. The curious palm sails of the Chinese and the lateen sails of the Italians give a picturesqueness to the scene, transporting the observer in imagination to far-distant countries.

The observations recorded on the following pages were made during the summers of 1892 and 1894; the first season covering the interval between June 20th and August 23d, and the second season between June 15th and August 28th. In June and July of 1892 my attention was partly directed to the land birds, but in 1894 my whole time was devoted to the water birds, and I was therefore able not only to verify but to extend the observations of the first season.

In 1892 I confined my efforts, so far as the water birds were concerned, chiefly to the bay, but in 1894 my work was carried on mainly on the ocean proper. The Hopkins Seaside Laboratory at Pacific Grove was my base of operations both seasons. The directors—Dr. Charles H. Gilbert and Dr. O. P. Jenkins—courteously placed at my disposal one of the rooms of the laboratory, where I prepared my specimens. Each morning I had my boatman row out to the ocean to a whistle buoy anchored in deep water about a third of a mile northwest from the outlying rocks at Pt. Pinos. Here I would spend an hour or two watching migratory birds round the Point. Afterward I would move down the coast in the boat, sometimes nearly as far as Pt. Cypress, or go out from three to ten miles from land, usually returning to the Seaside Laboratory by two o'clock. A trip was made, June 25th, by land, to Carmel Bay, where I visited a Cormorant rookery on a rocky islet at the extremity of Pt. Carmel, or Pt. Lobos, as it is locally known.

EARLY SOUTHWARD MIGRATIONS.

As I have already published* some account of the migrations witnessed during the summer of 1892, my present remarks will be restricted chiefly to the summer of 1894.

The sea offers peculiarly favorable opportunities for studying migratory movements. On land much of migration readily escapes observation. Often only the birds that stop in a locality are noted. The greater perils and the natural and other obstructions necessitate a greater elevation of flight. Further, migration over the ocean continues during the daytime to an extent not usualy observed on land, resembling, perhaps, more the night migration of land birds. The vegetation of the land also affords means of concealment, and stragglers escape notice that would readily be seen on the water.

The occurrence of stragglers on isolated islets, as American birds on Heligoland (see Seebohm, "Ibis," 6th ser., vol. iv, pp. 1–32) or of the Catbird on the Farallones (Townsend, "Auk," vol. ii, p. 215), illustrate in another way the favorableness of the sea for the study of migration. It is not to be supposed that estrays visit such islets more frequently than they do the adjacent mainland. Over miles of water they find but a single resting place, so the chances of meeting them are many, but on the mainland in an area of equal extent, where any spot may be a resting place, the chances of seeing them are extremely few. Some years ago in upper South Caro-

* "The Auk," vol. xi, pp. 27, 28, 29, 30, 95–98.

lina I tried the experiment, during the height of migration, of sending out each morning an assistant I had trained, to collect birds in a direction opposite from the one I would take myself. The results of our day's shooting were often very different—so different were they that I have since been fully convinced that a single observer, diligently spending each day in the field, can know but little of the rarer birds that happen to be in his neighborhood at the time of his observations, and how impossible it is for a single observer ever to exhaust a locality, even one of but a few square miles in extent.

The incipient stages of southward migration of the species that breed in a locality are not always very apparent. Daily records intelligently kept afford a key, however, to these indefinite movements, for they gradually develop into those that are unmistakable. The immediate vicinity of Monterey is an advantageous situation for the study of such migration in sea birds, for there are no suitable places for rookeries, which causes the southward movements of individuals of breeding species of the region, from rookeries further up the coast, to be like the movements of species that rear their young only in the high north. At the rookeries, migratory movements may not always be apparent at the outset, for departure from them after reproduction is over may be simply forsaking of the land for the water, the real home of sea birds.

June.—On June 16th, California Murres were moving down the coast. A number of individual birds, a few couples, and one small company, were seen flying southward, following the shore-line. This was apparently a migratory movement, for in the weeks that followed these Murres continued to pass south in increasing numbers, *with no return movements*. While no breeding places of this species were actually discovered, my observations later in the season led to the conviction that there must be a rookery a short distance to the northward of Pt. Santa Cruz. Also on the 16th, many Dark-bodied Shearwaters were seen. They were flying steadily northward several miles out from land. In an hour not less than a thousand passed my boat. The movements of Shearwaters on the days that immediately followed indicated, for they were southward ones, that this was probably a local movement, though it may have been the ending of the northward migration of the species in this vicinity. In two males and two females that were taken, the organs of reproduction exhibited no signs of recent erotic development, which was also true of all captured afterwards. This circumstance raises the question whether the breeding habitat of this Shearwater is not in the Southern Hemisphere, as is believed to be the case in Wilson's Petrel.

June 18th, many California Murres were heading southward. Some were also on the water. Viewed in the light of the after movements, the stragglers on the water were apparently birds that had temporarily paused in the southward migration. Two females were taken. The ovaries in both showed that they had recently bred. In 1892, no females were secured before July 11th. The first migratory waves, however, were not comprehended that season, nor was there opportunity for thorough study of them. As many Dark-bodied Sheawaters were noted as on the 16th, but all were moving southward.

June 19th and 20th the movements were about the same as on the previous days. On the 21st there was an increase in California Murres, individuals and small parties in single file appeared from the northward, and passing quickly by disappeared to the southward, keeping the course of the coast-line. Solitary ones were quite numerous on the water. At least a thousand Dark-bodied

Shearwaters were seen. They were several miles off shore, and followed two parallel lines of flight about a quarter of a mile apart. They came from the northward, singly, in little companies, and in straggling flocks, and passed rapidly down the coast.

On the 23d there was considerable southward movement in these two species. On the 30th no Shearwaters were observed, and only several California Murres. The scarcity of this latter species rendered still more prominent its extensive movements during the preceding fortnight.

About a dozen Scoters, all seemingly female *deglandi*, were found near the Del Monte beach at Monterey on the 22d. Like the specimens procured two years ago, three females that were captured were in very worn plumage, and had apparently not bred, the ova in each being very indistinct. During the rest of my stay this species was seen at intervals. Its periods of absence were apparently similar to those intervening between migratory movements. The Surf Scoter was found in July and August, and was likewise somewhat irregular in its occurrence.

On my arrival, Western Gulls seemed to greatly outnumber Heermann's Gulls, but toward the end of June the ranks of the latter were apparently re-enforced, for they became about as numerous as the Western Gulls. Adult Heermann's were very scarce in June and during the early part of July, those with mottled heads and birds of the year being almost the only kind met with.

It will be observed that the conspicuous movements of the latter half of June were confined to the California Murre and Dark-bodied Shearwater. Incipient movements apparently occurred in other species, and another season of study, with the light I now have, would probably enable me to define such movements with certainty.

July.—July 2d there was a great movement southward of Dark-bodied Shearwaters. The sea was very calm, although there was a heavy surf, and at midday the sky was clear. I was out on the water from eight in the morning until two in the afternoon, going as far down the coast as Pt. Cypress, keeping near the land on the way down and several miles off shore in returning. Only a few Shearwaters were seen before midday. These were heading rapidly down the coast. About noon many were seen at a distance off Pt. Cypress. They were also flying southward. On going out several miles from land it was found that there was almost a continuous stream of these birds coming from the northward and passing southward. They flew only a few feet above the water, flapping their wings a few times, then sailing for a few moments. The line of movement diverged somewhat to the seaward, as the birds sheared off to avoid the boat. After awhile, they began to pass on both sides of the boat, and it was discovered that there were two parallel lines of flight, as on the 21st of June. In returning up the coast to Monterey Bay, I kept in the path of movement, which was several miles distant from land, for about five miles. As it was clear, the birds could be seen a long way up and down the coast. It was fully determined that they followed the coast-line leading southward, conforming their course to the inward bend at Monterey Bay.

In rounding Pt. Pinos the Shearwaters approached much nearer to the shore than in the bay, a few stragglers even coming within several hundred yards of the surf. After passing the Point all shaped their course so as to regain their former distance from the land. The observations for the entire season in this and other species indicated that Pt. Pinos is a prominent landmark for water birds journeying southward.

At least three thousand of the Shearwaters were seen during the last two hours I was out on the ocean, and there was apparently no abatement in their movement before I returned to land. The two streams in which they moved were formed of straggling companies, varying from a few individuals to flocks of considerable size. Often there was a complete break, no birds being in sight for several minutes. None were seen on the water. Although a portion passed directly through a great gathering of Cormorants and Gulls that had been frightened from the water by the boat, they did not deviate from their course, apparently paying no attention to the great mass of birds flying in confusion about them. Neither did they decoy to wounded comrades, though Gulls were attracted to them, several alighting on the water close by the wounded birds.

With one of the large companies there was a white colored Fulmar. It was probably *Fulmarus glacialis rodgersii*, as that subspecies was secured a little later in the season.

Only two California Murres were seen. Both were flying southward, following the coast-line south of Pt. Pinos.

Two male Black Turnstones, with minute testes, were shot at some rocky islets—known locally as Seal Rocks—about a mile north of Pt. Cypress. There were perhaps a half dozen in all.

The occurrence of the Black Turnstone on the California coast in each of the summer months (as upon the Farallones, Bryant, *fide* Emerson, Proc. Cal. Acad, Sci., 2d ser., vol. i, p. 44) is not an exceptional circumstance, for the same thing happens on the Atlantic seaboard in other boreal Limicolæ—for example, on the Gulf Coast of Florida (Scott, "Auk," vol. vi, pp. 156–159).

It has long been held that the individuals of a species found during the summer months south of the breeding range, but not breeding, are actual summer residents, having failed to migrate northward, or at least failed to complete the migration, owing to barrenness or some accidental cause. Such cause might possibly exist in temporary sickness or wounds, or the way may have been lost, particularly if the loiterers were young birds. This view is not incompatible with the fact of early southward migration. Such stragglers may occur, and when the tide of migration sets southward they may join the ranks of the early migrants of other species. The Fulmar alluded to above may be an example. It may not have reached the breeding habitat, and have come from a locality to the southward of it, joining the Shearwaters as they passed by or falling in with them on the way down the coast. The fluctuations occurring in the Scoters may have been occasioned by the early departure southward of June birds and the arrival of others a little later from further north—such local movement being in advance of the migration from the boreal breeding grounds. While fully recognizing physical debility and accident as factors in this question, too great stress must not be laid upon them, for ample allowance must be made for late northbound migrants and early southbound migrants, as the two migratory movements nearly or quite bridge over the interval of summer. Movements of Black Turnstones from the southern frontier of their breeding range, if they consumed a fortnight and were as early as those of the California Murres, would reach the vicinity of Monterey about the 1st of July. It should be added that the young of this Turnstone are able to take wing in July and leave the flats of the Lower Yukon for the sea-coast (Nelson, Rep. Nat. Hist. Coll. Alaska, p. 130).

On the 5th of July, in the vicinity of Monterey, there was no evident migratory movement. Only one Dark-bodied Shearwater and four California Murres were noted; the latter were on the water.

July 6th. Four Western Grebes were found near the surf within the bay. With the exception of one shot on the 2d, these were the first observed of the season.

July 7th. Several California Murres were seen. All were going southward. No Shearwaters were met with, although it was foggy—a favorable state of the weather for their occurrence near land. A male Harlequin Duck was shot as it was rounding Pt. Pinos. The testes were very large—those of a breeding bird. Another followed shortly after. Both came from the nothward.

July 9th. I remarked in my journal on this day that the Tufted Puffin is apparently to be reckoned among the early migrants. While there had been no pronounced migration, individuals had been passing southward daily for some time, with no corresponding return movements, as in Brandt's Cormorants, for example, which were continually coming into the bay to fish and returning to their rookery south of Pt. Pinos. In the morning a young Marbled Murrelet was discovered at a patch of kelp near the Seaside Laboratory. This was the first instance, for the season, of the species coming under my observation. There was also a decided flight of Pigeon Guillemots. Previously no definite movements had been observed. All were adults. They appeared in twos and threes at frequent intervals, following the shore-line in a southward direction. Two males were taken. The testes were very large in both. But few California Murres were seen, and they were on wing, moving down the coast. Only one Dark-bodied Shearwater was noticed. It was going south.

July 10th and 11th but little migration seemingly took place. The 11th nevertheless was a notable day, for the Northern Phalarope made its appearance, affording additional evidence that the movements witnessed all along were truly migratory movements. At first two were seen coming from the northward, about a mile off shore and a mile south of Pt. Pinos. Then two were found on the water a little further up the coast. When forced to take wing, they continued their flight southward. Nearly opposite Pt. Pinos, still a mile off shore, another appeared from the northward and alighted on the water. Within the bay, fully a mile out from land, a sixth was seen. It also came from up the coast. The only specimen taken was a female, and it had evidently bred the present year.

On the 12th, a visit to Seal Rocks disclosed that large numbers of Western and Heermann's Gulls were congregated there. That a considerable influx of these Gulls had taken place had been manifest for several days at the kelp along the south shore of the bay. A female Fulmar, apparently Rodgers's, was shot on the water near the extremity of Pt. Pinos at midday. It must have arrived while I was at Seal Rocks, for I passed over the spot during the morning. The sea was like glass, so it could hardly have been overlooked. Its ovaries were not those of a bird that had lately bred, and its plumage was greatly abraded. Two more Black Turnstones were secured at Seal Rocks. They were both males, and showed no enlargement of the sexual organs.

The 13th was another day of no very obvious migratory movements.

The second week of July marked a decided increase in the number of California Brown Pelicans seen. Most were young of the year. Each forenoon they came into Monterey Bay from the direction of Pt. Cypress. It was not ascertained whether they were actually accessions to the locality (stragglers of a night or off-shore migration, visiting the bay from temporary roosting places), or whether they were simply birds from rookeries a little lower down the coast and were enlarging their food-area by making daily excursions into the bay. On the 14th an incident happened that seemed to indicate that migration from the north was actually in progress in this species. A little company of young, headed by a fine adult, rounded Pt. Pinos during the morning and moved south. At the Point the young seemed to show a disposition to break away from the leadership of the old one, attempting several times to return into the bay, but each time the old bird got them straightened out by heading them off, and finally disappeared with them in line at his tail in the direction of Pt. Cypress. Also on the 14th, quite a number of Pigeon Guillemots were observed on the water and passing in and out of the bay, but no migratory movement was apparent, there seemingly being a lull in the migration of the species. Three or four California Murres were seen; two were on wing going southward. A Wandering Tatler appeared on the rocks near the Seaside Laboratory. This species was not met with in the weeks that preceded. In 1892, the 23d of July was my earliest record.

On the 16th, 17th, and 18th migration was almost at a standstill. On the 16th an American Eared Grebe and two young Pigeon Guillemots were captured. The ovary of the Grebe was that of a breeding bird of the season. Another young Pigeon Guillemot was seen on the 17th, upon which day adults were scarce. The Grebe and the young Guillemots were new birds for the season, so far as determined by my observations. On the 17th, for the first time, Heermann's Gulls appeared to outnumber

Western Gulls on the kelp. A few California Murres passed south on each of the three days, but none were seen on the water. A small flock of Killdeers was found one morning on the kelp, and Long-billed Curlews, on wing, began to be conspicuous about the bay. Both circumstances seem to point to migration, for there is general dispersion during migration, and birds are stranded in all sorts of situations after the passage of a migratory wave. Two Dark-bodied Shearwaters—the only ones noticed since the 9th—were seen on the 16th, following the shore line south.

The lull in the migration of California Murres, which began at the end of June, was broken July 19th, when solitary individuals and companies of less than half a dozen passed down the coast at short intervals during two hours I spent in the morning off Pt. Pinos. They shaped their course so that it brought them near to Pt. Pinos, then they diverged from the land, taking a direction that would carry them several miles out from shore as they passed Pt. Cypress. Not a single one was seen on the water. On the 19th also, it was very apparent that the adult Heermann's Gulls were more numerous than the Western Gulls or the immature Heermann's, an extensive inroad having taken place within a few days. There was a similar intrusion of adult Heermann's Gulls about the middle of July, 1892. Both instances furnish examples of migration indicated solely by increased abundance, as in neither case were the birds observed *in transitu*. A Dark-bodied Shearwater was shot and another was seen. Both were moving southward.

From the 20th to the 25th I did not make any observations, but my boatman, who was out each day, informed me that on the 20th and 21st a good many California Murres went down the coast, and that Pigeon Guillemots were more numerous than at any time since he had been with me.

On the 26th a few adult Marbled Murrelets appeared—the first of the season. They came from the northward in couples, and passed with great rapidity down the coast. There was quite a movement southward of California Murres, chiefly of individual birds. A few scattered Northern Phalaropes were also seen going south. A female that was taken displayed no signs of recent oviposition. It had retained more of the breeding plumage, however, than the one shot on the 11th.

The passage of individual California Murres down the coast continued on the 27th as on the 26th. One Murre, frightened by being shot at, turned upon its course and flew northward for some distance. Then it rose higher in the air, apparently to get its bearings, altered its course so as to head south again, and finally descended to the ordinary level of flight. This maneuver was resorted to a second time before it got fully in line with the southerly trend of the shore. The whole circumstance seemed to indicate that the bird recognized the landmarks, and was able to determine the direction by them and regain its former course.

There was also considerable migration south in Northern Phalaropes on this day. In two hours during the forenoon seven small companies rounded Pt. Pinos—the largest one had nine birds in it. The testes of a male that was shot were those of a bird that had bred. A visit to Seal Rocks revealed that a large flock of Black Turnstones was occupying the place of the few individuals found there on previous visits (the last occasion being July 18th), proving an invasion from another locality, presumably from the north, as in the case of the Northern Phalaropes.

Very little migration was observed on the 28th. It was confined to the California Murre and Northern Phalarope.

On the 30th conspicuous movements took place, particularly in the Marbled Murrelet and California Murre. Adults of the former species passed down the coast, singly and in couples, all the forenoon. The height of their movement was during the morning. They flew swiftly, and rounded Pt. Pinos chiefly between the buoy and the shore. No young birds were seen. The movement of California Murres was greater than at any time before during the season. Single birds and little strings were continually passing south, near the Point and far out, the whole time I was on the water. There appeared also to be a steady migration southward of California Brown Pelicans, though not on so large a scale by far as in the Murres and Murrelets. Northern Phalaropes showed a slight increase over former numbers. One large company and several small ones were observed, *en route* to the south.

July 31st there was a dense fog that lasted until noon. It had the effect to deflect the path of migration so that the birds came near to the shore. During half an hour, at eight o'clock, when the fog was densest, a large number of Dark-bodied Shearwaters on their way down the coast passed within a few hundred yards of the surf in front of the Seaside Laboratory. Among them was a Pink-footed Shearwater. It appeared to be bewildered by the fog. It was the first one of the season met with. No Dark-bodied Shearwaters were seen on the 26th and only one on each of the following days up to the 31st. These stragglers were all going south.

Quite a number of Northern Phalaropes were flying about at random during the forenoon of the 31st, apparently lost in the fog. Marbled Murrelets were moving down the coast as on the 30th, but not in such large numbers. There were some young ones among them. With the exception of a single trio, two adults and a young bird, the Murrelets flew in couples. There was some migration of California Murres, but not nearly as much as on the 30th. After the fog cleared away there appeared to be a decided migratory movement southward of Heermann's Gulls, small bands passing down the coast in frequent succession. They flew near the shore and rather high in the air. There was a directness and steadiness of flight not observed before. They appeared to be bent upon a journey, not merely going and coming from feeding grounds. As already stated, previous indications of migration in this Gull had been manifested solely by increased abundance, particularly noticeable in the adults. Four Pomarine Jaegers were seen after the fog. They followed the same path of movement as the Heermann's Gulls. A solitary Cassin's Auklet was taken. It was a female, apparently a breeding bird of the season. This species had not been noted before during the summer.

Western Gulls gained in numbers during the month instead of diminishing as they appeared to do in July, 1892. The great increase in Heermann's Gulls, however, rendered them less prominent.

From the above detailed account, it will be seen that increase in abundance in some species of summer and extensive migratory waves in others, followed by intervals of scarcity, and the appearance of boreal species, were the chief features of migration in July.

Excluding the Scoters and Black Turnstone, the northern birds to appear were Rodgers's Fulmar, Harlequin Duck, Marbled Murrelet, Northern Phalarope, Pomarine Jaeger: all except the last one appearing during the first half of the month.

Conspicuous migratory waves occurred in the Dark-bodied Shearwater (on the 2d and 31st), in the California Murre (at intervals from the 19th to the 31st, high-water mark for June and July being reached on the 30th), in the Pigeon Guillemot (on the 9th, and probably on the 20th and 21st), in the Marbled Murrelet (on the 30th and 31st, especially on the 30th), in the Northern Phalarope (from the 27th to the 31st, forestalling the greater waves of August), in the California Brown Pelican (on the 30th), and in Heermann's Gull (on the 31st).

A notable feature in the migrations of July was the length of time when there were no migratory waves in the California Murre and Dark-bodied Shearwater, only passing stragglers occurring in either species.

August.—The greatest flight of Dark-bodied Shearwaters I observed during my stay occurred August 1st. There was no fog and they kept well away from the land. Three miles off shore they began to pass the boat in great numbers, and as far out as I could see at eight miles they were equally abundant. All passed rapidly down the coast as upon former occasions, except at midday when a few flew about at random, apparently temporarily halting in their migration. One was seen on the water. It was with a Black-footed Albatross. Accompanying the Dark-bodied Shearwaters were a white Fulmar, probably Rodgers's, and nine Pink-footed Shearwaters. The latter came straggling along at intervals, not more than two being seen together. Many Northern Phalaropes were observed two miles and outward from land. They flew up the coast as well as down, indicating that there was a pause in their migration also. Marble Murrelets were not numerous. Their path of migration was near the shore. California Murres were migrating as on previous days, but apparently in smaller numbers. They were all within three miles of the land. Three Pomarine Jaegers were seen, and an immature California Gull was taken, the first example of the season.

On the morning of the 2d two young Marbled Murrelets were found on the bay near the Seaside Laboratory, and quite a number of adults were seen on wing heading down the coast. Many young Pigeon Guillemots and some adults were on the water. They were very tame, and appeared to be tired birds resting after a night's migration.* Previously the adults had been very shy, usually taking wing out of reach of gun shot when the boat was turned toward them. Only three were seen off the water—an adult accompanied by two young. They were going south. Migration was slight in the California Murre on this day. Only a few were seen, and these *in transitu*. It was somewhat foggy and Dark-bodied Shearwaters, on their way down the coast, came within half a mile of the land. Not so many were noted as on the day before, and only two Pink-footed Shearwaters. An adult female Red Phalarope was captured alive. Northern Phalaropes were common, passing and repassing up and down the coast. Two years before on this day, instead of there being an eddy in their migration there was a great wave southward.

In my notes for Aug. 3d I find the following with regard to the Western Gull: "This species is now migrating in considerable numbers. At two o'clock many were moving south near the shore off the Laboratory. Adult birds are plentiful, but immature ones are more abundant. A good many were on the kelp, and some at Seal Rocks."

Heermann's Gulls swarmed on Seal Rocks. The sides toward the land were literally hidden by them. They were also scattered all along the kelp within the bay, and

* A similar instance occurred Aug. 6, 1892.

near Pt. Pinos there was a great gathering of them, evidently attracted by a school of "blue fish" that were leaping to the surface. As I came in from the ocean at two o'clock, a succession of small flocks passed the Seaside Laboratory, closely following the shore-line in the direction leading southward. On no previous day of the season was this species so abundant. In all situations the majority were dark birds, showing that the migration of the young was fully under headway. Young birds outnumbered the adults Aug. 1, 1892. Although it was foggy during the morning, the only Dark-bodied Shearwaters seen were one small flock and a few couples. Northern Phalaropes were common, but no rush took place. A female Surf Bird, apparently a bird of the year, was shot at Seal Rocks. Two others were with it. There was also a large flock of Black Turnstones on these rocks. The tide of migration was in the ascendency in the California Murre. Besides single birds and companies in indian file, one large wedge-shaped flock was seen, the first of the kind for the season. All came from the northward, passed quickly by, and disappeared to the southward, following the line of the coast in their flight.

On the 4th it was foggy during the morning, but afterward the fog retreated several miles out to sea, leaving a clear highway along the shore. Northern Phalaropes and California Murres were the only birds to appear in numbers to take advantage of it. Numerous small parties of the former species and some large ones came from the northward and passed down the coast. Migration in the Marbled Murrelet and Dark-bodied Shearwater was very feeble. Only a few individuals of either were noticed. Over two miles out from land several Surf Birds were seen flying southward. The Parasitic Jaeger was added to the list of the migrants from boreal regions, two individuals being noted. In 1892, one was taken and another seen on the first day of the month.

6th. When I reached the beach in the morning small flocks of Heermann's Gulls were passing down the coast at brief intervals, and also Western Gulls in fewer numbers. No movement of this kind had been observed before at so early an hour in the day. It was high tide on this day in the migration of Northern Phalaropes. There were large flocks, small companies, and single birds. Most of them were flying southward, following the shore-line, sweeping inward at Monterey Bay and outward after passing Pt. Pinos. Scarcely any were on the water, and comparatively few went up the coast. Many passed within a quarter of a mile of the shore, although there was no "low fog." Several small companies of Cassin's Auklets were found two or three miles out on the ocean. They were apparently the vanguard in the migration of this species in this vicinity, as but a single one had been met with before. Many solitary California Murres and little parties of half a dozen or less passed south. A bird of the year, under the charge of an adult, was captured on the ocean several miles north of the buoy. It was the first one I saw. As its wings were not sufficiently grown to enable it to fly, it was probably hatched not far north of Pt. Santa Cruz. Marbled Murrelets did not appear in any numbers. All that were seen were adults, flying southward in pairs. Two large straggling flocks of Dark-bodied Shearwaters, going south, were seen two or three miles north of the buoy. A Pink-footed Shearwater was obtained from among them. It was the only one noticed.

7th. Five miles north of the buoy numerous companies of Cassin's Auklets, varying in size from half a dozen to a score of individuals, were scattered about on the water.

They were apparently thoroughly tired out. Many were so weak of wing they struck the crests of the waves frequently in flying short distances to keep out of the way of the boat. Two young California Murres, each accompanied by an adult, were found five miles out from land in the direction of Pt. Santa Cruz. Quite a number of old birds were on the water, and many solitary ones and little parties in files were flying down the coast. A Pacific Fulmar—a female of the dark phase—was captured about three miles off shore. It was very lean and in very worn and faded plumage, and had apparently not bred during the season. Four or five miles out a good many Dark-bodied Shearwaters flew by the boat in a southward direction. Only one Pink-footed Shearwater was observed. Not so many Northern Phalaropes were encountered as on the day before.

There was almost a complete cessation of migration on the 8th, 9th, and 10th. Cassin's Auklets were not as abundant on the 8th as the day before. Between six and ten miles northwest of Pt. Pinos a good many little companies, however, were resting on the water. Fewer were seen on the 9th. They flew without difficulty. A considerable number of California Murres, in little groups, were on the ocean between Pt. Pinos and Seal Rocks on the 10th. A few Dark-bodied Shearwaters and one Pink-footed Sheawater were noted on the 8th. On the 9th, over ten miles northwest of Pt. Pinos, quite a number were seen flying eastward. A few others, nearer Pt. Pinos, were flying south. A few Northern Phalaropes were seen on the 8th and 9th. None of them were on the water, and as many went up the coast as down on the latter day. Only a small flock of Black Turnstones was at Seal Rocks on the 10th, the great flock that had been there having disappeared.

The 11th was one of the rare days when there was no fog, when the shore-line could be seen for miles. The sea too was calm. It was a day of great migration in Dark-bodied Shearwaters. They were passing Pt. Pinos all the forenoon about three miles off shore. The eye could follow them a long way as they came down the coast and disappeared to the southward. There were two almost continuous streams of them made up of straggling companies and loose flocks. At nine o'clock and again at eleven some of the latter were of such large size the two streams became merged into a single broad one at least an eighth of a mile in width. One of these flocks was estimated to be two miles in length. Leadership appeared to be exercised among them, for one of the birds, apparently seeing I was making havoc with my gun in a flock just in advance, left the flock he was in and flew back along the advancing column, and as he passed by the birds sheared off to the seaward, going past the boat out of range. The whole manœuvre was so obvious that my boatman, who had also been intently watching it, unconsciously to me, exclaimed, "that bird must be some sort of a general."

An adult Pacific Fulmar of the dark phase was shot as it was resting upon the water. It had apparently dropped out of the ranks of the Dark-bodied Shearwaters, for I had been over the spot where it was taken only an hour before and it was not there then. Its plumage was greatly worn, and its ovary had no appearance of recent functional enlargement. There was a good deal of migration in the California Murre and some in the Marbled Murrelet, but only one or two companies of Northern Phalaropes were seen.

13th. During the morning there was a low fog hanging over the bay and ocean. It seemed to arrest migration. After it had risen many Northern Phalaropes, chiefly in small flocks, passed down the coast. Dark-bodied Shearwaters were migrating in smaller numbers than on the 11th, and nearer to the land owing to the fog. There was quite a flight of California Murres, especially after the fog. A marked increase was noticeable in the size and number of the flocks of California Brown Pelicans coming into the bay in the morning from the direction of Pt. Cypress. A Pomarine Jaeger was shot, and four other Jaegers were seen.

14th. There had evidently been a migratory movement of adult Marbled Murrelets the night before or early on the morning of the 14th, for many pairs were found resting on the water between the Seaside Laboratory and the buoy, from half a mile to a mile off shore, during the forenoon. They did not attempt to fly, but dived to escape pursuit. Only one was observed on wing. On former occasions but very few had been seen upon the water. Several of the pairs were secured. The birds of each pair proved to be mated, one being a male and the other a female. They displayed strong attachment for each other. If one was shot the survivor would begin to call and look anxiously about for its mate, or if they became separated in diving, one would call and the other respond as soon as they came to the surface. A touching instance of fidelity occurred a few days before. A female had been shot and the male followed the boat as we returned to land, finally alighting near it and looking toward us in evident distress called piteously. Only three young Murrelets were noted. There was scarcely any migration in California Murres or Northern Phalaropes. Quite a number of the former, however, were on the water. But one Dark-bodied Shearwater was seen.

But little migration was in progress on the 15th and

16th. On the latter day more California Murres were found on the water than any time before during the summer. One flock numbered fully a score. Several young of the year were seen. Each one was under the guardianship of an adult. A male Pacific Fulmar of the dark phase was taken on the water well toward the eastern shore of the bay. Its plumage was bleached and worn, and new feathers were beginning to appear. The testes had apparently been dormant during the breeding season.

Migratory movement on the 17th was limited chiefly to the California Murre. Many were on the water, but the greater number were pursuing their way south. One flock of migrants had thirty in it. Four companies of White-winged Scoters in high black plumage came into the bay from the direction of Pt. Cypress during the forenoon. They were adjudged to be recent arrivals from the breeding grounds, because their general mien and their plumage was so entirely different from that of the ragged and faded birds found not far out from the surf along the sandy beaches earlier in the summer. In 1892 there was a similar appearance of these birds in high feather.

Migration in the California Murre was greater on the 18th than upon any previous day of the season. Not only did they appear in quicker succession, but large wedge-shaped flocks were numerous. A good many companies were on the water, but these were insignificant in numbers compared with those winging their way southward. There was no migration apparent in other species, except in the Northern Phalarope and Dark-bodied Shearwater. In both it was slight.

20th. There was a heavy fog during the forenoon until about eleven o'clock, when it lifted for awhile, settling down again between twelve and one. After one o'clock the fog wholly disappeared. Decidedly the greatest movement of California Murres during my sojourn took place on this day. During the first part of the forenoon great numbers were going down the coast within a few hundred yards of the surf. I spent several hours between the buoy and the outer rocks at Pt. Pinos watching them pass by. They moved chiefly in large wedge-shaped flocks, of greater size on an average than any previously seen. When the fog lifted we went out about three miles north of the buoy. Two miles from land and beyond great numbers of Murres were on the water, scattered about in large companies. They appeared to be very tired. Many tried to fly when approached, but most of these fell back upon the water after flying a few yards. Some of them did not rise high enough to keep from striking the water with their wings, and a high wave generally threw them back upon the water. One was discovered asleep with his bill tucked under his wing. He did not wake until the boat was almost upon him. A young bird with wings not developed sufficiently for flight, was taken. It was under the care of an adult. No other was seen.

Over two miles off shore, a male Rodgers's Fulmar was secured. It was flying about apparently searching for food. Its generative organs had the same degenerate appearance as those in the Fulmars previously taken. Its plumage was much worn. It was also moulting and new feathers were appearing. There was some migration in Dark-bodied Shearwaters. They appeared in twos, threes, and fours during the fog, passing within a few hundred yards of the surf between the Seaside Laboratory and buoy as they made their way down the coast. Two were resting on the water with the Murres. Many small companies of Northern Phalaropes were journeying southward. The fog also drove them inshore. They seemed to be confused by it. At the Point some hesitated and alighted on the water. They were apparently not tired, but afraid to venture out on the open ocean, for they took wing as soon as the boat approached them. Upon no previous occasion were so many seen on the water. Some flew back into the bay.

21st. No migration was observed in the California Murre. Only a very few were seen anywhere on the water, the great numbers of the day before having disappeared. Two California Gulls were taken—the second and third examples of the season. They were in the company of Western and Heermann's Gulls. Small parties of Dark-bodied Shearwaters, passing south, were found from two to five miles out from land, north of the buoy. There was quite a continuous flight of them. But one Pink-footed Shearwater was noticed. It was flying southward alone. Few Northern Phalaropes were seen.

22d. It was foggy in the morning and at intervals during the rest of the forenoon. Many Dark-bodied Shearwaters were following the shore-line south close to the land during the morning. When the sun broke through the mist their path of migration receded several miles from the shore. About midday the fog banked in the northern and western parts of the bay, leaving a partially clear strip along the south shore from a little north of the Del Monte beach to Pt. Pinos. Great numbers of Dark-bodied Shearwaters passed outward along this open highway, keeping just without the denser mist. There were some large flocks, but small ones greatly predominated. At times the flocks appeared so quickly one after the other that they formed an almost unbroken column. Three Pink-footed Shearwaters followed the path of the

Dark-bodied. They kept to themselves, however, appearing singly when there was a lull in the latter species. Northern Phalaropes did not occur in any numbers. There was a small flight of California Murres—of solitary birds and small squads. Some sixteen Pomarine and Parasitic Jaegers were seen. Three were taken at one spot. Quite a gathering of Gulls had collected there over several dead ones that had been thrown overboard as decoys. These in turn attracted a half-dozen Jaegers that were passing, an Arctic Tern, and two large Terns, probably the Royal. The fog had suddenly set in toward the south shore of the bay, deflecting migration, and putting the boat in the path of the Shearwaters and other birds migrating at the time. Among the Gulls that decoyed was a California Gull.

23d. A low fog closed down upon the bay and ocean at intervals during the entire forenoon, having the usual deflecting influence upon migration. Small flocks of Cassin's Auklets were going south all the forenoon. They rounded Pt. Pinos in the vicinity of the buoy. A few individuals were seen on the water. But little migration occurred in the California Murre. Several adults with young were on the water. There was considerable migration in Pomarine and Parasitic Jaegers. A single Long-tailed Jaeger was taken. Dark-bodied Shearwaters in small flocks were passing south during the whole forenoon. They were not nearly as numerous, however, as the day before. Many passed between the Point and buoy. There was a good deal of migration in Northern Phalaropes. When the fog was thickest, they showed a disposition to stop on the water. A notable event of the day was the capture of two Sabine's Gulls—an adult and a bird of the year. They were apparently migrating.

24th. I did not go out on the water until after nine

o'clock. There was a heavy cross sea and a strong west wind. The sky was clear. It was hard work for my boatman to row against the wind, and most of the forenoon was spent in getting out to the buoy. A great many Northern Phalaropes were trying to make their way southward. They were following the south shore of the bay and had to breast the full force of the wind. Many became tired out, making short flights, stopping to rest between. The spirit of migration was strong in them. The wind blew so hard that the Brandt's Cormorants returning to their rookery could not keep in any order, but struggled against it in confused straggling flocks. Numerous flocks of Phalaropes were on the water just inside of the bay off Pt. Pinos. They were as mindful as we were to face the waves. If a white cap suddenly developed in front of them they flew lightly over it, immediately settling again on the water. A fine illustration of migration retarded by a strong head wind was afforded in this incident. There was no migration in other species. Perhaps the wind checked it.

25th. There was but little migration on this day, although the sun rose in a clear sky—a rare occurrence in this region in the summer months. An adult male Red Phalarope was secured as it was resting on the water about a mile off shore. It was the second specimen of this Phalarope to be positively identified, although individuals were thought to have been seen occasionally after the capture of the one on the 2d. Northern Phalaropes—single birds, and in several instances little companies—were distributed about on the water between the Seaside Laboratory and the buoy.

27th. As soon as I got out on the water in the morning, I noticed that there was an extensive movement of Phalaropes in progress. As there was no "low fog," I was surprised to find them within a few hundred yards of the beach near the Seaside Laboratory, flying out toward Pt. Pinos. Their large size and light-colored backs soon made it apparent that they were not Northern, but Red Phalaropes. At the buoy they were seen to turn Pt. Pinos and head south in the manner observed all along in other migrants. The majority were in small flocks. Some individuals, however, were migrating alone. Toward noon the flight began to subside, but none apparently stopped on the water. There was a greater flight of Northern Phalaropes. It continued without abatement as long as I was out on the water. Few of them came nearer to Pt. Pinos than half a mile. Four solitary Black-vented Shearwaters and one couple passed the boat, as it was stationed near the buoy, on their way down the coast. Their advent, seemingly portending the beginning of a migration later than that of their dark-bodied congener, was not wholly unexpected, for one was captured August 10, 1892. There was some movement in Dark-bodied Shearwaters. Its inner edge reached the buoy. A second Arctic Tern and two California Gulls were taken. The latter species was apparently becoming common. Several Surf Birds, proceeding southward, were seen out on the ocean near the buoy. Passing individuals had been observed at different times along since the 4th.

Few white-headed Western Gulls were seen toward the close of the month. Birds of the year and older immature ones, however, were numerous. The same remarks apply about as well to Heermann's Gull, for dark birds were almost the only ones met with toward the end of my stay. The two species seemed to become equal again in numbers at the last.

It was not satisfactorily ascertained whether any migration occurred in the Brandt's Cormorants breeding in the vicinity. There were fluctuations in abundance that may have been due not to shifting of fishing grounds, but to departure of adult birds and arrival of others later from further north.

The peculiar features of migration in August, as compared with July, were the larger number of birds that temporarily halted by the way, the greatly increased size and frequency of the waves, and the greater prominence of boreal species.

The additional northern birds to arrive were the Red Phalarope, Surf Bird, Parasitic Jaeger, and Pacific Fulmar (dark phase), during the first half of the month, and the Arctic Tern, Long-tailed Jaeger, and Sabine's Gull, during the second half.

Conspicuous migratory waves were observed in the following species:—

Dark-bodied Shearwater, on the 1st and 2d, the wave beginning July 31st and reaching its height on the 1st; on the 6th and 7th; on the 11th and 13th, the height of the wave perhaps occurring on Sunday, the 12th; from the 20th to the 23d, the height being reached on the 22d; on the 27th.

Marbled Murrelet, on the 1st and 2d—the aftermath of the wave of the closing days of July; on the 14th, manifested solely by the presence of the birds on the water.

Northern Phalarope, from the 3d to the 7th, the height being on the 6th; on the 13th; on the 20th; on the 23d and 24th; on the 27th.

California Murre, from the 3d to the 7th; from the 11th to the 13th, the height probably being attained on the 12th, Sunday; from the 17th to the 20th, the height apparently being on the 20th.

Cassin's Auklet, on the 6th and 7th, manifested solely by birds on the water; on the 23d.

Pomarine and Parasitic Jaegers, on the 22d and 23d.

Red Phalarope, on the 27th, probably beginning on Sunday, the 26th.

The Western and Heermann's Gulls have been omitted from the list, as their fluctuations in abundance were not closely followed at the last.

Summary.—As has been shown by the facts presented, early southward migration was indicated by increased abundance in certain "summer species," by the appearance of species not previously observed, and by the passage of migratory waves, the birds being seen actually *in transitu*. As early as the latter half of June the tide of migration apparently began to set southward, migratory waves seemingly occurring in the California Murre and Dark-bodied Shearwater. During the first fortnight of July northern birds began to appear, the Northern Phalarope being the most notable example. There were extensive waves of "summer species," particularly during the latter part of the month. The closing days were signalized by conspicuous waves in the Marbled Murrelet and Northern Phalarope, and by the advent of the Pomarine Jaeger. During August the waves assumed much larger proportions and increased in frequency. Boreal birds became prominent, the waves of the Northern Phalaropes from the outset rivalling those of species summering in the region. The Sabine's Gulls, the Arctic Terns, and the wave of Red Phalaropes, appearing toward the close of the month, foreshadowed the great autumnal migrations that were to follow, as the Pomarine Jaeger, Marbled Murrelet, and Northern Phalarope, at the end of July, foreshadowed the movements of August.

While the general tendency was toward increase in size of the waves as the migrations advanced in a species, lesser waves also intervened between larger ones. Small waves were sometimes immediate forerunners of large ones. Individuals apparently preceded the first waves of northern species. Usually a large wave extended over several or more days, beginning gradually, reaching a day of maximum height, and then subsiding, being followed by a period when little migration took place in the species. With the progress of migration these lulls generally became of shorter duration, and were marked by increasing numbers of birds temporarily pausing by the way, resting upon the water or flying about at random. When a number of species were migrating at one time, it did not uniformly happen that the height of migration occurred upon the sâme day in each, for often the waves of some were waning while those of others were waxing. The California Murre typically exemplified early southward migration in species breeding in the region and the Northern Phalarope in species breeding in boreal regions.

"Barren birds" did not play an important part in the migrations, and young birds of the year did not precede the adults. In some instances young birds were found accompanying the adults, as in the Marbled Murrelet on July 31st. Such young birds, weak of wing, drop by the way, furnishing seeming instances of prior occurrence of young birds during the early movements of species into regions south of their breeding habitats.*

It has been seen that the Dark-bodied Shearwater, a highly pelagic species, followed the coast-line in migrating in the same manner as the Northern Phalarope. That the

* The *mere* occurrence of the young in a given locality before the presence of adults has been detected proves nothing beyond the bare fact that young birds were observed there earlier than adults. It does not prove that they left the region of their birth in advance of their parents, any more than the habitual absence in a locality of a species breeding to the northward and wintering to the southward of it, proves that the species does not migrate.

Shearwaters were guided on their journey by the land is shown by their conforming their course to the inward sweep of the shore-line at Monterey Bay, and by their deflecting their line of flight toward the land during dense fog.* The Brandt's Cormorants breeding in the neighborhood, and well acquainted with the surroundings, were not so dependent upon the land for guidance, finding their way readily in a fog over a placid sea from their fishing grounds to their rookery.†

Although the migrations in summer off Monterey are extensive, they are insignificant in comparison with the movements that follow in autumn.‡ Owing to the lack of proper situations for rookeries, there was not a large breeding population,§ and hence there could not be extensive movements in birds breeding in the immediate vicinity. The food-supply is temporarily very great in such a thinly populated region, being far in excess of the demands of the breeding colonies—a condition highly favorable for early southward migration. The area below the line of snow and ice in winter in North America is

* Shearwaters and other birds habitually flying near the surface of the water must from necessity migrate near the land if they desired to keep in sight of it, particularly in the region of such constant fogs as the vicinity of Monterey in summer. The coast at Pt. Santa Cruz, though mountainous and less than twenty-five miles away, was scarcely ever discernible from Pt. Pinos owing to the foggy state of the atmosphere.

† Pelagic migration of birds, especially in its relation to isolated oceanic islands, will be further considered in another paper, now in preparation.

‡ This is also the case in the smaller land birds in upper South Carolina. See "Auk," vol. ix, pp. 33-39.

§ Sea birds are necessarily very local on this coast during the season of reproduction owing to the isolated character of their breeding places. Certain species are found associated on some islets and not on others. Such distribution is probably due sometimes to lack of room, and not to actual scarcity in a species or to the physical conditions. All sea birds having the same breeding range cannot find habitation at one spot, no matter how favorable the situation may be.

comparatively small. The vast region above this line, abounding in summer with the means of supporting bird life, must be largely depopulated before winter. Hence there is southward migration of birds—migration that extends even into a land of summer in the Southern Hemisphere. Winter also enforces depopulation of Arctic seas. In the lapse of time sea birds like the California Murre have seemingly learned, after the cares of reproduction are over, to move further south in the sparsely populated region of great food store, making room for the countless hosts that must leave the region of their birth before the chilling breath of winter has turned it into a region of desolation and famine.*

GENERAL REMARKS ON THE BIRDS OBSERVED.

Only species of which specimens were taken are mentioned in the notes that follow. When no year is given, 1894 is to be understood. The determinations are based in most instances upon printed descriptions and are therefore in a measure provisional.

Æchmophorus occidentalis. WESTERN GREBE.—Toward the end of summer this Grebe became tolerably common. A male, taken July 2d, was the first individual of the season met with.

Colymbus nigricollis californicus. AMERICAN EARED GREBE.—An adult female was captured on the bay July 16th. This was the only example of this species observed during the summer.

Podilymbus podiceps. PIED-BILLED GREBE.—A single specimen that was found dead upon the water near the Seaside Laboratory, August 11th, was the only one seen.

*As there is early southward migration in temperate climates in breeding representatives of "resident" land birds, it is not surprising that an apparently similar migration exists in "resident" sea birds. In this connection, see "Auk," vol. ix, pp. 33-39; xi, pp. 100, 101, 103, 104, 108, 109.

Lunda cirrhata. TUFTED PUFFIN.—Individuals were quite common from the outset. Most of them appeared to be south bound migrants. I was told that a small breeding colony was located every year on an islet in Carmel Bay. The "Sea Parrots" apparently have a great deal of curiosity, for they were often observed to change their course when flying by so as to pass near the boat. One, on being shot at, flew back in the direction from which it came for a considerable distance; then it returned, passing close to the boat, and seemingly scrutinizing it.

In a male shot July 13th, the ear-tufts are very short, being worn off at the end. In another male, August 4th, one ear-tuft has entirely disappeared and several worn feathers alone remain of the other. Birds of the year, strong of wing, were observed as early as the first week of August.

Ptychoramphus aleuticus. CASSIN'S AUKLET.—So far as determined, Cassin's Auklet occurred only as a migrant. It was first noted July 31st.

Brachyramphus marmoratus. MARBLED MURRELET.—The Marbled Murrelet appeared early in July, but it did not become common until at the end of the month. With the exception of a single female taken August 15th, all the adults secured were in the "marbled" plumage. This female was nearly in complete winter garb. Its ovary did not have the appearance of recent functional enlargement, as was the case in the other females examined. This circumstance may account for its earlier assumption of the winter dress.

Brachyramphus hypoleucus. XANTUS'S MURRELET.—Of a Murrelet shot July 28th on the ocean three miles north of Pt. Pinos, Mr. Ridgway has written me: "—the bird is *Brachyramphus hypoleucus* with an unusually short bill. It is otherwise a normal *hypoleucus*." The tarsi of this specimen are scutellate in front—as distinctly scutellate as in *Synthliboramphus antiquus*. This record apparently extends the known range of this species, for I find no mention in the general literature of its occurrence so far north.

Cepphus columba. PIGEON GUILLEMOT.—This species was rather common at Monterey Bay upon my arrival. It increased in abundance with the progress of the migrations. A breeding colony was apparently established on the south side of Carmel Bay at the time of my visit, June 25th, for fifty or more adults suddenly appeared from the water's edge of the rocky islets along the shore when I fired my gun. Before not one had been in sight. A party of Stanford University students found a nest with young the day following in the same locality.

Uria troile californica. CALIFORNIA MURRE.—Although a common bird during the latter half of June, the California Murre was apparently only a migrant in the vicinity of Monterey in summer. As has already been stated, young birds, unable to fly and under the care of adults, appeared early in August, probably from a rookery somewhere in the vicinity of Pt. Santa Cruz. These young birds were expert divers. When an adult and its charge were approached, the young bird would dive first. If the two became separated, the old one would call loudly, and as soon as the young responded the old bird would dive, coming to the surface at the spot where the young one had taken refuge. I shot an adult, and quite unexpectedly it proved to be a male. California Murres, on being approached, have a curious habit of frequently dipping their bills into the water. They also have the habit of standing erect on the water and flapping their wings, apparently to free the plumage from water.

Stercorarius pomarinus. Pomarine Jaeger.—During August of both seasons this Jaeger became quite common. In 1892, it was first detected August 1st, and in 1894, July 31st.

Stercorarius parasiticus. Parasitic Jaeger.—Nearly the same remarks apply to this species as to the preceding one. It was not positively identified, however, the second season before August 4th.

Both Jaegers were very bold. Often they would decoy to Gulls thrown overboard to attract them. Several times individuals came and hovered over the boat for a moment, apparently drawn by the dead birds plainly in view in it. They did not linger, however, over wounded companions as did the Gulls. Upon one occasion three of them tried to capture a small bird that was flying over the bay not far out from the shore. Although their attack was a concerted one, the bird succeeded in dodging them and keeping above them, finally escaping to the land.

Stercorarius longicaudus. Long-tailed Jaeger.—August 23d an adult male Jaeger was obtained that is apparently this species. The tarsi were light bluish in life, in marked contrast with the black of the toes. In drying the color of the tarsi has become olivaceous. The slate-gray of the under tail-coverts prevails over the abdomen. The nasal shield and unguis are about equal in length.

So far as I am aware this species has not been previously reported from California.

Larus occidentalis. Western Gull.—With the exception of the Brandt's Cormorants, the Western Gulls are the most prominent birds of the bay during the early part of summer. Later, although increased in abundance, they are overshadowed, for some weeks at least, by the Heermann's Gulls. No Gull rookeries were discovered. The beds of kelp growing along the shore of the bay a short distance out from the surf were favorite resting-places for both Western and Heermann's Gulls. They freely associated on the kelp and elsewhere. On the rocks and on the open water, especially where there were schools of fish, Brandt's Cormorants and California Brown Pelicans were found in their society. Often great congregations of Gulls and Cormorants were formed where the fishing was good. Both Gulls were very unsuspicious as a rule, and allowed the boat to draw quite close before taking wing. They invariably decoyed when dead birds were thrown out on the water to lure them, large flocks as well as solitary birds being attracted.

Larus californicus. CALIFORNIA GULL.—The first one was noticed August 1st. Toward the close of the month they became somewhat common. They were found in company with other Gulls.

Larus heermanni. HEERMANN'S GULL.—Immature birds were common the latter half of June, but adults were scarce. In the middle of July adults became abundant, exceeding the immature birds or the Western Gulls. By August, birds in dark plumage were in the ascendency, a great inroad having occurred, adult birds, too, having diminished. Toward the last of the month Western and Heermann's Gulls for the second time appeared to be equal in abundance. It is significant that Mr. Henshaw found only adult Heermann's Gulls during the latter part of June at Santa Barbara (Ann. Rep. Chief Engineers. 1876, Appendix J J, p. 497), while I found almost exclusively immature birds at Monterey at the same season of the year.

Individuals began to moult as early as June. Most of the mottled-headed birds of the first part of August had only partially grown tails. Many of them appeared to be fork-tailed, two of the old outer feathers remaining.

Xema sabinii. Sabine's Gull.—An adult (apparently a male) and a female bird of the year were obtained August 23d. They were decoyed within range with dead Gulls. In the adult, some white feathers show in the plumbeous of the head and upper part of the neck. Otherwise it appears to retain the full summer plumage. Mr. Bryant has recorded a previous specimen from California in "Zoe," vol. iii, p. 165. This specimen is No. 379 of the collection of the California Academy of Sciences. It is a bird of the year, and is labelled "San Francisco Bay, Cal., Oct. (10?), 1889;" "From E. F. Lorquin."

Sterna paradisæa. Arctic Tern.—Two females were taken—one, August 22d, the other, August 27th. The former is apparently in summer plumage, but the latter has white mixed with the black on the top of the head.

Diomedea nigripes. Black-footed Albatross.—On the 1st of August two Black-footed Albatrosses were secured about eight miles north of the buoy. These were the first observed. During the rest of my sojourn individuals were seen every few days, one of them within half a mile of the buoy. They were very unsuspicious. One bird, sighting the boat a long way off, came directly toward us and alighted on the water about a hundred yards away. It sat there motionless until dispatched at very short range. Another, quite a distance off, changed its course immediately when a couple of Gulls were tossed into the air to attract it, and headed in a bee-line for the boat, only stopping in its career when cut down, scarcely fifty feet away.

Diomedea albatrus. SHORT-TAILED ALBATROSS.—June 18th an adult male was shot near the Chinese fishing village at Monterey. It was evidently a straggler that had sought safety within the bay on account of its disabled condition, for the outer primaries of the left wing were broken off and the feet bore the marks of gunshot wounds, long healed. It was very tame, and flew only a short distance when pressed.

Fulmarus glacialis glupischa. PACIFIC FULMAR.—As has been stated in a previous part of this paper, three examples of the dark phase were taken in August—on the 7th, 11th, 16th.

Fulmarus glacialis rodgersii. RODGERS'S FULMAR.—A Fulmar with plumage greatly bleached, shot July 12th, appears to be this subspecies, and not a faded example of the light phase of *glupischa*. A specimen, procured August 20th, is undoubtedly typical *rodgersii*, it having enough fresh fall plumage for satisfactory determination. July 2d and August 1st and 17th, three other white Fulmars were seen.

Puffinus creatopus. PINK-FOOTED SHEARWATER.—They were observed both seasons during August as passing migrants, but were not abundant. In 1894, the first one was seen July 31st. The flight of these Shearwaters when migrating is not as direct as that of the Black-vented and Dark-bodied. They circle frequently and cross their track, much as Swallows are wont to do when migrating singly or in small companies.

Puffinus gavia. BLACK-VENTED SHEARWATER.—One was shot August 10, 1892, and six were seen August 27, 1894.

Puffinus griseus. DARK-BODIED SHEARWATER.—After the 16th of June this Shearwater apparently occurred only as a south bound migrant. During the passage of its migratory waves it was very abundant. It was observed both years. A series of forty-seven specimens was secured the second season.

Phalacrocorax penicillatus. BRANDT'S CORMORANT.—The going and coming of the "Shags" in their fishing excursions into the bay from their rookery at Seal Rocks is the most striking feature in the bird life of the vicinity of Monterey during summer. They were abundant in 1892 and still more abundant in 1894. The latter year, also, wedge-shaped flocks were formed earlier and more birds were found at the outset on the rocks along the shore.

Sometimes solitary Cormorants returning to their rookery joined the files of migrating California Murres, and frequently single Murres were observed bringing up the rear of strings of outgoing Cormorants. On one occasion a California Brown Pelican was seen at the end of a line of Cormorants.

Great rafts of these Cormorants collected on the bay whenever "the feed came in." At the distance these gatherings present a very peculiar appearance. The water seems to be thickly set with black sticks, often covering an area of several acres. Gulls, particularly, congregate with the Cormorants upon such occasions.

Two rookeries were discovered; one at Pt. Carmel, and the other at Seal Rocks. June 25th I visited the former, which is situated on a rock, or little islet, in the ocean at the extremity of Pt. Carmel, about fifteen yards from the mainland. This rock rises perpendicularly some forty or more feet above the water. At first sight it does not seem that it can be scaled, but closer inspection reveals that a foothold may be had in the seams and protuberances on its water-worn sides. Only on days when the sea is very calm can the rock be landed upon, and then only from the sheltered channel separating it from the mainland. Fortunately, it happened that the sea was quiet the day of my visit. The following day a party of Stanford University students were unable to land on account of the heavy surf.

We first took a view of the rookery from the mainland. The Cormorants were very tame, remaining on their nests while we clambered down the sloping rocks, and while we stood watching them, on the same level, only a few yards away. They were safe, however, from nearer approach, the deep though narrow channel with its precipitous walls of rock, effectually cutting off further advance. They were equally tame when the boat drew near, as we approached from the water.

The clefts in the sides of the rock were occupied by Baird's Cormorants and the top by Brandt's. There were comparatively few of the former, but of the Brandt's Cormorants there were upwards of two hundred pairs. Their nests covered the top of the rock, every available situation being occupied. The surface was so uneven that all the nests could not be seen from one spot. Standing in one place I counted one hundred and eighteen.

All the nests of the Brandt's Cormorants on the rock contained eggs (apparently in an advanced state of incubation), with the exception of eleven, which had young birds in them. In ten, the young were just out of the shell. In the remaining one, they were as large as "spring chickens." The eggs in seventy-seven nests were counted by a companion. Twenty-one contained four eggs each; thirty-six, three eggs; fourteen, two eggs; three, five eggs; three, one egg. The most frequent numbers were therefore three and four, probably the ordinary clutches.

"Sardines"* were lying in little bunches near the nests, apparently placed there as food for the birds that were setting.

The smell from the accumulated excrement was sickening. The sides of the rock were so daubed that it appeared to be white toward the top. Flies swarmed about the rookery.

It was not until I fired my gun that the brooding birds began to desert their eggs. The Baird's Cormorants were the first to go. Many of the Brandt's Cormorants lingered on the edge of the rock while I walked about among the nests, only a few steps away. Finally all were driven to the water, where they formed a great raft. They began to return as soon as I left the top of the rock.

The rookery at Seal Rocks was much larger than the one at Pt. Carmel. The rocky islet upon which it was located is considerably greater in size and much lower in elevation than the Pt. Carmel islet. From the mainland, less than a hundred yards distant, no nests were in sight, all being on the side toward the ocean, hidden from view by a sort of dividing ridge. The Del Monte drive passes along the shore directly opposite the Rocks. It is a much frequented roadway, and the summer visitors have greatly persecuted the birds with firearms, forcing them to seek shelter for their nests behind the protecting rock.

My first visit to the rookery was made July 2d. As at Pt. Carmel, a landing could be effected only on the shore side of the islet. The resident population was composed

---

*Dr. Charles H. Gilbert kindly identified the "sardines" taken from the gullets of the Brandt's Cormorants during the summer of 1892. They proved to be a species of Rock Cod—*Sebastodes paucispinis*.

exclusively of Brandt's Cormorants. Their nests were crowded so closely together on the uneven surface of the rock that room to place the foot was not always readily found. Some of the nests were on little points of rock, others in crevices, every available spot being utilized. Most of the eggs had hatched. The young were in different stages of growth, varying in size from those just out of the shell to half-grown ones. The larger left the nests when approached, and huddled together on the edge of the islet well above the reach of the surf. There was such a complete mixing up of babies that the old birds must have had some trouble in sorting them out when they returned, for immediately after I landed most of the adults retreated to the water, congregating in a great raft a short distance away. A few of the bolder remained behind for awhile. Several, apparently females, kept close by their young until I approached within ten feet of them, when their courage failed and they took flight, leaving the young to shift for themselves. Two of the larger young birds sought refuge on an outlying rock, separated from the islet by a little channel. They had apparently never been in the water before. They succeeded, nevertheless, in swimming across the channel and climbing up the steep sides of the rock, although a number of times they were buried out of sight by incoming waves.

A vibratory movement of the gular sac, apparently occasioned by fear, was noticed in a number of adults and half-grown young. Most of the adults observed on the rookery appeared to have lost the nuptial filaments.

The general form of the nests was circular, except where wedged in between rocks. They appeared to be constructed entirely of eel grass (*Zostera*).* Those con-

*I am indebted to Mr. H. L. Kimball, an investigator at the Hopkins Seaside Laboratory, for the determination of this plant.

taining the larger young were trampled down. Two typical, untrampled nests yielded the following measurements, in inches:

Outer diameter	22	19
Inner diameter	10	10
Depth	4	4
Height	5½	7

Not many fish were lying about the nests. There were too many hungry mouths to be filled for a store to accumulate as at Pt. Carmel rookery.

It was evident that sanitary measures were not in vogue, for the decaying bodies of several birds were suffered to remain and add to the almost intolerable stench of the excrement deposits. Quantities of feathers were scattered about and there were myriads of flies. Some of the flies accompanied us in the boat most of the way to Pt. Pinos, much to our annoyance.

On the 27th of July all the young observed during previous visits were apparently still on the islet. There were also a few eggs and a few young recently hatched. When I drew near them, the older of the young birds crowded to the edge of the islet and many of them tumbled into the water, where they seemed to be as much at home as their parents.

Phalacrocorax pelagicus resplendens. BAIRD'S CORMORANT.—Save a solitary one seen July 5th about a mile south of Pt. Pinos, the only Baird's Cormorants observed were those at the Pt. Carmel rookery. There were some twenty pairs nesting in the crevices on the sheltered sides of the rock facing the mainland. They were inclined to keep apart from the Brandt's Cormorants, most of them retreating to a rocky point on the mainland instead of joining the raft of Brandt's Cormorants on the water. They were shyer than their larger congener. All had the white patch on the flanks, but in some it was larger than in others. No young birds were discovered. The eggs examined, however, appeared to be well advanced toward the hatching point.

Pelecanus californicus. CALIFORNIA BROWN PELICAN.—In June it was rather common. Through July and August it increased steadily in abundance, toward the last becoming one of the most conspicuous birds of the bay.

Histrionicus histrionicus. HARLEQUIN DUCK.—On July 7th an adult male, on its way down the coast, was captured as it was rounding Pt. Pinos. It was followed a short time after by a second individual.

Oidemia deglandi. WHITE-WINGED SCOTER.—Both seasons White-winged Scoters were quite common off the sandy beaches during the latter half of June. They were present at intervals during July and August. During the closing half of August birds apparently began to arrive from the breeding grounds.

Oidemia perspicillata. SURF SCOTER.—Surf Scoters were also present during summer, occurring in the same manner as the White-winged, except there was no marked inroad toward the end of August.

Ardea herodias. GREAT BLUE HERON.—After the 1st of July individuals were occasionally seen flying over the bay and ocean or sitting upon the rocks along the shore. July 13th a fine adult male was secured on the kelp, about half a mile out on the bay. Another was seen the same day on drifting kelp on the ocean at Pt. Pinos, the sea being very still.

Fulica americana. AMERICAN COOT.—A few apparently bred at a lagoon behind the dunes, about a mile north of Monterey, for they were found there the middle of June.

Crymophilus fulicarius. RED PHALAROPE.—One was handed to me alive on the morning of August 2d. I was told that it swam into the little cove at the Seaside Laboratory and climbed upon a rock, where it was struck with an oar and captured. It was an adult female, and retained much of the nuptial plumage. It was apparently in good health. An adult male was taken on the 25th. On the 27th an extensive wave of these Phalaropes passed the vicinity of Monterey.

Phalaropus lobatus. NORTHERN PHALAROPE.—Appearing July 11th, the Northern Phalarope became very abundant during August as a transient migrant.

Ereunetes occidentalis. WESTERN SANDPIPER.—A male in high plumage was taken July 18, 1892, on the beach north of Monterey. It was with a company of Snowy Plovers.

Symphemia semipalmata inornata. WESTERN WILLET.—About two miles north of the buoy a female was secured August 17th from a company of three that were flying toward Pt. Santa Cruz. A small flock going in the same direction was seen on the 20th near the same spot.

Heteractitis incanus. WANDERING TATLER.—This Sandpiper was not discovered breeding either season. The first were observed about the middle of July. They became very common at the end of that month, frequenting the rocks just above the surf.

Numenius longirostris. Long-billed Curlew.—In July, Long-billed Curlews on wing began to be conspicuous about the bay, evidencing that the nesting season was drawing to a close and migration was under headway.

Ægialitis vocifera. Killdeer.—Killdeers were quite common both summers in marshy situations.

Ægialitis nivosa. Snowy Plover.—In 1892, Snowy Plovers were found commonly on the sandy beaches. The last of June of that season fully fledged birds of the year were abroad. A female taken July 4th contained ova nearly ready for the shell.

Aphriza virgata. Surf Bird.—Besides three found at Seal Rocks on the 3d, a few passing migrants were noted at intervals during August.

Arenaria interpres. Turnstone.—A female was captured July 18, 1892, on the beach about a mile to the northward of Del Monte.

Arenaria melanocephala. Black Turnstone.—Black Turnstones were first met with on the 2d of July, about a half a dozen being found at Seal Rocks. On the 27th, in the same situation, there was a large flock. Early in August there was a marked diminution in the number of these birds found at Seal Rocks. This scarcity was attributed to migration.

Hæmatopus bachmani. Black Oyster-catcher.—They probably bred at Pt. Carmel, for several were seen there June 26th, by a party of students from the Seaside Laboratory. Two pairs were taken by myself at Seal Rocks; one on the 2d and the other on the 18th of July.

COLEOPTERA OF BAJA CALIFORNIA.

(SUPPLEMENT I.)

BY GEORGE H. HORN.

[With Plate xx.]

The following additional account of Coleoptera from the Cape Region of Baja California is the result of the examination of new collections received from the California Academy of Sciences since the publication of my last paper. The result is the addition of fifty-two species, half of which are new. The collections were made by the expedition sent out by the Academy to the Cape Region and the Pacific mainland of Mexico during the month of September, 1894.

One of the most interesting in the matter of distribution is a specimen which I am compelled to refer to *Calosoma Sayi*, unfortunately a female and without legs. The discovery of the male may prove it to be a distinct species.

This as well as the former collections from the Cape Region of Baja California contains a few species identical with those from the Pacific mainland of Mexico, indicating the varied origin of the Cape Region fauna. The types will be found in the collection of the Academy of Sciences in San Francisco.

CARABIDÆ.

*CALOSOMA SAYI Dej. The unique female is rather smaller than those found in the States. Pennsylvania to Texas. San José del Cabo.

DYSCHIRIUS TRUNCATUS Lec. Colorado, Nevada, California, Arizona. San José del Cabo.

*On the day after the proof-sheets of these pages left me a male of the *Calosoma Sayi* was received, enabling me to state with certainty that it agrees in all details with the forms from the Atlantic region.

Bembidium jucundum n. sp. San José del Cabo, Sierra San Lazaro.

Pristonychus complanatus Dej. More decidedly black than the specimens from California. Probably introduced through commerce. Europe, California. San José del Cabo.

Euphorticus pubescens Dej. The unique example is of the Central American type mentioned by Bates in which the surface is decidedly brassy and the thorax slightly broader with more arcuate sides. Occurs from North Carolina southward through Mexico to Santa Catharina, Brazil. San José del Cabo.

Tetragonoderus pallidus Horn. San Diego, Cal., Tucson, Ariz. Sierra El Taste.

Tetragonoderus fasciatus Hald. Eastern United States from Michigan southward, Texas, Arizona, southern California. San José del Cabo.

Apenes lucidula Dej. The specimen referred to this name is duller than usual and the surface sculpture better marked. From the method of variation of the species it seems unwise to describe the unique as a distinct species. There are no Mexican forms allied to it. Eastern United States. Sierra San Lazaro.

DYTISCIDÆ.

Celina angustata Aubé. Atlantic region and Florida to Texas. San José del Cabo.

HYDROPHILIDÆ.

Cercyon rufescens n. sp. Sierra San Lazaro.

STAPHYLINIDÆ.

Tachyporus chrysomelinus Linn. Europe, Atlantic States. San José del Cabo.

SILPHIDÆ.

Silpha truncata Say. Kansas, New Mexico. Sierra San Lazaro.

SCYDMÆNIDÆ.

Eumicrus lucanus n. sp. Arizona. San José del Cabo.

COCCINELLIDÆ.

Scymnus collaris Mels. Canada to Texas. San José del Cabo.

Scymnus cinctus Lec. (*suturalis* Lec.) Louisiana westward to Los Angeles. San José del Cabo.

Scymnus ardelio Horn. Separated by me from the series formerly called *marginicollis* from which it differs in usually red legs and the absence of tubercle on first ventral of male. California, Arizona. San José del Cabo.

CRYPTOPHAGIDÆ.

Tomarus bisignatus n. sp. San José del Cabo.

DERMESTIDÆ.

Attagenus piceus Oliv. Cosmopolitan. Sierra San Lazaro.

HISTERIDÆ.

Paromalus (Carcinops) viridicollis Mars. To this is referred *P. mimeticus* Horn. Arizona, Mexico. Sierra Laguna.

Saprinus oregonensis Lec. The unique differs from typical forms in having the punctures of the apex less extended toward the humeri. Oregon, California. Sierra San Lazaro.

DASCYLLIDÆ.

Scirtes humeralis n. sp. San José del Cabo.

ELATERIDÆ.

HORISTONOTUS DENSUS Lec. Sierra San Lazaro and San José del Cabo.

LAMPYRIDÆ.

MALTHODES LUCANUS n. sp. San José del Cabo.

TELEGEUSIS DEBILIS n. g., n. sp. Sierra San Lazaro.

CLERIDÆ.

CREGYA VETUSTA Spin. Pennsylvania to Missouri. San José del Cabo.

HYDNOCERA DISCOIDEA Lec. Very variable in its coloration. Texas, Arizona. Calmalli Mines and Sierra San Lazaro.

PTINIDÆ.

GIBBIUM SCOTIAS Scop. Europe, Atlantic States. San José del Cabo.

PITNUS PYGMÆUS Ghm. By error this appears in the previous list as a Ptinus.

PASSALIDÆ.

NELEUS TLASCALA Perch. Mexico. San José del Cabo. Identified through a series sent to the National Museum by the editors of Biologia Centrali-Americana.

SCARABÆIDÆ.

OCHODÆUS PENINSULARIS n. sp. Sierra San Lazaro, Sierra El Taste.

CERAMBYCIDÆ.

RHOPALOPHORA BICINCTA n. sp. San José del Cabo.

DECTES SPINOSUS Lec. Middle States to Kansas, Texas. Mexico. Sierra San Lazaro.

LEPTOSTYLUS BIUSTUS Lec. Widely distributed in the Atlantic region, Texas. San José del Cabo.

DYSPHAGA DEBILIS n. sp. San José del Cabo.

CHRYSOMELIDÆ.

LEMA TRILINEATA Oliv. Atlantic States, Texas, Arizona, Mexico. San José del Cabo.

LEMA OMOGERA Horn. This species varies with the elytra entirely yellow.

GRIBURIUS MONTEZUMA Suff. The generic name Scolochrus is adopted by European authors, although three years posterior to Haldeman's name. Suffrian (Linn. Ent., vii, p. 104) suppresses the older name for no valid reason. Arizona, Mexico. San José del Cabo.

METACOLASPIS CONSPERSA n. sp. San José del Cabo.

COLASPIS MŒSTA n. sp. San José del Cabo.

BLEPHARIDA ATRIPENNIS n. sp. San José del Cabo.

CALLIGRAPHA ANCORALIS Stal. Mexico, Sierra San Lazaro.

CREPIDODERA PENINSULARIS n. sp. San José del Cabo.

PSYLLIODES CONVEXIOR Lec. Atlantic States, Texas, Nevada and California. San José del Cabo.

BRUCHIDÆ.

BRUCHUS DISTINGUENDUS Horn. Florida, Arizona. Sierra San Lazaro.

TENEBRIONIDÆ.

TRIPHALUS CRIBRICOLLIS n. sp. Miraflores.

EURYMETOPON PUNCTULATUM Lec. By an oversight, this species appears twice on page 347 of my previous paper.

ANTHICIDÆ.

XYLOPHILUS BRUNNIPENNIS Lec. Canada, Georgia, Arizona, California. El Taste, La Paz.

MELOIDÆ.

MACROBASIS LANGUIDA n. sp. San José del Cabo.

CURCULIONIDÆ.

CLEONUS QUADRILINEATUS Chev. Kansas, Texas, New Mexico, Arizona. Sierra San Lazaro.

OTIDOCEPHALUS ALTERNATUS n. sp. La Paz.

OTIDOCEPHALUS SPARSUS n. sp. Cape Region, probably.

CONOTRACHELUS LUCANUS n. sp. San José del Cabo.

ZASCELIS OBLONGA n. sp. Arizona. Sierra San Lazaro.

COPTURUS SOBRINUS n. sp. La Paz, Sierra San Lazaro, Sierra El Taste.

SCOLYTIDÆ.

XYLEBORUS sp. Closely related to *xylographus*, and a little larger. One specimen. San José del Cabo.

DESCRIPTIONS OF NEW SPECIES.

The following pages contain descriptions of those indicated as new in the preceding list, together with notes on some previously recorded. I have availed myself of the kind permission previously accorded by the publication committee and have added descriptions of a few species from related regions.

BEMBIDIUM JUCUNDUM n. sp.

Form of *axillaris*, but somewhat more robust, piceous, shining, head and thorax slightly bluish-green, elytra with a broad pale vitta with irregular sides extending from base to apex between the third and seventh striæ. Antennæ and palpi testaceous. Head smooth, polished, frontal striæ double, convergent, eyes large and prominent. Thorax nearly twice as wide as long, sides moderately strongly arcuate and narrowing posteriorly, hind angles scarcely evident, not carinate, base with short peduncle; disc convex, polished, median line extremely fine. Elytra oblong oval, one-third longer than wide, humeri obtuse, disc feebly convex, striæ composed of moderate punctures not closely placed, obliterated at apical third, sutural stria at apical half and the eighth impressed, the latter distant from the margin, third interval finely bipunctate, punctures near the third stria. Body beneath smooth and shining. Legs and coxæ yellowish testaceous. Length, .14 inch; 3.5 mm.

This pretty little species belongs to the group xv, as defined by Dr. LeConte (Proc. Acad. Phil., 1857, p. 5), containing *axillare*, *quadrimaculatum et al.*, to which group *B. sphærodcrum* and *cyclodes* Bates (Biol. 1, pp. 147 and 290) probably belong. In fact, the present species seems closely allied to *cyclodes*.

Two specimens. San José del Cabo and Sierra San Lazaro.

CYMINDIS CALIFORNICA n. sp.

Piceous feebly shining, elytra paler, legs testaceous. Antennæ brownish, basal joint paler. Head not closely punctate. Thorax distinctly broader than long, trapezoidal, sides arcuate in front, slightly sinuate posteriorly, hind angles distinct, not prominent, side margin not translucent nor reflexed, disc very feebly convex, coarsely punctate at sides and base, more sparsely at middle, median line fine, sutural margin with three setæ, the middle one short. Elytra very finely striate, striæ with fine indistinct punctures, intervals flat, indistinctly alutaceous and with one series of irregularly placed punctures, each with a very short hair. Body beneath piceous, very sparsely punctate. Length, .30 inch; 7.5 mm.

A small species without translucent side-margin to the thorax and with the thorax very distinctly wider than long. It is therefore more allied to *unicolor* than any other in our fauna. The latter has, however, but one lateral setigerous puncture and the side margin of thorax distinctly reflexed and the disc convex and coarsely closely punctate. The elytral intervals are also closely punctate. In the present species the side margin of the thorax is not at all reflexed and the disc as flat as in *Apenes nebulosa*.

One specimen. San Luis Obispo. (Chas. Fuchs.) This is the first instance of the occurrence of Cymindis in California.

APENES LIMBATA n. sp.

Form of *nebulosa*, head and thorax rufo-testaceous, elytra piceous with a broad lateral pale margin, a humeral lunule and a somewhat sinuous band near the apex testaceous. Antennæ pale. Head finely rugulose and sparsely finely punctate. Thorax alutaceous, faintly wrinkled and very sparsely punctulate, median line distinct from base to apex. Elytra with fine striæ, finely punctulate, intervals distinctly alutaceous, sparsely minutely punctulate. Body beneath and legs testaceous, abdomen piceous, paler at middle. Length, .30 inch; 7.5 mm.

The pale lateral border covers the outer three intervals. The humeral lunule is like that of many Cicindelæ. The apical band is very near the apex and is composed of short lines on the intervals, the innermost one being at the sutural angle.

This species is entirely unlike any in our fauna in its paler color and especially the pale sides of the elytra. This is the first indication of the occurrence of Apenes in the California fauna. One specimen collected at San Luis Obispo and given me by Mr. Chas. Fuchs.

Having had occasion to refer to an essay of Baron Chaudoir on some "Aberrant genera of the group Cymindides" (Bull. Mosc., 1875. ii. pp. 1–61 in sep.), there was observed a species described as *A. opaca*, the name having been used by LeConte in 1866.

HYDROCHARIS RICKSECKERI n. sp.

Oblong oval, slightly broader behind the middle, form of *castus*, beneath black, above piceous with olivaceous tinge, margin of thorax and elytra yellow. Antennæ, legs and palpi yellow, the trochanters piceous. Head finely punctulate. Thorax finely punctulate with three groups of coarse punctures, one near the front angle, a second oblique line each side of middle in front, a third more numerous from the middle of the side obliquely backward. Elytra minutely punctulate with the usual series of coarse punctures. Prosternum not prolonged at apex. Tarsal claws abruptly bent, with a broad lobe at base. Length, .46 inch; 11.5 mm.

The yellow lateral border of the thorax is narrow and well defined, that of the elytra broader but not sharply limited.

One specimen. Harris' Pond, near Santa Rosa, Cal. Ricksecker.

The species of Hydrocharis, four in number, equally divided between the Atlantic and Pacific regions. They are as follows:

Prosternum prolonged in point behind the coxæ.
  Maxillary palpi long. *castus*.
Prosternum not prolonged.
  Piceous with entire border yellow. *Rickseckeri*.
  Piceous slightly bronzed. *obtusatus*.
  Surface pale glaucous green. *glaucus*.

CERCYON RUFESCENS n. sp.

Oval, convex, form of *tristis*, rufescent moderately shining. Head and thorax sparsely finely punctate. Thorax without basal marginal line. Elytra without impressed striæ but with rows of moderate punctures not closely placed, the eighth, ninth and tenth rows with the punctures deeper and closer for part of their length; intervals flat, equal in width, irregularly biseriately punctate. Body beneath colored as above. Metasternal area not well defined, moderately coarsely punctate. Mesosternum oval, acute in front, coarsely punctured. Prosternum strongly carinate. Length, .08 inch; 2 mm.

From its form and the fact that the head is vertical the species seems best placed near *tristis* and *floridanus*, but it differs from all of that series by its pale color, which is not due to immaturity. At first glance it would be taken for an Olibrus.

One specimen. Sierra San Lazaro.

ANISOTOMA MERKELIANA n. sp.

Oval, slightly oblong, moderately robust, piceous black, shining. Antennæ rufo-testaceous, the outer three joints piceous. Head sparsely punctate but with four large vertical punctures in a transverse row. Thorax more than twice as wide as long, sides strongly arcuate, hind angles broadly rounded, disc sparsely punctate with a few coarse punctures along the base. Elytra moderately coarsely striato-punctate, intervals flat, minutely sparsely punctulate, the intervals 3–5–7–9 with the usual series of very coarse distant punctures. Body beneath piceous black, shining, sides of metasternum coarsely punctured, abdomen less coarsely punctured. Legs piceous black. Length, .16 inch; 4 mm. Pl. xx, fig. 7.

This species so closely resembles *valida* and agrees with it in most of its characters so closely that it is probable the females cannot be separated. In the present species the posterior femora has a large tooth near the outer condyle, the distal edge oblique, the tibiæ are slightly sinuate not curved as in *valida*. In the latter the posterior femur is strongly angulate at its middle.

One male from the State of Washington, kindly given me by Mr. Merkel, to whom I take pleasure in dedicating it.

Eumicrus lucanus n. sp.

Reddish brown, sparsely clothed with paler, semi-erect hair. Antennæ slender longer than half the body, fifth joint as long as the preceding two and slightly longer than the two following. Head shining, very minutely sparsely punctate. Thorax longer than wide, apex narrower than base, sides regularly arcuate, widest one-third from apex, disc convex, sparsely and minutely punctulate, at base moderately quadrifoveolate. Elytra not wider at base than the thorax, a slight depression within the humeri, widest at middle, disc convex, surface indistinctly punctulate. Tarsi slender, longer than half the tibiæ. Length, .06 inch; 1.5 mm.

The male has the anterior tarsi slightly dilated.

This species is evidently allied to *commilitonis*, as described and figured by Dr. Sharp (Biol. 11, pt. 1, p. 67, pl. 2, fig. 22), but is of more elongate form and with more slender antennæ.

One specimen. San José del Cabo, with which I associate two from Arizona in my cabinet.

There are now six species of this genus known in our faunal limits—*grossus*, *Motschulskii*, *vestalis*, *Caseyi*, *lucanus* and *Zimmermanni*. The latter species has for a synonym *punctatus* Casey. The genus Cholerus has, in Mexico, half more species than Eumicrus, but none are yet known to me from our fauna. *E. Caseyi* Brend., which is unknown to me, is remarkable in its polished surface.

Piestus extimus Sharp (Biol. 1, pt. 2, p. 713).

Through the kindness of Mr. Ricksecker I have a specimen which I refer to this species, collected in Arizona. The type is from Chihuahua. *P. erythropus* Er. is said by Fauvel (Bull. Soc. Linn., Norm., ix, p. 29) to have occurred at Opelousas.

ZALOBIUS Lec. Trans. Am. Ent. Soc. v, p. 49, March, 1874.

This genus is introduced with the desire to give sketches of the head and thorax of the two known species. At the same time some additional characters will be made known and a new form allied to Zalobius described.

The two species differ greatly in the form of the thorax, as will be seen by reference to the plate. On both the disc is quadricostate, the costæ abbreviated and indistinct in *spinicollis*, entire and well marked in *serricollis*.

The terminal joint of the maxillary palpi is scarcely twice as long as the third joint, although Dr. LeConte gives it greater length. The gular sutures coalesce at their middle, forming but one, as will be observed in all the genera in our fauna of the sub-tribe Coprophilini, of which Zalobius is a member, as stated by LeConte. The outer five joints of the antennæ are described as rather abruptly wider (Class. Col. N. A., ed. 1883, p. 103), while in Syntomium but three joints are so. While this character is fairly good, it will be observed that the structure of the joints themselves affords a more certain character. In Zalobius the last four joints have the dense punctuation indicating their sensitive character; the joints one to seven are smooth, and in *spinicollis* without hairs, in *serricollis* somewhat hairy. In Syntomium the last three joints have the dense punctuation, the other joints glabrous but hairy.

It will also be observed that the head is considerably prolonged behind the eyes in both species of Zalobius, suddenly narrowed to a neck, while the head is not capable of retraction on account of the elevation of the back of the head above the neck. Of the genera of Coprophilini, Coprophilus is the only one with spinules on the outer edge of the tibiæ.

These characters have been passed in review with the view of suggesting a modification of the arrangement of the genera of Coprophilini as represented in our fauna. In the table of the classification above quoted the "Oxytelini genuini" and "Coprophilini," as defined by Erichson (Staph., p. 30), have been included in one group, Oxyteli, but it now seems to me better to revert to the Erichson method and separate the genera with five-jointed tarsi from those with but three.

The genera may be arranged in the following manner:

Tibiæ spinulose on the outer edge.

Antennæ gradually thicker, the outer five joints with sensitive punctuation and pubescence. *Coprophilus.*

Tibiæ not spinulose.

Maxillary palpi rather short and stout, the terminal joint shorter than the third. Antennæ with three outer joints obviously wider. *Syntomium.*

Maxillary palpi rather slender and elongate, the terminal joint longer than third.

Gular sutures confluent at middle.

Middle coxæ cavities confluent. Antennæ not stouter externally. Elytra not costate. *Deleaster.*

Middle coxæ cavities distinctly separated.

Antennæ with outer five joints stouter. Elytra distinctly costate. *Zalobius.*

Gular sutures separated their entire length. Middle coxæ cavities separated.

Maxillary palpi very slender. Outer four joints of antennæ stouter. Elytra costate. *Asemobius.*

Zalobius is represented by two species.

ZALOBIUS SPINICOLLIS Lec. Trans. Am. Ent. Soc., v, 1874, p. 49.

Occurs from Vancouver southward to Santa Clara (Ricksecker). Pl. xx, fig. 8.

ZALOBIUS SERRICOLLIS Lec. Loc. and cit., 1875, p. 170.

Collected by me at Fort Tejon, not Owens Valley, as stated by LeConte. Pl. xx, fig. 9.

ASEMOBIUS n. g.

Form elongate, much depressed, surface moderately shining. Head moderate in size, scarcely longer than wide, slightly prolonged behind the eyes, very abruptly constricted, the occiput not elevated transversely at the constriction. Mandibles moderately prominent, acute at tip, toothed at middle. Labrum transverse, truncate. Maxillary palpi long and slender, the last joint a little longer than the third. Antennæ slender, first joint cylindrical, stout, second shorter and more slender, third very slender and nearly as long as the first two, joints four to seven equal in length, slender, slightly thickened at their distal end, joints one to seven glabrous with few hairs, joints eight to ten quadrate, broader, eleventh more elongate, joints eight to eleven densely punctulate and pubescent. Prothorax with sides explanate and serrate. Elytra quadrate, emarginate at base, apex truncate, acutely notched each side, disc costate. Abdomen depressed, gradually narrowed to apex, sides widely margined. Middle coxæ separated. Legs slender, tibiæ not spinulose. Tarsi slender, joints one to four nearly equal, fifth as long as the three preceding united.

ASEMOBIUS CÆLATUS n. sp.

Pale castaneous, moderately shining. Head slightly concave each side, surface coarsely reticulate. Thorax broader than long, the disc convex along the middle, sides laminate lateral margin broadly arcuate, subangulate and obliquely narrowed behind the middle, edge coarsely serrulate, disc with two feeble costæ near the middle which are bifurcate posteriorly, surface with very coarse rather closely placed umbilicate punctures. Elytra conjointly a little wider than long, side margin acute and crenulate, disc with striæ of coarse punctures wider than the intervals, the third and seventh intervals elevated the former two-thirds, the latter the entire length of elytra. Abdomen longer than the elytra, sparsely punctate, with short golden hairs. Prothorax beneath with few very coarse punctures at middle, polished at the sides, abdomen sparsely punctate gradually more closely to apex. Length, .17 inch; 4.5 mm. Pl. xx, fig. 10.

Male. Last ventral with a deep oval emargination, the face of the segment slightly flattened.

The genus Asemobius proposed above is allied to Zalobius, but differs in having the gular sutures separated and by the terminal joint of the maxillary palpus scarcely longer than the preceding joint. The occiput may be retracted under the apical margin of the thorax, but in Zalobius the occiput is elevated above the neck, so that it applies against the edge of the thorax.

One specimen. California without special locality, but I think given me by Mr. L. E. Ricksecker.

TOMARUS BISIGNATUS n. sp.

Pale fusco-testaceous, each elytron with a small piceous spot centrally placed, form of *pulchellus* but somewhat more acute behind, surface with moderately coarse pubescence. Head moderately coarsely, not closely punctate. Thorax twice as wide as long, slightly narrower at apex, sides moderately arcuate, disc convex more coarsely and less densely punctate than the head, basal foveæ well marked. Elytra punctate, the punctures somewhat larger and more distant than those of the thorax, very much finer and nearly obliterated at apex. Body beneath darker than above, sparsely finely punctulate. Legs pale. Length, .06 inch; 1.5 mm.

Camp Grant, Arizona; San José del Cabo.

The species of Tomarus are not numerous and may be distinguished in the following manner:

Elytra coarsely punctate, pubescence coarse.
- Margin of thorax continuous; each elytron with a small piceous spot at center. *bisignatus.*
- Margin of thorax irregular; each elytron with a piceous incomplete band at middle, a piceous area near the apex. *hirtellus.*

Elytra finely punctate or nearly smooth; pubescence fine.
- Thorax not wider in front, anterior angles obtuse; elytral punctures distinct. *pulchellus.*
- Thorax distinctly wider in front, anterior angles truncate; elytral punctuation almost entirely obliterated; testaceous, each elytron with a transverse piceous spot at middle of the side margin. *acutus.*

Specimens of *pulchellus* have been collected in Arizona and at Los Angeles (Coquillett).

SCIRTES HUMERALIS n. sp.

Oval, slightly oblong, piceous, moderately shining, head and thorax rufescent, a reddish band along the side margin from the humerus one-third to apex, each elytron with an indefinite paler area at apex, surface sparsely fulvo-pubescent. Antennæ testaceous, gradually darker to tip. Head rufescent, sparsely punctate, occipital region narrowly piceous. Thorax sparsely punctate. Scutellum rufescent, piceous at middle, punctate. Elytra slightly more coarsely punctate than the thorax. Body beneath piceous; metapectus smooth; abdomen moderately closely punctulate. Legs rufo-testaceous, the femora darker. Length, .12 inch; 3 mm.

This species is more oblong and convex than the others of our fauna. It is most nearly related to *orbiculatus* by the presence of a distinct infra-orbital ridge.

One specimen. San José del Cabo.

MALTHODES LUCANUS n. sp.

Slender, piceous, thorax yellow. Antennæ slender longer than the body, joints two and three equal, each a little shorter than the fourth. Head shining with few punctures. Thorax broader than long, disc irregular with few scattered punctures. Elytra as long as the body, scabrous, sparsely pubescent. Length, .10 inch; 2.5 mm.

Male. Last ventral large parallel, curved upward, channeled its entire length, acutely notched at tip. Last dorsal oval without processes. Pl. xx, fig. 2.

Allied most closely to *curvatus* Lec., but differs in the male characters, its color and by the shorter third joint of the antennæ.

One specimen. San José del Cabo.

MASTINOCERUS OPACULUS n. sp.

Slender, luteous yellow, abdomen piceous the last two segments yellow, surface very sparsely pubescent. Head scabro-punctate, but indistinctly. Thorax similar in form to that of *texanus* but less arcuately produced at apex, surface opaque very finely granular, with sparsely placed indistinct punctures. Elytra opaque, the disc very flat, surface minutely scabrous. Venter shining, sparsely punctate and pubescent. Length, .18 inch; 4.5 mm.

Closely resembles *texanus* but is much smaller, the dorsum more flat and the surface opaque without very decided sculpture.

Arizona, southern, but locality unknown.

Mr. Gorham (Biol. Cent. Am., iii, pt. 2, p. 106) seems to have the opinion that our species are not congeneric with the Chilian form described by Solier. The antennæ of the latter are said to have the eleventh joint biramose. While it is possible that such a character may be found, all instances heretofore mentioned have been found erroneous on second examination. Mr. Gorham admits that Cenophengus is related to Mastinocerus, because the second and third joints of the antennæ are short and without rami. This is equally true of both the Mastinocerus of our fauna.

The most remarkable opinion of Mr. Gorham is the placing of these genera together with Pterotus (which he does not know) in the Lymexylonidæ. To carry the idea to its legitimate extent, Phengodes, Zarhipis and Tytthonyx must also take the same course.

TELEGEUSIS n. g.

Facies of a Malthinus, slender depressed. Head oval, abruptly prolonged in front of the eye in a short muzzle, the clypeal region abruptly deflexed but not distinctly separated from the front, behind the eyes the head is feebly arcuately narrowed. Eyes prominent, finely granulated, distant from the thorax. Antennæ in front of the eyes inserted in the angle of the muzzle formed by the deflexed clypeus, eleven jointed (?*female*) first joint stout, conical, longer than the next two together, second joint short, stout, shorter than the third, fourth longer than third, five to ten very slightly shorter, eleventh longer and fusiform. Labrum short and broad, anterior edge bisinuate. Mandibles falciform but not prominent, acute at tip. Mentum small, pentagonal. Labial palpi three, maxillary palpi four-jointed, the basal joints in each case very small, the terminal joint spathuliform and nearly as long as half the entire body. Thorax quadrate, slightly narrower at base. Scutellum longer than wide, parallel-sided, truncate at apex, longitudinally impressed. Elytra about half the length of the abdomen, dehiscent at their apical half, without distinct epipleuræ. Anterior and middle coxæ conical, prominent and contiguous, the mesocoxæ at apex only, the anterior with large trochantin. Abdomen flat, composed of eight segments, the first very short and lateral, the joints 2–7 nearly equal in length, eighth elongate oval, narrower. Legs moderate, femora not stout, tibiæ slender with distinct spurs. Hind tarsi longer than half the tibiæ, slender, first joint as long as the next two, fourth scarcely bilobed. Claws simple. Body with well developed under-wings.

TELEGEUSIS DEBILIS n. sp.

Form slender, resembling Mathinus, testaceo-piceous, moderately shining, sparsely clothed with short semi-erect yellowish hairs. Head sparsely indistinctly punctate. Thorax quadrate with obtuse angles, apex truncate, base arcuate, sides slightly sinuate at middle, disc feebly convex with scarcely evident sparse punctuation, in each angle a distinct depression. Elytra half the length of abdomen, broader at base than the thorax, humeri rounded, surface sparsely punctate becoming gradually granulate and at apex densely granulate. Abdomen above sparsely punctate, the three basal segments almost entirely smooth, beneath more closely punctate. Legs and coxæ yellowish. Length, .22 inch; 5.5 mm. Pl. xx, fig. 1.

In this genus we doubtless have our first representative of the tribe Drilini. It is, however, remarkable in the length of the terminal joint of the palpi, nothing at all approaching it is known to me in the entire order of Coleoptera. The insect seems most nearly allied to Drilus, but in the absence of the male nothing positive can be asserted.

Sierra San Lazaro.

ATTALUS SETOSUS Horn, Proc. Cal. Acad. Sci., iv, p. 381.

This species was described from an unusually pale colored specimen. Others recently received show that the color may be entirely piceous, faintly bronzed. Others have simply a pale humeral spot, which extends and widens, covering nearly the entire elytral surface, except narrow sutural and marginal piceous stripes. It is easily known, however, in all its varieties by the short erect setæ.

In addition to San José del Cabo, it occurs at Miraflores, Sierra El Chinche, Santo Domingo del Taste and Sierra San Lazaro.

OCHODÆUS PENINSULARIS n. sp.

Oval, slightly oblong, fulvo-testaceous, sparsely clothed with short fulvous hair. Antennæ, except the first two points, piceous. Head very coarsely punctate, occiput slightly piceous, clypeus oval at middle, slightly sinuate each side, the margin narrowly reflexed, front not tuberculate. Thorax fully twice as wide as long, disc convex, not closely granulate, a vague median depression posteriorly. Elytra about one-fourth longer than broad, the striæ feebly impressed, moderately coarsely and closely punctate, intervals flat, very irregularly biseriately punctate. Length, .16-.22 inch.; 4-5.5 mm.

The mentum is a little longer than wide, deeply impressed its whole length and emarginate in front.

In the six specimens examined, in which both sexes are undoubtedly represented, I find no armature either of the femora or tibiæ. The male hind tibiæ are, however, somewhat shorter and more hairy. This species is most closely related to *biarmatus*, and more especially the female, which has not the clypeal tubercles of the male. In both sexes of *biarmatus* the condyle of the hind femur forms a narrow plate along the posterior edge, terminating in a small tooth, one-third from the knee. No such structure exists in this species.

Sierra San Lazaro and Sierra El Taste.

OCHODÆUS CALIFORNICUS n. sp.

Broadly oval, piceous black, elytra dull brown, paler near the humeri. Head moderately closely punctate, without tubercles or ridges, clypeus arcuate, the margin not reflexed. Thorax twice as wide as long, moderately densely tuberculate, a vague median impression posteriorly. Elytra vaguely striate, punctures of striæ coarse but not close, intervals flat, more closely and finely granulate than the thorax. Length, .18 inch.; 4.5 mm.

Mentum broader than long, deeply concave, the apex emarginate.

The unique before me is a female, and shows no armature of the legs.

This species, by the form of mentum and simple clypeus, is not especially related to any in our fauna. In its color it is unique, and might be mistaken for an Onthophagus.

Pomona, Cal., from H. C. Fall.

OCHODÆUS FRONTALIS Lec.

This occurs in Coahuila, Mex. It evidently escaped Mr. Bates' observation, unless O—4 (Lamellic. p. 107) is it.

RHOPALOPHORA BICINCTA n. sp.

Black, rather dull, elytra with a basal and a post-median slightly arcuate narrow band of silvery white pubescence. Head closely punctate, occiput between the eyes smoother. Antennæ (*female*) yellow, not longer than the body, joints 5–10, gradually shorter, 11 slightly longer than 10, joints 8–10 slightly serrate. Thorax as in *R. longipes*, sparsely punctate, more closely on the flanks, surface alutaceous, beneath more shining, slightly transversely wrinkled and sparsely punctate. Elytra parallel, scarcely wider than the thorax, apices obtuse 3 or 4 dentate, surface closely cribrate punctate. Body beneath black, shining, sparsely punctate and with sparse hairs, but with denser patches of silvery white pubescence at the sides of mesosternum, the hind angles of metasternum and at the sides of the first ventral segment along the suture. Legs piceo-testaceous, the tibiæ paler, the femora at base darker. Length .20 inch.; 5 mm.

This species is placed temporarily in the genus from its very close resemblance in all the more important structural details, although the antennæ suggest a relationship with the Cleomenides, as defined by Lacordaire. The antennæ are not setaceous as in Rhopalophora, but rather filiform, the three penultimate joints slightly flattened and subseviate.

The specimen at my disposal is a female and unique, from which the useful parts of the mouth have disappeared. The ornamentation of the elytra is not in line with Rhopalophorus, and it seems very probable that the discovery of the male will show the necessity for suggesting a new generic name.

One specimen. San José del Cabo.

Dysphaga debilis n. sp.

Slender, elongate, pale yellowish testaceous, abdomen piceous. Front not punctate, neck obsoletely punctate. Thorax a little wider at middle than long, sides slightly arcuate, surface dull, not distinctly punctate, sparcely pubescent. Elytra one-third longer than the head and thorax, extending beyond the middle of the abdomen, surface shining, slightly wrinkled and vaguely bicostulate. Abdomen piceous, the last ventral segment yellow at apical half. Length, .18 inch; 4.5 mm.

The single specimen has the last ventral deeply triangularly incised. This is thought by Dr. LeConte to be a female character.

San José del Cabo.

The species of Dysphaga are few in number and specimens are always rare. They may be separated as fol lows:

Head and thorax piceous black.

Thorax distinctly punctate; elytra not costulate. *tenuipes.*

Thorax smoother bicallous at base. *lævis.*

Head and thorax reddish yellow; elytra piceous and distinctly bicostulate. *bicolor.*

Pale yellowish testaceous, elytra bicostulate. *debilis.*

METACOLASPIS n. g.

Head oval, not deeply inserted, eyes free, entire, not very convex, oval, finely granulated; front feebly sculptured, clypeus not distinctly separated, labrum short, transverse. Last joint of maxillary palpi slender, attenuate at tip, longer than the preceding joint. Antennæ slender, three-fourths the length of body, first joint thickened, second one shorter than the third, joints 3–11 gradually slightly longer. Prothorax transverse, the front angles not dentiform, hind angles very obtuse, lateral margin entire. Scutellum oval obtuse at apex. Elytra very irregularly substriately punctate. Prosternum between the coxæ twice as long as wide. Legs moderate in length, femora slightly fusiform, the tibiæ slender, the four posterior feebly emarginate at outer angle of the tip. Tarsi moderate in length, claws deeply bifid, the inner division long and slender.

The insect for which the above generic name is proposed has the general aspect of *Colaspis brunnea* except the color. It is certainly allied to Metachroma by the structure of the prothoracic episterna, the bifid claws and the emarginate tibiæ.

From Metachroma it differs in having the second antennal joint much shorter than the third. The other two genera are from Malacca and the Philippine Islands. Pyropida has a broad prosternum and less deeply bifid claws, Chrysopida has dentate femora.

METACOLASPIS CONSPERSA n. sp.

Form moderately robust, general color pale yellowish-white, antennæ, metapectus, tibiæ and tarsi piceous; elytra brown with numerous yellowish-white spots usually small, irregularly placed, surface glabrous. Head smooth. Thorax one-half wider than long, sides feebly arcuate, front angles distinct, hind angles obtuse, surface polished with numerous moderately coarse punctures at the sides. Scutellum smooth. Elytra wider at base than the thorax, about one-third longer than wide, sides feebly arcuate, disc convex with closely placed deep punctures, substriately arranged with three faintly indicated costæ. Body beneath smooth, sides of metapectus finely punctate and pubescent. Length, .16 inch; 4 mm.

San José del Cabo.

COLASPIS MŒSTA n. sp.

Form robust, piceous, upper surface feebly dark-bronzed, moderately shining. Antennæ piceous, the basal five joints gradually paler. Head punctate, less closely on the occiput. Thorax twice as wide as long, narrower in front, sides broadly arcuate, scarcely undulate, margin narrowly reflexed, disc convex, finely punctate, but more coarsely at sides. Elytra very little wider than the thorax, disc vaguely subcostate near the apex, surface moderately coarsely, irregularly, not densely punctate. Body beneath very sparsely punctate. Legs piceous, the front and middle femora and all the coxæ slightly paler. Length, .25 inch; 6.5 mm.

Similar in form to *Rhabdopterus picipes*, but larger and more robust. It seems most nearly related in our fauna to *C. nigrocyanea*, and like that species has the prosternum rather wider than in the other Colaspis.

San José del Cabo.

BLEPHARIDA ATRIPENNIS n. sp.

Form moderately robust, rufo-testaceous, antennæ, elytra, tibiæ and tarsi piceous black. Basal joint of antennæ testaceous in front. Head smooth. Thorax fully twice as wide as long, slightly narrowed in front, front angle slightly nodiform, hind angle obtuse, disc convex, smooth, sparsely finely punctate. Scutellum smooth, rufo-testaceous. Elytra slightly wider than the thorax, oval, one-third longer than wide, convex, surface dull, very finely alutaceous, irregularly geminately striato-punctate, the punctures not coarse nor close. Body beneath sparsely punctate, slightly pubescent on abdomen and sides of metapectus. Length, .22 inch; 5.5 mm.

While the species is placed in Blepharida as a matter of expediency, there does not seem to be any valid character for its separation from that genus. The only other species at present in hand is our *rhois*, which differs in some characters from those given by Chapuis. The elytral sculpture differs notably from that of *rhois*, but several Mexican species resemble it in this respect.

San José del Cabo.

CREPIDODERA PENINSULARIS n. sp.

Oval, moderately robust, rufo-testaceous, elytra piceo-rufous with distinctly æneous surface. Head smooth, antennæ darker at tip. Thorax twice as wide as long, slightly narrowed in front, anterior angles truncate, hind angles acute, sides moderately arcuate, disc convex, with few coarse irregularly placed punctures, transverse impression moderately deep and with punctures. Elytra wider at base than the thorax, humeri obtuse, surface with rows of coarse and deep but not very closely placed punctures, intervals not convex, smooth. Body beneath smooth, abdomen indistinctly punctate and alutaceous. Legs paler. Length, .10 inch; 2.5 mm.

This species is not particularly related to any in the Boreal fauna, but is probably nearer *chiriquensis*, although from description abundantly distinct.

One specimen. San José del Cabo.

Luperodes histrio n. sp.

Form oblong, color variable above, beneath yellowish, metapectus piceous. Antennæ rather longer than half the body, testaceous, the apex and front edge of each joint piceous. Head smooth, blue or green, the front yellow. Thorax slightly wider than long, not narrowed in front, sides slightly arcuate anteriorly, a feeble sinuation posteriorly, hind angles acute, the anterior slightly nodiform, disc moderately convex, scarcely perceptibly punctulate. Elytra wider than the thorax, twice as long as wide, sides nearly parallel, disc moderately convex, sparsely regularly, finely punctate. Body beneath yellow, sides of metapectus piceous, surface sparsely puuctate with short hairs. Femora yellow, with upper edge and apex piceous, tibiæ and tarsi piceous. Length, .16–.22 inch; 4–5.5 mm.

This insect is very variable in color.

Typical form: Head bicolored; this is constant in all the varieties. Thorax yellow. Elytra violet-blue to green, with an elongate yellow spot from base to middle, narrowing behind, an oval spot near the apex.

Variety: Thorax piceous with greenish surface lustre, the borders very narrowly pale. Elytra greenish, a basal elongate oval pale spot.

Variety: Thorax entirely yellow. Elytra almost entirely yellow, with a very narrow greenish margin all around and along the suture.

This species is placed in Luperodes, or rather the generic name is adopted for reasons given by me in a synopsis of Galerucini of Boreal America (Trans. Am. Ent. Soc. 1893, p. 108). It is, however, quite certain that some other use or arrangement of characters must be adopted for the genera of the group Luperites as defined by Chapuis.

Sierra San Lazaro, Sierra El Chinche.

SCELOLYPERUS CYANELLUS n. sp.

Form elongate, piceous black, moderately shining, elytra violaceous, blue or bluish-green. Head smooth. Thorax quadrate, very little longer than wide, sides very feebly arcuate, anterior angles slightly nodiform, hind angles sharply rectangular, disc moderately convex, smooth. Elytra one-half wider than the thorax, twice as wide as long, disc moderately convex, sparsely punctulate. Body beneath piceous or blue-black, sparsely punctulate and pubescent. Legs black, femora bluish, alutaceous, the anterior and middle coxæ and all the trochanters yellow. Length, .12–.16 inch; 3–4 mm.

The males have the posterior tibiæ straight. Variations occur which seem to be merely due to immaturity—the antennæ may be brownish-yellow and the under side of prothorax, tibiæ and tarsi yellowish-testaceous.

This species resembles a diminutive form of that variety of *maculicollis* with black thorax.

El Taste and San José del Cabo.

TRIPHALUS CRIBRICOLLIS n. sp.

Piceous black, feebly shining. Head moderately coarsely and closely punctate with a slight tendency to become strigose; mandibles more densely and finely punctate. Thorax quadrate, slightly broader than long and slightly narrower in front, sides very slightly arcuate, base arcuate, disc convex coarsely and closely punctate, each puncture (as on the head) bearing a short yellow hair. Elytra regularly oval, broadest at middle, one-half longer than wide, base not wider than base of thorax, disc convex with the striæ of coarse punctures not closely placed, intervals slightly convex and with two series of finer punctures irregularly placed each bearing a short yellow hair. Body beneath more shining than above, coarsely but not closely punctate. Length, .22 inch; 5.5 mm.

Similar in form to *perforatus*, but with the striæ of the elytra more closely and the thorax more densely punctured. The finer punctures of the intervals in *perforatus* are scarcely evident and the hairs very short.

One specimen. Miraflores.

Cerenopus angustatus Horn. Proc. Cal. Acad. Sci. 1894, p. 426.

A recent invoice with a dozen specimens shows that my description of the female was erroneous. The female at hand was merely a feeble specimen of *concolor*.

The female of *angustatus* has the clypeus very different from the male, which may be described as truncate with an abrupt median notch either oval or subtriangular. In the female there is a deep sinuation each side of the notch and the angles are obtusely produced so that the front has somewhat a quadridentate appearance. The posterior femora have on the inner side of the lower edge a series of well marked tubercles. Pl. xx, figs. 4, 5.

In addition to the locality at San José del Cabo it also occurs at Pescadero, Santo Domingo del Taste, Sierra El Chinche, San Lazaro and Miraflores.

Macrobasis languida n. sp.

Form slender, general color pale luteous clothed with slightly paler pubescence, antennæ and tarsi black. Thorax longer than wide, a feebly impressed median line.

Tarsi black with a ring of whitish pubescence at the base of each joint. Length, .40 inch; 10 mm.

Male. Antennæ setaceous the first joint as long as the next two, not sinuate at apex, second three-fourths as long and equal to the next three joints together. Anterior tibiæ with two terminal spurs, the first joint of tarsi longer than the second. Last ventral segment incised.

This species resembles *linearis* very closely and it is probable that the females will be difficult to separate.

San José del Cabo.

At the time of my study of Macrobasis (Proc. Amer. Philos. Soc. 1873) the male of *linearis* was unknown. The following are the sexual characters which will separate it from *languida:*

M. linearis ♂ Lec. Antennæ setaceous, the first joint as long as the next five not sinuate at tip, second joint scarcely a third of the first. Anterior tibiæ with a single spur, first joint of tarsi much shorter than the second and strongly compressed at base.

Otidocephalus alternatus n. sp.

Form of *vittatus*, piceous black, shining, clothed with recumbent and not dense gray hairs the alternate intervals of the elytra naked, without erect setæ. Rostrum smooth in front, not carinate, punctate at the sides. Thorax elongate oval, one-half longer than wide, equally narrowed at apex and base, regularly convex, coarsely not closely punctate. Scutellum densely clothed with white pubescence. Elytra elongate oval, widest at middle, humeri well marked. Punctures in striæ rather coarse and closely placed, intervals slightly convex, intervals 2–4–6–8 moderately densely pubescent, the others sparsely biseriately punctate, each puncture with a hair. Body beneath sparsely hairy, a denser line at the sides of meso- and metapectus. Legs brownish, sparsely punctate and hairy. The femora with a small tooth. Length, sine rost., .18 inch; 4.5 mm.

Related to *vittatus* and with similar vestiture but less dense, the hairs are simple and not tufted. It differs also in the absence of smooth thoracic line and is without erect setæ.

One specimen. La Paz.

OTIDOCEPHALUS SPARSUS n. sp.

Form of *scrobicollis*, piceous black shining, surface clothed with tufted pubescence not closely placed, thorax and elytra with erect white hairs. Rostrum coarsely punctate, smooth at middle in front. Thorax slightly oval, equally narrowed at apex and base, one-half longer than wide, sides feebly arcuate, disc convex, very coarsely and deeply, moderately closely punctate, tufted hairs not closely placed, erect hairs directed toward the front. Elytra oblong oval, widest slightly behind the middle, humeri distinct, surface with striæ of moderately coarse punctures, not deep nor closely placed, intervals flat sparsely punctate. Body beneath sparsely clothed with tufted hairs and erect setæ. Legs piceous, tarsi pale. Femora with small tooth. Claws with broad rectangular tooth at base. Length, .18 inch; 45 mm.

This species does not particularly resemble any in our fauna; it has vestiture similar to *Ulkei*, but less dense.

One specimen in my cabinet, probably from the Cape Region.

OTIDOCEPHALUS CARINICOLLIS, n. sp.

Form of *scrobicollis*, piceous black, surface very dark bronze, shining, clothed with grayish pubescence with smooth spaces, the intervals with semi-erect black setæ. Rostrum vaguely bisulcate each side. Thorax elongate oval, one-third longer than wide, widest slightly in fro of middle, base and apex equal, disc convex, very coarsely deeply and moderately closely punctate, the median line carinate from base two-thirds to apex. Elytra oval, widest slightly behind the middle, humeri distinct, disc with disdinctly impressed striæ having coarse, deeply impressed and rather closely placed punctures, intervals flat. Body beneath clothed with much finer pubescence than the upper surface. Tibiæ and the femora at base rufescent. Femora with a small tooth. Claws with quadrangular tooth at base. Length, .21 inch; 5.25 mm.

This species viewed in profile is more convex than any of our hairy species, resembling in this respect *Chevrolati*. The vestiture of the upper surface is rather formed of very large narrow scales than simple hairs. They are however not at all tufted as in *sparsus* or *Ulkei*. The carinate thorax will at once separate it from any of our species.

Occurs in Texas. One specimen was given me by Dr. Dohrn as from California, but I believe the locality erroneous.

In the Annals of the N. Y. Acad. Sc., 1892, Capt. Casey reviews the species of Otidocephalus describing several new ones. *O. nivosus* from intermediate specimens from Utah does not seem specifically separable from *vittatus*.

O. cavirostris should be compared with *Poeyi*, the inflated and excavated beak on which so much stress is laid is purely a sexual character. (See Lac. Genera, vi, p. 569, note, also Schoenh. Curcul. vii, 2, p. 208. Neither author, however, recognizes that it is a male character.)

O. perforatus Horn. has been separated as a distinct genus, *Oopterinus*, based on the absence of scutellum. The genus is not valid, as a distinct scutellum exists although less evident than in the species with distinct humeri.

COPTURUS SOBRINUS n. sp.

Piceous, elytra, legs and beak castaneous. Rostrum and sides of head with broad white scales, eyes contiguous on the front. Thorax a little broader at base than long, slightly narrower in front, feebly constricted at apex, sides feebly arcuate, disc slightly convex, coarsely, deeply and closely punctate, at sides with broad white scales, at middle a short line from the base with narrower scales. Elytra deeply striate with elongate punctures, intervals subcostiform and imbricately sculptured, color castaneous ornamented with short lines of white scales on the intervals forming in a general way a basal band, a humeral lunule extending inward at middle, a white area on the suture near the apex, another at sides near apex, the scales otherwise are brownish. Body beneath and legs densely clothed with white scales, broad on the body, narrow on the legs. Length, .08 inch; 2 mm.

Closely related to *minutus* in color, form and markings, but it is readily known by the form of elytral intervals. In *minutus* the intervals are flat and decidedly wider than the striæ, the intervals of *sobrinus* are subcostiform and narrower than the striæ. The antennæ are similar in that the second joint of the funicle is scarcely visibly longer than the third.

La Paz, Sierra San Lazaro, Sierra El Taste.

ZASCELIS OBLONGA n. sp.

Oblong, form of *irrorata*, dark brown, dull, sparsely clothed with dull white, slim, erect hairs, with more erect short black setæ intermixed. Rostrum very coarsely cribrate, at base slightly expanded, forming alæ over the scrobes. Head coarsely and deeply perforate punctate. Thorax a little wider at base than long, sides oblique convergent at apical third, posteriorly parallel, disc mo erately coarsely, deeply and closely punctate, with faintly indicated median carina. Elytra wider at base than the thorax, disc not depressed, striæ with deep quadrate punctures, intervals flat, the third and fifth finely carinate, rather roughly punctate. Body beneath coarsely, closely and equally punctate. Femora not toothed, tibiæ serrate. Length, .18–.22 inch.; 4.5–5.5 mm.

Allied in its vestiture to *serripes*, but in form to *irrorata*. Is much less robust and with less coarsely toothed tibiæ than the first, and differs from the second in vestiture and the absence of femoral tooth.

Two specimens. Arizona, Sierra San Lazaro.

CONOTRACHELUS LUCANUS n. sp.

Oblong oval, similar in form to *naso*, piceous, dull, elytra marmorate with brownish and dirty white hairs, the latter more evident at the sides and with four more conspicuous white spots at the base of the third and sixth intervals. Antennæ rufescent, inserted less than a third from the tip of the rostrum, which is rather deeply trisulcate on each side, the sulci coarsely punctured, surface glabrous, a few hairs near the base. Head densely punctate, with fulvous hairs. Thorax quadrate, a little wider than long, abruptly constricted in front, sides feebly arcuate, disc convex, very coarsely cribrate, with a median carina from apex, not quite reaching the base, surface with sparse erect hairs. Elytra more than half wider at base than the thorax, humeri not acute, disc with striæ of quadrate punctures not densely placed, the intervals 3–5–7–9 finely carinate, and with a row of short erect hairs. Body beneath very coarsely and deeply punctate. Mesosternum protuberant. Femora unidentate, claws divergent and acutely toothed. Length, .20 inch.; 5 mm.

This species belongs to the group defined by Dr.

LeConte as 1—B (Rhynchophora, p. 239), and by the moderate length of beak, which is strongly striate, and by the protuberant mesosternum, is allied to *posticatus*. It resembles the latter species in the vestiture, and differs especially in the very coarse sculpture of the thorax, which approaches that of *cribricollis*.

One specimen. San José del Cabo.

CONOTRACHELUS ECHINATUS n. sp.

Form of *erinaceus*, piceous, clothed with luteous and dark brown scales, indiscriminately intermixed on the thorax, on the elytra at declivity an indefinite band of the paler color, on the intervals are moderately long black hairs, in the punctures of the striæ a slender pale hair. Beak longer than head and thorax, sulcate at the sides, in front rather densely scaly and with erect hairs, antennæ inserted close to the tip. Thorax wider than long, not greatly constricted in front, sides feebly arcuate, disc convex, coarsely, deeply and not closely punctate. Elytra nearly twice as wide at base as the thorax, one-third longer than wide, humeri prominent but not dentiform, disc striate, striæ punctate, intervals flat. Abdomen coarsely and deeply punctate. Legs piceous, tibiæ and tarsi paler, densely scaly and with whitish hairs. Femora not toothed. Claws cleft. Length, .10 inch; 2.5 mm.

This species belongs to a small group containing but two species, characterized by the cleft claws and the presence of erect setæ on the intervals. It differs from either of the two at present known by the style of vestiture, the longer erect hairs and the entire absence of femoral tooth. As is usual with species of bicolored scales the surface is variegated in color and not purely piceous.

Two specimens. Southern Arizona.

EXPLANATION OF PLATE XX.

Fig. 1. Telegeusis debilis *Horn*, greatly enlarged.
Fig. 2. Malthodes lucanus *Horn*, terminal ventral segments of male, underneath view.
Fig. 3. Malthodes lucanus *Horn*, terminal ventral segments of male, lateral view.
Fig. 4. Head of male Cerenopus angustatus *Horn*.
Fig. 5. Head of female of same.
Fig. 6. Anisotoma valida *Horn*, hind leg of male.
Fig. 7. Anisotoma Merkeliana *Horn*, hind leg of male.
Fig. 8. Zalobius spinicollis *Lec*.
Fig. 9. Zalobius serricollis *Lec*.
Fig. 10. Asemobius cælatus *Horn*.

THIRD REPORT ON SOME MEXICAN HYMENOPTERA, PRINCIPALLY FROM LOWER CALIFORNIA.

BY WILLIAM J. FOX.

[With Plate xxi.]

The present paper is based on collections made in the fall of 1894, by Messrs. Eisen and Vaslit. By far the greater portion was gathered at San José del Cabo in September, and contains a number of species not mentioned in my former paper. The remainder, from the vicinity of Tepic, in the Territory of Tepic, on the Pacific Coast of Mexico, is of interest as it demonstrates the difference of the fauna of the main land from that of the peninsula, as the species are in nearly every case different from those of the latter region. The ants and parasitic hymenoptera are not included herein, but will form the subjects of other papers by Messrs. Pergande and Ashmead respectively.

CHRYSIDIDÆ.

Chrysis selenia Costa. Numerous specimens. San José del Cabo.

MUTILLIDÆ.

Mutilla ornativentris Cress. San José del Cabo. One specimen. Apparently a widely distributed species, as I have seen specimens from the eastern United States and Canada.

Sphærophthalma Nortoni Bl. Tepic. Two specimens, differing from the type in having the orange spots on anterior portion of second dorsal abdominal segment connected with the orange on posterior portion; otherwise the same, but a little larger.

Sphærophthalma Sackenii Cress. San José del Cabo. Five ♂ specimens, varying from 9–15 mm. in length.

SPHÆROPHTHALMA MAGNA Cress. San José del Cabo. Thirteen specimens.

SPHÆROPHTHALMA GLORIOSA Sauss. San José del Cabo. Fifteen specimens. The ground color of this species varies considerably; usually entirely red, yet forms occur with it entirely black, but forms intermediate are commoner. The color of the unusually long, white pubescence with which the species is clothed is invariable. The size varies from 8–17 mm.

SPHÆROPHTHALMA XALAPA Bl. San José del Cabo. One specimen.

SCOLIIDÆ.

MYZINE HAMATA Say. Three females. San José del Cabo and Tepic. The markings of the abdomen show considerably variation in this species. The form occurring in the eastern United States usually has the third and fourth dorsal segments, in the ♀, banded with yellow; in the two specimens from San José these bands are interrupted medially, but are more prolonged than the lateral spots on the other segments; in the Tepic specimen the third segment only is banded.

MYZINE HYALINA Cress. One ♂. San José del Cabo.

MYZINE SEXCINCTA Fabr. What I take to be a variety of this species is represented in the collection by a single ♀ from Tepic. Dorsal abdominal segments 1–4 are each marked by a broad yellow band; the coxæ, trochanters and basal two-thirds of the femora black; flagellum blackish. I considered this specimen at first as representing a new species, but the absence of substantial structural characters together with the fact that forms, more or less intermediate with the typical one, occur in the United States, led me to consider it but a variety.

Engycystis* gen. nov.

Plate xxi, fig. 1, female; fig. 2, head from front; fig. 3, abdomen of ♂; fig. 4, hind leg of ♀; fig. 5, fore leg of same.

Form elongate, *slender in both sexes*. Eyes tolerably large, reaching the base of mandibles, *not emarginate in either sex*. Mandibles *armed internally with a strong tooth between middle and apex*. Ocelli distinct. Insertions of antennæ protected by a prominent lobe, the distance between the insertions less than that between them and the eye margin. Antennæ rather long and slender, tapering to apex, in length about equal to the thorax, joints long and easily distinguished, 12 in ♀, 13 in ♂. *Maxillary palpi* 5-, *labial palpi* 4-*jointed*, the joints of the former unusually *long and slender*, of the later short and stout. Thorax elongate-oblong, narrower than head. Pronotum much longer than dorsulum and shorter than middle-segment, the dorsulum small, if anything, shorter than scutellum, the latter somewhat narrowed posteriorly and longer than broad. Mesopleuræ *ridged anteriorly, and beneath separated from the mesosterum by an indistinct furrow*. Metapleuræ less deeply sunken than in *Myzine*. Middle-segment in the only known species coarsely rugose or rugoso-reticulate. *Legs slender in both sexes*, the femora and tibiæ *not flattened or broadened*. Hind femora at apex beneath *produced into a rather sharp tooth*. Fore tibiæ with one, medial and hind pair with two spurs, the hind pair being also strongly serrated in the ♀, less strongly in ♂, and with the tarsi well spined. Anterior metatarsus *with a feeble comb*, composed of short, straight spines. Middle coxæ a little separated. Abdomen about as in ♂ *Myzine*, but not so strongly constricted at base and apex of the segments. Last dorsal segment of ♀

* ἐγγύς close by + κύστις = cell.

with a well developed pygidial area *which is smooth and bounded laterally by a sharp ridge*, in the ♂ this segment is emarginate at apex into which the huge, curved spine fits, which extends up from beneath. Wings ample, *the stigma large, lanceolate*, at least half the length of the marginal cell; *the latter lanceolate, firmly united to the costal margin its entire length*. Three distinct submarginals and an indication of a fourth, *the second shorter than either the first or third*. Recurrent nervures received by the second and third submarginals. *Basal vein confluent with the transverso-medial nervure in both sexes*. Cubital vein of hind wings originating a little *before* the apex of the submedian cell *in both sexes*, rarely confluent with the transverso-medial nervure.

Type, *E. rufiventris*. This genus is allied to *Myzine*, but is very distinct, as will be seen by the characters given above. It is erected for *Myzine rufiventris* Cress., and it is surprising that observers like Cresson and Cameron (Biol. Centr. Amer. Hym., ii, p. 258) should have overlooked such prominent generic characters as this species possesses. Moreover, the coloration—head, thorax and first abdominal segment black, with the remainder of abdomen bright red, is strikingly different from any of the species of *Myzine* that I am aware of.

The ♀ which has not before been noticed, is very similar to the ♂; the head is much more sparsely punctured, and the thorax in general more coarsely sculptured, except the mesopleuræ, which are very densely punctured, while in the ♂ they have the punctures large and well separated; abdomen above almost impunctate, else the strong scattered punctures evident in the ♂ are very feeble; wings in the middle and at base hyaline, otherwise fuscous, so that they appear bifasciate, hind pair except apex hyaline; first and second joints of flagellum about equal in length; in the ♂ the first is shorter than the second; colored alike in both sexes.

Two ♀ specimens. San José del Cabo. I am indebted to Mr. Samuel Henshaw for the loan of Cresson's type, which is a ♂.

In my opinion this genus differs sufficiently from the rest of the Scoliidæ, particularly *Tiphia* and *Myzine*, to be constituted a tribe of that family.

Scolia badia Sauss. San José del Cabo. One ♀.

Scolia guttata Burm. Two specimens, ♀ ♂. Tepic. Thoracic markings in these specimens are entirely absent. The only pale color to be found on them exists on the second and third dorsal abdominal segments in the shape of a large spot on each side.

Elis dorsata Fabr. San José del Cabo and Tepic.

Elis tricincta Fabr. Two specimens. Tepic.

Elis trifasciata Fabr. One ♀. Tepic.

Elis plumipes Dr. San José del Cabo. Numerous ♂ specimens. These differ from the more northern specimens by the clypeus being spotted on either side, as in specimens from the United States the clypeus is entirely black. The maculated clypeus is a characteristic of *E. limosa*, but *plumipes* differs in having the fore and medial tibiæ more or less yellowish, whereas in the first mentioned the medial tibiæ are entirely black.

Elis xantiana Sauss. San José del Cabo.

POMPILIDÆ.

Pompilus philadelphicus St. Farg. San José del Cabo and Tepic. Five females. These are much bluer than the form common in the United States.

Pompilus æthiops Cress. San José del Cabo. One ♂.

POMPILUS TENEBROSUS Cress. One ♀. San José del Cabo.

POMPILUS FULGIDUS Cress. One ♀. San José del Cabo. This specimen is smaller and less blue than the typical form, but otherwise agrees with it.

POMPILUS CORRUSCUS var. JUXTA Cress. San José del Cabo.

POMPILUS INTERRUPTUS Say. A variety of this species from San José del Cabo.

PRIOCNEMIS FLAMMIPENNIS Sm. Tepic. One female.

AGENIA MEXICANA Cress. Tepic. One female.

PEPSIS RUBRA Dr. Two females, one male. San José del Cabo.

PEPSIS TERMINATA Dhlb. One ♂ from San José del Cabo.

SPHEGIDÆ.

SCELIPHRON LUCÆ Sauss. Three specimens of this handsome species. San José del Cabo.

SCELIPHRON HISTRIO St. Farg. Tepic. One specimen.

SCELIPHRON *(Chalybion)* CÆRULEUM Linné. Twelve males. San José del Cabo.

SCELIPHRON *(Chalybion)* ZIMMERMANNII Dhlb. Five females. San José del Cabo. *Sc. aztecum* Sauss is probably the ♂ of this species.

SPHEX HABENA Say. One specimen. San José del Cabo.

SPHEX FLAVIPES Sm. Tepic. Two females. This is the form described by Saussure as *hirsutus*.

SPHEX ABDOMINALIS Cress. San José del Cabo. One ♀.

SPHEX THOMÆ Fahr. Tepic and San José del Cabo. Two females.

AMMOPHILA MICANS Cam. One ♀. Tepic.

AMMOPHILA ZANTHOPTERA Cam. Tepic. One female. Seems to be close to *mediata* Cress.

AMMOPHILA FEMUR-RUBRA Fox. San José del Cabo. One specimen.

AMMOPHILA PICTIPENNIS Walsh. Three females. San José del Cabo. These differ only from more northern specimens in their smaller size.

AMMOPHILA YARROWI Cress. Eight specimens. San José del Cabo.

CERCERIS GRAPHICA Sm.

I am not acquainted with the female of this species, the sex described by Smith. A ♂ specimen from Tepic agrees so well with Smith's description that I refer it to *graphica* without hesitancy. The flagellum is reddish beneath to the apex and above, at base the scape being entirely of that color; clypeus entirely, sides of face broadly, yellow; other markings on head and those on thorax as described in the female; as are also those on the abdomen, except that the sixth segment is banded and the pygidium is more or less yellow; inclosure at base of middle segment coarsely punctured basally, polished at apex.

The reddish legs with yellow tarsi and the peculiar combination of reddish, yellow and black of abdomen are characteristic of this species.

SPHECIUS CONVALLIS Patt. San José del Cabo. Two specimens.

STIZUS GODMANI Cam. Two specimens, ♀ ♂. San José del Cabo.

Bembex monodonta Say. Four males. San José del Cabo.

Trypoxylon albitarse Fabr. Tepic. One female.

Trypxoylon aztecum Sauss. San José del Cabo. Five ♀ and thirteen ♂ specimens. In the ♂ the color of the first tarsal joint varies from white to black. The reddish color existing on the basal joints of flagellum beneath in the ♂ does not occur in the female.

Trypoxylon spinosum Cam. Two specimens. Tepic and San José del Cabo.

Trypoxylon centrale n. sp.

♀.—Head higher than broad, finely and closely punctured; fore ocellus separated from the posteriors by a slightly greater distance than the latter are from each other, a distinct space between the posteriors and each eye; front just above the antennæ with a prominent, longitudinal ridge or carina, which extends less than half way to the fore ocellus; face depressed on each side of this ridge, above it the front convex, parted by a not strong impressed line running from the ridge to fore ocellus; occipital notch indistinct; clypeus flat, not carinated, its anterior margin rounded out, subangular in the middle; space between eyes at the vertex greater than the combined length of antennal joints 2 and 3, at the base of clypeus this space is about one-third less than at the vertex; first joint of flagellum about one-third longer than the second; pronotum strongly and bluntly shouldered at each side; dorsulum finely and very closely punctured, the scutellum less so and not impressed; mesopleuræ less closely punctured than the dorsulum; middle segment above distinctly striated, obliquely and transversely, with a slight, rather indistinct medial impression, posterior face with the sculpture hidden by dense, pale pubescence, strongly furrowed down the middle, however, sides shining, delicately striato-punctate; abdomen lông and slender, first segment subnodose at apex, fully one-third longer than the second, second and third segments united about one-quarter longer than the first. Black; mandibles, darker apically, greater part of fore tibiæ, their tarsi and a ring at base of middle tibiæ, not always present, however, reddish-yellow, sides of the second, third and fourth abdominal segments basally reddish; tegulæ pale testaceous; calcaria whitish; wings hyaline throughout, iridescent, nervures and stigma black, space between the recurrent nervure and the transverso-cubital nervure on the cubital vein distinctly greater than the width of the submarginal cell at apex; clypeus, head in front, including the eye-emarginations, cheeks, and thorax beneath less densely, with silvery pubescence. Length, 9 mm.

San José del Cabo. Two specimens. Distinct by the longitudinal ridge of front, sculpture of middle segment above, clear wings, shape of clypeus, coloration, and position of ocelli, which are not connected by a furrow. Its form and coloration is similar to those species with the middle segment trisulcate above, but in this there is but one sulcus. It is evidently more clearly related to *frigidum* and *bidentatam*.

EUMENIDÆ.

ODYNERUS DORSALIS Fabr. Tepic. One specimen representing the red variety of this species.

ODYNERUS FUSUS Cress. San José del Cabo. One specimen. Differs from Texan examples by having the truncation of the clypeus a little narrowed, and abdomen more suffused with yellow.

ODYNERUS LACUNUS Fox. San José del Cabo. One ♀ specimen.

VESPIDÆ.

POLISTES FUSCATUS Fabr. Ten specimens. Tepic.

POLISTES CARNIFEX Fabr. Ten specimens. Tepic. All very large.

POLISTES BELLICOSUS Cress. Ten specimens of a form which I consider a variety of this species, differing from Texan individuals by the scutellum and metanotum (postscutellum) being entirely yellow, and the markings in general, particularly those of the abdomen, more extended; the middle segment lacks the parallel yellow lines which occur in the Texan form. The specimens are from San José del Cabo.

POLISTES MINOR Pal. Bve. San José del Cabo. Three specimens.

POLISTES LINEATUS Fabr. Numerous specimens. San José del Cabo.

POLYBIA DECEPTA n. sp.

Entirely of a very dark, almost blackish brown, except the mandibles, which are ferruginous, and the antennæ, which are black, clothed with a sericeous pile; wings: basal two-thirds or more of anteriors black, as are also the nervures, on the remainder whitish, the nervures yellowish, posteriors blackish throughout; middle-segment with a deep, convex, depression, extending from base to apex; hind coxæ very large; medial and hind tarsi with joints 3 and 4 greatly extended at apex within, particularly the fourth joint; petiole of abdomen robust, most convex in middle above, before the apex with a transverse impression. Length, 15 mm.

One specimen, either a ♀ or ☿. Easily distinguished by the color of wings, which is exactly as in *Chartergus apicalis*, as in the rest of the insect. It evidently belongs to Saussure's Division Iota, and seems to come closest to *P. socialis* Sauss., from Brazil. Its similarity in color to *Chartergus apicalis* is really remarkable.

MISCHOCYTTARUS LABIATUS Sauss. Tepic. One specimen.

APIDÆ.

AGAPOSTEMON MELLIVENTRIS Cress. San José del Cabo. One ♀ specimen. Another unidentified species having the abdomen entirely black, is in the collection, from the same locality.

COLLETES sp. A specimen of a species related to *aethiops* Cress. Tepic.

CALLIOPSIS MEXICANA Cress. Tepic. One specimen.

PERDITA sp. One specimen. Tepic. Related to *8-maculatus* Say, from the United States.

ANTHIDIUM NOTATUM Latr. Two ♂ specimens from San José del Cabo, which represent a variety of this species, differing in the paler markings, entirely black femora, and by having greater part of tibiæ black; structurally there is apparently no difference from specimens from the eastern United States.

MEGACHILE sp. San José del Cabo. One ♀.

CERATINA MEXICANA Cress. Tepic. One ♀ specimen.

MEGACILISSA THORACICA n. sp.

♀.—Head black, the vertex with black pubescence front, face, clypeus at the sides and cheeks with pale pubescence, more or less intermingled with black, except on cheeks; clypeus with strong, sparse punctures; labrum reddish, strongly furrowed down the middle; mandibles reddish medially, strongly furrowed, the outer margin with a fringe of long, golden hairs; antennæ black, the flagellum beneath fuscous, the apical joint beneath reddish testaceous; thorax black, its greater part clothed with dense, black pubescence, which gradually shades from hind part of dorsulum into the fulvous pubescence with which the middle segment is clothed; pubescence of mesopleuræ posteriorly colored as on the front; legs blackish brown, fore femora with black pubescence, that of medial femora a mixture of pale and black, while on the hind pair it is longest and fulvous, tibiæ and first joint of tarsi with short, dark hair on outer portion, internally reddish, apical tarsal joints reddish, with pubescence of a darker color; wings pale fulvo-hyaline, nervures and stigma black; abdomen above with the base of segments 2–4 and the fifth entirely bluish, the apical portion of segments 2–4 greenish, first segment and extreme sides with long fulvous pubescence, the fifth and sixth except base, with long black pubescence, the remaining dorsal segments clothed with short, thin, appressed, yellowish pubescence, which is most conspicuous on apical or greenish portion, beneath, the abdomen pale testaceous, the apical margins of the segments with a fringe of long, fulvous hair, which is, however, blackish medially on the fourth and fifth segments, the sixth ventral with black pubescence only. Length 18 mm.

Tepic. One specimen. Distinguished by the color of the thoracic pubescence and of antennæ. Related to *mexicana* Cress. by color of abdomen.

MEGACILISSA MEXICANA Cress. Tepic. One specimen.

XYLOCOPA FIMBRIATA Fabr. Tepic. Two specimens.

Xylocopa varipunctata Patt. Ten ♀, one ♂, specimens. San José del Cabo.

Centris flavifrons Fabr. Tepic. One specimen.

Eulema fasciata St. Farg. One specimen. Tepic. Smith (Ann. Mag. N. H., (4) xiii, p. 442) unites *fasciata* and *cajennensis* under the latter name, notwithstanding that *fasciata* is the first described.

Bombus diligens Sm. Tepic. One ☿.

Bombus sp. From Tepic are two ☿'s and one ♀ not structurally distinct from *B. medius* Cress., although the thorax lacks the broad black band so prominent in *medius*, and is entirely yellow above. It is probably but a variety of *medius*.

Trigona bipartita St. Farg. One specimen. Tepic.

Apis mellifica Linné. Numerous specimens, ♂ and ☿ from both localities.

THE NEOCENE STRATIGRAPHY OF THE SANTA CRUZ MOUNTAINS OF CALIFORNIA.

BY GEORGE H. ASHLEY.

[With Plates xxii-xxv.]

INTRODUCTION.

The following paper gives the results of a preliminary study of the Tertiary stratigraphy of the Santa Cruz Mountains, but such additional notes upon the earlier and later rocks are given as may be of value to subsequent observers.

The data from which the stratigraphic column herein given has been worked out were obtained chiefly in the cliffs along the sea coast, backed up by a reconnoissance of the mountains themselves. A large number of fossils were collected and determined, and a beginning was made upon a detailed geological map of the mountains, chart No. 3055 of the U. S. Coast and Geodetic Survey being used as a foundation.

To determine how far the conditions found in the Santa Cruz Mountains hold good in other parts of the Coast Ranges, short excursions were made into the Mount Hamilton Range, the Gavilan Range, and to a number of points in Los Angeles county.

Altogether about five months were spent in the field.

TOPOGRAPHY AND PHYSICAL GEOGRAPHY.

Santa Cruz Mountains—This name has been given to the series of parallel ridges extending from the Bay of Monterey northward to Point San Pedro. The ridges have a strike of nearly northwest and southeast. They reach their highest point in Mount Bache, 3780 feet above sea-level; further to the north Black Mountain reaches a height of nearly 3000 feet. Further to the northwest, and south of Pilarcitos Lake and Creek, the

Cumbre de las Auros rises in high bare ridges 2000 to 3000 feet high, and again near Point San Pedro this same ridge attains a height of 1940 feet in Mount Montara.

The ridge which starts in at Mussel Rock, north of Point San Pedro, reaches its highest point in Black Mountain, and to the southeast runs out gradually into the plain of Santa Clara Valley. The granite ridge which rises at Point San Pedro breaks down before reaching Pilarcitos Creek.

West of the high ridge, of which Mount Bache is a part, is a granite ridge running from near Santa Cruz nearly to Pescadero Creek. In this region considerable study would be required to systematize the ridges. The ridges and their narrow intervening valleys occupy a region about sixty-five miles long by twenty-five wide as a maximum. On the ocean side the mountains in some places reach the sea; in other places they are separated from the ocean by a narrow strip of flat land. From Lake Merced to Mussel Rock the foothills are cut by the ocean in bluffs from 200 to 700 feet in height. At Point San Pedro the prominent ridge, of which Mount Montara is the highest point, is cut by the ocean in bluffs from 1000 to 1500 feet high. Between Mussel Rock and Point San Pedro occurs the nearly level Quarternary. Starting at sea level at the mouths of San Pedro, Calera, Salt Lake and Milagra valleys, it extends back into these valleys and northward toward Mussel Rock. In the latter direction the ocean has cut a perfect section of the deposit, which at the Rock is over two hundred feet thick. The surface of the deposit rises gently as it recedes from the ocean. From Point Montara the Quaternary extends southward to Lobetus. At Spanish Town it is about two miles broad, becoming narrower to the south. At Pillar Point a long hill of the Merced series rises from the Quaternary, and is cut by the ocean. Much of the distance from Point Montara to Lobetus, the Quaternary and underlying Miocene form low cliffs.

From Lobetus to Pescadero Creek the ocean cuts a number of ridges forming high bluffs, the Merced series being exposed, with a small amount of Quaternary overlying. From Pescadero Creek to Point New Year the flat land occupies a narrow belt from a few hundred feet to a half a mile broad. Between Point New Year and El Jarro Point the mountains are cut in high bluffs, the rock being the White Miocene Shale.

At El Jarro Point the level belt begins again and continues to the Bay of Monterey, broadening out at Santa Cruz, and extending up the valley of the Pajaro River as a broad plain. At Santa Cruz this horizontal land appears at four distinct levels, the lower two being very noticeable, the highest standing over 700 feet above sea-level.

The level land, which extends up the Pajaro River and Arroyo de las Llagas, cuts off the range from the Gavilan Range to the south and merges almost imperceptably into the broad level of the Santa Clara Valley. At San José the Santa Clara Valley has a breadth of about twelve miles, but to the north is largely occupied by San Francisco Bay, so that at San Carlos and to the north the foothills are close to the Bay.

From South San Francisco to Lake Merced, a low gap cuts off the Santa Cruz Mountains from the San Bruno Mountains and other hills in and about San Francisco.

The main axis of the range is indicated by the name "Santa Cruz Mountains" on the sketch-map. At present only the northern end of the range has been mapped topographically. But in general it may be said that the whole area, within the limits indicated above, is occupied

forms the valley of San Francisquito Creek, finally ending in Coal Mine Cañon in the flank of the Black Mounain.

The next salient topographic feature is the main ridge of the mountains. This starts at Mussel Rock and follows the same general direction, nearly southeast. At its northern end it can hardly be distinguished topographically from the foothills previously described; but near San Andreas Lake, it reached a height of 1300 feet and is distinctly marked off by the valley just described. In width it extends here from the Crystal Springs Valley to Pilarcitos Valley. Its northwestern end is serrated by the San Pedro, Calera, Salt Lake, Milagra and other valleys into a number of long projecting points running down to or nearly to the shore. Near San Andreas Lake this main ridge is somewhat broken by San Mateo Creek, and, after being much reduced between San Mateo Creek and Pilarcitos Lake, rises and continues southeastwards with very even crest as far as the San Mateo-Spanish Town road. South of this road it becomes higher, with narrow, slightly uneven crest, until it swings around the head of the long valley described, and joins Black Mountain.

The end of another ridge is met near Point San Pedro. This ridge, largely of granite, rises abruptly from the ocean at Point San Pedro, quickly attaining a height of 1940 feet in Mount Montara. Continuing southeastward, where it is known as the Cumbre de las Auros, it rises into bare, precipitous ridges, which break down before reaching the San Mateo-Spanish Town road.

On the ocean side of this ridge and the main ridge described in the preceding paragraph, high, irregular ridges extend seaward between the streams shown on the sketch-map. These end more or less abruptly, or else run out at the coast line or a short distance inland.

PREVIOUS WRITERS.

The following will show in a general way the development of our present knowledge on the Coast Ranges, and the Santa Cruz Mountains in particular. Only the more important points are mentioned here, as these and other works will be referred to in detail in subsequent chapters.

Beechey.—In 1825 to 1828, Captain Beechey* made a voyage to the Pacific and Behring Straits. His notes and collections on the geology of the vicinity of San Francisco were worked up by Professor Buckland. A map is given of the headlands about the Golden Gate, upon which serpentine, sandstone and jasper rock are represented.

Tyson.—In 1849, Mr. P. T. Tyson† visited California. He notes the presence of sandstones containing a big oyster (*Ostrea titan*, Conrad) near Martinez and in Livermore Valley, and assigns the strata to Eocene or Miocene age. He finds strata which he thinks may be quite recent or late Tertiary. He describes quite accurately the cherts, jaspers, etc. (phthanites). He notes the presence of hypogene and metamorphic rocks of many kinds, which have been twisted about and mixed together in the most confused manner, and mentions having found in a small space near Bodega Point, gneiss, mica slate, indurated talcose slate, hornblende slate and serpentine, the last containing chromiferous iron. Near San Diego he notes the extensive diffusion of diluvial drift.

---

* Zoology of Captain Beechey's Voyage to the Pacific and Behring's Straits in 1825–1828. London. 1839.

† Report of the Secretary of War, Communicating Information in Relation to the Geology and Topography of California. Senate, Ex. Doc. No. 47. 1850, pp. 15 *et seq.*

isk.—Dr. John B. Trask*, as State Geologist, made ort in 1855, in which he defines the Coast Mountains, the local name of Santa Cruz Range. He divides rocks of the Coast Ranges into three groups: Ter- ry, Primitive and Volcanic. The volcanic rocks he inks to be late Miocene in age. His Primitive Group ncluded the syenites, mica schist, granite, gneiss, porphyries and the older greenstone, including also the serpentine rocks; he also groups with these a crystalline limestone, quite common from Santa Cruz southward, and makes it the same in age as the group which extends through three hundred miles of the Sierra Nevada, noting that it is older than the igneous rocks. In the Tertiary he calls attention to the widespread existence of bituminous or Monterey shale, calling it, by way of distinction, the "infusorial group," and its age the "Infusorial Period," but he makes them horizontal, and their exposure due to simple uplift. Above this he places "sandstones and slates," the former predominating. These upper beds are very fossiliferous.

Dana.—In the report of the U. S. Exploring Expedition, Prof. J. D. Dana† calls attention to terraces on all rivers of Oregon and northern California as evidence of recent lifting, and to the fiords of British America as evidence of subsidence in that region.

Blake.—In 1856 Prof. Blake‡ made a report in whic he gives a general map of the geology about San Fra cisco; he also gives sections on Yerba Buena Island,

* Report of the Geology of the Coast Mountains, etc., by Dr. John Trask. Senate, Doc. No. 14, session 1855.

† U. S. Exploring Expedition, etc., under command of Charles W N. Vol. x, Geology, by James D. Dana, 1849, pp. 659–678.

‡ Explorations and Surveys, etc., for a railroad fr ... Pacific Ocean. 1856. Senate, Ex. Doc.

Navy Point, Benicia, and from San Francisco to the Pacific. He names and describes the San Francisco sandstone, abundant at San Francisco and elsewhere. He assigns it to the Tertiary, though a portion of the Upper Cretaceous may be represented. He describes alluvial deposits about San Francisco Bay, drift deposits in low passes in San Francisco and sand dunes in the same place. From finding "Post-Pliocene deposits" at Monterey, Santa Barbara, San Pedro and San Diego, he argues for a very recent uplift of that region.

Whitney.—During the survey under Prof. Whitney,* the Santa Cruz Mountains were crossed in several directions. His report in a very general way points out the distribution of the different formations in those mountains. He makes the metamorphic rocks and San Francisco sandstone Cretaceous, thinks certain shales in the valley near Searsville are Cretaceous, but assingns most of the later rocks of the mountains to the Miocene. He calls the strata between Lake Merced and Mussel Rock Pliocene, overlaid unconformably by post-Pliocene.

In 1880, in his "Contributions to American Geology," Whitney† calls attention to the fact that in the Coast Range the movement has produced crushing and breaking, rather than folding and uplifting. He notes Pliocene near San Diego, at north end of San Fernando Valley, unconformable on Miocene, and subaerial Pliocene gravels all about the Santa Clara Valley. The Miocene he divides into two groups; one a fine grained slate or shale often highly bituminous and the other a rather coarse grained sandstone, the latter being the lower member. He makes

* Geological Survey of California, J. D. Whitney, State Geologist. Geology, vol. i, 1865, pp. 61 *et seq.*

† Auriferous Gravels, J. D. Whitney. Memoirs of Museum of Comparative Zoology, 1880, pp. 15 *et seq.*

the general uplift post-Miocene with many local movements since Pliocene, the Miocene and Pliocene being conformable in some places, in others unconformable.

Becker.—In 1888, Dr. Becker.* in his report on the "Quicksilver Deposits of the Pacific Slope," also makes the metamorphic rocks Cretaceous, but thinks the limestone is the lowest member and probably older. He thinks the first upheaval took place in lower Cretaceous. The cherts before mentioned he calls phthanites and argues that the serpentine is derived from sandstone.

Cooper.—In the Seventh Annual Report of the State Mineralogist, Dr. J. G. Cooper † gives a list of the fossils of California with their geographical and geological range. This is a good index of the ages to which some of the more fossiliferous horizons were assigned at that time and practically up to the present. The following ages are given to some of the beds to be discussed later. The horizontal beds from Santa Barbara to San Diego are all called Quaternary. The beds of Seven Mile Beach and their continuation to the southeast are called Pliocene. The fossiliferous beds along the coast from Half Moon Bay to Soquel are generally called Pliocene, but in a few places are referred to as Miocene.

Fairbanks.—Mr. Harold W. Fairbanks ‡ argues for the pre-Cretaceous age of the metamorphic rocks and makes the early upheaval post-Jurassic. He mentions the intrusion of granite into the metamorphics of the Gavilan Range.

Lawson.—In the Geology of Carmelo Bay, Prof. Law-

* U. S. Geological Survey, Monograph XIII, 1888.

† Cal State Mining Bureau, Seventh Annual Report of the State Mineralogist, 1888, pp 223 *et seq.*

‡ American Geologist, vol. IX, Mar. 1892, p. 133.

son* suggests that the White Miocene shale or Monterey shale is of volcanic origin. Gravels which he finds there overlying the granite he thinks are of Eocene age. He also differentiates the Pliocene and Quaternary. In a later paper† he summarizes the evidence of post-Pliocene uplift between San Diego and San Francisco. He describes in a general way the beds on Seven Mile Beach and their structure, calling them all Pliocene and naming them the Merced series. He makes them continuous with the beds in the cliffs at Pillar Point, giving a section from Lake Merced to Pillar Point.

A number of other geologists have touched the Santa Cruz Mountains, without their reports giving us any new information on the main features of the geology. As for example Dr. J. S. Newberry‡ and Dr. Thos. Antisell.§

GENERAL GEOLOGY.

Formations Represented.—The Pleistocene is well represented in belts of nearly level ground, which skirt the mountains on all sides, and form benches up many of the streams.

The Pliocene is recognized on Seven-Mile Beach and to the southeast. Merced series.

The Miocene is the formation which predominates. Three facies are recognized. The lower part of the Merced series, very fossiliferous, yellow and drab sands more or less consolidated; the Monterey series, principally a light colored bituminous shale, containing few fossils; the Pescadero series (in part), represented by wide spread yellow sandstone, fossiliferous in places.

The Eocene is thought to be represented, but no faunal evidence of its presence has been found.

* Univ. of Cal., Bull. Dept. of Geol., vol. i, pp. 1 *et seq.*, May, 1893.

† Univ. of Cal., Bull. Dept. of Geol., vol. i, pp. 115 *et seq.* Dec. 1893.

‡ Pacific Railroad Survey, vol. vi, part ii.

§ Pacific Railroad Survey, vol. vii, part ii.

The Cretaceous is thought to be represented. The strata containing lower Miocene fossils form the upper part of a series several thousand feet thick. (Pescadero series.) No fossils were found in the lower part of the series, but it seems probable that the Inoceramus reported by Prof. Whitney came from these beds, and partly upon that as a basis they have been referred to the Cretaceous.

In addition to the above are areas of granite and limestone that are probably pre-Cretaceous.

There are large areas of metamorphosed sandstones, phthanites, serpentines and associated rocks, shales and older eruptives, of which the age is as yet in question.

*Distribution of the Formations**—Granite.—While from the position and character of its outcrops it seems to be evident that the granite has considerable body underneath, its outcrops are not very conspicuous in the Santa Cruz Mountains. It occupies the ridge running from a little south of Point San Pedro nearly to the San Mateo-Spanish Town road; also the ridge on the west of the San Lorenzo River, which extends from near Santa Cruz nearly to Pescadero Creek. A little granite appears at a few places along the summit of the main ridge.

Metamorphics.—This may include the Gavilan limestone, the metamorphic sandstone, the phthanite or radiolarian chert and the older eruptives. They make up most of the hills running diagonally across San Francisco from Fort Point and Point Lobos to Hunter's Point and Visitacion Valley. The western side of the San Bruno Mountains. The main ridge, starting from Mussel Rock, as far as the San Mateo-Spanish Town road. The eastern foothills from Milbrae to the Redwood City-Searsville road. From Searsville southeast to Black Mountain and

*The distribution is given only for the regions visited by the writer.

southeastward. The main peaks shown upon the sketch-map and adjacent region.

Pescadero Series.—This series makes up the hills of northeastern San Francisco. The eastern side of the San Bruno Mountains and hills northwest of Ocean View. At Point San Pedro. Eastern foothills, from Redwood-Searsville road to south of Mayfield. Main ridge, from San Mateo-Spanish Town road to headwaters of Pescadero Creek. Pescadero Point to Pigeon Point.

Monterey Series.—Western flank of main ridge, from near Spanish Town, southeastward. Coast, from Point New Year to Santa Cruz.

Merced Series.—Lake Merced to Mussel Rock on the coast, and extending southeast to Milbrae, forming foothills. Foothills near Stanford University and Mayfield. Foot of main ridge, west side of Coal Mine Cañon. Coast, from Point Montara to Pescadero Creek. At Point New Year. Santa Cruz to south of Capitola.

Pleistocene and Recent.—Most of San Francisco. Valley of Lake Merced. Between foothills and Bay of San Francisco. Santa Clara Valley. Small areas in foothills and in stream valleys. Mussel Rock to Point San Pedro. Raised beaches on coast, from Point Montara to Pájaro River. Valley of Pajaro River.

Stratigraphical Relations of the Formations.—There are three, and probably four, marked periods during each of which sedimentation was more or less continuous, each being followed by upheaval and folding, and each being laid down unconformably upon the preceding. The formations belonging to the first of these have been grouped together and have been called for want of a better name,

The Metamorphics of the Coast Ranges.—These include the Gavilan limestone, the phthanite or radiolarian chert and the metamorphic sandstone. Little is known accurately of these formations. The phthanite and metamorphic sandstone appear to be conformable, the sandstone certainly underlying and probably overlying the phthanite. Beyond that, the relations of the formations to each other are not known. There appear to be two beds of the Gavilan limestone, one of which at least is several hundred feet thick. The limestone is found associated with the metamorphic sandstone and metamorphic slate, also with younger formations where brought up by faulting. Dr. Becker,* in treating the rocks of the Gavilan Range, calls the limestone the lowest member. Its age is unknown. The phthanites, upon the evidence of Radiolaria found in them, have been thought to be of Jurassic or Cretaceous age.† In general, all the metamorphics are thought to be pre-Cretaceous by Mr. H. W. Fairbanks.‡

Above the metamorphics is a great thickness of sandstones and shale, frequently thin bedded, topped by heavy beds of conglomerate, which, to distinguish, we have called the

Pescadero Series.—This series comprises all the upper portion of what has been known as the San Francisco sandstone, also rocks thought to be Eocene, and at the top heavy beds of conglomerate which contain Miocene fossils. The relation between this series and the metamorphic rocks below was not made out, all or nearly all the evidence seeming to indicate that there was no break between this series and the metamorphic sandstone and phthanite below. However, as nearly all the geologists at present actively engaged in studying the metamorphic rocks agree in placing them unconformably below the lowest Cretaceous and as the Pescadero series

* U. S. Geol. Surv. Monograph xiii, p. 181.

† Bull. Dept. of Geol., U. of Cal., vol. i, p. 237, Oct., 1894.

‡ American Geologist, vol. ix, Mar. 1892, p. 163.

has been found to extend upward to the Tertiary, it will be assumed here, purely upon their evidence, that such a break does exist here between the metamorphics and the Pescadero series. The series has a thickness of several thousand feet.

During the next period of sedimentation there were laid down the two formations known as the

Monterey and Merced Series.—The Monterey series is a considerable thickness of white, or light colored, frequently bituminous, shale, with some sandstone, and is overlain conformably by soft sandstones, shales and conglomerates or grits, the whole, known as the Merced series, having locally a thickness of nearly a mile. The Monterey series contains but few fossils, those being of Miocene age. The Merced series is very fossiliferous, at the bottom containing many Miocene forms and at the top appearing to run nearly if not quite into the Pleistocene.

The two formations appear to grade into each other so that locally sedimentation seems to have been continuous from the beginning of the Monterey period at least through the Pliocene. On the other hand there is evidence of many minor movements during this long period, many parts of the area now covered by the formations having evidently been land areas during part of the time.

These two formations were laid down unconformably upon the preceding Pescadero series.

Unconformably upon all the preceding formations there are all around the edge of the Santa Cruz Mountains unconsolidated deposits of Pleistocene or Recent age. Locally, as just south of Mussel Rock and one or two other places, these reach a thickness of over two hundred feet; but generally they average from five to fifty feet in thickness. An interesting feature of these deposits is the part that wind-blown sands play in them, appearing at the bottom of the water deposits, and at the top covering a major part of San Francisco county.

General Structure.—As stated above, each of these periods of deposition has been followed by upheaval and erosion. These upheavals have been progressively gentler. On the San Francisco peninsula no distinction could be found between a movement which followed the laying down of the Pescadero series and a post-Jurassic movement which it is claimed followed the laying down of the radiolarian cherts and associated rocks.*

Both formations, whether distinct or not, were subjected to a powerful upheaval, being folded, faulted, crushed until a structureless mass has frequently resulted. The lower or so called metamorphic rocks have generally lost all trace of bedding, the phthanites excepted. In some places they show secondary silicification. The upper beds usually show bedding, though in many places this is lost and secondary silicification appears similar to that more common in the lower metamorphic rocks. High and vertical dips prevail in both formations.

Movement seems to have taken place largely by faulting, though the structure taken as a whole appears to be that of an anticline. The bedding of the Pescadero series along the top of the main ridge is generally more or less nearly horizontal, on either flank becoming vertical or nearly so. The Merced and Monterey series were evidently laid down upon and around this anticlinal mountain of early Miocene age, and in late Pliocene or in Pleistocene time came another movement which elevated the mountains nearly to their present position. As before, faulting predominates and is the controlling factor in most of those details of the topography due to structure. Many of these faults have a downthrow of hundreds of feet or

* American Geologist, vol. ix, Mar. 1892, p. 133 *et seq.*

even running into the thousands. These faults and nearly all the structural features have a strike of about northwest and southeast.

As a rule the Monterey-Merced series are only gently folded, the strata usually having a dip lower than 45°, but locally the dip is frequently found to be as high as 75° or 80°. The axis of greatest disturbance seems to have been in about the same line as in the early Miocene upheaval. During the Merced period a volcanic outflow of andesite took place in the area of the foot-hills near Stanford University. Just above this is a layer of barnacles, cidaris spines, etc., and in places Pholas borings in it, with other evidence showing that the outflow formed later a part of the shore line. Other evidences of slight movements during the Monterey-Merced period are found in the distribution of the two formations. In places the Monterey series lies upon the granite, in others the granite or Pescadero series underlie the Merced series.

The folding, as a rule, has been sharper on the side of the mountains toward the bay. The fossils there are generally distorted and the strata are firmer and more consolidated. The most noticeable faults are also all on the northeast flank of the mountains. Along the exposure on the Seven Mile Beach, faults of from a few feet or inches up to ten or twenty feet are frequent; but aside from those only major faults have been recognized, and these partly by the distribution of the formations, partly by topography. There appear to be two main directions of faulting, parallel with the strike of the mountain, and in a nearly due east and west direction. The strata are generally tilted up at a high angle, often being perpendicular, but usually having lower dips as we recede from the line of greatest disturbance.

The Quaternary was a period of oscillation. During

this period there was no folding but simple uplift with the axis of uplift as before in the axis of the range. The shore lines show that these movements of subsidence and uplift rested at a number of levels, the lowest subsidence being at least 1500 feet* below the present level. Prof. Lawson proposes as an hypothesis to differentiate the Pliocene and Quaternary; "that the Pliocene corresponds to the period of more or less continuous depression of the coast; and that the Quaternary corresponds to the more or less continuous uplift which has affected the coast since the maximum depression was reached."† This hypothesis would seem to meet the conditions along the southern coast, but will not those north of Monterey Bay, where the Pliocene has been upturned and eroded before the Quaternary was laid down and where land deposits form the bottom of the Quaternary. In that region the difference in the character of the uplift must be used, the Quaternary being only gently upraised and nowhere dipping more than a few degrees, while the Pliocene is lifted to angles of from 20° to 80°.

The movements of the Quaternary may be summarized as follows:

1st. The post-Merced uplift folded the strata of the Santa Cruz Mountains. Subsequent erosion reduced them to base level. Movements aggregating 1000 to 1200 feet raise this plane, and again erosion produces marked benches.

2d. Uplift to a level nearly 400 feet above present level. Land period. Abundant æolian and sub-aerial deposits. San Francisco Bay a valley.

3d. Submergence bringing shore line of San Francisco Bay to the foothills, producing benches on coast. Pres-

* Univ. of Cal., Bull. Dept. of Geol., vol. i, p. 157.

† Univ. of Cal., Bull. Dept. of Geol., vol. i, p. 57.

ent site of San Francisco an island. Bay filling Santa Clara Valley.

4th. Recent uplift to present conditions.

COLUMNAR SECTION OF THE SANTA CRUZ MOUNTAINS.

Period	Series	Formation	Fossils	Thickness
Quaternary.		Æolian deposits. Freshwater deposits. Marine deposits. Land deposits.	Elephas, conifers.	Thickness, 200 ft.+
Pliocene.	Merced Series.	Fossiliferous beds of sandstone and conglomerate. Transitional from the Miocene through the Pliocene.	Living shells. "Venus pajaroensis" (a Mactra). Large Pectens, Arcas, etc. Cetacean bones.	Thickness, 4700 ft.+
Miocene.	Monterey Series.	Bituminous or White Miocene shale; with some sandstone.	Infusoria. Pecten peckhami. Tellina congesta.	Thickness, 1000 ft.
Eocene?		Sandstones and conglomerates: Carmelo Series(?) Pescadero sandstones and shales.	Turritella hoffmanni. Ostrea titan. Liropecten estrellanus. Dosinia, etc.	Thickness, 2000–10,000 ft. (?)
Cretaceous		(The "San Francisco sandstone" in part.) Aucella beds.	Plant remains. Inoceramus. Aucella.	?
?		Radiolarian chert or phthanite. Metamorphic sandstone.	Radiolaria.	Thickness, 400 ft.+
		Gavilan limestone.	Foraminifera.	Thickness, 300–1100 ft. (?)

DETAILED STRATIGRAPHY.

In a more detailed study of the stratigraphy, the formations will be taken up from earlier to later. The metamorphic sandstone will be considered as at least a more or less marked facies of the pre-Miocene sandstones. The future will decide whether it be simply a metamorphic facies of the lowest bed of the great series which have been known as the San Francisco sandstone, or whether this term has been applied to beds differing not only in age, as it is quite probable that they do, but in all their structural relations. The evidence observed will be given under the relations of the Pescadero series.

LIMESTONE.

Lithology.—This bed of limestone is in places highly crystalline, in other places it does not appear to be so much metamorphosed, and contains long lenticular masses of chert through it in the direction of dip. Prof. Whitney describes it in one place as follows*: "The upper layers are thin bedded, and some strata are light colored, others dark; below the stratification is less distinct, the layers heavier and the rock more crystalline."

The evidence would seem to indicate that there are two layers, both of considerable thickness. One of the layers is probably not less than 300 feet thick, a section on Permanenta Creek giving much more than that, but with such poor exposures over part of the outcrop that the section was not considered reliable. Whitney estimated at one place that it "must be over 1000 feet in thickness." †

Distribution.—This limestone was observed in large masses in only a few places: one in the Calera Valley, where it caps some small knolls; and again on the ridge

* Geol. Surv. of Cal., Geology, vol. I, p. 75.

† Geol. Surv. of Cal., Geology, vol. I, p. 75.

between the San Andreas and Crystal Springs Valley and Pilarcitos Valley, from the San Mateo-Spanish Town road, where it used to be quarried, northward a few miles. Near the head of Stevens Creek it outcrops and is well exposed along the hill on the east of the valley; also further east, evidently having been faulted or folded.

Blocks and small outcrops were found at a number of places, but from their occurrence were generally thought to be only fragments brought up by faulting.

Relations and Age.—Aside from some indications that this limestone occurred low down in the series, no definite relations could be made out between it and other formations exposed. In the Gavilan Range, Becker makes it "the lowest sedimentary formation encountered."* From the association with it there of rocks of the Archæan gneiss type, he thinks it may be very old, though he suggests the possibility of its being a member of the Knoxville series of the Lower Cretaceous.

In the section up Permanenta Creek there seems to be some evidence that it is a conformable member of the series containing the phthanite and metamorphic sandstone, and would seem to underlie the phthanite by some distance. No good outcrop of the phthanite occurs on the creek, but its position is indicated by fragments in plenty. Between it and the limestone the bedding in the sandstone and shale is very indistinct, but in several places indicated a dip and strike, which is found to agree with the dip and strike of the limestone when that formation is reached further up stream.

PHTHANITES, OR RADIOLARIAN CHERT.

Lithology.—Nearly every writer on the geology of California has noted or described a series of thin-bedded, cherty strata, ranging in color from red or green to brown

* U. S. Geol. Surv., Monograph xiii, p. 181.

or black. They are usually very thin-bedded, averaging about an inch, but ranging from a fraction of an inch to several inches. They are very hard, and break up into little parallelopipedons. Almost everywhere they are exposed, the strata are very much and very characteristically contorted, yet maintaining their bedding almost unbroken, like the edges of a bale of cloth which has been crumpled up.

It has recently been shown that in places the phthanite contains an abundance of Radiolaria, though in poor preservation.*

From this it has been thought that the phthanites were originally silicious deposits, the silica having been derived from organic remains.

Occurrence.—Small outcrops of phthanite are abundant, and it frequently appears to be all mixed up with the sandstone, serpentine, etc., as though in the crushing to which the metamorphics had been subjected the phthanite bed had been torn into small masses which had been thoroughly scattered among the other beds. That we do not observe the same thing for the other beds is doubtless due to the fact that, as a rule, we have no way of noting the lack of relation between adjacent outcrops or masses. These small outcrops are scattered all along the foothills from Milbrae to Haakerville, more especially on the edges of the large outcrops of serpentine. The low hill at Point Coyote and all the hills close to Belmont and San Carlos are phthanite, and on most of them may be found good exposures. There are besides a few scattered exposures—on the ridge between Calera and San Pedro Valleys close to the road, on the San Francisquito Creek near Searsville, etc. There are many good exposures of phthanite about San Francisco. It is the predominating

* Bull. Dept. of Geol., Univ. of Cal., vol. i, pp. 199, 200.

rock in the line of hills from Point Lobos southeastward past the almshouse to the hills just north of Visitacion Valley on the bay side. This belt includes Bernal Heights. It is not confined to this belt, and usually may be found wherever the serpentine or metamorphic sandstone occurs. As typical exposures readily accessible from San Francisco might be mentioned exposures in the small hill in the northeast corner of Golden Gate Park; also in the hill near Strawberry Hill, in the same park, upon which the Prayer Book Cross has been erected. In numerous cuts in the group of hills surrounding the almshouse, and in fact nearly everywhere that it occurs in any quantity, it is much quarried for road material and hence the abundance of excellent exposures.

Relations and Age.—The evidence, both on the San Francisco peninsula and elsewhere, makes it appear that the phthanite is conformably interbedded with the older sandstone. Mr. F. Leslie Ransome reports this relation to exist just north of the Golden Gate and on Angel Island.* It may therefore be considered as of the same age. By a comparison of its radiolarian fauna with that of certain cherts of Europe, it has been thought that the phthanite might be of Jurassic or Cretaceous age.†

THE METAMORPHIC SANDSTONE.

Lithology.—In the area studied a metamorphic sandstone is very abundant in which bedding is visible in only a few places, and then can be followed, as a rule, but a few feet. It is usually light brown or gray in color, rather fine grained, but variable in that respect, breaking along joint planes which seem to form a fine network through it and which are generally stained brown or black.

* Univ. of Cal., Bull. Dept. of Geol., vol. i, pp. 73–74, and pp. 198, 199.

† Univ. of Cal., Bull. Dept. of Geol., vol. i, pp. 237–238.

In many places over considerable areas hand specimens show a schistose structure full of slickensides and evidently bearing no relation to its original structure and character, sometimes appearing as though it had been reduced to a pasty mass and afterward consolidated.

In many places this old sandstone is further distinguished by secondary silicification, the rock being filled with a net work of fine quartz veins.

In some regions the metamorphic sandstone can be recognized at once. In others the resemblance to the sandstone of the Pescadero series is so great that it is impossible to draw the line between them. For example, in the San Bruno Mountains the northeastern side is clearly made up of the Pescadero sandstones showing the bedding and characteristic features. As the southern end of the mountain is rounded the bedding becomes more obscure and the dip less regular; this continues, the beds being found at all angles until finally the bedding only shows here and there, phthanite occurs to some extent, secondary silicification is quite marked in some places, and the rock as a whole on the west side of the mountain would be pronounced the metamorphic sandstone. Yet the occasional presence of characteristic Pescadero sandstone, the gradual transition from the one sandstone to the other, together with the structure of the mountain, would seem to suggest that the metamorphic sandstone and accompanying phthanites simply made up the lower part of the Pescadero series.

This relation is suggested by the exposures on many of the hills in and around San Francisco. Indeed, it is not to be wondered at that all the early geologists and down to within a few years have placed the metamorphic sandstones, the phthanites, and the younger sandstones in one series, for such a relationship appears to be the true one on the San Francisco peninsula.

As, however, most of the recent observers agree in placing a marked non-conformity between the metamorphic rocks and the Cretaceous rocks, with the latter of which the Pescadero series has been thought to, in part, agree, the writer has preferred in this paper to accept their conclusions.

Occurrence.—The metamorphic sandstone occurs in connection with phthanite or serpentine in most of hills included in the broad belt which extends from Fort Point and Point Lobos southeastward to the bay between Mission Bay and Visitacion Valley. It also forms the southwestern flank of the San Bruno Mountains.

It is found abundantly on the edge of the foothills facing San Francisco Bay from Milbrae to Redwood City; on the ridge which starting from Mussel Rock extends southeast on the west side of San Andreas and Crystal Springs Valleys. Also in Black Mountain and southward, and in general wherever the metamorphic rocks are found. In the Santa Cruz Mountains much of the eastern slope and the highest peaks are made up of the metamorphic sandstone and associated rocks.

Relations.—From the lack of bedding or structure it is seldom that any relation can be made out between the metamorphic sandstone and the other formations. The only relation so far observed is that it lies conformably under the phthanite and probably also above. This appears in several of the phthanite hills in San Francisco, notably on Castro Heights. Mr. F. Leslie Ransome reports the same relation to exist north of the Golden Gate* and on Angel Island.†

Correlation.—No fossils have ever been found in the metamorphic sandstone of the San Francisco peninsula.

* Univ. of Cal., Bull. Geol. Dept., vol. i, pp. 73-74.

† Univ. of Cal., Bull. Geol. Dept., vol. i, p. 198.

Its age is therefore not known. From its association with the phthanite it has been thought to be of about the same age, and so is called Jurassic by some. Assuming the existence of a nonconformity between these rocks and the Cretaceous, it may be best for the present simply to consider them as pre-Cretaceous.*

THE PESCADERO SERIES.

The San Francisco sandstone was one of the earliest formations recognized and described in this State. Of late, however, as intimated above, it has been thought that the formation originally described under the name San Francisco sandstone, instead of being one formation represents at least two of very differing ages. In most, if not all, the recent papers the California or San Francisco sandstone has been assumed to belong to the pre-Cretaceous series. But the writer has found that a large part of what was originally described as the San Francisco sandstone belongs to a series which certainly is in part Miocene, as at first described by Blake, probably in part Tejon and possibly in part Cretaceous. The best exposure of the series was found in the low bluffs near Pescadero, extending from Pescadero Point nearly to Pigeon Point. It was the only place found appearing to give anything like a complete section. The beds are practically perpendicular and near one end of the section are fossiliferous. As a matter of convenience, the formation will be distinguished in this paper as the Pescadero series.

The Pescadero series has a thickness estimated at from 2000 to over 10,000 (?) feet. Though apparently overlooked by some of the recent writers on California geology, all the earlier workers recognized the existence below the White Miocene Shale of sandstones containing older Mi-

* American Geologist, vol. ix, 1892, pp. 153 *et seq.*; also vol. xi, 1893, pp. 69-84

ocene fossils. It is now believed that Blake was right in placing these with the San Francisco sandstone.

The sandstone and shale, which is quite typically developed in Telegraph Hill and other hills in the northeastern part of San Francisco, is found making up a large part of the San Bruno Mountains, just south of the city. It is then found typically developed at Point San Pedro and again at Pescadero Point. The section there gave a thickness calculated at 10,800 feet. Near the southern end of this section are several hundred feet of conglomerate, and in this conglomerate were found *Turritella hoffmanni* Gabb, and a few other forms, not yet identified. In the headwaters of Stevens' Creek and Coal Mine Cañon occurs a conglomerate indistinguishable from the conglomerate near Pigeon Point, and also containing *Turritella hoffmanni* Gabb, together with *Liropecten estrellanus* Conrad, Ostrea and Dosina. Again, in Alum Rock Cañon in the Mt. Hamilton Range, near San José, are found similar sandstones and conglomerate with a similar fauna, and containing also the characteristic *Ostrea titan* Conrad, and some other forms.

The Pescadero section has thus served as a key, showing as it does in continuous section rock very typical of the old San Francisco sandstone, as developed in San Francisco, and conglomerate identical in appearance and fossils with the somewhat abundant fossiliferous sandstones and conglomerates underlying the White Miocene shale.

The writer visited the original locality from which Professor Lawson described his "Carmelo Series," and is inclined to the belief that the Carmelo series will be found to be the equivalent of the conglomerate of the Pescadero series.

In the Coast Range Mountains of Southern California, Dr. Antisell studied the lower Miocene strata in some detail. It is believed that the strata described by him will be found to belong to the Pescadero series.

Petrography and Stratigraphy.—This formation may be described under three facies, the first two of which everywhere grade into each other, the last being more distinct.

The first is typically developed at Point San Pedro and the neighboring bluffs. The exposure consists of dark and black shales and slates, or shaly sandstones which, while showing finer bedding, is distinctly bedded in layers of from one to three inches or upwards. Many layers of coarse grained, sometimes conglomeritic, grey or white hard sandstone or quartzite occur interstratified in that series. This sandstone, while sometimes heavily bedded, is not in continuous layers, but constantly varying in thickness and sometimes disappearing altogether. At the top of the point, where the nearly vertical shales have been deeply weathered, they have become the same in appearance as the soft yellow sandstones which in other regions have been assigned to the Miocene age, as, for example, the sandstone exposed along the top of the ridge west of Woodside and Searsville. The bedding runs from the shale into the sandstone without break, leaving no doubt that the one is only the weathered product of the other. In places the joints have been filled with a white mineral like barite. In San Francisco this facies is well shown by a deep cut on Second street, several blocks from Market, also at the corner of Jones and Washington streets.

The second facies, which is quite common, consists of heavy bedded sandstone, generally of a yellow or brown color where weathered, but a grayish blue in the interior of the strata where not weathered. In some places the beds will average a foot or two thick, the stratification being distinct. In other places the strata are very heavily bedded, and frequently so intersected with joint planes that it is difficult to make out the real bedding. In many places these sandstones were observed to be peculiarly and quite characteristically weathered, so as to give the surface a pitted or honeycombed appearance, the cells being from a fraction of an inch up to five or six inches across. The resisting cell walls have probably been hardened by iron which has infiltrated into the network of joint planes. At Point San Pedro these strata lie upon the granite, and are largely made up of granite boulders and fragments. The proportion of granite increases as the main granitic mass is approached, until it becomes difficult to recognize the separating line.

These two facies were not found occupying definite horizons, but, as in the Pescadero section, grade into each other, and succeed each other irregularly in the vertical section.

The third facies is more characteristic. It is a conglomerate, heavy bedded, and apparently made up of the rocks of the metamorphics, fragments of phthanite being particularly abundant. This conglomerate is usually quite hard, the included fragments breaking across where a piece of the rock is fractured. The rock is usually brown or dark colored, except where relieved by the red or other colors of the phthanite. The pebbles are usually not very large, varying from an inch or under to three or four inches in diameter. They are frequently found to be distinctly faulted.

The conglomerate of Carmelo Bay is laid down upon granite and contains considerable granite, which the conglomerates of the Santa Cruz Mountains do not, as far as observed. This is, however, due to local conditions and may not affect the main question of their original continuity.

As previously stated, the best section observed was that running from Pescadero Point almost to Pigeon Point. (Plate xxiv). The strata are almost perpendicular for the whole distance of over five miles; at the northern end disappearing under some gently dipping strata, and at the southern end, where the dip of the beds becomes lower, being cut off by a fault. The section was made by pacing and abundant observations upon the dip and strike, corrected by the U. S. Coast and Geodetic Survey chart of that region. The section was not examined in great detail, and the frequently varying strike gave room for the presence of faults; but though carefully watched for, none were found of any magnitude except the one at the south end of the section. The section gave a thickness of nearly 11,000 feet, composed, with the exception of 600 feet of conglomerate at the southern end, of the thin bedded and heavy bedded sandstone, changing rapidly from one to the other all through the section. The predominating layers average from three inches to a foot in thickness. Over a large part of the exposure the dip is as high as 85°, but near the southern end quickly becomes lower, at the fault being only about 30°. South of the fault the dip is still lower, becoming nearly horizontal just north of Pigeon Point.

One question of vital importance remains undecided. That is, which is the bottom and which is the top of the section? Other things being equal it would generally be assumed that the conglomerate was at the bottom of the series. But the low dips at the south end of the section point to the conglomerate as the uppermost member. If the conglomerate is the lowest member then the whole series belongs to the Miocene Tertiary, which hardly seems probable. If the conglomerate is the uppermost member, then the series may be the result of continuous sedimentation through part or all of the Cretaceous and Eocene. It is not easy on the latter theory to explain the presence of such large quantities of the metamorphic rocks, which presumably had just been buried beneath the rest of the section, unless we assume considerable erosion previous to their laying down. In the Pescadero section no evidence of a break was observed; the conglomerate seems to run into the thin-bedded sandstone, the strike and nearly vertical dip being the same in both.

The conglomerate of the Carmelo series was laid down upon granite. If it is rightly correlated with the conglomerate of the Pescadero series, then the conglomerate of Pescadero must represent the bottom of the series or else there must have been uplift, erosion, and exposure of the metamorphic rocks between the laying down of the conglomerate and the rest of the series, notwithstanding the evidence to the contrary at Pescadero. This uplift and exposure may have been local. As the evidence seems to favor the latter theory it will be taken tentatively here.

Professor Lawson estimated the Carmelo series to have a thickness of at least 800 feet.* The conglomerate at Pescadero was estimated to have a thickness of 720 feet at least. Blake gave the San Francisco sandstone a thickness of 2000 or 3000 feet.†

Dr. Antisell, working south of the Santa Cruz Mountains, obtained the following section of the strata below the White Miocene shale.

---

* Univ. of Cal., Bull. Geol. Dept., vol. i, p. 19.

† Rep. Geol. Recon. in Cal., 1858, p. 153.

"Grits and calcareous sandstones, as at Panza and Santa Margarita	360 feet.
San Antonia sandstones, with Dosina, etc	250 feet.
Gypseous and ferruginous sandstones of Santa Inez, Panza and Gavilan, containing Ostrea, Turritella, etc."*	1,200 feet.
Or thickness below White shale	1,810 feet.

Occurrence.—The Pescadero series is well exposed at a number of points about San Francisco Bay. At Benicia, Blake found in rocks of this series a Trochus, Turritella and shark's tooth. In San Francisco it makes up the three hills in the northeast, or business quarter, of the city. The hills just north and west of Ocean View and most of the San Bruno Mountains belong to this series. It is finely exposed at Point San Pedro and at Pescadero. It appears to be one of the most abundant formations of the Santa Cruz Mountains. It occurs abundantly in the Mount Hamilton Range and to some extent in the Mount Diablo Range.

If it be correctly correlated with the lower Miocene strata worked up by Mr. Antisell, it will probably be found that this formation is widespread all through the Coast Ranges.

Relations.—The relation to the underlying metamorphic rocks has already been discussed under that head. The existence of water-worn metamorphic fragments in the Pescadero series would argue for a nonconformity between the two formations, unless, as above, we assume uplift and extensive erosion during one part of the period of deposition of the Pescadero series.

At Pescadero Point the Pescadero series is overlain unconformably by strata having only a small dip, and in turn both are overlain by about six feet of horizontal

* Pacific R. R. Report, vol. 7, p. 197

Quaternary. The age of the overlying strata was not determined, as they differed petrographically from any rock found elsewhere. Their position and structure would seem to indicate that they belong to the Monterey series. In San Francisquito Creek, above the Mayfield-Searsville road, the Pescadero series with a high dip is overlain by the gently dipping Merced series; and as the Merced and Monterey series are found to be conformable, that would place the Pescadero series unconformably below the Monterey series.

At Carmelo Bay Dr. Lawson believes a nonconformity exists between the Carmelo series and the White Miocene shale or Monterey series. In the same place the Carmelo series is seen resting upon the granite.* At Point San Pedro the formation also rests upon the granite,

Correlation.—The fossiliferous formations just underlying the White shale that are described by Dr. Antisell and Prof. Whitney are believed to be correctly correlated with part of the Pescadero series. Their Miocene age can hardly be questioned.

An attempt to work out the faunas of the three divisions made by Dr. Antisell for the strata below the White shale was only partly successful. The following is a partial list:

c. The grits and calcareous sandstones.
 Ostrea titan Conrad.
 Ostrea panzana Conrad.
 Hinites crassa Conrad.
 Pallium estrellanus Conrad.
 Cyclas permacra Conrad.
 Balanus estrellanus Conrad.
 Asterodapis antiselli Conrad.

b. The San Antonio sandstones.
 Dosinia alta Conrad.
 Dosinia montereyana Conrad.

---

* Univ. of Cal., Bull. Geol. Dept., vol. i, p. 7.

Dosinia montana Conrad.
Dosinia longula Conrad.
Dosinia suboblique Conrad.

a. The gypseous and ferruginous sandstones.
Mytilus inezensis Conrad.
Pachydesma inezensis Conrad.
Turritella variata Conrad.
Ostrea.

The gypseous and ferruginous sandstones do not seem to have been very fossiliferous and are associated with conglomerate and lignitic layers. They would thus seem to have some similarity to the Carmelo series of Lawson.

Among the specimens found at the head of Stevens' Creek were the following:

Ostrea sp. ind.
Liropecten estrellanus Conrad sp.
Turritella hoffmanni Gabb.
Dosinia sp. ind.

Ostrea titan has been found near this point.

In Alum Rock Cañon in what are thought to be the same beds were found:

Natica sp.
Ostrea titan Conrad.
Liropecten estrellanus Conrad sp.
Glycimeris generosa? Gould.
Nummulites?

Assuming that the younger part at least of the series is of Miocene age there seems to be some evidence that part of the series is of Tejon (Eocene) age. There is claimed to be a striking resemblance between the sandstone of Searsville Valley and the Tejon sandstones.

In a well near Haakerville light colored sandstone, full of the stems of some plant, was found. Mr. Gabb reports the same stems, and identifies them with fucoids in the "Cretaceous rocks overlying the coal at Mount Diablo."* These beds have since been shown to be

* Geol. Surv. of Cal., Geology, vol. 1, p. 71.

Eocene at Mount Diablo. In Alum Rock Cañon, near San José, a piece of float rock was found to contain what were thought to be Nummulites, though only cross sections could be obtained. The genus Nummulina, though known to exist from the Carboniferous to the present, is only known to have attained a considerable size and to have been of any geological importance in one age, the Eocene. The evidence of the fucoids near Haakerville is of only minor importance, but must be taken into consideration.

If, as the structure suggests, the conglomerate in the Pescadero section is near the top of the series, it would seem possible that the great thickness of strata below represent not only the deposition of the Miocene, but possibly also of the Eocene.

In conclusion, it may be stated that below the White Miocene shale or Monterey series of the Miocene, there has been found to lie, unconformably, a series of sandstones, shales or shaly sandstones and conglomerates, having a thickness of several thousand feet, in part, at least, of older Miocene age, and possibly extending back through the Eocene.

THE MONTEREY-MERCED PERIOD.

The Monterey series was among the earliest of formations described in California. It was assigned to the Miocene, and later investigations have not modified that decision.

The first mention of the Merced series was by Whitney, who merely mentions the finding by Gabb and Remond of Pliocene strata on Seven-Mile Beach.* These beds seem to have commanded little attention until recently. They were recently described more in detail by Professor

* Geol. Surv. of Cal., Geology, vol. i, p. 79.

Lawson, by whom they were called the Merced series.* They have been called Pliocene by all who have worked with them. From Half Moon Bay southward, at a number of places, occur very fossiliferous beds, which have often been mentioned or referred to, but of which little has been written. The fossils obtained from these beds have been referred by some to the Miocene, by some to the Pliocene. The same beds have been referred to the Pliocene where little disturbed, and to the Miocene where much disturbed.

The field work done by the writer has seemed to show:

1. That, though minor oscillations have occurred, there has been continuous sedimentation from the beginning of the Monterey series to the end of the Merced. (Due to minor oscillations, this is not always true locally.)

2. That the two are similar in structure, that structure having been received from the movement which took place at the end of the Merced period.

3. That the fossiliferous beds south of Half Moon Bay are conformable with the Monterey series below them and the Merced series above them.

4. That the Monterey series is Miocene, the Merced series on Seven-Mile Beach principally Pliocene, and the fossiliferous beds transitional between the two, containing a mixture of Miocene and Pliocene forms. That means that if a line were drawn between the Miocene and Pliocene it would not come at the top of the Monterey series, as usually defined, but from one to several hundred feet higher, somewhere in the period of the fossiliferous beds.

It would be difficult, if not impossible, to draw the line between the two ages, as it would be largely governed by individual inclinations. Accordingly the writer prefers

* Univ. of Cal., Bull. Dept. of Geol., vol. 1, p. 143.

to simply call them the "transition beds" of the Merced series, grouping them with the Merced series because petrographically they are similar to the predominating rocks of that series.

THE MONTEREY SERIES.

Lithology.—The Monterey series or formation, or, as it is variously known, the White Miocene shale, or Bituminous shale, is the most characteristic formation of the Miocene. It is for the most part nearly white or light buff in color, of a shaly, porous character. It is quite soft and without grit, yet resisting weathering to a remarkable degree, loose fragments or artificial exposures maintaining their sharp edges. It is usually quite thin bedded, the bedding being very distinct. It is generally cut up by joint planes which determine the surfaces of fracture. The porosity has been found in some cases to be due to the leaching out of minute shells, probably foraminifera. This shale has distributed through it some carbonaceous material, which, though not generally showing, except occasionally as small black specks, gives rise to the bituminous springs and deposits so common along the coast.

Near Santa Cruz, apparently underlying but probably part of the same formation, is a black bituminous rock, like a coarse grained sandstone, in which the matrix is bitumen. Large quantities of this rock are shipped to San Francisco.

At places, as on Bald Knob near the road down Tunitas Creek, the shale appears to be silicified into chert or chalcedony or sometimes resembling opal. It maintains its white color and bedding.

At Monterey and other places it is very rich in infusorial forms, diatoms, sponges, etc., so that it has been considered as a vast deposit of such forms. Recently Professor Lawson* has advanced the suggestion, upon

* Univ. of Cal., Bull. Dep. Geol., vol. i, p. 24.

chemical and microscopic examination, that the White shale is of volcanic origin. He estimates its thickness at over 1000 feet.

There is, however, one feature which would seem to contradict that theory. At Santa Cruz and for a short distance south, the bluffs are formed of the fossiliferous Transition beds. The structure is a shallow syncline. Plate xxiv. Going north from the lighthouse the lower of the Transition beds are crossed, as shown in the section, plate xxiv, and the top of the White Miocene shale reached. As the bottom of the Transition beds is approached, the beds take on more and more the character of the White Miocene shale until before the parting is reached they become indistinguishable from it, showing that the conditions held over from the one period to the other. Further, as mentioned below, its apparent existence in the middle of the Merced series on Seven Mile Beach.

Occurrence.—The White shale is almost wanting on the northeastern slope of the mountains. A small exposure occurs on the Menlo Park-Searsville road, a mile or two from Searsville; also, on the road on top of the ridge, half a mile south of the Searsville-Pescadero road. But on the west side of the main ridge it forms one of the principal rocks. One of the best exposures in the Santa Cruz Mountains is along the coast from Scott's Creek south to Santa Cruz. Here it occurs dipping gently to the ocean and can be well studied in the ravines which have cut down through it. On the south point, at the mouth of Wood's Gulch on Seven Mile Beach, two or three hundred feet above the beach, there occurs an outcrop of rock identical in appearance with the White shale. Its presence here in the middle of the Merced series is difficult to explain, but will be touched upon later. Considering the thickness and extent of this formation on the coast side of the mountains, the question naturally arises, what has become of this formation on the bay side, where only one small outcrop near Searsville has been noted? The most acceptable answer is that it has been eroded to that extent. As we shall see under the discussion of the Merced series, that formation in places rests upon the granite and older strata; the fragmentary character of the outcrops of White shale on the east side of the mountains is strengthened by the further evidence of great erosion. Faulting may also play an important part in its present distribution.

Outside of the Santa Cruz Mountains the White shale is found abundantly over a large area, occurring for several hundred miles along the coast.

Relations.—The relation to the underlying rocks has already been discussed. In the Santa Cruz Mountains it is believe to overlie unconformably all the older formations, in some places lying upon the granite, in others upon the Pescadero and older series. In the sections given by Dr. Antisell from the southern part of the State, the White shale is uniformly represented as conformable with the underlying Miocene.*

The relation to the Merced series will be treated under the relations of the Merced series.

Correlation.—The White Miocene shale is generally very unfossiliferous, but characteristic fossils have been found at a number of places, notably at Monterey.

A few fossils were found by Dr. J. P. Smith near the summit south of Searsville. They were identified as follows:

Pecten peckhami Gabb.
Lucina borealis Linnaeus.
Pandora conf. scapha Gabb.
Nucula sp. ind.

* Report on Pacific Railroad Survey, vol. vii, pls. i *et seq.*

Blake reports* the following from this horizon:

Tellina congesta Conrad.
Meretrix traskii Conrad sp.
Mercenaria perlaminosa Conrad.

In addition Professor Lawson notes† the following forms:

Arca sp. (Nov.?)
Saxidomus sp.
Leda sp. (Nov.?)
Lucina like L. crenulata.
Clementia? sp.
Young Cardium, or small Venericardia.
Macoma sp. (Nov. ?)

Among those mentioned by Gabb in vol. i of the Paleontology of Cal. is *Turritella hoffmanni* Gabb.

Though none of these species except *Lucina borealis* Linn. occur on the east coast, by a comparison of similar forms Conrad concluded that this formation was Miocene. And as we have both above and below species almost identical with Miocene species of the east coast, we may accept his determination.

THE MERCED SERIES.

The beds which will be assigned to this division have alternately been called Miocene and Pliocene, but of recent years have come to be considered Pliocene. The field-work of the writer seems to show that they are Pliocene, though at the bottom probably transitional from the Miocene. This formation is of considerable thickness and is very fossiliferous. Its location is very favorable for the exposure of fine sections, so that along the seaboard it is exposed almost continuously the whole length of the Santa Cruz Mountains, and in cliffs averaging perhaps 75 feet high (see plate xxiv), but on Seven

* Geol. Recon. of Cal., 1858, pp. 182, 179.

† Univ. of Cal., Bull. Dept. Geol., vol. i, p. 27.

Mile Beach becoming over 700 feet high. It is doubtful if there is anywhere in America a Tertiary formation presenting so many advantages and attractions for study. That it is intimately concerned in the structure of the mountain, that it is cut and interbedded by igneous eruptions, and that during the deposition of its nearly a mile of sediment a gradual but marked change in the fauna can be traced, increases the interest it must have for the student.

Petrography.—The Merced series of the Santa Cruz Mountains is composed of a great thickness of partly consolidated sands, clays, argillaceous sands and hard, fine conglomerates. On the bay side of the range the strata have felt the mountain-making forces more, and are usually harder and, as shown by the distortion of the fossils, more crushed. Local metamorphism and minor variations will be mentioned later. The most abundant and characteristic rock is a dark drab or slate colored argillaceous sand, breaking into small fragments of about half an inch cube, which in many places are bright red on the joint faces. In places it shows a yellow ochre-like deposit. It varies in hardness from that which crushes easily in the hand to tough and more argillaceous varieties which are like a hard clay. Locally it sometimes forms very hard nodules or layers, generally due to the lime from inclosed shells, many of the shells collected being obtained by splitting open these nodules. In places this sandstone is thin-bedded, but it is more apt not to show many bedding planes in a thickness of 100–200 feet. This rock makes up the most of the long hill at Point Pillar, and likewise much of the bluffs to the south and on Seven-Mile Beach.

The next most abundant rock is a yellow or buff colored sand, generally quite soft; sometimes it weathers in the bluffs until it seems to be filled with great pot-holes.

Just below the sheet of basalt south of Stanford University the sandstone is coarse grained, nearly white and very hard. Above the sheet of basalt it has been leached out until the fragments on the surface are almost as porous as a sponge.

Along the Seven-Mile Beach in particular there are many thin layers of hard conglomerate. The pebbles are usually water-worn fragments of phthanite and the other metamorphics, usually less than an inch in diameter. These layers are, as a rule, very hard and seldom more than a foot or two thick, though sometimes there are many layers close together. At several places along the Seven-Mile Beach section they resist weathering and stand out very prominently from the softer surrounding strata, one layer at lowest tide being traceable several hundred feet out into the ocean. Most of these fine conglomerates contain many commuted fragments of shells, which may account for their hardness. In some the proportion of shell fragments becomes so great that they would more properly be called shell breccias or brecciated limestone. Such a breccia outcrops quite prominently in the foothills west of San Bruno; also along the contact running southeast from Mussel Rock between the Miocene and metamorphic or igneous rocks. Just above the sheet of basalt south of Stanford University occurs a similar brecciated limestone, largely composed of fragments of Balanus, and rather soft and friable. In all the ravines running into San Francisquito Creek above Searsville a ledge of soft limestone is crossed. It is more strictly a calcareous sandstone, and is full of fossils which undoubtedly furnished the lime.

A number of layers of lignite from an inch or two thick up to a foot in thickness, occur in the section along Seven-Mile Beach. In some cases the structure of the wood still shows, but in most cases it is invisible macroscopically.

At the top of the section along Seven-Mile Beach, the sandy strata become quite soft—there are found to be no hard strata. As we rise in the series the strata become more and more unconsolidated and sandy, the upper layers consisting almost entirely of yellow and orange sand.

Fine gravel occurs in some abundance, and there is a little coarse gravel, but in no case is it consolidated into a conglomerate.

The most noticeable layer in this upper series is a white, chalky layer, which Dr. Lawson* considers a volcanic ash. On the beach it appears as a bed having a uniform thickness of about one foot, interbedded with the other strata. At one point on the northeast slope of the bluffs, near the head of Lake Merced, it has a thickness of six feet. The Spring Valley Water Works Company mine it here for "chalk." Under the microscope it shows no crystalline structure, nor could any diatoms or foraminifera be made out; a chemical analysis would probably throw much light on its composition and origin.

Stratigraphy.—The best section obtained was that along Seven-Mile Beach. This section gives probably a nearly complete series at the top. The bottom of the section, however, is about 670 yards north of the contact with the igneous rocks of Mussel Rock, so that the section is not complete; but the dip at this point becomes low, and there is abundant evidence of faulting, so that it is not thought that more than a few hundred feet are omitted.

A study of the contact between the Merced series and the igneous rocks from Mussel Rock southeastward shows, however, that the bottom of the series has not been ex-

* Univ. of Cal., Bull. Dept. Geol., vol. i,p. 144.

)osed on Seven-Mile Beach; in other words, that the Merced series was not laid down upon the igneous rock of Mussel Rock, the contact between the two formations being plainly a fault contact. The absence of a large part of the Transitional beds is evidence in the same direction.

This section does not include the fossiliferous Transition beds, of which no section was obtained. They would probably carry the strata downward several hundred feet, the rock being similar to the drab and pink argillaceous sand so common above.

The following is the

Section of Merced Series along Seven-Mile Beach:

	Feet.
72 Variable layers of soft, yellow and orange sand	200
71 Light colored argillaceous sand	6
70 Hard ferruginous layer of brown sand	1
69 Yellow and orange sand	8
68 Hard layer light yellow sand	1
67 Yellow sand	8
66 Sand and fine gravel, cross bedded	12
65 Layer of "Volcanic ash" of Lawson	1
64 Yellow and orange sand	15
63 Coarse gravel scattered in orange sand	4
62 Yellow and orange sand	75
61 Yellow sand containing recent fossils	10
60 Drab argillaceous sand with red joint faces, containing plant remains and a few lamelli branches	40
59 Upper bed, upper gasteropod bed, light brown sand, full of shell and fragments	1-2
58 Yellow sand	[illegible]
57 Lower bed, upper gasteropod bed, dark gray sand	1
Drab argillaceous sand with red joint faces	3
56 Drab argillaceous sand, red jointing, with plants (?)	[illegible]
55 Light drab sand slightly cross bedded	
54 Layers fine conglomerate and shell fragments, very hard	1
53 Bluish argillaceous sand	
52 Drab argillaceous sand, quite firm, with many scattered standella	
51 Soft sandstone	

		Feet.
50	Sandstone with fossils	40
49	Many thin layers of conglomerate and shell fragments	20
48	Sandy layers	125
47	Conglomerate	2
46	Sand	10
45	Conglomerate with *Scutella interlineata* Stimp	5
44	Yellow sandstone	30
43	Yellow sandstone with many layers of fine conglomerate	20
42	Yellow sandstone with *Scutella interlineata* Stimp.	15
41	Yellow sandstone	55
40	Thin layers of shells	10
39	Yellow sandstone	20
38	Lignite, several thin layers	
37	Yellow sandstone	25
36	Conglomerate with *Scutella interlineata* Stimp	20
35	Sandstone, containing many layers a few inches thick almost made up of *Scutella*	55
34	Bright yellow sandstone	20
33	Same with hard sandstone layers or nodules a few inches thick	10
32	Bright yellow sandstone with scattered shells and full of pot-like holes from a foot to several feet in diameter	145
31	Layers of fine gravel and conglomerate	20
30	Yellow sandstone with *Lucina*, etc	110
29	Drab sandstone with thin layers of lignite	10
28	Yellow sandstone	25
27	Lignite.	
26	Buff sandstone, showing a few scattered fossils	530
25	Same, showing layers of fine and coarse conglomerate with *Crypta*, *Olivella*, *Nassa*, etc.	55
24	Drab clay with bright red jointing	6
23	Buff sandstone	235
22	Gravel and conglomerate with fossils, *Crepidula grandis* Con., *Standella californica* Con., *Scutella*, etc	10
21	Buff sandstone	60
20	Yellow sandstone with some layers of conglomerate	105
19	Layer of lignite from one to eight inches thick	
18	Yellow sandstone	65
17	Conglomerate, very hard, with *Scutella*	2
16	Light drab sand with many hard projecting layers of conglomerate	150
16	Same as above	195

		Feet.
15	Buff sandstone and drab clay with hard layers or nodules of sandstone, containing *Venus pajaroensis* Con., *Siliqua patula* Dixon, etc. Very similar to beds at mouth of Purisima Creek	220
	Conglomerate and shell fragments.	
14	Several layers full of *Crepidula grandis* Midd.	15
13	Buff sandstone with *Venus* (Mactra)	10
12	Beds with Crepidula grandis Midd. every few feet	40
	Seam of lignite.	
11	Light drab sandstone, containing beds of conglomerate and shell fragments	100
10	Buff colored sandstone	250
9	Numerous scattered lamellibranchs.	
8	Buff colored sandstone	160
7	Lignite	1
6	Drab argillaceous sandstone with pink jointing, contains *Venus*, etc. Same rock as a Pillar Point	200
5	Same with few fossils. *Tellina*, etc	120
4	Drab argillaceous rock	192
3	Lower gasteropod beds. Several hard layers from 1-2 feet thick containing a great abundance of gasteropods, also *Scutella*, *Standella*, etc	18
2	Buff sandstone with few fossils	170
1	Buff sandstone	100
1	Buff sandstone	235

Total thickness of the section, 4740 feet.

Some of the upper beds are exposed in a long landslide, which has given a little uncertainty to a few of the thicknesses. It is thought that the upper part of the section as given here extends into the Pleistocene, but no way was found of drawing the line. The upper beds are practically horizontal and contained no fossils as far as could be discovered.

The long landslide mentioned gives a good opportunity to study the variation in the layers. It is about a mile long and the course agrees with the strike of the strata, so that the beds are exposed horizontally the whole distance. It was found that in that distance there was quite a variation. The two beds of the strata called the upper gasteropod beds run all the distance and quite uniformly three feet apart. Their own thickness, however, varies from one to five feet, while the beds overlying them are entirely different in different parts of the course.

The section was made by pacing (corrected by map), the minor layers being estimated by rough measurement or by eye. Frequent landslides at the foot of the bluff make it difficult to get accurate measurements of many of the layers. At Wood's Gulch there was found to be a fault of 825 feet, downthrow on the south side. This was the only fault of any magnitude discovered, except those at the south end which could not be measured. Professor Lawson estimated the same strata to have a thicknoss of 5626 feet.* These beds appear to thin out to the southeast.

This section was the only one obtained, since the strata in the sections south of Point San Pedro have, as a rule, low dips and would require instrumental surveys to assure any degree of accuracy.

A little south of Point Montara this formation lies upon the granite. The lowest bed where it lies close to the granite is almost made up of pebbles and boulders of granite. In a short distance the proportion of granite rapidly decreases and the layer becomes very fossiliferous, gasteropods predominating, and from their resemblance to those of the lower gasteropod beds of the section on Seven-Mile Beach it is thought the two belong to the same horizon.

In the long hill at Point Pillar, the strata are excellently exposed dipping south, the dip being as high as 40° part of the distance. The same rock, the drab argillaceous sandstone with red joint faces, runs through the whole

---

* Univ. of Cal., Bull. Dept. Geol., vol. i, p. 147.

exposure. *Lucina borealis* Linn. is very abundant here and a few other Miocene forms were found. At Point Pillar the dip becomes horizontal or a little north and the stacks and long reefs seen at low tide show almost complete connection with the bluffs a mile south of Spanish Town. Here is the same rock containing *Crepidula grandis* Midd. From here to Pescadero Creek the same sandstone makes up the cliffs under the Quaternary wherever it was examined. In a low anticline south of San Gregoria Creek it was estimated that about 500 feet of strata were exposed. Layers of pectens were especially abundant at the center of this anticline. At Capitola the thickness actually exposed is small, the rock varying from very soft and friable drab sandstone to the hard layers of the same thing, lime from the contained shells probably causing the difference; the beds here are very fossiliferous and the fossils generally fairly well preserved. See plate xxiv.

These bluffs from Point Pillar to Capitola will doubtless yield excellent sections when a detailed study of them is made.

No exposure of sufficient continuity to yield sections of any value were found in the mountains on either the bay or coast side.

In the mountains over the San Fernando tunnel in Los Angeles county and the foothills in that neighborhood, and in the foothills in the city of Los Angeles, the great thickness of unfossiliferous White Miocene shale is overlaid conformably by a few hundred feet of fossiliferous gravels and conglomerates whose fauna would seem to place them in this Merced series. At San Fernando there appears to be considerable thickness, but the strata are folded and faulted so that it would require some detailed study to estimate it. In going up the cañon leading to the tunnel from the south, the structure is seen to be anticlinal, so that the beds that make up the foothills run down on the north side of the anticline in the ridge through which the tunnel is cut. The gravelly deposits in Los Angeles give good evidence of being shore deposits, for the gravel seems to to be largely derived from the underlying formation and the boulders are pierced and sometimes almost honey-combed with borings of one of the rock boring mollusks.

Distribution.—The Merced series, antedating as it did the upheaval which gives the Santa Cruz Mountains its present topographic position, was probably originally laid down over all or most of the region now occupied by those mountains. The upheaval, however, seems to have operated largely by faulting, which resulted in some parts having been elevated and exposed to erosion more than others, so that the beds left now were probably under water during the Quaternary and thus preserved, and it is only the post-Quaternary uplift which has exposed them. As we might expect if this theory is true, we only find these Merced series beds on the lower flanks of the mountains probably not over 1000 feet above sea-level. They are exposed in the bluffs along the sea-coast from two miles north of Mussel Rock to Capitola at least, not continuously, however, for between Mussel Rock and Point Montara, at Spanish Town, and along the stretch north of Santa Cruz this formation is lacking, erosion having exposed the underlying strata or cut down the bluffs.

How far back into the mountains the Merced series extends was not accurately determined. In Purissima, Lobetas, Tunitas, San Gregoria and Pomponia creeks it was found to extend back at least four or five miles from the coast.

From Seven-Mile Beach the Merced series extends southeast in the direction of strike to Milbrae. From there to Redwood City the underlying metamorphics are exposed. South of Stanford University the foothills are fossiliferous and show the presence of this formation. It is also shown above Searsville in the lower part of the ravines running into the valley of San Francisquito Creek from the west and south.

Whitney* notes the presence of several hundred feet of this formation on the north flank of the Palo Scrito Hills.

It seems possible that most of the beds from which older Pliocene fossils have been reported will be found to correspond with this series. Dr. Lawson believes that the strata which he recently described as the "Wild Cat Series"† in Humboldt county may be correlated with the Merced series. Thus the fauna reported from Kirker's Pass and Green Valley, Contra Costa county; Santa Rosa and Russian River, Sonoma county; would suggest the presence of the Merced series.

Relations.—The relation between the Merced and Monterey series was best seen where the strata of the Soquel basin rest upon the Monterey series so abundant north of Santa Cruz. A little north of Surfside, a suburb of Santa Cruz, the contact is well exposed, the line being marked by a line of scattered pebbles of older igneous and chert rock. The beds have a low dip south and are conformable. Going toward the lighthouse the beds are crossed in ascending order. Just above what has been taken as the line of contact the beds are indistinguishable from the most characteristic white shale. Ascending the series the rock changes imperceptably from the characteristic,

* Geol. Surv. of Cal., Geology, vol. i, p. 154.

† Univ. of Cal., Bull. Dept. Geol., vol. i, p. 255.

creamy, fine-grained shale to the light or dark drab sandy strata characteristic of the Merced, and which can be traced to the fossiliferous beds of Soquel and Capitola. (See plate xxiv.)

At Point Montara the Merced lies upon the granite, and at several places in the San Francisquito basin the Merced series lies unconformably upon the Pescadero series. There is thus some evidence of erosion or transgression between the two periods.

South of Half Moon Bay the same relation seems to exist as at Santa Crúz, as indicated in certain well borings. On Purissima and Tunitas creeks wells have been sunk for oil to depths of from 600 to 800 feet. On Purissima Creek, about a mile from the village of that name, a well was sunk to a depth of 770 feet; oil was struck at 240 feet and from that depth downward. At that depth they also struck "some fossil clam shells."* Calling to mind the scarcity and character of the fossils of the white or bituminous shale it seems probable that these clams belong to the horizon of the fossiliferous beds in the bluffs on the coast. The suggestion is made that at that depth is the top of the bituminous or white shale. In one of the Tunitas Creek wells oil was reported at a depth of 350 feet. The well is 130 feet above sea-level and the dip of the shaly strata southwest. The above facts would indicate, if our interpretation is right, that the fossiliferous beds at Purissima, etc., are quite near the bottom of the series.

Some of the early writers grouped the fossiliferous or transition beds with the Monterey series, but it has seemed best for the following reasons to group them with the overlying Merced series:

* Seventh Ann. Rep. of State Geologist for 1887, p. 101.

1st. Petrographically the beds are similar to the Merced series.

2d. Faunally there is a strong resemblance, many of the species, if not a majority, being common to both.

3d. They and the Monterey series differ petrographically, the White shale being a white, thin bedded, silicious shale, while the upper formation is largely an argillaceous sandstone, sometimes becoming an arenaceous clay, and sometimes becoming very sandy. The latter is seldom thin bedded, generally showing but few bedding planes in a considerable thickness. The thin beds of conglomerate of the latter have not been noted in the White Miocene shales.

4th. They differ greatly in their faunas. The White shale having a small fauna but sparsely represented, while the Transition beds as well as the Merced series have an abundant fauna, widely and abundantly represented. Further, of the species quoted as found in the White shale, only one species, the *Lucina borealis* Linn., has been noted among the Transition beds.

5th. The position of strata at San Fernando and Los Angeles, where beds whose fauna seem to place them in the same horizon as the Merced series are to be seen overlying the White shale. Also the finding by Professor Whitney of "A group of rocks, newer in age than the bituminous shale,"* east of Monterey, whose fossils ally them to the Transition beds of the Santa Cruz Mountains. He quotes the following species from this locality:

Neptunea recurva Gabb.
Modiola recta Conrad.
Modiola capax Conrad.
Arca canalis Conrad.

Structure.—Plate xxiv gives a section of the strata as

* Geol. Surv. of Cal., Geology, vol. 1, p. 154.

exposed along Seven-Mile Beach. The section is not at right angles to the strike, so that the dip does not appear as high as it is. Some of the dips were 30° N. 9° E., 65° N. 9° E., 75° N. 14° E., 65° N. 24° E., 60° N. 14° E., 65° N., 65° N. 4° E., 67° N. 4° E., 35° N. 4° E., 30° N. 9° E., the line of section varying from S. to a few degrees E. of S. The bearings refer to the true meridian. The section gives a good idea of the upheaval to which these beds with the underlying beds were subjected.

Plate xxiv gives also a section from Lake Merced to Purissima. From this figure it is evident, if our conclusions are correct, that the uplift along Seven-Mile Beach is not a local uplift, but is intimately connected with the main recent upheaval, and that we must assign the mountains to an age later than the deposition of these beds.

Where the Santa Cruz Range is cut by the Pacific its structure would appear to be a simple anticline, rising probably more by faulting than by folding, and the different ridges due merely to erosion as influenced by this faulting. Thus, as already pointed out, there is a fault of 825 feet downthrow at Wood's Gulch, and evidence of this fault is found in the ravine which heads against Wood's Gulch and flows to the bay. About in this same line there is evidence of a fault in the north fork of Twelve-Mile Creek, and in a cut near the Happy Valley House. Just north of Mussel Rock a fault zone commences and continues to Black Mountain. This is evidently not a single fault, but seems to be a line of fracturing, The contact between the Tertiary and older formations which run southeast from Mussel Rock follows this fault line at least to San Andreas Lake. The bluffs along the northeast side of San Andreas, Crystal Springs and San Francisquito valleys appears to be due to faulting.

If the streams running from the foothills to the bay be

examined it will be noticed that quite a number of them in the upper part of their course run nearly due east. It will also be noticed that in many of these cases the rocks exposed upon the two sides of the stream differ. This has led to the suggestion that a system of east and west faulting exists in the foothills, and that the streams have followed these faults.

All through the region, but especially at the northern end, are found small undrained basins, many of them containing standing water all the year. Several of these have been cut into in comparatively recent times by backward cutting streams, and fault lines exposed. This is very finely illustrated at the head of Wood's Gulch, where the faulting appears to have produced such an undrained basin, the fault scarp forming a perpendicular or overhanging cliff. Gradually the basin filled in against this face, partly by washing from the surrounding hilltops, partly by wind deposits, partly by fragments from the face of the cliff, until all trace of the fault is covered up, only to be exposed when erosion eats its way into the basin. Judging from those we can examine, these little basins are the result of faulting, and by an examination of the map we can get some idea of the amount of faulting that has taken place.

It would thus seem, from what has been said, that the structure which has resulted from the last uplift is essentially fault structure, the area having been cut up with fault lines which follow two main directions, and probably others which were not discovered.

Correlation.—For convenience we may consider separately the area from Point Montara to Capitola, that from Seven Mile-Beach to Milbrae, and that from Redwood City southward.

The first division, from Point Montara to Capitola, has yielded the following fauna:

	Pillar Point	Purisima	Lobetus	Tunitas Creek	San Gregoria Creek	San Gregoria	Capitola	Extinct	Living only Elsewhere
GASTEROPODA.									
Astyris gausopata Gould		*					*		
Calyptraea inornata Gabb?		*						*	
Cancillaria tritonidea Gabb							*	*	
Chorus belcheri Hinds							*		
Crepidula grandis Middendorf		*			*	*	*	*	
Cryptochiton c. f. stelleri Middendorf		*							
Lunatia lewisii Gould		*		*	*	*	*		
Nassa californica Conrad						*	*		
Nassa perpinguis Hinds						*			
Natica [illegible] Broderip and Sowerby?							*		
Neptunea humerosa? Gabb						*		*	
Neptunea tabulata Baird		*				*			
Purpura crispata Chemnitz	*								
Purpura saxicola Valenciennes	*								
Surcula carpentaria Gabb							*		
Volutilites indurata Conrad		*						*	
LAMELLIBRANCHIATA.									
Acila castrensis Hinds		*		*					
Arca canalis Conrad		*						*	
Arca microdonta Conrad		*				*	*	*	
Arca sulcicosta Gabb		*					*	*	
Cardium corbis Martyn							*		
Cardium meekianum Gabb							*	*	
Chione simillima Sowerby							*		s.
Crytomya californica Conrad							*		
Cyrena californica Gabb.							*	*	
Glycimeris generosa Gou									
Lucina borealis Linnæus	*	*	*				*		n. s.
Macoma edulis Nuttall							*		
Macoma nasuta Conrad		*				*	*		
Meretrix traskii Conrad ?							*	*	
Modiola flabellata Gould						*	*		s.
Mya truncata ?		*							
Pachydesma ineziana Conrad?							*	*	
Pecten caurinum Gould	*					*	*	*	
Pecten pabloensis Conrad			*	*				*	
Pecten propatulus Conrad		*	*	*				*	
Psamnobia rubroradiata Conrad ?							*		
Sanguinolaria nuttalliana Conrad?							*		
Saxidomus gibbosus Gabb			*	*		*		*	
Schizothœrus nuttalli Conrad			*				*		

	Pillar Point	Purisima	Lobitos	Tunitas Creek	San Gregorio Creek	San Gregorio	Capitola	[illegible]	Living only Elsewhere
Siliqua patula Dixon sp							•		
Solen sicarius Gould		•		•			•		
Standella californica Conrad			•						
Standella falcata Gould						•	•		
Standella nasuta Gould ?						•	•		
Tapes staminea Conrad							•		
Tapes tenerrima Carpenter				•					
*Venus pajaroensis Conrad	•	•	•	•		•	•	•	
Yoldia cooperi Gabb						•			s.
Zirphoea crispata Linnæus	•	•							
ECHINODERMATA.									
Scutella gibbsi Remond							•	•	
Scutella interlineata Stimpson	•							•	

Relations of the Fauna.—We have fifty-two species, of which eighteen are not known living and four are not known in the present fauna of the same region. Or, using the old method of percentages, we find 56% of the fossil fauna in the living fauna. It is found that twenty-two of the above list have been found in strata whose Miocene age is not questioned, of which number five are strictly Miocene. This would place these strata in the upper Miocene or according to some authorities in the lower Pliocene. But fortunately we can here use the more modern method of comparison with known faunas. We do not as yet feel safe in asserting the identity of any of these species with those found in the Atlantic Miocene. In many cases, however, the resemblance is so strong that for all practical purposes we may assume them to be of the same type and use them as though we felt sure of their specific identity.

* Recently shown by Dr. Merriam to be a Mactra

These strata, like the Atlantic Miocene, are characterized by many huge Pectens, large Arcas, and other forms which have no representatives in the present waters of the coast. Thus, there is on the coast of California one very small species of Arca, found at San Diego. In these strata we find great numbers of several species of Arca, some of which are over four inches broad. The most common of these, the *Arca microdonta* Conrad, will fit the figure and description of *Arca arata* Say of the Maryland Miocene just as well as it does Conrad's figure and description of the west coast species. The presence of the large Pectens, six or seven inches across, gives the fauna a strong resemblance to the Atlantic Miocene of Virginia and Maryland. Aside from the above localities, these Pectens have been previously quoted only from strata generally acknowledged to be Miocene. The *Crepidula grandis* Midd. is another form about four times as large as any of its living representatives. The *Cardium meekianum* Gabb, *Saxidomus gibbosus* Gabb, *Mactra* (not *Venus) pajaroensis* Con. and the Echinii, *Scutella Gibbsi* Rem. and *S. interlineata* Stimp. are among those which have no living representatives. A number of others are only known now living in distant seas. In several cases it is found that certain characteristics of a species have changed.

We have seen that there is good evidence that the beds just north of Mussel Rock and those between Point Montara and Pillar Point are the same as those further south at Purissima, etc. The fauna at Point Pillar though small did not give a single form living in the present fauna there, though the lower beds toward Point Montara did; thus seeming to support the structural evidence.

The fauna, while closely related to living faunas, as shown by the percentages given above, is found to have quite a number of species closely resembling species which in eastern America are typical of the Miocene. Two other facts of interest come in here; first, as already pointed out, the evidence of practically continuous sedimentation from the White Miocene shale to the top of the Pliocene or a little beyond; and second, the interesting way in which, as the series is ascended, the older forms drop out one by one. Thus, the large pectens are found only near the very bottom of the series, none of them having been found in the main body of the section as exposed on Seven-Mile Beach. In the same way many of the other species can be traced part way up the column, when they disappear, as the *Crepidula grandis* and large Arcas, thus showing finely a gradual dying out of one fauna and replacement by another.

It would therefore seem that the lowest, or what might be called the Pecten beds, are more closely related to the Miocene, but a rapidly changing fauna soon gives the beds a Pliocene aspect which is maintained through most of the section. The writer has, therefore, thought it best to call the lower beds, as exposed along the coast south of Half Moon Bay and at Capitola, "Transition Beds."

On Seven-Mile Beach the great thickness of strata gives a splendid opportunity for the study of faunal changes. On account of the friable nature of most of the specimens, the lists given are very incomplete. The suggestion has been made that the top of the Merced series on Seven-Mile Beach extends into the Pleistocene. To show the ground for such a belief the fauna of the strata from the "upper gasteropod bed" upward is given separately, the fossiliferous strata above the upper gasteropod bed especially having a Pleistocene aspect, all the forms of which still live on the coast.

Of the forms given, the *Neptunea tabulata*, *Calyptræa*

filosa and *Crepidula prærupta* occur only at the bottom of the section, the *Crepidula grandis*, *Arca microdona*, *Cardium meekianum*, *Saxodomus gibbosus*, *Mactra* (not *Venus) pajaroensis* and *Scutella interlineata* disappear at different horizons and are replaced in the uppermost layers by living forms. Thus for example the living *Echinarachinus excentricus* replaces *Scutella interlineata*, *Cardium corbis* replaces *C. meekianum*, from which it probably descended, etc.

The fauna of the main body of the section is as follows:

LIST OF FOSSILS.	Seven-Mile Beach	Twelve-Mile Creek	Foothills	Extinct	Living Elsewhere
GASTEROPODA.					
Astyris gausapata Gould	●				
Calyptræa filosa Gabb?	●			●	
Chemnitzia tenuicula Gould	●			●	
Chorus belcheri Hinds?	●				s.
Crepidula grandis Middendorf	●			●	
Crepidula prærupta Conrad	●			●	
Drillia incisa Carpenter	●	●	●		
Lunatia lewisii Gould	●				
Nassa californica Conrad. Var.	●	●			?
Nassa fossata Gould	●				
Neptuneata bulata Baird	●				
Olivella biplicata Sowerby	●	●			
Purpura crispata Chemnitz	●	●			
Purpura saxicola Valenciennes	●				
Turritella sp. ind	●			?	s.
LAMELLIBRANCHIATA.					
Arca microdonta Conrad	●	●		●	
Cardium meekianum Gabb	●	●		●	
Cardium quadragenarium Conrad			●		s.
Chione succincta Valenciennes	●				s.
Cryptomya californica Conrad	●		● ?		
Macoma edulis Nuttall?	●				n.
Macoma nasuta Conrad	●				
Mytilus californianus Conrad	●	●			
Saxidomus gibbosus Gabb	●			●	
Schizothærus nuttali Conrad	●		●		
Siliqua patula Dixon	●		●		
Solen siccarius Gould	●		●		
Standella californica Conrad	●	●	●		
Tapes staminea Conrad	●				
Mactra (not Venus) pajaroensis Conrad	●			●	
ECHINODERMATA.					
Scutella interlineata Stimpson	●		●	●	

We have here only a small fauna, thirty-two species having been identified among those collected. Of these nine are not known to be living and five are found only in some other district. This gives out of thirty-two species in the fossil fauna eighteen in the present fauna of the same region, or 57%.

The southern character of the fauna is very noticeable. Several of the living species are known only in more southern waters, and some of the extinct forms, as the *Arca microdonta* Conrad, are related to forms found only south of this region.

TABLE OF FOSSILS OF UPPERMOST BEDS OF MERCED SERIES.	Living	Not Represented in Present Fauna	Seven-Mile Beach, Upper Gasteropod Beds	Seven-Mile Beach, above Upper Gasteropod Beds.	Twelve-Mile Creek	Near San Bruno
Astyris gausapata Gould	*		*			*
Bittium armillatum Carpenter	*				*	
Cardium corbis Martyn	*		*	*		
Cardium quadragenarium Conrad?	*	*	*			
Chione succincto Valenciennes	*	*	*			
Columbella richthofeni Gabb		*	*			
Crepidula navicelloides Nuttall	*		*			
Crytomya californica Conrad	*		*			
Drillia incisa Carpenter	*					*
Drillia penicillata Carpenter	*	*	*			
Macoma edulis Nuttall	*	*	*			
Macoma nasuta Conrad	*			*		
Monoceras engonatum Conrad	*		*			
Mytilus edulis Linnæus	*		*	*?		
Nassa californica Conrad, var.			*		*	
Nassa fossata Gould	*		*			
Nassa mendica Gould	*		*	*		
Neptunea recurva Gabb?		*				*
Olivella biplicata Sowerby	*		*		*	*
Olivella intorta Carpenter	*				*	
Ostrea sp. ind.	*?		*			
Placunanomia macroschisma Deshayes	*		*			
Psephis lordi Baird	*		*			
Purpura canaliculata Duclos?	*		*			
Purpura crispata Chemnitz	*		*	*		*
Purpura saxicola Valenciennes	*		*			
var. ostrina? Gould	*					
Saxidomus gracilis Gould	*		*	*?		
Schizothœris nuttalli Conrad	*		*			
Standella californica Conrad	*		*			
Tapes staminea Conrad	*		*			
Echinarachinus excentricus Escholtz	*			*		

To sum up, it will be seen that out of thirty-one species that were specifically identified, six are not known to the fauna of this region (from San Francisco to Santa Cruz), leaving twenty-five species known in the present fauna, or 81% of the fossil fauna are living in that region. This by the old method would make the beds Pliocene. But considering the smallness of the fauna and the questionableness of several of the specific identifications, that method can have but little value here. Nor have any satisfactory results been obtained from a comparison with known Pliocene faunas, for the species are all peculiar to the west coast, or are not known in the fossil state elsewhere.

The arguments from the structural side are, that while, as shown by the fauna, this formation is possibly later than Pliocene, it is found to be overlaid by horizontal or nearly horizontal strata containing Elephas bones, with evidence of a land period between. Since these beds were laid down, therefore, there has been a movement which has tilted them at angles of from 5° to 40°, followed by their being exposed to subaerial erosion, and later by being submerged and covered by deposits at one place over 200 feet thick and then the whole subjected to a more or less general elevation to its present position. It will thus be seen that our assignment of these strata to the Pliocene is only in a homotaxial sense.

Not only does the fauna suggest that these upper beds might be considered by themselves, but the structural relation to the lower beds is just obscure enough to prevent a positive assertion that they are conformable. The first writers on the subject made them unconformable.

The presence of Pliocene on the Seven-Mile Beach was noted by Gabb and Remond* at the time of the first

* Geol Sur of Cal., Geology, vol. 1, p 79

survey; but the beds we are here considering the top of the Pliocene they called "Post-Pliocene." They noted the presence of two extinct forms—*Scutella interlineata* Stimp. and *Crepidula grandis* Midd.—among several species still living on this coast. It was on the finding of these two forms that the main section was judged to be Pliocene. In the upper two beds they found only shells of recent species, among which the genera Tellina, Mytilus and Buccinum are represented. Of these two beds, the lower, according to Mr. Remond, have a northwest strike, and dip to the northeast at an angle of 35°, while the upper ones have an inclination of only 10°. We have already seen that over most of the distance the lower beds really have a dip of from 65° to 75°, and we find that these upper beds have a dip of from 5° to 40°.

Structure and Relations of Uppermost Beds of Merced Series.—The present position of the strata of this formation as exposed along Seven-Mile Beach is shown on the left of plate xxiv. The dip over most of the section is from 20° to 30°, but rises as high as 40°, and at one point was 50°, but was probably local, due to recent movement. At the northern end the strata becomes practically horizontal. Except along the stretch of landslide, the strata have nearly the same strike as the beds underlying, ranging from N. 50° W. to nearly due N.

The contact between the Pliocene and underlying beds is obscured by its occurring just at the south end of the long landslide. A detailed examination failed to reveal any definite evidence in proof of or against nonconformity at this point.

In favor of conformity is—a similarity between certain of the strata lithologically, a similarity in their structure, and the lack of definite proof of nonconformity, which factor must be taken as favorable to conformity.

In favor of nonconformity is the difference in the fauna, several of the extinct forms of the lower beds, as *Scutella interlineata* Stimp., *Crepidula grandis* Midd., *Venus pajaroensis* Con., and others, not occurring in these upper beds. Also differences in the strata, the entire absence in the upper formation of the conglomerates and sandstones which ring when struck with a hammer, and which are abundant to the very top of the lower formation. Slight structural differences, for while the lower beds range in dip from 35° to 78° and over most of the exposure are between 65° and 75° the upper beds will only range from 5° to 30°, though locally rising to 40° and 50° in one or two places. Further, while in some places the strike of the strata is nearly the same, at the point of contact there is a marked change of nearly 90°. The strike which along the landslide is practically the same as the course along the beach, just south becomes nearly at right angles to the beach. This difference may be influenced by but is not due to the landsliding, the change being very marked in the cliff back of the slide. It was found impossible to correlate strata of the upper and lower formations. The manner, for example, in which the upper gasteropod beds run under the beach and do not appear again can only be explained by a nonconformity or a fault. The same thing is true of the bed exposed in the bluffs back of the slide. There seems, therefore, to be good evidence of either a fault or nonconformity or both at the south end of the long landslide. Some of the evidence is best explained by a fault and some by a nonconformity. In Twelve-Mile Creek the contact appears to be near a fault.

The third area, from Redwood City southward, gives the following list of fossils:

	Foothills, above Basalt.	Foothills, below Basalt.	Coal Mine Canon.	Near Searsville.	Extinct.	Living Elsewhere.
Arca microdonta Conrad			*	*	*	
Cardium meekianum Gabb		*	*		*	
Cidaris sp. ind	*				?	?
Crassatella collina Conrad?			*		*	
Crepidula grandis Middendorf			*		*	
Dosinia ponderosa Gray	*		*			s.
Glycimeris generosa Gould		*	*			
Lucina borealis Linnæus	*	*				s.n.
Lucina nuttalli Conrad			*			
Lunatia lewisii Gould			*	*		
Macoma nasuta Conrad				*		
Nassa californica Conrad				*		
Neptunea recurva Gabb				*	*	
Pecten islandicus Müller	*	*				n.
Pecten latiauritas Conrad	*	*			?	s.
Saxidomus gibbosus Gabb?				*	*	
Schizothœris nuttalli Conrad				*		
Solen siccarius Gould		*	*			
Standella californica Conrad				*		
Standella planulata Conrad?			*			
Tapes staminea Conrad						
Tapes tenerrima Carpenter			*			
Terabratella caurina?	*					
Mactra (not Venus) pajarœnsis Conrad			*		*	
Yoldia impressa Conrad		*				s.

Out of the twenty-five species given seven are extinct, not counting the questionable cases, and six are known living only to the north or south. That is, out of twenty-five fossil species eleven are found in the present fauna of this part of the coast or 44%. The fossils in this area though abundant are very poor, seldom showing sculpture and generally much distorted; the determinations are therefore in many cases somewhat questionable.

Merced Series at San Fernando Pass.—At the San Fernando tunnel in Los Angeles county, the beds that have been considered as Miocene of the Monterey series are overlaid conformably by a series of calcareous sandstones and conglomerates, which are quite fossiliferous. As this series occupies structurally the same position that has been assigned to the Transition beds of the Santa Cruz Mountains it will be of interest to compare their faunas.

Among the fossils collected in the ridge over the San Fernando tunnel the following have been specifically determined:

Amusium caurinum Gould.
Calyptra filosa Gabb?
Cancellaria c. f. **vetusta** Gabb.
Cardium meekianum Gabb.
Chione simillima Sowerby?
Crepidula rogosa Nuttall?
Dentalium hexagonum Sowerby.
Dosina ponderosa Gray.
Drillia torosa Carpenter.
Lunatia lewisii Gould.
Macoma nasuta Conrad.
Myurella simplex Carpenter
Nassa Californica Conrad.
Neptunea humerosa Gabb.
Ostrea veatchii Gabb
Pachopoma gibberosum Chemnitz.
Liropecten estrellanum Conrad.
Pisania fortis Carpenter.
Saxidomus gibbosus Gabb?
Solen sicarius Gould.
Turritella cooperi Carpenter.
Turritella jewetti Carpenter.
Venericardia venticosa Gould.

In this list of twenty-three species, fourteen are living in the present fauna of that region or sixty per cent.

Summary.—Based upon the above data the following conclusions seem justified:

I. A series of beds of considersble thickness and importance and quite fossiliferous overlies the White Miocene shale or Monterey series.

II. The age of this series is Pliocene, but at the bottom is transitional from the Miocene and at the top probably transitional in the Pleistocene.

III. This series was laid down before the main uplift which has given the Santa Cruz Mountains their present structure took place.

IV. The beginning of the period of their deposition was marked by minor movements of the earth's surface. Evidence of this does not always appear.

A better knowledge of both the fossil and living fauna would unquestionably modify these results, but it is believed that it would not materially alter them. Had all the species collected been identified it is thought the percentages given would be somewhat lower.

The field work done by the writer in Los Angeles county has shown that the Coast Ranges are not all of one age. The Santa Cruz Mountains were certainly elavated at a later date than the mountains of Los Angeles county. The suggestion is made that the parallel ranges of the Central California coast line agree in age with the Santa Cruz Mountains; that the east and west ranges of Santa Barbara county, and to the south, are of one age, having been raised during the Merced Period and near the close of the Miocene. Attention has been called above to the presence at San Fernando of fossiliferous strata conformably above the Monterey series. The fauna of these strata has also been shown to agree closely in its character with the fossiliferous or transitional beds at the bottom of the Merced series. The mountains of that region were evidently raised soon after and *before* the deposition of the uppermost beds of the Merced series in the Santa Cruz Mountains. This is shown by the existence at San Pedro and elsewhere along the coast of horizontal strata, evidently of later age than the neighboring mountains, which contain an abundant fauna, believed to correspond with the fauna of the top of the Merced series in the Santa Cruz Mountains.

THE PLIOCENE OF SAN PEDRO.

The plain of Los Angeles, stretching from the city to the coast, is broken at the harbor of San Pedro by a long hill, 1475 feet high, according to the Coast Survey, known as San Pedro Hill. It is smooth and bare of timber, but interesting, both on account of its terraces, of which

that basis assigns the middle layer on San Pedro Hill to the Pliocene.*

It is hoped that the few hours spent there while working up this paper will put the geology at that point in a little clearer light and on a more substantial basis, and it is thought that Dr. Dall's conclusions will be substantiated.

Surrounding the hill on the east and north is a very broad terrace, from 20 to 40 feet above tide, most of the town of San Pedro being on this terrace. Near the depot this terrace is still flanked by soft deposits. At the top is a layer, two feet thick, of black sandy soil, containing many *Pecten æquisulcatus* Cpr., *Chione simillima* Sby., and other living shells. This bed will be treated more fully under the Quaternary. Below that bed are three to four feet of dark sand, then another thin layer, which is quite fossiliferous, and is the middle layer referred to above. Below that is a sandy deposit that has all the appearance of sanddune structure. At the entrance to the harbor this lower deposit is cut away and a recent sea-cliff runs out to Point Fermin and around the island. The middle layer or Pliocene is exposed at the top of this cliff lying uncomformably on the disturbed Miocene. At one point, where a short drain has cut into the cliff, the Pliocene layer is especially fossiliferous.

Altogether 125–150 species were collected in this layer, of which number 104 species have been determined specifically.

In the following list of Pliocene fossils collected at San Pedro the first column gives those which have been reported† from the Miocene; the second gives those previously reported from the Pliocene, most of which are

* Univ. of Cal., Bull. Dept. Geol., vol. i, p. 128.

† From Dr. Cooper's list, 7th Ann. Rep. State Mineralogist, 1888, pp. 223 *et seq.*

from the list of fossils from the San Diego well, which Dall referred to the Pliocene.* The third column notes those previously reported from the Quaternary, but which in reality mostly came from these beds, believed now to be Pliocene, from Santa Barbara, San Pedro and San Diego. The fourth column gives those known to be living, the (o) indicating those which have not been found in the living state. The fifth column indicates those which are not known in the present fauna of San Pedro, but which are known to the north (n) or to the south (s).

The last column will be referred to beyond:

NAMES OF FOSSILS.	Miocene.	Pliocene	Quaternary	Living	Living Elsewhere	Living on Catalina Is.
GASTEROPODA.						
Acmea mitra Escholtz			•	•		
Acmea patina Escholtz?				•		
Amphissa corrugata Reeve		•	•	•		
Astyris carinata Hinds				•		
Astyris tuberosa Carpenter		•	•	•		
Bittium armillatum Carpenter		•	•	•		
Bittium asperum Gabb		•	•	•		
Bulla nebulosa Gould		•	•	•		
Calliostoma canaliculatum Martyn			•	•		
Cerithidea californica Haidemann	•	•	•	•		
Chemnitzia tenuicula Gould			•	0		
Chlorostoma aureotinctum Forbes			•	•		
Chlorostoma brunneum Philippi			•	•		
Chlorostoma funebrale A. Adams			•	•		
Chlorostoma montereyi Kiener				•		
Chlorostoma p[illegible]			•	?		
C[illegible] tab[illegible] Baird		•		•	n.	•
C[illegible] H[illegible]		•	•	•		
C[illegible] Sowerby			•	•	n.	
C[illegible] Broderip				•		
C[illegible] Broderip			•	•	n.	

* Proc. Cal. Acad. Sci., vol. v, p. 296.

NAMES OF FOSSILS.	Miocene	Pliocene	Quaternary	Living	Living Elsewhere	Living on Catalina Is.
Crepidula navicelloides Nuttall		*	*	*		
Crepidula rugosa Nuttall			*	*		
Crucibulum spinosum Sowerby		*	*	*		
Dentalium hexagonum Sowerby		*	*	*		
Drillia c. f. aurantia Carpenter				*		
Drillia torosa Carpenter			*	*	n.	
Fusus robustus Trask?			*	0		
Fusus rugosus Trask?			*	0		
Glyphis aspera Escholtz			*	*		
Hipponyx antiquatus Linnæus			*	*		
Ischnochiton regularis Carpenter				*		
Lacuna solidula Loven			*	*		
Leptonyx sanguineus Linnæus?				*	?	
Lottia gigantea Gray				*	?	
Lunatia lewisii Gould		*	*	*		
Lunatia pallida Broderip and Sowerby		*		*	n.	*
Mangelia variegata Carpenter?			*	*		
Margarita pupilla Gould			*	*	n.	*
Monoceras engonatum Conrad			*	*		
Nacella incessa Hinds				*	?	
Nassa fossata Gould	*	*	*	*		
Nassa mendica Gould		*		*		
Nassa perpinguis Hinds	*	*		*		
Neverita recluziana Petit	*	*		*		
Ocinebra lurida Middendorf		*	*	*		
Ocinebra interfossa Carpenter			*	*		
Odostomia gravida Gould			*	*		
Olivella biplicata Sowerby		*	*	*		
Olivella intorta Carpenter		*	*	*	n.	
Olivella pedroana Conrad		*	*	*		
Pleurotoma perversa Gabb			*	*	n.	*
Priene oregonensis Redfield			*	*	n.	
Purpura crispata Chemnitz		*	*	*	n.	
Ranella californica Hinds				*	?	
Scalaria hindsii Carpenter				*	?	
Surcula carpentariana Gabb		*	*	*		
Thalotia caffea Gabb			*	*	n.	
Thylocodes squamigerus Carpenter				*		
Tornatina cerealis Gould			*	*		
Trochiscus norrisii Sowerby			*	*		
Trophon orpheus Gould		*	*	*	n.	
Turritella cooperi Carpenter		*	*	*		
Turritella jewetti Carpenter		*	*	*	n.	
Turritella variata Conrad	*			0		

NAMES OF FOSSILS.	Miocene	Pliocene	Quaternary	Living	Living Elsewhere	Living on Catalina Is.
LAMELLIBRANCHIATA.						
Acila castrensis Hinds	•	•	•	•		
Cardium centifilosum Carpenter	•	•	•	•	n	
Cardium corbis Martyn			•	•	n	
Cardium procerum Sowerby		•		•	s	
Cardium quadragenarium Conrad		•	•	•		
Cryptomya californica Conrad	•	•	•	•		
Donax californicus Conrad			•	•		
Hinnites giganteus Gray	•	•	•	•		
Liocardium substriatum Conrad			•	•		
Lucina borealis Linnæus	•	•	•	•	n	•
Lucina californica Conrad			•	•		
Lucina nuttalli Conrad			•	•		
Lutricola alta Conrad	•	•	•	•		
Macoma inquinata Deshayes		•	•	•		
Macoma nasuta Conrad	•	•	•	•		
Macoma secta Conrad		•	•	•		
Ostrea lurida Carpenter			•	•		
Pachydesma crassatelloides Conrad			•	•		
Pecten hastatus Sowerby	•	•	•	•		
Pecten islandicus Muller	•	•	•	•	n	
Pecten latiauritus Conrad			•	•?		
Pecten paucicostatus Carpenter			•	•		
Pecten ventricosus Sowerby		•	•	•	s	
Pholididea ovoidea Gould			•	•		
Placunanomia macroshisma Deshayes		•	•?	•		
Platyodon cancellatum Conrad			•?	•		
Psephis lordi Baird?			•	•		
Saxidomus gracilis Gould	•	•	•	•		
Saxidomus c. f. nuttalli Conrad			•	•		
Schizothærus nuttalli Conrad		•	•	•		
Solecurtis californianus Conrad		•	•	•		
Standella californica Conrad	•	•	•	•		
Tapes staminea Conrad	•	•	•	•		
Tapes tenerrima Carpenter		•	•	•		
Tellina bodegensis Hinds	•		•	•		
Venericardia ventricosa Gould	•	•	•	•	n	•
Zirphoea crispata Linnæus				•	?	
Zirphoea gabbi Tryon	•?		•	0		

In summary, it will be seen that of 104 species identified 99 or 95% are living. But many of these living are only known now in, for example, the Arctic fauna; thus it is found that of the 104 species, 26 have not been reported from San Pedro, leaving 78 species of 104 fossils species known to be living or 75%. Even supposing that a better knowledge of the fauna should reduce by one-half the number not known there at present, we should still have only 87% of the fossil fauna represented in the living. Again notice that of the species which have migrated all but two have gone northward, a number being known only in Arctic waters at present. As indicated in the last column a number of these northern forms are still found on Catalina Island. But as has been shown by Professor Lawson,* Catalina Island did not share in the subsidence of which we have such abundant evidence at San Pedro and all along the coast. Having in mind then that these beds lie on what are apparently wind deposits, and are overlaid by beds whose fauna differ markedly from that of these beds, but agrees very closely with the present fauna of the coast, also that the hill presents evidence of having been almost or completely submerged, we seem justified in drawing the following conclusions:

1st. The deposition of these beds has been followed by a submergence of at least 1200 to 1400 feet, and later has come uplift to the present level.

2d. That there has been a change in the climate from cold to warmer.

3d. That these changes have occupied a long time, as judged by the fact that so few of the migrating forms have become extinct; and by the extent of wave erosion exposed on the hill.

Based on the above, the following suggestion is made:

---

* Univ. of Cal., Bull. Dept. Geol., vol. i, p. 138.

that these beds antedate the close of the ice age and may have been deposited early in the Pleistocene, or in the Pliocene.

It is a matter of judgment and precedent as to which of those two periods we shall assign these beds. On account of the distinct character of the fauna, we cannot compare it with known Pliocene areas, and therefore our only resource is to fall back on the old method of percentages and say that as all the beds of known Quaternary age have a higher percentage of forms living in the present fauna, we may conditionally assign these beds to the Pliocene. On the other hand the arctic character of the fauna would suggest that these deposits were laid down during the early part of the Pleistocene in what is known as the Ice age.

It may be of interest to note that the above list as compared with the latest published list of California fossils (1888)* shows thirty-three species not noticed before as fossils at San Pedro, eleven species not previously noticed as fossil and fifty-four not before known as Pliocene, while one species has only been known as Miocene.

Santa Barbara and San Diego.—A study of the species recorded as Quaternary from Santa Barbara and San Diego in Dr. Cooper's "List of California Fossils" reveals the fact that at those places even a larger percentage of the Quaternary fauna is not found in the present fauna, and suggests that there exist at those places similar conditions as to those found at San Pedro, that is, two distinct horizons that have not been differentiated, the lower of which may be correlated homotaxially with the Pliocene.

Japan.—Dr. David Brauns+ in his study of the geology

---

* 7th Ann. Rep. State Mineralogist, pp. 223 et seq.

\+ Mem. [illegible] of Sci. Dept., Univ. of Tokio, No. 4, 1881, p. 77.

of Japan finds similar beds with a similar fauna along that coast, and reasoning partly from the changes and migrations which have taken place in the fauna, and partly on the non-conformities which the beds there show in places with overlying beds, he assigns them to the Pliocene period.

4. THE QUATERNARY.

The Santa Cruz Mountains are fringed around the base with a belt of nearly horizontal deposits. Topographically these extend from the edge of the foothills to sea-level with a very gentle slope. On the ocean side these deposits and the older deposits upon which they rest unconformably have been much eroded by the waves, so that much of the distance the lower part of the slope has been cut away and a vertical cliff is left to mark the advance of wave erosion. See plate xxii. As erosion is strongest at about mean tide level, there are usually numerous reefs exposed at lowest tide. At a few places stacks, sea caves and natural bridges attest the rapidity with which the erosion is progressing. At Point San Pedro erosion has destroyed all trace of this old bench or terrace. The upper edge of this bench, where it meets the mountains or foothills, varies somewhat in height. A study of this variation shows in the case of the Santa Cruz Mountains that it is due to the character of the uplift, that is, the uplift has not been a general vertical rise of so many feet, but has been strongest in the old axes of uplift, the main ridges of the range. The result of this is that the line of contact with the foothills has an elevation inversely proportional to its distance from the lines of uplift. Thus at the Mussel Rock, the terrace is over 220 feet above sea-level.

Quaternary of Mussel Rock.—From Mussel Rock

southward the sea has cut a fine section of these beds. The deposits here are at least 150–200 feet thick. At the bottom is a heavy bed of sand having the peculiar structure of sanddunes. This rests upon the surface of the old eruptive rock which makes up Mussel Rock. Above the æolian deposit are marine deposits of sand and gravel to the top of the terrace, the gravel being largely fragments of the old igneous rock. See plate xxv.

In the ravine a short distance south of Mussel Rock, the deposit is a difficult one to understand. Between the wind deposit and igneous rock are several argillaceous sandy layers which are not strictly conformable among themselves, but appear to have been laid down during a comparatively rapid rising of the shore. Over these beds come the wind deposits, and above those the marine deposits of sand and gravel. A short distance south the sand loses its gravel, and, except at a few places where the irregular surface of the igneous rock is exposed above the beach, the deposit continues to be sand to the end of the section at sea level. Near each exposure of igneous rock, the deposit is largely made up of fragments of the rock.

On the north side of Mussel Rock, just above the igneous rock, fragments of wood and cones of a conifer are found. Dr. Lawson reports these on the authority of Professor E. L. Greene to be *Pinus insignis*, or Monterey pine. Dr. Lawson,* however, placed the beds containing these cones below the Merced series. My observations make them a part of the nearly level strata lying on the surface of the igneous rock from which the Merced series had been eroded. These are the same as the nearly level strata better exposed on the south of Mussel Rock.

* Univ. of Cal., Bull. Dept Geol , vol 1, p 143

South of Point San Pedro.—Wave erosion has formed bluffs nearly the whole distance from Point Montara to Capitola. The Quaternary forms the top of these bluffs most of the way. Thus, at Purisima we have about thirty feet of horizontal sand and gravel overlying about fifty feet of the upper Miocene, the lower beds dipping to the north. See plate xxiv. The gravel referred to proves upon examination to be fragments of rock exactly similar to the layer of volcanic ash in the Pliocene of Seven-Mile Beach. These fragments of volcanic ash occur abundantly not only in the Quartenary all along the coast from Half Moon Bay to Capitola, but also in the Quaternary deposits in the little valleys opening out toward the Bay of San Francisco. This leads to the suggestion that the layer of volcanic ash on Seven-Mile Beach originally extended all around the mountains, and that these fragments tell what has become of it and explain why the Pliocene is not found over a broader area. These fragments argue strongly for a land period with subaereal erosion between the Pliocene and Quaternary and support the evidence of the sanddunes at Mussel Rock. These fragments of volcanic ash were used to some extent in recognizing the Quaternary.

Fossils are not abundant in the Quaternary along the coast, though at places Haliotis and some other shells are quite plentiful.

At Santa Cruz, besides the lower broad terrace already mentioned, several others are observable. Dr. Lawson* counted nine, the highest one 1201 feet above sea-level. The terraces have their sea-cliffs, from the foot of which the ground slopes gently to the next lower sea-cliff.

Along the Bay of San Francisco.—Along the bay the Quaternary deposits have not been cut by the waves, but

* Univ. of Cal., Bull. Dept. Geol., vol. i, p. 141.

form a continuous slope from the foothills to the center of the bay itself. In some areas, as between San Mateo and Coyote Point, the layer forming the surface is a black earth full of commuted fragments of shells. Small areas of apparently similar deposits occur at the surface of the Quaternary near the mouths of nearly all the little valleys which open out upon the main valley of the Bay of San Francisco. Smaller patches of a few acres occur at many points on the flanks of the foothills, sometimes several hundred feet above the bay.

The most abundant shells among these fragments are:

- Cardium corbis Martyn.
- Cerithidea californica Haldemann.
- Macoma nasuta Conrad.
- Mytilus californianus Conrad.
- Ostrea lurida Carpenter.

The study of the Quaternary on the side of the mountains toward the bay is not easy, on account of the difficulty of distinguishing between subaerial and marine deposits. Much of the flat land of the valley appears to be of subaerial origin, as shown in cuts made by streams, but this seems to be overlain by bay deposits.

Deposits in Hills near Seven-Mile Beach.—At a number of places recent erosion has exposed fresh water and wind deposits. These have already been noticed under the Merced series in the paragraph on structure. In some of the drainless basins these deposits are at present forming, in some cases, according to those living in the vicinity, gaining an inch or two a year. During the rainy season the deposit is a water deposit, and during the long dry summer the wind carries off the lighter sand or carries in sand from the surrounding higher ground.

The short distance that these secondary deposits have been transported has produced a marked similarity in appearance between these beds and some of the upturned beds upon which they lie. But though so similar in appearance, erosion discloses a marked difference in the solidity of the two formations. In the deposit at the head of Wood's Gulch, already described, the end of a tusk, presumably of elephant or mastodon, was found about 75 feet below the top of the deposit.

Though these deposits for the most part appear to be of fresh water and wind origin, there are some facts that suggest that they are, in part, at least, of marine origin. In some cases the top of the deposit is so little below the elevation of the surrounding land, its area so large, as compared with the area of the hills from which it might derive material, that we are led to suppose, either, that the surrounding points are almost reduced to base level (in this case the top of the deposit), or else that submergence has permitted the deposition of marine deposits. There is one feature that strongly favors the latter theory. Nearly all the streams running to the bay show a marked terrace almost to their head. In most cases the streams have cut down through this terrace, revealing deposits from a few feet to twenty-five or thirty feet in depth. In many cases these cuts are very recent, as traces of wagon roads are still visible at their heads, or are shown upon the U. S. Coast and Geodetic Survey map of 1869. These terraces follow about the present inclination of the stream bed, and, though they can seldom be traced continuously, would appear to join the general level land which slopes from the foothills to the bay. The character of these stream terraces, like the isolated deposits described in the preceding paragraph, is varied. In places they appear to be beyond a doubt subaerial stream deposits, containing trunks of spruce or redwoods, sometimes three or four feet in diameter. The wood in these trees is still in good preservation. Above the trees, which lie horizontally, the strata are suggestive of marine origin, strengthening the evidence presented above of a very recent submergence. Dr. Lawson ascribes to these upper beds a Pliocene age.*

As some question has been raised as to whether these terraces are not the result of erosion rather than of sedimentation, it may be well to call attention to the evidence more in detail. In the first place, these terraces and fillings show only horizontal bedding: further, the bedding of the strata forming the hillsides is in nearly, if not quite, every case where found, more or less highly inclined, and finally, in nearly every ravine one or more contacts were found where the horizontal strata can be clearly seen lying upon the highly inclined strata of the Merced series. A few of these might be mentioned. The formation at the head of Wood's Gulch has already been described. In the ravine which heads up against Wood's Gulch, a few yards below the Old San Pedro-Colma road, the Merced sandstones and thin bedded gravels are well exposed, having an almost perpendicular dip. On the edges of these perpendicular beds lie the horizontal Quaternary strata. These horizontal strata may be traced continuously down the ravine to where they contain quite a number of horizontal pines near the crossing of the New San Pedro-Colma road. In a branch of this same ravine occurs a recent cutting from fifty to seventy-five feet deep. The inclined stratification of the lower beds was not seen in the cut, but was found only a hundred or two feet away.

In the cut just north of the Happy Valley House the horizontal Quaternary overlies strata of the Merced series having a dip of 35° N. [illegible] E.

Dr. Lawson has pointed out that the whole coast has

* [illegible] Bull. Dept. Geol. vol. 1, p. [illegible]

recenty st ood for a considerable time at an elevation of 1600 to 2100 feet below its present level.* Evidence of this can be seen in the level summit of the main ridge to the north of Black Mountain.

Standing upon the hills near South San Francisco station where a comprehensive view can be obtained of the line of hills extending from Seven Mile-Beach to Redwood City, between Crystal Springs Valley and the bay, their summits can be seen to present a remarkably even horizontal line. Examined on the ground the top of this line of hills has the aspect of a plateau, from which rise a few sharply conical knobs. These have the appearance of remnants left by the eroding waters which planed off the top of the hills leaving the plateau. Further evidence of such a submergence and erosion is found in the fact that a little further south, where these hills are not separated from the main mountain ridge by the broad and deep valley, their summits are more or less covered with boulders of metamorphic rocks well water worn.

Altogether the evidence seems quite strong to show that the Santa Cruz Mountains have very recently been submerged to a depth near San Francisco Bay of at least 600 feet. In a later uplift the mountains seem to have stood for some time at a level about 100 feet below their present elevation. This has resulted in the marked shore line where the rounded foothills meet the nearly level floor of the valley surrounding the bay. This level floor seems to have been the result partly of erosion and partly of deposition. It seems quite possible that this upward movement is still in progress.

What preceded this recent submergence? The evidence, from two standpoints, would indicate a long land

* Univ. of Cal., Bull. Dept. Geol., vol. i, pp. 115-160.

period. The first is the presence of land and fresh water deposits beneath the recent marine deposits and above the upturned and eroded Merced series. The other is the evidence gained from a topographic study of San Francisco Bay and neighborhood. Such a study as recently pointed out by Prof. Lawson* shows the strong resemblance of San Francisco Bay, of Rodeo Lagoon, Tomales Bay, Walker's Creek, Drake's Bay, Bolinas Bay, the valley of Lake Merced, etc., to sunken and submerged valleys.

The question of the order of the above events is an open one. Prof. Lawson makes the land period followed by a slight submergence the last events. The fact that the marine deposits overlie land deposits has led the writer to place the recent submergence as subsequent to the land period, and in its turn it to precede the still more recent uplift to present conditions.

It is possible that a double movement would explain all the evidence. Aside from the water-worn boulders capping the foothills south of San Francisquito Creek, the marine deposits noticed are confined to the lower levels, about one hundred feet on the bay side, up to over two hundred feet on the ocean side, and higher still in the bench fillings of the streams running into the bay. This theory would give the following record of events: First, submergence to a depth of from 1600 to 1800 feet. Second, uplift of about 1200 feet. Third, uplift to nearly 400 feet above the present level, according to Prof. Lawson,† followed by long period of subaerial erosion. Fourth, submergence to topographic shore-line of present valley. Fifth, slow uplift to present height. (See fig. 1.)

This theory would account for some of the flooded

---

* Univ. of Cal., Bull. Dept. Geol., vol. i, p. 263.

† Univ. of Cal., Bull. Dept. of Geol., vol. i, p. 267.

valleys appearing to have been cut out of the eroded pene-plane; it places the tree trunks found in so many of the ravines, and the mastodon bones in the third period.

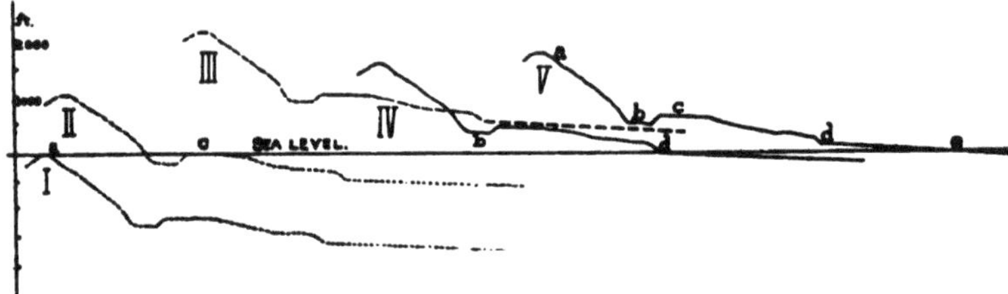

Fig. 1.—Diagram showing movements of the Santa Cruz Mountains during the Quaternary, and the development of the present topographic features. I-V—Successive positions of the mountains. Dotted line—Undeveloped features. Broken line—Partially developed features. Full line—Developed features. a—Main ridge. b—Spring Valley. c—Summit of foot-hills. d—Edge of present valley. e—Shore of San Francisco Bay. NE.-SW. section through Belmont. Vertical scale five times horizontal.

In Los Angeles County.—In the south part of the State the evidence of recent submergence is in many places very striking. As at Santa Cruz, wave cut terraces and sea-cliffs indicate the various levels at which movement has rested and given time for wave erosion. At San Pedro Hill eleven of these terraces have been counted. In the San Fernando Valley nine were counted, and several noticed in San Gabriel Valley. These terraces have been noticed by nearly all the previous writers who visited the coast of Southern California. Prof. Lawson has given us more accurate knowledge of these terraces at a number of localities, and concludes that the movement has been epeirogenic in its character.*

At San Pedro the most recent deposit, that of the lowest terrace, is quite fossiliferous in a few places. It has already been described in the discussion of the Pliocene of San Pedro. The following shells were collected from this layer:

---

* Univ. of Cal., Bull. Dept. Geol., vol. i, p. 157.

Bulla nebulosa Gould.
Cerithidea californica Haldemann.
Cerostoma nuttalli Conrad.
Chione simillima Sowerby.
Chione succincta Valenciennes.
Chlorostoma gallina.
Crepidula excavata Broderip.
Haliotis cracherodii Leach.
Liocardium substriatum Conrad.
Ostrea lurida Carpenter.
Pachydesma crassatelloides Conrad.
Pecten æquesulcatus Carpenter.
Schizothœrus nuttalli Conrad.

This fauna is very similar to that on the beach below at the present time.

With the exception of one point no trace of a nonconformity was found between the Pliocene and Quaternary at San Pedro Hill. At one point a V-shaped bed of fine gravel appears in the underlying Pliocene. At first glance it looks like an old stream filling; but examined closely, the edges of the gravel bed are not as sharply defined from the rest of the layer as that theory would seem to require. Though the transition is made in a distance of two or three inches, the two deposits seem to blend along that line as though they were local variations of deposition. The resemblance to a stream cut filling is so strong, however, that the writer believes that a more careful examination is required before we can accept Prof. Lawson's theory of the relation of the Pliocene and Quaternary. In brief, that theory is, that the Pliocene was a period of subsidence and the Quaternary a period of elevation.*

IGNEOUS ROCKS.

Granite.—The existence of granite just south of the area of the detail map near Point San Pedro has already been mentioned. Until the opportunity has been afforded to make more careful observations on the granite, the writer prefers not to commit himself to any theory as to its age or relations to the other formations.

Old Eruptives.—Over much of the country where the

* Univ. of Cal., Bull. Dept. Geol., vol. i, p. 57.

metamorphic series is exposed an eruptive rock is very abundant. It is well exposed at Mussel Rock and to the southeast of there, making much of the ridge which runs southeast to San Andreas Lake. Since this paper is concerned chiefly with the Tertiary rocks no attempt was made to map the igneous rock. Its relation to the limestone is shown on the shore at the limestone outcrop in Calera Valley. It is undoubtedly younger. On the east side of San Andreas Lake small exposures show it to be also younger than the phthanites and metamorphic sandstone. It is thus suggested that it may be contemparaneous with the post-Jurassic upheaval.

Serpentines.—The abundance and location of the serpentine has been described. The question of its origin remains unanswered. Owing to the interest which has attached to that question the writer gave to the subject some study, but beyond coming to the belief that the bronzite rock so abundant here represents an older form of the serpentine, the serpentine being, according to that theory, simply an alteration product of the old basic eruptive, he has left the problem where he found it. See Dr. Charles Palashe's paper on "The Lherzolite-Serpentine and Associated Rocks of the Potrero, San Francisco."*

Merced Eruptives.—West and south of Stanford University a large sheet of andesite is exposed. It extends from San Francisquito Creek over the foothills nearest the Bay to beyond the Page Mill Road, which runs up Matadero Creek. At its most northern exposure near San Francisquito Creek it is charactized by columnar structure, the columns being vertical where best exposed, the sheet of andesite being horizontal at that point.

On the Page Mill Road it appears to have been cut by later dykes. These dykes, however, preceded the depo-

*Univ. of Cal., Bull. Dept. Geol., vol. i, No. 5.

sition of the overlying calcareous beds. At the same point an old shore line shows finely, with its Pholas borings, its beach strewn with rolled fragments from the cliff of andesite, and the overlying deposit consisting almost entirely of fragments of barnacles.

The rock under the andesite appears somewhat metamorphosed. From fossils collected from above and below the andesite its age is evidently in the Merced period as that has been defined in this paper. The fossils do not show any marked change in the fauna and seem to indicate that the outflow took place during the Merced and not at the end.

Other late eruptives require further study before being reported upon.

PALEONTOLOGY.

Distribution.

The Californian Province.— In the present faunas the California province, extending from the Straits of Fuca to Cape San Lucas at the southern end of Lower California, is a well marked province. The provinces north and south overlap to some extent, but taken as a whole the fauna is quite distinct. At the present time this fauna is distinguished by the abundance of Chitonidæ, Patellidæ, Haliotidæ, Trochidæ and others.

Aside from the introduced species, the species which are found in both this province and in some Atlantic province are found on the Atlantic side only in northern waters. About fifty species have been noted as occurring in the California province, which have been found in the North Atlantic or northern Europe. Complete lists of the California fauna would probably largely increase that number. A very few of these, as the *Pecten islandicus* Müller, so common in the Merced period near Stanford University, and in the Pliocene elsewhere, are at present found in the North Atlantic, but are not known living in the Pacific.

Many species are found in Japan and Kamtschatka, which are common on the west coast of North America. Among these may be mentioned: *Amusium caurinum* Gld., *Cardium corbis* Mart., *Crepidula aculeata* Gmel., *Chrysodomus carinatus* Dunker, *Cryptochiton stelleri* Midd, *Glycimeris generosa* Gld., *Laqueus Californicus* Koch., *Leptothyra sanguinea* Cpr., *Lima dehiscens* Con., *Lucina borealis* Linn., *Siliqua patula* Dixon, *Macoma edulis* Nutt., *M. secta* Con., *M. nasuta* Con., *Mytilus edulis* Linn., *Natica clausa* Brod. & Sby., *Placunanomia macrochisma* Desh., *Priene oregonensis* Redf., *Saxidomus nuttalli* Con., *Siphonalia kelletii* Fbs., *Solen siccariens* Gld., *Tellina bodegensis* Hds.

Local Divisions.—On the coast of California the distribution of species, so far as known, suggests several local divisions of the California province. One extending from San Diego to Santa Barbara county. From a list of the Mollusca of Santa Barbara county, by Dr. S. G. Yates,* it is found that Santa Barbara county is the northern limit of about seventy species, aside from the strictly local fauna. It is also the southern limit of about sixty species. It is next found that the Bay of Monterey marks the northern limit of a large number of species, and suggests a division extending from the Bay of Monterey to Santa Barbara county. It is probable that Santa Catalina Island should be included in this division as well as the Santa Barbara Islands.

From the Bay of Monterey to Bodega Bay seems to mark another division, but the nothern limit may be such simply because of lack of observations further north. Little data is at hand on the faunas between Bodega Bay and Washington.

* Bull. Sta. Barbara Soc. Nat. Hist., vol. i, No. 2, p. 37.

Santa Catalina Island.—Prof. Lawson * has called attention to the fact, that Santa Catalina Island did not apparently share in the Quaternary subsidence. The fauna of this island presents some interesting features, which would seem to confirm Prof. Lawson's deduction.

Thus, by a study of Dr. Cooper's List of California Fossils, 1888, we find five otherwise extinct species from the Pliocene and Quaternary of Santa Barbara, San Pedro and San Diego are living on the island: *Amycla undata* Carpenter, *Daphnella clathrata* Gabb, *Nassa insculpta* Carpenter, *Psephis salmonea* Carpenter, *Solarellia peramabilis* Carpenter. There are also found living on the island and fossil on the coast of the mainland one species, *Crytodon flexuosus* Montagn, only known elsewhere in the North Atlantic; one, *Lucina borealis* Linneas, known elsewhere only in Arctic waters; one, *Laqueus californicus* Koch, known elsewhere only in the North Pacific. In addition to these, thirteen species, *Bittium asperum* Gabb, *Callista newcombiana* Gabb, *Cardium centifilosum* Carpenter, *Chrysodomus tabulatus* Baird, *Diala acuta* Carpenter, *Leptothyra bacula* Carpenter, *Lucina trunisculpta* Carpenter, *Lunatia pallida* Broderip and Sowerby, *Margarita pupilla* Gould, which are found on or about Santa Catalina Island, are only known elsewhere on the coast to the north; while it forms the northern limit of four species, *Chorus belcheri* Hinds, *Nucula exigua* Sowerby, *Omphalius fuscescens* Philippi, *Ostrea conchaphila* Carpenter.

These facts indicate that the fauna of Catalina Island has been little affected during a time when many species on the mainland have become extinct and others forced to migrate.

The fauna is largely northern though possessing a few

* Univ. of Cal., Bull. Dept. Geol., vol. 1, p. 138.

southern forms. It thus resembles quite strongly the fauna studied at San Pedro and called in this paper Pliocene.

This case is interesting, not alone from showing that the fauna considered Pliocene on the coast of the mainland is still living on or about the island, but from the way the fauna supports the deductions made from the topography and vice versa.

Quaternary faunas.—The faunas of the beds recognized as Quaternary are so similar to existing faunas that what has been said about the present distribution of species along the coast will hold for the Quaternary species.

Pliocene faunas.—The Pliocene faunas bear quite a striking resemblance to the existing forms of the same region, except that there are present a large number of northern forms with some forms which have since become extinct. Until the Pliocene has been differentiated at Santa Barbara, San Diego and intermediate points, as the writer believes they will be ere long, it will be unsafe to generalize on the local distribution of species.

Miocene fauna.—The Miocene fauna, while showing some regional relation to existing faunas, is principally distinguished by its uniformity over the State and its southern character.

Climatic changes.—The facts given indicate that during Miocene times a tropical or subtropical climate existed in California. This is followed by a colder climate and the introduction of boreal forms of life. The end of the Ice age, presumably, brings a return of warmer conditions and the northern forms return to the North Pacific, some species suffering extinction under the changed conditions.

The question arises whether the species found in the Pliocene of California, which are at present living in the North Atlantic, have migrated from the Pacific, or whether they represent southward migrations in both oceans at a time preceding the Ice age. From the occurrence of many of the Pliocene forms of California in the British Crag (Pliocene), it would seem that their distribution is due primarily to pre-glacial migrations.

A comparison of these results with those obtained by a study of fossil vertebrates of California would seem to show a disagreement. Thus Dr. Cooper concludes from the finding of a large species of lion, a llama, a third larger than the living camels, also one smaller, a Megalomeryx, another of the camel family, a Protohippus, etc., which have been thought to have been of Pliocene age, that the Pliocene of California had a tropical climate.* Were the Pliocene age of the deposits from which these fossils were obtained determined beyond question, the disagreement would be vital, and the beds described in this paper as Pliocene would probably have to be considered post-Pliocene.

In the first place, if the deduction upon the age of the lower beds of the Merced series made in this paper is correct, it seems probable that many other beds in the State which have been considered Pliocene may prove to be Miocene. Some of the beds from which the vertebrate fossils have been described may be among the number.

Again, the few vertebrate remains found by the writer were none of them in the beds described here as Pliocene. Elephant remains were found at two places in the Quaternary. Sharks teeth were found at two places in the beds assigned to the top of the Miocene; those previously reported have been called Pliocene. Whale bones from several localities were found in the Miocene, these also have previously been quoted only from the Pliocene.

* Proc Cal. Acad of Sci., vol v, p. 390.

Another argument of little value, but suggestive, is the possible contemporaneity of the basaltic outflow near the top of the Miocene in the Santa Cruz Mountains with some of the volcanic outflows in the northern and eastern part of the State. Several of the Pliocene vertebrates are described as from under the lava.

There seems to be some ground, therefore, for suspecting that many, if not all, of these tropical vertebrates may ultimately prove to be of Miocene age.

RELATIONS BETWEEN LIVING AND TERTIARY FAUNAS.

Notwithstanding all the evidence of repeated earth movements and climatic changes, molluscan forms have been very persistent in the California Tertiary and Quaternary. In Dr. Cooper's list of California fossils of 1888, thirty-two species are quoted as running from the Miocene to the present, and the determinations of this paper have largely increased that number by making many species Miocene which had previously been thought to go back not farther than the Pliocene. In the same list 118 species are reported from the Pliocene to the present; this also would be largely increased by recent data. From the same source 331 are quoted as found in the Quaternary and living.

If to these figures be added the species which range from the Miocene to the Pliocene or to the Quaternary and those ranging from the Pliocene through the Quaternary, it will be evident that the dividing of the Tertiary and post-Tertiary strata into zones or minor groups will be a difficult, if not impossible, work from the paleontological standpoint. It is probably that fact more than any other that has kept the Tertiary stratigraphy so long in confusion.

Changes in species.—One of the most interesting features of the study of the paleontology of the Tertiary and post-Tertiary of California is in studying the changes which have taken place in species which did not migrate, or in tracing the relation between living and old or extinct species of the same type. Lack of time and lack of material have prevented studying this subject as the writer had hoped to do. The few notes given are those taken down in determining the fossils. These notes will be given under the descriptive paleontology, but a few instances of such changes are noted at this point.

Thus, *Cardium corbis* Martyn and *Cardium meekianum* Gabb, appear to grade into each other in such a way as to suggest that *C. corbis* is simply the living representative of *C. meekianum*, no break coming between them. In the Pliocene at San Pedro is a *Chlorostoma* which agrees perfectly with *C. funebrale* A. Adams, except that the last coil is sculptured with a number of strong revolving ribs, the recent species usually having only two, the anterior one rather faint, and a few obsolete ribs. A large number of recent and fossil forms were examined without showing any intermediate specimens.

A comparative study of *Crepidula grandis* Middendorf with *C. rugosa* Nuttall and *C. excavata* Broderip may show a very close connection. The Miocene forms of *Lunatia lewisii* Gould are found to differ from the living in uniformly lacking the constriction near the suture, which is so marked in large living specimens and upon which Gould lays great stress in his description of the species. The Pliocene specimens of *Monoceras engonatum* Conrad differ from those living along the coast to-day in being from one and one-half to double the size. The same thing is true of some of the Purpuras.

SUMMARY.

The salient points of this paper are:

1. The description of the conglomerate and sandstone formation underlying the Monterey series.

2. The description of transition beds between the Monterey and Merced series.

3. Fixing the ages of the Merced and Post-Merced uplifts in southern and northern California.

4. Outlining the Quaternary history of the region about San Francisco Bay.

5. Studies in the Neocene changes in the faunal geography of the California coast.

TABLE OF CONTENTS.

CHANGES IN FAUNA AND FLORA OF CALIFORNIA.—ON THE POWER OF ADAPTATION IN INSECTS.

BY H. H. BEHR.

The power of adaptation to new circumstances plays a most prominent part in the changes that take place in the fauna and flora of newly settled countries.

It is chiefly the want of this power that causes the disappearance of types which were characteristic to districts before the harmony of organic life was disturbed by the interference of man.

To illustrate the disappearance of such characteristic features, it is only necessary to mention two instances, which all old inhabitants will confirm as soon as their attention is directed to them.

Up to the year 1856 a considerable part of the neighborhood of San Francisco was covered by a chaparral consisting almost exclusively of *Ceanothus thyrsiflorus*. Part of this formerly impenetrable thicket has been removed by human agency, and to a great extent has been replanted by our California *Cupressus macrocarpa*, *Pinus insignis*, Australian Acaciæ and Eucalypti. The more remote part of this thicket, where human interference was not directly at work, still exists, but in another shape, the Ceanothus being replaced by *Silybum Marianum*, a thistle with large blotched leaves, originally at home in Mediterranean Europe.

Another instance of similar nature is the striking change that has taken place in our aquatic vegetation. Our brooks and pools, as far as sewerage does not disable them to support phanerogamic vegetation, at present are covered by the luxuriant growth of an African weed, the *ula coronopifolia*, the round, yellow heads of which now familiar to the most superficial observer. Old itants will recollect the beautifully varied carpet produced by a graceful waterfern, the *Azolla Caroliniana*, that covered the water now monopolized by the luxuriant but coarse weed—the Cotula.

In both instances it was not so much direct interference that changed the character of the vegetation, but a certain inability of the native vegetation to conform to altered conditions of things.

There is a series of phenomena in the complex system of changes that follows the occupation of new territories which practically as well as theoretically is of great importance. It is a series of changes that gradually establish themselves in the relations between the vegetable kingdom and the insect world.

In new countries we find a certain harmony in these relations. Undisturbed nature characterizes itself by a greater variety of species than those exhibited in the agricultural stage, when the battle for existence has begun to thin out the original inhabitants of the soil. Another peculiarity of this undisturbed state is a certain equilibrium in the number of individuals of the different species, amongst whom there is no preponderation of species, otherwise than a very rare and even then a very transitory one.

The exclusive cultivation of agricultural plants and domestic animals proves a first cause of long series of effects, of complications, modifying each other. For instance, the planting of extensive orchards has favored an increase of those insect species that live on different varieties of fruit trees, and that formerly had a more or less precarious existence on wild species of the same order, mixed up with a forest vegetation of orders that do not favor the multiplication of insect species, depending exclusively on species related to certain fruit trees.

This, of course, is changed after some time, when such

fruit trees are grown to the exclusion of other species, forming, so to say, a forest of their own.

We will at present refer to a single group of insects, the Coccides, and especially the species of Lecanium and its allied genera, which in common life are comprised under the name "scalebug." All these insects produce a sticky exudation, which partly hardens into the protecting scale from which the group received its vernacular name, partly it covers leaf and branch in form of a kind of viscosity.

This viscosity again retains the spores of minute parasitic fungi of different varieties, one of the most common forms being the Capnodium, and we soon will observe a sootlike substance covering many leaves in our orchards, as a concomitant of the scalebug. Under ordinary circumstances but few spores out of a million reach their destination, that is a spot favorable to their development. But with the facilities offered by the sticky surface of leaves that are infested by the scalebug and the numbers of individuals that have already developed and have matured their spores, the number of germs floating in the air becomes such that the chances for the development of the parasites become more and more favorable; and as the Capnodium, the vegetable parasite, is but little dependent on the species of plant on which it develops, it soon begins to infest the forests, as well as orchards and hedges.

As most of our forest trees, being evergreen, never shed their leaves at once, the foliation of these evergreens has ample time to foster and breed on their surfaces, roughened by Lecanium and Capnodium, other fungoid growths, more detrimental to vegetation than the unsightly but comparatively harmless Capnodium.

Forms of Uredo will pass their dimorphic stage there,

to attack afterwards, in their Puccinia stage, grass and cereals; Rhytisma and Dædalea spores carried in the feathers of birds will be carried from tree to tree and by a slow but sure process will kill the giants of the forest. This is one of the causes that but seldom you find a healthy Madroña tree in the neighborhood of cultivated land and that the Sycamores at Niles Station are dying away.

Up to this point it is easy enough to classify the phenomena and derive them from a common source, but further on the effects become modified by the consequences of other changes that have taken place in the relation of the different forms of organic life. The agency still exists and marches on although we loose sight of the Ariadne thread, which we followed into the labyrinth of co-operating and antagonistic causes. We soon loose sight of the wave of the streamlet after its having joined the waters of the river bed.

The circumstances mentioned here form only part of a general system of changes that have taken place in organic life since the settlement of this country.

One of the most remarkable features of these changes is the degree of ability in the different species that form part of our flora and fauna to adapt themselves to altered circumstances, and the methods by which they adapt themselves.

We will consider here a few cases that were easy to follow up:

1. *Danais Plexippus*, a butterfly, common nearly through this whole continent and lately emigrated into the Sandwich Islands, New Zealand and Queensland. The insect was rare in the vicinity of San Francisco up to the year 1856, when suddenly a great number of this showy butterfly appeared in the streets of San Francisco. The caterpillar feeds exclusively on different Asclepiadace-

ous plants (called milkweed), which type does not grow on this side of the bay. In the following year I found a great number of the larva on *Asclepias fascicularis* growing on marshy grounds near Brooklyn, then called San Antonio. Some years afterwards I found the caterpillar in a garden in San Francisco on an exotic plant of the Asclepias family, *Gomphocarpus Curassavicus.*

From that time the butterfly has visited our streets every fall, and swarms of this insect working against the western current of air, peculiar to our summer months, fly out to the lighthouse, where they disappear, probably drowned in the ocean.

Since the year 1880 they have not visited the city, which omission easily could be accounted for by the circumstance that the marsh where their food plant, the Asclepias formerly grew, was converted into fields and orchards.

In this instance the insect has but little power of adaptation to new food, because it has shown itself dependent upon plants of the milkweed family and became locally extinct, at least temporarily in a district where the ground containing the milkweed was ploughed over. But, on the other hand, it has shown wonderful powers of adaptation to different climates; proof of it, its wide geographical range and the colonies formed in countries beyond the sea, where it probably has been carried in its chrysalis state in the ballast of vessels.

2. *Pyrameis Cardui*, the most cosmopolitan butterfly in existence, of an almost unlimited power of adaptation, because with the sole exception of the real tropics, Australia and the regions beyond the Arctic circle, it exists everywhere. Even the Australian species, *Pyrameis Kershawii*, by some authorities is considered identical and not merely related.

In regard to food the larva prefers thistles, but will feed on all other Compositæ, Malvaceæ and Urticaceæ, in our Golden Gate Park even on Lupinus. It conforms to most climates and very different circumstances. It has a single generation in Lapland and I have counted five generations in California. With all this the insect is less common in California than its near relation,

3. *Pyrameis Carye*, which is the most common of the series, but seems to be restricted to our coast as the Chilean specimens that I have seen, exhibit constant and well marked differences. The larva proves the same power of adaptation as its congener by feeding on Urticaceæ, Malvaceæ, Compositæ, etc., but its limited geographical distribution speaks against a facility of the species to adapt itself to changed climatic circumstances or inconveniences of travel.

4. *Pyrameis Hunteri* or *Virginiensis* feeds in California exclusively on Compositæ. It is here the rarest of the three congeners, but is common in the Atlantic States from where it extends as far as Buenos Ayres. So its adaptive power to new food plants is less than that to climatic changes.

In general the power of adaptation is greater amongst the Heterocera than amongst the Rhopalocera, of which several species have been lost to our neighborhood by ploughing or building of grounds formerly grown over by *Lupinus Chamissonis*.

On the other hand one of the East India silkworms, *Saturnia Cynthia*, has shown great adaptation powers in its conforming itself to different climates and no power whatever in adapting itself to a new variety of food. The insect has escaped from zoological gardens and experimental stations and established itself in different countries like an indigenous species, but, as far as I could ascertain, it feeds exclusively on *Ailanthus glandulosa*, which by some strange coincidence has been introduced as an ornamental tree into America and into Europe. Without this coincidence the insect would have perished for want of food in its larva state and would not have derived any advantage from its adaptability to different climates.

Deilephila lineata possesses the advantage of easy adaptation to climate and to food, combined with the enormous power of locomotion peculiar to its class. It is generally considered indigenous to the old and the new world, but as its organization excludes it entirely from the circum-polar regions it is evident that there must exist another center from which the species has spread. The insect is considerably more common im America than in Europe. Our specimens are larger and show their greater vitality by their power of adaptation in the larva state.

I infer from these circumstances that the original center from which this showy species spread is on our continent and not in Europe, where the species is comparatively rare, when on our continent it has occasionally developed into what was supposed to be an insect pest, but owing to the very transitory nature of its devastations, can be considered only a fright. The original food of the larva are probably Onagraceæ, which group of plants is much affected by the congeners of the Deilephila in other climates. In California it shows a predeliction for foreign plants of the same order. It prefers for instance the Fuchsia cultivated in our gardens to our native Œnothera, Godetia, Boisduvalia, etc., without neglecting them entirely. It has also adapted itself to Rumex and Polygonum, to Purslane and its relations, Claytonia and Calandrinia. It has been discovered in great numbers by Mrs. Brandegee in the Gila desert, feeding on a species of Lupine, and has occasionally frightened our winegrowers by attacking, in company with the larva of *Chærocampa Achæmon*, the grapevine. In regard to the latter circumstance, I have to take the part of the Deilephila larva, as much more mischief is done by the Chœrocampa, which feeds exclusively on the grapevine, when Deilephila generally prefers the weeds of the vineyard.

The caterpillar is frequently mistaken for the army-worm, which term is correctly applied to the larvæ of different Agrotides, which are much smaller but considerably more dangerous than the formidably looking Deilephila, whose devastations are exceedingly transitory.

The Indians eat the caterpillar, and that to my opinion is another proof of its American birthright, as all savages avoid eating animals which they do not know for generations.

I have here to mention another insect, which, originally very rare, gradually has developed into an insect pest, without possessing any advantages in regard to locomotion, nor powers for adaptation in regard to food or climate.

It is *Phryganidia Californica*, a type most interesting to the student of systematic entomology, but without any attraction as to beauty.

This insect feeds on oaks, formerly exclusively on live-oaks, but since these oaks have diminished in number, in proportion to the insect having multiplied the larva has begun to invade *lobata* and *Kelloggii*, but there its power of adaptation to new food stops.

I do not know how many generations in a year are produced by the insect. I have counted four, but am certain there is at least one more.

In consequence of these many generations the destruction of leaves extends over the whole year, and the poor tree has scarcely produced a new crop of leaves, when it is stripped again by a new generation, so that many liveoaks in our surroundings die by exhaustion.

Phryganidia Californica was formerly very rare, and it was not before the year 1859 that I could add the female to my collection. But from that year the Phryganidia constantly has gained ground, which circumstance has to be accounted for chiefly by the decrease in the number of insect-feeding birds. The group of the warblers (Muscicapa) was formerly well represented on the peninsula and around the bay, but at present most species of these birds have become very rare, some of them being entirely lost to our neighborhood.

Wherever the small bird is protected, as for instance in the Golden Gate Park, the liveoak is free from the Phryganidia and grows well, only being plagued by füngoid growths like the Capnodium, where the groves by too dense growth shut out air and light.

The introduction of the English sparrow has made matters worse, because the sparrow is himself a very poor insect-feeder, but, by pugnaciousness and a certain social organization of his own, he manages to drive away the real insect-feeders, warblers, titmice and swallows, and in this way has become the protector of several of our insect pests. He is a bird of great power of adaptation, but of no utility whatever, and in very short time will serve as another living proof how easy it is to disturb the natural relation of things and how difficult to restore it.

THE FISHES OF SINALOA.*

BY DAVID STARR JORDAN,

Assisted by

EDWIN CHAPIN STARKS, GEORGE BLISS CULVER AND THOMAS MARION WILLIAMS.

[With Plates xxvi-lv.]

The Mexican State of Sinaloa lies along the east shore of the Gulf of California, mostly to the north of the Tropic of Cancer, extending from Rio Fuerte on the north, which separates it from Sonora, to the northwest boundary of Jalisco. The greatest length of the State along the coast is about 325 miles. The land forms an irregular and broken slope from the high table-lands and cliffs of the Sierra Madre on the east downward to the coast. Down this slope flow several streams of clear water, which acquire great volume in the rainy season (June to November) and which dwindle rapidly in the dry season of the winter. The coast line is very irregular, being formed of rocky islands, mostly of volcanic origin, and of abrupt cliffs or "rincones," the terminations of hills or spurs from the Sierra Madre. Between these are long curved sand beaches, and occasionally sand-spits across the mouth of some estuary which is thus converted into a lagoon. The water of the sea off the coast is very clear. The bottom is very irregular, as is the contour of the shore.

The chief port of Sinaloa is Mazatlan. This city of about 20,000 inhabitants lies on a peninsula between the Estuary or Astillero de Mazatlan on the south and a curving bay known as the Puerto Viejo on the north. On this peninsula are two considerable headlands, Nevería on the north and Vijía on the southwest, between which is a sand beach, facing the west, noted for its high surf, for

* Contributions to Biology from the Hopkins Seaside Laboratory of the Leland Stanford Jr. University. No. 1.

which it is named las Olas Altas. North of Puerto Viejo, at a distance of about seven miles, are three large rocky islands, very much alike, close together and in a right line, known as the three Venados. Opposite them on the shore is a similar headland, Camarron. About all these headlands and islands are many rock-pools and basins left filled with water by receding tides. Beyond the extremity of Vijía is a tall conical island, over 500 feet in height, known as Creston. This is surmounted by a lighthouse and is the most conspicuous land mark of the harbor of Mazatlan. North of Creston lie a number of large barren rocks of white volcanic rock, known collectively as Islas Blancas. The scanty harbor of Mazatlan lies to the south of Vijía and Creston, between these and the Isla de los Chivos and Isla de las Piedras. It ends in a long deep winding channel, known as the Astillero or Estuary, which extends around the south side of the city, with many muddy arms lined with Mangrove bushes, then turns to the south, forming for some ten miles the narrow channel between Isla de las Piedras and the mainland. No fresh waters of importance flow into the Astillero and the tides form strong currents as the waters pass in and out.

At Altata, in the northern part of Sinaloa, is a small harbor, the port of the capital City of Culiacan.

Of the several rivers in the State, only one, Rio Presidio or Rio de Mazatlan, was visited by us. This is a swift clear stream, rising in the mountains. At Presidio and Villa Union, where it was visited by us, it flows rapidly over gravel, being in January some three rods wide and rarely more than two feet deep.

The fishes of Sinaloa are known chiefly from the collections made by Dr. Charles H. Gilbert in the winter of 1881. Under the auspices of the U. S. Fish Commission,

Mr. Gilbert spent six weeks at Mazatlan where he secured a collection of about 180 species, of which number about fifty were new to science. These were described by Jordan and Gilbert in the Proceedings of the U. S. National Museum in 1881, the typical specimens being deposited in the Museum at Washington. Previous to this time a number of specimens had been sent, by collectors who had visited Mazatlan and Altata, to the Museum at Vienna, where they were described by Dr. Franz Steindachner, and to Berlin where they were recorded by Professor Peters.

Subsequent to the visit of Dr. Gilbert, collections were made at Mazatlan and Presidio by Mr. Alphonse Forrer, now of Santa Cruz, California. Most of these were sent to the U. S. National Museum, where they were described by the present writer. A few specimens were also sent to the British Museum.

In December, 1894, through the kindly interest of Mr. Timothy Hopkins of Menlo Park, California, and under the auspices of the Hopkins Seaside Laboratory, a branch of the Leland Stanford Jr. University, an expedition was sent to Mazatlan for the purpose of collecting fishes. This was in charge of David S. Jordan, assisted by George B. Culver and Edwin C. Starks. In addition, Mr. Thomas M. Williams, Mr. Norman B. Scofield and Mr. James A. Richardson accompanied the expedition as volunteer assistants, with Frank H. Lamb as botanist, and Mr. George B. Seward as herpetologist.

One month, Dec. 24, 1894 to Jan. 25, 1895, was spent at Mazatlan in the collection of fishes. One hundred and eighty-five species were obtained, of which twenty-nine seem to be new to science, besides two species from La Paz. A full series of the specimens obtained is in the Museum of Leland Stanford Jr. University. Other series nearly complete have been sent to the British Museum and to the Museums at Vienna and Berlin. Partial sets are in the Academy of Sciences at San Francisco and in the U. S. National Museum. It is evident that the list here given is by no means a complete record of the fishes of Sinaloa. Doubtless all the species enumerated from Sonora by Gilbert, Jenkins and Evermann, and by Gilbert and others from Lower California, will ultimately be found in this region. Every day spent at Mazatlan either by Dr. Gilbert or by ourselves brought some addition to the list, and the deep water fishes have not been studied at all.

Besides our obligations to Mr. Hopkins, and to the volunteer assistants above named, the writers wish to express their especial indebtedness for local assistance to Dr. George Warren Rogers, a scholarly physician resident at Mazatlan; to Señor Ygnacio Moreno, the leading fisherman of the port, whose efforts in aiding our work were unwearying. We also owe many favors to Messrs. William W. Felton, Bert L. Smith, John L. Kendall and J. Rippey, American residents in Mazatlan. From Dr. Charles H. Gilbert, in whose laboratory the present paper has been written, we have received much valuable aid in many ways.

The plates accompanying this paper have been drawn by Miss Anna L. Brown, artist of the Hopkins Laboratory.

The following species are here described as new to science. The numbers after each name are those borne by the type specimens on the register of the Museum of Leland Stanford Jr. University.

Pristis zephyreus Jordan & Starks. (Skin.)
Narcine entemedor Jordan & Starks. 1699.
Urolophus rogersi Jordan & Starks. 1700.
Urolophus umbrifer Jordan & Starks.
Pteroplatea rava Jordan & Starks. 1587.

Galeichthys gilberti Jordan & Williams. 1666, 1667, 1668.
Galeichthys azureus Jordan & Williams. 1575.
Stolephorus scofieldi Jordan & Culver. 2941.
Pœcilia presidionis Jordan & Culver. 2687.
Siphostoma starksii Jordan & Culver. 2686.
Mugil hospes Jordan & Culver. 2890, 2954, 1695.
Thyrina evermanni Jordan & Culver. 2688.
Thyrina crystallina Jordan & Culver. 2685.
Scomberomorus sinaloæ Jordan & Starks. 1720.
Caranx medusicola Jordan & Starks. 2645.
Hynnis hopkinsi Jordan & Starks. 1563.
Trachinotus paloma Jordan & Starks. 2690.
Trachinotus culveri Jordan & Starks. 2691.
Mycteroperca venadorum Jordan & Starks. (British Museum.)
Mycteroperca boulengeri Jordan & Starks. 1621.
Lythrulon opalescens Jordan & Starks. 2963.
Microspathodon azurissimus Jordan & Starks. 1636, 2895, 1610.
Teuthis crestonis Jordan & Starks. 2899.
Balistes naufragium Jordan & Starks. 1656.
Aboma etheostoma Jordan & Starks.
Gobius manglicola Jordan & Starks. 3095.
Scorpæna mystes Jordan & Starks. 1616, 1617, 2919, 1501.
Symphurus williamsi Jordan & Culver. 2943.
Orthopristis reddingi Jordan & Richardson.
Alexurus armiger Jordan & Richardson.

Family GINGLYMOSTOMIDÆ.

1. **Ginglymostoma cirratum** (Gmelin.) GATA.

Two large specimens, respectively five and six feet in length, were taken. These agree fairly with published descriptions, except that the black spots scattered over the body are very small and pepper-like. It is possible that these spots vanish with age, and that *Ginglymostoma fulvum* Poey, the unspotted form, is the adult of the other.

This species was obtained by Dr. Gilbert, at Mazatlan and Panama.

Family GALEIDÆ.

2. **Galeus lunulatus** (Jordan & Gilbert). Gato.

Rather common at Mazatlan, where the original types were obtained by Dr. Gilbert.

3. **Galeocerdo tigrinus** Müller & Henle.

Recorded by Dr. Gilbert, from Mazatlan and from San José de Guatemala; not seen by us. It has not been compared with the Brazilian type of the species.

4. **Scoliodon longurio** (Jordan & Gilbert).

Rather common in the harbor at Mazatlan, where the original types were taken by Dr. Gilbert, who also found the species at Panama.

5. **Carcharhinus æthalorus** Jordan & Gilbert.

Original described from Mazatlan; not seen by us. Also recorded by Dr. Gilbert, from Panama. It is not likely that *Carcharhinus limbatus* occurs on the Pacific Coast. Probably this related species has been mistaken for it.

6. **Carcharhinus lamiella** (Jordan & Gilbert).

A very young specimen with a deformed tail was obtained by us at Mazatlan, the first record of the species from that port.

7. **Carcharhinus fronto** Jordan & Gilbert.

This large shark is not uncommon about Mazatlan, where the original types were taken by Dr. Gilbert. No specimens were seen by us, but the species is said to be common in the surf about the Olas Altas. It is said that during the time that Mazatlan was occupied by French soldiers a number of these were killed by the sharks while bathing in the surf.

Family SPHYRNIDÆ.

8. **Sphyrna tiburo** (Linnæus). CORNUDA.

One specimen obtained by us at Mazatlan. It was not secured by Dr. Gilbert; this being the first record on the Pacific Coast of America of this common Atlantic species. Our specimen seems to agree fully with an example from Florida.

9. **Sphyrna tudes** (Cuvier). CORNUDA.

Not rare at Mazatlan, where specimens were obtained by Dr. Gilbert, and one by the Hopkins expedition.

10. **Sphyrna zygæna** Linnæus. CORNUDA.

Common in the sea about Mazatlan. Three young specimens taken by us. Also recorded by Dr. Gilbert from Mazatlan and Panama.

Family PRISTIDIDÆ.

11. **Pristis zephyreus** Jordan & Starks n. sp. PEZ DE ESPADA.

Snout to nostrils, 3 in length to base of caudal; breadth of saw at anterior end between first two pairs of teeth half breadth of its base behind the last pairs; teeth on saw trenchant behind, arranged in 22 pairs; hinder teeth wide apart, the interspaces 5 times their base; posterior teeth turned slightly backward, a groove on their posterior edge; front teeth not quite half as long as the saw is broad at their base; distance between first and second tooth three times base of first. (Other specimens examined for us by Dr. G. W. Rogers show 18 to 21 pairs of teeth.) Eye equal to spiracle, contained 3 times in base of saw just behind last pair of teeth; width of mouth a little greater than base of saw; mouth with about 65 series of blunt teeth; slant height of pectoral in front, a little more than half distance from tip of snout to mouth. Dorsals sub-equal; first dorsal inserted in advance of ventrals; about half its base over ventrals; caudal, with a lower lobe, which is equal to slant height of pectoral; tail with a keel on side.

Color, plain olive grey above, light below.

Measurements—Length, 50 inches; caudal, 7 inches; pectoral, 7 inches; dorsal front, 5½ inches; snout without nostril, 11 inches.

Type—A skin in L. S. Jr. Univ. Museum.

Common in brackish waters at the mouth of the Rio Presidio, where one fine specimen was obtained. The species is also recorded (as *Pristis perroteti*) by Dr. Gilbert from Mazatlan, and by Dr. Günther from Chiapam. Dr. Günther identifies this species with *Pristis perroteti* described by Müller & Henle, from the Senegal River. In view of the great difference in the fauna of the Gulf of California from that of Equatorial Africa, this identification may be questioned, especially as there are several details in which the description of *Pristis perotteti* differs from our fish.

We append the description of Müller & Henle, as also the descriptions given by Latham of his *Pristis antiquorum* and *Pristis pectinatus*, together with our account of the common saw fish of the Gulf of Mexico, usually and probably correctly identified as *Pristis pectinatus* Latham.

The following is the original description of

"Spec. 4. Pristis Perotteti, N.

Kopf. "Die Form des Kopfes und der Naslöcher wie Pristis antiquorum. Die Säge läuft nach vorn sehr allmählig spitz zu. Sie ist an der Basis 1 Zoll 7 Linien, an der Spitze zwischen den beiden letzten Zähnen 10 Linien breit, 19 Zähne jederseits. Die hintersten Zähne sind kurz, wahrscheinlich abgenutzt. Die vordersten sind etwas breiter als die Hälfte der Breite der Säge, alle am hintern Rande gerinnt. Die hintern Zähne stehen weit aus einander, um 5-6 Mal die Breite des Zahns. Die vordersten sind einander etwas mehr genähert. Die Distanz zwischen den beiden letzten ist nicht ganz 3 Mal so breit als die Basis des Zahns. Alle Zahne nur wenig nach hinten geneigt.

"Die obere Nasenklappe reicht mit ihrem innern Rande bis zum innern Nasenwinkel. Die Zahne sind grösser als bei Pristis antiquorum, 60–70 in einer Reihe.

Flossen. Die Brustflossen vom Kopf scharf abgesetzt. Erste Rückenflosse mit der Halfte ikrer Basis vor den Bauchflossen. Schwanzflosse mit kurzem aber deutlichem untern Lappen.

Farbe. Farbe wie *Pristis antiquorum.*

Maasse. Von der Spitze der Sage zur Mitte zwischen den

"aussern Naswinkeln	11″	6‴
Von den Naslöchern zum Maul	1″	10‴
Vom Maul zum After	11″	
Vom After zur Schwanzflosse	8″	
Lange der Schwanzflosse	5″	6‴
Breite der Sage in der Mitte	1″	2‴
Lange des Langsten Zahns		6‴
Breite desselben		1½‴
Distanz der Naslöcher	2″	
Breite des Maules	2″	

Fundort. Aus dem Senegal. Soll nur im süssen Wasser leben. Ein Exampler ♂ trocken in Paris durch Perottet."

(Müller & Henle, Plagiostomen, p. 108.)

From the work of Latham we take the following description of his

"Pristis antiquorum:

Pr. rostro spinis validis utrinque 18–24. Tab. 26, f. i.

Squalus pristis, Lin. Syst. Nat. I., p. 401, 15. Faun. Suec. 297. Mus. Ad Fr. I., p. 52. Mull. Lin. Th. 3, Tab. ii, f. 2 (spin. 18). Gmel. Lin. I., p. 1494, 15. Fab. Fn. Groenl., 130, 91, Mull. Prodr., p. 38, 319. Klein. Miss. Pisc. 3, p. 12, No. ii, tab. 3, f. 1, 2. (pullus.)

Plin. Nat. Hist., lib. 9, cap. 2. Clus. Ex., tab. 14, p. 136 (spin. 20). Aldrov. Cet., p. 692. Will. Icth., p. 61, Tab. B. 9, fig. 5 (fig. Clusii). Raii, Syn. Pisc., p. 23. Olear. Mus., p. 41, t. 26, f. 1. Rondel. Pisc. 487.

Bell. de Aq., t. in p. 66 (Langue de Serpent).

Valent. Amboin, p. 33, t. 19, f. 52. Bloch, Fisch. Deutsch., p. 37, t. 120. Du Tertre Ant., p. 207 (Spadon). Bonann. Mus. Kirch., t. 288, t. 21. Cabinet de Ste. Genev., t. p. 100. Brouss. Act. Par. 1780, p. 671. (La Scie.) Pis. Ind. Occ., p. 51. Marcgr. Bras., p. 158 (Araguagua). Gronov. Zooph., p. 33. Arted. Syn. 66, Id. Syn. 93. Brown. Jam. 458, I.

HABITAT IN OCEANO.

"Totum corpus ad 15 pedes longum, supra nigricans, seu leucophæo-griseum, abdomine albicante. Caput antice planum. Rostrum ad 5 pedes longum, spinis validis numero utrinque 18–24. Os dentibus granulatis instructum. Oculi magni iride aurea. Pone oculos foramina duo oblonga. Spiracula quinque. Pinna dorsalis prima ventralibus opposita, altera inter primam & caudæ apicem media. Pectorales latæ longæque. Caudalis brevior quam congeribus.

This species and the following grow to the largest size of any which have yet come under the inspection of the naturalist, some specimens measuring 15 feet in length.

The head is rather flat at top, the eyes large, with yellow irides, behind which is a hole, which some have supposed may lead to an organ of hearing.*

* Nos foramina hæc meatus auditorios esse credimus. Willughb.

The mouth is well furnished with teeth, but they are blunt, serving rather to bruise its prey than to divide it by cutting. Before the mouth are two other foramina, supposed to be the nostrils. The rostrum, beak or snout, is in general about one-third of the total length of the fish, and contains in some eighteen, in others as far as twenty-three or twenty-four spines on each side; these are very stout, much thicker at the back part, and channelled, inclining to an edge forwards. The fins are seven in number, viz.: two dorsal, placed at some distance from each other; two pectoral, taking rise just behind the breathing-holes, which are five in number; two ventral, situated almost underneath the first dorsal; and lastly the caudal, occupying the tail both above and beneath, but longest on the upper part. The general color of the body is a dull grey, or brownish, growing paler as it approaches the belly, where it is nearly white." (*Latham*, Trans. Linn. Soc., 1794, p. 277.)

Mr. Latham thus describes his

"**Pristis pectinatus:**

Pr. rostro spinis augustioribus utriuque ad 34. Tab. 26, fig. 2.

Pristis seu Serra, Gesner Aq., fig. in p. 728 (spin. 34), Id. Ic. An., p. 171. Mus. Besler, tab. 17, f. 3 (spin. 28). Id. f. i (caput, spinis, 25) Aldr. Cet. f., p. 692. Johnst. Pisc., p. 8, t. iii (spin. 28). Blas. Anat., p. 466, t. 49, f. 13. Bloch, Deutsch. p. 37, t. 120 rostr. ar[illegible] Knorr. Delic., p. 56, t. H. 4. Olear. Kunst., p. 38, t. 25. f. [illegible] P[illegible] p. Hist. Norv. ii., p. 240 (spin. 25).

HABITAT IN OCEANO.

[illegible] differt. Rostri spinæ longiores, & minus validæ, numero variant a 25 usque ad 34. Pinnæ posticæ magis excavatæ.

This and the former species have been confounded hitherto by naturalists, nor are we certain that any others have been observed by them; and if we may judge by their figures of each, it should seem that the first described was the most plentiful. That figured in Gesner is far from a bad representation, and the one engraved by Knorr in his Deliciæ is sufficiently accurate. This species differs from the first, in having the snout more narrow in proportion at the base, and the whole of it more slender in all its parts; whereas the first is very broad at the base, and tapers considerably from thence to the point. The spines on each side also are longer and more slender, and vary from 24 to 34 in the different specimens; we have indeed been informed of one which contained no less than 35 spines on each side of the snout; but we must confess that we have never been fortunate enough to have seen such a specimen. This is supposed to grow to as great a size as the former, and in the general make and shape of the body does not materially differ." (*Latham*, Trans. Linn. Soc., 1794, p. 278.)

The following description of *Pristis pectinatus* Latham (*Pristis granulosa* Bloch & Schneider) is taken from a specimen two feet long, from Key West, Fla.:

Snout to nasal-lobes, 3 in length of body to base of caudal; width of anterior end of saw between first two pairs of teeth, equal to the internasal space, ½ the base behind last pair of teeth; saw with 26 teeth on a side; eye larger than spiracle, half interorbital space; width of mouth equal to its distance to front of nostril; teeth in mouth in about 70 series; width across outer angle of pectoral fins, 2½ in length from eyes to base of caudal; width of body behind pectorals, 7. Height of pectoral slant in front, 3 in snout to mouth; dorsals subequal; caudal, with no lower lobe, equal to pectoral slant.

Color, uniform brown above, below light.

Family RHINOBATIDÆ.

12. **Rhinobatus glaucostigma** Jordan & Gilbert. GUITARRO.

Very common on sandy bottoms in the estuary or Astillero at Mazatlan, where the species was originally found by Dr. Gilbert.

Family NARCOBATIDÆ.

13. **Narcine entemedor** Jordan & Starks, n. sp. ENTEMEDOR.

Two specimens taken in the estuary at Mazatlan, and

a third procured by Mr. James A. Richardson in the harbor of La Paz. Specimens had also been obtained by Dr. Gilbert, at Panama, in 1883, but having been destroyed by fire, the species has remained undescribed until the present time.

Snout 3¼ in length of disk: preocular part of snout equals preoral; interocular space in snout. 1½; width of mouth, 2¼. Eye much smaller than spiracle; spiracles edged with small tubercles. Length of disk equal to its width; disk equal to length of tail, without caudal fin; tail with a loose fold of skin on each side. First and second dorsals equal, rounded behind; ventrals large, ending midway between posterior edge of disk and caudal fin. Color: Pale olive brown, a little clouded with darker; second dorsal edged with pale; dots on head dusky.

Length of largest specimen, 20 inches. Type, No. 1699, L. S. Jr. Univ. Mus.

The Spanish name *Entemedor* seems to be equivalent to *Intimidator*.

Family DASYATIDÆ.

14. **Urolophus asterias** Jordan & Gilbert. RAIA.

Very common in the surf and on the sandy beaches about Mazatlan. Spinules on back and tail 18 to 32 in number. The upper side of the disk is marked with round dusky spots, faint, as if washed or faded out.

15. **Urolophus rogersi** Jordan & Starks, n. sp.

Disk broader than long by a distance 2½ times the interorbital width: anterior margins of disk nearly straight, the tip of snout projecting: snout from eye, 3¾ in length of disk: eyes little smaller than spiracles: width of mouth, 2½ times in preoral part of snout: caudal spine inserted in front of middle of tail. Skin with minute prickles on margin of pectorals and on middle of back, leaving smooth areas near middle of pectorals and over branchial arches; 16 to 20 large spinules along median line of back and tail.

Color, plain brown; caudal fin darker, edged with white.

This species differs from *Urolophus asterias*, in having a wider disk, more acute snout, much smaller prickles, and fewer spinules on back and tail.

Three specimens obtained in the Astillero, the longest 18 inches in entire length. Type, No. 1700, L. S. Jr. U. Museum.

This species is named for Dr. George Warren Rogers, a scholarly physician, native of Vermont, but long resident in Mazatlan.

16. **Urolophus umbrifer** Jordan & Starks n. sp.

Occasionally taken with *Urolophus asterias*, but much less common.

Disk round, not wider than long, its length greater than tail; snout pointed, not exserted. Snout from eye, 4½ in disk; eyes equal to spiracles; mouth 2 in distance to tip of snout; caudal spine inserted in front of middle of tail; skin perfectly smooth.

Color, brown above, with blackish cross-shades or bars, radiating from the shoulder; a dark band behind eyes, and one from eyes; caudal fin dark.

One adult female specimen, the uterus containing four young.

This is probably not identical with Garman's *Urolophus nebulosus*, being perfectly smooth and different in color.

17. **Dasyatis longus** Garman.

Rather common at Mazatlan, where specimens were also taken by Dr. Gilbert; also recorded by Mr. Garman from Acapulco and from Panama, and by **Evermann &** Jenkins from Guaymas.

18. Pteroplatea crebripunctata Peters. MANTARAIA.

Very common on sandy shores everywhere about Mazatlan. from which locality it was originally described; also taken by Dr. Gilbert.

Width of disk twice length to posterior end of anal slit; snout forming a regular curve from a little in front of middle of pectorals, a very small blunt projection at tip; anterior margin of disk convex near snout and lateral angles, pectorals concave medially; posterior margin weakly convex; posterior angle broadly rounded; lateral angle sharply rounded; distance from snout to a line drawn through lateral angles, 2½ times in distance to tip of tail.

Interorbital a little wider than its distance to tip of snout; eyes twice spiracles; mouth equals snout, 6½ in disk. Tail rat-like, with a scarcely perceptible fold of skin on its dorsal side.

Ground color olive brown, everywhere with small dark points, not so close set as in *Pteroplatea rava*, indistinct greyish spots, half as large as iris, scattered over the body among the dark points, these spots are more distinct on anterior edge of disk; tail mottled with darker; lower parts light. Markings nowhere so distinct as in the next species.

Several specimens, the largest 15 inches long.

19. Pteroplatea rava Jordan & Starks, n. sp. MANTARAIA COLORADA.

One specimen taken in the Astillero at Mazatlan.

Length of disk 1⅔ width: snout forming an angle which is almost a right angle: pectorals slightly concave medially: posterior margin of disk weakly convex; posterior angle not broadly rounded. but curved in somewhat suddenly: lateral angles acute.

A line drawn through lateral angles would bisect a line from snout to tip of tail. Interorbital 1⅓ in snout; eye 1½ in spiracles; mouth 7 in disk, 1½ in snout; tail straight and slender, with a very slight fold on dorsal side.

Ground color light olive brown, thickly set with sharp cut black points; conspicuous grey or white spots, half as large as iris, scattered over the body, around which the black spots form rings; brighter yellowish spots and half spots around anterior edge of disk; tail mottled above with darker; lower parts chiefly light orange red or rust colored in life.

All the markings are very distinct and clear cut, the reddish of the belly conspicuous.

One specimen, 12 inches long. Type No. 1587, L. S. Jr. Univ. Mus.

20. **Ætobatus narinari** (Euphrasen). Gavilan.

Rather common in the harbor of Mazatlan, where it was also taken by Gilbert; a beautifully colored species reaching a large size.

Length of disk 1⅔ in width; proximal half of anterior margin of pectoral fins straight, distal half convex; posterior margin concave, the end of each ray forming a small scallop; lateral angle sharp.

Snout forming an angle, from its tip to division of nasal-lobes, 1⅓ times breadth of head; width of snout 1⅛ times distance from its tip to the division of nasal-lobes; nasal-lobes projecting back over the mouth; width of mouth 1½ its distance to tip of snout; numerous blunt buccal papillæ around upper dental plate and on ridge between nostrils; interorbital 4¾ in disk; eyes smaller than spiracles, which are as long as base of dorsal. Ventrals well rounded, 3⅔ in length of disk; tail 3½ times disk. First caudal spine equals base of dorsal, which is half second spine.

Color bluish black with many round yellowish spots scattered equally over the back and ventral fins; spots about as large as eye on back, smaller on head, sometimes two spots run together forming an elliptical spot, about sixteen spots from eye along anterior margin of pectoral to lateral angle; posterior margin of pectoral very narrowly margined with white; ventral side pearly white.

From the description of *Ætobatus laticeps* this species differs in the following respects: disk not so broad; tail not so long; width of head and snout less; ventrals not truncated behind; pectorals not margined with blackish; spots on ventrals not assuming the form of ocelli.

Five large specimens obtained; length of disk in each, 15 inches.

This description has been compared by Dr. Barton W. Evermann, with specimens of *Ætobatus narinari* from Brazil. No difference of any importance appears, and in his judgment the Atlantic and Pacific Coast American forms are identical.

NOTE.—This species has been several times obtained by Dr. Gilbert and others in the Gulf of California, having been identified as *Ætobatus laticeps* of Gill. It does not, however, agree with Dr. Gill's description and there is no evidence that his specimen came from Mexico. *Ætobatus laticeps* was described from an example from unknown locality received from San Francisco. It is therefore quite as likely to have come from Honolulu or from China, as from the Gulf of California.

The following is Dr. Gill's description:

"*Aëtobatis laticeps* Gill.

"The greatest width is rather *more* than twice as great as distance from snout to front of anus. The head is broad and nearly equals the distance from snout to division of nasal lobes. The snout is obtusely angulated in front, and at its sides is convex and scarcely angulated;

its width at a line in front of the nostril is as great as the distance from its point to interlobular nasal emargination, The rostro-frontal fontanelle is constricted at its anterior third; the interval between the crests of the anterior portion enters about 2⅔ times in the interorbital area; at the constriction, about 4 times; at the posterior portion, about 2⅔ times; the posterior portion gradually expands backwards and terminates with an oval contour behind. The nasal lobes are about twice as long as wide, their length externally exceeding half the length or breadth of the rostral area.

"The dental plate has a triangular contour; its anterior angle obtusely rounded.

"The dorsal commences immediately behind the pectoral fins. The ventral fins almost truncated behind, between the well rounded angles; their breadth 2½ times their length. The tail is four or five times as long as the body.

"The color is bluish-black above, relieved on the head by numerous, but rather distinct, whitish or yellowish spots, smaller than eye, much larger on the body and behind towards the sides, and on the ventrols sometimes assuming the form of ocelli; below white; pectorals margined with blackish.

"This species is closely related to *A. narinari* and its allies, and especially *A. latirostris* A. Dum., but is apparently distinguished by the combination of characters given in the diagnosis. It belongs to the genus *Goniobatis* Ag., proposed for a species with a more angular lower dental plate than in *A. narinari*, and is related to the *Goniobatis meleagris* Ag.* of the Sandwich Islands,

* "This species has not been characterized, but a dried Aëtobatine obtained at the Sandwich Islands by the Wilkes Exploring Expedition probably belongs to it."

but is distinguished by the more declivous forehead and the shape of the rostro-frontal fontanelle.

"A single specimen was forwarded to the Smithsonian Institution by S. E. Hubbard, Esq., of San Francisco, Cal." (Gill.)

21. **Manta birostris** (Walbaum).

Said to be frequently seen in the open sea about Mazatlan; not obtained by us.

Family SILURIDÆ.

22. **Felichthys pinnimaculatus** (Steindachner).

Occasionally taken in the estuary. Recorded by Gilbert from Mazatlan and Panama, by Steindachner from Altata, Costa Rica and Panama. Two specimens obtained by us.

23. **Felichthys panamensis** (Gill).

Not rare in the estuary, reaching a considerable size. Obtained by Gilbert at Mazatlan, Libertad, Punta Arenas; by Gill and Günther at Panama; and by Steindachner at Magdalena Bay, Altata and Panama. One specimen obtained by us.

24. **Galeichthys peruvianus** Lütken. PANAMA.

Recorded by Steindachner from Altata; not seen by us, and taken by Dr. Gilbert only at Panama; apparently not common.

The so-called genus *Galeichthys* is distinguished from *Hexanematichthys* only by having the bones of the head covered by skin. In several species of other genera (notably *platypogon*, *dasycephalus*, *gilberti*), the skin on the head is thickened in females, obscuring the outline and granulation of many of the bones. It may be that the species called *Galeichthys* represent only the extreme of this condition, and that the species referred to it should be arranged in other groups.

As the dentition of the typical species of *Galeichthys* agrees in essential respects with that of *Hexanematichthys*, we unite the two groups under the earlier name, *Galeichthys*.

25. Galeichthys gilberti Jordan & Williams, n. sp. BAGRE BLANCO. Plate xxvi.

Extremely abundant in the upper part of the Astillero, along sandy bottoms, exceeding by far in numbers all other cat fishes. Also found by Gilbert at Mazatlan, whence it was erroneously recorded by Jordan & Gilbert as *Arius assimilis* Günther. Large numbers of this species are left on the beach after seining, and the various sea birds, pelicans, man-of-war birds, gulls and the like, come down to take possession of them. In two cases specimens of this cat-fish were swallowed by pelicans; the spines were erected after the fish was partly engorged, and these spines entering the skin of the sack of the pelican, made it impossible for the bird to swallow them or to dislodge them. Considerable numbers of pelicans are doubtless destroyed every year by attempting to swallow living cat-fish which have been left by the fishermen.

The following description is essentially that of Jordan & Gilbert, Bull. U. S. Nat. Mus., 1882, under the name of *Arius assimilis*. The type of that description, 29,213 U. S. N. M., from Mazatlan, coll., Gilbert, may be taken as the special type of the species, numerous co-types (numbered 1666, 1667 and 1668, L. S. Jr. Mus.), having been sent by us to different museums:

Head, $3\frac{4}{7}$ to 4; width of head, $5\frac{1}{5}$: depth, 5; D. I. 7; A. 4, 14.

Body comparatively elongate, the head depressed but not very broad, somewhat broader than high; eye rather large, 5 to 6 in length of head; width of interorbital space, $2\frac{1}{4}$ in head; breadth of mouth, $2\frac{3}{5}$; length of snout, 3.

Teeth all villiform; bands of vomerine teeth separated by a rather wide interval, each small, roundish, confluent with the neighboring palatine band, the junction marked by a slight constriction; palatine bands ovate, broad behind, varying considerably in size and somewhat in form, the width ranging from one-third diameter of eye to two-thirds, being generally largest in adults; band of palatine teeth without backward prolongation; band of maxillary teeth rather broad and short, its length about five times its breadth. Maxillary barbel broad and flattened at base, reaching a little past base of pectoral in the young, scarcely to the gill opening in the adult; outer mental barbels, 2 in head, inner 3. Gill-rakers, 4+12.

Dorsal shield very short, narrowly crescent-shaped, its length on the median line not more than half that of one of its sides. Occipital process subtriangular, not quite as long as broad at base, with a strong median keel, its edges slightly curved. A short distance in front of the beginning of the keel is the end of the very narrow groove-like fontanelle, which is somewhat widened anteriorly, finally merging into the broad, flat, smooth interorbital area, the boundaries of which are not well defined; shields of head usually smooth, all finely and very sparsely granular, the granules not forming distinct lines.

Gill membranes forming a rather broad fold across isthmus.

Dorsal spine long, usually, but not always, shorter than the pectoral spine, about $1\frac{3}{4}$ in head; axillary pore absent. Humeral process rather broadly triangular, not much produced backward, less than half length of pectoral spine, its surface not granular, covered by skin. Adipose fin half length of anal, its posterior margin little free. Upper lobe of caudal longest and somewhat falcate, about as long as head. Ventrals unusually long about reaching anal in females, shorter in the males. Vent much nearer base of ventrals than anal.

Color olive green, with bluish luster, white below; upper fins dusky olivaceous; caudal yellowish dusky at tip; anal yellowish with a median dusky shade; ventral yellowish, the basal half of the upper side abruptly black; pectorals similarly colored, the black area rather smaller; maxillary barbel blackish; other barbels pale.

Length, 12 to 18 inches.

The following specimens from Dr. Gilbert's Mazatlan collections are registered in the United States National Museum:

28,161, 28,189, 28,210, 28,213 (2), 28,221, 28,232, 28,276, 28,304.

This species is nearest allied to *Galeichthys seemanni* (Günther), a Panama species. *Galeichthys jordani* (Eigenmann) from Panama differs in the gill rakers and in other regards. *Galeichthys assimilis* is an Atlantic species, not yet known from the Pacific Coast. With each of these *Galeichthys gilberti* has been at one time or another confounded. *Galeichthys gilberti* differs from *Galeichthys seemanni*, as described by Dr. Eigenmann, in the absence of pectoral pore, in the shorter spines and in the fontanelle not quite reaching occipital process; ventrals unusually long, no dark specks on side of belly, barbel short, compressed. As noted below, *Galeichthys gilberti* bears a superficial resemblance to *Netuma platypogon*. Its teeth are different, the ventrals are much longer, and the adipose dorsal much larger. *Netuma platypogon* has the sides of belly much soiled by dark specks.

26. **Galeichthys azureus** Jordan & Williams, n. sp. BAGRE AZUL. Plate xxvii.

Head 3¼; width of head 4⅝, depth 9. Length from tip of snout to tip of upper lobe of caudal fin 19¼ inches. D. I, 7: P. I, 10. A. 4, 14. Gill rakers 6+13.

Body robust, its width anteriorly greater than its depth; caudal peduncle short, stout; distance from end of anal fin to base of median caudal rays about one-half length of head. Head flat, very broad; its depth at posterior angle of jaw about one-half its width; interorbital region flat, smooth anteriorly and granulated posteriorly; fontanelle almost obsolete, wide anteriorly and ending in a short groove posteriorly at a point one-half distance from tip of snout to posterior end of occipital process; top of head, occipital process and dorsal shield finely granular, granulations mostly arranged in radiating striæ and extending forward to a line with the pupils, nostrils very large and close together; posterior one with a broad valve.

Occipital process pentagonal, its length 4½ in head, about as long as wide, with a very low ridge; dorsal shield crescent shaped with points extending back on each side of fin, its median length about one-half the length of its side. Eye small, about 9 in head; interorbital width almost 2 in head; snout 3 in head: breadth of mouth 2$\frac{3}{10}$ in head.

Maxillary barbel slender, thick at base, 1⅘ in head; outer mental barbel reaches to posterior angle of jaw, about 2⅓ in head; inner mental barbel about 4 in head.

Teeth all villiform: premaxillary band narrow, about one-eighth as wide as long, vomerine and palatine bands of teeth fully confluent on each side, forming together a crescent-shaped patch, narrowly divided on the median line of the vomer; form of vomerine bands similar to that of the palatine bands but smaller. Palatine band of teeth without backward prolongations.

Opercle with radiating ridges; humeral process granular, triangular, lower posterior corner prominent; axillary pore very small. Gill membranes forming a broad fold across isthmus.

Dorsal fin short, base not including spine equal to base of adipose dorsal; dorsal spine robust, but little shorter than pectoral spine, about two in head; its anterior serræ small and tubercle-like; its posterior edge, as well as that of pectoral, retrosely serrate; soft rays of dorsal extending but little beyond spine, the longest about three-fifths length of head. Adipose dorsal about one-half as high as long. Caudal lobes unequal, the upper lobe about one-third longer than lower lobe. Anal short, of medium height. Distance from vent to base of ventrals one-half distance from origin of anal. Pectoral spine very strong, its anterior margin with serræ towards the tip, which become small tubercles towards base; soft rays but little longer than spine, which reaches slightly beyond one-half distance from its origin to base of ventrals.

Color dark blue with silvery reflections on sides; belly pale, mental barbels dusky; maxillary barbels light below and black above; paired fins darkest on inner side; other fins almost uniformly dusky.

One specimen, 19¼ inches long, was taken by the Hopkins expedition in the estuary at Mazatlan. It is numbered 1575 in the collection of the Leland Stanford Jr. University.

27. **Galeichthys guatemalensis** (Günther).

Taken by Dr. Gilbert at Mazatlan; not seen by us. Also recorded from Chiapam (Günther), and the coast of Colima (Xantus).

28. **Netuma platypogon** (Günther).

Very common at Mazatlan; several specimens taken in Astillero, where it is scarcely less abundant than *Galeichthys gilberti*. Also recorded by Dr. Gilbert from Mazatlan, Libertad and Punta Arenas; by Günther from San José; and by Steindachner from Magdalena Bay and Callao. To the southward it is very abundant.

In some specimens, perhaps females, granulations are visible on the occipital process only, the other bones being covered by smooth skin, as in the subgenus called *Galeichthys*. This species much resembles *Galeichthys gilberti*. It is, however, readily known by the short, pale ventrals, as well as by the generic character of the backward extension of the palatine bands of teeth.

29. **Netuma kessleri** (Steindachner).

Recorded by Steindachner from Altata; recorded from Panama both by Gilbert and Steindachner. Not taken by us.

30. **Sciadeichthys troscheli** (Gill). BAGRE COLORADO.

Rather common in the Astillero at Mazatlan, reaching a considerable size. Also taken at Mazatlan by Gilbert, at Altata by Steindachner; found by Gilbert and Steindachner at Panama, and by Gilbert at Punta Arenas. Its general coloration is decidedly reddish or coppery. The sculpture of the large dorsal shield and of the occipital process is subject to considerable variation, and possibly more than one species of this type exists.

We follow Dr. Eigenmann in referring the short description of *Sciades troscheli* Gill to the species called *Arius brandtii* by Steindachner. Dr. Gill does not fully describe the dorsal shield and the type of his description is lost. In recalling the matter to his memory, he is, however, positive that the type of *troscheli* had the large dorsal buckler shown in Steindachner's figure of *brandtii*. In that case *troscheli* and *brandtii* must be the same.

Family MURÆNIDÆ.

31. **Muræna lentiginosa** Jenyns. ANGUILA PINTA.

Not rare in the rocky places about the islands at Mazatlan, where a few specimens were taken by us. Numerous others, the types of *Muræna pinta*, were found by Dr. Gilbert. The species is widely distributed, having been recorded from Cape San Lucas (Xantus), Colima (Xantus), Panama (Rowell) and San Josef Island (Nichols).

32. **Lycodontis dovii** (Günther). ANGUILA PINTITA.

Not seen by us at Mazatlan. The original types of *Muræna pintita* (which we now identify with *dovii*) were taken at Mazatlan by Dr. Gilbert. Specimens which we have elsewhere referred to this species have been recorded from Espiritu Santo (Belding), Galapagos Islands (Herendeen) and from Panama (Günther).

The name *Gymnothorax* as originally proposed by Bloch, is an exact synonym of *Muræna* as understood by us. Of the many later names applied to this type, *Lycodontis* of McClelland seems to claim priority.

33. **Lycodontis castaneus** (Jordan & Gilbert).

This enormous eel is very common about the islands near Mazatlan, where numerous specimens were obtained both by Dr. Gilbert and by us. Our largest specimen is 5½ feet in length. The species is very close to the West Indian *Lycodontis funebris* (Ranzani), but is apparently distinct from the latter. The colors are not the same, *funebris* being of a greenish black and *casteneus* bordering upon purplish chestnut. This species and its congener *(funebris)* reach a larger size than any other American morays.

Family OPHICHTHYIDÆ.

34. **Myrichthys tigrinus** Girard. CULEVRA.

Not uncommon in the harbor of Mazatlan, where several specimens (types of *Ophichthys xysturus* Jordan & Gilbert) were taken by Dr. Gilbert. Several specimens were also obtained by us. It has been recorded also from Acapulco and Panama. The original types of *Myrichthys tyrinus* were said to come from Adair Bay in Oregon. It has, however, not yet been taken north of the Gulf of California, and the locality assigned to the type is very doubtful. We have not been able to find a bay of this name on any map of Oregon.

35. **Ophichthus triserialis** (Kaup.) (*Ophisurus californiensis* Garrett; *Herpetoichthys callisoma* Abbott.)

Recorded by Gilbert from Mazatlan; not seen by us. A specimen certainly belonging to this species has been lately obtained by Dr. Gilbert in the Bay of Monterey. The only other definitely known localities are Cape San Lucas and the Galapagos Islands, whence it was described as *Ophichthus rugifer* Jordan & Bollman.

36. **Ophichthus zophochir** Jordan & Gilbert.

Rather common in the Bay of Mazatlan, where it was also taken by Dr. Gilbert. We have examined specimens from Acapulco.

Olive brown, abruptly paler olive below middle of side. Dorsal with a black edge, which shades toward olive at base of fin; anal similar, paler. Pectoral uniformly dusky, the base paler. Teeth 2-rowed above and below, canines small. Pectoral $2\frac{2}{5}$ in head; snout $5\frac{1}{2}$; eye $1\frac{2}{3}$ in snout; gape $2\frac{3}{5}$ in head; head and body $1\frac{2}{3}$ in the long tail.

Family MURÆNESOCIDÆ.

37. **Muræenesox coniceps** Jordan & Gilbert. CULEVRA BLANCA, ANGUILA BLANCA.

Very common about the islands in the neighborhood of Mazatlan. It reaches an enormous size, a specimen obtained by us being 6 feet and 10 inches long and having a girth of 22 inches.

Family CHANIDÆ.

38. **Chanos chanos** (Forskål). SÁBALO.

Very common on the sandy shores of the bay, reaching length of about 5 feet. The flesh is poor, and the fish is seldom brought into the market, but is frequently used as bait. The hard enamelled scales are used for ornamental work by the Indians. We are unable to see any difference between our specimens and others brought by Dr. Jenkins from the Hawaiian Islands. We have no doubt that our species is identical with the common East Indian form.

Head $4\frac{2}{5}$; depth 4; D. 2, 12; A. 2, 9; V. 12; scales 12–70–14; snout $3\frac{1}{2}$ in head; eye $3\frac{1}{2}$; maxillary $4\frac{1}{3}$; pectoral $1\frac{3}{5}$; ventral $1\frac{4}{5}$; caudal $\frac{1}{3}$ longer than head; dorsal $1\frac{1}{4}$ in head.

Body elliptical, moderately compressed, the caudal peduncle slender. Head pointed, rounded above. Eye and side of head covered by a large transparent, imperforate adipose eyelid. Mouth small, terminal, toothless, transverse, the lower jaw included: maxillary broad, slipping under the adipose preorbital, without supplemental bone. Branchiostegals 4. Opercle truncate behind. Pseudbranchiæ very large. Gill-rakers fine and flexible, very close set, rather long, the gill-rakers of all the arches bound together so as to form a perfect strainer. Bones of gill-rakers flexible. Scales firm, enamelled at base, with strongly marked longitudinal striæ, becoming bony when dry; used by the Indians for ornamental work. Lateral line well developed. Dorsal somewhat nearer snout than base of caudal, before ventrals, its first ray falcate, its last produced in a short filament, longer than pupil. Base of fin with a large scaly sheath; pectoral and ventral with scaly axillary appendage. Anal similar to dorsal, but much smaller. Pectorals and ventrals rather small; caudal very long, forked to the base, its lobes subequal, straight; base of fin with small scales. Ventrals somewhat falcate.

Brilliant silvery in color, greenish above; fins more or less darker; inside of pectoral and ventral blackish.

Stomach forming a muscular crop. Pyloric cæca many. Intestinal canal long, filled only with remains of plants.

The skeletal peculiarities of *Chanos* are numerous and remarkable, many archaic characters persisting. The following account of the skeleton has been prepared by Mr. Starks:

SKELETON OF CHANOS CHANOS.

a. Cranium.

The frontals are very large, covering nearly the whole top of the head, and extending over the dorso-anterior part of the parietals, supra-occipital and the parotic process.

On the side of the skull there is an area bounded by the supra-occipital, the opisthotic and the sphenotic, which is not ossified but is composed of cartilage.

Between the frontals, at about their middle, there is a place in which the bone is fibrous and largely cartilaginous; it is easily broken through.

The basal cavity under the brain cavity is large.

On the upper part of the operculum is a large scale-like bone.

The suborbitals are well developed and plate-like, extending back nearly to the posterior edge of the preopercle.

b. Vertebral Column.

There are forty-two vertebræ in the spinal column.

The first vertebra is co-ossified to the skull, and apparently bears no ribs; the second vertebra supports a pair of very small, slender ribs, which articulate directly with the sides of the vertebra; the third vertebra supports the first pair of large ribs; they are articulated with the transverse processes.

The first fourteen or fifteen neural spines and pairs of transverse processes are articulated with the vertebræ by sutures, they are easily separated from the vertebræ by boiling or maceration.

The vertebræ gradually increase in size and reach their largest size about two-thirds of the distance from the anterior to the posterior end of the spinal column, where they are three or four times the size of the anterior ones. This character is more marked in the adult than in the young.

c. Shoulder Girdle.

The shoulder girdle is exceedingly well braced, the post-temporal is widely forked, and strongly articulated to the epiotic processes of the skull.

The supra-clavicle is long and slender, its posterior face is hollowed out and attached some distance from the upper end of the clavicle, which projects upward.

This projecting upper end of the clavicle is braced to the skull by two long bones.* The first bone is very slender, at its anterior end it is connected to the exocci-

* See Dr. R. W. Shufeldt's report on the osteology of *Amia calva:* Bull. U. S. F. C., 1883, page 59.

pital; near its middle it is connected with the posterior end of the post-temporal, at which point it turns at a sharp angle and runs to the clavicle. The second bone is much larger, it is articulated to the basioccipital. Its posterior edge is nearly straight for its whole length, but its anterior edge is produced and much swollen near its middle, and joins the post-temporal over the first bone, then runs to the upper end of the clavicle.

The inner part of the clavicle and the coracoid are thin and pierced by many holes, so that the bone in places is little more than network.

The hypercoracoid has a very large foramen; at its posterior edge is a projection which supports a thin bone, probably a dermal bone.

The mesocoracoid is well developed.

There are four actinosts; the first is long, but they rapidly decrease in size to the fourth, which is short and triangular.

The first ray of the pectoral is large at the basal end, and hollowed out; it works directly on the hypercoracoid.

d. Branchial Apparatus.

The branchial apparatus is peculiar in the adult, in having gill-rakers somewhat resembling the filaments of a feather, on both sides of each arch and on the basibranchial. They meet in a middle line between the arches and unite forming a continuous lattice-work screen, through which nothing but the very smallest bodies can pass. The pharyngeals have no teeth, but have gill-rakers similar to those on the arches; they are enclosed in sac-like projections on each side.

This description is taken from the skeleton of a large specimen 4 feet long. The gill-rakers are not united in young specimens.

e. Other Parts.

The septæ between the myotomes are ossified about half an inch under the skin, forming long, slender rays of bone.

There is an upper series running from the middle of the sides up on the back, and a lower series from the sides down on the belly, they form a sort of a basket around the body. Those below have a single branch near the middle of each, the ones above have two branches each, these branches are lost towards the posterior end.

These bones are not present in the young.

The large caudal fin is attached very firmly to the hypural, the long rays of each lobe join the hypural at about the same oblique angle, the base of each ray is deeply divided and articulated immovably with the hypural. The middle short rays are all nearly horizontal and are much less firmly fastened.

The first interspinal ray of the anal is hollow and cone-shaped, the posterior end of the air-bladder runs into it as in the genera *Eucinostomus* and *Calamus*. The scales are very thick and closely imbricated; the skin anteriorly is a quarter of an inch thick.

Family ELOPIDÆ.

39. **Elops saurus** Linnæus. CHIRO.

Very common in the estuary, ascending into brackish mud puddles at high tide; not valued as food. Also found by Gilbert at Mazatlan.

Family ALBULIDÆ.

40. **Albula vulpes** (Linnæus). SANDUCHA.

Very common in the estuary at Mazatlan; not valued as a food fish. Also found by Gilbert at Panama and Mazatlan. The band-shaped young, which Dr. Gilbert has shown to be the larvæ of this species, were obtained in abundance.

Family CLUPEIDÆ.

41. **Sardinella stolifera** (Jordan & Gilbert). SARDINA DE ACEITE. Plate xxviii.

Exceedingly abundant in the Astillero at Mazatlan, where many specimens were taken by Dr. Gilbert, as well as by the Hopkins expedition. This species is also recorded by Gilbert from Panama, and has been found in several other localities. The flesh of this sardine is very rich and delicate, quite equal to that of the European Pilchard (*Clupanodon pilchardus*), and it is therefore a most excellent pan fish. It is, however, not eaten by the Mexicans, no fish having less than one-half pound weight being salable in the market at Mazatlan. The art of properly cooking delicate fish like this is unknown to the people of this region.

42. **Opisthonema libertate** (Günther). SARDINA MACHETE.

Common in shallow water, in the surf and in the harbor at Mazatlan, where it was also taken by Dr. Gilbert.

43. **Opisthopterus lutipinnis** (Jordan & Gilbert).

Extremely common in the surf outside the bay, where great numbers are taken with the seine; a delicate fish which, probably, is of excellent quality as food.

Our specimens are all smaller than the single one taken by Dr. Gilbert at Mazatlan, and they differ in some minor details. Doubtless all belong to the same species.

Head $4\frac{2}{5}$; depth $3\frac{5}{8}$: scales 48–13; D. 14; A. 54; snout 4 in head; eye $3\frac{1}{3}$: maxillary 2; pectoral $1\frac{1}{6}$; anal base $2\frac{1}{4}$ in body; scutes 27.

Gill-rakers moderate, slender, about x+15.

Body strongly compressed, translucent, the belly much compressed, with sharp scutes: vent midway between tip of snout and base of caudal. Front of dorsal midway between preopercle and base of caudal. Teeth strong, sharp, unequal in both jaws; small teeth in patches on palate and tongue. Maxillary pointed behind, reaching middle of eye.

Color bright silvery, bluish above; a very distinct black spot at shoulder on level of eye, two-thirds diameter of eye; chin and nose black. Fins all pale, with no yellow; a trace of a broad diffuse, lateral streak of silvery, most distinct in young. Upper ray of pectoral dusky, some pale olive spots on back, very faint.

Very many specimens taken, the longest 5½ inches in length.

Family ENGRAULIDIDÆ.

44. **Stolephorus miarchus** Jordan & Gilbert.

Obtained by Dr. Gilbert in the open water about Mazatlan; not found by us. These translucent type speciments are apparently immature, but the small number of anal rays would indicate that it is a species distinct from any other now known.

The immature or larval specimens obtained by us in the open sea have the fin-rays of *Stolephorus ischanus* and must belong to that species.

45. **Stolephorus exiguus** Jordan & Gilbert.

Originally found by Dr. Gilbert in the Astillero at Mazatlan; not seen by us.

46. **Stolephorus curtus** Jordan & Gilbert.

Rather common in the Astillero at Mazatlan, where it was originally found by Dr. Gilbert. Numerous specimens taken by us.

47. **Stolephorus ischanus** Jordan & Gilbert.

Very common in the Astillero at Mazatlan, where it was originally found by Dr. Gilbert. Many specimens obtained.

In the open sea many slender larvæ, similar in form to *Stolephorus miarchus* were obtained by the use of dynamite. The number of anal rays shows that these larvæ belong to the present species.

48. Stolephorus lucidus Jordan & Gilbert.

Originally found by Dr. Gilbert in the Astillero at Mazatlan; not obtained by us.

49. Stolephorus scofieldi Jordan & Culver, n. sp.

* Head 3¾ to 3$\frac{9}{10}$ in length to base of caudal; depth 4½ to 5; eye 3¾ to 4 in head; dorsal 12; anal 25 or 26; scales 41 or 42.

Close to *Stolephorus delicatissimus*, but with larger head, wider lateral band, and greater number of dorsal and anal rays.

Body somewhat compressed and elevated, the belly not carinated or serrated. Teeth in both jaws, and on palatines; a few on vomer. Maxillary covered with teeth its entire length and reaching beyond base of mandible, but not to opercular margin.

Gill-rakers 10+12, the longest a little more than half the eye.

Origin of dorsal midway between base of median caudal rays and center of eye; anal not quite as long as head, its origin below the middle of dorsal. Lower caudal lobe longer than upper; longest ray equaling length of

* The following are the measurements, etc., of seven specimens:

Anal rays.	Dorsal rays.	Head in length.	Depth in length.	Eye in head.	Scales.
26	12	3 9/10	4¾	4	42
26	12	3 9/10	4½	4	41
26	12	3 9/10	4¾	3¾	42
25	12	3 4/5	4¾	3¾	41
26	12	3¾	4¾	3¾	42
26	12	3¾	4¾	3¾	42
25	12	3¾	5	3¾	41

the head; shortest caudal ray 2½ in longest. Pectorals not reaching ventrals, 1¾ in head. Both anal and dorsal fins preceded by a rudimentary spine, not half length of first true ray.

Color translucent, with a distinct broad silvery stripe as wide as the eye, growing more diffuse at lower anterior edge, narrowing on caudal peduncle, and becoming fan-shaped on the base of caudal. Tip of snout black; a distinct median band of black specks extending from tip of snout to base of caudal. No distinct black markings on fins.

Length, 3 inches. Type, No. 2941, L. S. Jr. Univ. Mus.

Found in the Astillero at Mazatlan, not very abundant.

Named for Mr. Norman Bishop Scofield, a member of the Hopkins expedition to Sinaloa.

50. **Anchovia*** **macrolepidota** (Kner & Steindachner).

Originally described from the neighborhood of Panama; recorded by Dr. Gilbert from the Bay of Mazatlan, but not seen by us there; apparently rare.

Family SYNODONTIDÆ.

51. **Synodus scituliceps** Jordan & Gilbert. CAIMAN.

Not very common, on sandy bottoms in the Bay, where the species was originally found by Dr. Gilbert; also recorded from Panama.

Color brown, with markings of pale bluish green. No yellow anywhere.

52. **Synodus jenkinsi** Jordan & Bollman.

Not rare, occurring in deeper water than the preceding and reaching a much larger size. The two species are very closely related, but seem to be distinct. In *Synodus*

* *Anchovia* (Jordan & Evermann, Fishes of North America), is a new generic name applied to this species, distinguished from *Stolephorus* by its robust form and the absence of teeth in the adult.

jenkinsi, the head is much larger and the form more robust, besides slight differences in the scales. The specimens obtained were sent to us by Señor Ygnacio Moreno after our departure from Mazatlan.

Family PŒCILIIDÆ.

53. **Pœcilia butleri** Jordan.

Common in the fresh waters of the Rio Presidio below the village of Presidio, where the species was originally taken by Mr. Alphonse Forrer.

Head 3½; depth 2¾ to 3⅓; dorsal 9; anal 6; scales 26–9; eye 3 in head, equal to snout; interorbital 2; pectoral 1¼ in head; caudal equal to head. Longest dorsal ray 1¼ in head in male; 1⅔ in female.

Body much deeper and more compressed than in *Pœcilia presidionis*, the profile rather steeply rising to front of dorsal. Dorsal and ventral outlines of head meeting at mouth and forming a somewhat sharp point; snout as viewed from above, truncate. Teeth in two series, the inner smaller, more close set, not trifid, the two series well separated. Interorbital space wide and flat, about twice as wide as eye.

The sexes differ greatly in the position of the anal fin, it is under or rather behind dorsal in females, much in front in males, the tips of ventrals reaching much past the base of fin. The sexes similar in size, not very unlike in coloration; both with traces of faint olive cross-bands, especially on caudal peduncle; a dark curved streak behind eye on opercle bounding a roundish silvery area on opercle and breast.

Male green with pale blue spots on each scale surrounded by pale bronze shades: no bars. Dorsal and caudal pale orange, with many small black spots. Lower fins pale. Female similar, paler, without cross-bands, with a dark spot behind pectoral; lower fins bright orange, caudal nearly plain; dorsal speckled as in male. Form similar to that of male, deeper than in *Pœcilia presidionis.*

Alcoholic specimens show no dark spot behind pectoral and only a few specimens show traces of orange coloration on fins.

The following is a list of the species of fishes found in the fresh waters of Rio Presidio about Presidio and Villa Union:

Sardinella stolifera. Scarce.
Pœcilia butleri. Rather common.
Pœcilia presidionis. Very common.
Thyrina crystallina. Rather common.
Agonostomus nasutus. Very common in ripples.
Siphostoma starksii. Common in algæ in sluggish water.
Centropomus ensiferus. Common in cut-offs of rivers.
Centropomus pedimacula. Scarce.
Eucinostomus gracilis. Common.
Xystæma cinereum. Not rare.
Heros beani. Common in deep places.
Philypnus lateralis. Common (young very common).
Eleotris æquidens. Scarce.
Dormitator latifrons. Common.
Awaous taiasica. Common.
Citharichthys gilberti. Not rare in river; colors very bright.
Achirus mazatlanus. Very common.
Achirus fonsecensis. Scarce.

54. **Pœcilia presidionis** Jordan & Culver, n. sp. Plate xxix.

In the clear waters of the Rio Presidio, about Presidio; with the preceding, and still more abundant.

Head $4\frac{1}{5}$; depth $3\frac{1}{5}$ to $4\frac{1}{5}$; D. 7 or 8; anal 7; scales 28–9; eye equal to snout, $3\frac{1}{2}$ in head; interorbital 2; caudal 1 to $1\frac{1}{5}$; pectoral $1\frac{1}{4}$. Body rather elongate, shaped as in a *Fundulus*, the profile scarcely rising to dorsal.

Teeth much as in *Pœcilia butleri*, the outer smaller

than in *butleri;* broad and movable, apparently in two well separated series, the inner row similar to the outer, but smaller.

Fins all low and short, except anal in male, in which the first one or two rays are produced and extend back nearly to the caudal fin.

Dorsal in female inserted over middle of anal, behind anal in male; caudal truncate.

Female greenish above, sides with violet sheen; three or four black cross bars, sometimes obsolete in adult, but very distinct in young; one or two blackish oblong spots before the anterior bar, representing other bars; a dark pencil-like streak on sides of body below the scales; a dark blotch on opercle; a trace of a dark ocellus on last ray of dorsal at base. Fins without spots; lower fins plain; a dark streak along edge of caudal peduncle; faint traces of black markings on edge of dorsal and caudal.

Male much smaller, reddish, with the lower fins yellowish; the coloration generally similar; both sexes rather dull.

Type, No. 2687, L. S. Jr. Univ. Mus.

Family ESOCIDÆ.

55. **Tylosurus fodiator** Jordan & Gilbert. AGUJON.

Common in the harbor at Mazatlan, where numerous specimens, large and small, were taken; the largest of these is about four feet long.

It reaches a length of five feet. Greatly valued as food in Acapulco; but not at Mazatlan, the people disliking it on account of the green bones. It often leaps at lights in boats, and is regarded as a species dangerous to fishermen, as its sharp beak readily pierces their scanty clothing.

56. **Tylosurus stolzmanni** (Steindachner). SIERRITA.

Occasionally taken in the harbor of Mazatlan, where specimens, the types of *Tylosurus sierrita*, were taken by Dr. Gilbert. One large specimen obtained by us. Its measurements differ somewhat from those given in the type of *Tylosurus sierrita*. The distance between the eyes is 8¾ in head. The maxillary reaches beyond the vertical from front of pupil. The eye is 3 in postorbital part of head. Head not quite 2 in length. D. 1.15; A. 1.17. Pectorals with dusky specks, but not notably black at tip.

This fish is probably identical with *Tylosurus stolzmanni*, described by Steindachner from Tumbez, Peru. The snout in our specimen, as in the type of *sierrita*, is shorter than in *Tylosurus stolzmanni*.

Family HEMIRAMPHIDÆ.

57. **Hyporhamphus roberti** (Cuvier & Valenciennes). PAJARITO.

Exceedingly common about Mazatlan, swimming in schools in open water, especially numerous in the bay; those of the same age and size go together. Schools of adults and schools of half grown specimens will be found, each moving about independently of the other. It is highly valued as a food fish, although distinctly inferior to *Sardinella stolifera*.

Lower jaw, measured from tip of upper, two times length of rest of head. Snout, 2½ in head.

This species is found along the whole Pacific Coast of tropical America, and from Cape Cod to the mouth of the Rio Grande, being everywhere common southward. We have seen no specimens from the West Indies.

The type of *Hemirhamphus roberti* Cuvier & Valenciennes, came from Cayenne, coll. Poiteau. Through the kindness of our friend, Dr. F. Bocourt, of the Museum at Paris, we have received a drawing of this specimen. In the drawing the lower jaw, from tip of upper, is 1¾ times length of head. The head, with lower jaw, is 1¼ times in length from tip of upper jaw to base of caudal. The ventral is midway between front of eye and base of caudal. The name *roberti* belongs, therefore, to the common long-jawed form: the short-jawed West Indian form being *Hyporhamphus unifasciatus*.

Family SYNGNATHIDÆ.

58. **Siphostoma starksii** Jordan & Culver, n. sp. Culevra de Rio. Plate xxx.

Common in the Rio Presidio in sluggish water, on the bottom, about a mile below the village of Presidio. The species is probably found in brackish and fresh waters rather than in the sea.

Head 10½; depth 21; dorsal 38, on 0+10 or 11 rings. Rings 13 or 14—37 or 38. Head and body in tail 2. Snout 2⅓ in head. Dorsal half longer than head.

Body rather stout. Head scarcely carinate above. Snout with a slight smooth carina. Two lateral keels, confluent into one behind.

Belly slightly keeled; no keel on opercle.

Color, dark olive, much mottled with darker but without distinct markings; yellow below.

Male and female common in the fresh waters of Rio Presidio among algæ; not seen in salt or brackish water. The pouch of the male teeming with eggs in January.

Length [illegible] inches.

Type, No. [illegible], L. S. Jr. Univ. Mus.

59. **Siphostoma arctum** [illegible]

[illegible] specimens taken [illegible] Mazatlan, both [illegible] inches. [illegible] This species resembles the preceding, but its dorsal fin has but 20 rays, being placed on 0+5 rings.

60. Hippocampus ingens Girard. CABALLITO DE MAR.

Rare in the harbor at Mazatlan. Three male specimens and one female, each about six inches long, obtained. Also recorded by Dr. Gilbert.

D. 19. Rings about 11+36; dorsal on 3+2 plates.

Spines on head and body high, with large fringed flaps and with many small papillæ. Every 3d to 5th tubercle of dorsal series enlarged.

Greatest depth $1\frac{1}{10}$ to $1\frac{1}{3}$ in head. Tail longer than rest of body. Snout moderate, $2\frac{1}{8}$ to $2\frac{1}{4}$ in head, rather longer than opercle, $2\frac{1}{3}$ times eye. Shoulder girdle with three tubercles; anterior spine on frontal triangle much smaller than the others.

Color blackish, unspotted, faintly barred with darker; dorsal speckled with black and edged with white; papillæ on body pale, giving an appearance of scattered whitish dots everywhere; a white speck before eye; a faint trace of radiating streaks behind it; one specimen further dotted with black on body, the radiating streaks behind eye distinct.

Here described from an adult male, 6 inches long. The female is entirely similar except that the body is much more slender, the depth $1\frac{3}{8}$ in head; the snout is longer, as long as rest of head.

The male specimens agree fairly with the description of *Hippocampus ingens*. The female evidently corresponds to *Hippocampus gracilis* Gill.

Family FISTULARIIDÆ.

61. Fistularia depressa Günther. CORNETA.

Common in the Bay at Mazatlan; many specimens taken with the seine in shallow water. Also, found in abundance by Dr. Gilbert; not yet recorded from localities further south.

Family ATHERINIDÆ.

62. **Eurystole eriarcha** (Jordan & Gilbert). Plate xxxii.

One specimen found in a rocky pool by Dr. Gilbert; a second one taken by us with a seine on the sandy beach just south of Mazatlan. Only these two specimens are known, and the species is probably rare. This species is allied to the genus *Menidia* rather than to *Atherina.* It differs from the species of *Menidia* chiefly in the extremely long anal fin and in the smallness of its dorsal, which is unusually far backward. These characters have been used by Jordan and Evermann to define the genu *Eurystole*, of which this species is type. The mouth is shorter than in *Menidia*, but its structure is exactly the same.

Head 5; depth 5; dorsal III–I, 11 or 12, anal I, 27; scales about 48.*

Body short, deep, much compressed; head short, deep, about ¼ longer than deep, rather broad above; opercles, truncate behind, the interorbital space about equal to eye. Mouth very small, terminal, very oblique, with curved cleft as in *Menidia;* the premaxillary very short, wide behind, with curved edge, slipping under the narrower maxillary; the premaxillary protractile, but not much movable; jaws subequal, the lower slightly included. Maxillary scarcely as long as eye, not quite reaching front of eye. Teeth rather large, hooked backward. Snout short, 3½ in head. Eye large, 2¾ in head. Gill-rakers numerous, long and slender. Scales smooth, caducous, not easily counted, 21 before dorsal. Pectoral moderate,

*Not to be exactly counted; the number (36-7) stated in our original description is an error.

not falcate, inserted high, $1\frac{1}{3}$ in head, 6 in body, reaching to the middle of the small ventral. Belly not especially compressed, not cultrate. First dorsal very small, slightly nearer snout than base of caudal, over first ray of anal; last ray of dorsal much before last of anal. Anal very long, somewhat elevated in front, its base 3 times in length of body. Soft dorsal and anal scaleless.

Color translucent green, very pale; back, lips and bases of vertical fins faintly dotted; lateral band very broad and highly silvery, about two-thirds as broad as eye; lower fins pale; air-bladder not visible through the flesh.

One specimen, $2\frac{3}{4}$ inches long.

63. **Thyrina evermanni** Jordan & Culver, n. g. and n. sp. Plate xxxiii.

Common in the estuary. In this species the structure of the mouth is exactly as in *Thyrina crystallina*. It differs from that seen in *Menidia* only in having the upper jaw shorter. It is apparently closely related to the genus *Atherinella* of Steindachner, but it has not the toothed scales of the type of that genus, *Atherinella panamensis*. The other characters of *Atherinella*—the great length of the pectoral fin, the great compression of the breast and the long anal fin—are shared by this species which we have made the type of a new genus, *Thyrina*. The name (θύρις, window) refers to the translucent sides. Both *Eurystole* and *Thyrina* are intermediate between *Menidia* and *Atherinella*.

Thyrina evermanni differs from *Thyrina crystallina* in the longer anal, the more falcate pectoral, the smaller scales, more compressed breast and the absence of black on the fins.

Head $4\frac{3}{4}$; depth $4\frac{2}{3}$ to 5; dorsal IV, I, 7; anal I, 23 to I, 25; scales 36–9; eye $2\frac{2}{3}$ in head; snout $3\frac{2}{5}$ in head; maxillary $3\frac{2}{5}$ in head; lower jaw $2\frac{2}{3}$ in head;

pectoral 1¼ longer than head, 3½ in body; caudal slightly longer than head; interorbital space broad, nearly equal to eye.

Body much compressed, the belly sharp edged, concave on each side below pectorals, as if pinched together between the fingers, the ribs reaching the edge, the scales passing around it; the edge almost carinate. Back narrow. Scales smooth, none on dorsal or anal. Mouth small, terminal, the short jaws curved, the structure precisely as in *Menidia*, the teeth moderate, curved, those in the upper jaw longer; opercles oblique behind, not vertically truncate. Gill-rakers numerous, long and slender. Pectorals very long and falcate, reaching to front of anal and beyond tips of the short ventrals, their posterior margin concave; spinous dorsal small, inserted midway between edge of preopercle and base of caudal, about over sixth ray of anal; last ray of dorsal considerably before last of anal; base of anal 1⅔ times length of head, 2¾ in body.

Color, light green, much dotted above, translucent below; a black streak of dots along base of anal; some on sides of head; median line of back dusky; fins all pale; no black on spinous dorsal, ventral or pectoral; lateral stripe ⅔ width of eye, underlaid by black; a large, perfectly transparent, space above front of anal, marking the posterior portion of the air-bladder.

Length, 2½ to 3 inches. Rather common in the estuary at Mazatlan.

About twelve specimens obtained, numbered 2688 in the L. S. Jr. Univ. Mus.

64. **Thyrina crystallina** Jordan & Culver, n. sp.

Rather common in the Rio Presidio in fresh water: not seen elsewhere. It is apparently not found in the sea, but confined to fresh or brackish waters.

Head 4¾; depth 4½ to 5; dorsal IV–I, 8; anal I, 21; scales 40–11; pectoral $\frac{1}{8}$ longer than head, 4¼ in body; anal base more than half longer than head, 3 in body; eye 2¾ in head; snout 3¼; maxillary 2⅔; lower jaw 2½.

Body rather deep and compressed; snout shortish; opercle shortish, rounded behind; mouth small, the upper jaw very protractile, the premaxillary strongly curved; jaws equal; teeth rather strong, the outer curved, those in upper jaw largest; eyes very large, silvery; breast compressed, as in *Thyrina evermanni*, but less sharp at edge, appearing as if pinched between thumb and finger; pectoral long, pointed, not truly falcate, reaching more or less past the middle of the short ventrals, its posterior margin not concave, the middle rays considerably more than half length of upper rays; dorsal and anal naked; gill-rakers numerous, long and slender; first dorsal small, behind front of the long anal, midway between gill opening and base of caudal; first ray of soft dorsal over about fourth of anal; last rays of soft dorsal considerably before last of anal. Caudal lunate, the lower lobe the longer and broader, as long as head. Color, translucent green, with considerable dusky dottings, no yellow; fins dotted; ventrals black, as are lobes of second dorsal and anal; silvery stripe narrow, little more than half diameter of the eye; first dorsal and base of anal dusky; air-bladder evident through the translucent sides of body, but less clearly so than in *Thyrina evermanni*.

In fresh water, very common in the lower Presidio; many specimens taken; the longest 3¼ inches long. Type, No. 2685, L. S. Jr. Univ. Mus.

Family MUGILIDÆ.

65. **Mugil cephalus** Linnæus. LISA MACHO. LISA CABEZUDA.

Very common in the bay of Mazatlan; a fish of almost universal distribution on both coasts of tropical America, and extending to Europe. We are unable to distinguish the specimens from the two coasts one from another, and find no permanent difference between these and specimens from the Mediterranean. This species is largely used as food, and often enters lagoons and sheltered places.

66. **Mugil curema** Cuvier & Valenciennes. LISA BLANCA.

Excessively common everywhere, especially in the harbor and estuary. This species is also valued as a food, but reaches a considerably smaller size than the other. In life the iris is tinged with orange, and there is an orange spot on the side of the head behind the eye. This species, like the preceding, is very widely distributed, being found on both coasts of tropical America.

67. **Mugil hospes** Jordan & Culver, n. sp. LISITA. Plate xxxi.

Rather scarce in the harbor at Mazatlan, where it occurs in company with schools of the preceding species; some eight specimens obtained by us. According to Dr. Gilbert, it is quite common at Panama, but the specimens obtained there by him in 1883 were destroyed by fire, so that the species has not thus far received a name. Most specimens of this species have in the mouth or about the branchial cavity a small Crustacean allied to *Oniscus* or *Cymothoa*, the condition being similar to that seen in the eastern Menhaden (*Brevoortia tyrannus*). This Crustacean is found in none of the other species of mullet and its presence is a distinctive character of the present one, which is also readily known at sight by the much greater length of its pectoral fins as compared with *Mugil curema*. The Crustacean is also common and characteristic of the same species at Panama.

Head 3⅔ to 4; depth 4 to 4⅓; D. IV–8; A. III, 9; scales 38–13; eye 4½ in head; snout 4; maxillary 4.

Body a little slenderer and more compressed than in *Mugil curema*, the back considerably more arched, the profile evenly curved from tip of snout to soft dorsal. Eye moderate, with a large adipose eyelid. Head broad and round above; interorbital width 2⅗ in head. Teeth very small, perceptible with a lens. Tip of lower jaw forming about a right angle. Space between dentaries club-shaped, very much larger than in *Mugil curema*, the subopercles barely touching below. First dorsal inserted above middle of body nearly over tip of ventral spine. Second dorsal moderate; its edge incised. Upper lobe of caudal a little longer than lower, as long as head. Anal rather high. Ventral inserted before middle of pectorals. Pectoral very much longer and more pointed than in *curema*, 1¼ in head.

Soft dorsal and anal covered with small scales.

Color much as in *curema*, rather greener above, sides silvery, with less trace of longitudinal streaks. Fins pale; base of pectoral with a round black spot. Upper edge of pectoral and end of caudal dusky. No golden on head. Iris with a little brown, green above eye.

Types, Nos. 1695, 2890, 2954, L. S. Jr. Univ. Mus.

68. **Mugil setosus** Gilbert.

Four young specimens taken in a rock pool. The pectoral is as long as in *Mugil hospes*, reaching the first dorsal, and there is a distinct dark blue spot at its base. Color bluish above, much as in *Mugil curema*; much darker than in the original types of the species, with which our specimens have been compared. The original specimens came from a bottom of volcanic ashes.

69. **Chænomugil proboscideus** (Günther). LISITA.

Very common in rocky places, reaching a length of about 6 inches; not found by us in open water.

70. **Querimana harengus** (Günther). VERDE.

Very common in the bay and estuary; often seen swimming in schools on the surface after the fashion of whirligig beetles; occasionally taken in rock pools. Back bright green, in life with a large, shining, silvery spot on each side of the back. This spot becomes inconspicuous when the fish is taken out of the water, but is a prominent recognition mark while the fish is swimming.

71. **Agonostomus nasutus** Günther. TRUCHA.

Extremely abundant in the fresh waters of the Rio Presidio, especially in the swift places or ripples. It reaches a length of over a foot, but most of the specimens are much smaller.

Head 4 to $4\frac{1}{4}$; depth $4\frac{1}{3}$ to $4\frac{1}{2}$; dorsal IV–1, 8; anal usually II, 10, very rarely II, 9; scales 43–13; maxillary $3\frac{1}{3}$ to $3\frac{1}{10}$; eye $3\frac{2}{3}$ to $4\frac{1}{3}$; snout $3\frac{2}{3}$ to 4; pectoral $1\frac{1}{3}$ to $1\frac{1}{2}$; caudal equal to head.

Body moderately elongate, not much compressed, nape prominent, rounded. Interorbital much rounded, 3 in head. Preorbital narrow, as wide as pupil. Mouth rather small; maxillary reaching front of pupil; lower jaw included. Eye large without adipose eyelid. Teeth small, in villiform bands. Gill-rakers slender, short, close set. Pectoral short, not reaching first dorsal. Ventrals under middle of its length, each with a small axillary scale. Anal and soft dorsal with the free edge concave; caudal well forked. First spine of anal very short, almost rudimentary; second $3\frac{1}{3}$ in longest soft ray. First soft ray slender, but articulate, half length of longest ray.*

Olivaceous, sides creamy, white. Many scales on sides punctate so that black scales seem scattered among the others. A conspicuous black bar at base of pectoral, followed by a white streak; a narrow black rim around lower half of eye. Fins all creamy yellow, the upper ones blotched and dotted with blackish. Young with a black blotch surrounded by orange on first dorsal. Spot on pectoral distinct at all ages.

Family SPHYRÆNIDÆ.

72. **Sphyræna ensis** Jordan & Gilbert. VICUDA.

Rather common in the harbor, where it was found by Dr. Gilbert; also recorded from Panama by Gilbert, and from San Bartholomé Bay and Panama by Steindachner. An excellent food fish, but reaching a smaller size than most species of the group.

Family POLYNEMIDÆ.

73. **Polydactylus approximans** (Lay & Bennett). RATON.

Very common, especially on sandy beaches; many specimens taken by us; also recorded by Gilbert from Mazatlan and from other localities. Used as food.

74. **Polydactylus opercularis** (Gill.)

Obtained by Dr. Gilbert from Mazatlan and Panama; not seen by us.

Family HOLOCENTRIDÆ.

75. **Holocentrus suborbitalis** Gill. MOJARRA CARDENAL.

Very abundant in all rocky pools about Mazatlan. It reaches only a small size, barely exceeding six inches,

* Apparently taken for a spine by Dr. Günther, who counts A, III, 9.

and its coloration is less red than that of the Atlantic species of the genus.

Head 3; depth 2⅔; D. XI, 12; A. IV, 8; scales 3–36–7; longest dorsal spine 1¾ in head; longest dorsal ray 1¾; caudal lobes 1⅓; third anal spine 1⅔; pectoral 1⅗; ventral rays 1, 7. Seven scales on cheek. Maxillary slipping under preorbital. Ventral with accessory scale. Dorsal lying in a groove.

Body short and deep, compressed, with slender caudal peduncle; anterior profile rounded. Mouth small; upper jaw protractile. Teeth in villiform bands on jaws, vomer and palatines. Maxillary moderate, slipping under the very narrow preorbital, which, like rest of suborbital ring, is armed with close-set sharp teeth, turned backwards. Preopercle, opercle, subopercle, interopercle and postemporal armed with similar teeth. Preopercular spine nearly as long as pupil; nearly as long as eye on large specimens. Two spines on opercle.

Steel gray, underlaid by bright coppery red, which becomes brighter after death. Everywhere much punctate with black, the dots coarse. Sides, and especially back, with purple reflections. Top and side of head coppery; a curved bright silvery streak from tip of snout, below eye and around it, ceasing opposite middle of pupil. A vertical silver streak on edge of opercle and extending out on spine. Head yellowish, upper lip reddish; lower with throat silvery. Dorsal brown, clouded with reddish and dark; dark brown near edge, then a series of grayish clouds; roundish, irregular, whitish spots at its base. Second dorsal reddish, its rays pale, its first two black; the caudal red, base pale; the upper and lower rays dark yellowish, darkest in young, the dark extending on peduncle above and below. Anal spines whitish, the soft rays bright red, the last ones pale, the first soft rays dark. Ventral reddish, the spine and first soft ray whitish, the first ray dark red; when the fin is closed it seems reddish, edged with whitish or yellowish, and with a blackish line. The dark is fainter in larger specimens.

It is not impossible that *Rhamphoberyx pœcilopus* Gill is the very young of this species. *Rhamphoberyx leucopus* may be the young of *Myripristis occidentalis*, which has the ventrals plain.

Family MULLIDÆ.

76. **Upenus grandisquamis** Gill. Chivo. (*Upenus tetraspilus* Günther.)

This small species, rarely exceeding a foot in length, is generally common in the harbor and estuary at Mazatlan, where it was found also by Dr. Gilbert. It seems to be everywhere common on the coast.

Color evanescent, olive with two rows of light bluish green spots toward back, then a bronze band, then a blue streak on level of pupil; 2 or 3 yellowish streaks below it. Sides of head golden, with a light green streak forward from eye and some blue behind eye. A large black blotch below last dorsal spine. First dorsal reddish, clouded with dark. Second mesially black, edged with orange. Caudal and anal red. Ventral and pectoral pale.

In alcohol much red appears. In life, sides with curved light yellowish brown, cross bands most distinct on the silvery lower parts.

Family SCOMBRIDÆ.

77. **Germo alalunga** (Gmelin).

Recorded from near Mazatlan by Lay and Bennett; not seen by us, it being probably a migratory fish coming in the spring or fall.

78. Scomberomorus sierra Jordan & Starks, n. sp.
SIERRA.

Rather common in the harbor at Mazatlan, numerous specimens being taken: also found by Dr. Gilbert at Panama. This is not valued as a food fish, little attention being paid to it by fishermen. This, however, may be due to the lack of appreciation of good fishes by the people of Mazatlan, who have not learned the art of properly cooking any fish.

This species is very closely allied to its Atlantic cognate, *Scomberomorus maculatus*. It differs in the slightly more backward insertion of its soft dorsal, in its coloration, the spots in *maculatus* being elliptical and fewer in number, and perhaps in the fewer pores in the lateral line (175 in *maculatus*). In *Scomberomorus maculatus* the soft dorsal is inserted one eye's diameter before anal.

Head $4\frac{3}{4}$: depth equal head: dorsal XVIII–15–IX; anal II–15–IX: maxillary $1\frac{3}{4}$ in head: eye 5 in head; pectoral $1\frac{3}{4}$: ventral $3\frac{1}{2}$: dorsal and anal lobes equal, $1\frac{3}{4}$ in head.

Body elongate, its dorsal and ventral outlines about equal: profile straight from snout to dorsal: head small and pointed: mouth large, oblique: jaws equal: maxillary reaching to posterior edge of orbit. Teeth large, compressed and sharp, 20 to 32 in each jaw: gill-rakers $4+11$. Soft dorsal inserted almost directly over front of anal: lateral line undulating, about 165 pores.

Silvery, above bluish, sides with numerous round brownish spots: three rows of spots below lateral line and one above. Spinous dorsal white at base, black above: soft dorsal tinged with yellowish, its margins black: anal white: posterior face of pectoral entirely black, anterior face yellowish with blackish borders: caudal black.

Another example supposed to be a male has five rows

of spots below the lateral line, these spots decrease in size towards the belly, covering both sides nearly to level of pectoral.

Types, 1720, L. S. Jr. Univ. Mus.; the largest 24 inches long.

Family CARANGIDÆ.

79. **Oligoplites altus** (Günther). MONDA.

One large specimen taken by us. Recorded by Dr. Gilbert from Mazatlan and Panama.

80. **Oligoplites saurus** (Bloch & Schneider). MONDA.

Common in the harbor of Mazatlan, where it was also taken by Gilbert. On comparison of specimens from Mazatlan with others from Havana we are unable to find any difference whatever. The species called *inornatus* is therefore fully identical with *saurus*.

81. **Trachurops crumenophthalmus** (Bloch).

Common in the harbor at Mazatlan, where numerous specimens were taken; not recorded by Dr. Gilbert. Specimens have been compared with others from Havana and no difference of any kind is observable. *Trachurops brachychirus* must therefore be regarded as an exact synonym of *Trachurops crumenophthalmus*.

82. **Caranx vinctus** Jordan & Gilbert.

Rather common in the estuary, where numerous specimens were taken. The original types were found by Gilbert at Mazatlan, and the species has been recorded from San Blas and Punta Arenas.

83. **Caranx caballus** Günther. COJINERO.

Extremely common in the harbor; also found in abundance by Dr. Gilbert.

84. **Caranx medusicola** Jordan & Starks, n. sp. Plate xxxiv.

Rather common in the surf outside the harbor. Not found in the Astillero. The young from 1 to 2 inches long live in the body cavity of the large white jelly fish, which is very abundant about the Venados Islands in January. Sometimes two or three specimens will be found in the body cavity of one jelly fish.

Head $3\frac{3}{8}$; depth $2\frac{1}{6}$; D. VII–1, 22 or 23; A. II, 1, 19 or 18; scutes 30 to 32; pectoral $\frac{1}{6}$ longer than head; dorsal lobe $1\frac{2}{3}$ in head; caudal lobe, as long as head; curve of lateral line $1\frac{1}{2}$ in straight part; height in chord 4; eye 4 in head; snout 3; maxillary 3; ventral $2\frac{1}{3}$.

Body unusually deep and compressed, the back elevated, the belly similarly arched; head moderate, deep, the nape arched. Mouth small, maxillary broad, with broad supplemental bone. Teeth in moderate bands, the outer enlarged but not canine-like; upper teeth rather larger and in broader bands. Villiform bands on vomer, palatines and tongue. Eye moderate; preorbital rather narrow. Gill-rakers rather long and slender, about 12 below angle of arch. Soft dorsal and anal with falcate lobes. Caudal well forked, the lobes equal. Pectoral very long and falcate; ventrals short. Lateral line rather strongly curved, with moderate armature. Breast entirely scaly.

Clear blue above, silvery below; no bands or spots anywhere, except a small black axillary spot and a blue green patch on back of caudal peduncle; pectoral bright yellow; anal yellow, the lobe blackish; caudal grayish, the lobes black with whitish posterior edge; ventrals yellow.

Length of largest specimens, 6 inches. Type, No. 2645, L. S. Jr. Univ. Mus.

Another example was, in life, blue above, silvery below; no dark spots on opercle or pectoral; pectoral bright yellow, very long. D. and A. and C. lobes, all tipped with black. Base of dorsal bright blue. Anal and dorsal largely blue. Base of caudal peduncle green above. No trace of bands; a slight dusky shade on axil.

The very young, taken from the body of a *Medusa*, may be thus described:

Head 3 in length; depth $2\frac{5}{8}$; dorsal IX, 24; anal II, I, 18 or 19; ventral with a sheath; scales minute; caudal keel scarcely appreciable; lower jaw projecting; mouth oblique; body deep, compressed; caudal peduncle slender, the fin short, moderately forked; pectoral short, not falcate, shorter than head; maxillary broad, reaching pupil; preorbital narrow; dorsal and anal not falcate; lateral line arched before, then straight; jaws with teeth; preopercle with flexible spines.

Clear white, fins all pale, a bright violet blue area above and behind eye, fading in spirits; dark dots above; dorsals both dusky at tip.

85. **Caranx marginatus** (Gill).

Not rare in the Astillero, where several specimens were taken by us. This species is well distinguished from *Caranx latus*, with which it has hitherto been confounded, since it was originally described by Dr. Gill. The following are its characters:

Head $3\frac{1}{3}$; depth $2\frac{2}{3}$; dorsal VIII–1, 19; anal II–1, 15; eye $3\frac{2}{3}$ in head; pectoral $3\frac{1}{3}$ in length, equal to head; ventral $7\frac{1}{4}$; dorsal lobe $5\frac{1}{4}$; caudal $3\frac{2}{3}$.

Dorsal outline of body evenly curved from snout to caudal peduncle; ventral outline straight from gill openings to anal spine, behind which it is curved like the dorsal portion.

ıy difference between specimens from the west d specimens from Havana.

ıathanodon speciosus (Forskål). MOJARRA RADA.

common in the harbor and estuary, being one of ɛ valuable food fishes, the flesh being firm and We have compared specimens with others taken Jenkins at Honolulu and find no difference. We ꞉refore, no hesitation in continuing to identify our (*Caranx panamensis* Gill) with this common East ısh, of which the oldest name is *speciosus*.

ɛ, everywhere deep golden yellow, with black nds.

:ula dorsalis (Gill). PÁMPANO.

:r common in the estuary. Three specimens y us, one half-grown and the others adult, the in form being strikingly marked, as will appear : following descriptions:

'orsalis (half grown):

$3\frac{1}{8}$; depth $1\frac{3}{4}$; D. VI–I, 19; A. II, 1, 17; eye :ad, the orbit $3\frac{1}{2}$; snout $2\frac{2}{3}$; pectorals $2\frac{1}{2}$ in body, than head; ventrals 3; caudal lobe equal to head; /ith one long filament, as long as body, reaching ɔf caudal; anal with one filament; caudal mod- elongate, the lobes equal; pectoral very long, reaching tenth anal ray; ventrals small, reach- past vent.

deep, compressed, rather ovate than angular; traight from the vertical truncate snout to nape, ınded, then straight to front of dorsal. A nearly line from chin to front of anal. Eye rather small, al deep. Mouth large, the lower jaw included. nall, in broad bands on jaws, vomer and palatines,

Top of head, snout, lower jaw, orbitals, maxillary, lower two-thirds of opercle and preopercle naked; cheeks scaled; eye large, with membranous eyelid to posterior edge of pupil in specimens six or eight inches long, not conspicuous in young examples. Snout equal to eye, twice width of preorbital; lower jaw entering profile; maxillary reaching to posterior edge of orbit. Teeth strong, in a single row; lower teeth close together, with two canines in front; upper teeth larger, the distance between them irregular, not much enlarged anteriorly; vomer, palatines and tongue with exceedingly small villiform teeth. Gill-rakers hardly half eye, 4–13. Breast scaled; curved part of lateral line, 1½ in straight part; scutes large, about 30; scales, 80.

Color, silvery, bluish above with golden reflections below; a dark band along plates of lateral line; fins largely yellow, dorsal, anal and caudal, broadly edged with black; a distinct small black spot at upper end of gill-opening; a dark blotch on opercle, and one behind pectoral.

Body more elongate than in *Caranx latus*, the fin rays fewer, the eye larger and the coloration more yellow, with more black on the fins.

86. Caranx latus Agassiz.

Occasionally taken in the bay at Mazatlan, and generally distributed throughout the waters of the tropical Pacific and West Indies. We are unable to distinguish the specimens from the west coast of Mexico from the common West Indian form.

87. Caranx hippos (Linnaeus). Toro.

Very common in the sea about Mazatlan, occasionally entering the estuary. A food fish of some importance, reaching the length of two or three feet. We are unable to see any difference between specimens from the west coast and specimens from Havana.

88. Gnathanodon speciosus (Forskål). MOJARRA DORADA.

Very common in the harbor and estuary, being one of the more valuable food fishes, the flesh being firm and delicate. We have compared specimens with others taken by Dr. Jenkins at Honolulu and find no difference. We have, therefore, no hesitation in continuing to identify our species (*Caranx panamensis* Gill) with this common East Indian fish, of which the oldest name is *speciosus*.

In life, everywhere deep golden yellow, with black cross bands.

89. Citula dorsalis (Gill). PÁMPANO.

Rather common in the estuary. Three specimens taken by us, one half-grown and the others adult, the change in form being strikingly marked, as will appear from the following descriptions:

Citula dorsalis (half grown):

Head $3\frac{1}{8}$; depth $1\frac{3}{4}$; D. VI–I, 19; A. II, 1, 17; eye $4\frac{3}{4}$ in head, the orbit $3\frac{1}{2}$: snout $2\frac{2}{3}$; pectorals $2\frac{1}{2}$ in body, $\frac{1}{8}$ longer than head; ventrals 3; caudal lobe equal to head; dorsal with one long filament, as long as body, reaching middle of caudal; anal with one filament; caudal moderately elongate, the lobes equal; pectoral very long, falcate, reaching tenth anal ray; ventrals small, reaching just past vent.

Body deep, compressed, rather ovate than angular; profile straight from the vertical truncate snout to nape, then rounded, then straight to front of dorsal. A nearly straight line from chin to front of anal. Eye rather small, preorbital deep. Mouth large, the lower jaw included. Teeth small, in broad bands on jaws, vomer and palatines, maxillary reaching pupil. Cheek entirely scaly, some scales on opercle above. Breast naked, body well scaled. Body with small scales, the nuchal region naked, scarcely carinate. Gill-rakers rather long, 2+15.

Lateral line evenly curved, the curve high, equal to straight part. Scutes small, eighteen with keels; the total number of scales on straight part 58.

Steel blue above, silvery below, with golden reflections and shades; fins all pale, tinged with yellowish, none of them dusky; no black on pectorals. Axil jet black; opercle slightly dusky, blackish within; a dark spot on orbit above.

Specimen described, ten inches long.

Citula dorsalis (adult):

Length 24 inches; head 3¼; depth 2¼; D. 18; A. 17. About 25 scutes developed. Body moderately compressed, with angular outlines. Profile of head rounded, of belly somewhat concave, forming an angle at anal similar to one at front of dorsal. Eye 5 in head. Maxillary 2⅙; lower jaw included. Teeth in broad villiform bands on both jaws and on vomer and palatines. Nostrils large, equal, close together. Gill-rakers 3+14, rather stout, shorter than eye. Dorsal spines nearly obsolete, three of them present; first dorsal ray filamentous, 1¾ in body. Long anal ray 2¾ in body. Caudal keel considerably elevated, with a small keel above and below it; scutes not sharp. Caudal lobes subequal, about as long as head. Pectoral falcate, ⅐ longer than head. Ventral short, 3⅔ in head. Curve of lateral line low, 1⅛ times in straight part, its height ¼ its chord. Maxillary broad, with very broad supplemental bone, its width ⅔ eye.

Color, silvery, strongly tinged with golden, olive on upper parts, pearly reflections below. A large black spot in axil, nearly as large as eye. Fins pale.

90. Alectis ciliaris (Bloch). Pámpano.

Obtained by Dr. Gilbert; not seen by us. We have hitherto been unable to distinguish the specimens of this species from the two coasts of Mexico. We are furthermore unable to find any distinction between the American form called *crinitus*, and the East Indian species, *Alectis ciliaris*. We do not believe that any distinction exists, and therefore find ourselves compelled to believe that this species, like *Caranx hippos* and *Caranx latus*, is almost cosmopolitan in the tropical seas, ranging from the coast of Arabia to the West Indies. None of the three are found in the Mediterranean.

91. Hynnis hopkinsi Jordan & Starks, n. sp. Pámpano. Plate xxxv.

One large specimen taken with the seine in the harbor at Mazatlan.

Head 3½; depth 2⅛; D. VI-1, 18; A, II, 1, 15; snout 2¾; eye 3⅝ in head; maxillary 2¾; pectoral, 3⅛ in body; ventral, 2¼ in head; dorsal lobes 2⅛ in head; caudal lobes 1⅝ in head; anal lobe, 2¼; preorbital, 4¼ in head.

Body oblong, compressed, elevated, with angular outlines, ventrals outline sharp. Top of head sharply carinate; profile nearly straight from snout to nape, there boldly convex, then nearly straight to elevated front of soft dorsal; a concavity in profile before soft dorsal and before anal. Mouth oblique, rather large, the jaws equal. Broad bands of small sharp teeth on jaws, vomer and palatines. Eye very large. Dorsal and anal lobes low. Lateral line with a long arch, as long as straight part, which has about twelve elevated scutes and thirty-seven scales in all from end of curve; curved part of lateral line undulating behind. Gill-rakers short rather few, twelve in all, those above angle obsolete. Body minutely scaly. Belly and lower parts largely naked, a large patch of scales on cheeks; head otherwise naked.

Bright blue above, with bright reflections, sides bright silvery; no golden; a narrow brownish streak not quite so wide as pupil from upper part of gill opening to middle of base of soft dorsal. Pectoral tipped with black; axil of pectoral dusky. Upper fins rather dusky, lower white. Dusky on opercle inside and out but without definite spot.

More elongate than *Citula dorsalis*, the anterior profile more convex, the base of dorsal and anal more elevated, the caudal scutes stronger and fewer, the ventrals longer though the specimen is larger. Gill-rakers fewer. Pectoral long and falcate, reaching seventh anal ray. Ventrals not short, reaching vent. Caudal moderate.

One specimen obtained, twenty-six inches long, No. 1563, L. S. Jr. Univ. Mus.

We take great pleasure in naming this interesting fish for Mr. Timothy Hopkins, in recognition of his great interest in scientific research.

We provisionally admit *Citula* and *Hynnis* as genera distinct from *Alectis*. No structural characters of importance distinguish this group, and all these genera are merely form variations from *Caranx*.

92. **Vomer setipinnis** (Mitchill).

Recorded by Dr. Gilbert as common at Mazatlan and Panama; no specimens, however, were seen by us. It is not unlikely that this species disappears from the coast with the end of the rainy season.

93. **Selene œrstedi** Lütken.

Recorded by Dr. Gilbert as frequently found both at Mazatlan and Panama. One specimen, sixteen inches long, taken by Ygnacio Moreno and sent to us.

Head 3; depth 2; dorsal V–I, 15; anal (II) I–14; eye 4 in head; snout 1¾; maxillary 2¾; ventral 3⅓; caudal lobes equal to head; pectoral one-eighth longer than head.

Body compressed and elevated; profile oblique, concave over snout then straight to occiput, which is well rounded; line of back straight to soft dorsal, then lightly curved to caudal peduncle; ventral outline rounded on breast to ventrals, then straight to anal, forming an angle at first ray, then straight to caudal peduncle. Mouth projecting, with minute teeth on jaws, vomer, palatines, and tongue; gill-rakers thick and blunt, many of them knobbed at tip—in old examples at least, one above angle with 3 or 4 rudimentary ones, and 13 below. A large bony knob at occiput, conspicuous in adult, the thickened supraoccipital crest.

Pectoral falcate, reaching to tenth anal ray; dorsal and anal lobes filamentous, reaching past tips of caudal lobes; lateral line strongly arched; curve equal to straight part. Color silvery, with bluish reflections above, dorsal and caudal dark, pectoral, ventral and anal white; axil dusky.

94. **Selene vomer** (Linnæus).

One large specimen obtained by us. Recorded by Dr. Gilbert as common at Mazatlan and Panama. It perhaps disappears with the end of the autumn, going farther south.

95. **Trachinotus paloma** Jordan & Starks, n. sp. PALOMA.

A few small specimens taken in the surf at Puerto Viejo, just north of Mazatlan; other specimens were taken by Mr. Xantus on Cape San Lucas, and still others were obtained by Dr. Gilbert in San Juan Lagoon. The species is apparently not common, and it is not known to the fishermen. On the Atlantic coast, the very closely related Pámpano. *Trachinotus carolinus*, is one of the most valued food fishes. We are unable to see any difference of any importance between the present species and the Pámpano of the gulf other than the fact that in the Sinaloan form the head seems to be larger and longer. On this difference we have ventured to give a new specific name to our specimens from Mazatlan. We shall not, however, be surprised if the species proves inseparable from *Trachinotus carolinus*.

Allied to *Trachinotus carolinus*, but with the head larger.

Head 3, depth $2\frac{1}{2}$: D. VI–I. 24. A. II. I, 23; eye $3\frac{1}{2}$ in head; snout $3\frac{2}{3}$; maxillary $2\frac{1}{3}$: dorsal lobe $1\frac{2}{3}$; caudal $1\frac{1}{10}$.

Body rather elongate, the back moderately and regularly arched; snout bluntish. Mouth large, horizontal, the lower jaw included, maxillary reaching past pupil. Lateral line little arched, its curve $1\frac{1}{6}$ in straight part. Teeth well developed. Caudal not widely forked.

Silvery without spot or band: anal creamy orange, its tip whitish. Other fins pale, except dorsal lobe which is dusky. Axil silvery.

A few specimens taken in the surf, the largest $2\frac{1}{2}$ inches long. Type No. 2690 L. S. Jr. Univ. Mus. Other specimens taken by the Albatross in San Juan Lagoon examined; some of these are five inches in length.

96. **Trachinotus rhodopus** Gill. (*Trachynotus fasciatus* Gill: *Trachynotus nasutus* Gill.)

Very common on sandy shores about Mazatlan, reaching the length of about a foot: not much valued as food. Readily distinguished at all ages by the reddish color of the lobes of the dorsal, anal and caudal. These lobes become considerably elevated with age, but at all times they are marked by shades of brownish red or maroon color. There seems to be little doubt that the *Trachynotus rhodopus* Gill is the young of the species which he called at the same time *Trachynotus fasciatus*. The very young specimens to which Gill gave the name *Trachynotus nasutus* were probably also the young of the same species, but it may be that they were the young of *Trachinotus kennedyi*. Dr. Jordan's identification of the great Pámpano of the Florida Keys with Gill's *Trachynotus rhodopus* is doubtless incorrect. There is at present no evidence that any species of *Trachinotus* is common to both coasts of Mexico.

Young specimens, 2½ inches long. Blue above, white below, no bars. Dorsal and caudal lobes black, with strong orange shade. Lobes of caudal orange brown, verging on black. Pectoral and ventral white.

Specimens 6 or 7 inches long, have from 3 to 5 narrow dark cross-bars, not quite so wide as pupil, running from a point on a level with pectoral fin to within a short distance of the dorsal line of the back, but never quite to it; these bars vary in number and position; posterior face of pectoral fin dusky. Otherwise colored as the younger ones.

97. **Trachinotus culveri** Jordan & Starks n. sp. PALOMETA. Plate xxxvi.

Five specimens, each 7 inches long, obtained in the market at Mazatlan; no others seen. This species is related to *Trachinotus falcatus* of the Atlantic, but its fins are lower and different in coloration. It is also allied to *Trachinotus kennedyi*, but the body is much deeper and there is no black axillary spot. It does not seem possible that with age *culveri* should become transformed into *kennedyi*.

Head $3\frac{2}{5}$; depth 1½; D. VI–I, 17; A. II–I, 17; maxillary 3 in head; eye $3\frac{2}{3}$; snout $4\frac{1}{3}$; dorsal lobe $1\frac{1}{11}$ in head; pectoral $1\frac{1}{3}$ in head; caudal $\frac{1}{7}$ longer than head.

Body very deep, compressed, the back much elevated. Snout very blunt and convex, the rest of profile straight and steep; base of dorsal and anal very oblique. Dorsal and anal lobes rather low. Caudal long. Lateral line little elevated in front, the curve $1\frac{1}{4}$ in straight part. Gill-rakers very short, about 5—9. Teeth persistent, in specimens 7 inches long.

Bluish gray, silvery below, tinged with yellow, everywhere much soiled with blackish spots, no distinct markings anywhere, the axil only slightly dusky; fins all dusky except middle of caudal and lobe of anal, and the ventrals which are whitish.

Types, No. 2691, L. S. Jr. Univ. Mus.

98. **Trachinotus kennedyi** Steindachner. PALOMETA.

Two large specimens obtained in the surf. This species was originally described by Steindachner from Magdalena Bay, and has been recorded by Dr. Gilbert from Mazatlan and from Panama.

Head $3\frac{2}{3}$; depth at vent $2\frac{1}{3}$; at anal $2\frac{1}{10}$; D. VI–I, 19; A. II, I, 16. Curve of lateral line $1\frac{2}{3}$ in straight part. Eye 5 in head; maxillary $2\frac{2}{3}$; dorsal lobe $1\frac{1}{3}$; caudal $\frac{1}{4}$ longer than head; pectoral $1\frac{1}{4}$ in head; snout $3\frac{2}{3}$; least depth of caudal peduncle $3\frac{1}{2}$ in head.

Body oblong, compressed, and elevated at bases of dorsal and anal. Anterior profile of head an even curve, the snout blunt and convex; line straight from nape to dorsal. Mouth moderate, very oblique, subinferior, the lower jaw much shorter than upper, the maxillary reaching to posterior border of pupil. Teeth obsolete. Tail widely forked, the lobes equal. Lobes of dorsal and anal low, not sharp.

Gray above, with deep green reflections, lower half silvery, with strong golden tinge. Axil jet black, the color covering base of fin and extending behind for a distance nearly equal to eye, so that the fin does not cover it; upper fins dusky, the caudal edged with paler, anal dusky with golden tinge, ventrals purplish white. Pec torals dusky; maxillary with a black streak.

99. **Seriola mazatlana** Steindachner.

Originally described from Mazatlan by Steindachner, but not seen by Dr. Gilbert or by us; probably a migratory species.

Family NEMATISTIIDÆ.

100. **Nematistius pectoralis** Gill. PAPAGALLO.

Very common in all the waters about Mazatlan; specimens reaching the length of about three or four feet found about the islands of Venados, Isla Blanca and Creston.

Color silvery, iridescent bluish above, with black bands; the first across tip of snout; the second across interorbital, involving the top of membranous eyelid; the third from nape across opercle; the fourth including the first dorsal spine and running obliquely down on the belly, where it fades out at about the tip of the pectoral fin; the fifth running from middle of first dorsal obliquely to lateral line, then backwards along lateral line to upper lobe of caudal, including the whole upper half of caudal peduncle; a sixth indistinct band, following the line of the back for a short distance, under the soft dorsal; upper part of maxillary dusky; long spines of dorsal with alternate bands of yellow and black, and much slaty-bluish at base; soft dorsal and caudal uniform dusky; pectoral with a black spot on lower rays, not involving the axil; ventrals white; anal slightly dusky.

Described from a specimen sixteen inches long.

The two anal spines united with rest of the fin. No free anal spines. Ventral ray really I, 5, the inner ray very wide, made up of four branches so that the rays seem more numerous; ventral spine obscure. Anal fin short. Pectoral fin falcate. Both dorsal and ventral with sheath. Soft dorsal and anal low, the last ray slightly lengthened.

Dr. Gill is probably right in regarding *Nematistius* as type of a family distinct from the *Carangidæ*.

Family STROMATEIDÆ.

101. **Rhombus medius** (Peters).

Originally described by Dr. Peters from Mazatlan; not seen by Dr. Gilbert or by us. Only the original type in the museum at Berlin seems to be yet definitely known.

Family CHEILODIPTERIDÆ.

102. **Apogon dovii** Günther.

This species was found by Dr. Gilbert at Mazatlan, but was not seen by us.

103. **Apogon retrosella** Gill. CARDENAL. Plate xxxvii.

Two specimens of this most beautiful little fish were obtained by us with dynamite off the Isla Blanca and Creston Islands. Only the very young, found by Mr. John Xantus, at Cape San Lucas have been hitherto known.

Head $2\frac{6}{7}$; depth $2\frac{7}{8}$; scales 3–26–9; dorsal VI–1, 10; anal II, 9; eye $2\frac{7}{8}$ in head; maxillary $1\frac{3}{4}$; snout $4\frac{1}{2}$; interorbital 4; first dorsal $2\frac{1}{4}$; second dorsal $1\frac{2}{5}$; caudal $1\frac{1}{3}$; pectoral $1\frac{1}{2}$; ventral $1\frac{3}{5}$.

Body rather plump, not much compressed, the profile rising steeply from snout to first dorsal. Caudal peduncle long and strong; eye very large; mouth large, oblique, the maxillary opposite posterior margin of pupil. Teeth small, the outer scarcely enlarged. Premaxillary protractile; no supplemental maxillary.

Bright scarlet much dotted with black, cheek with many dark points, a diffuse dark blotch on opercle; a diffuse black blotch at base of caudal. First dorsal with triangular red area in front. Second dorsal red at base, the anterior half jet black above the red, the posterior half translucent. From black anterior rays, a rather faint black saddle falls to middle of side. Caudal red at base, upper and lower lobes black, the middle pale. Anal red at base, the anterior rays black, the posterior pale. Pectoral white, the base deep scarlet. Ventral white, red at base, blackish at tip. Opercle reddish within, with some dusky. Preopercle minutely serrulate on its vertical margin only, these serrulations soft and easily rubbed off.

A younger specimen was, in life, scarlet, deeper below and on tail, fading on fins; second dorsal, anal, and caudal tipped with blackish. An oblong inky spot at middle of base of caudal. An inky bar below soft dorsal extending to level of pectoral and spreading on base of soft dorsal. A black bar from snout through eye to gill opening, broader and clearer behind, overlaid by reddish, a fainter dusky band below parallel with it.

Family SERRANIDÆ.

104. **Alphestes multiguttatus** (Günther).

This species is found in rocky places along the coast, having been taken by Gilbert at Mazatlan and Panama. But one small specimen was obtained by us.

105. **Epinephelus labriformis** (Jenyns). Cabrilla Pinta.

This species is generally common about the islands on the coast of Mexico all the way from Cape San Lucas to the Galapagos Islands. Only young specimens were seen by us.

Inside of mouth salmon yellow; pectoral with salmon color, its edge pale; caudal with a maroon band above and below; dorsal edged with blackish red, spots on belly nearly white; dorsal with white on membranes.

106. **Epinephelus analogus** Gill.

This species is also common in rocky places along the coast from Mazatlan to Panama. Several specimens were obtained by us.

107. **Promicrops guttatus** (Linnæus). MERO.

Rather common about the islands and in deep water, reaching an enormous size, greater than that of any other bony fish found in the region. The largest seen by us weighed some seventy pounds, but it is said to attain the weight at times of 500 or 600 pounds. Only one specimen was obtained in a condition for preservation. This was a small one 20 inches long. The species was found by Dr. Gilbert at Mazatlan, Panama and Punta Arenas; the type of *quinquefasciatus* were obtained by Dr. Bocourt at Tauesco.

This species seems to agree fully with the account of *Promicrops guttatus*, given by Gilbert & Swain, in 1884. There is not much doubt of the identity of the Pacific Coast *Promicrops quinquefasciatus* with *Promicrops guttatus* of the Atlantic.

108. **Dermatolepis punctatus** Gill.

This species seems to be rare along the coast. The type was found by Mr. Xantus at Cape San Lucas, another specimen was brought by Lieut. Nichols from Socorro Island, and a third was found by Dr. Gilbert about the islands near Mazatlan. It was found in abundance by Dr. Gilbert about the Revillagigedos.

109. Mycteroperca boulengeri Jordan & Starks, n. sp. CABRILLA RAIZER. "MANGROVE GROUPER." Plate xxxviii.

This species is found with *Mycteroperca jordani* Jenkins & Evermann in about equal abundance. It reaches a much smaller size than any other species of *Mycteroperca*. It is in many ways an aberrant form, showing affinities with *Epinephelus*. The anal fin is short, as in *Epinephelus*, while the general appearance and coloration is that of *Mycteroperca*. The structure of the skull shows that its affinities are with the latter.

Head $2\frac{4}{8}$ in length; depth $2\frac{5}{8}$. Dorsal XI–14 or 15; anal III–9 or 10; scales about 90, 20 above and 42 below; snout $3\frac{1}{2}$ in head; maxillary $2\frac{1}{8}$; eye $5\frac{1}{2}$; pectoral $1\frac{3}{4}$; ventral $1\frac{5}{8}$; longest anal ray $1\frac{2}{3}$; caudal $1\frac{3}{8}$; longest dorsal spine $2\frac{1}{2}$; gill-rakers short, about 6+17, the longest about $\frac{3}{8}$ eye; longest dorsal ray 2 in head; length 10 inches.

Body short and deep, compressed. Head moderate, compressed, its profile not steep, nearly straight, a depression before eye. Upper canines moderate, the lower quite small. Nostrils small, well separated, the anterior slightly larger. Lower jaw very strongly projecting. Maxillary reaching opposite posterior edge of pupil. Preopercle slighily notched, the angle slightly salient, with enlarged teeth. Dorsal not deeply notched, the fourth spine not much elevated. Second dorsal high, not long, its angle not rounded. Caudal scarcely lunate, the upper lobe long, the lower truncate. Anal very high, strongly elevated; its posterior border incised, the anterior rounded. Pectoral and ventral moderate. Scales smoothish, not very small.

Color olive gray, covered everywhere with oblong irregular markings of black, between which the ground color forms rivulations. Gray lines radiating from the eye. A black blotch below maxillary. Pectoral olive yellow. Other fins blackish, clouded with pale. First dorsal with faint small black spots.

The supraoccipital and temporal crests are high, the supraoccipital crest extending to the posterior margin of orbit; the temporal crests are parallel to each other, and extending to pupil; interorbital space concave.

Several specimens, the largest (No. 1621, L. S. Jr. Univ. Mus.) one foot in length, taken in the Astillero at Mazatlan.

We take pleasure in naming this interesting species for Dr. George Albert Boulenger of the British Museum, in recognition of his excellent work on the *Serranidæ*, in the first volume of his Catalogue of the Fishes of the British Museum, the proof sheets of which have been kindly placed in our hands.

110. **Mycteroperca rosacea** (Streets). CABRILLA CALAMARIA.

Occasionally taken at Mazatlan in rather deep water. Three specimens only of this species have been preserved; one of them from Mazatlan, collected by Gilbert; one, the original type, obtained by Dr. Streets at some point further northward in the Gulf of California, and the third sent to us by Señor Ygnacio Moreno after our return from Mazatlan. In all of these the life color seems to be bright orange.

111. **Mycteroperca venadorum** Jordan & Starks, n. sp. GARLOPA.

A very large species found in some abundance about the islands along the coast, in rather deep water. But a single specimen, weighing 75 pounds, was obtained by us, this specimen being a type of the species. We are told by Dr. George W. Rogers and others that specimens weighing 150 pounds are not uncommon. The specimen from which the species is described was taken by the explosion of dynamite outside in the deep water not far from the island called Isla Blanca.

Head $3\frac{1}{8}$ in length; depth $3\frac{1}{4}$. Scales, small, smoothish, about 130. Dorsal XI, 16; anal III, 11. Snout 3 in head; maxillary 2; eye 8. Gill-rakers 3+8; pectorals $1\frac{9}{10}$; 4th dorsal spine $3\frac{3}{8}$; longest dorsal rays 3; longest anal ray $2\frac{1}{8}$; caudal lobe $1\frac{3}{4}$; ventrals $2\frac{1}{4}$.

Body robust, not strongly compressed, the head large. Lower jaw much projecting. Posterior nostril three times diameter of anterior. Preopercle scarcely notched, its angle scarcely salient, its teeth a little enlarged. Gill-rakers short, thick, few in number. Dorsal deep notched, 2d spine a little lower than the 4th. Soft dorsal high, slightly angulated. Anal very high, with exserted rays. Caudal well forked, lobes unequal.

Color olive brown, almost uniform; no spots or bands. Dorsal, anal and caudal with broad black margin narrowly edged with whitish. Pectoral and ventral darker behind. Pectoral with pale edge.

The type, a specimen weighing in life seventy-five lbs., has been sent as a skin to the British Museum. Its length was 40 inches to base of caudal fin.

112. **Mycteroperca pardalis** Gilbert. CABRILLA PINTITA.

This species is said to be rather common at the Venados and other islands in the neighborhood of Mazatlan. A single specimen was obtained by us; a head was also found in the market. Dr. Gilbert tells us that he has seen salted specimens apparently of this species preserved by the fishermen at Guaymas, together with specimens of a very large species, probably our *Mycteroperca venadorum*.

Head 3 in length; depth $3\frac{1}{10}$; dorsal XI, 16; anal III, 11. Scales 100, small, smooth, imbedded, difficult to count. Eye $6\frac{2}{3}$ in head; maxillary $2\frac{1}{2}$; pectoral $1\frac{2}{3}$; longest anal ray $1\frac{3}{4}$; longest dorsal 2; longest dorsal spine $3\frac{1}{2}$. Caudal upper lobe $1\frac{1}{3}$; ventrals 2.

Body deep, robust; anterior profile rather steep and straight; lower jaw moderately projecting. Small canines in both jaws; preopercle with notch and a salient angle. Gill-rakers about 15+25, rather stout, the longest about $7\frac{1}{2}$ in head; snout $3\frac{1}{2}$. Posterior nostril oblong, 4 times as long as anterior. Dorsal spines low, the third and fourth but little longer than the last. Dorsal fin pointed behind; anal very high, triangular in form; anterior margin convex, posterior concave. Sixth soft ray very high, reaching far beyond tip of last, which is short; spines graduated. Caudal fin broad, on a broad peduncle, unequally lunate; upper lobe longer and broader than lower. Pectorals rounded.

Color olive gray, paler below, clouded with dark above. Everywhere covered with small roundish dark olive or bronzed spots so thick as to obscure the ground color; very close set on head and back, small and distinct, not larger than anterior nostril, growing larger and less thick-set below; posteriorly still larger, often half diameter of pupil, and tending to run together forming elongated blotches and vermiculations. Dorsal similarly spotted with spots which grow faint on soft rays; pectoral, anal and caudal like soft dorsal. All soft fins growing dusky toward margin. Soft dorsal, anal and caudal very narrowly edged with pale. Pectoral with broader pale margin; ventral like pectoral, pale edge narrower. When seen from back an appearance of about 10 very faint dusky cross-shades, probably very conspicuous in young.

113. **Mycteroperca jordani** (Jenkins & Evermann).
CABRILLA DE ASTILLERO.

Common in the Astillero at Mazatlan, reaching a much smaller size than any of the three preceding, the largest among them not being more than two pounds in weight. It is not found about the rocks, but lives in abundance in the branches of the Astillero on the muddy bottoms below a growth of the mangrove bushes.

Head 2¾, depth 3⅛. D. XI, 15. A. III, 10. Scales 23–125–43. Gill-rakers 3+10, short, barely longer than pupil. Eye 6½ in head; snout 3⅔; maxillary 2⅓. P. 1⅘. V. 2. 4th D. spine 3. Longest soft ray 2¾. A. 2⅛. C. 1¾.

Body moderately elongate, compressed; profile anteriorly a little convex, depressed before eye. Mouth moderate, the lower jaw longer. Nostrils well separated, subequal. Preopercle scarcely notched, the teeth at angle scarcely enlarged. First dorsal low, scarcely notched, the fourth spine not elongate. Soft dorsal low and rounded. Caudal truncate or very slightly rounded. Anal high but not rounded, its posterior border not incised. Pectorals and ventrals moderate.

Color olive gray, with very obscure marks of darker olive in the form of diffuse dark clouds; lower parts pale olive. Pectorals yellowish green; other fins blackish, the soft dorsal and caudal narrowly edged with whitish. Sides of head with wavy blackish streaks; a black mustache behind maxillary; lower side of head clouded, lower lip greenish.

Several specimens, each about a foot long.

An adult specimen of the same species shows the following characters:

Head 2⅔ in length; depth 3¾. Dorsal XI, 17; anal III, 11. Scales 120. Snout 3¼ in head; maxillary 2;

eye $7\frac{1}{3}$; pectoral $1\frac{3}{4}$; ventral $2\frac{1}{5}$; anal ray $2\frac{1}{5}$; caudal $1\frac{4}{5}$. Longest dorsal spine $2\frac{3}{5}$; longest dorsal ray $2\frac{3}{5}$. Gill-rakers short 3+8, not longer than pupil.

Body robust, rather elongate. Head large, low, its profile not steep, a depression before eye. Canines in both jaws, rather strong. Nostrils well separated, the posterior scarcely longer than anterior. Lower jaw strongly projecting. Preopercle slightly notched, the angle little salient. Dorsal rather deeply notched, the fourth spine not especially elevated. Second dorsal high and long, with rounded angles. Caudal slightly lunate. Anal high, but not falcate, its middle rays much elevated but not exserted; both outlines nearly straight.

Color olive almost black above, with four series of oblong blackish, cloud-like blotches along sides; these irregular in size, the largest twice length of eye. Fins all dark, clouded with darker. A little dark red on pectoral and on the lower edge of anal and caudal. Pale edge on dorsal, anal, and caudal very slight; none on pectoral. Cheeks and opercles clouded, the cheeks faintly reticulate, the lower parts grayish, faintly mottled. Inside of mouth pale.

114. **Mycteroperca xenarcha** Jordan.

One specimen, 22 inches long, from the Venados Islands.

Head $2\frac{2}{3}$; depth 3. Dorsal XI, 16. Anal III, 11. Scales 25–110 to 115–50.

Body rather deep and compressed; head compressed, with rather short, sharp snout, which is 4 in head; profile steep and nearly straight. Mouth large, the maxillary reaching scarcely beyond eye, 2 in head. Lower canines small; upper canines (two in number) strong, scarcely directed forward. Eye small, 7 in head. Preorbital narrow, ¾ width of eye. Interorbital area convex, its width 4½ in head. Nostrils small, the posterior scarcely the larger, separated from the anterior by one diameter. Angle of preopercle scarcely salient, but provided with coarser teeth; a small sharp notch above it. Opercular spine flat and divided into about six teeth at the end. Gill-rakers moderate 9+18. Scales moderate, scarcely ctenoid. Dorsal spines low, the outline of the spinous dorsal gently convex, the fourth spine longest, 3 in head. Soft dorsal high, its outline angular, the tenth ray produced, $1\frac{5}{6}$ in head. Anal fin formed as in *Mycteroperca falcata*, its seventh ray produced and falcate, $1\frac{4}{5}$ in head, its posterior outline concave. Caudal subtruncate, the outer rays slightly produced. Pectoral 1¾ in head.

Color plain dark olivaceous, the edges of the fins scarcely darker; no evident markings on body.

115. **Paralabrax maculatofasciatus** (Steindachner). CABRILLA PINTA.

Rather common at Mazatlan. This is one of the very few northern species which extends its range thus far to the southward. It is found in some abundance about San Diego, and its center of distribution is probably between Mazatlan and San Diego, these two places being the limits of its range, so far as now known.

116. **Diplectrum radiale** (Quoy & Gaimard).

This small species is abòut a foot in length and is generally common on the Coast. It is apparently not very abundant at Mazatlan, the few specimens seen by us being all taken in the Astillero.

Much cherry red on head and fins in life, sides salmon color, streaks on head greenish.

117. **Prionodes fasciatus** Jenyns.

Generally common in rocky islands on the Coast. Obtained by Gilbert trom the islands about Mazatlan, whence it was described as *Serranus calopteryx*. Not taken by us.

Serranus bulleri, lately described by Dr. Boulenger from Las Peñas, Jalisco, seems to be identical with *Prionodes fasciatus*.

118. **Rypticus xanti** Gill. Jabon.

This species was found by Gilbert in some abundance at Mazatlan. It was not seen by us.

Family CENTROPOMIDÆ.

119. **Centropomus viridis** Lockington. Robalo.

A common and valued food fish at Mazatlan, where it was also taken by Dr. Gilbert.

This Pacific Coast fish seems to be really a species distinct from *Centropomus undecimalis*, with which it has hitherto been identified. The only differences we find are these: In *Centropomus viridis* the anterior appendages to the air-bladder are two to three times diameter of orbit (in *C. undecimalis* not longer than orbit), and the third anal spine projects beyond second. In *C. undecimalis* the second spine is the longer.

Color in life olivaceous, the sides dull silvery, a very little yellow on ventral, none elsewhere; ventrals not black.

120. **Centropomus nigrescens** Günther. Robalo Prieto.

Rather common: a food fish of some importance, reaching a length of about two feet, less common than *Centropomus viridis*. Recorded from Chiapam by Günther, and from Mazatlan, Panama and Punta Arenas by Gilbert.

121. **Centropomus pedimacula** Poey. ROBALITO, OR CONSTANTINO DE LAS ALETAS PRIETAS. (*Centropomus medius* Günther.)

Rather common, reaching a length of a little more than a foot; found at Chiapam (Günther), San Blas (Nichols) and Punta Arenas (Gilbert).

We find but one difference between the Pacific form called *Centropomus medius* and its Atlantic analogue, *Centropomus pedimacula* Poey. In the Pacific specimens, *Centropomus medius*, the second anal spine is curved and $1\frac{1}{2}$ to $1\frac{3}{8}$ times in head. In *Centropomus pedimacula* it is straightish and longer, $1\frac{1}{4}$ to $1\frac{1}{3}$ in head. This difference is of very doubtful value, and for the present we place *medius* in the synonymy of *pedimacula*.

Color greenish, the sides bright silvery. Ventral pale yellow, black at tip, a little yellow on anal, none elsewhere. Upper fins dusky; dusky on anal behind the spine.

122. **Centropomus robalito** Jordan & Gilbert. CONSTANTINO, OR ROBALITO DE LAS ALETAS AMARILLAS.

Rather common in the estuary and freely ascending the fresh waters, numerous specimens being taken by us in various places in the Rio Presidio. The species was found by Gilbert at Mazatlan and at Panama; it is probably generally common along the coast.

At our request, Dr. Evermann has compared specimens of the Pacific form called *Centropomus robalito* with *Centropomus ensiferus* from Cuba. He is unable to find any differences, and probably the two are identical. *Centropomus armatus* Gill from Panama is, however, distinct from *ensiferus* or *robalito*.

Olivaceous with bluish reflections; sides silvery, brightest above; ventrals bright yellow, not black at tip. Anal more or less bright yellow; upper fins dusky.

Family LUTIANIDÆ.

123. **Hoplopagrus guntheri** Gill. PARGO COCONACO.

This beautiful and most interesting species is very common about Mazatlan in deep water among the islands. It reaches a considerable size, the largest specimen seen by us having a length of 26 inches. There is considerable difference between the young and the old in coloration, the bands so conspicuous disappearing with age. The species has been found in abundance at Cape San Lucas, Altata and Guaymas, but has not been noticed further south.

Adult greenish above, belly coppery pink; head olive, sides with eight cross bands of warm brown, unequally placed; fins dusky olive shaded with pinkish and brown; ventrals black tipped. A dark crescent at base of pectoral.

124. **Lutianus novemfasciatus** Gill. PARGO PRIETO. PARGO MAREÑO.

This species reaches a much larger size than any other members of the genus on the Pacific Coast, those specimens obtained by us with dynamite among the Venados Islands having a weight of about twenty-five pounds. It is a food fish of some importance. It undergoes very considerable changes with age, as the notes below will show. The young are dark in color, the bodies banded and the amount of red very slight. The adult becomes uniformly colored with much red, and with increased age there is a progressive lengthening of the snout and widening of the preorbital.

Description of adult of 30 inches: Head 3: depth 3 ($3\frac{1}{3}$ in young): dorsal X, 14: anal III, 18: scales 6 (4)–50–13: eye $6\frac{1}{3}$ in head: snout $2\frac{1}{2}$: maxillary $2\frac{2}{5}$. Pectoral $1\frac{1}{4}$. Ventral 2. Anal 3: 3d anal spine $5\frac{3}{4}$:

caudal $1\frac{3}{4}$; preorbital $3\frac{5}{8}$ ($4\frac{1}{3}$ in smaller specimens 20 inches long; 5 in those of one foot long).

Body very robust, not much compressed, the back not sharp. Head very large, the mouth very large, reaching middle of eye. Canines very strong, in front of jaw and on sides of lower. Vomerine teeth in a V-shaped patch, not prolonged behind. Gill-rakers 7, very small, the longest less than pupil. Posterior nostril oblong, much longer than anterior. Preopercle slightly notched; 7 or 8 rows of scales on cheeks.

Dorsal deeply notched, rather low. Soft dorsal low and rounded. Anal low and rounded. Pectoral long and pointed. Caudal short, scarcely concave. Anal spines short, graduated. Scales above lateral line not in a parallel series.

Maroon color above, copper red below, becoming salmon color before. Fins blackish, tinged with maroon. Pectoral dull yellow olive, blackish at tip; a blackish cross spot on base of pectoral, growing faint with age. Inside of the mouth salmon. Ventral quite dark, the tips black. Iris salmon color; no blue spots or line below eye.

Young with spinous dorsal edged with black; anal and caudal black; ventrals black tipped. A black crescent on upper part of base of pectoral.

Young of one foot, black with progressively less red and narrow preorbital. Color largely blackish, tinged with copper on belly and lower parts.

The young are called Pargo Negro; the half grown, Pargo Prieto; the adult Pargo Mareño, or Maroon Snapper.

125. **Lutianus argentiventris** (Peters). PARGO AMARILLO.

Very abundant everywhere about Mazatlan, and probably common all the way from Guaymas to Panama. It reaches a weight of about five pounds, and is a food fish of some importance.

Back olivaceous, anterior parts washed with maroon red, bright on sides of head, becoming more orange posteriorly; posterior half of body bright yellow; some pale streaks on scales. Pectoral light orange red. Other fins mostly bright yellow. A row of round blue spots below eye. Belly silvery, slightly washed with red; inside of mouth white; iris white.

126. **Lutianus colorado** Jordan & Gilbert. PARGO COLORADO.

This large, handsomely colored species, is one of the staple food fishes at Mazatlan, being brought into the market every day, both from the estuary and from the deep water about the islands. It reaches a weight of about ten pounds. Thus far it has been recorded only from Mazatlan and Punta Arenas, all the known specimens having been collected by Dr. Gilbert.

127. **Lutianus guttatus** (Steindachner). PARGO FLAMENCO.

This small, beautifully colored species, is generally common about Mazatlan, and probably in all the localities along the coast; it is found both in the estuary and in the neighborhood of the rocks. It rarely reaches a pound in weight.

Light olivaceous above, the markings bronze olive; sides pale crimson, the marks more yellow. Belly golden yellow. Scarlet on iris, yellow about eye: first dorsal reddish, second with reddish brown markings; caudal deep rich red: lower fins golden: pectoral nearly colorless: side of head pink with golden stripes.

128. Lutianus aratus (Günther). PARGO RAIZERO.

This beautiful species is not very abundant about Mazatlan, specimens being only occasionally taken. It rarely reaches five pounds in weight. It is generally distributed along the coast, having been recorded from Punta Arenas by Gilbert, and from Chiapam and Panama by Günther.

Dark green, the dark stripes on sides dark brown, the interspaces yellowish white; belly coppery red; some bluish on cheek; pectoral maroon red; ventrals salmon red, the first ray white; anal creamy red; caudal dark red, blackish towards tip; dorsals dusky; throat silvery.

129. Rabirubia inermis (Peters). Plate xxxix.

The original type of this species in the museum at Berlin was said to have been brought from Mazatlan. A single specimen from Panama is in the museum of Stanford University. In this species the supra-occipital crest is continued forward on the head to the ethmoid region, as in the genus *Ocyurus*. This character widely separates *inermis* from the genus *Lutianus*. The genus *Rabirubia* Jordan & Fesler, of which it is the type, is separated from *Ocyurus* chiefly by the small number of the gill-rakers.

Family HÆMULIDÆ.

130. Hæmulon sexfasciatum Gill. RONCADOR ALMEJERO.

This species reaches a larger size than any other of the group, none that were found by us being less than two feet in length. It is not very common, living mainly about the islands. It was obtained by Peters and Gilbert at Mazatlan, and ranges from Cape San Lucas to Panama.

131. **Hæmulon scudderi** Gill. RONCADOR PRIETO.

This species reaches a length of about fifteen inches, and is very common at Mazatlan; more so than any other member of the group. Large specimens were taken by dynamite in the deep water about the Venados, and the young are rather common in the estuary. The species seems to have indifferently eleven or twelve dorsal spines, and there is a greater variation than usual in the form of the body and in the shade of coloration. There seems to be no doubt, however, that all the forms usually referred to this species belong to a single one. The species is found from Cape San Lucas to Panama.

Back bright yellow-olive to opposite front of soft dorsal, the posterior half, more or less abruptly, steel blue black. The vertical fins all blackish; in some the whole back is greenish, in others only half; lower parts all gray; most of the large ones show no traces of spots on scales, some show a few spots; fins silvery, with golden above and below; mouth red within; black under preopercle.

132. **Hæmulon steindachneri** (Jordan & Gilbert). RONCADOR RAIADO.

This small species, not reaching a length of more than eight inches, and too small to be regarded as a food fish, is very abundant in the harbor at Mazatlan, especially about the wharf and in the quiet waters in the estuary. It is generally distributed along the coast from Guaymas to Panama. It seems to be indistinguishable from a species found along the Brazilian coast and north to St. Lucia. For this species we have formerly taken the name of *Hæmulon schranki* Agassiz. This identification is probably an error. *Hæmulon schranki* is probably based on a faded example of *Hæmulon melanurum*. Apparently the appropriate name of *Hæmulon steindachneri* should stand.

Fins all golden yellow; body dark bronze, with rows of pearly blue spots; a large black blotch at base of caudal.

133. **Lythrulon flaviguttatum** (Gill). (*Hæmulon margaritiferum* Günther.)

This species is not very common in the estuary at Mazatlan, a few specimens having been taken by Dr. Gilbert. It is widely distributed along the coast from Guaymas to Panama.

134. **Lythrulon opalescens** Jordan & Starks, n. sp. Plate xl.

Rather common in the estuary at Mazatlan, not yet noticed elsewhere; all the specimens of *Lythrulon* from other localities examined by us being referable to *Lythrulon flaviguttatum*.

Head 3½; depth 2⅔; dorsal XII, 16; anal III, 9; snout 3⅔ in head; maxillary reaching slightly past front of pupil, 2½ in head; orbit $2\frac{5}{8}$; interorbital 3⅔; longest dorsal spine 2; longest dorsal ray 4; second anal spine 2½; pectoral $1\frac{1}{18}$; ventrals 1½; scales 7–54–13.

Body deep, compressed, the back well elevated, the dorsal outline nearly uniformly curved from tip of snout to caudal peduncle; ventral outline curved from chin to breast, thence straight to anal spine, and slanting obliquely upwards to caudal peduncle.

Snout small and pointed; mouth small and oblique, the lower jaw slightly projecting; teeth all small, the outer scarcely enlarged; preopercle finely serrate, the posterior limb somewhat concave, the angle broadly rounded.

Gill-rakers short and slender, about half the diameter of pupil, 8+15; scales above lateral line arranged in oblique series; tip of snout, chin and maxillary naked; scales on head small and crowded; soft fins scaled.

Pectoral reaching to vent; ventrals reaching half way to second anal ray; second anal spine a little longer and stronger than third; upper lobe of caudal the longer, about equal to head.

Color as in *Lythrulon flaviguttatum*, in spirits, dark steel gray; a small very distinct pale spot on each scale of back and sides, surrounded by darker. This spot is, in spirits, light yellowish; in life of a pearly blue. Head plain; a small dusky blotch under angle of preopercle. Fins plain bright yellow in life. Young with a large black blotch at base of caudal, as in *Hæmulon steindachneri* and *Orthostœchus maculicauda*, and without the dusky horizontal streaks seen in most of the other species.

This species differs from *Lythrulon flaviguttatum* in having fewer gill-rakers, the depth and arch of the back greater.

Described from a specimen (No. 2003, L. S. Jr. Univ. Mus.) 9 inches long. Two others were obtained.

135. Orthostœchus maculicauda Gill.

This small species was not found at Mazatlan either by Dr. Gilbert or by the Hopkins expedition. Specimens from Mazatlan and from Acapulco have been recorded by Steindachner. It was obtained by Xantus at Cape San Lucas and Colima, and by Dr. Gilbert at La Paz and Panama.

136. Anisotremus interruptus (Gill). Mojarron.

This large species occurs in great abundance about the islands near Mazatlan. [illegible] specimens, the largest [illegible] by dynamite. [illegible] dis- [illegible] obtained by [illegible] before.

yellow on posterior half; the back tinged with brassy olive, which grows darker behind, the posterior parts pretty distinctly yellow; fin spines gray, the soft fins olive, the fins growing dusky at tip; scales on back and sides each with a distinct black spot; iris yellow; scales above lateral line much enlarged, 4 in number, 7 in an oblique series; 52 pores.

The generally larger size of the scales above the lateral line may possibly separate this species from the common Atlantic form, *Anisotremus surinamensis.*

137. **Anisotremus cæsius** (Jordan & Gilbert).

This species is known only from two or three specimens obtained by Dr. Gilbert in 1881 from Mazatlan. It was not seen by us, and is doubtless rare.

138. **Anisotremus dovii** (Günther).

This species was found by Gilbert at Mazatlan and Panama, but no specimens were obtained by us.

139. **Anisotremus tæniatus** Gill. CATALINA.

This species is rather common about the islands. It reaches a length of about 18 inches, and in life is very brilliant in color. It is seldom found in shallow water. It ranges from Magdalena Bay to Panama.

140. **Pomadasis macracanthus** (Günther). BURRO.

This species is extremely common everywhere about Mazatlan. It is a food fish of some importance, but the flesh is rather coarse. It reaches a length of about 18 inches. When taken from the water it makes a loud and singular noise extremely similiar to the noise made by the donkey or burro, from which this species receives its common name. Every species of the genus makes some noise, but in no case is it so loud as in this one.

141. **Pomadasis branicki** (Steindachner).

This small species, rarely exceeding six inches in length, was found by us in some abundance in the Astillero at Mazatlan. It was obtained by Gilbert both at Mazatlan and Panama. Steindachner described it from Tumbez on the coast of Peru.

142. **Pomadasis panamensis** (Steindachner).

This species is generally common along the west coast, but it was not seen by us. Dr. Gilbert found it both at Mazatlan and Panama.

143. **Pomadasis axillaris** (Steindachner). Burro Blanco.

This species reaches the length of about a foot, and is occasionally taken at Panama; a single specimen being found by us at Mazatlan. Both Steindachner and Gilbert also record it from Mazatlan, and a single specimen has been found by us in the collection of Dr. Streets from the coast of Lower California. It has not been noticed from any other locality.

144. **Pomadasis nitidus** (Steindachner).

This species was found at Mazatlan by both Steindachner and Gilbert, but it was not seen by us. Gilbert records it also from Panama.

145. **Pomadasis leuciscus** (Günther). Burrito.

This small species seldom exceeds a length of six inches, and is generally common in the bay at Mazatlan, and on sandy bottoms where the water is shallow. We found large variations in the depth of body, in the width of the preorbital and in the length of the anal spines, but in no case have we been able to make these variations agree exactly with any of the differences by which we have hitherto distinguished *Pomadasis elongatus* (Steindachner) from *Pomadasis leuciscus* (Günther). We have reached the conclusion that all of these forms belong to one species, and that *elongatus*, as we have understood it, cannot be maintained as a separate species. The two supposed forms have been recorded from various places between Guaymas and Panama. The name *elongatus* was first applied to a Peruvian specimen, which is possibly different from *leuciscus*, as we have seen none exactly like Steindachner's figure.

The young show yellowish shades on fins. Second dorsal mottled with blackish; a diffuse dusky blotch on opercular angle, and evident dark streaks, three or four, along middle of sides.

146. **Orthopristis chalceus** (Günther).

This species is generally common along the coast from Guaymas to Panama. It was obtained by Steindachner and Gilbert at Mazatlan, but no specimens were secured by us.

147. **Isaciella brevipinnis** (Steindachner).

The original type of this species was obtained by Dr. Steindachner at Mazatlan. A specimen from Panama, now in the museum of Yale University, was obtained by Prof. Bradley. The species seems to be rare, and no specimens were secured by us.

148. **Microlepidotus inornatus** Gill. JOPATON.

Five specimens of this rare species, the largest about fifteen inches in length, were obtained by us with dynamite off the shore of the southernmost of the three Venados Islands.

In life, steel-blue, with stripes of bright bronze; upper fins with golden; caudal partly dusky; preorbital with vertically oblong spots.

Family SPARIDÆ.

149. Calamus brachysomus (Lockington). Mojarra Garabata.

This species is very common about Mazatlan, being a food fish of some importance and reaching a length of about fifteen inches. It was also obtained by Dr. Gilbert. Its range southward is not certain, but it is generally common in the Gulf of California.

Family KYPHOSIDÆ.

150. Kyphosus analogus (Gill). Salema.

This beautiful species is rather common about Mazatlan, both in the estuary and in deep water in the neighborhood of the islands. It was not found by Dr. Gilbert, and its range along the coast is not definitely distinguished from that of the following species, the two having been recorded as identical by authors who had seen but one. They were first properly distinguished by Jenkins and Evermann, who obtained both at Guaymas. The marked difference in color, however, does not appear in the description of Jenkins and Evermann, which was drawn from specimens preserved in alcohol.

Head 4; depth $2\frac{1}{4}$; dorsal XI, 14; anal III, 12; eye $4\frac{1}{2}$ in head; snout 3; maxillary $3\frac{1}{3}$; pectoral $1\frac{3}{4}$, equal to ventrals; longest ray of soft dorsal $3\frac{1}{2}$; longest dorsal spine $2\frac{1}{3}$; upper lobe of caudal as long as head.

Body compressed, elliptical; profile in some specimens evenly curved from tip of snout to dorsal, in others slightly produced before eyes and concave over snout.

Mouth small, horizontal; jaws equal; teeth in a single series, from 22 to 28 in each jaw; maxillary extending to the vertical from the front of eye. Snout, lower jaw and preorbital naked, head elsewhere with scales; 12 to 15 rows of scales on opercle; scales on body much crowded anteriorly; scales 13–76–20; all the fins, with the exception of spinous dorsal, entirely scaled.

Tip of pectoral sharply rounded; front of anal not greatly elevated, its longest ray 3 in base of fin, which is about equal to head; spinous dorsal higher than soft dorsal; upper lobe of caudal the longer.

Color, steel blue, brighter than in *elegans*, with bronze streaks along the edges or rows of scales, much brighter than in *elegans*. A broader gray streak bordered with bronze at base of soft dorsal. A large brassy spot in the axil, extending along shoulder girdle; a deep bronze stripe through eye, another back from angle of mouth; the two separated by steel blue; fins all blue black, with some bronze, especially on pectoral. Body more elongate than in *elegans;* the form more elliptical; the mouth less blunt, with fewer teeth; the scales smaller and more crowded anteriorly; the fins lower, especially the anal. Well separated from *Kyphosus elegans*, living chiefly in the rocks outside; rare in the bay. Largest specimen eighteen inches long.

151. **Kyphosus elegans** (Peters). CHOPA.

This species is rather common about Mazatlan, especially in the sluggish waters of the Astillero. Like the preceding, it reaches a length of about fifteen inches.

Head 3⅔; depth 2; dorsal XI, 12; anal III, 11; eye 4 in head; snout 3¼; maxillary $3\frac{1}{8}$; pectoral $1\frac{3}{8}$, equals ventral; longest ray of soft dorsal 2½; longest dorsal spine 2½; longest anal ray 2; upper lobe of caudal equals head.

Body ovate, compressed; profile rounded, slightly produced before eyes; concave over snout in some specimens, straight in others; a gentle curve from eyes to dorsal. Mouth small, horizontal, the jaws equal; teeth in a single series, about 36 in each jaw; maxillary extending to the vertical from anterior edge of orbit; snout, lower jaw and preorbital naked, head everywhere else scaled; opercles with 8 or 9 rows of scales: scales on body large, somewhat crowded anteriorly; scales 11–63–17; all the fins, except spinous dorsal, with scales to their edges, those on caudal exceedingly small.

Tip of pectoral sharply rounded, not reaching to tips of ventrals; ventral spine half as long as soft rays; anal spines short and stout, graduated; anal elevated in front and higher than soft dorsal; middle spines of dorsal the longest, about equal to highest rays of soft dorsal; upper lobe of caudal the longer.

Color grayish black, with paler centers to the scales; sides with large faint diffuse yellowish white spots; a little bluish and yellowish on sides of head; a yellow streak below lower part of eye. Vertebræ 9+16 or 10+15.

Family SCIÆNIDÆ.

152. **Cynoscion reticulatus** (Günther). Corvina.

Generally common on the sandy bottoms about Mazatlan. An excellent food fish, very often brought into the markets, and reaching a length of nearly 3 feet. It was found by Dr. Gilbert at Mazatlan and is common south to Panama.

Caudal fin yellowish orange in life; inside of mouth deep orange yellow.

153. **Cynoscion xanthulum** Jordan & Gilbert. Corvina Aletas Amarillas.

Found in company with *Cynoscion reticulatus*, but rather less abundant and perhaps reaching a smaller size. It is also a food fish. It has thus far been recorded only from Mazatlan, where the original types were taken by Dr. Gilbert.

154. Larimus argenteus (Gill).

One large specimen obtained; also found in the Gulf of California and southward on sandy shores to Panama.

155. Larimus breviceps Cuvier & Valenciennes.

Specimens of this species were obtained by Dr. Gilbert at Mazatlan, Punta Arenas and Panama. None were seen by us.

156. Corvula macrops (Steindachner). VACUOCUA.

One fine specimen from the Astillero at Mazatlan.

Head 3½; depth 3; dorsal XI, I, 25; anal II, 9; eye 3½ in head; snout 4¾; maxillary 2⅙; longest dorsal spine 1¾; longest dorsal ray 2¼; second anal spine 2⅓; ventrals 1½; pectoral $1\frac{2}{5}$; caudal fin 1½.

Body oblong, moderately compressed, not much elevated; dorsal outline uniform from tip of snout to caudal peduncle; ventral outline rounded from chin to breast, then straight to anal spine, then slanting obliquely upward to caudal peduncle.

Snout blunt, shorter than large eye; upper jaw slightly projecting, teeth small and sharp, in one or two irregular series in lower jaw, in several series in upper jaw, the outer row slightly enlarged; maxillary extending to posterior edge of pupil; chin with four large pores; edge of preopercle covered with skin, which is serrated on the edge.

Gill-rakers slender, 9+13; scales ctenoid on the body, cycloid on the head; scales 8–56–11.

Spinous dorsal a little higher than soft dorsal; first dorsal spine very short, second about 5 times longer, third twice as long as second, third, fourth, fifth and sixth subequal, the others rapidly shorter; first anal spine very small, the second many times longer and stouter, but shorter than soft rays; ventrals inserted behind pectorals and reaching beyond them; caudal truncate.

131. **Hæmulon scudderi** Gill. RONCADOR PRIETO.

This species reaches a length of about fifteen inches, and is very common at Mazatlan, more so than any other member of the group. Large specimens were taken by dynamite in the deep water about the Venados, and the young are rather common in the estuary. The species seems to have indifferently eleven or twelve dorsal spines, and there is a greater variation than usual in the form of the body and in the shade of coloration. There seems to be no doubt, however, that all the forms usually referred to this species belong to a single one. The species is found from Cape San Lucas to Panama.

Back bright yellow-olive to opposite front of soft dorsal, the posterior half, more or less abruptly, steel blue black. The vertical fins all blackish; in some the whole back is greenish, in others only half; lower parts all gray; most of the large ones show no traces of spots on scales, some show a few spots; fins silvery, with golden above and below; mouth red within; black under preopercle.

132. **Hæmulon steindachneri** (Jordan & Gilbert). RONCADOR RAIADO.

This small species, not reaching a length of more than eight inches, and too small to be regarded as a food fish, is very abundant in the harbor at Mazatlan, especially about the wharf and in the quiet waters in the estuary. It is generally distributed along the coast from Guaymas to Panama. It seems to be indistinguishable from a species found along the Brazilian coast and north to St. Lucia. For this species we have formerly taken the name of *Hæmulon schranki* Agassiz. This identification is probably an error. *Hæmulon schranki* is probably based on a faded example of *Hæmulon melanurum*. Apparently the appropriate name of *Hæmulon steindachneri* should stand.

Fins all golden yellow; body dark bronze, with rows of pearly blue spots; a large black blotch at base of caudal.

133. Lythrulon flaviguttatum (Gill). (*Hæmulon margaritiferum* Günther.)

This species is not very common in the estuary at Mazatlan, a few specimens having been taken by Dr. Gilbert. It is widely distributed along the coast from Guaymas to Panama.

134. Lythrulon opalescens Jordan & Starks, n. sp. Plate xl.

Rather common in the estuary at Mazatlan, not yet noticed elsewhere; all the specimens of *Lythrulon* from other localities examined by us being referable to *Lythrulon flaviguttatum*.

Head $3\frac{1}{2}$; depth $2\frac{2}{3}$; dorsal XII, 16; anal III, 9; snout $3\frac{2}{3}$ in head; maxillary reaching slightly past front of pupil, $2\frac{1}{2}$ in head; orbit $2\frac{5}{8}$; interorbital $3\frac{2}{3}$; longest dorsal spine 2; longest dorsal ray 4; second anal spine $2\frac{1}{2}$; pectoral $1\frac{1}{18}$; ventrals $1\frac{1}{2}$; scales 7–54–13.

Body deep, compressed, the back well elevated, the dorsal outline nearly uniformly curved from tip of snout to caudal peduncle; ventral outline curved from chin to breast, thence straight to anal spine, and slanting obliquely upwards to caudal peduncle.

Snout small and pointed; mouth small and oblique, the lower jaw slightly projecting; teeth all small, the outer scarcely enlarged; preopercle finely serrate, the posterior limb somewhat concave, the angle broadly rounded.

Gill-rakers short and slender, about half the diameter of pupil, 8+15; scales above lateral line arranged in oblique series; tip of snout, chin and maxillary naked; scales on head small and crowded; soft fins scaled.

long and spear-shaped, very much more slender in proportion to its length, not hollow and not receiving any of the air bladder. This structure is seen in *Gerres cinereus* (Walbaum), in *Gerres peruvianus* Cuvier & Valenciennes, and in *Gerres lineatus* Humboldt, as also in several West Indian species.

Eucinostomus californiensis is generally common along the west coast of Mexico, from Guaymas to Panama. It is probably, however, not found in the West Indies, the closely related *Eucinostomus harengulus* being apparently a different species. The specimens called *californiensis* by Gill, having the premaxillary groove semi-oval or ∩-shaped, seem to represent the adult of this species. Those called *gracilis*, with the premaxillary groove linear, are the young or half-grown. Still others, especially adults, have the premaxillary groove round, forming a pit, and every intermediate character may be found.

At first we thought it possible to separate *californiensis* and *gracilis* as distinct species. The careful re-examination of some 200 specimens leaves us wholly unable to separate them, as all grades of variation occur. Apparently the premaxillary groove is linear in the young, growing broader with age, but the changes very irregular. The name *Eucinostomus californiensis* has priority over *E. gracilis*.

NOTE.—The genus *Gerres* was established by Cuvier in the second edition of the Règne Animal, the name being based on seven species as enumerated by him, *rhombeus*, *oyena*, *aprion*, *poieti*, *lineatus*, *argyreus* and *filamentosus*. One of these species must, therefore, be chosen as the type of *Gerres*. In 1842, Ranzani established the genus *Diapterus* on *auratus*, a species closely related to *rhombeus*, or rather to the allied *olisthostoma*. In 1850, the name *Catochænum* was proposed by Cantor as a substitute for *Gerres*, regarded as preoccupied by the earlier name *Gerris*, applied by Fabricius to a genus of insects. The name *Catochænum* can only be used if *Gerres* is regarded as ineligible. By the rules followed by us, *Gerres* must be retained, being spelled differently from *Gerris*. In different publications of Poey, *plumieri* is made the type of *Gerres*, although it is not one of Cuvier's original species. Bleeker substitutes *Diapterus* for *Gerres* and *Catochænum*, specifying *plumieri* as its type, while Gill and Poey have used the name *Diapterus* for the allies of *gula*, to which the name *Eucinostomus* had been applied in 1855 by Baird and Girard. Although *plumieri* cannot be made the type of *Gerres*, it seems to us that the cognate species *lineatus* can be so regarded. If this view is adopted, the restricted *Gerres* of the present paper would correspond exactly with the restricted *Gerres* of Poey and Gill. This fact certainly justifies us in choosing *lineatus* as the type of the genus.

There can be no doubt of the generic value of *Eucinostomus (gula)* and of *Ulæma* Jordan & Evermann MS. *(lefroyi)*, as distinguished from *Gerres*. Of the other groups represented in American waters, *Xystæma* Jordan & Evermann MS. *(cinereus)* seems to be a valid genus, while *Diapterus (auratus)* should stand rather as a subgenus of *Gerres*. *Diapterus* differs from *Gerres* chiefly in the entire preorbital. *Xystæma* has the preopercle as well as preorbital entire, while *Ulæma* has the second interhæmal very short, and the two spines of the anal are themselves scarcely enlarged.

Moharra Poey *(rhombeus)* differs from *Diapterus* only in the presence of two anal spines instead of three, a character of low importance, as the relation of the species included in the two groups is very close.

The exotic genera of this group have not been studied by us.

The specimens recorded by Eigenmann from San Diego Bay as *Gerres cinereus* var. (Amer. Nat., 1891, 156) seem to be *Eucinostomus californiensis*.

166. **Xystæma cinereum** (Walbaum). MOJARRA BLANCA.

Very abundant at Mazatlan, being one of the staple food fishes, and reaching a length of nearly two feet; its flesh is of an excellent quality. The species was found by Dr. Gilbert at Mazatlan and Panama, and seems to be generally common along the coast. Like the rest of the genus, it occurs in shallow water on sandy bottoms, away from the surf.

167. **Gerres peruvianus** Cuvier & Valenciennes. MOJARRA DE LAS ALETAS AMARILLAS.

This small species is abundant at Mazatlan, although less common than *Eucinostomus californiensis*, and *Xystæma cinereum*. It rarely exceeds six inches in length.

Gerres brevirostris Sauvage, from Rio Guayas, near Guayaquil, is not evidently different from this species.

168. **Gerres lineatus** (Humboldt). MOJARRA CHINA. (*Gerres axillaris* Günther).

Rather common at Mazatlan, with the preceding, but reaching a rather larger size, from eight to twelve inches, and frequently used as food. It was found by Dr. Gilbert at Mazatlan, and has been recorded from Acapulco by Humboldt and Bradley, from San Blas by Nichols, and from Chiapam by Günther.

Family CIRRHITIDÆ.

169. **Cirrhites betaurus** Gill.

The young of this species, from two to six inches in length, are very abundant in rock pools about Mazatlan, where numerous specimens were obtained by us, as well as by Dr. Gilbert. These small specimens are identical with those obtained by Xantus at Cape San Lucas, the types of *Cirrhites betaurus*. It has been supposed that these are the young of *Cirrhites rivulatus* Valenciennes, abundant about the Galapagos and Revillagigedos, as no differences except those of color appear. The color differences are, however, strongly marked, and we are disposed to let *Cirrhites betaurus* stand provisionally as a distinct species. The coloration of *betaurus* has been well described by Dr. Gill; that of *rivulatus* is well figured by Dr. Günther.

First dorsal fin bright orange red in life; second reddish; cross bands on body black.

Family CICHLIDÆ.

170. **Heros beani** Jordan. MOJARRA VERDE.

Common in the deeper and more quiet places in the Rio Presidio, especially just below the village of Presidio. It reaches a length of about eight inches, and is occasionally taken by the hook, its habits being very similar to those of the abundant sun fishes as seen in the more northern waters.

Adult light olive, banded with darker; black spots on each scale. First dorsal edged with dark red, the two black blotches and black bars obsolete. Young with the bars distinct; no blue, yellow or red in life.

Family POMACENTRIDÆ.

171. **Eupomacentrus rectifrænum** (Gill). PESCADO AZUL. (*Pomacentrus analigutta* Gill.)

This beautiful fish is very abundant in the rock pools about Mazatlan. It is excessively wary and hard to catch. Great changes in coloration, due to age, have been noticed by Dr. Günther and others. The chief peculiarity is in the greater uniformity in coloration of the adult, in which the blue shades become obscure, and the ocelli, so conspicuous in the young, are more or less lost.

This species is exceedingly close to *Eupomacentrus fuscus* (Cuvier & Valenciennes), a species found on the Brazilian coast. Comparing specimens from Bahia with ours from Mazatlan, we note that in *E. rectifrænum* the blue markings persist longer and that the scales on the head are smaller, more crowded and more mixed with small scales in *E. rectifrænum* than in *Eupomacentrus fuscus*.

Head $3\frac{1}{3}$; depth 2; D. XII, 13; A. II, 11; scales 3–28–9; eye 4 in head; snout $2\frac{3}{5}$; D. lobe $1\frac{3}{5}$; C. upper lobe $1\frac{2}{5}$; V. $1\frac{1}{3}$; P. $1\frac{1}{8}$.

Preorbital and preopercle strongly serrate. Teeth firm, flattened, not notched. Lateral line ending under ninth dorsal ray. Caudal lunate, the upper lobe the longer. Dorsal and anal rounded, ventral filamentous. Gill-rakers short, slender, weak, numerous.

Color of adult ($5\frac{1}{2}$ inches) nearly uniform blackish olive, darker on head, back and fins, paler on pectoral and on axil, where is a yellowish area below the small axillary spot.

The coloration of the young and partly grown has been well described by Dr. Gill. Dr. Gill's last account (Proc. Ac. Nat. Sci. Phila., 1863) of this and related species is most excellent. The only error of importance contained in it is the failure to examine the teeth of "*Pomataprion*" *bairdii* and *dorsalis*. *Pomataprion* is identical with *Microspathodon*.

172. **Eupomacentrus flavilatus** (Gill). PESCADO AZUL DE DOS COLORES. Plate xlii.

This little fish is equally abundant with the preceding in rock pools. It seems to reach a smaller size. The differences between the two are comparatively slight but very persistent, and we believe that the two species are fully distinct from each other. In life *Eupomacentrus flavilatus* is the most beautiful fish found on the coast of Mexico, showing a most intense shade in the blue of its back and the orange of its sides. Both this species and the preceding were found at Cape San Lucas, but only *Eupomacentrus rectifrænum* has been taken at Panama.

An irregular line from snout below eye to soft dorsal divides the fish into two parts; below this line all is brilliant yellow with an orange shade, deepest on anal; above all is the brightest sky blue. Scales darker, but all edged with sky blue, six sky blue stripes on upper part of head. An indigo spot on base of first soft dorsal and last dorsal spines extending on back, this surrounded by a ring of sky blue; a similar smaller ocellated spot on back of caudal peduncle.

173. **Abudefduf* saxatilis** (Linnæus).

Common in rock pools about Mazatlan, where it was obtained in abundance by Dr. Gilbert and by us. The largest specimens were taken by dynamite off the Venados Islands.

Careful comparison of these specimens with others from the West Indies shows no difference whatever. *Glyphisodon troscheli* Gill, the name given to the Pacific Coast form, is therefore fully synonymus with *Abudefduf* (or *Glyphisodon*) *saxatilis*.

In life, bright greenish yellow above with steel blue bands. Dorsal like back; other fins dusky; axillary spot faint.

In alcohol, the color is a slaty brown tinged with red-

* *Abudefduf* Forskål seems to be identical with *Glyphisodon* and is entitled to priority, notwithstanding its barbarous form.

dish brown below, showing faint dark cross bars, with no bright color anywhere, the yellowish green of the back being last to fade; behind the pectoral each scale has a white spot, these form white lines that run back to a little past the tip of pectoral. All fins dark except pectoral, which is colorless.

174. **Abudefduf declivifrons** (Gill).

This species occurs in rock pools in abundance everywhere about Maxatlan, in company with *Abudefduf saxatilis*, from which its duller color readily distinguishes it.

In life, dusky brownish with many pale spots on edge of scales; these vary a good deal; cross bands blackish; no bright colors. Black spot at base of pectoral conspicuous, a good mark, varying in size, larger in older specimens.

175. **Microspathodon bairdii** (Gill). Plate xliii.

Numerous small specimens taken in the rock pools in company with *Eupomacentrus flavilatus*, a species which the present one closely resembles in color, and which scarcely excels it in brilliancy. This species has been well described by Dr. Gill. It seems to reach only a small size, none of ours being more than two inches long.

It differs from the other species of *Microspathodon* in its low fins and in color. The latter may be a matter of age only, but this does not seem likely, as the young of *Microspathodon dorsalis* (called by Dr. Gill *quadrigutta*) has essentially the coloration of the adult. Apparently four species of *Microspathodon* exist on the west coast of Mexico, but it is possible that all are forms of one protean species, for which the earliest specific name is *dorsalis*.

Head 3; depth 2; dorsal XII, 16; anal II, 13; eye 2⅔ in head; pectoral 1¼; anal ⅛ longer than head; soft dorsal and anal lobes equal 1⅓ in head; caudal lobe 1¼ in head.

Body compressed, ovate; profile convex; mouth wide, lower jaw included; teeth in a single row and movable; gill-rakers small and numerous; head entirely scaled; scales on body large 3–30–9; scales running well up on fins; lateral line high, ending under last dorsal ray.

Color: Body divided into two parts by a line from the opercular flap to posterior end of soft dorsal, below this line it is rich, bright yellow, above it is sky blue, darker on head, with brilliant sky blue spots; a chain of these spots following the suborbitals below eye; a spot at angle of mouth, two converging lines of spots more or less run together from tip of snout to upper edge of orbit, each scale on nape with a spot and a few scattering spots on opercle; scales on upper part of body edged with dark; a dark spot on caudal peduncle anteriorly edged with sky blue; fins all more or less dusky except anal and ventrals, which are white and edged with black.

176. **Microspathodon dorsalis** (Gill). (*Pomacentrus quadrigutta* Gill.)

A single specimen 4 inches in length was obtained in a rock pool on the Peninsula called Vijía, by Mr. George B. Culver.

This specimen corresponds almost perfectly to Dr. Gill's account of *Pomatoprion dorsalis*. A smaller specimen entirely similar was also obtained. The distinctions between this species and *Microspathodon bairdii* are constant though slight.

Head 3; depth $1\frac{5}{8}$; D. XII, 16; A. II, 12; scales 3–28–10; eye $2\frac{1}{2}$ in head; snout 4; D. lobe 1; C. lobe equals head; P. $1\frac{1}{6}$; V. equals head.

Body compressed, the profile rounded, depressed before eye so that snout projects. Gill-rakers numerous, very short, slender, close set. Preorbital deep. Preorbital and preopercle entire. Teeth in a single row, movable.

Family LABRIDÆ.

178. Harpe diplotænia Gill.

A single young female specimen was obtained by us at Mazatlan. This species is rare in collections, but is apparently not uncommon around the rocky islands. It has been recorded from Cape San Lucas by Xantus, and numerous specimens from the Revillagigedos have been taken by Dr. Gilbert. The form called *Harpe pectoralis* Gill is the male of the same species of which *Harpe diplotænia* Gill is the female.

179. Pseudojulis notospilus Günther.

This small species is common in rock pools about Mazatlan, where numerous examples, the largest about six inches long, were obtained by us. It was found in these pools by Gilbert, and has been recorded from Panama by Günther.

Coloration of adult blue green; bar across base of pectoral very bright; no dark spot behind eye; corners and tip of caudal pale, as in young. Each scale of posterior part of body with a small sky blue spot at tip; edges of scales bluish, the base olivaceous. Axil blue, golden behind. Breast and throat pale salmon color, with bluish streaks and shades; cheeks yellowish, snout blue. Young with blue spots more distinct, especially one behind eye. Adult with four dark shades on back extending on dorsal, the largest at front of soft dorsal; blackish spot diffuse, not ocellated. Caudal with faint bluish cross-streaks on faint bronze ground color, the angles broadly whitish; anal bronze with three bluish streaks, the tip pale. Ventrals dusky edged.

Young colored like adult but brighter, a paler olive streak from mouth across opercle above pectoral to base of caudal, this obsolete in adult. Dorsal unlike that of adult. First dorsal bronze with bluish cross-streaks, the large black blotch ocellated with blue and with a patch of bright yellow before and behind it. Interspaces between this and the two other, smaller black spots also bright light yellow.

180. Halichœres dispilus (Günther). Plate xlv.

This beautiful little fish was found to be rather abundant in the branches of the Astillero which cross Isla de las Piedras south of Mazatlan. Unlike most species of the group, it lives on the muddy bottoms, and is abundant about the roots of the mangrove, which border the muddy branches of the Astillero. It reaches a length of about six inches. A few specimens were also obtained in tide pools with sandy bottom.

Head $3\frac{1}{3}$; depth 4; dorsal IX, 11; anal III, 12; eye 6 in head; snout $3\frac{1}{3}$; maxillary $4\frac{1}{4}$; pectoral $1\frac{2}{3}$; anal 3; caudal fin 2.

Body slender and compressed; dorsal and ventral outlines similar; head pointed, the profile slightly convex; mouth small, the jaws equal; teeth in a single row: canines $\frac{2}{4}$ in front of jaws; at the posterior end of the premaxillary is a single strong, sharp tooth, pointing forward, and entirely below the angle of mouth. Lateral line high, following the curve of the back to the eighth dorsal ray, where it curves sharply down through two rows of scales, and then runs straight through middle of caudal peduncle to tail; pores of lateral line simple: scales large 2–27–10; head entirely naked; gill-rakers very small and pointed 6+7. Dorsal spines slender but pungent: caudal slightly rounded, the upper angle slightly acute: ventrals short not filamentous: scales before dorsal in about six rows, not covering middle line.

Length of specimen described, five inches. Number 2904, L. S. Jr. Univ. Mus.

In life olive green, a bright blue streak, narrow and somewhat interrupted, from eye to base of caudal; a broader dark bronze streak just below it, containing a series of small dark spots, mostly arranged in threes, the last one darkest, at base of caudal, just above middle line, these obsolete in adult; below the bronze band, a faint blue streak, then a broad brown one, then a short one, bright sky blue bounding the belly, ending over the middle of anal; belly and throat pearl white. Head cherry red and bronze anteriorly, becoming olive in all specimens behind, mottled with blue; a dark blue edged spot behind eye; a large black spot smaller than eye below fifth dorsal spine, the spot crescent shaped, bordered with yellow behind, mostly on one scale. Iris red. A golden crescent at base of pectoral. Dorsal bright orange, bluish below. Caudal cherry red. Anal bright orange. No spots on fins. Larger specimens deeper in color, the head cherry red, a dark spot bordered with blue behind eye. Pectoral not black. In alcoholic specimens pearly streaks appear on sides of head and behind pectoral.

Found by Dr. Gilbert at Mazatlan. Specimens have also been obtained at Panama by Günther, and at Acapulco by Steindachner.

Our specimens differ somewhat in color from those described by Dr. Günther, especially in the hue of the head and caudal and in the presence of a black spot behind eye. They are, however, probably not specifically distinct.

181. **Thalassoma lucasanum** (Gill).

Obtained by Dr. Gilbert at Mazatlan; not seen by us. Also recorded by Mr. Forrer from Tres Marias, the original types taken by Xantus at Cape San Lucas.

Family SCARIDÆ.

182. **Scarus perrico** Jordan & Gilbert. PERRICO.

This large parrot-fish is rather common about the rocky islands near Mazatlan. A single specimen was obtained by us. The original type was found by Dr. Gilbert at the same locality. The fins of another specimen were found on the beach at La Paz by Mr. James A. Richardson.

Body olive brown. The markings, fins, teeth and spots on head all bright blue green.

Family EPHIPPIDÆ.

183. **Chætodipterus zonatus** (Girard).

Occasionally seen at Mazatlan, several specimens being taken by us in the Astillero. It was found by Dr. Gilbert at Mazatlan and Panama. The original type of the species came from San Diego, where no author subsequent to Girard has seen it. It is probably generally diffused along the coast, although less abundant than the corresponding species (*Chætodipterus faber* L.) is in the Atlantic.

Chætodipterus zonatus agrees with *Chætodipterus faber* in nearly all respects. The chief differences arè that behind the great band from soft dorsal to anal in *Ch. zonatus* there are two other bands; one under middle of soft dorsal, the other at base of caudal, both distinct complete rings; no other bands. The third dorsal spine is not very high, being only about half length of head, and about twice height of the fourth. Dorsal VIII–1, 18; anal II, 16; scales 70. Long rays of soft dorsal and anal ¼ longer than head.

Family CHÆTODONTIDÆ.

184. **Chætodon humeralis** Günther. MUÑECA.

Exceedingly common in the Astillero, especially on rock bottom. It reaches a length of about six inches, and is seldom used as food, although its striking color, which has suggested the name of *Muñeca* or doll, makes it an object of attention.

185. **Pomacanthus zonipectus** (Gill). MOJARRA DE LAS PIEDRAS. (*Pomacanthus crescentalis* Jordan & Gilbert.)

Not uncommon in rocky places about Mazatlan. Two specimens were obtained by us with dynamite about the wreck of a French man-of-war in the Astillero. Smaller specimens, very different in color from the adult, and hence taken by us to be a distinct species (*Pomacanthus crescentalis*), were obtained by Dr. Gilbert at Mazatlan and Panama. The original type of the species was taken at San Salvador by Capt. Dow.

Description of the adult of *Pomacanthus zonipectus:*

Head $3\frac{2}{3}$: depth $1\frac{1}{4}$: D. XI, 23: A. III, 20. Preopercular spine longer than eye, $3\frac{1}{3}$ in head. Last dorsal spine $1\frac{1}{2}$ in head. Longest dorsal ray $\frac{1}{3}$ longer than head, falcate. Anal rounded. Caudal short, truncate, $1\frac{1}{4}$ in head. Pectoral moderate. Ventral very long, $\frac{1}{4}$ longer than head. Preorbital equals maxillary, $1\frac{1}{3}$ in head. Eye $3\frac{1}{2}$ in head. Interopercle with one stoutish spine. Preopercle very finely serrate. A large hump at nape in adult.

Dark gray, blackish posteriorly, most scales with black centers; edges of scales, bright sky blue in life, especially posteriorly; a triangular bronze yellow patch in front of line connecting pectorals with ventrals, then a diffuse blackish bar from front of dorsal along region behind pectorals to ventrals, then a broad curved bar of yellow, obscured by blackish centers of scales; behind this a diffuse blackish area; breast vermiculated with blue and yellowish; a blackish bar covering most of head, behind which the opercles and nape are yellowish; jaws pale bluish; dorsal orange, vermiculate with sky blue, the edge bright sky blue, below which is orange; caudal orange, vermiculated with sky blue, the edge orange, the very margin blackish. Anal blackish, vermiculated with sky blue; pectorals light orange, marked with grayish blue. Ventrals largely blue-black, tipped with orange, the spine bluish.

Family TEUTHIDIDÆ.

186. Teuthis crestonis Jordan & Starks n. sp. BARBERO NEGRO. Plate xlvii.

Common in the Astillero and in rocky places about the islands. Also obtained by Dr. Gilbert in 1881 at Mazatlan and Panama. These specimens having been destroyed by fire, have never been described, and were provisionally and incorrectly referred to the West Indian species *Teuthis tractus (bahianus)*, from which this species differs in a few respects.

Head $3\frac{1}{3}$: depth $1\frac{5}{6}$; D. IX, 26; A. III, 24; snout $1\frac{2}{3}$ in head; eye $3\frac{1}{3}$; pectoral equal to head; caudal $\frac{1}{8}$ longer than head; longest dorsal spine equal longest soft ray, $1\frac{1}{2}$ in head; ventral $1\frac{1}{8}$ in head.

Body deep and compressed, the anterior profile steep, convex before eye; caudal lunate, the upper ray $\frac{1}{3}$ longer than middle one, ventrals very long.

Body slaty brown, mottled with gray but without bands; dorsal with a bluish gray band at base, then a bronze one, forking on soft dorsal inclosing a bluish gray band; five gray bands and four bronze ones on dorsal more or less distinct, especially in young; anal with five bluish gray and five bronze bands more oblique than those on dorsal and hence not continuous the whole length of fin; caudal peduncle black, a whitish yellow cross-band behind spine, faint in adult, the anterior margin vertical, the posterior concave; rest of caudal black. Pectoral yellowish; ventrals dusky, the spine black.

Adult with the pectoral quite yellow; pale band at base of caudal growing faint with age; a blue streak along base of dorsal.

Numerous specimens, the largest about six inches in length, numbered 2899, in the L. S. Jr. Univ. Mus.

187. **Xesurus punctatus** (Gill). COCHINITO. Plate xlvi.

Young specimens very abundant in rock pools about Mazatlan, hitherto known only from Cape San Lucas. It was not found by Dr. Gilbert at Mazatlan. Most of our specimens were secured by the use of the fish poison called gervo. By pouring this liquid into the rock pools at low tide this and several other species were obtained in numbers. This gervo or gerbo is the milky juice of a tree called *hava*, abundant in the forests about Mazatlan, and apparently allied to the *Strychnos nux-vomica*. In rock pools no specimens exceeding two inches in length were found. Several very large specimens were obtained with dynamite about the islands of Creston and Isla Blanca, where the species reaches a length of 16½ inches.

Description of adult:

Head 4; depth 2; dorsal VII, 26; anal II, 23; snout 1⅓ in head; eye 5⅓; pectoral long as head: ventral 1⅔; caudal 1⅙; second dorsal spine 2.

Body deep, compressed, covered with fine velvet. Caudal with three stout compressed blunt spines, with broad bases, the tips turned upward. Some specimens with no other spines; others with many spines, similar in form but much smaller, scattered over posterior half of body; most numerous about the other spines. Gill-rakers extremely small and weak. Caudal evenly lunate. Pectoral not falcate; anterior profile concave before eye then convex, the short conic snout projecting; lower jaw included. Preopercle obliquely placed, its bony edge slightly roughened.

Color in life olive green, slightly paler below, everywhere evenly covered with small round black spots, close-set and not confluent, the largest about equal to nostril. Caudal peduncle and fin abruptly bright yellow, unspotted. Other fins colored like the body and similarly spotted, the spots more sparse, the edges dusky with few spots. Large caudal spines whitish, their bases black; other spines all black.

Among the young two different styles of coloration were noticed, but all probably belong to the same species:

1. Specimens with the caudal yellow are more dusky, the dark spots much smaller and more distinct than in the others. Ground color of back light steel blue gray, lighter below head. Caudal canary yellow, clouded with dark at base, the yellow running forward on caudal peduncle.

2. Specimens with the caudal white have ground color lighter, more milky in general, much more silvery below eye, the silvery forming an irregular triangular patch on breast and opercle; caudal gray and white, black at base, white running forward slightly on caudal peduncle; dark spots on body forming pale reticulations, above lateral line white patches. Body deeper than in yellow-tailed specimens.

Both have the first dorsal and anal black at base, otherwise mostly white: white line bounding the back; dark bar from nape to eye; snout dusky; breast and opercles silvery.

This species is the type of the genus *Xesurus* Jordan & Evermann (MS.), distinguished from *Prionurus* by the armature of the caudal peduncle, as above described.

Family BALISTIDÆ.

188. Balistes polylepis Steindachner. PEZ PUERCO.

Generally common in rocky places on the coast from Magdalena Bay to Panama. Many specimens were obtained by us, the largest of them sixteen inches in length. It was found at Mazatlan also by Gilbert and by Steindachner.

189. Balistes naufragium Jordan & Starks n. sp. PEZ PUERCO DE PIEDRA.

Four specimens obtained with dynamite, about the wreck of a French man-of-war in the Astillero at Mazatlan, in company with *Pomacanthus zonipectus*. The largest of these was fourteen inches in length.

Allied to *Balistes carolinensis*.

Head 3; depth [illegible]; D. III–27; A. 24; scales 50; 12 rows on cheek; snout [illegible] in head; eye 5; 1st D. spine [illegible]; longest ray [illegible]; longest anal ray [illegible]; upper caudal lobe [illegible]; pectoral [illegible].

Body very plump, not strongly compressed; no streaks on cheeks; no spinules on caudal peduncle; a few larger scutes behind gill-openings; groove before eye, slight not naked. Lateral line traceable for most of its length. First dorsal spine very stout, [illegible] moderate. [illegible] rounded. Caudal [illegible] longer than [illegible].

Dark [illegible] of scales

largely pale blue, especially toward the tail; faint traces of numerous dark cross-bands. Fins dusky olive, the pectoral and first dorsal paler, base of pectoral dusky.

Type No. 1656 L. S. Jr. Univ. Mus.

190. Pachynathus capistratus (Shaw). COCHE.

Common in rocky places about the islands of the Ven ados, Creston and Isla Blanca; many specimens obtained. This species was found by Gilbert at Mazatlan, and by Steindachner at Cape San Lucas. We have thus far been unable to find any distinction between the American form and the common East Indian species, to which the name *capistratus* was first given. Two markedly different types of coloration were obtained, supposed by us to be of the two sexes, since no other difference except that of coloration is noticeable. In all specimens obtained, however, the sexual organs were so immature that the sexes could not be distinguished thereby.

Specimens supposed to be female dull olive with darker clouds; no yellow on posterior parts which are scarcely paler behind; fins all plain olive blackish; streak behind mouth light bluish, very faint, soon fading after death; lower lip blue, then golden, then a blue ring, then yellow., then bluish; upper lip livid, bluish above.

Others supposed to be male are in life dark olive clouded with darker; posterior part of body deep yellow, below median line; fins blackish; first dorsal bright olive yellow on membranes; green on caudal membranes, the rays black. Anal reddish. Streak behind mouth bright red in one specimen, whitish in another. Upper lip livid blue then orange, then golden, then livid blue or purplish, then orange, then crimson, then dark.

Still other specimens were marked with whitish shades instead of red.

Family TETRAODONTIDÆ.

191. **Spheroides annulatus** (Jenyns) var. *politus* Girard. Tambor.

Very common everywhere in the Astillero. Specimens entirely smooth, and those variously prickly, were obtained; prickly ones, both young and old, were found, but no very young which were smooth. There seems to be no specific difference recognizable among these. All of them, however, differ from specimens taken farther south in the larger size of the dark spots and in a somewhat greater tendency to smoothness of the body. All of these, smooth or rough, seem to belong to the form called *politus*, which is probably the northern form or representative of *Spheroides annulatus*.

192. **Spheroides lobatus** (Steindachner). Botete.

Rather common in the estuary with the preceding, reaching a smaller size, the largest seen not over six inches in length. The species was first described by Steindachner from Altata, but until its recent discovery in the Albatross collections it was confounded with *Spheroides angusticeps* (Jenyns), from which it is probably distinct, although the latter, entirely smooth and uniform dusky in color, may prove to be the adult form. In both species the two small black flaps on the shoulder are present, and in both the interorbital space is very narrow and concave. Specimens taken at La Paz by Mr. James A. Richardson are intermediate in color, but retain the prickles.

In life grass green, with maroon colored spots and markings.

Family DIODONTIDÆ.

193. **Diodon hystrix** Linnæus. PUERCO ESPINO.

Very common about rocky places, especially among the islands, where it was also found by Dr. Gilbert. All specimens taken belong to the typical *Diodon hystrix*. *Diodon holocanthus*, if different, is unrepresented in our Mazatlan collections.

Family MOLIDÆ.

194. **Mola mola** (Linnæus). PEZ MOLA.

Found in the open sea from San Francisco to Mazatlan. It was seen at the latter locality by Dr. Gilbert, but not by us.

Family SCORPÆNIDÆ.

195. **Scorpæna mystes** Jordan & Starks, n. sp. LAPON. Plate lii.

Common in the Astillero, on the bottom. Very tenacious of life, and much dreaded by the fishermen from the poisonous sting of its dorsal spines.

Allied to *Scorpæna plumieri* Bloch, which species it represents on the Pacific Coast.

Head 2⅓; depth 3⅓; dorsal XII, 10; anal III, 5; scales about 30; orbit 6½ in head; maxillary 2; pectoral 2; highest dorsal spine 3½; second anal spine 3; caudal 2.

Body robust, not much compressed; interorbital space wide, not deeply concave, ½ wider than orbit; a pit between preorbital and eye, and a broad depression behind coronal spines; membranous flaps on preorbital, edge of preopercle, over nostrils and above eyes; preocular, supraocular, tympanic, coronal, occipital, nuchal and exoccipital spines present. Maxillary reaching to behind eye; lower jaw included; gill-rakers short and thick, about

3+6; head naked, with the exception of a few imbedded scales on preopercle and posterior part of opercle; scales on body large, many of them with membranous flaps.

Olive-brown almost black, marbled with light drab; opercular flap with pale edge; the fins much spotted and marbled, all except spinous dorsal, with white margin, more distinct in the young; caudal fin showing three indistinct cross-bars; axil jet black, with white spots.

Largest specimen fourteen inches long.

This species differs from *Scorpæna plumieri* in having a wider and flatter interorbital area; the lower jaw wider and more rounded in front; the knob at symphysis not so sharp and projecting; the pit behind coronal spines broader and not so deep, and the color darker.

This species was also obtained at Mazatlan by Dr. Gilbert, who identified it provisionally as *Scorpæna plumieri*.

Types numbered 1501, 1616, 1617, 2919 on the L. S. Jr. Univ. Mus. register.

196. Scorpæna sonoræ (Jenkins & Evermann).

This small species is not uncommon in the Astillero, where numerous specimens, none of them over three inches in length, were obtained. It has hitherto been recorded only by Jenkins & Evermann from Guaymas.

Gray above, the flaps pinkish, the bars blackish; lower parts pink, bright on ventrals and anal; axil orange, mottled with dusky; ventrals and pectorals black at tip, edged with pale. Middle rays of pectoral slightly divided at tip, not all of them being strictly simple.

Family TRIGLIDÆ.

197. **Prionotus horrens** Richardson.

Two small specimens, each about two inches long, obtained in the Astillero.

Family GOBIIDÆ.

198. **Philypnus lateralis** Gill. ABOMA DE MAR.

Common in the Rio Presidio and occasionally taken in the Astillero, especially where the fresh water soaks into it. The species is common in fresh waters along the coast, but has not hitherto been noticed at Mazatlan.

199. **Dormitator maculatus** (Bloch). PUÑECA.

Rather common in the Rio Presidio and also in the brackish waters about the estuary. The young occur in considerable abundance in the mud puddles left by the winter rains or by the high tides. It reaches in the river a considerable size, and is a food fish of some importance, said to be the most valuable in the Rio Presidio. It is generally common along the coast, as well as everywhere along the Atlantic side.

200. **Eleotris æquidens** Jordan & Gilbert. GUAVINA.

Rather scarce in the Rio Presidio, where only one young specimen was obtained by us. A few others were found in brackish waters or muddy places about the estuary.

Blackish everywhere, sides with faint whitish streaks, along rows of scales; a broad blackish lateral band occupying whole of side; back and belly paler; traces of faint dark cross-bands; caudal black, with a pale margin and some dark cross-shades; pectorals, dorsals and ventrals more or less barred with black; a whitish bar at base of caudal with a darker one before it. Scales 68; preopercular spine well developed.

201. **Cotylopus gymnogaster** (Ogilvie-Grant).

Recorded from streams about Mazatlan; not seen by us.

202. Awaous tajasica (Lichtenstein). ABOMA DE RIO.

Found in company with *Philypnus lateralis*, from which most fishermen scarcely distinguish it. It is rather less abundant in the river, and was not noticed by us in the Astillero. Elsewhere on the coast it has been recorded only from the river at San José del Cabo in Lower California, where it was found by Mr. Lyman Belding and more recently by Dr. Gustav Eisen.

Comparison with specimens from Havana shows no differences.

203. Gobius soporator Cuvier & Valenciennes. CAIMAN.

Found in abundance in all rock pools, ascending farther above the low-tide mark than any other marine species. It does not occur in fresh water. There seems to be no difference between these specimens and those from the Gulf of Mexico, where it is found everywhere in water not exceeding two feet in depth.

204. Gobius sagittula (Günther).

A few small specimens, not over four inches in length, found in the Astillero on muddy bottoms. It was also taken by Dr. Gilbert at La Paz, Mazatlan and at Panama. *Gobius longicauda*, described by Jenkins & Evermann from Guaymas, is no doubt the adult of the same species, as Dr. Gilbert has already indicated.

Head $4\frac{1}{3}$; depth $6\frac{1}{4}$; caudal $\frac{2}{3}$ longer than head; eye $3\frac{1}{2}$ in head; maxillary $2\frac{2}{3}$; snout $3\frac{1}{2}$; scales about 52, the first 37 very small; dorsal VI–13; anal 14; skull with a median lengthwise ridge, interorbital space narrow, channelled, skull somewhat broader behind; scales before dorsal minute, head naked, scales ctenoid, much reduced anteriorly, lower jaw short, included; no flaps on shoulder girdle maxillary reaching to pupil; dorsal spines slender, some filamentous; caudal lanceolate; teeth sharp, rather small, the outer larger; lower jaw thin and flat, its acutish tip elevated.

Olive, speckled and marbled; side with five oblong black spots, the smallest at base of caudal; a black blotch on opercle; dark cross-bars under soft dorsal; head much mottled; dorsal speckled; caudal with ten zigzag cross-bars of dark specks; pectoral faintly barred; anal and ventral plain; a dark curved streak about yellowish base of pectoral; lower lip dusky; a blackish cross-blotch above gill opening.

In the adult, called *Gobius longicauda*, the caudal is much longer, but there is no other difference of importance.

205. **Gobius manglicola** Jordan & Starks n. sp.

One specimen found in the mud of the Astillero among the roots of mangrove bushes *(Rhizophora mangle)*.

Head 4¼; depth 5⅔; D. VI–12; A. 12; scales about 35, not to be exactly counted; caudal lanceolate, 2⅔ in body; pectoral about equal to head; dorsal spine slender, not filamentous, 1⅔ in head; eyes large, close together, the range partly vertical, the narrow interorbital deeply furrowed; no flaps on shoulder girdle; scales moderate, ctenoid anteriorly, becoming smooth behind; median keel on head slight; head naked.

Body long, compressed, the head depressed, the cheeks tumid; snout bluntly truncate; mouth large, the maxillary reaching the middle of eye, not produced backward, truncated behind, somewhat oblique, the lower jaw a little the longer; lower jaw flat; teeth strong, the outer in both jaws enlarged; cranium without median crest, abruptly widened behind eyes.

Color light olive mottled with darker; six oblong blotches of blackish on sides as in *Gobius boleosoma*, the bar at base of caudal; dorsals and caudal finely checkered and barred with dark brownish orange and blackish; anal mottled; a dark shoulder spot; a dark bar before eye and one below eye; ventrals dusky, the edge pale.

The species seems nearest allied to *Gobius sagittula*.

One specimen, $1\frac{1}{2}$ inches long, numbered 3095 on the L. S. Jr. Univ. Mus. register.

206. Garmannia paradoxa (Günther). Plate xlix.

A single specimen found on muddy bottom among the mangroves lining the estuary.

Head $3\frac{1}{2}$; depth $4\frac{1}{3}$; D. VI–11; A. 9; eye 4 in head; snout $4\frac{1}{4}$; pectoral $1\frac{1}{2}$ in head; dorsal spine $1\frac{1}{3}$.

Form of *Gobiosoma bosci*. Body compressed; head broad and depressed, with tumid cheeks; snout not very blunt, short, oblique-truncate; eyes rather large, high, the maxillary not produced, extending to their posterior margin; mouth large, oblique; lower jaw heavy, slightly projecting; teeth strong; gill-openings narrow, not wider than base of pectoral. First dorsal rather high, the first spine filamentous, reaching past soft dorsal; other fins low. Head and anterior half of body to front of soft dorsal naked; scattering scales coming in above, twelve rows of imbricated slightly ctenoid scales along median line of caudal peduncle and forward to middle of soft dorsal, the scaled area about as long as head, the upper parts better scaled than lower. No flaps on shoulder girdle.

Olivaceous with seven or eight dark cross-shades—two on head, one across gill-openings, one behind pectoral, and a broad one below soft dorsal; dorsals dusky, the filamentous ray pink; lower half of soft dorsal yellowish, upper dusky; lower fins black; caudal dusky; a dark speck at angle of opercle; skin everywhere punctate with black; a pale olive bar at base of caudal.

Skull without median crest. Interorbital space not concave. Head not very abruptly widened behind eyes.

One specimen 1½ inches long obtained. This specimen differs but slightly from Günther's account of *Gobius paradoxus*, a species which is the type of the genus *Garmannia* of Jordan & Evermann (MS.), distinguished from *Gobius* by the half-naked body. The genus is named for Mr. Samuel Garman, the accomplished ichthyologist of the Museum of Comparative Zoology at Cambridge, Massachusetts, in recognition of his important contributions to ichthyology.

207. **Aboma etheostoma** Jordan & Starks, n. gen. and n. sp. Plate I.

A single small specimen found in the mud on a shallow bottom in the Astillero.

Aboma, new genus, allied to *Microgobius* Poey, distinguished by the large, ctenoid scales, which cover the body; head naked, rather long, pointed in profile, the mouth moderate, not very oblique; teeth rather strong. Dorsal spines more than six, none of them filamentous; soft dorsal and anal short; no flaps on shoulder girdle. Cranium with a slight median crest. The name *Aboma* is used by the Mexicans in Sinaloa as synonymous with goby. Besides the new species, *Aboma etheostoma*, which is the type of this genus, probably *Gobius chiquita* Jenkins & Evermann, and *Gobius lucretiæ* Eigenmann & Eigenmann, will be referable to it.

Head 3⅓; depth 5; D. VIII–11; A. 10; scales 26; longest dorsal spine 1¾ in head; eye 3; snout 4; maxillary 3.

Body long and low, moderately depressed and pointed forward. Scales large, ctenoid behind, none on head, those on nape and belly much reduced. Mouth moderate, terminal, moderately oblique; the maxillary reaching middle of pupil. jaws subequal or the lower a little the longer: teeth rather strong. No flaps on shoulder girdle. Cranium with a slight median crest. Interorbital ridge not hollowed out: skull not abruptly widened behind.

Color olivaceous. side with a very broad jet black lateral band. three times interrupted by silvery. Caudal white with four ‹-shaped bands. growing progressively fainter behind. Pectoral mottled gray. with a jet black oblique crescent towards its base. surrounding a large yellow spot. side of head with four round gray spots separated by black. the largest below eye. with a black streak before it. First dorsal jet black: second mottled: the produced spine with yellowish. Ventrals and anal pale.

One specimen. 1¹⁄₄ inches long. in the Museum of the Leland Stanford Jr. University.

208. Evermannia zosterura (Jordan & Gilbert). Plate ii.

Very common on sandy bottoms everywhere about the estuary. numerous specimens being dug out of the sand by Mr. Williams. It is seldom found much if any below the mark of low tide. It is a very handsomely colored species. the male being more strikingly marked than any other of our Gobies. The species has hitherto been known only from a single specimen taken by Dr. Gilbert at Mazatlan.

Head [illegible]; depth [illegible]; dorsal IV–[illegible]; anal [illegible]; eye equals snout, [illegible] in head; pectoral [illegible]; caudal [illegible].

Body compressed [illegible] not very [illegible] posterior [illegible]

[illegible]

Male sometimes with traces of eight olive cross-bands. Fins very ornate, the dorsal and anal yellowish at base, then a broad median band of jet black, then a broad white margin. Middle of caudal yellow to the tip, with a black band above and below, and a white edge above and below this as in dorsal and anal; no bands on tail.

Female with dorsal filament short, reaching about to first soft ray. Dorsals and anal checkered with blackish; caudal faintly barred; all vertical fins with pale edgings, but without the black stripe of the males.

Family GOBIESOCIDÆ.

209. **Gobiesox adustus** Jordan & Gilbert.

Obtained by Dr. Gilbert in rock pools at Mazatlan. Rare and not found by us.

210. **Gobiesox erythrops** Jordan & Gilbert.

Found rare in rock pools at Mazatlan by Dr. Gilbert, who also records a specimen from Tres Marias. Not seen by us.

211. **Gobiesox zebra** Jordan & Gilbert.

Very abundant in rocky places at Mazatlan, especially among sea urchins. Numerous specimens were obtained by us, as also by Dr. Gilbert.

The coloration is quite variable, although the markings are rather constant. In general, light pink with markings of gray, blackish and olive; a distinct dusky blotch behind eye and a dark bar across caudal.

212. **Gobiesox eos** Jordan & Gilbert.

Found in rock pools at Mazatlan by Dr. Gilbert. Not recorded from any other locality.

Two specimens obtained by us from rock pools among echini. The bright cherry red coloration is distinctive and persists in alcohol.

Family OPISTOGNATHIDÆ.

213. Opistognathus punctata Peters.

The original type of this species was described by Dr. Peters from Mazatlan. It was also found by Dr. Gilbert at Panama, the two specimens mentioned being as yet the only ones known.

Family BLENNIIDÆ.

214. Isesthes brevipinnis (Günther).

This species was found to be rather common in rock pools at Mazatlan both by Dr. Gilbert and by us.

215. Rupiscartes atlanticus (Cuvier & Valenciennes.)

This species is very common in rock pools about Mazatlan, where it reaches a length of about six inches. It was found in numbers by Dr. Gilbert at Mazatlan, but has not been recorded from localities farther south. Mr. Charles H. Townsend found it at San Cristobal Bay, and Mr. John Xantus at Cape San Lucas. Thus far no difference has been found between these specimens and those from the West Indies.

Body liver brown, paler below. Fins mostly blackish; an orange area on upper edge of caudal; a yellow one tinged reddish below. Eye red posteriorly.

216. Rupiscartes chiostictus (Jordan & Gilbert).

Only the original types of this species found by Dr. Gilbert in the tide pools at Mazatlan have been recorded. It was not seen by us.

Family CLINIDÆ.

217. Labrosomus xanti (Gill).

Very common at Mazatlan in rock pools with *Rupiscartes atlanticus* (Cuvier & Valenciennes), and reaching about the same size. It was also found by Richardson at La Paz and by Gilbert at Mazatlan. It has been recorded from Cape San Lucas by Xantus and from San Cristobal Bay by Townsend. The Pacific form called *Labrosomus xanti* seems to be scarcely if at all distinguished from the West Indian form, *nuchipinnis*, cognate to it. The only difference we have found is in the dentition of the vomer, and this may not be constant.

218. **Labrosomus delalandi** (Cuvier & Valenciennes).

Extremely common in rock pools at Mazatlan, where it was also found by Dr. Gilbert. It has not been noticed from any other locality on the Pacific Coast. Thus far we have not been able to distinguish it from *Labrosomus delalandi* of the coast of Brazil.

219. **Enneanectes carminalis** (Jordan & Gilbert) n. gen. Plate liii.

Four specimens, types of the species, were found by Dr. Gilbert in a rock pool at Mazatlan. A single small example was obtained by us.

The short chubby body, large rough-ctenoid scales, little rounded profile, and short fins distinguish this species sufficiently from *Tripterygion* Risso, and characterize the new genus *Enneanectes*, framed for it by Jordan & Evermann.

220. **Auchenopterus monophthalmus** Günther.

Several specimens taken in rock pools at Mazatlan. At low tide it is often left by the recession of the water, in which case it creeps about in the Corallina.

In this species the first dorsal is higher and better separated from the rest of the fin than in the California species, *Auchenopterus integripinnis*, and there are some constant differences in coloration.

Family FIERASFERIDÆ.

221. **Fierasfer arenicola** Jordan & Gilbert.

A single specimen found in the sand at Mazatlan by Dr. Gilbert. At first described as a new species, *Fierasfer arenicola* Jordan & Gilbert, and subsequently identified with the species which occurs in more or less abundance in the shells of the pearl oyster. It was not found by the Hopkins Expedition. According to Prof. Putnam, the West Coast species, *Fierasfer arenicola*, is not distinct from *Fierasfer dubius* Putnam, of the Florida Keys. We may, however, retain the former as distinct until comparison of specimens can be made.

Family BROTULIDÆ.

222. **Dinematichthys ventralis** Gill. Plate liv.

Found abundant in rock pools at Mazatlan, where specimens were taken reaching a length of about four inches. This fish has hitherto been recorded as extremely rare, and very few were obtained by Dr. Gilbert. This is one of the species that were brought from their hiding places by the introduction of the poisonous juice of the Hava tree into the water. It has been recorded from Cape San Lucas and Mazatlan.

Color in life, everywhere liver brown, the fins edged with whitish or pinkish.

Family PLEURONECTIDÆ.

223. **Syacium ovale** (Günther).

Occasionally taken in the Astillero at Mazatlan, where specimens were found by Dr. Gilbert and by us. It is more abundant at Panama. The broad-headed form called *Syacium latifrons* (Jordan & Gilbert), which has been supposed, perhaps wrongly, to be the male of this species, has been seen only at Panama.

224. Citharichthys gilberti Jenkins & Evermann. LENGUADO.

Very common everywhere in the Astillero, and also ascending the Rio Presidio in the fresh waters nearly as far as the village of Presidio. In fresh water the color is considerably brighter than in the sea, and these fresh water specimens correspond to those described by Jordan & Goss as *Citharichthys sumichrasti*. These seem to be, however, of the same species.

225. Azevia panamensis (Steindachner).

Common in the Astillero, reaching a length of about eight inches. The following is a count of the fin rays of nine specimens: D. 95, A. 75; D. 89, A. 67; D. 92, A. 71; D. 89, A. 71; D. 94, A. 74; D. 89, A. 71; D. 90, A. 72; D. 92, A. 71; D. 91, A. 72.

These specimens seem to be inseparable from *Azevia panamensis*.

226. Etropus crossotus Jordan & Gilbert.

Rather common in the Astillero with the preceding species, but reaching a smaller size, rarely exceeding four inches. On careful comparison of our specimens with others from Beaufort, Pensacola, Panama, and other localities, we are unable to find any differences. The color varies with the bottom, some being plain light brown, others are much mottled with lighter or with darker.

227. Hippoglossina macrops Steindachner.

This species was described by Steindachner from a specimen obtained at Mazatlan. We have not seen it.

228. Paralichthys adspersus (Steindachner).

Very common in the bay and Astillero at Mazatlan, and in fact everywhere on the coast from Guaymas and

La Paz to Panama and Callao. It reaches a length of about three feet, and is a food fish of some importance, most specimens, however, being much smaller.

Head $3\frac{1}{2}$; depth about 2 in length of body; D. 73 (70 to 76); A. 57 (53 to 60); P. 12; V. 6; scales on lateral line about 106–8 with 35 dorsally and 36 ventrally.

Flesh firm. Body oblong, moderately compressed; mouth large, oblique, the mandible very heavy, slightly projecting; 4 canine teeth on each side of lower jaw in adult specimens, 8 in young, the two anterior teeth long; anterior teeth of upper jaw strong, but smaller than those in the lower jaw; the lateral teeth very small and close set. Eye small, shorter than snout, about 7 (6 to 8) in length of head; interorbital area, smooth, flattish, $\frac{2}{3}$ width of eye. Scales cycloid, small anteriorly and larger posteriorly. Lateral line strongly arched anteriorly, arch about $3\frac{1}{3}$ in straight part.

Gill-rakers of medium length, broad, retrose-serrate on inner side, longest about $\frac{2}{3}$ length of eye, from 4+13 to 5+14 in number, counted in eight specimens; pectoral fin about as long as mandible, slightly more than half length of head. Dorsal low, anterior origin opposite anterior margin of eye; caudal barely double concave; caudal peduncle very strong. Anal spine obsolete; ventral fins small, inserted symmetrically. Fins all scaly.

Color—Large specimens are dark brown, with blotches on fins; small specimens are covered with pearly white and very dark brown blotches. The brown blotches are almost circular, larger and with less definite outlines near the center of the body, very dark and distinct on caudal.

Seven specimens were taken by the Hopkins Expedition in the estuary at Mazatlan, where they reach a length of 44 cm. Several specimens were also taken at La Paz.

These specimens seem to be identical with *Paralichthys adspersus*, described from Callao by Steindachner. The original types have on the average more gill-rakers than we find on our Mazatlan specimens, but this character is subject to variation, and no other distinction appears.

In one of Dr. Steindachner's types from Callao (11,417, Mus. Comp. Zool.) we find the gill-rakers longer, 6+17; depth 2½ in length; D. 67; A. 51; scales 120; arch of lateral line barely twice as long as high, nearly 5 in straight part; maxillary 2⅙ in head.

Mr. Garman has kindly examined for us six other specimens, with the following results:

"*Paralichthys adspersus* from Callao has gill-rakers—

7 above, as long as the eye;
17 below.
$\frac{6}{15}$ about ⅔ as long as the eye.
$\frac{7}{18}$ nearly as long as the eye.
$\frac{3}{14}$ about ⅔ as long as the eye.
$\frac{5}{15}$ about ⅔ as long as the eye.
$\frac{6}{17}$ near ¾ as long as eye."

—(Garman, in lit., May 3, 1895.)

Family SOLEIDÆ.

229. **Achirus mazatlanus** (Steindachner). LENGUADO DE RIO. (*Solea pilosa* Peters.)

Very abundant in the fresh waters of the Rio Presidio below the village, varying considerably in color, and somewhat in form. One specimen was taken in the brackish waters of the estuary.

230. **Achirus fonsecensis** (Richardson).

Two specimens found in the Rio Presidio with *Achirus mazatlanus;* not seen at Mazatlan.

231. **Symphurus williamsi** Jordan & Culver, n. sp. Plate lv.

Two specimens, the largest about 1⅛ inches long, were obtained by Mr. Thomas Marion Williams in tide pools with sandy bottom, in very shallow water, near the estuary at Mazatlan.

Head $4\frac{4}{5}$; depth 3⅔; D. 93; A. 73; scales 92. Body slenderer than in *Symphurus plagiusa*, which it much resembles, but not so slender as in *Symphurus elongatus*, and the caudal fin not black. Upper eye slightly in advance of lower.

Sand color in life; light gray, everywhere finely mottled with light and dark, with traces of a few very narrow dark-cross bands. Fins all mottled; the caudal and posterior part of dorsal and anal not black, scarcely darker than anterior part.

Type numbered 2943, in the register of L. S. Jr. Univ. Mus.

Family ONCOCEPHALIDÆ.

232. **Oncocephalus elater** (Jordan & Gilbert).

One specimen, the type of the species, presented to Dr. Gilbert by Dr. Bastow, then a resident of Mazatlan. It is found in deep water, and was not seen by us, but numerous specimens have been since dredged by the Albatross in localities further to the south, so that the species is now well known.

SUPPLEMENTARY NOTE ON THE FISHES OF LA PAZ HARBOR.

Mr. James A. Richardson, a member of the Hopkins Expedition, spent two days at La Paz, the chief city of Baja California, where he made a small collection of fishes. The work was done under very unfavorable conditions, as La Paz has no fish market and its fish supply is obtained by the spear and the hook and line. There is but one seine at La Paz, a very old and rotten one, which was rented by Mr. Richardson, as was also a parachute seine and a small dip-net. Considering all the difficulties encountered, the list here given shows that the locality is well worthy of a detailed exploration.

Concerning the harbor of La Paz, Mr. Richardson has the following notes:

"The approach to La Paz estuary is guarded by several large islands, uninhabited, wild and precipitous. The entrance to the estuary is very wide, apparently ten or fifteen miles, the general direction being north and south and the length of the estuary about fifteen miles. The estuary gradually narrows to about one mile at ten miles from the entrance. As the steamer proceeds up the estuary it is noticed that she hugs the left bank closely. I was told that in all that breadth of water there is but a very narrow channel, the balance of the space in the estuary being of a sand formation, the sand bars coming very near the surface of the water so that they can be seen from the deck of the steamer. The steamer in following the channel nearly doubles on itself occasionally, and in the darkness of the night a boat is lowered and a search is made for certain buoys. The left bank is made up alternately of gravel beach and abrupt cliffs all the way to La Paz. The country behind La Paz is hilly and mountainous, of no value, covered with rocks and cactus. The right bank opposite La Paz, as far as one could see, is one vast stretch of sand and mangrove bushes lying a little above tide water. This is considered to be fine soil for cocoanut trees, but it is uninhabited and uncultivated. The sand beach is very fine; one could ride a bicycle here for fifty miles following the shore line."

1. **Narcine entemedor** Jordan & Starks.

Common. One specimen somewhat decayed found on the beach.

2. **Opisthonema libertate** (Günther).

Two specimens obtained (1¾ in. long).

3. **Stolephorus ischanus** Jordan & Gilbert.

Two small specimens.

4. **Stolephons curtus** Jordan & Gilbert.

One specimen.

5. **Mugil cephalus** Linnæus.

Very common.

6. **Mugil curema** Cuvier & Valenciennes.

Very common.

7. **Querimana harengus** (Günther).

Very abundant in the lagoons and small estuaries.

8. **Holocentrus suborbitalis** Gill.

Common in rock pools.

9. **Paralabrax maculatofasciatus** (Steindachner).

Common.

10. **Lutianus novemfasciatus** Gill.

Two specimens.

11. **Lutianus argentiventris** (Peters).

One specimen obtained.

12. **Xenistius californiensis** (Steindachner).

Several young specimens obtained.

Silvery, with continuous streaks of bright warm brown along the rows of scales.

13. **Pomadasis macracanthus** (Günther).

Common.

14. **Orthopristis reddingi** Jordan & Richardson, n. sp. Plate xli.

Allied to *Orthopristis ruber* (Cuv. & Val.)

Head $3\frac{1}{6}$; depth 3; dorsal XII, 15; anal III, 10; scales 8–52–15; 53 pores.

Eye $4\frac{1}{4}$ in head; maxillary $3\frac{1}{4}$; preorbital $4\frac{1}{4}$ in snout; pectoral $1\frac{2}{5}$ in head; longest dorsal spine $2\frac{5}{6}$; longest soft ray $3\frac{3}{5}$; second anal spine $4\frac{2}{3}$; ventral $1\frac{2}{3}$; upper caudal lobe $1\frac{1}{2}$; base of soft dorsal in spinous $1\frac{3}{4}$.

Body oblong, the back not much elevated; the anterior profile straightish, slightly depressed above the eye; mouth small, low, the maxillary reaching to opposite the nostril; teeth subequal, in broad bands; lower jaw included; nostrils both oblong, the anterior the larger; eye rather large, about as wide as the broad preorbital; preopercle very finely serrated on its posterior margin only, the serrations very weak; gill-rakers short and small, about 12; scales moderate, the rows above lateral line very oblique, those below nearly horizontal, the series from the scapular scale reaching middle of spinous dorsal. Spinous dorsal moderate, not deeply notched, the median spines injured in youth in the type specimen; soft dorsal low, free from scales; anal spines low, the second a little longer than third; soft rays scaleless; caudal lunate, the lobes unequal, the upper longer than lower, which is more obtuse. Ventrals rather long, inserted just behind axil of pectoral. Pectoral rather short, not quite reaching tips of ventrals.

Color pearly gray, darker above; each scale of back and sides, with a bright bronze spot behind its center; these forming nearly continuous streaks along the rows of scales. These streaks run upward and backward anteriorly and nearly horizontally on sides, when they are more or less interrupted or transposed. Head plain gray, dorsal with some streaks and clouds: outer fins plain; ventrals somewhat dusky.

One specimen, 8¾ inches long, was taken by Mr. Richardson.

This species is very closely allied to the Atlantic species, *Orthopristis ruber* (Cuv. & Val.), but has the body a little more slender and the head larger.

The specimen from Guaymas provisionally referred to *Orthopristis cantharinus* (see Jordan & Fesler. Rept. U. S. Fish Com. for 1889 to 1891, 500, 1893), is perhaps a second specimen of *Orthopristis reddingi*.

This species is named in honor of Hon. Benjamin B. Redding, first Fish Commissioner of California, a man deeply interested in scientific research, to whom Mr. Richardson has been indebted for many favors, in his former capacity of Superintendent of the California Fish Hatching Station at Sisson.

15. **Microlepidotus inornatus** Gill.

One specimen, 10 inches long, obtained. Common.

16. **Umbrina xanti** Gill.

Common.

17. **Micropogon ectenes** Jordan & Gilbert.

One specimen.

18. Eucinostomus gracilis (Gill).

[illegible]

19. **Xystæma cinereum** (Walbaum.)

Common. About twenty specimens obtained.

20. **Gerres lineatus** (Humboldt).

Common.

21. **Scarus perrico** Jordan & Gilbert.

One specimen, found dead on the beach.

22. **Spheroides lobatus** (Steindachner).

Common. Two specimens obtained. In color these approach *Spheroides angusticeps* (Jenyns). It may be that *lobatus* is, after all, the young of *angusticeps*, as was supposed by Jordan and Gilbert.

23. **Diodon holacanthus** Linnæus.

Common. One specimen, 11 inches long, was obtained. D. 12; A. 12; back and sides covered with spots; no spots on fins or tail; back very dark; a dark band between eyes; frontal spines nearly as long as pectoral spines which are longest.

24. **Alexurus armiger** Jordan, n. g. and sp. GOBIIDÆ. Plate xlviii.

Head 4⅔; depth 8; dorsal VI–13; anal 11; V. I, 5; scales about 102–30; eye 8 in head; maxillary 2⅔; mandible 2½; snout 5⅔; interorbital 4⅓; pectoral 1⅘; caudal equals head; ventral 2; last dorsal ray 1⅗.

Body long and low, compressed posteriorly, depressed in front. Head flattish and broad above, the cheeks moderately tumid. Eyes small, high up, separated by a broad flattish interorbital space; snout short; mouth moderate, very oblique, the maxillary ceasing below the center of pupil; lower jaw very heavy, oblique, projecting beyond upper, its outline horseshoe-shaped, obtuse in front. Teeth in rather broad bands, the outer enlarged below, but scarcely so above: none of them canine-like. Top of head with very small scales. Cheeks and opercles with rudimentary scales above. Preopercle with a concealed antrose hook below as in *Eleotris*. Scales on body very small, perfectly smooth, partially imbedded: scales on nape and throat minute. Gill membranes extending a little forward below, so that the branchiostegals are free from the isthmus.

Insertion of dorsal twice as far from middle of base of caudal as from tip of snout: the fin low, its slender rays slightly filamentous. Soft dorsal low, its last ray highest. Anal similar, beginning under second dorsal ray. Caudal long, bluntly pointed behind, with strongly procurrent base above and below, the base above two-fifth length of head formed of fourteen short rays, that below a little shorter, of twelve rays, this procurrent portion forming an angle with the caudal proper where it joins it. Pectoral and ventrals short, the ventrals inserted under pectorals.

Color olive green, dusky above, paler below, but everywhere covered with fine black dots. Both dorsals with the membranes pale, the rays each barred with black. Caudal mesially blackish, all the rays barred or chequered in fine pattern. Pectoral and anal pale, similarly speckled; base of pectoral dusky: ventral finely speckled.

One specimen, 6½ inches long, taken by Mr. James A. Richardson in the harbor of La Paz.

This species seems to be the type of a distinct genus allied to *Eleotris* and *Erotelis*, distinguished from *Eleotris* by its very small cycloid scales, from *Erotelis* by its concealed preopercular hook, and from both by the procurrent caudal fin. The generic name is from [illegible], to [illegible] tail.

25. **Gobius sagittula** (Günther).

Two large specimens, each six to eight inches long, besides one very young example, corresponding to the form called *Gobius longicauda* of Jenkins & Evermann. As Dr. Gilbert has noticed, this is the adult form of the species called by Dr. Günther *Euctenogobius sagittula*, of which specimens were found by us at Mazatlan.

The species is very similar to *Gobius oceanicus* of the Atlantic.

26. **Gobius soporator** Cuvier & Valenciennes.

Very common.

27. **Scorpæna mystes** Jordan & Starks.

Common.

28. **Labrosomus xanti** Gill.

Common in rock pools.

29. **Labrosomus delalandi** (Cuvier & Valenciennes).

Common in rock pools.

30. **Auchenopterus monophthalmus** Günther.

Not rare; in rock pools.

31. **Paralichthys adspersus** (Steindachner).

Very common; about ten specimens taken.

LIST OF PLATES.

SOME MEXICAN NEUROPTERA.

BY NATHAN BANKS.

The following sixteen species of Neuroptera were collected by Dr. Gustav Eisen and Mr. Frank H. Vaslit mostly in the Cape Region of Baja California; a few, however, are from Tepic, Territory of Tepic.

Although the collection is too small for generalization, it may be noted that the described species from Baja California were previously known from California; and there is not a single species of general distribution in the United States. The three species of Chrysopa which appear to be new have the wings less veined than usual, the cells being larger than in our eastern species. Of the three species from Tepic one appears to be new, though it would seem strange that such a prominent insect had escaped observation. All of the specimens are alcoholic.

Sialidæ.

Raphidia australis nov. sp.

Length 13 mm. Face yellowish; palpi fuscous, annulate with white; antennæ yellowish, darker toward tips; head granulate, above and behind shining greenish, with a smooth, elongate, rufous spot reaching nearly to the ocelli; sides of head behind the eyes straight, not convex; prothorax slender, rufous, contracted before middle, granulate, each granule giving rise to a short hair, an indistinct black fork on the basal portion; legs pale yellow; abdomen black, with many longitudinal rows of yellow spots, those on sides are larger and more or less connected, a black median line on the yellow venter, ovipositor nearly as long as the abdomen, upcurved. Wings quite short, but six or seven costal transverse veinlets, subcosta runs into the costa at a point before the pterostigma twice the length of the latter; pterostigma short, barely twice as long as broad, four short cells beneath it in fore wings, in hind wings two short cells and then a longer one.

Described from two specimens from San Lazaro, Baja California.

MANTISPIDÆ.

SYMPHASIS SIGNATA Hagen.

Hagen, Stett. Entom. Zeit. 1877, p. 208.

Two specimens, one from San Lazaro, the other, San José del Cabo, Baja California. September.

CHRYSOPIDÆ.

CHRYSOPA EXTERNA Hagen.

Hagen, Syn. Neuropt. N. Am. 1861, p. 221.

A single specimen from Tepic. October.

CHRYSOPA PERFECTA nov. sp.

Length 15 mm. Pale yellowish, a black stripe from each eye to the mouth; palpi black; antennæ shorter than the wings, pubescent, whitish, slightly darker toward the tips; prothorax nearly as long as broad, narrowed in front, a large, elongate, rufous spot each side starting from the anterior edge, but not reaching the posterior margin; anterior lobe of mesothorax with a rufous spot each side; meso- and metathorax sometimes with a whitish median stripe; legs whitish, white haired; abdomen yellowish. Wings moderately slender, anterior pair scarcely pointed at tips, the posterior pair more distinctly so; costal transversals black; most of the other transverse veinlets are fuscous at ends, and faintly margined with fuscous; the veinlet connecting subcosta and radius is beyond [illegible] costal transversal; second cubital cell about [illegible] as long as broad, distinctly swollen on anterior margin [illegible] third cubital cell with the division ending [illegible] the first transversal [illegible] the radial sector:

cells large, four gradate veins in the first series, five or six in the second series; pterostigma not distinct.

Several specimens, San Lazaro and El Taste, Baja California.

CHRYSOPA VALIDA nov. sp.

Length 14 mm. Pale yellowish, no spots on head; tips of palpi fuscous; antennæ much shorter than wings, whitish, a red line above on basal joint; prothorax much shorter than broad, narrowed in front, barely rufous each side; rest of thorax and abdomen yellowish; legs short, whitish, with white hairs. Wings moderately slender, anterior pair barely pointed, the posterior pair very distinctly so; nearly all the transversals and gradate veins in anterior pair are black or fuscous, and some gradate veins in hind wing; usually but two black transversals before the veinlet which connects subcosta and radius; second cubital cell more than twice as long as broad, barely swollen on the anterior side; division of third cubital cell ends just beyond the first transversal to radial sector; cells quite large, three gradate veins of the first series, and six in second series; pterostigma very distinct, opaque whitish.

A few specimens, El Taste and San José del Cabo, Baja California. September.

CHRYSOPA INCERTA nov. sp.

Length 14 mm. Brownish yellow, face yellow, without spots; antennæ shorter than wings, black, except the basal joint which is yellow and inflated; tips of palpi fuscous; prothorax much broader than long, much narrowed in front, a fuscous stripe on each side, meso- and metathorax with a few fuscous spots; abdomen with many fuscous dots and streaks; legs pale with black hair, tarsi darker, an elongate fuscous spot above near tip of femur.

Wings moderate, anterior pair scarcely pointed at tip, posterior pair a little more so; in anterior pair the transversals, gradate veinlets, many of the marginal veinlets, and most of radial sector, black or fuscous; in posterior pair the transversals and gradate veinlets fuscous; connecting veinlet between subcosta and radius is beyond the second costal transversal; second cubital cell more than twice as long as broad, barely swollen on anterior side; division of third cubital cell ends beyond the first transversal to radial sectors; two or three gradate veins in first series, five or six in second series; pterostigma moderately distinct.

Two specimens, El Taste, Baja California.

MYRMELEONIDÆ.

ACANTHACLISIS FALLAX Rambr.

Rambr., Neuropt., p. 385.
Hagen, Syn. Neuropt. N. Am., p. 223

Several specimens from San Lazaro and San José del Cabo, Baja California, September. It is distributed throughout tropical America, from California to Argentine.

ACANTHACLISIS CONGENER Hagen.

Hagen, Syn. Neuropt. N. Am., p. 224.

One male, San José del Cabo, Baja California, September. It is more common northward on the Pacific coast.

BRACHYNEMURUS PEREGRINUS Hagen.

Hagen, Syn. Neuropt. N. Am., p. 234 (sub Myrmeleon).
H[illegible] Ent. [illegible] 1888, p. [illegible]

Several from San José del Cabo, September, and one from Sierra El Taste, Baja California. This is found as far westward as Colorado and Nebraska, but is more abundant in the southwest.

Brachynemurus sackeni Hagen.

Hagen, Can. Entom. 1888, p. 94.

Many specimens from San José del Cabo, Baja California. September. Previously known from Arizona, Texas, California and Mexico.

Brachynemurus californicus nov. sp.

Length ♀ 26 mm., ♂ 35 mm., alar expanse 50–55 mm. Head yellowish, a broad fuscous band through antennal sockets from eye to eye, vertex with two transverse rows of fuscous spots, and behind on each side a brown and a black spot; antennæ with two basal joints pale, a band in front and behind on first one, rest annulate with fuscous; prothorax with a broad fuscous stripe each side, which is forked in front and contains a pale spot behind; on each lower side is a shorter and less definite fuscous stripe; many fuscous spots on each upper side of meso- and metathorax, forming a more or less maculose stripe; other spots on middle, some of them shining black; a broad black stripe below wings, containing an interrupted pale line; legs thickly dotted with fuscous, black on tips of tibiæ, third and fourth tarsal joints wholly, and the tip of the fifth joint; spurs about as long as the basal joint. Abdomen of ♂ longer than the wings, of the ♀ shorter; pale, lineated with fuscous, near the tip often broken into spots more or less connected; second segment mostly fuscous, the others with a median and side stripes, the former narrow in the middle, usually wanting on basal part of basal joints, last segment almost wholly fuscous. Nearly everywhere scantily clothed with white hair and black bristles. The appendages of the ♂ pale, and dotted with fuscous, twice as long as last segment, slender, bowed and slightly upcurved, a comb of stiff bristles above near tip. Wings narrow, hyaline; veins fuscous, pointed with white; small fuscous spots, most distinct on radius and along cubitus; basal part of hind margin of wings with a long fringe; the costal area biareolated from near the base to the pterostigma; the space behind the radius, before the origin of the radial sector, is also biareolated; the furcation of cubitus is considerably beyond the origin of the radial sector.

Numerous specimens from San Lazaro, Sierra El Taste and San José del Cabo, Baja California. September. By its biareolated costal space it is evidently related to *B. inscriptus* Hagen, but the markings are quite different.

Brachynemurus fraternus nov. sp.

Length ♀ 18 mm., ♂ 26 mm., alar expanse 35–40 mm. This is so extremely similar in structure and in coloration to the preceding species (*B. californicus*) that it may be but a small variety of it. But the specimens were taken at the same time and place, and show no intermediate sizes, so I shall consider it, for the present at least, as distinct. It differs from *B. californicus*, besides the size, only in that the appendages of the ♂ are scarcely twice the length of the last segment of the abdomen; and the markings on the sides of the abdomen of the ♂, and on the prothorax are usually much less distinct.

Several specimens from San Lazaro and San José del Cabo, Baja California. September.

Brachynemurus mexicanus nov. sp.

Length ♀ 28 mm., ♂ 42 mm., alar expanse 55–60 mm. Head yellowish, a broad black band across antennal region, vertex with two comma-shaped black spots, sometimes connected behind to the prothoracic stripes; antennae brown, the basal joint pale above, annulated with [illegible] yellowish, with a dark brown stripe each [illegible] branch reaching anterior margin, outer branch falling short, sometimes, however, uniting with the inner branch; lower margin each side fuscous; meso- and metathorax above interruptedly lined with fuscous, sides with many fuscous spots usually connected; legs yellowish, with scattered brown dots, tips of tarsal joints blackish; abdomen blackish, a yellowish line above on each side of the basal joints; clothed throughout with black bristly hair. Wings hyaline, veins pale; both pairs evenly spread with many small fuscous spots, of about equal size, usually having for their center the base of a transverse veinlet; pterostigma large, distinct, cream yellow. Wings broad, costal space not biareolate till near middle of wings, space behind radius simple, cubitus forks beyond the origin of radial sector, but before the forking of radial sector. Antennæ of ♂ longer than thorax, a little shorter in the ♀; spurs longer than basal tarsal joint; abdomen in the ♂ longer than wings, in the ♀ shorter; appendages of the ♂ shorter than last segment of abdomen, nearly parallel, upcurved, their tips divaricate, a comb of stiff bristles near the tip above.

A number of specimens from Tepic. November.

MYRMELEON RUSTICUS Hagen.

Hagen, Syn. Neuropt. N. Am., p. 233.
Hagen, Can. Entom. 1887, p. 210.

Several specimens from San José del Cabo, Baja California. September.

ASCALAPHIDÆ.

ULULA MEXICANA McLach.

McLachlan, Journ. Linn. Soc., Zool. 1871, p. 248.

A single specimen from Tepic, Mexico. October.

ULULA BICOLOR nov. sp.

Length 30 mm., alar expanse 55 mm. Hair on face white, more grayish on head and thorax; white on coxæ

and femora, rest of legs pale yellowish, with black bristles, abdomen marked with brown and yellowish in the usual pattern, pruinose beneath; antennæ much shorter than wings, yellowish, tips of joints fuscous, club pale brownish; pterostigma of fore wings cream white, of hind wings more fuscous; a curved dark cloud below pterostigma of hind wings in some specimens. Venation like *U. hyalina*.

Several specimens from San José del Cabo, September, and Sierra El Taste, Baja California. Differs from *U. hyalina* principally in the shorter antennæ and light colored pterostigma.

THE SPECIES OF THE GENUS XANTUSIA.

BY JOHN VAN DENBURGH,

Curator of the Department of Herpetology.

A large number of specimens of *Xantusia vigilis* and several of *Xantusia henshawi*, which I have recently collected, tempt me to present revised descriptions of the species of this genus of lizards.*

KEY TO THE SPECIES.

a¹. One series of small plates (superciliaries) ever eye.
  b¹. Ventral plates in twelve longitudinal series.
    c¹. A single frontal, eye large. *X. vigilis.*
    c². A pair of frontals, eye small. *X. gilberti.*
  b². Ventral plates in fourteen longitudinal series. *X. henshawi.*
a². Two series of small plates (superciliaries and supraoculars) over eye. Ventral plates in sixteen longitudinal series. *X. riversiana.*

XANTUSIA VIGILIS Baird.

Xantusia vigilis.
1859, Baird, Proc. Ac. Nat. Sci. Phila., 1858, p. 255.
1893, Stejneger, N. A. Fauna, No. 7, p. 198, pl. iii, figs. 1a-1c.

Description.—The body is subcylindrical, with very short limbs. The upper surface of the head is flattened, curving towards the snout. There are three folds on the throat, the anterior connecting the ears and encircling the head. The nostril is pierced at the junction of the rostral, internasal, postnasal and first labial plates. The rostral is in contact with the first labial and internasal plates. The two internasals are followed by a large sub-

* Since this paper was written, two new genera of Xantusidæ—Zablepsis and Amœbopsis—have been proposed (Am. Nat., xxix, Aug. 1895, p. 757), to contain *Xantusia henshawi* Stejn. and *Xantusia gilberti* Van D. Neither appears to me well founded. None of the characters alleged to be distinctive of Zablepsis is constant, even as a specific character of *X. henshawi;* and the general scutellation and coloring of *X. gilberti* are so like those of *X. vigilis* as to indicate close relationship. Why these separations should have been made and *X. riversiana* still have been left in the original genus, I cannot comprehend.

hexagonal frontonasal. Behind this are two prefontals (in contact), bordered posteriorly by the single broad frontal and the first superciliary plates. Each of the two frontoparietal plates forms sutures with the frontal, second, third and fourth superciliaries, first supratemporal, parietal, interparietal and its fellow of the opposite side. The parietals and the very large interparietal are bordered behind by the two large occipitals. A row of small supratemporal scutes lies along the outer edge of the occipital and parietal plates. The two large loreals are in contact below with the superior labials, and above with the frontonasal and prefontal plates. In front of the first loreal is a large postnasal. A series of small plates, the upper of which are the superciliaries, usually surrounds the eye. Between this ring and the larger loreal are two or three small plates. There are four or five superior and three or four inferior labials to a point below the pupil. The eye is large, without lids, and with vertical pupil. Its diameter is contained about twice in the distance from the end of the snout to the orbit. The oblique ear opening has a very weak anterior denticulation. The inferior labials are in contact with the large sublabials. The first pair of the latter are in contact on the median line. The back, sides, upper and posterior surfaces of the limbs, and the gular regions, are covered with subhexagonal granules. There is a series of large plates along the edge of the last gular fold. The quadrate ventrals are in twelve longitudinal and twenty-seven to thirty transverse rows. The large preanal plates are arranged in two series of two each, sometimes surrounded by a few smaller scales or granules. The tail is conical, and covered with whorls of smooth scales, which are very narrow and transversely convex. Its length is very variable. Six to ten femoral pores form a series along each thigh.

The ground color in different specimens varies from smoke gray, through many shades of yellow and brown, to clove brown. Scattered granules are dark brown or black. At times these dark granules are so numerous as to become confluent, with a tendency to form longitudinal lines. In other individuals they are scarcely visible. Some specimens have heavy dotting on a very pale ground; in others the dotting is heavy on a dark ground; many show faint dots on a light ground; and several have few dots on a dark ground. A yellowish line usually runs back on the neck from the outer edge of each occipital plate. Two similar lines are sometimes present on the nape. The lower parts are creamy white, sometimes clouded with brown towards the sides. The young average much darker than the adults.*

	mm.	mm.	mm.	mm.	mm.
Snout to vent	47	44	42	37	22
Tail	40†	47	61	41	24
Hind limb	17	16	15½	15	9½
Fore limb	12	11	11	10¾	7
Shielded part of head	9¾	9	9	9	6
Snout to ear	9	8½	8	8	5½
Snout to anterior gular fold	9	8½	8	8	5½
Snout to posterior gular fold	15	15	14	13	9
Base of fifth to end of fourth toe	6½	6	5¾	5½	4

History.—The first representatives of *Xantusia vigilis* were found at Fort Tejon, California, by Mr. John Xantus, who furnished the three specimens upon which Professor Baird based his original description, published in the Proceedings of the Academy of Natural Sciences of Philadelphia for 1858. Nothing more concerning it appeared until May, 1893, when Dr. Stejneger recorded two specimens secured by the Death Valley Expedition in 1891. Nothing has been known about its habits, and

* See remarks under *Habits*, p. 527.

† Regrown.

this very interesting species has been considered one of the rarest of our reptiles.

Distribution.—In reality, *X. vigilis* is the most abundant lizard in the territory it has chosen for its home. It seems to be peculiarly dependent upon the presence of tree yuccas. A glance at Dr. Merriams' map* shows that these weird plants grow in each of the localities from which the species has been recorded, viz.: Fort Tejon in the Cañada de las Uvas, and Hesperia, in California, and Pahrump Valley, in Nevada.

Dr. Charles H. Gilbert and the writer collected specimens near Mojave, and found a portion of a cast skin at Victor, California, in November, 1893. In September of the following year, the writer found this species common at Mojave and Hesperia, and secured a single specimen near Cabazon on the eastern slope of San Gorgonio Pass, California. The first three of these localities are situated in the great *Yucca arborescens* belt, which extends along the southwestern edge of the Mojave desert. The last is in a small and apparently isolated grove of smaller tree yuccas, seemingly of another species.

Habits.—Mojave, California, Nov. 4, 1893. About a mile from the station, there is a considerable forest of *Yucca arborescens*. The many trees and wind-broken branches, which lie decaying on the ground, afford a home to numerous colonies of white ants, scorpions, viscious-looking black spiders, and several species of beetles. In a deep crack of one of these branches a small lizard was discovered which, when caught, proved to be a young [illegible]. Probably it had not yet learned how to [illegible] undis- [illegible]

[illegible] the collection

of a large series of these lizards was merely a matter of physical exertion. Every fourth or fifth stem that was examined gave up its Xantusia, and in one instance five, as many as were previously known to collections, were found under a single tree.

Most of the lizards were found between the bark and the ground, but many had hidden in the thick clusters of dead leaves, from which it was very difficult to dislodge them. When first exposed to the light, they were dark colored, and seemed dazzled for a moment, during which they made no attempt to escape. They were not at all sluggish, however, and, if not caught immediately, made for the nearest cover as fast as their very short legs would permit. This cover was often the collector, and the little lizards either hid under his shoes, or climbed his legs, sometimes even reaching his shoulders. They showed no desire to enter the numerous holes in the ground about them, or to escape by burrowing. Put into a glass bottle they became very light colored in a few minutes, but began to turn dark again immediately after sundown. Young specimens were numerous, and remained dark longer than adults. Many fragments of cast skins were found, but never a whole skin in one place. The stomachs of several individuals contained the wings of some small dipterous insect, the elytra of a little brown beetle, and some very small white bodies which resembled spiders' eggs.

Several specimens were taken alive to The Leland Stanford Junior University, and kept for some months in a large glass jar in which some fine sand and pieces of wood and bark had been placed. At first, they ventured out from their retreat only at dusk unless disturbed, but after a few days they seemed to become more restless, and, urged perhaps by hunger, showed themselves many times each day. At night, when they were always more active, they often climbed to the top of a piece of yucca stem placed upright in the middle of their cage. No desire to burrow was observed. All declined to show any interest in the small beetles and flies, both dead and living, which were placed in the jar, and finally became greatly emaciated. They were chloroformed in March, 1894.

Mojave, Cal., Sept. 17–18, 1894. As it was not practicable to learn by actual investigation whether or not *X. vigilis* hid, during the day, among the thick-growing leaves of the living yuccas, the localities examined in 1893, still clearly marked by the displaced rubbish, were again searched with great care. The fact that very few specimens were now secured in this previously-worked area while the species was very common just outside its limits, is evidence that the specimens found on the ground under the dead branches were in their true diurnal home, and not mere stragglers from the living yuccas.

The specimens were all caught alive and put into a large glass bottle, but were soon killed by the heat, although care was taken to keep them in the shade as much as possible. Count was kept as the lizards were put in the bottle, and showed later that several more were taken out than had been put in. This may have been due to a mistake in the record, but was more probably caused by the birth of young after capture. The adults were afterwards carefully examined and three were found to contain young, showing that the species is ovoviparous. One of the three contains two fetuses, the others have one each. These fetal specimens are about the size of the young found under the dead branches.

Hesperia, Cal., Sept. 26, 1894. Xantusias were very abundant. Young were as numerous as at Mojave, Sept.

17–18, 1894, and the habits observed were the same as recorded there.

Cabazon, Cal., Sept. 28, 1894. A single specimen, secured after several hours searching, was shaken from the dry leaves of a dead but still standing yucca about two feet high.

XANTUSIA GILBERTI Van D.

Xantusia gilberti.

1895, Van Denburgh, Proc. Cal. Acad. Sci., vol. v, p. 121, pl. xi.

Description.—The body is subcylindrical, with very short limbs. The upper surface of the head is flattened, curving towards the snout. There are three folds on the throat, the anterior connecting the ears and encircling the head. The nostril is pierced at the junction of the rostral, internasal, postnasal, and first labial plates. The rostral is in contact with the first labial and internasal plates. The two internasals are followed by a large frontonasal, which separates the prefrontal plates. Behind the latter are two large frontals. Each of the two frontoparietal plates forms sutures with one of the frontals, the second and third superciliaries, first supratemporal, parietal, interparietal, and its fellow of the opposite side. The parietal and the very large interparietal are bordered behind by the two large occipitals. A row of small supratemporal scutes lies along the outer edge of the occipital and parietal plates. The two large loreals are in contact below with the superior labials, and above with the frontonasal and prefrontal plates. In front of the first loreal is a large postnasal. A series of small plates, the upper of which are the superciliaries, surrounds the eye.* Between this ring and the larger loreal are two small plates. There are five superior and four inferior

*Most of these plates are united on the side of the head shown in the figure, pl. xi.

labials to a point below the pupil. The eye is small, without lids, and with vertical pupil. Its diameter is contained about two and one-half times in the distance from the end of the snout to the orbit. The oblique ear opening has a weak anterior denticulation. The inferior labials are in contact with the large sublabials. The back, sides, posterior surfaces of the limbs, and the gular regions, are covered with smooth subhexagonal granules. These are flattened on the gular region, but convex on the back and sides. There is a series of large plates along the edge of the last gular fold. The quadrate ventrals are in twelve longitudinal and thirty-two transverse rows. The tail is conical, and covered with whorls of smooth scales, which are very narrow and transversely convex. There are eight and nine femoral pores.

The color above is dark brownish clay, dotted with black on single granules. A pale yellowish line, two granules wide, runs posteriorly from each occipital plate, but is soon lost on the back to reappear over the thigh. The lower surfaces are pale yellowish white.

Snout to vent (about) 39 mm. Tail (about) 38 mm. Hind limb 14 mm. Fore limb 10 mm. Shielded part of head 8⅜ mm. Snout to ear 8 mm. Snout to anterior gular fold 7¼ mm. Snout to posterior gular fold 12¾ mm. Base of fifth to end of fourth toe 4½ mm.

Distribution.—*Xantusia gilberti* is known from a single specimen taken at San Francisquito, Sierra Laguna, Lower California, Mexico.

XANTUSIA HENSHAWI Stejn.

Xantusia henshawi.
1893, Stejneger, Proc. U. S. Nat. Mus., p. 467.

Description.—The body is greatly depressed, with very short limbs. The upper surface of the head is very flat. There are three folds on the throat. The nostril is pierced in a small scute at the junction of the rostral, internasal, postnasal, and first labial plates. The rostral is broad and rather low, bounded by the first labial, nasal and internasal plates. The two internasals are followed by a large subquadrate frontonasal, which is sometimes divided longitudinally, behind this are two prefrontals, bordered posteriorly by the broad frontal and the first superciliary plates. Each of the two frontoparietal plates is in contact with the frontal, second third and fourth superciliaries, first supratemporal, parietal, interparietal and its fellow of the opposite side. The parietals and interparietal are bordered behind by the two large occipitals. One or more interoccipitals are sometimes present. There is a row of small supratemporals along the outer edge of the occipital and parietal plates. The two large loreals are in contact below with the superior labials, and above with the frontonasal and prefontal plates. The eye is surrounded by a series of small plates, the upper five of which are the superciliaries. Between this ring and the larger loreal are two small plates. There are five superior and three inferior labials to a point below the pupil. The eye is large, without lids, and with vertical pupil. Its diameter is contained about twice in the distance from the end of the snout to the orbit. The ear opening has a very weak anterior denticulation. The symphysial plate is very long. The inferior labials are in contact with the large sublabials. The first pair of the latter are in contact on the median line. The back, sides, upper and posterior surfaces of the limbs, and the gular regions, are covered with subhexagonal granular scales. There is a series of large quadrate plates along the edge of the last gular fold. The quadrate ventrals are in fourteen longitudinal and thirty-three or thirty-four transverse rows. The preanal plates are arranged in three or four rows, the two median ones of the posterior series being largest. The conical tail is somewhat depressed at its base and is covered with whorls of smooth scales, which are very narrow and transversely convex. Eight or ten femoral pores form a series along each thigh.

The ground color above is broccoli brown. On this are numerous large irregular rounded blotches of very dark seal brown, between which run more or less continuous lines of pale yellow. The upper surfaces of the limbs and head are similarly, but less distinctly, marked. The tail is yellow with irregular blotches and half rings of blackish seal brown. The lower surfaces are uniform yellowish white.

	mm.	mm.	mm.
Snout to vent	65	63	57
Tail	83	69	66
Hind limb	—	27	26
Fore limb	—	16	10
Shielded part of head	13	14	12½
Snout to ear	—	13	12
Snout to anterior gular fold	—	13	12
Snout to posterior fold	—	21	20
Base of fifth to end of fourth toe	—	10	9½

Distribution.—*Xantusia henshawi* has been found only at Witch Creek, San Diego County, California. This locality is in the chaparral belt, at an "altitude of about 2,700 feet."

Habits.—Here this species lives among the numerous granite boulders, and comes out into the narrower crevices between them a few minutes before dark. It is, therefore, practicable to hunt for it only about fifteen or twenty minutes each day. If a bit of string or a straw be introduced into the domain of one of these lizards it will often be seized, the reptile apparently mistaking it for some stray insect.

XANTUSIA RIVERSIANA Cope.

Xantusia riversiana.

1883, Cope, Proc. Ac. Nat. Sci. Phila., p. 29.
1889, Rivers, Am. Nat., xxiii, p. 1100.
1889, Cope, Proc. U. S. Nat. Mus., p. 147.

Description.—The limbs are very short, and the body is somewhat depressed. The upper surface of the head is very flat. The nostril is pierced in a small scute at the junction of the rostral, internasal, postnasal, and first labial plates. The rostral is broad and rather low, bounded by the first labial, nasal, and internasal plates. The two internasals are followed by a large hexagonal frontonasal. Behind this are two prefrontals, bordered posteriorly by the broad frontal and the first superciliary and first supraocular plates. Each of the two frontoparietal plates is in contact with the frontal, second, third and fourth supraoculars, parietal, interparietal, and its fellow of the opposite side. The interparietal is bordered behind by the two large occipitals, which are separated from the parietals by two small scutes. There is a row of large supratemporals along the outer edge of the occipital and parietal plates. The two large loreals are in contact below with the supralabials, and above with the frontonasal and prefrontal plates. The eye is surrounded by a series of small plates, the upper five of which are the superciliaries. Between this ring and the posterior loreal are two or three small plates. A series of four supraoculars separates the superciliaries from the frontal and frontoparietal plates. There are five superior and four or five inferior labials to a point below the pupil. The eye is large, without lids, and with vertical pupil. The ear has a weak anterior denticulation. The inferior labials are in contact with the large sublabials. The first pair of the latter are in contact on the median line. The back, sides, upper and posterior surfaces of the limbs, and the gular regions, are covered with flattened granules. There is a series of large plates along the edge of the last gular fold. The quadrate ventrals are in sixteen longitudinal and thirty-two to thirty-five transverse rows. The large preanal plates are arranged in two or three series, edged by smaller scales and granules. The conical tail is covered with whorls of smooth scales which are very narrow and transversely convex. There is a series of from ten to twelve femoral pores along each thigh.

The ground color is smoke gray or cinnamon, with numerous irregular maculations of dark brown or black. These markings are much smaller and less numerous on the lower surfaces. There is considerable variation in the color pattern. One specimen has two narrow parallel black lines, originating at the posterior edge of each occipital plate, and running the whole length of the back. The space between each pair of these lines is unmarked, but the rest of the upper surface is irregularly spotted. Other specimens offer an almost perfect imitation of coarse granitic rock.

Snout to vent 106 mm. Tail (injured) 73 mm. Hind limb 38 mm. Fore limb 30 mm. Shielded part of head 24 mm. Snout to ear 24 mm. Snout to anterior gular fold 20 mm. Snout to posterior fold 34 mm. Base of fifth to end of fourth toe 14 mm.

Distribution.—This largest species of the group has been recorded from San Nicolas, Santa Catalina, and San Clemente Islands, California.

A LIST OF LICHENS COLLECTED BY MR. ROBERT REULEAUX IN THE WESTERN PARTS OF NORTH AMERICA.

BY DR. STIZENBERGER, KONSTANZ, GERMANY.

Mr. Reuleaux having kindly favored me with a small collection of lichens gathered during his last year's travels through the United States, and comprising specimens from Yellowstone Park and Monterey, Cal., as well as from Sitka, Alaska—some of them never found in North America before—I made the above-mentioned cryptogams the subject of my special study, the results of which are laid down in the following list, intended to serve as a supplement to the late Professor Tuckerman's Synopsis of North American Lichens:

1. SPHÆROPHORON CORALLOIDES Pers., Tuck. New Engl. 82. Sitka.
2. BÆOMYCES ICMADOPHILUS (Ehrh.) Nyl., Tuck. Syn. ii, 7, 8; on dead wood. Sitka.
3. CLADONIA FIMBRIATA f. TUBÆFORMIS (Hffm.) Nyl., Tuck. Syn. i, 241. Sitka.
4. CLADONIA BELLIDIFLORA (Ach.) Schær., Tuck. Syn. i, 252. Sitka.
5. CLADINA RANGIFERINA (L.) Nyl., Tuck. Syn. i, 242. Sitka.
6. RAMALINA CERUCHIS (Ach.) DN. var. CEPHALOTA Tuck. Syn. i, 21; on dead twigs of shrubs. Monterey.
7. RAMALINA RETICULATA (Noehd.) Krmplh., Tuck. Syn. i, 22. Sitka, Monterey.
8. RAMALINA FARINACEA (L.) Ach., Tuck. Syn. i, 25. Sitka.
9. RAMALINA POLLINARIELLA Nyl.; sterile. Sitka.

10. ALECTORIA SARMENTOSA Ach., Tuck. Syn. i, 45; sterile. Sitka.

11. ALECTORIA PROLIXA (Ach.) Nyl., *A. jubata* c. *implexa* Tuck. Syn. i, 44; fertile. Sitka.

12. CHLOREA VULPINA (L.) Nyl., Tuck. Syn. i, 28; sterile on dead wood. Yellowstone Park.

13. PLATYSMA LACUNOSUM (Ach.) Nyl., Tuck. Syn. i, 35; sterile. Sitka.

14. PARMELIA SULCATA Tayl., Nyl., Tuck. Syn. i, 59; on trees. Sitka.

15. PARMELIA VITTATA (Ach.) Nyl., Tuck. Syn. i, 60; fertile on twigs of trees. Sitka.

16. PARMELIA ENTEROMORPHA Ach., Tuck. Syn. i, 60; evernioid, fertile. Sitka.

17. STICTINA SCROBICULATA (Scop.) Nyl., Tuck. Syn. i, 102; fertile. Sitka.

18. STICTA PULMONARIA (L.) Ach., Tuck. Syn. i, 94; fertile. Sitka.

19. NEPHROMA ARCTICUM (L.) Fr., Tuck. Syn. i, 103; sterile. Sitka.

20. PHYSCIA LYCHNEA (Ach.) Nyl., Tuck. Syn. i, 50; on bark. Sitka.

21. PHYSCIA LYCHNEA (Ach.) var. PYGMÆA (Bory) Nyl., Tuck. Syn. i, 51; upon granitic rocks. Sitka.

22. PHYSCIA LYCHNEA (Ach.) var. LACINIOSA (Schær.) Stzb. Helv. No. 305. Thallus red on application of hydrate of potassa; on bark. Sitka.

23. LECANORA ELEGANS (Link) Ach., Tuck. Syn. i, 170; on rocks. Sitka. Yellowstone Park.

24. LECANORA CERINELLA Nyl. Luxb. 370; on thin twigs of coniferous trees. Monterey.

25. LECANORA LACINIOSA (Dut.) Nyl., *Theloschistes concolor* Tuck. Syn. i, 51 p. p. Thallus without reaction on application of hydrate of potassa; on bark. Sitka.

26. LECANORA POLYTROPA (Ehrh.) var. ILLUSORIA Ach., Tuck. Syn. i, 192 p. p. (*Lecanora varia* var. *polytropa*); on rocks. Yellowstone Park.
27. LECANORA SYMMICTA (Ach.) Nyl., Tuck. Syn. i, 192 p. p. Thallus red on application of hypochlorite of lime; on dead wood. Sitka.
28. LECANORA HYPOPTOIDES Nyl. in Flora 1867, 371; on dead wood. Yellowstone Park.
29. LECANORA PALLESCENS (L.) Ach., Tuck. Syn. i, 196; on bark. Sitka.
30. LECANORA COARCTATA (Sm.) Ach., Tuck. Syn. ii, 15; on rocks. Yellowstone Park.
31. CŒNOGONIUM INTERPOSITUM Nyl. Coen. 91, Tuck. Syn. i, 258; on thin twigs. Monterey.
32. LECIDEA MEIOCARPA Nyl. in Flora 1876, 577; on cones of cypress. Monterey.
33. LECIDEA SANGUINEO-ATRA (Ach.) Nyl., Tuck. Syn. ii, 21, f. *corticola;* on thin twigs of coniferous trees. Monterey.
34. LECIDEA MYRIOCARPA (DC.) Nyl., Tuck. Syn. ii, 97; on cones of cypress. Monterey.
35. OPEGRAPHA ATRORIMALIS Nyl. in Flora 1864, 488; on cones of cypress and on thin twigs. Monterey.

It still remains to add here the diagnosis of a new western lichen, kindly sent me by Mr. Henry Willey, New Bedford, Mass.

ALECTORIA PACIFICA Stzb. n. sp.

Thallus fruticulous, prostrate, rigid, terete, smooth, brown and shining, from 1 to 1.5 cm. in length, 1–1.5 mm. in width, very much divaricately branched, the branches flexuous, densely intertangled, 0.25 mm. in diameter, at the ends forked with very short branchlets, scarcely 0.05 mm. in width, apothecia and spermogonia unknown.

The anatomical structure perfectly agreeing with *Alectoria;* no traces of an orthogonal-trajectoric direction of hyphæ (as it is found in *Cetraria aculeata).* Cortical and medullary layer with equal, nearly longitudinally running filamentous elements. No central cavity; medullary layer cottony, very loose, sprinkled with heaps of gonidia (these 0.004–8 mm. in diameter). Thin sections of the thallus bordered with a very thin light-brown line. The cortical layer neither thickened nor interrupted by larger cavities (which are frequent in the older cortical tissue of *Cetraria).* No reactions on application of hydrate of potassa and hypochlorite of lime.

Found in the Island of Guadalupe (Pacific Ocean), on humous earth, by Dr. Palmer.

SOME PARASITIC HYMENOPTERA FROM BAJA CALIFORNIA AND TEPIC, MEXICO.

BY WILLIAM H. ASHMEAD.

Through the kindness of Mr. Wm. J. Fox, of the Philadelphia Academy of Sciences, I have been enabled to examine and report upon another interesting collection of parasitic Hymenoptera, made in Baja California and Mexico, in the fall of 1894, by Messrs. Eisen and Vaslit, members of the California Academy of Sciences.

The collection, although small in numbers, represents thirty-eight distinct species, distributed in seven families, and many of which, especially among the microscopic forms, prove to be new to science, and are briefly characterized below.

Family PROCTOTRYPIDÆ.

Mesitius Spinola.

1. Mesitius nigripilosus sp. n.

♀.—Length 4.5 mm. Black, shining, with sparse black hairs, more especially apparent on the head and the apical half of the abdomen. Scape, pedicel, mandibles and legs, except coxæ and posterior femora, reddish yellow; palpi white; flagellum dark brown; wings subfuscous, the veins brownish yellow.

The head is scarcely longer than wide across the eyes, alutaceously sculptured, with some sparse, shallow, thimble-like punctures scattered over its surface. Antennæ 13-jointed, filiform, tapering toward tips and extending a little beyond the tegulæ; the scape is obconical, slightly curved, about four times as long as thick at apex, while the flagellar joints are all longer than thick, averaging from 1½ to 2 times as long as wide. The pronotum is long, subtrapezoidal, as long as the mesonotum and scutellum united; mesonotum with two complete furrows; scutellum triangular, with a punctured frenum; metathorax quadrate, with several longitudinal raised lines on its disk. Abdomen conic-ovate, polished, longer than the thorax, with the fourth and following segments sparsely fimbriate with black hairs.

Described from one ♀ specimen from Tepic.

GONIOZUS Förster.

2. GONIOZUS MEXICANUS sp. n.

♀.—Length 2 to 2.1 mm. Allied to *G. cellaris* Say, agreeing with it in colorational detail, and in having a small, closed, triangular discoidal cell, but it is readily separated from it, as well as *G. palliditarsis* Cam., by its smaller size, much longer head, black mandibles and longer abdomen.

The head in this species is nearly twice as long as wide, not narrowed behind the eyes, the space behind the eyes being fully as long as the eye itself, while in *G. cellaris* the head is only a little longer than wide, rounded behind the eyes, the space being much shorter than the length of the eye.

The abdomen is as long as the head and thorax united or slightly longer. The antennæ, except the 7 apical joints, as well as the tibiæ and tarsi, are yellow or brownish yellow, although the middle and posterior tibiæ are sometimes dusky.

Decribed from two ♀ specimens from Tepic.

3. GONIOZUS TEPICENSIS sp. n.

♂.—Length 2.6 mm. Agrees in color and sculpture with *G. mexicanus*, but the head is a little longer, fully twice as long as wide, the space behind the eye being twice as long as the eye, while the anterior wings have no closed triangular discoidal cell, the basal vein having only a slight stump of a vein present.

Described from one ♂ specimen from Tepic.

The long oblong head readily separates this species from all other described species in our fauna, having no triangular closed discoidal cell in the front wings.

Family CYNIPIDÆ.

Eucœla Westwood.

4. Eucœla mexicana sp. n.

♀.—Length 1.6 mm. Polished black; mandibles and legs, including all the coxæ, rufous; antennæ black, the 5 or 6 basal joints beneath, rufo-piceous; wings hyaline, the veins pallid, with a slight yellowish tinge. Antennæ 13-jointed, reaching to the base of the abdomen, the scape longer than the pedicel, the first joint of the flagellum slender, but as long as the scape, the following joints, 2–5, gradually shortening, but increasing in thickness, joints 6–10 oblong-moniliform, equal, the last joint ovate, longer than the preceding, the joints 7 to 11 are all delicately fluted.

The scutellum is rugulose, its cup oval, connected with the hind margin of the mesonotum by a short carina, the disk flat, with four punctures and a small fovea on its posterior margin. Front wings with the marginal cell closed, the first abscissa of radius about two-thirds the length of the second. Metapleura with a small tuft of wool just above the hind coxæ. Abdomen a little longer than the head and thorax united, with a narrow but dense woolly girdle at base.

Described from 1 ♀ specimen from San Lazaro.

Hexaplasta Förster.

5. Hexaplasta californica sp. n.

♀.—Length 1.1 mm. Polished black; mandibles and legs, including coxæ, reddish-yellow; wings hyaline, strongly ciliated, the veins dark brown; the marginal cell except at basal one-third, open along fore margin.

Antennæ 13-jointed, black; scape not quite twice as long as the pedicel, the latter oval; funicle 5-jointed, slender, the first joint 2½ times as long as the second, joints 2–3, moniliform, scarcely longer than thick, joints 4–5 a little longer; club 6-jointed, the joints, except the last which is ovate, oblong-moniliform. Cup of scutellum elliptic, with a small fovea posteriorly and four punctures on its disk. Front wings with the second abcissa of radius stouter and a little longer than the first. The metapleura have a small tuft of wool just over the base of coxæ, while the abdomen is not longer than the head and thorax united, with the usual woolly girdle at base.

Described from one ♀ specimen from San Lazaro.

Family TENTHREDINIDÆ.

Hylotoma Latreille.

6. Hylotoma pœciloides sp. n.

♂.—Length 8 mm. Testaceous; head, antennæ, three spots on anterior margin of pronotum, the lateral lobes of the mesonotum, the tegulæ, the wings (except a subhyaline streak medially), the prosternum, a line on each side of the mesosternum, the anterior tarsi (except the first joint beneath), middle and hind tibiæ and their tarsi, apex of the sixth abdominal segment and the following segments, all black.

Described from one ♂ specimen from San José del Cabo.

The species is allied to *Hylotoma pœcila* Klug and *H. intermedia* Cam., but is readily recognized by its color: From the former in having the stigma and veins black, not yellow, the wings being less distinctly banded at the middle, while the anterior tibiæ are pale not black. From *H. intermedia* it differs in its smaller size, by the black lateral lobes of the mesonotum and by the different color of the legs.

Family BRACONIDÆ.

Iphiaulax Förster.

7. Iphiaulax megaptera Cam.

Biol. Centr.-Am. Hym., p. 358, Tab. xv, f. 5. ♀

Four ♀ specimens from San José del Cabo.

Bracon Fabr.

8. Bracon excelsus Cam.

Biol. Centr.-Am. Hym., p. 321. ♀

The male of this species has never been described, but to it I refer a single ♂ from San José del Cabo, which differs in no ways from the ♀, except in its smaller size and in the usual sexual differences.

9. Bracon foxii sp. n.

♀.—Length 6.5 mm.; ovipositor about twice as long as the abdomen. Reddish-yellow; the stemmaticum, antennæ, palpi and legs, except the hind coxæ and basal two-thirds of hind femora, black. Wings smoky black, subhyaline at base, the tegulæ, stigma and veins black.

The surface is smooth, highly polished, impunctate; head subquadrate, the face feebly punctate and clothed with glittering white pile; the under surface of the thorax and the legs are also clothed with pile, but more sparsely so. Abdomen smooth, but the second segment has a triangular elevation at the basal middle, with broad depression on each side of it, while the third has two oblique lateral furrows, and is separated from the second by a deep slightly arcuate but smooth furrow.

Described from one ♀ specimen from San José del Cabo. This lovely species is dedicated to Wm. J. Fox, as a slight appreciation of the many favors he has shown me.

Hedysomus Förster.

10. Hedysomus quadriceps sp. n.

♀.—Length 9 mm.; ovipositor a little longer than the abdomen. Black, with the middle lobe of the mesonotum, the metathorax and the abdomen red.

Head quadrate, polished, the face below antennæ rugulose, the clypeus fimbriate; mandibles broad without teeth within.

Antennæ longer than the body, the scape large, stout, thicker and longer than the pedicel and first joint of flagellum; the first joint of the flagellum is twice as long as the pedicel. The thorax, except the metanotum is smooth and polished, the metanotum being reticulate with large, coarse punctures. The anterior wings have the median and submedian cells of an equal length, with the second abcissa of the radius twice as long as the first; hind wings with the radial cell divided by a cross-vein. Hind femora short, much swollen.

Abdomen much wider than the thorax, the first and second segments irregularly longitudinally striated, the second being divided into three parts by a semicircular grooved furrow.

Described from one ♀ specimen from San José del Cabo.

This genus closely resembles Doryctes, but the relative lengths of the three basal joints of antennæ, basal cells of anterior wings, and the characteristics of the abdomen, as pointed out by Förster, readily separate it.

Hormius Nees.

11. Hormius albipes sp. n.

♀.—Length 2 mm.; ovipositor very short. Ferruginous; stemmaticum, lobes of mesonotum, scutellum, metanotum and first segment of abdomen dusky or black;

palpi and legs whitish. Antennæ 20-jointed, a little longer than the body, the joints of the flagellum more than three times as long as thick. Wings hyaline, the second abscissa of the radius not or scarcely longer than the first.

Described from two ♀ specimens from San José del Cabo.

CHELONUS Jurine.

12. CHELONUS ALBOBASILARIS Ashm.

Proc. Cal. Acad. Sci. (2) IV, p. 123.

One ♂ and one ♀ from San Lazaro. The ♂ agrees with the female except that the abdomen has a transverse fissure at the apex, as in *C. fissus* Prov., *C. minimus* Cr., etc. Some authors have described these as females, but I have ascertained by a careful examination that they are really males.

APANTELES Förster.

13. APANTELES MEXICANUS sp. n.

♂ ♀.—Length 2–2.2 mm. Black, shining, the mesothorax above punctate; the head, disk of mesopleura and episterna of metathorax smooth, impunctate; the scutellum is almost smooth but with some sparse punctures; palpi white; scape, pedicel beneath and legs (except hind coxæ, tips of hind femora and tibiæ and their tarsi which are black or fuscous) reddish-yellow; the abdomen beneath and sometimes the suture between segments 3 and 4, brownish-yellow; flagellum brown-black, paler beneath. Wings hyaline, the stigma and veins brown, the inner vein of the open areolet is a little longer than the recurrent nervure. Metathorax short, rugulose with a median carina. Abdomen with the plate of first segment trapezoidal, about 1½ times as long as wide, the hind angles slightly rounded; the second segment in ♀ is a little shorter than the third, in the ♂ a little longer than the third; segments 1 and 2 are feebly sculptured, the following all smooth and polished: the ovipositor not prominent.

This species spin their cocoons in large masses covered with a white woolly secretion, resembling a cotton-ball, and it is probably parasitic on the larva of some large sphinx moth.

In appearance it resembles *A. congregatus* Say, but is readily separated by its metathoracic and abdominal characters.

Described from many specimens from Tepic and San José del Cabo.

TOXONEURON Say.

14. TOXONEURON SEMINIGRUM Cr.

Tenthredoides seminigrum Cr. Proc. Ent. Soc. Phil. iv, p. 291.
Toxoneuron seminigrum Cr. Can. Ent. v, p. 69.
Toxoneura seminigra Cr. Syn. Hym. p. 239; Ashm. Proc. Ent. Soc. Wash. iii, p. 52.

One ♀ specimen from San José del Cabo.

I have restored the original spelling of this genus, since there is a Dipterous genus Toxoneura.

OPIUS Wesmael.

15. OPIUS BRUNNEIVENTRIS Cr.

Trans. Am. Ent. Soc. iv, p. 178.

One ♀ specimen from San José del Cabo.

The species was originally described from Texas, but it also occurs in the Western States and in Canada.

PHÆNOCARPA Förster.

16. PHÆNOCARPA MEXICANA sp. n.

♀.—Length 2 mm.: ovipositor as long as the abdomen. Polished black: prosternum rufous: three basal joints of antennæ, mandibles and legs brownish-yellow; palpi white: flagellum much longer than the body, brown-black, pubescent, the second joint one-third longer than the first. The mesonotum has a small fovea just in front of the large fovea at base of the scutellum. Metathorax rugulose, with two smooth areas at base. Wings hyaline, the stigma and veins light brown, the second submarginal cell twice as long as the first transverse cubital vein, the second transverse cubital vein only about two-thirds as long as the first. Abdomen scarcely longer than the thorax, with the first segment longitudinally striated.

Described from one ♀ specimen from San Lazaro.

Family ICHNEUMONIDÆ.

Subfamily II, OPHIONINÆ.

Enicospilus Curtis.

17. Enicospilus mexicanus Cr.

Ophion mexicanus Cr. Proc. Phil. Acad. Sci., 1873, p. 374.
Enicospilus mexicanus Cam. Biol. Centr.-Am. Hym., p. 290.

One ♂ specimen from San José del Cabo.

18. Enicospilus maculipennis Cam.

Biol. Centr.-Am. Hym., p. 292.

One ♂ specimen from San José del Cabo.

Ophion Fabr.

19. Ophion subfuliginosus Ashm.

Proc. Cal. Acad. Sci. (2), iv, p. 126.

Two specimens, one ♂, one ♀, from San José del Cabo.

Exetastes Grav.

20. Exetastes fascipennis Cr.

Proc. Ent. Soc. Phil., iv, p. 278.

One ♀ specimen from San José del Cabo.

Subfamily IV, ICHNEUMONINÆ.

Trogus Grav.

21. Trogus pulcherrimus sp. n.

♂.—Length 18 mm. Yellow-fulvous; the upper half of the head, the antennæ, upper part of pronotum, metanotum, a spot beneath, the tegulæ, hind legs, apical one-fourth of anterior wings and a band across before the stigma, apical two-thirds of hind wings and the three terminal segments of abdomen black. The head is smooth, polished; the mesonotum punctate, becoming very finely and closely punctate toward the lateral margins, the pronotum punctate only along the upper hind margin; the scutellum conically elevated, sparsely punctate; metanotum with a deep transverse furrow at base, the posterior face with shallow punctures, clothed with a sparse black pubescense; abdomen, except the basal half of the petiole, longitudinally shagreened.

Described from one ♂ specimen from San José del Cabo. A most lovely species, imitating some of the forms found in the genus Joppa.

Œdicephalus Cresson.

22. Œdicephalus albomaculatus sp. n.

♀.—Length 9 mm. Black, shining; the antennal joints 10–20, orbits, face, mandibles, palpi, lateral margins of pronotum, two short lines on disk of mesonotum, spot on the scutellar ridges, the scutellum, except the depression at base, the post-scutellum, a small spot before each metathoracic spiracle, two large spots on the posterior face of metathorax, a large spot beneath the tegulæ, the anterior coxæ and trochanters and a spot on the hind coxæ above at base, middle coxæ except behind, and a spot on the middle of mesopleura and metapleura, all white: legs mostly, apical margins of all the abdominal segments and the venter, yellow. The antennæ, except the broad white annulus, the middle coxæ behind, the hind coxæ except as noted, the hind trochanters above, their femora above, tips of their tibiæ and last joint of tarsi, black: the middle chanters behind and their femora above, as well as their tibiæ at tips, are more or less brown or dusky. Wings hyaline, the tegulæ except a white spot at base, the stigma and the veins black; the areolet is pentagonal, but the lateral veins strongly converge toward each other above, so that the portion of the radius which forms its upper side is very short. The head is polished, impunctate, the mesonotum polished, but finely, although not closely, punctured; the metathorax punctured and finely rugulose, distinctly areolated, the median area divided into two by a central longitudinal carina, while the spiracular and middle pleural areas are confluent. The abdomen has the apex of the petiole and the second segment closely punctate, otherwise smooth and polished, with the gastrocœli distinct, but widely separated.

Described from one ♀ specimen from San José del Cabo.

Subfamily V, CRYPTINÆ.

JOPPIDIUM Walsh.

23. JOPPIDIUM ANNULICORNE sp. n.

♂.—Length 13 mm. Dark rufous, closely punctate; the flagellum, except joints 10–19, prosternum, lower part of mesosternum, surroundings of scutellums, a broad band on the metanotum, all coxæ and trochanters, the hind legs, except tarsi, and the petiole of abdomen, black. Antennal joints 10–19, anterior legs, tips of middle tibiæ and all tarsi, yellow. Wings smoky black, except a yellowish streak at base of the stigma and along the stigma of hind wings; the areolet is large, subquadrate.

Described from one ♂ specimen from San José del Cabo.

This species approaches nearest to *Joppidium ardeus* Cr.

POLYCYRTUS Spinola.

24. POLYCYRTUS ALBOANNULARIS sp. n.

♀.—Length 20 mm. Head and thorax, except scutellums, black; joints 7–14 of antennæ, labium, palpi, anterior coxæ beneath, scutellums, spots on three apical segments of abdomen and the apical half of basal joint of hind tarsi, as well as joints 2 and 3, white; abdomen and legs, except hind tibiæ, basal half of first joint of their tarsi, as well as the two last joints, red; hind tibiæ and tarsi, except as already noted, black or dark fuscous; wings subhyaline, dusky at tips, the stigma and veins fusco-black.

The head and thorax are closely punctate, except a smooth shining space at sides of collar, on upper middle of mesopleura, and that portion of metanotum enclosed by the first transverse carina; the metanotum behind this ridge is transversely rugulose, the lower part of the mesopleura, the surface beneath the insertion of hind wings and the metapleura being clothed with an appressed pubescence.

Described from one ♀ specimen from San José del Cabo.

Subfamily VI, PIMPLINÆ.

PIMPLA Fabr.

25. PIMPLA FERALIS Cr.

Proc. Acad. Sci. Phil., 1873, p. 399.

One ♂ specimen from San José del Cabo.

Family EVANIIDÆ.

EVANIA Fabr.

26. EVANIA APPENDIGASTER Linn.

Ichneumon appendigaster Linn. Syst. Nat., ed. xi, p. 566.

One ♀ specimen from San José del Cabo.

For a list of the extensive synonymy of this species consult August Schletterer's "Die Hymenopteren-Gruppe Evaniiden, Wien, 1889–90."

Family CHALCIDIDÆ.

SMICRA Spinola.

27. SMICRA DELIRA Cr.

Trans. Am. Ent. Soc., iv, p. 41.

One ♂ specimen from Tepic. The species was originally described from Texas, but is widely distributed over the Southern and Western States.

HALTICHELLA Spinola.

28. HALTICHELLA XANTICLES Walk.

Hockeria xanticles Walk., Ann. Soc. Ent. Fr. (2), i, p. 147.
Haltichella xanticles Cr., Syn. Hym. N. A., p. 234.
Haltichella americana How., Bull. No. 5 U. S. Dept. of Agric., p. 9.

Two ♂ specimens from Tepic. This species is also widely distributed over the United States and is very variable in size.

EURYTOMA Illiger.

29. EURYTOMA SEMINATRIX Walsh.

Am. Ent., ii, p. 299.

Three ♀ and two ♂ specimens from Tepic. The species infest woolly cynipid galls and I fail to find any difference between those from Tepic and those bred by myself from galls in Florida. Walsh considered the species only a variety of his *E. auriceps*, but from a close study of many specimens I fail to find intermediate grades, and, as the specific characters are constant, I believe it should be elevated to a distinct species.

30. EURYTOMA TEPICENSIS sp. n.

♀.—Length 2.2 mm. Black; scape and legs, except coxæ and the hind femora medially brownish-yellow; tegulæ black; wings hyaline, the veins brownish-yellow.

Head and thorax, flagellum, legs and apex of abdomen clothed with a sparse, glittering white pubescence. The flagellum is not quite three times as long as the scape, the pedicel is half as long as the first joint of funicle very little longer than thick at apex; funicle 5-jointed, without counting the single ring-joint, the first joint being the longest, or fully twice as long as thick, the following very slightly decreasing in length so that the fifth joint is only slightly longer than thick; club 3-jointed, a little shorter than the scape. The pronotum seen from above is almost as long as the mesonotum, the scutellum to its tip being considerably longer than the mesonotum. Metanotum medially sulcate, the sulcus having two delicate parallel carina, the space between them being filled with delicate transverse raised lines. Marginal vein of front wings rather stout, 1½ times as long as the stigmal, the latter a little shorter than the postmarginal.

Abdomen subsessile, not longer than the thorax, blunt at apex, the fifth segment the longest, about 2½ times as long as the fourth, the sixth about half the length of the fifth, the seventh a little longer than sixth, bearded with white hairs and bearing spiracles, the eighth segment retracted.

Described from one ♀ specimen from Tepic.

EVOXYSOMA Ashmead.

31. EVOXYSOMA DECATOMOIDES sp. n.

♂.—Length 2.5 mm. Black; face, orbits, streak behind ocelli, prosternum, middle lobe of mesonotum posteriorly the axillæ and the scutellum, brownish-yellow; basal half of scape, palpi and legs, white; flagellum black: wings hyaline, the veins pale brownish, the postmarginal vein a little longer than the marginal, or more than twice as long as the stigmal. The head is wider than the thorax, with large prominent eyes, the occiput deeply roundedly emarginate, while the face has a deep antennal emargination. The flagellum is long, filiform, the joints binodose, with whorls of long, white hairs. The abdomen is polished black, ovate, attached to the thorax by a long, punctate petiole.

Described from one ♂ specimen from Tepic.

PERILAMPUS Latreille.

32. PERILAMPUS TRIANGULARIS Say.

LeConte Ed. Say's Works, i, p. 381.

Six specimens from San Lazaro.

ORASEMA Cameron.

33. ORASEMA VIRIDIS sp. n.

♀.—Length 3 mm. Head, thorax, scape of antennæ, femora and petiole of abdomen, all metallic green; flagellum black; mandibles rufous, the right with two teeth, the left with only one within; tibiæ and tarsi brownish-yellow; abdomen æneous black; wings hyaline, the veins dark brown.

Described from one ♂ specimen from Tepic.

METAPON Walker.

34. METAPON MEXICANUM sp. n.

♀.—Length 2 mm. Head and thorax æneous black, closely punctate; mandibles rufous; scape, pedicel and legs, except the coxæ and anterior and middle femora, ferruginous; anterior and middle femora metallic brown; hind coxæ blue. Abdomen pointed ovate, polished black, as long as the thorax and keeled beneath. Flagellum incrassated toward tip, the joints, after the first, increasing in width and wider than long. Wings hyaline, the veins brown, the marginal vein twice as long as the stigmal. The fourth abdominal segment is not quite half as long as the third, the fifth only a little shorter than the third, while the sixth and seventh are a little longer.

Described from one ♀ specimen from Tepic.

The species comes nearest to *M.* [illegible] Walk., but its smaller size, the metallic colored anterior and middle femora and the shape and relative length of the abdominal segments readily distinguish it.

Catolaccus Thomson.

55. Catolaccus tepicensis sp. n.

♀.—Length 2.8 mm. Aeneous black, the dorsum of mesothorax and the scutellum bronzy-green, finely closely punctate and clothed with a fine whitish pubescence. Scape, knees, tibiae and tarsi, brownish-yellow, the tibiae medially more or less infuscated; mandibles rufous, the teeth black; wings hyaline, the veins light brown. Head transverse, wider than the thorax. Ocelli whitish. Antennae 13-jointed, the pedicel not quite as long as the first joint of funicle, the last two joints of funicle a little wider than long. Marginal vein of front wings 1½ times as long as the stigmal, the latter a little shorter than the postmarginal. Abdomen conic-ovate, a little longer than the head and thorax united, aeneous black, clothed with a sparse white pubescence beneath.

Described from one ♀ specimen from Tepic.

Elasmus Westwood.

56. Elasmus sp.

Of this interesting genus there is a single ♀ specimen from Tepic, but with the abdomen and hind legs gone and in too poor a condition to satisfactorily describe.

[illegible]

Wings hyaline, the veins pale yellowish. Antennæ 10-jointed, with two ring joints, the funicle 3-jointed, the first two joints subequal, more than twice longer than thick, the third only about twice as long as thick, the club ovate, 3-jointed. Abdomen ovate, polished black, with a short finely rugose petiole, the body of abdomen at apex is clothed with sparse black hairs.

Described from one ♀ specimen.

This genus has only been characterized recently by the writer, the types coming from St. Vincent, West Indies. The present species approaches nearest to *C. petiolata*, but it is larger, not so smooth, with the facial striæ coarser, while the joints of the funicle are proportionately longer. For description of the genus and the other two species see Journal of the Linnean Society, Zoology, vol. xxv, 1894, pp. 178–179.

TETRASTICHUS Haliday.

38. TETRASTICHUS ORBITALIS sp. n.

♀.—Length 1.1 mm. Shining black: face, orbits, scape, pedicel beneath, anterior margin of pronotum, inner margins of the scapulæ, tegulæ, base of abdomen and legs, except the coxæ and the middle and hind femora, brownish-yellow. Flagellum light brown, the three funicle joints gradually shortening, but also thickening, the first the longest, a little more than twice as long as thick, the last only about 1½ times as long as thick, the club large, stout, ovate, 3-jointed, fully twice as thick as the first joint of the funicle. Wings hyaline, ciliated, broadly rounded at tips. Abdomen sessile, ovate, as long as the head and thorax united and much broader than the thorax, with the sheaths of ovipositor somewhat prominent.

Described from one ♀ specimen from San Lazaro.

A REVIEW OF THE HERPETOLOGY OF LOWER CALIFORNIA. PART II—BATRACHIANS.

BY JOHN VAN DENBURGH,

Curator of the Department of Herpetology.

The long peninsula of Lower California, parched and barren except where some stream, escaping from the sheltering shadows of the upland oaks and pines, winds down to the ocean or sinks almost immediately into the panting soil, has few attractions to offer the batrachia. In consequence, few representatives of this class have been found within its limits. Those that do occur either live in the moister mountainous areas or are of wide distribution and comparatively great adaptation for life in a land arid and desolate. The *Bufo* and the *Scaphiopus* range as far east as Texas; the *Batrachoseps* and *Hyla regilla* occupy a considerable area along the Pacific; while the *Plethodon* has been taken, elsewhere, only in southern California.

HYLA REGILLA B. and G.

Hyla regilla.

(1852, Baird and Girard, Proc. Ac. Nat. Sci. Phila., vi, p. 174.)
1866, Cope, Proc. Ac. Nat. Sci. Phila., p. 313.
1877, Streets, Bull. U. S. Nat. Mus., No. 7, p. 35.
1883, Yarrow, Bull. U. S. Nat. Mus., No. 24; p. 171.
(1887, Belding, West Am. Scientist, iii, 24, p. 99.)

Hyla curta.

(? 1883, Yarrow, Bull. U. S. Nat. Mus., No. 24, p. 171.)
(? 1887, Belding, West Am. Scientist, iii, 24, p. 99.)

Hyla regilla var. *laticeps.*

1889, Cope, Bull. U. S. Nat. Mus., No. 34, p. 359.

I fail to find any constant difference between specimens of this species from various parts of its range. The characters which have been claimed to be distinctive seem to be purely individual, and to occur wherever a series of specimens has been secured.

This *Hyla* has been recorded from Cerros Island, La Paz, and Cape San Lucas, in Lower California.

List of specimens of Hyla regilla.

Cal. Acad. Sci. No.	Locality.	Date.	Collector.
403	San Francisquito, Sierra Laguna, L. C.	Mar. 27, 1892	Gustav Eisen.
404 to 408	Sierra Laguna, L. C.	"	"
427 to 431		"	"
600	San Rafael Valley, L. C.	Apr. 29, 1893	A. W. Anthony.
601	San Pedro Martir Mt., L. C.	May 19, 1893	"
682 to 697	San Ignacio, L. C.	April, 1889	W. E. Bryant.
698 to 702	Comondu, L. C.	Mar., 1889	"
988	Miraflores, L. C.	Oct., 1893	Gustav Eisen.
989	"	"	"
997	San José del Cabo, L. C.	Sept., 1893	"
2250	San Francisquito, Sierra Laguna, L. C.	Mar. 27, 1892	"
2256	Sierra Laguna, L. C.	Oct., 1893	"
2257	"	"	"

HYLA CURTA Cope.

Hyla curta.

1866, Cope, Proc. Ac. Nat. Sci. Phila., p. 313.
(1875, Cope, Bull. U. S. Nat. Mus., No. 1, pp. 30, 92.)
(1881, Brocchi, Miss. Sci. au Mex., Batraciens, p. 39.)
(1884, S. Garman, Bull. Essex Inst., xvi, 1, p. 45.)
(1887, Cope, Bull. U. S. Nat. Mus., No. 32, p. 15.)
(1889, Cope, Bull. U. S. Nat. Mus., No. 34, pp. 351, 360.)

There seems to be some confusion in the published references to this species. At the end of the original description of *Hyla curta* is the statement, "No. 5293, 19 specimens half ♂), Cape St. Lucas. Jno. Xantus." Later, in the list of specimens of *H. regilla* in "The Batrachia of North America," is written, "No. 5293—19 specimens—Fort Tejon, Cal.—John Xantus," which is also the entry made by Dr. Yarrow in his Check List. Thus, apparently, the types of *H. curta* are referred to *H. regilla*, with a change in the statement of locality. But *H. curta* is still recognized as a distinct species (Bull. U. S. N. M., No. 34, pp. 351, 360). It would be interesting to know definitely which locality is the correct one, and whether two species of *Hyla* really exist in Lower California.

SCAPHIOPUS COUCHII Baird.

Scaphiopus couchii.
(1854, Baird, Proc. Ac. Nat. Sci. Phila., vii, p. 62.)
(1889, Cope, Bull. U. S. Nat. Mus., No. 34, p. 301.)
Scaphiopus varius.
1864, Cope, Proc. Ac. Nat. Sci. Phila., 1863, p. 52.
(1881, Brocchi, Miss. Sci. au Mex., Batraciens, p. 27.)
Scaphiopus couchii (var. *varius*).
(1866, Cope, Proc. Ac. Nat. Sci. Phila., p. 313.)
Scaphiopus varius varius.
(1875, Cope, Bull. U. S. Nat. Mus., No. 1, p. 31.)
(1883, Yarrow, Bull. U. S. Nat. Mus., No. 24, p. 177.)
Scaphiopus couchi.
1883, Yarrow, Bull. U. S. Nat. Mus., No. 24, p. 177.
(1887, Belding, West Am. Scientist, iii, 24, p. 99.)
Scaphiopus couchii varius.
(1884, S. Garman, Bull. Essex Inst., xvi, 1, p. 46.)
(1887, Cope, Bull. U. S. Nat. Mus., No. 32, p. 12.)
1889, Cope, Bull. U. S. Nat. Mus., No. 34. fig. 75.

This species appears to be much less abundant in Lower California than *Bufo punctatus*. It has been recorded only from the San Lucas Fauna, where Mr. Xantus collected it at Cape San Lucas, and Mr. Belding at La Paz.

List of specimens of Scaphiopus couchii.

Cal. Acad. Sci. No.	Locality.	Date.	Collector.
533	San José del Cabo, L. C.	Sept. 23, 1890	W. E. Bryant.
759	"	Sept. 27, 1890	"
763		"	"
930 to 972		Sept., 1893	Gustav Eisen.
1360 to 1363	Miraflores, L. C.	Sept., 1894	Eisen and Vaslit.
2440 to 2444	San José del Cabo, L. C.	"	"
2512	Miraflores, L. C.	"	"

Bufo punctatus B. & G.

Bufo punctatus.

(1852, Baird & Girard, Proc. Ac. Nat. Sci. Phila., vi, p. 173.)
1866, Cope, Proc. Ac. Nat. Sci. Phila., p. 313.
(1875, Cope, Bull. U. S. Nat. Mus., No. 1, p. 29.)
(1883, Yarrow, Bull. U. S. Nat. Mus., No. 24, p. 162.)
(1887, Cope, Bull. U. S. Nat. Mus., No. 32. p. 10.)
(1887, Belding, West Am. Scientist, iii, 24, p. 99.)
(1889, Cope, Bull. U. S. Nat. Mus., No. 34, p. 262.)

Bufo beldingi.

1882, Yarrow, Proc. U. S. Nat. Mus., p. 441.
(1883, Yarrow, Bull. U. S. Nat. Mus., No. 24, p. 163.)
(1887, Belding, West Am. Scientist, iii, 24, p. 99.)

Judging from the large series of specimens secured by the Academy's collectors, toads of this species must be very numerous in the "Cape Region" of Lower California. Several which were collected in September contain eggs nearly ready for deposit.

Mr. Xantus found this toad at Cape San Lucas, and Mr. Belding at La Paz.

List of specimens of Bufo punctatus.

Cal. Acad. Sci. No.	Locality.	Date.	Collector.
432	Santa Anita, L. C.	Apr. 4, 1892	Gustav Eisen.
634 to 636	San Ignacio, L. C.	April, 1889	W. E. Bryant.
745 to 747	Agua Caliente, L. C.	Oct., 1890	
895 to 900	San José del Cabo, L. C.	Sept., 1893	Gustav Eisen.
927 to 929		"	"
977	Corral de Piedras, Sierra El Taste, L. C.	"	
978	"		
1364 to 1386	Miraflores, L. C.	Sept., 1894	Eisen and Vaslit.
2129 to 2190	San José del Cabo, L. C.	"	"
2401 to 2439			
2445 to 2511	Miraflores, L. C.	"	"

BATRACHOSEPS ATTENUATUS (Esch.)

Salamandrina attenuata.

"1833, Eschscholtz, Zool. Atl., pt. v, 1, pl. 21, figs. 1-14."

Batrachoseps attenuatus.

1880, Lockington, Am. Nat., xiv, p. 295.

Mr. Lockington has reported *Batrachoseps* from La Paz, where it was secured by Mr. W. J. Fisher. The Academy has a specimen (No. 619) collected by Mr. T. S. Brandegee on San Pedro Martir Mt., Lower California.

PLETHODON CROCEATER Cope.

Plethodon croceater.

(1867, Cope, Proc. Ac. Nat. Sci., Phila., p. 210.)
1869, Cope, Proc. Ac. Nat. Sci. Phila., p. 100.
(1875, Cope, Bull. U. S. Nat. Mus., No. 1, pp. 27, 92.)
1880, Lockington, Am. Nat., xiv, p. 295.
(1883, Yarrow, Bull. U. S. Nat. Mus.. No. 24, p. 192.)
(1887, Cope, Bull. U. S. Nat. Mus., No. 32, p. 9.)
1889, Cope, Bull. U. S. Nat. Mus., No. 34, p. 150.

Mr. Xantus is said to have collected a *Plethodon* of this species at Cape San Lucas, and Mr. Lockington has recorded one from the northern part of the peninsula, seventy-five miles southeast of San Diego, California.

THE CALIFORNIA PHRYGANIDIAN (PHRYGANIDIA CALIFORNICA PACK.)

BY VERNON L. KELLOGG AND F. J. JACK.

With Plate IV.

Last fall and this spring the oak trees, especially the live-oaks (*Quercus agrifolia*), in the vicinity of the Leland Stanford Jr. University and of Palo Alto suffered serious defoliation by the attacks of the larva of *Phryganidia californica*. The pest is not remembered to have been so abundant here before. The caterpillars appeared in astounding numbers, the continuous dropping of frass from the infested trees attracting common attention. The caterpillars were conspicuous, also, on the tree trunks and on fences and the walls of buildings near trees, often massing in a way suggesting the well-known *Datana* masses, although never forming such compact and isolated bunches. The special interest attaching to this insect, because of its systematic isolation among the Heterocera, its limited geographical range, and its capacity, abundantly shown last fall and this spring, for damage, led us to make the observations recorded in the following notes.

The larvæ, mostly full-grown, were noticed in great numbers on September 10, 1894. They were feeding singly, although crowded together by numerical abundance. The massing already referred to was especially noticeable in the crotches of the trees and on the trunks and large branches. The larvæ in these groups maintained an irregular jerking of the head, much less pronounced than the jerking of *Datana* larvæ but very like it. The caudal extremity of the body is commonly elevated when the larva is at rest, although the anal feet are used when the caterpillar is walking.

A few chrysalids were seen on September 10. By October 1 most of the larvæ had pupated, the naked chrysalids being conspicuous objects on the tree-trunks, on fences and the walls of buildings near trees. When the larva is ready to pupate, it lies along the bark head downward with body contracted longitudinally and a little curved. It then spins a thin irregular net of silk (showing well when the larvæ pupated on the glass sides of breeding jars), covering very thinly the surface against which it lies. The chrysalid is attached to the supporting object only by the projecting caudal process (see *d*, plate lvi), although the body of the chrysalid rests against the thin silken net. The fresh chrysalid is fleshy pinkish-yellow, and it retains a considerable sensitiveness and mobility up to the time of the issuance of the imago.

By the middle of October many moths were flying. In the laboratory the pupal stage was uniformly of ten days' duration. The moths were very abundant all through the latter half of October, fluttering with a pretty, wavering flight through the foliage of the oaks. The eggs are laid in patches commonly on the under side of the oak leaves (occasionally on the upper side), from two dozen (rarely fewer) to four dozen being laid together. They lie in a single layer, almost or barely touching one another, and often in irregular lines. In the laboratory the egg stage lasted twelve days in all instances noted. But out of doors the eggs did not begin hatching until about December; and then they hatched irregularly, unhatched eggs being found up to January 1. As late as February 14 larvæ in the first stage were found. The eggs show in a couple of days a shallow polar depression, and surrounding it a zone of pinkish-brown. This zone in eggs five days old is a striking cherry red. Just before hatching the egg becomes pink and ashy mottled all over. The larvæ display a singular slowness of growth.

in the laboratory, from eggs hatched December 15, the larvae first moulted December 29, or fourteen days after hatching. The second moult occurred January 11, the third January [illegible]—at this time larvae in the trees out of doors were in the second stage, *i. e.*, had moulted but once—the fourth moult occurred February 22, the fifth, [illegible] and pupation March 27. The imagines issued [illegible] ten or eleven days after pupation, agreeing with the duration of the pupal stages in the fall before. Thus we have the winter passed by the insect in [illegible] larval stage, the evergreen condition of the live-oak lending itself to such a life history. Out of doors the larval stages were passed more slowly than in the laboratory. At the time of the issuance of the moths in the indoor breeding cages on April 6, larvae just making the third moult were found in the live-oaks. The out of doors larvae began pupation by the middle of May, the moths appearing at the end of the month and during the first part of June. The newly hatched larvae and those in the second stage, in feeding, merely skeletonize the leaves, the soft parenchymatous tissue being eaten, and the firmer vascular tissue being left unattacked. After the second moult, however, all the leaf substance is eaten.

The larvae of the winter and spring were, although abundant, less numerous in many localities than those of the late summer and fall brood. This disparity in numbers was largely due to the commendable zeal of a particular parasite,* *Pimpla behrensii* Cresson. To a discerning observer the abundance of this parasitic ichneumon about the oaks in September was as apparent as the hordes of caterpillars. A resident of Palo Alto complained of the large number of "small wasps" (the ichneumons) which entered his house, and buzzed in the

*Determined by Mr. L. O. Howard.

windows. The effectiveness of this parasite wherever it occurred is shown by the fact that of 100 chrysalids examined on November 1, 67 were parasitized by this ichneumon; and from 144 chrysalids gathered and kept in the laboratory but 11 moths issued, 99 of the chrysalids being parasitized by *Pimpla behrendsii* Cress., 7 chrysalids containing other parasites, and the others being dead from various causes.

Although most abundant on the live-oaks *(Q. agrifolia)* the larvæ attack other oaks. We have found them on *Quercus lobata*, *Q. kelloggii*, *Q. dumosa* and *Q. douglassii*. The live-oaks in this vicinity begin to put out new leaves about January 1, but in the case of many of the trees badly defoliated by the larvæ in the autumn, new leaves appeared much earlier than the first of January. The wintering of the insect in a larval condition is only possible in the evergreen oaks, and they are thus the natural and usual host of the pest. At the time of the hatching of the first of the autumn brood of eggs (latter part of November) the leaves of the deciduous oaks begin to fall. But, oddly, the eggs were found to be deposited on the leaves of both the white oak and Douglas's oak (deciduous oaks), and the larvæ hatched only to die of starvation. By this suicidal means the pest aids in depleting its own numbers. The new leaves of the deciduous oaks appear about April 1, before the eggs for the summer brood of larvæ are deposited. These eggs, therefore, can safely be laid on the leaves of these trees, but the eggs laid by the fall moths on the foliage of these trees give up their young to certain destruction.

As to the number of generations which appear annually of this insect, Henry Edwards (quoted by Packard in Hayden's Report of the U. S. Geological Survey of the Territories for 1875, and in Forest Insects, Fifth Report of the U. S. Entomological Commission, 1890) states that there are two. Packard, in Forest Insects, quotes Mr. Behrens of San Francisco as saying that three generations appear annually. In our year of observation but two generations appeared. Moths were flying in May, 1894, the larvæ from whose eggs became full-grown in September, and produced the October moths. From the slowly and irregularly hatching eggs of these moths came the slow-growing winter brood of larvæ which became full-grown in the laboratory in the last of March, but out of doors not until the middle of May. From these came again a late May and early June brood of moths. It is to be noted that the occasional appearance of moths, as, for example, two specimens captured on February 20, 1895, resulting from the more rapid growth and transformation of a few individuals of a brood, is not an unusual phenomenon in the life history of this insect. It may explain, too, some of the unwarranted statements occasionally heard concerning this pest, crediting it with five or six annual generations.*

The descriptions of the egg, larval stages and chrysalid follow. The only illustration of the larva of this species we have seen, that of Stretch, after which the figure for Packard's Forest Insects was made, is a case of mistaken identity, the conspicuous tufts of hair on the figured caterpillar having no counterparts on the Phryganidia larva.

Egg (see *a*, plate lvi)—Smooth, spherical, becoming slightly depressed at one pole soon after exclusion, this depression becoming conspicuous in a few days. Diam-

* Dr. H. H. Behr in an article entitled "On the power of adaptation in insects," Proc. Cal. Acad. Sci. ser. 2, vol. 5, Aug., 1895, states that there are four or five generations annually, but after considering again the data at command concludes (as expressed in a private letter) that "Phryganidia has but two regular generations, but under certain circumstances there must develop at least one more."

eter .85 mm. Color, shining yellowish-white. In two days a circular zone about the depressed apex is brownish pink, which in five days after exclusion becomes cherry-red. Just before hatching the whole egg is mottled with dark-pinkish. Laid on the lower (rarely upper) surface of the leaves of *Quercus agrifolia* and of other species of *Quercus*, in patches of from thirty to fifty.

Larva, first stage (see *b*, plate lvi)—Head large, rounded, bilobed, the mesal line distinct, bearing a few prominent hairs; color when excluded from the egg, pearl-gray, soon changing to light brown, shining, width .68 mm. Body cylindrical, with conspicuous setigerous tubercles, arranged as follows: a row on each side of the dorsimeson, composed of two large approximated tubercles on each segment, the caudal tubercle of each segment lying slightly laterad of the cephalic one. These dorsal rows bend a little laterad on the second and third segments. Two rows on each side, the upper row consisting of large tubercles one on each segment, the lower row of smaller ones two on each segment, the caudal tubercle of the lower pair being a little above the cephalic one. Cervical shield, broad and widest at mesal part, brown. Anal shield, distinct, brown. Color of the body ashy at exclusion, changing in four hours to bright yellowish green above and ashy below, tubercles brown to black, and there is a narrow, subdorsal, interrupted reddish line extending whole length of the body; legs concolorous with venter, and marked with brown blotches. Triangular, brownish blotches occur in mesal line on segments 4, 6, 8 and 11, and may or may not extend across the dorsum of segment. Length (at exclusion from egg) 2.05 mm.; width .26 mm.

Second stage—Head shining brown, ocelli and mouth

parts dark reddish brown; width 1.14 mm. The brown coloration of the tubercles has disappeared. Two prominent, reddish, continuous, subdorsal lines, also a faint interrupted median dorsal line. On the pleurum, reddish markings on segments 4–11 forming two interrupted rows of short, sinuous lines. Prominent reddish-brown blotches on segments 8 and 11. Length 7 mm.; width .65 mm.

Third stage—Width of head 1.45 mm. Color of body bright yellow. Subdorsal lines more pronounced than in previous stages and continuous; mesal line present or not; on segments 2 and 3 are two parallel reddish lines extending ventrad from the subdorsal line a short distance. Cervical shield black. Length 10 mm.; width .78 mm.

Fourth stage—Width of head 1.88 mm. There may or may not be faint interrupted lines between the median dorsal and subdorsal lines. A more or less interrupted, ill-defined reddish supra-stigmatic line appears, with narrow uneven lines running longitudinally between this supra-stigmatic line and the subdorsal lines. (All these lines unite in the last larval stage to form the broad subdorsal band, in which the composing lines may be partly traced.) An interrupted reddish stigmatic line is apparent on segments 1–9. General color of body bright to dirty yellow, the cervical and anal shield and dorsal blotches black. Length 18 mm.; width 1.5 mm.

Fifth stage—Width of head 2.22 mm. Markings of body as in previous stage, but usually more pronounced. Length 22 mm.; width 2 mm.

Sixth stage (see *c*, plate lvi)—Head large, rounded, bilobed, mesal line pronounced, bearing a few conspicuous hairs; smooth, dark brown, shining; ocelli and adjacent region black; clypeus prominent, ashy-gray;

proximal segment of antennæ whitish, distal parts of remaining segments black; width of head 2.57 mm. Body cylindrical, 11th and 12th segments humped, smooth, shining, tubercles and hairs not noticeable; general color light yellow. An alternative body color is black above and on sides, and ashy-gray on venter. The majority of the fall brood were yellow; the majority of the spring brood black. Cervical shield prominent, black; anal shield small, black. Thoracic legs black. The reddish median dorsal line widens or becomes a large blotch on the 8th segment. There is a conspicuous transverse blotch on the humped 11th segment. Smaller blotches occur also sometimes on the 4th and 6th segments. There is a narrow, uneven, black line on each side of the median dorsal line parallel with it and continuous from 1st to 12th segments. Laterad of these narrow lines, there is a conspicuous broad black subdorsal band composed of several contiguous narrow lines, the composing lines frequently blending. On segments 6–11 just above the bases of the prolegs, which are yellowish white with reddish markings, there are two short sinuous reddish lines, the lower one of each pair being the broader and more distinct and the space between them being pearly-white in color. Connecting the upper one of these two lines with the subdorsal band there is on segments 6–9 a short sinuous vertical reddish line. When the general body color of the larva is black the spaces between the narrow dorsal lines remain yellowish appearing as four narrow parallel dorsal lines running the whole length of the body. Length 23–27 mm.; width 3 mm.

At end of this stage, the larvæ let themselves down from the tree to the ground by a silken thread, and then crawl up on the side of an adjacent building or upon a fence, or upon the trunk of the tree, or they crawl down

from the foliage to the tree trunk, and form a naked chrysalid.

Pupa (see *d*, plate lvi)—Naked, suspended by cremaster, greenish-white with yellow suffusion and black markings. Wing cases pearly bluish-white, with many black lines of different lengths. On the dorsum of body an interrupted median black line frequently expanding blotch-like. On either side of it on abdomen, and separated from it by distinct yellowish markings, an interrupted broad black band extending to base of cremaster. Spiracles black, surrounded by yellow. Ventrad of the line of spiracles a faint narrow longitudinal line of pinkish, and on the venter two submedian bands of black composed of narrow lateral blotches. Cremaster single, strong, length 1.5 mm. Chrysalid concave on dorsum and convex on venter, length 12 mm.; width 4 mm.

Imagines (see *e*, plate lvi)—Males easily distinguished from females by longer pectinations of antennæ, and indistinct yellowish patch just beyond apex of discal cell of fore wing.

DESCRIPTION OF A NEW SPECIES OF GOBIESOX FROM MONTEREY BAY, CALIFORNIA.

BY SETH EUGENE MEEK AND CHARLES J. PIERSON.

[With Plate lxxi]

Gobiesox muscarum n. sp.

Head $3\frac{4}{5}$ in length; depth $8\frac{3}{4}$; dorsal 6; anal 5.

Body elongate, slender, depressed anteriorly, very narrow but slightly compressed posteriorly, the greatest width of body immediately behind head, 7 in length.

Head narrow, much depressed, wider posteriorly. Eye small, its diameter $2\frac{1}{2}$ in interorbital width, 5 in head. Maxillary reaching to the front of the eye, its length less than 3 in head. Teeth in upper jaw conical, acute, curved, forming a crescent-shaped patch, those of the anterior row enlarged. In the lower jaw is an anterior row of about five broad, entire incisors, placed nearly horizontally; behind these a crescent-shaped patch of teeth, similar to those in the upper jaw, becoming canine-like laterally. No evident opercular spine.

Ventral disk longer than broad; its length $1\frac{1}{2}$ in head, $6\frac{1}{4}$ in length. Distance from vent to front of anal $2\frac{1}{2}$ in the distance from vent to disk.

Pectoral fin broad, short, $2\frac{1}{3}$ in head. Dorsal and anal fins small, the anal slightly in advance. Caudal fin rounded.

Ground color—in alcohol—light yellowish, paler below. Above, everywhere sparsely covered with distinct brownish-red spots about as large as pupil. A lateral band of the same color begins on the front of the snout, where it joins the band of the opposite side, extends through the eye across the opercle to the caudal, becoming very indistinct posteriorly. This lateral stripe is in strong contrast with the uniform pale ventral surface.

Two speciments were dredged in January, 1895, in

Monterey Bay, at a depth of about eight fathoms. One of these, the type (1½ in. long), is numbered 3030 on the Register of the L. S. Jr. Univ. Mus. The second specimen (1 1/10 in. long) resembles the type, but has the dorsal spots confined to the top of the head and nuchal region, the lateral stripe disappearing slightly behind middle of body, and the ventral surface marked posteriorly with brownish red spots like the dorsal surface.

ON THE CRANIAL CHARACTERS OF THE GENUS SEBASTODES (ROCK-FISH).*

(With Plates lvii-lxx.)

BY FRANK CRAMER.

The rock-fishes of the Pacific, commonly but erroneously called "rock-cod," constitute a large section of the Scorpænidæ, a family of the mail-cheeked fishes, and present extremely interesting problems in distribution and classification. Fifty or more species have been described during the past forty years from the west coast of North America, between the southern boundary of the United States and Bering Strait. Quite a large number of species also, distinct from the foregoing, have been discovered on the coast of Japan, and all the indications point to many more that are still undescribed. To the southward of the United States the group abruptly disappears, but reappears again in the temperate and cold waters of western South America, which undoubtedly still hold out a rich field for investigation of this group.

The rock-fishes of American waters are characterized by having 13 dorsal spines, while their nearest allies, the rose-fishes *(Sebastes)*, have a larger number. Some of the Japanese forms, however, are described as varying in the number of dorsal spines from 13 to 14. If this is so, the further study of the rock-fishes of the Japanese coast will furnish new and interesting material upon which to base the systematic arrangement of the group, for no such variation is found in all the fifty or more species of the western coast of North America.

* I wish to thank Prof. Charles H. Gilbert for putting at my disposal the material on which this paper is based, and for generously sacrificing valuable specimens, in order that the series might be made as complete as possible. The collection of skulls is now in the Museum of the Leland Stanford Jr. University.

There has hitherto been no agreement among ichthyologists as to the boundaries of the genera of rose- and rock-fishes. European writers, believing that the difference in the number of dorsal spines is not a sufficient basis for a generic separation of the Pacific forms, include them all in the old Cuvierian genus *Sebastes*. American writers, however, lay greater stress on this difference, which they have shown to be connected with a constant difference in the number of vertebræ. They are also prompted by the desirability of breaking up so large and unwieldy a genus into smaller natural groups, and have thus not only segregated the Pacific forms with 13 dorsal spines and 12+15 vertebræ in the genus *Sebastodes*, but have made several efforts to break up the latter genus into several smaller ones. Between 1854 and 1861 W. O. Ayres[1] described numerous species from the Pacific Coast of California, including them all under the old genus *Sebastes*. In 1861 Gill[2] proposed the genus *Sebastodes* for the *Sebastes paucispinis* of Ayres. In 1862 he placed all the remaining rock-fish of the West Coast in a new genus, *Sebastichthys*, but all the generic characters which he assigned have proved worthless.

Ayres accepted the genus *Sebastodes*, but redefined it so as to include the species *ovalis*, *flavidus*, *melanops* and *pinniger*. It will be seen that this was a natural group, the characters which he selected being correlated with others of which he knew nothing. He retained all the remaining West Coast rock-fish in the genus *Sebastes* "with the characters of *Sebastes* as given by Cuvier, except that the top of the head is always marked by spinous ridges, the orbits being commonly crested, so as to leave a depression between them."

[1] Ayres Proc Cal Acad Sci., 1854-1862

[2] Gill Proc Acad Nat Sci Phil., 1861, p 165; 1862, p. 329.

In 1864 Gill[3] separated the then known rock-fishes of the Pacific Coast into four genera: *Sebastodes, Sebastichthys, Sebastosomus and Sebastomus*. The groups which he thus indicated form natural assemblages of species, but thus far he has never defined them satisfactorily. The genera proposed by him have generally been accepted as of subgeneric value by later workers in the group, but with a knowledge of the early known species which Gill was unable to examine, together with many others discovered since, they have found it impossible to draw the lines of generic separation indicated by him.

In 1880 Jordan and Gilbert[4] discovered and described fifteen or more new species, and adopted a more definite terminology for the spinous ridges of the cranium, which seemed to them to furnish the most reliable characters. The arrangement adopted by them on the basis of these characters agreed in the main with the generic grouping already proposed by Gill. Since, therefore, the characters furnished by the top of the head had been most relied upon for the grouping of the species, and it was still a mooted question whether they should all be included in one genus or distributed among several, it seemed to the writer desirable to make a detailed examination of a series of skulls in order to determine what other cranial characters, if any, were correlated with those of the top of the head, and whether there were any gaps in the series which would serve as points of separation into genera.

As will be seen later, the writer has been unable to discover a basis for such generic separation and is convinced that the cranial characters fail to indicate such. Since the present investigation was completed, however, an at-

[3] Gill: Proc. Acad. Nat. Sci. Phil. 1864, p. 145.

[4] Jordan & Gibert: Proc. U. S. National Museum, 1880, p. 287.

tempt has been made by Eigenmann and Beeson * along the same lines, and with opposite results. It therefore becomes necessary to examine their conclusions in some detail. As a basis for the primary division of the group, they have selected the condition of the parietals, classifying the species according as their parietals meet or do not meet above the supraoccipital. The character is elsewhere described as the "union or non-union of the parietals," and the statement made that "the value placed on such a character .... need not be defended here." During the course of his investigation the writer also attempted to make use of this variation in the extent of the parietals, but came to the conclusion that it had little, if any, taxonomic value. The inner edges of the parietals are strictly superficial in position, overlapping the supraoccipital. Their inner margins are irregular, and the extent of the lap somewhat variable within the limits of each species, depending both on original individual variations and on the extent to which the thin edges of the bones have been absorbed. Taking a series of species, we have presented every degree of approximation of these margins, from the condition where they are wide apart and leave exposed a broad strip of the supraoccipital, to that in which they touch, meet, or overlap. Union is never effected between the parietals and it is misleading to speak of such. The manner in which the parietals reach or pass over the middle line is so variable as to suggest anything but genetic relationship. In a few species the inner edges of the parietals are parallel and seem to abut against each other in the middle line, in others the inner outlines are curved and the left parietal overlaps the right. In some cases

* Preliminary Note on the Relationship of the Species Usually United Under the Generic Name Sebastodes. C. H. Eigenmann and C. H. Beeson. American Naturalist, vol. XXVII, pp. 668-671, July, 1893. For convenience of reference this paper is given in full in the appendix, which see.

one of the parietals reaches the middle line and the other does not; in other cases the posterior part of one parietal and the anterior part of the other reach the middle line, and yet a wide strip of supraoccipital separates the two bones throughout their length. All of these conditions are evident in the accompanying figures. There is no more reason why that condition of the parietals in which they barely meet should be chosen as the line of separation between two groups of species than that any other degree of approximation or overlapping should be chosen. The character is unfitted *a priori* to serve as a primary character. The kind of difficulties into which its adoption leads is illustrated, among other instances, by the fact that *S. elongatus* and *S. levis* are placed in the group with separated parietals, although in some individuals the parietals plainly meet.

Not only is the condition of the parietals, by the nature of the character, unsuited for the purpose which it is made to serve, but it is not correlated with a single other important cranial character. After it is adopted as the primary character it does not serve in the slightest degree as a key to the rest of the structure. The degree of development of the cranial spines and ridges, the condition of the interorbital space, the curvature of the base of the skull, the condition of the ventral process of the basisphenoid and the direction of the mesethmoid processes are all closely correlated with each other and all lead to the same arrangement of species. The condition of the parietals not being correlated with a single other character, its use as a primary character is bound to rupture all the correlations that do exist; and that is what it does. To select a single illustration from among a host of them, the genus *Sebastomus*, as made up by Eigenmann and Beeson, includes species from all parts of the group:

rosaceus, *ruber*, *constellatus*, etc., with concave interorbital space, straight base of skull, and strong spines and ridges; and *miniatus* and *pinniger* with convex interorbital space, curved base of skull, weak spines and ridges and depressed mesethmoid processes. In every point of structure and conformation of skull the last two species are most closely related to the species placed in the genera *Primospina*, *Sebastosomus* and *Acutomentum;* and are widely separated from the other species of the genus *Sebastomus*.

The condition of the parietals was the first character selected by the writer as a basis for the arrangement of the species, but it was soon found unreliable from every point of view and had to be rejected; and the further the investigation proceeded the more clearly was its rejection justified. An examination of all the cranial characters in a large number of species will invariably lead to the same result.

Of the fifty or more species recognized from the Pacific Coast of America, the following thirty-two have been examined by me: *S. paucispinis*, *goodei*, *mystinus*, *melanops*, *flavidus*, *entomelas*, *ovalis*, *atrovirens*, *pinniger*, *miniatus*, *introniger*, *aurora*, *chlorostictus*, *rosaceus*, *constellatus*, *rhodochloris*, *ruberrimus*,* *saxicola*, *diploproa*, *elongatus*, *rubrivinctus*, *levis*, *serriceps*, *rastrelliger*, *auriculatus*, *vexillaris*, *caurinus*, *maliger*, *carnatus*, *chrysomelas* and *nebulosus;* besides two or three unidentified

* The specific name *ruberrimus* is here proposed as a substitute for the *ruber* of recent authors, not of Ayres, which latter must be regarded as a synonym of *auriculatus*. That the specimens to which the name *ruber* was first applied belonged to the species *auriculatus* is clearly shown by the careful description of the spines on the top of the head. The statements concerning color and size do not apply to *auriculatus*, but apply equally well to each of the three species *ruberrimus*, *pinniger* or *miniatus*. (Ayres, Proceedings California Academy of Sciences, vol. 1, p. 7, 1854.)

skulls. The following West Coast species were not available: *ciliatus, proriger, brevispinis, umbrosus, nigrocinctus, alutus, serranoides, rufus, melanostomus, rupestris, eos, ærcus, gilli, zacentrus, sinensis.*

The series upon which the following conclusions are based consisted of fifty-one skulls of thirty-two different species. Although many skulls could not be procured, the series is essentially complete, containing representatives from all parts of the group.

The cranial characters that have hitherto proved useful relate to the cranial ridges and the spines in which they end. The characteristic spines and ridges are: the preocular on the anterior superior border of the orbit; the supraocular, near the edge of the frontal bone above the middle of the orbit; the postocular, behind the supraocular, and the tympanic, behind the postocular on the frontal bone near the superior posterior angle of the orbit; and the parietal, present in all the species, a longitudinal ridge on the middle of the parietal bone. Of these ridges all may be absent except the parietal,* and in the different species in which they are present differ exceedingly in the degree of their development.

In a comparison of the crania some characters which it was at first supposed would furnish good marks by which to subdivide the genus into groups, proved otherwise. The thickness of the bones of the skull is generally correlated with other characters, rather thin papery skulls bearing strongly developed bony ridges, while thicker and more bony skulls have the ridges low or obsolete. But there are several exceptions to the rule. Other characters at first seem important, but as they occur

*Prof. Eigenmann has changed the name of this ridge and its spine from "occipital" to "parietal," and I have adopted his name for it, because it seems much more appropriate.

in a few species only, far apart in the series, they must be regarded as sporadic; thus nuchal spines are present in *S. levis*, *chlorostictus*, *aurora* and *constellatus* (in the last species connected with a tendency of the ridges to break up into spines and tubercles), but they are inconstant even in the species in which they occur; so that it is doubtful whether they are always present in any species. The coronal spines, likewise inconstant, are usually present in *S. aurora*, and nearly always present in *S. auriculatus*.

In some species in which pairs of spines are normally absent, these are sometimes present in a rudimentary or distorted form, either singly or in pairs. Although the *paucispinis* group is characterized by the absence of the usual pairs of spines in adults, two adult *paucispinis* skulls had a rudimentary supra- or postocular on the left side, and a very young skull of this division had rudimentary tympanic spines on both sides and a postocular on the left side: a medium-sized *melanops* had a rudimentary right tympanic; and a large one had a pair of postoculars and a deformed left supraocular; a young *flavidus* had a rudimentary right postocular; in an *elongatus*, in which the supraoculars are normally absent, the spines were still present in the form of low humps on the ridge; in another specimen the supraocular spine was sharp and perfectly distinct.

Hilgendorf expressed the belief that when one of the three pairs of spines (supraocular, postocular and tympanic) is absent, it is the supraocular and not the postocular that has disappeared.* This is proved by several

*Hilgendorf Uebersicht uber die japanischen Sebastes-Arten, Sitzungs-Bericht der Gesellschaft Naturforschenden Freunde zu Berlin, 21. Dec., 1880, p. 168. "Das maximum von Dornen am Oberkopf kommt bei *S. marmoratus* vor, namlich einer in der Nasengegend, der nasaldorn, drei am Augenrand, Orbital-dornen, von denen der mittlere bei den andern zuerst verschwindet."

series of facts. When the three spines are present together, the distance from the base of the tympanic to the base of the supraocular on the one hand, and the distance from the supraocular to the preocular on the other hand, are to each other in many species as 1 to 1, varying from this ratio to 1 to 3 in *rosaceus;* while where one of the spines is absent, the relative distances vary from 3 to 10 to 3 to 15 (except *nebulosus*, 2 to 5). These measurements give the all but invariable rule that, when one of the spines is absent the so-called supraocular occupies the position of the postocular. When both the supraocular and postocular are present and differ in size (which is usually the case), the supraocular is invariably weaker than the postocular. The depression between the tympanic and postocular is always deep, while between the postocular and supraocular there is frequently a well-marked ridge (*chlorostictus*, *rhodochloris*, *ruberrimus*). In *levis* the true supraocular is usually present; in the skull at hand it was absent, but on one side a blunt knob occupied the position required by the rule of relative distances, and just behind this point, on both sides, there was a depression in the otherwise continuous ridge, marking the depression between the supra- and postoculars. In the skull of *elongatus*, in which one of the pairs of spines is normally absent, there is a low, conical rudimentary spine on the left side, occupying the position of the supraocular, as required by the rule of relative distances. These facts, taken together, seem to establish the conclusion that when one of the trio of pairs of spines is absent, the supraocular spine has disappeared, and the supraocular ridge merged with the postocular.

A source of error that had to be studiously avoided in the comparison of species is that due to the changes that take place with increasing age. Of these, the following are among the most constant: The bones of the skull grow thicker and in very large specimens become spongy. The processes of the mesethmoid become depressed; and the ventral process of the basisphenoid, when present at all, sometimes suffers complete, and always partial absorption. The interorbital space grows relatively wider, this being one of the most striking and constant variations. In the present paper the width of this space is always given as measured at its narrowest part (which usually falls immediately behind the preocular spines), and compared with the total length of the base of the skull. In a young *vexillaris*, the ratio of interorbital width into the length of the base of the skull is $5\frac{1}{4}$, in a medium-sized one $4\frac{8}{9}$, and in a large one 4. In a young *maliger* it is $4\frac{4}{5}$, in an old one $4\frac{1}{3}$; in a young *miniatus* $3\frac{4}{7}$, in an old one $3\frac{1}{11}$; in a young *flavidus* $3\frac{7}{13}$, in an old one 3. In a very young *ruberrimus* it is $6\frac{4}{7}$, in one two or three times as large $5\frac{2}{7}$, in one in which the cranial ridges are almost competely serrated 5, and in a very large, old specimen $4\frac{5}{22}$.

The degree of approximation of the parietals seemed at first to be a valuable character, and it will be seen from the key given below that in several parts of the group closely related species have the parietals in contact; but while it serves well as a character of subordinate importance, the mere fact that any two species have parietals which meet or overlap is no proof of affinity unless it is supported by other agreements.

The most reliable cranial characters for the purpose of classification of the species are: the degree of curvature of the base of the skull; the convexity or concavity of the interorbital space and its relative width; the direction of the mesethmoid processes; the degree of development of the ventral process of the basisphenoid; and the strength or weakness of the cranial ridges. These characters are closely correlated, and furnish the only basis for the arrangement of the species within the genus. In the *paucispinis, melanops* and *pinniger* groups (see classification below) the base of the skull is strikingly curved; the interorbital space is always convex (at most flat, never concave) and relatively wide, its width never being more than 3½ in the length of the base of the skull; the mesethmoid processes are never directed upward; the ventral process of the basisphenoid is absent, or reduced to a mere point or at most occasionally present in very young specimens; the cranial ridges are poorly or not at all developed and the spines are delicate or absent. In the *rosaceus-nebulosus* groups the base of the skull is straight or nearly so; the interorbital space is always concave and narrow, its ratio in the base of the skull varying from 4¼ to $6\frac{4}{7}$; the mesethmoid processes are always directed more or less upward; and the ventral process of the basisphenoid, the cranial ridges and the spines are strongly developed.

These two groups of characters would furnish an ample basis for the division of the genus into two, if the species mentioned were alone to be considered. But between the two groups distinguished by these characters lies another (*introniger-aurora*) in which the base of the skull is somewhat curved (approaching straightness), the interorbital space is flat or slightly concave, of medium width, 4 to 4½ into the base of the skull, the processes of the mesethmoid are directed but little upward and the ventral process of the basisphenoid is poorly developed. By the interposition of this group it is possible to arrange a series from *paucispinis* to *rosaceus* in which there is an almost perfect gradation of all the above-mentioned characters, from strikingly curved to straight base of skull, from convex and broad to concave and narrow interorbital space, from mesethmoid processes depressed to those directed forty-five degrees above the dorsal plane of the skull, from a rudimentary to a fully developed ventral process of the basisphenoid and from nearly obsolete to strongly developed cranial ridges.

The single species *ruberrimus* furnishes at different stages in its development a series of characters that parallel in a striking way the series just described. The very young skull is so much like those of *rosaceus* and *rhodochloris* that, if it were the only *ruberrimus* at hand, it might easily be put between them in a series. The width of the interorbital space is $6\frac{4}{7}$ into the base of the skull, relatively narrower than that of any other skull in the collection of fifty, and deeply concave; the mesethmoid processes are directed upward and the ventral process of the basisphenoid is well developed. The very large skull of the same species is almost exactly adapted to the description of the *aurora-introniger* group. The interorbital space is perfectly flat and $4\frac{5}{22}$ into the base of the skull, the mesethmoid processes extend forward nearly horizontally and the ventral process of the basisphenoid is rudimentary. The gap between these two extremes is completely closed by skulls of intermediate age.

S. saxicola and *diploproa* constitute another intermediate group with the base of the skull markedly curved, the interorbital space slightly convex or flat, of medium width, $3\frac{3}{4}$ to $4\frac{1}{4}$ into the base of the skull, mesethmoid processes directed but little upward, and the ventral process of the basisphenoid rudimentary or fairly developed. This intermediate group, unlike the other, lacks the supraocular spine and probably forms one of the links between the *entomelas-pinniger* group and the other rockfish in which the supraocular is wanting.

The following classification, based exclusively on cranial characters, summarizes what has been said and includes some details not hitherto mentioned:

A. Base of skull markedly curved. Interorbital space convex or flat, broad, less than 3½ in the base of the skull. Processes of mesethmoid not directed upward. Ventral process of basisphenoid rudimentary. Cranial ridges obsolete or weak, spines absent or delicate.

a. Cranial ridges (except parietal) obsolete or very slightly developed. Cranial spines absent or very inconstant and weakly developed.

b. Parietals not meeting; mesethmoid processes weak and depressed; skull moderately thick; parietal ridges weak, with minute spines or none; other ridges none.

c. Interorbital space plainly convex, *paucispinis*.

cc. Interorbital space nearly flat, *goodei*.

bb. Parietals meeting in the middle line, but separated posteriorly by a wedge-shaped exposure of the supraoccipital. Mesethmoid processes better developed, straight and horizontal; skull thick; the bones striated; parietal ridges low, spineless, other ridges none.

d. Preocular spines none, *flavidus*, *melanops*.

dd. Preocular spines present, *mystinus*.

aa. Cranial ridges somewhat developed; preocular, supraocular, postocular, tympanic and parietal spines present, all delicate; ventral process of basisphenoid sometimes present in young. (Tympanic spines usually absent or imperfect in *atrovirens*.)

e. Parietals not meeting; interorbital space usually plainly convex; bones thick, more or less striated.

f. Supraocular spine present.

g. Base of skull strikingly curved; parietals nearly meeting, *entomelas*, *ovalis*.

gg. Base of skull less strikingly curved; parietals well separated.

h. Interorbital space plainly convex, *pinniger*.

hh. Interorbital space flat or nearly so, *miniatus*.

ff. Supraocular spine absent; parietals well separated; interorbital space but little convex; mesethmoid processes directed somewhat upward, *atrovirens*.

B. Base of skull markedly curved. Interorbital space flat or slightly concave, of medium width, 3¾ to 4½ in base of skull. Processes of mesethmoid directed but little upward. Ventral process of basisphenoid rudimentary or fairly developed.

h. Cranial ridges fairly developed, supraocular spines absent, skull thin, papery, mesethmoid processes horizontal.

i. Parietals not meeting. *saxicola*.

ii. Parietals meeting. *diploproa*.

C. Base of skull nearly straight (slightly curved). Interorbital space flat or slightly concave, of medium width, 4 to 4½ in base of skull. Processes of mesethmoid directed but little upward. Ventral process of basisphenoid rudimentary or poorly developed. Cranial ridges and spines quite strong.

j. Cranial ridges well developed. Preocular, supraocular, postocular, tympanic, parietal and nuchal spines present. Coronal spines usually present. *introniger*, *aurora*.

D. Base of skull straight or nearly so. Interorbital space concave and narrow, 4½ to 6⅔ in base of skull. Processes of mesethmoid directed upward. Ventral process of basisphenoid well developed. Cranial ridges high and strong.

k. Supraocular spine present. Parietals not meeting.

l. Skull thick; cranial ridges broken into tubercles and spines; interorbital space flat; mesethmoid processes horizontal; ventral process of basisphenoid rudimentary in adult (the skull of young almost exactly as in *rosaceus;* see below). *ruberrimus*.

ll. Skulls somewhat papery; ridges smooth; interorbital space concave; mesethmoid processes directed upward; ventral process of basisphenoid well developed in both young and old. *constellatus*, *rosaceus*, *rhodochloris*, *chlorostictus*.

kk. Supraocular spine absent.

m. Interorbital space not widening markedly backward.

n. Parietals not meeting; skull papery. *elongatus*.

nn. Parietals meeting; skull bony.

o. Nuchal spines none. *rubrivinctus*, *levis*.

oo. Nuchal spines present; ridges thick and high. *serriceps*.

mm. Interorbital space widening markedly backwards; parietals not meeting.

p. Coronal spines present, skull bony. *auriculatus*.

pp. Coronal spines none.

q. Skull thick; bones striated; interorbital space slightly convex. *rastrelliger*.

qq. Interorbital space concave and the cranial ridges strong and high. *vexillaris*, *maliger*, *carnatus*, *chrysomelas*, *nebulosus*.

The interorbital space becoming more concave and narrower and the ridges stronger and higher from the beginning to the end of the series.

It has been impracticable in some cases to separate closely related species in the above classification according to cranial characters, some of them agreeing even in color patterns and differing only in colors and other details, and showing no tangible differences in the skulls.

S. serriceps is probably placed a little too high up in the series, as its other characters indicate closer connections with the last group. It is evident that the cranial characters do not furnish a basis for the division of the rockfishes of the West Coast into several genera. All the characters that are at all available for purposes of classification serve remarkably well for arranging the species in series, but the changes which those characters undergo in the successive species are so perfectly graduated that they cannot be used to break up the genus. Jordan and Gilbert[5] first grouped the species in 1883, using the number and degree of development of the cranial ridges and spines as principal characters. Their arrangement not only remains, but is more firmly established, with one or two doubtful exceptions, by the remaining cranial characters.

Connected with this series of cranial characters and their modifications are a number of other characters. Although the correlations are not always exact, an arrangement of species based on these external characters would differ but little from that given above.

Ayres long ago pointed out that "the border of the caudal fin changes insensibly in the successive species from the slight emargination of *paucispinis* to the slight rounding of *nigrocinctus*." In *paucispinis* the anal spines are graduated, but this feature gradually changes in the series until in the *rosaceus* group the second anal spine is longer than the third. In the group represented by *paucispinis* and *pinniger* the longest rakers on the anterior limb of the first arch are relatively much longer than in the group represented by *rosaceus*, etc. The decrease in length is gradual in the series and is quite closely cor-

[5] Jordan & Gilbert: Synopsis of the Fishes of North America, 1883, pp. 652–678.

related with the decrease in the number of rakers on the anterior limb.

The scales also become successively larger, from very small ones in *paucispinis* to large scales in *introniger*. But it is impossible to use the size of the scales for the purpose of generic distinction. In the whole genus the transverse rows of scales corresponding in number with the pores are very oblique (making an angle of about forty-five degrees with the vertical) and have rarely, if ever, been counted as the "transverse rows of scales." Besides these there is a series that is actually vertical, making an angle of about forty-five degrees with the former. For each "oblique transverse" row there are two plainly visible vertical rows, and as a scale for each of the latter rows lies upon or nearly upon the lateral line these have been depended on for the determination of the "transverse rows of scales." Occasionally the scale of a vertical row lies far enough above or below the row of pores to be left out of the count, although the row to which it belongs is continuous above and below the line. This counting of the scales on the lateral line instead of the vertical rows to which they belong has led to confusion, because no two specimens of the same species give similar results.

It is an easy matter to arrange the species in a probably natural order; but, even with the fine series of graduated characters described above, it has been impossible to construct a "genealogical tree." The genus is prob- [illegible] extensive comparisons [illegible] species with the adult [illegible] species with each [illegible] the later embry- [illegible] together with the [illegible] measure solve the problem of genetic relationship in this interesting group.

I include below a diagnosis of the genus, with an analysis of the North American species.

SEBASTODES* Gill.

Rock-Fish; "Rock-Cod."

(*Sebastosomus*, *Sebastomus*, *Sebastichthys* Gill; *Acutomentum*, *Primospina*, *Pteropodus*, *Auctospina* Eigenmann and Beeson.)

(Gill, Proc. Acad. Nat. Sci. Phila., 165, 1861: type *Sebastes paucispinis* Ayres; Jordan and Gilbert, Synopsis of Fishes of North America, 652, 1883.)

Body and head somewhat compressed; head large, $2\frac{3}{5}$ to $3\frac{2}{3}$ in length of body†; depth $2\frac{1}{4}$ to $3\frac{3}{4}$ in length of body; mouth moderate or large, with the jaws equal or the lower more or less projecting; the maxillary reaching middle of eye or beyond, sometimes beyond posterior edge of orbit, its length from $1\frac{3}{4}$ to 3 in length of head; teeth in villiform bands on jaws, vomer and palatines. Head more or less evenly scaled, without dermal flaps; interorbital space convex or concave, widening markedly with age; cranial ridges‡ more or less developed, one or more of the following pairs always present, usually ending in spines: preocular, supraocular, postocular, tym-

---

* A very doubtful species, which may be the young of *Sebastes marinus*, with an abnormal number of spines, is accredited to the Atlantic Coast, viz.: *S.? fasciatus* (Storer). "Body elongated, not convex in front of dorsal fin as in *Sebastes norvegicus;* four distinct dark brown transverse bands upon the sides, the broadest at the posterior portion of the body." D. XIII-14; A. III, 7. Provincetown, Mass. (Storer). (*Sebastes fasciatus* Storer, Proc. Bost. Soc. Nat. Hist., v, 31, 1854.)

An equally doubtful fossil species is referred to this genus, viz.: *Sebastodes (?) rosæ* Eigenmann. It is known only from a fragment, the horizontal limb of a preopercle, which was found at Port Harford, Cal., among various tertiary fossils, thirty feet above the sea; but the finder himself thinks it *may have been left* there by the Indians. (*Sebastodes (?) rosæ* Eigenmann, Zoe, i, 16, 1890.)

†Length of body is measured from tip of snout to base of caudal fin.

‡For illustrations of cranial ridges and spines, see explanation of plates.

ANALYSIS OF NORTH AMERICAN SPECIES OF SEBASTODES.

a. Interorbital space convex (never concave), broad, less than $3\frac{1}{2}$ in base of skull, cranial ridges very low or obsolete, the spines when present, delicate; base of skull strongly curved, mesethmoid processes not elevated (not directed upward), ventral process of basisphenoid rudimentary (or fairly developed only in young); skull usually thick; anal rays III, 9 to III, 6; gill-rakers usually long and slender; snout, preorbitals and jaws more or less scaly.

b. Cranial ridges (except parietal) all obsolete or very slightly developed, cranial spines absent or very inconstant and minute (regularly present only in young), (preocular spines usually present in *mystinus*); lower jaw much projecting.

c. Parietal bones not meeting, mesethmoid processes weak and depressed; scales small, 90–100 transverse series of scales above lateral line, peritoneum white, lower jaw much projecting, entering profile, a large symphyseal knob, directed forward. A. III, 8–III, 9.

d. Head $2\frac{2}{3}$; depth $3\frac{2}{3}$; D. XIII–13; A. III, 9; lat. l. tubes 65–80, transverse rows of scales about 100. Maxillary reaching beyond eye (in adult), $1\frac{2}{3}$ in head; lower jaw much projecting, with a large symphyseal knob, eye large, 4–6 in head. Scales very small, irregular. Anal spines small, graduated. Pale dull orange red, dark brown above; young olivaceous. Peritoneum white. San Diego to San Francisco, abundant, and to British Columbia (Bean). *paucispinis.**

dd. Head, $2\frac{4}{5}$; depth, $3\frac{3}{5}$. D. XIII–14; A. III, 8; lat. l. 55 (pores), transverse rows of scales above lat. l. about 90. Maxillary reaching little beyond middle of orbit, $2\frac{1}{3}$ in head; lower jaw much projecting, with large symphyseal knob. Eye $3\frac{3}{8}$ in head; nasal spine obsolete; interorbital width $4\frac{1}{2}$ in head; anal spines, short, strong, graduated. Scales rough-ctenoid. Dusky olivaceous, silvery below, flushed with red. Peritoneum white. San Diego to San Francisco. *goodei.*†

cc. Parietal bones usually meeting, mesethmoid processes better developed, straight, not elevated.

e. Peritoneum white; dorsal fin deeply emarginate.

f. A. III, 9.

---

* *Sebastes paucispinis* Ayres, Proc. Cal. Acad. Sci., i, 6, 1854.

† *Sebastodes goodei* Eigenmann and Eigenmann, Proc. Cal. Acad. Sci., 12, 1890.

g. **Pectorals broad reaching tips of ventrals, but not quite to vent. Head 3; depth 3. D. XIII-15; A. III, 9; lat. l. about 60. Maxillary reaching nearly to posterior margin of eye, 2 in head; lower jaw much projecting, with large symphyseal knob. Scales medium. Anal spines low, graduated, second as long as eye; olivaceous; caudal yellowish or greenish. Peritoneum white. San Diego to San Francisco, abundant.** ***flavidus.****

gg. **Pectorals not reaching tips of ventrals, not nearly to vent. Head 3; depth about 3⅗. D. XIII-15 or 16, A. III, 9; lat. l. 60 (pores). Elongate. Lower jaw projecting, entering profile. Eye 4½ in head, 1⅓ in interorbital space. Scales large, those of head greatly reduced. Anal spines slender, graduated. Gray of varying shades, back darker; a series of large white blotches along sides of back, much more marked in some than in others; fins yellowish. Cortes Banks to San Francisco.** ***serranoides.*†**

ff. **A. III, 8; head 3; depth 2⅔. D. XIII-16; lat. l. 53, transverse rows of scales 60-70. Maxillary nearly reaching posterior margin of orbit, a little less than two in head; lower jaw projecting, its tip entering profile. Eye large. Scales moderate, accessory scales numerous. Anal spines small, graduated. Olive brown, dark above; upper part of sides thickly marked with small slaty-black spots; caudal dark. Peritoneum white. Monterey to Sitka, abundant northward.** ***melanops.*‡**

ee. **Peritoneum black, colors dusky, fins blackish, dorsal fin not very deeply emarginate.**

h. **Head 3⅓; depth 3½. D. XIII-15; A. III, 8; lat. l. 66. Maxillary reaching posterior margin of pupil; lower jaw somewhat projecting, without prominent knob. Scales all ctenoid. Second anal not longer than third. Preocular ridges obsolete; frontal region not specially convex. Blackish green, sides rather pale. Peritoneum black. Coast of Alaska.** ***ciliatus.*§**

---

* *Sebastodes flavidus* Ayres, Proc. Cal. Acad. Sci., 209, fig. 64, 1862.

† *Sebastodes serranoides* Eigenmann & Eigenmann, Proc. Cal. Acad. Sci., 36, 1890.

‡ *Sebastes melanops* Girard, Proc. Acad. Nat. Sci. Phila., VIII, 135, 1854, and U. S. Pac. R. R. Surv. Fish, 81.

§ *Epinephelus ciliatus* Tilesius, Mem. Acad. Sci. St. Petersb., IV, 474, 1810.

hh. Head 3⅓; depth 2⅘. D. XIII-15; A. III, 9; lat. l. 66; 50–55 tubes. Maxillary dilated behind, reaching posterior margin of pupil, 2⅛ in head; lower jaw protruding. Anal spines graduated, the second 3⅓ in head. Preocular ridges present, usually ending in spines, frontal region between them bulging. Slaty black; paler below lateral line. Peritoneum black. Puget Sound to San Diego, abundant. *mystinus*.*

bb. Cranial ridges somewhat developed, preocular, postocular, tympanic and parietal spines usually all present, delicate (supraocular also present in some species; tympanic usually absent in *atrovirens*); lower jaw projecting, parietal bones usually not meeting.

i. Lower jaw much projecting, scales rather small; lat. l. 50–75: anal rays III, 7–III, 9; dorsal fin not deeply emarginate, soft dorsal low.

j. Second anal spine scarcely or not longer, usually shorter than third.

k. Supraocular spine wanting. Head 3¼; depth 3⅓. D. XIII-15; A. III, 8; lat. l. 65. Maxillary reaching middle of eye, 2⅓ in head; lower jaw protruding, its tip entering profile. Eye less than interorbital space, 4 in head. Anal spines graduated. Olive green; creamy below; fins dusky. Peritoneum black. Port Harford to Monterey, rare. *entomelas*.†

kk. Supraocular spine usually present.

l. Peritoneum black.

m. Maxillary reaching middle of eye. Head 3; depth 3½. D. XIII-14½; A. III, 8½; 56 pores in lateral line. Compressed, elongate; mandible with prominent symphyseal knob. Orbit 3½ to 4 in head. Head entirely covered with moderate-sized scales; those of body larger. Anal spines graduated. Rufous; variously marked with brown; caudal dusky. Peritoneum jet black. Cortes Banks, San Diego. *rufus*.‡

mm. Maxillary reaching posterior margin of eye, 2 in head. Head 3 in total length; depth 3⅘. D. XIII-13½; A. III, 7½. Elongate; head pointed; lower jaw projecting. Mandible, maxillaries and snout scaled; scales of head small, ctenoid, those of body larger. Anal

* *Sebastichthys mystinus* Jordan & Gilbert, Proc. U. S. Nat. Mus., 455, 1880; 56, 70, 1881.

† *Sebastichthys entomelas* Jordan & Gilbert, Proc. U. S. Nat. Mus., 142, 1880.

‡ *Sebastodes rufus* Eigenmann & Eigenmann, Proc. Cal. Acad. Sci. 13, 1890.

spines graduated. Mostly black above, lat. l. vermillion; a black band below it. Peritoneum black. San Diego. *macdonaldi.**

ll. Peritoneum white. Closely allied to *Sebastodes proriger*, but larger in size and more uniform in color; anal spines graduated. Coast of Alaska. *brevispinis.*†

jj. Second anal spine notably longer than third. Peritoneum black.

n. Supraocular spines usually present.

o. Head 3; depth 2⅗. D. XIII–14; A. III, 8; lat. l. about 70. Body ovate. Maxillary reaching posterior edge of pupil, 2⅓ in head; lower jaw considerably protruding. Eye slightly longer than snout. Maxillary and mandible scaly. Second anal spine longer and stronger than third, 2⅓ in head. Creamy olivaceous; upper fins greenish, lower yellowish, mostly dark edged. Young more green. Peritoneum black. San Diego to San Francisco, rare. *ovalis.*‡

oo. Body elongate; depth more than 3; pores of lat. l. 50–52.

p. Head 3; depth 3⅗; D. XIII–14; A. III, 7; transverse rows of scales about 52; pores of lat. l. about 51. Maxillary reaching center of pupil, about 3 in head; lower jaw much projecting, with prominent symphyseal knob. Orbit 3⅓ in head. Scales everywhere strongly ctenoid, rather small; accessory scales not very numerous. Pectorals not reaching vent. Cranial spines very weak, often absent. Colored more or less like *ovalis*. Peritoneum black. Pacific Grove, Cal.; rare. *hopkinsi.*§

pp. Head 2⅘; depth 3⅓. D. XIII–15; A. III, 8; lat. l. 50 (tubes). Maxillary reaching middle of pupil, 2⅓ in head. Eye 3⅓ in head; interorbital space 1⅓ in orbit; scales small, rough, much smaller above lateral line, irregular; scales smooth on breast, snout, maxillary and mandible. Second anal spine much stronger and longer than third, 2⅓ in head. Pectorals reaching vent. Dusky above, with faint traces of darker blotches along back. Santa Barbara Islands. *alutus.*‖

---

* *Sebastodes proriger* Eigenmann & Eigenmann (not of Jordan & Gilbert), Proc. Cal. Acad. Sci., 15, 1890, and *Acutomentum macdonaldi* Eigenmann & Beeson, Amer. Naturalist, 669, 1893.

† *Sebastichthys proriger* var. *brevispinis* Bean, Proc. U. S. Nat. Mus., 359, 1883.

‡ *Sebastodes ovalis* Ayres, Proc. Cal. Acad. Sci., 209, 212, fig. 65, 1862.

§ sp. nov. A full description will soon be published elsewhere.

‖ *Sebastichthys alutus* Gilbert, Proc. U. S. Nat. Mus., 76, 1890.

nn. Supraocular spines absent. Head 3; depth 3½. D. XIII-13; A. III, 7; lat. l. 75. Maxillary short, broad, reaching beyond middle of eye, 2¼ in head; lower jaw much projecting, with large symphyseal knob. Eye very large, longer than snout. Body rather elongate. Second anal spine much longer and stronger than third, 2¼ in head. Color chiefly red; lateral line running in a continuous red stripe; iris red. Peritoneum black. San Diego to San Francisco, not rare. *proriger.**

ii. Lower jaw little projecting; scales moderate; lat. l. 45–55; A. III, 7–III, 6.

q. Supraocular spine present; A. III, 7; color red.

r. Color chiefly orange; head 2¾; depth 2¾; D. XIII-14; A. III, 7; lat. l. 48. Maxillary reaching posterior margin of eye, 2 in head; lower jaw somewhat projecting, with a symphyseal knob; eye 4 in head. Accessory scales numerous; scales on mandible smooth. Anal spines graduated, the second 3 in head. Peritoneum pale. San Diego to Puget Sound, abundant. *pinniger.*†

rr. Color chiefly brick red. Head 2¾; depth 3; D. XIII-14; A. III, 7; lat. l. 47. Maxillary reaching past pupil, 2 in head; lower jaw somewhat projecting, with a moderate symphyseal knob. Scales rough-ctenoid; those on mandible rough. Second anal spine equal to third, about 3 in head. Back and sides everywhere with clusters of black dots. San Francisco to San Diego; not rare. *miniatus.*‡

qq. Supraocular spine wanting. A. III, 6. Olivaceous, marbled with darker. Head 3; depth 2¾. D. XIII-14; lat. l. 52. Maxillary extending beyond posterior border of pupil, 2 in head; lower jaw somewhat projecting. Eye 3½ in head. Interorbital space but little convex. Scales large; mandible with a few smooth scales. Tympanic spine usually absent; anal spines graduated, the second 2½ in head. San Diego to San Francisco, abundant. *atrovirens.*§

---

* *Sebastichthys proriger* Jordan & Gilbert, Proc. U. S. Nat. Mus., 327, 1880.

† *Sebastodes rosaceus* Ayres, Proc. Cal. Acad. Sci., ii, 216, fig. 62, 1862, not *Sebastes rosaceus* Grd.; *Sebastosomus pinniger* Gill, Proc. Acad. Nat. Sci. Phila., 147, 1864.

‡ *Sebastichthys miniatus* Jorban & Gilbert, Proc. U. S. Nat. Mus., 70, 1880.

§ *Sebastichthys atrovirens* Jordan & Gilbert, Proc. U. S. Nat. Mus., 289, 1880.

aa. **Interorbital space flat or slightly concave, of medium width, mesethmoid processes but little or not at all elevated, ventral process of basisphenoid rudimentary. Cranial ridges and spines moderately strong. Lower jaw usually not much, sometimes not at all, projecting; gill-rakers usually long and slender; A. III, 6, to III, 8. Deep water fishes.**

s. **Base of skull strongly curved, supraocular spine absent.**

t. **Parietal bones not meeting. Olivaceous above, silvery below. Head 2¾ to 2⅞; D. XIII-12 or 13; A. III, 7; lat. l. 45 (pores). Maxillary nearly reaching posterior margin of pupil, 2¼ in head; lower jaw somewhat projecting, with a conspicuous knob. Scales rough-ctenoid, present on maxillary, mandible and snout. Second anal spine longer and stronger than third, 2 to 2½ in head. Peritoneum black. Santa Barbara Islands.** ***saxicola.****

tt. **Parietal bones meeting. Uniform rose-red above, bright silvery below. Head 2½; depth 2¾. D. XIII-12 or 13; A. III, 7; lat. l. 35 (tubes). Maxillary reaching beyond middle of pupil, 2⅜ in head; premaxillaries with prominent dentigerous knobs, between which the tip of lower jaw fits. Eye 3 to 3⅓ in head; interorbital space 1⅓ in orbit. Scales large, minutely spinous, readily deciduous; very small and cycloid on maxillary, mandible and breast. Second anal spine longer and stronger than third, 2½ to 3 in head. Peritoneum jet black. Coronado Islands.** ***diploproa.***†

ss. **Base of skull nearly straight; supraocular spine present, quite strong. Coronal and nuchal spines usually present (except in *melanostomus*).**

u. **Second anal spine much longer and stronger than third, 2½ in head. A. III, 6; head 2½; depth 2¾. D. XIII-13 or 14; lat. l. 29 (pores). Maxillary nearly reaching posterior margin of orbit, 2¼ in head; mandible included. Eye large, 3⅓ in head, much longer than snout or interorbital space. Scales everywhere very rough-ctenoid, covering branchiostegal rays, mandible and maxillary. Uniform red, light below. Peritoneum black. Santa Barbara Islands.** ***aurora.***‡

uu. **Second anal spine little or not at all longer than third. A. III, 7.**

v. **Lower jaw projecting; longest dorsal spine 3½ or more in head; mouth and gill cavities black.**

w. **Anal spines graduated. Head 3¼ in total length; depth about 3. D. XIII-13½; A. III, 7½; lat. l. 43. Body short, deep. Maxillary reaching posterior border of pupil; lower jaw pro-**

---

* *Sebastichthys saxicola* Gilbert, Proc. U. S. Nat. Mus., 78, 1890.

† *Sebastichthys diploproa* Gilbert, Proc. U. S. Nat. Mus., 79, 1890.

‡ *Sebastichthys aurora* Gilbert, Proc. U. S. Nat. Mus., 80, 1890.

jecting. Orbit one in snout, 3⅗ in head. Cranial spines covered with skin. Coronal spines absent. Scales very large, but few accessory scales. Body scarlet, dark above; mouth and gill-cavity black. Peritoneum black. San Diego. *melanostomus*.*

ww. Second anal spine equal to third. Head 2⅓; depth 2⅘. D. XIII-13; A. III, 7; lat. l. 30 to 35 (pores); about 55 vertical series of scales. Maxillary nearly reaching posterior margin of pupil 2⅛ in head. Eye large, 3⅓ in head; interorbital width 5⅓ in head. Lower jaw projecting, with prominent symphyseal knob. Cranial spines quite strong. Scales large, everywhere strongly ctenoid; accessory scales numerous; highest dorsal 3⅓ in head. Red; axils black; mouth and gill-cavities largely black. Peritoneum jet black. Santa Barbara Islands. *introniger*.†

vv. Lower jaw scarcely projecting. Longest dorsal spine 2⅗ in head. Chiefly red; mouth and gill cavities and peritoneum dusky. D. XIII-14; A. III, 7. Nuchal and coronal spines present; maxillary reaching posterior border of eye, 1⅞ in head. Interorbital space a little less than eye. Both jaws covered with rough ctenoid scales; highest dorsal 2⅓ in head. Second anal spine scarcely longer than third. Yeso; Aleutian Islands. *matzubaræ*.‡

aaa. Base of skull straight, or nearly so; interorbital space as a rule concave and narrow; the cranial ridges and spines well developed. Mesethmoid processes directed upward; ventral process of basisphenoid well developed; skull comparatively thin. Gill-rakers usually short.

x. Supraocular spine present; interorbital space concave.

y. Cranial ridges broken and armed with accessory spines, and interorbital space nearly flat in adult (ridges smooth, interorbital space concave in young, as in *Sebastodes rosaceus*). Second anal spine scarcely longer than third. Head 3; depth 2⅗; D. XIII-14; A. III, 7; lat. l. about 50; maxillary reaching nearly posterior edge of eye, 2 in head; lower jaw a little projecting. Eye 4⅓ in head. Scales on head and body rough; accessory scales numerous, Color red, nearly plain. Peritoneum white. San Diego to Puget Sound; Alaska (Bean). *ruberrimus*.§

---

* *Sebastodes melanostomus* Eigenmann & Eigenmann, Proc. Cal. Acad. Sci., 17, 1890.

† *Sebastichthys introniger* Gilbert, Proc. U. S. Nat. Mus., 81, 1890.

‡ *Sebastes matzubaræ* Hilgendorf, Sitzber. Gesellschaft Naturforschender Freunde, Berlin, 170, 1880.

§ *Sebastodes ruber* Jordan & Gilbert (not of Ayres), Synopsis of Fishes of North America, 665, 1883.

yy. Cranial ridges smooth; second anal spine much longer, usually stronger than third.

z. Color more or less rosy, with three to five round blotches of pink on sides of back.

A. Dorsal spines usually low, the highest less than half the length of head; no small green spots on sides of back.

B. Head and body everywhere with many small roundish pale spots. Head 2½; depth 2½. D. XIII-13; A. III, 6; lat. l. 53. Maxillary very broad, extending beyond pupil, 2 in head; lower jaw slightly projecting. Eye 4 in head. Scales strongly ctenoid, accessory scales numerous; head densely covered with small scales. Second anal spine considerably longer than third, 2⅗ in head. Orange red, back olive-shaded. Peritoneum white. San Diego to San Francisco. *constellatus*.*

BB. Body without stellate spots.

C. Second anal spine longer than third.

D. The five large pink blotches washed with orange, general color light orange, overlaid with blackish. Head 2⅗; depth 2⅗. D. XIII-12; A. III, 6; lat. l. 40 (tubes). Maxillary reaching posterior margin of pupil, 2 in head; lower jaw scarcely projecting. Eye large, 4 in head; interorbital space rather broad. Scales moderate; many accessory scales; both jaws with small smoothish scales. Second anal spine 2½ in head. Santa Barbara. *umbrosus*.†

DD. Bright orange red; the pale blotches on sides surrounded by purple shades; head with purplish above. Head 2⅗; depth 3. D. XIII-13; A. III, 6; lat. l. 48. Maxillary not reaching posterior border of eye, 2 in head; jaws equal, eye very large, 3½ in head. Scales moderate; accessory scales numerous. Second anal spine much longer and stronger than third, 2½ in head, curved. Mandible naked. Peritoneum blackish. San Diego to San Francisco, abundant. *rosaceus*.‡

DDD. General color, bright clear rose-red; pale blotches on sides surrounded by green shades; no purple. Head 2½; depth 3. D. XIII-14; A. III, 6; lat. l. 58. Maxillary reaching beyond pupil, 2⅓ in head; jaws about equal. Cranial ridges very sharp;

**Sebastichthys constellatus* Jordan & Gilbert, Proc. U. S. Nat. Mus. 285, 1880.

†*Sebastichthys umbrosus* Jordan & Gilbert, Proc. U. S. Nat. Mus. 410, 1882; *Sebastodes æreus* Eigenmann & Eigenmann, Proc. Cal. Acad. Sci. 20, 1890.

‡*Sebastes rosaceus* Girard, Proc. Acad. Nat. Sci. Phila. viii, 146, 1854; and in U. S. Pacific R. R. Surv. Fish. 78, plate 21 (poor figure).

eye very large, 3½ in head; accessory scales very numerous, mandible partly scaly; second anal spine very long, longer than maxillary, 2 in head. Peritoneum dusky. Off Monterey and San Francisco, rather rare. *rhodochloris*.*

DDDD. Body and head intense rose-pink, color marks washed or faded. Head 2½; depth 3. D. XIII-13½; A. III, 6½; lat. l. 37. Maxillary reaching beyond eye, 2 in head; lower jaw included, symphyseal knob strong. Eye one in snout, slightly more than 4 in head. Maxillary and mandible scaly; accessory scales numerous on cheeks and opercles. Second anal spine 2¾ to 3 in head. Interorbital space flattish, with deep median groove. Peritoneum white or more or less dusky. San Diego. *eos*.†

CC. Second anal spine about as long as third. Head 3; depth 3. D. XIII-13½; A. III, 7½; lat. l. (pores) 44-45. Maxillary reaching posterior edge of pupil, 2 in head; lower jaw projecting, entering profile, without knob. Orbit one in snout, 4⅛ to 4½ in head, a little greater than interorbital width. Scales strongly ctenoid, accessory scales very numerous everywhere; mandible naked. Dorsal surface closely covered with small, bronze, roundish spots; ventral surface light geranium red. Peritoneum white, sparsely dotted with black. San Diego. *gillii*.‡

AA. Dorsal spines very high, the highest half the length of head. Body above with many small round green spots. Head 2½; depth 2¾. D. XIII-13; A. III, 6; lat. l. 50. Maxillary reaching to beyond pupil, 2⅛ in head; jaws equal, a conspicuous symphyseal knob. Eye 3½ in head. Mandible naked. Second anal spine much longer and stronger than third, 2½ in head. Olivaceous above, sides pinkish or golden; the pink spots less distinct than in *Sebastodes rosaceus*. San Diego to San Francisco, abundant. *chlorostictus*.§

zz. Color nearly as in *Sebastodes zacentrus*; no round pink blotches on sides of back. Head 2½; depth 2⅝ to 3. D. XIII-13; A. III, 7; lat. l. 31 (pores), about 60 vertical series of scales above lateral line. Maxillary reaching beyond middle of pupil, 2½ in head; jaws equal. Eye 2½ in head, longer than snout or interorbital space. Nuchal spines present. Scales rough-ctenoid; those on maxillary and mandible minute and smooth. Second anal spine longer and stronger than third, 2¾ in head. Peritoneum black. Santa Barbara Islands. *rupestris*.||

---

* *Sebastichthys rhodochloris* Jordan & Gilbert, Proc. U. S. Nat. Mus. 144, 1880.

† *Sebastodes eos* Eigenmann & Eigenmann, Proc. Cal. Acad. Sci. 18, 1890.

‡ *Sebastodes gillii* Eigenmann & Eigenmann, Amer. Naturalist, 154, 1891.

§ *Sebastichthys chlorostictus* Jordan & Gilbert, Proc. U. S. Nat. Mus. 294, 1880.

|| *Sebastichthys rupestris* Gilbert, Proc. U. S. Nat. Mus. 76, 1890.

xx. Supraocular spine wanting.

E. Mandible scaly, peritoneum dusky or black.

F. Lower jaw only slightly or not all projecting; peritoneum jet-black.

G. Head 2½; depth 3. D. XIII-12; A. III, 5; lat. l. (tubes) 40–45. Body short deep. Maxillary reaching beyond pupil, 2¼ in head; jaws about equal. Eye very large, 2¾ to 3 in head; interorbital space 6 in head. Scales small, mostly smooth and cycloid, irregular. Mandible and maxillary partly scaled. Second anal spine longer and stronger than third, 2 in head. Pale below, dusky above, blotched with reddish and black; mouth and gill cavities and peritoneum jet-black. Gulf of California. *sinensis.**

GG. Head 2¾; depth 3½. D. XIII-14 or 15; A. III, 7 or 8; lat. l. (tubes) about 42, 70 vertical series of scales above lat. l. Maxillary reaching middle of pupil, 2½ in head; lower jaw slightly the longer, with small knob. Eye much longer than snout, 3 to 3½ in head; interorbital space 1½ in orbit. Scales large, rough-ctenoid, those on maxillary and mandible smoother. Second anal spine usually longer and stronger than third, 1½ to 1¾ in head. Five vaguely defined black bars on back; some red on the sides. Roof of mouth posteriorly dusky; mouth and branchial cavities otherwise white. Peritoneum jet-black. Santa Barbara Islands. *zacentrus.*†

FF. Lower jaw strongly projecting. Peritoneum dusky. Head 2¾; depth 3½. D. XIII-13; A. III, 6; lat. l. 58. Maxillary reaching posterior margin of pupil, 2½ in head. Eye very large, longer than snout, 3½ in head. Scales large, not very rough; accessory scales numerous; maxillary and mandible scaly. Second anal spine much longer than third, 2 in head. Light red; sides above with irregular horizontal, interrupted olive-green bands. San Diego to San Francisco, abundant. *elongatus.*‡

EE. Mandible naked, peritoneum pale or white. Body usually deep.

H. Scales on head mostly cycloid; lower jaw projecting; head large, pointed.

I. Pink, with 4 interrupted cross-bars of black; back sometimes dusky. Head 2¾; depth 3. D. XIII-13½; A. III, 7½; lat. l. 50. Head very large; maxillary reaching posterior margin of pupil, greatly dilated behind, its width about equal to diameter of eye; lower jaw projecting, with a well developed symphyseal knob.

* *Sebastichthys sinensis* Gilbert, Proc. U. S. Nat. Mus. 81, 1890.

† *Sebastichthys zacentrus* Gilbert, Proc. U. S. Nat. Mus. 77, 1890.

‡ *Sebastes elongatus* Ayres, Proc. Cal. Acad. Sci. ii, 26, 1859, fig. 9.

Eye 5½ in head, one in interorbital space. Scales of body weakly ctenoid; those on head cycloid; accessory scales numerous; mandible and maxillary naked. Second anal spine 4½ in head. Peritoneum white. San Diego to Monterey. *levis.**

II. Pinkish white, banded with deep crimson. Head 2½; depth 2½. D. XIII-12; A. III, 7; lat. l. 55. Maxillary broad, reaching middle of eye, 2½ in head; lower jaw projecting. Eye very large, 3⅓ in head. Scales of body rather smooth; those of head thin, mostly cycloid; accessory scales very numerous; mandible naked; maxillary with a few scales. Second anal spine much longer and stronger than third, 2½ in head. Peritoneum white. San Diego to Monterey, rare. *rubrivinctus.*†

HH. Scales on head ctenoid; lower jaw usually included. Second anal spine little enlarged.

J. Nuchal spines usually present, sometimes coalescent with the parietals. Head 3; depth 2½. D. XIII-13; A. III, 5; lat. l. 50. Maxillary reaching middle of eye, 2½ in head; jaws equal. Eye small, 5 in head. Interorbital space closely scaled; jaws naked. Second anal spine scarcely longer than third, 2½ in head. Dark olive, blackish above, yellowish below; sides with about 7 oblique black cross-bands. Peritoneum pale. San Francisco to Cerros Island. *serriceps.*‡

JJ. Nuchal spines none. Interorbital space widening markedly from before backward.

K. Coronal spines usually present. Head 3½; depth, 2½. D. XIII-13; A. III, 7; lat. l. 45. Maxillary reaching beyond eye, 2½ in head; jaws nearly equal. Scales on body large, ctenoid; accessory scales not numerous; mandible naked. Second anal spine longer and stronger than third, 2½ in head. Blackish brown, mottled with light brown. Cerros Island to Vancouver's Island; very abundant. *auriculatus.**

KK. Coronal spines none.

L. Cranial ridges with entire edges.

---

* *Sebastichthys levis* Eigenmann & Eigenmann, Notes from San Diego Biol. Lab. i, 6, 1889; West American Scientist, 129, 1889.

† *Sebastichthys rubrivinctus* Jordan & Gilbert, Proc. U. S. Nat. Mus. 291, 1880.

‡ *Sebastichthys serriceps* Jordan & Gilbert, Proc. U. S. Nat. Mus. 38, 1880.

§ *Sebastes auriculatus* Girard, Proc. Acad. Nat. Sci. Phila. 131, 146, 1854, and U. S. Pac. R. R. Surv. Fish, 80; *Pteropodus dallii* Eigenmann & Beeson, Amer. Naturalist, 66, 1894. This last is probably a young *Sebastodes auriculatus* with coronal spines obsolete.

M. Head 3; depth 2⅘. D. XIII-13; A. III, 6; lat. l. 47. Maxillary reaching posterior margin of eye, 2½ in head; jaws equal, no symphyseal knob. Eye small, anterior, 4½ in head; scales on body large; accessory scales few. Gill-rakers extremely short, most of them as wide as high. Second anal as long as third, 3 in head. Dusky green, with paler mottlings. Peritoneum brownish. San Diego to Humboldt Bay; abundant southward. *rastrelliger.**

N. Highest dorsal spine notably more than half length of head.

O. Head and upper parts not speckled with orange; membrane of spinous dorsal not very deeply incised.

P. Color dark brown, varied with light brown; armature of head, fin-rays, gill-rakers and scales as in *Sebastodes vexillaris;* but the lower jaw is more projecting, the pale shades are better defined, and the dorsal spines are slender and much lower than in *Sebastodes vexillaris*. Puget Sound to Sitka; abundant. *caurinus.*†

PP. Reddish, varied with yellowish. D. XIII-16; A. III, 6; lat. l. 55. Maxillary extending behind orbit, 2 in head; lower jaw a little projecting, without knob. Eye high up, 4-4½ in head. Jaws naked. Dorsal spines higher than in *Sebastodes caurinus*. Second anal scarcely longer than third, 3 in head. Peritoneum white. San Diego and northward. *vexillaris.*‡

OO. Head and upper parts everywhere speckled with orange. Dorsal spines extremely high, their membranes deeply incised. Head 2⅗; depth 2½. D. XIII-13; A. III, 6; lat. l. 47. Maxillary reaching posterior margin of eye, about 2 in head; jaws nearly equal. Scales rough, jaws naked. Second anal spine little higher than third, 2½ in head; front of back yellowish; soft fins black. Peritoneum pale. Monterey to Sitka, very abundant northward. *maliger.*§

---

* *Sebastichthys rastrelliger* Jordan & Gilbert, Proc. U. S. Nat. Mus. 296, 1880

† *Sebastes caurinus* Richardson, Voy. Sulphur, Ichth. 77, pl. 41, fig. 1, 1845

‡ *Sebastichthys vexillaris* Jordan & Gilbert. Proc. U. S Nat. Mus. 292, 1880.

§ *Sebastichthys maliger* Jordan & Gilbert, Proc. U. S. Nat. Mus., 322, 1880.

NN. Highest dorsal spine little if any more than half the length of head.

Q. Pale blotches on sides not forming a continuous lateral band; parietal ridges moderate.

R. Pale markings flesh color, dark markings olivaceous. Head $2\frac{3}{5}$, depth $2\frac{3}{5}$. D. XIII-13; A. III; 6; lat. l. 43. Maxillary extending a little beyond posterior margin of eye, 2 in head; jaws about equal, no symphyseal knob, scales on head rougher than in *Sebastodes chrysomelas*, mandible and maxillary naked. Second anal spine slightly longer than third, $2\frac{3}{4}$ in head. Peritoneum white. San Diego to San Francisco, abundant. *carnatus*.*

RR. Pale markings yellow, dark markings blackish; pattern of coloration exactly as in *Sebastodes carnatus*. Head $2\frac{3}{4}$; depth $2\frac{3}{4}$. D. XIII-13; A. III, 6; lat. l. 45. Maxillary reaching posterior margin of eye, 2 in head; lower jaw slightly included. Scales moderate, rough; accessory scales few; mandible, maxillary and nasal region naked. Second anal strong, equal to third, $2\frac{3}{4}$ in head. Peritoneum pale. San Diego to San Francisco; abundant. *chrysomelas*.†

QQ. Pale blotches on sides forming a continuous lateral band. Head 3; depth $2\frac{3}{4}$; D. XIII-13; A. III, 7, lat. l. 49. Maxillary extending beyond pupil, 2 in head; jaws equal. Eye large. Scales rough, accessory scales numerous; jaws naked. Second anal spine slightly longer than third, $2\frac{1}{2}$ in head; parietal ridges very strong. Ground color blue black. Body and fins profusely speckled with pale; pale markings yellow. Peritoneum pale. Vancouver's Island to Port Harford, abundant. *nebulosus*.‡

LL. Cranial ridges with the surface broken, spinous. Head $2\frac{4}{5}$; depth $2\frac{3}{5}$. D. XIII-15, A. III, 7; lat. l. 50. Maxillary reaching beyond pupil, 2 in head; lower jaw very slightly projecting. Eye large, $4\frac{1}{4}$ in head. Scales rough; jaws naked. Second anal spine longer and much stronger than

---

* *Sebastichthys carnatus* Jordan & Gilbert, Proc. U. S. Nat. Mus., 73, 1880.

† *Sebastichthys chrysomelas* Jordan & Gilbert, Proc. U. S. Nat. Mus., 455, 465, 1880.

‡ *Sebastes fasciatus* Girard, Proc. Acad. Nat. Sci. Phila., 146, 1854, etc. (not of Storer); *Sebastes nebulosus* Ayres, Proc. Cal. Acad. Sci., i, 5, 1854.

third, 2¼ in head. Frontal ridges elevated. Bright red, with black bands. Peritoneum white. Monterey to Vancouver's Island, rare southward. *nigrocinctus*.*

APPENDIX.

For convenience of reference, I add in full the article by Eigenmann & Beeson, including their proposed rearrangement of the group based on a study of its cranial characters.

PRELIMINARY NOTE ON THE RELATIONSHIP OF THE SPECIES USUALLY UNITED UNDER THE GENERIC NAME SEBASTODES.

(Eigenmann & Beeson, American Naturalist, 668–671, July, 1893.)

On the Pacific coast of temperate North America, a large number of species of viviparous Scorpænidæ are found. They range all the way from tide water to a depth of 1600 feet, from Cerros Island to Alaska. They are most abundant on the coast of California, about 30 species being known from San Diego, and a like number from Monterey. In size they vary from 1 lb. to 30 lbs.

The species have been variously grouped as forming one genus by Jordan & Gilbert, as forming two by Jordan, and as forming four by Gill. Jordan & Gilbert, in their Synopsis, arranged the species known to them according to the greater or less prominence of the spiniferous ridges of the skull. In examining the skulls of a number of them, one of us, several years ago, noticed that in a number of species the parietals meet over the supra-occipitals, while in others they are separated, and the supra-occipital is exposed above for its whole length. A more recent examination of a larger series of skulls, tended to show that, if we admit the relationships pointed out by Jordan & Gilbert, this greater or less development

* *Sebastes nigrocinctus* Ayres, Proc. Cal. Acad. Sci., II, 25 and 217, fig. 6. 1859.

of the parietals is of no significance. A more thorough study has, however, convinced us that the species with united parietals are related, and that the relationships pointed out by Jordan & Gilbert are at fault.

The value placed on such a cranial character as the union or nonunion of the parietals need not be defended here. It may only be mentioned that in *mystinus*, which for other reasons we considered the hub to which the other groups proposed here are related as spokes, the parietals are united in 8 out of 10 specimens. The variation of this character in *mystinus* but confirmed our view that it is the radiating point.

Leaving the parietals, the next prominent characters are the development or nondevelopment of certain cranial spines and ridges. These spines are found in all stages, from minute points to comparatively huge spines. The variation in size for this reason, if there were no other objections, cannot be utilized for determining generic relationship. The spines are very regularly arranged, and in any given species certain ones are always present. (Individual variations should of course be expected in this character, as in every other, if a sufficient number of specimens are examined.) The *constancy* of the presence of certain spines in a given species warrants the use of the presence or absence of these spines in the different species in determining their true relationship. This relationship is usually borne out by a number of subsidiary characters. Considering the constancy of the spines, reinforced by subsidiary characters, we have divided the species usually united under the generic name *Sebastodes* as follows:

a. Parietals meeting above the supra-occipital.

b. Jaws equal; head narrow above; high and prominent cranial ridges ending in spines; preocular, supraocular, tympanic and parietals present. Scales usually very strongly ctenoid; accessory scales numerous; suborbital stay directed obliquely downward and backward; second anal spine much heavier than, and at least as long as the third; body short and deep, back arched; mouth very large; head heavy. All known species with cross bands.

SEBASTICHTHYS Gill.

*nigrocinctus, serriceps, rubrivinctus, diploproa.**

bb. Lower jaw much projecting; head broad, the skull usually convex; cranial ridges, when present, low; gill-rakers very long and slender; scales usually smooth, few if any accessory scales. Suborbital stay little if at all oblique.

c. Parietal ridges ending in spines; preocular, supraocular and tympanic spines well developed. Peritoneum black.

d. Postocular spine present. Second anal spine usually stronger and longer than third. Symphyseal knob strong, projecting forward. Dorsal low. (Peritoneum black, mandibles and maxillary scaled.)

ACUTOMENTUM[1] E. & B.

[1]Type *A. ovalis* (Ayres).

melanostomus, ovalis, rufus, **alutus, macdonaldi* n. sp. nov. = *S. proriger* E. & E., not of J. & G.

dd. Postocular spine not developed.

We have not been able to examine the two species (entomelas and atrovirens) and cannot vouch for their position.

cc. Parietal ridges not ending in spines.

e. Preocular spines well developed. Supraocular and tympanic spines sometimes present. Interorbital wide, convex. Peritoneum black. Approximated edges of sub-opercle and inter-opercle frequently ending in spines. PRIMOSPINA[2] E. & B.

[2]Type *P. Mystinus* (J. & G.)

The only species (*mystinus*) is the most variable species of the group.

ee. Preocular without spine, skull smooth, without spines. Peritoneum usually white.

SEBASTOSOMUS Gill.

flavidus, serranoides, melanops, **ciliatus.*

aa. Parietals separated by the supra-occipital.

f. Cranium with parietal ridges only. Lower jaw much projecting, entering the profile; a prominent symphyseal knob directed forward. Head broad, convex. Interobital convex, nearly smooth.

SEBASTODES Gill.

paucispinis, goodei.

*Species marked with an asterisk have not been examined in reference to the characters utilized.

ff'. Cranium with many ridges, all ending in spines.

g. Postocular and tympanic spines both present. Interopercle and subopercle without spines. Lower pectoral rays normal.

h. Coronal spines; nuchal spines, a spine below, another in front of eye. * *matzubarae* with this species we are not acquainted.

hh. No coronal spines. SEBASTOMUS Gill.

*miniatus, pinniger, levis, aereus**, *constellatus, umbrosus**, *rosaceus, rhodochloris**, *gilli**, *rupestris**, *eos, chlorostictus**, *ruber**, *rufus.*

gg. Postocular spine wanting.

i. Coronal spines none.

PTEROPODUS E. & B.[1]

Species with normal pectoral rays, (living off the bottom) *saxicola**, *proriger*†*, *brevispinis**, *elongatus, sinensis.*

Species with lower pectoral rays thick (living on the bottom) *zacentrus**, *maliger, caurinus, vexillaris, rastrelliger, nebulosus, carnatus, chrysomelas.*

ii. Coronal spines present.

AUCTOSPINA E. & B.[2]

*aurora**, *auriculatus.*

† The specimen described by E. & E., Proc. Cal. Acad. Sci. (2) III, 15, 1890, is a species distinct from *proriger.*

The inter-relationship of these genera is complex. It may be represented by the following diagram, where the genera with the united parietals are followed by an asterisk.

[1]Type *P. maliger* (J. & G.)

[2]Type *A. auriculatus* (Girard).

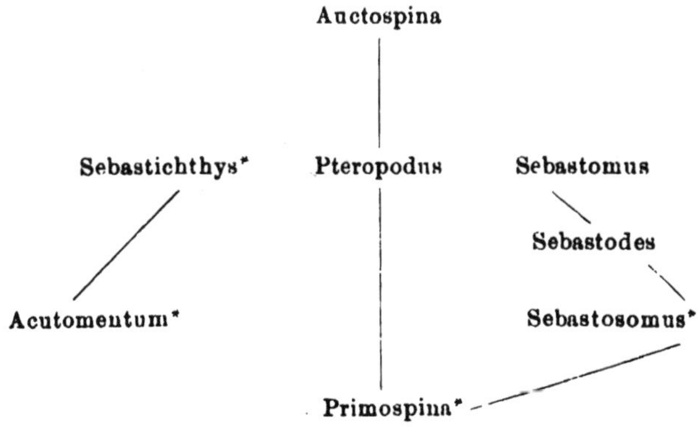

EXPLANATION OF PLATES.

PLATE LVII.

1. **Sebastodes paucispinis** × $1\frac{1}{8}$; **Monterey, Cal.** No. **1461, Mus. L. S. Jr. Univ.**
2. **Sebastodes goodei** × $1\frac{1}{3}$; **Monterey, Cal.** No. **1462, Mus. L. S. Jr. Univ.**
3. **Sebastodes flavidus** × $1\frac{1}{8}$; **Monterey, Cal.** No. **1471, Mus. L. S. Jr. Univ.**

PLATE LVIII.

4. **Sebastodes melanops** × 1; **San Francisco market.** No. **1466, Mus. L. S. Jr. Univ.**
5. **Sebastodes entomelas** × $2\frac{1}{8}$; **San Francisco market.** No. **1472, Mus. L. S. Jr. Univ.**
6. **Sebastodes ovalis** × $2\frac{1}{8}$; **San Francisco market,** No. **1474, Mus. L. S. Jr. Univ.**

PLATE LIX.

7. **Sebastodes pinniger** × 1; **Monterey, Cal.** No. **1469, Mus. L. S. Jr. Univ.**
8. **Sebastodes miniatus** × $\frac{3}{8}$; **Cortez Banks, Cal.** No. **1467, Mus. L. S. Jr. Univ.**
9. **Sebastodes miniatus** × $2\frac{3}{8}$; **San Francisco market.** No. **1468, Mus. L. S. Jr. Univ.**

PLATE LX.

10. **Sebastodes atrovirens** × $1\frac{1}{3}$; **San Francisco market.** No. **1493, Mus. L. S. Jr. Univ.**
11. **Sebastodes sp. incog.** × $1\frac{1}{3}$; **North Pacific.** No. **1473, Mus. L. S. Jr. Univ.**
12. **Sebastodes saxicola** × 2; **Santa Barbara Channel.** No. **1477, Mus. L. S. Jr. Univ.**

PLATE LXI.

13. **Sebastodes diploproa** × $1\frac{1}{8}$; **Santa Barbara Channel.** No. **1476, Mus. L. S. Jr. Univ.**
14. **Sebastodes introniger** × 1; **North Pacific.** No. **1479, Mus. L. S. Jr. Univ.**
15. **Sebastodes aurora** × $1\frac{1}{3}$; **Santa Barbara Channel.** No. **1478, Mus. L. S. Jr. Univ.**

PLATE LXII.

16. Sebastodes ruberrimus × $2\frac{3}{5}$; Monterey, Cal. No. **1484, Mus. L. S. Jr. Univ.**
17. Sebastodes ruberrimus × $1\frac{3}{8}$; San Francisco market. No. **1483, Mus. L. S. Jr. Univ.**
18. Sebastodes ruberrimus × $1\frac{1}{5}$; Monterey, Cal. No. **1482, Mus. L. S. Jr. Univ.**

PLATE LXIII.

19. **Sebastodes ruberrimus × $\frac{3}{8}$; San Francisco market. No. 1481, Mus. L. S. Jr. Univ.**
20. **Sebastodes rosaceus × $2\frac{3}{8}$; Monterey, Cal. No. 1485, Mus. L. S. Jr. Univ.**
21. **Sebastodes chlorostictus × $1\frac{1}{8}$; Monterey, Cal. No. 1487, Mus. L. S. Jr. Univ.**

PLATE LXIV.

22. **Sebastodes elongatus × 2; Monterey, Cal. No. 1490, Mus. L. S. Jr. Univ.**
23. **Sebastodes rubrivinctus × $1\frac{1}{8}$; Monterey, Cal. No. 1491, Mus. L. S. Jr. Univ.**
24. **Sebastodes levis × $\frac{3}{8}$; Monterey, Cal. No. 1492, Mus. L. S. Jr. Univ.**

PLATE LXV.

25. **Sebastodes serriceps × $1\frac{1}{8}$; San Pedro, Cal. No. 1501, Mus. L. S. Jr. Univ.**
26. **Sebastodes auriculatus × 1_6; Monterey, Cal. No. 1500, Mus. L. S. Jr. Univ.**
27. **Sebastodes rastrelliger × $1\frac{1}{3}$; Monterey, Cal. No. 1494, Mus. L. S. Jr. Univ.**

PLATE LXVI.

28. **Sebastodes vexillaris; Monterey, Cal. No. 1502, Mus. L. S. Jr. Univ.**
29. **Sebastodes vexillaris × 1, San Francisco market. No. 1503, Mus. L. S. Jr. Univ.**
30. Sebastodes maliger × $2\frac{1}{8}$; San Francisco market. **No. 1498, Mus. L. S. Jr. Univ.**

PLATE LXVII.

31. Sebastodes maliger × $1\frac{1}{3}$; Monterey, Cal. **No. 1497, Mus. L. S. Jr. Univ.**
32. Sebastodes nebulosus × $1\frac{2}{3}$; Monterey, Cal. **No. 1499, Mus. L. S. Jr. Univ.**
33. Sebastodes paucispinis × $1\frac{1}{6}$; Monterey, Cal. **No. 1461, Mus. L. S. Jr. Univ.**

PLATE LXVIII.

34. Sebastodes flavidus × $1\frac{1}{6}$; Monterey, Cal. No. 1471, Mus. L. S. Jr. Univ.
35. Sebastodes miniatus × $\frac{7}{6}$; Cortez Banks, Cal. No. 1467, Mus. L. S. Jr. Univ.
36. Sebastodes atrovirens × $1\frac{1}{3}$; San Francisco market. No. 1493, Mus. L. S. Jr. Univ.

PLATE LXIX.

37. Sebastodes saxicola $\times 2\frac{1}{3}$; Santa Barbara Channel. No. 1477, Mus. L. S. Jr. Univ.
38. Sebastodes ruberrimus $\times 1$; Monterey, Cal. No. 1482, Mus. L. S. Jr. Univ.
39. Sebastodes chlorostictus $\times 1\frac{1}{8}$; Monterey, Cal. No. 1487, Mus. L. S. Jr. Univ.

PLATE LXX.

40. Sebastodes elongatus $\times 1\frac{3}{8}$; Monterey, Cal. No. 1490, Mus. L. S. Jr. Univ.
41. Sebastodes rastrelliger $\times 1\frac{1}{3}$; Monterey, Cal. No. 1494, Mus. L. S. Jr. Univ.
42. Sebastodes nebulosus $\times 2$; Monterey, Cal. No. 1499, Mus. L. S. Jr. Univ.

CONTRIBUTIONS TO WESTERN BOTANY. NO. VII.

BY MARCUS E. JONES, A. M.

Having had an opportunity to examine the material in the National Herbarium I have been able to correct many errors, and possibly to make a few more. The following notes and new species are the result of those studies, and are only such as have come to light from the identification of my collection of 1894 made under the auspices of the U. S. Dept. of Agriculture as Special Field Agent. The long delay in the publication of the report necessitates the early publication of the new species. The number of new species and varieties discovered by me in 1894 and described here are 104, those collected and named but also found before by others are 29, a total of 133 new species and varieties in the collection. Other new species and varieties described from other sources are 57, a total of 190.

The collection of 1894 consisted of about 50,000 specimens and 1700 species, there being 1106 species and varieties in the sets, to these I have added in the sets which I will distribute about 50 others collected in 1893 and 1895. A very large amount of material was collected as the basis of an extended report on geographical distribution and plant adaptation; this material it is my intention to work up at some future date in connection with similar material gathered in the Great Basin since 1879.

The types of the new species are in my herbarium unless otherwise stated. Duplicate types of all species in the collection of 1894 are also in the National Herbarium.

New species signed "T. & E." are by Tracy and Earl.

New species signed "E. & E." are by Ellis and Everhart.

All numbers above 5000 belong to the collection of 1894.

In nomenclature I have tried to follow the recent circular signed by the representative body of American botanists, and not the peculiar nomenclature of "A Committee of the Botanical Club, American Association for the Advancement of Science." My reasons for doing so are that the circular is the first and only agreement of American botanists on nomenclature since the death of Dr. Gray; that it represents my own views with the exception of some unimportant particulars; that the "List of Pteridophyta and Spermophyta" prepared by "A Committee" represents the views of only a portion even of that "Committee;" that it is not representative; that it when published was not sanctioned by the "Botanical Club" or the American Association; that it is the product of a few practically self-appointed individuals; that it does not meet the views of American botanists; that it degrades the rank of species and genera and opens wide the flood gates for the indefinite multiplication of fictitious species and genera by those who have practically no field knowledge, who have of late been manufacturing species in the herbarium; that it destroys the relative standing of genera, species and varieties by elevating the latter to specific rank, by making genera of intimately related natural groups of species and thus destroying the subgeric relationship while leaving nothing in its place but a multitude of fictitious genera of no apparent relationship; that in the use of names it professes that priority whether varietal, specific or generic shall rule, while at the same time repudiating it by insisting that no earlier date shall be used than the Species Plantarum of Linnæus; that in "Once a synonym always a synonym" the bibliography of species, etc., is to be loaded up with a mass of new names nine-tenths of which are wholly useless, which will require thousands of changes of well known names and thousands of dollars expense in renaming herbaria with absolutely no compensating advantage. I have used double authorities when there are any because I believe it more just. I have followed the order of Gray rather than that of the Pflanzenfamilien for convenience and not because I believe it better, for I prefer the latter.

In three cases I have deviated from the letter of the circular referred to above, because I believe ultimately some date near 1885 will be fixed in which varietal names shall supersede specific ones in the same genus. Should this not be done the following will be the nomenclature of the three species: *Astragalus ceramicus* Sheldon for *A. angustus* and its varieties (the name *foliosus* Gray used as a variety, *foliolosus* Sheldon not Gray, having been given prior to 1885); *Astragalus salinus* Howell for *A. latus* Jones; *Hedysarum boreale* var. *flavescens* (Coult. & Fisher) for *H. boreale* var. *leucanthum* Greene.

CLEMATIS VERTICILLARIS var. COLUMBIANA (Nutt. Jour. Phil. Acad., vii, p. 7).

Clematis Columbiana, Nutt. l. c.

No. 5571. July 3, Provo, Utah, in Slate Cañon, 8000° alt., on moist slopes, among oak brush.

A comparison of many specimens in the National Herbarium, from various localities, together with my own throughout the Plateau region, makes it clear that this is a well marked variety, being characterized by the usually entire or crenulate (rarely toothed) leaflets and different fruit. I have seen but one transition specimen, from Minnesota. The typical species ranges from the Atlantic to the Rocky Mountains, and the variety ranges thence westward to the coast.

Akenes obovate, about 1″ long and ¾″ wide, with apex crowned by the long tail, which is very slender, and scarcely enlarged below: akenes usually obtuse at apex, never acuminate; sparsely pubescent below, much pubescent above; tail 2′ long; mature peduncles about 5′ long; leaflets thin, shortly acuminate.

Clematis verticillaris DC. has coarsely and sparsely dentate, ovate, shortly acuminate leaflets, which are often trifid: fruiting peduncles 2–3′ long, reflexed; akenes 1–1½″ long, obovate, with a broad and flat border, which tapers abruptly into a flat wide tail, which gradually narrows upward; akenes nearly equally hairy throughout, and sparsely so, nearly twice the size of those of the var. *Columbiana*.

CLEMATIS DOUGLASII var. BIGELOVII (Torrey, Pac. R. R. Rep., 4, 61).

Clematis Bigelovii Torrey l. c.

I can find no characters which separate this from *C. Douglasii*. Palmer's specimen from New Mexico has the long peduncle and flower of *C. Douglasii*, has three distinct pairs of leaflets besides the terminal one; the lower pairs are again 3-parted, and the divisions stalked (lateral stalks 2″ and terminal one 6″ long), making the leaf 2-ternate, the leaflets are again cut-toothed or parted into acute segments, whose general outline is ovate to lanceolate, 6–12″ long; petioles of the main pairs of leaflets 1′ long: leaflets nearly glabrous: petioles, etc., decidedly pubescent: inner sepals woolly.

Newberry's specimen from McComb's Expedition has leaflets of *Douglasii*, but broader, the fruit is that of *Douglasii*.

Lemmon's specimen from Arizona has filiform segments, but in other respects is *C. Douglasii*.

Shufeldt's specimen from Fort Wingate, New Mexico, has oblong-ovate leaflets, 6″ or less long, mostly entire, shortly acute, and giving a wholly different appearance.

but is manifestly only a form of *C. Douglasii*. These specimens are in the National Herbarium.

Anemone Tetonensis Porter.

No. 5763. August 7, Fish Lake, Utah, 10,800° alt., on gravelly knoll, devoid of trees, in the midst of a heavy forest of spruces and firs, along with *Anemone multifida*, but never showing any tendency toward intergrading with it.

This does not accord with Mr. Britton's description, but does accord with a part of the material on which Mr. Britton founded his description, *i. e.*, Bailey's specimen. It would seem that Coulter's specimen is either a form of *A. multifida* or is a transition form between the two. There was every facility for hybridization, and yet no trace of it where I collected my material, though I hunted for it for nearly an hour and saw hundreds of specimens. All specimens have long, not "short" styles, and akenes barely pubescent on the backs, flowers white; segments of the leaves narrower and half longer, tapering and very acute at both ends; petioles of bracts shorter for the blade; whole plant more strict and not weak. The other characters given for *A. Tetonensis* do not hold. The sepals are bluish outside and pilose, especially below, oval, 3″ long, not open much, anthers oval and apiculate; styles 2″ long, glabrous above, curved but not reflexed; heads oval, 3″ high; plants 6–8′ high, erect and not slender, tufted; stems 2–5 from the apex of an upwardly bent rootstock; dead leaf sheaths present.

RANUNCULUS CUSICKII. Referred provisionally by Watson to *alismæfolius* var. *alismellus*, and by Holzinger to *hydrocharoides*. Root leaves nearly round to ovate, cordate or truncate at base, obtuse, entire, 1′ long, edges barely sinuate, on petioles 1–3′ long, with dilated bases; stem leaves 1–2, similar and short petioled or nearly sessile; bracts lanceolate, entire, 6″ long; flowers on long (2–3′) slender peduncles, yellow; petals obovate-oval, veiny; akenes in a small globular head ⅙″ long, inflated, very obtuse, as in var. *alismellus*, with minute beak; whole plant glabrous, erect, 6–8′ high, slender, but stems rather thick below; roots fleshy-fibrous. The type is No. 1161, Cusick, Eagle Co., Or., 1884, 6000° alt. I also refer to this Watson's specimen from below Lobo Divide, Idaho, Aug. 20, 1880.

RANUNCULUS JUNIPERINUS.

No. 5011. April 4, at Copper Mine, 18 miles west of St. George, Utah, in Beaverdam Mts., 5000° alt., among junipers, in loose gravelly soil.

No. 5139x. April 30, at the head of the west branch of Santa Clara valley in the Beaverdam Mts., Utah. 5000° alt., in loose soil on rocks, among junipers.

Perennial, with many long, rather fleshy fascicled roots from the crown and when the crown is prolonged then there are many roots growing out from the axils of the old leaf-sheaths; plants densely tufted with many crowns which are covered with dark, long, narrow and rather fibrous leaf-sheaths 1½′ long; stems about 8–12′ long and generally erect; leaves rather fleshy and doubly-pedately-parted, the lobes variously toothed or lobed; petals white at first, veined on the outside with pink, when old becoming purple and enlarging, 5–8″ long, oval to orbicular, with rather uneven margin, veins very prominent, often rotate-spreading, usually cup-shaped; stamens many, with minute round anthers; sepals concave, veined, greenish, almost hyaline, only the claw reflexed, broadly-ovate, obtuse, enlarging with age and closed over the fruit, both petals and sepals persistent; flowers 2–3, long-peduncled, always with a leafly bract at the base of the peduncle, but with no other stem leaves: fruit flat with a thick marginal nerve, not at all inflated, 1–1½″ long, nearly 1″ wide, broadly obovoid and obliquely so, minutely apiculate; style minute; seed curved, nearly 1″ long, oblong-obovate, very much smaller than the akene; akenes many in a head. This is *R. Andersoni* var. *tenellus* Wats., but Mr. Watson seems to have failed to recognize the great difference in the akene.

This grows on gravelly mountain sides among the rocks and junipers, and is the earliest bloomer of all flowers.

DELPHINIUM SCOPULORUM var. ATTENUATUM.

No. 5893d. August 23, Head of Bullion Creek, Utah, near Marysvale, 11,500° alt., on gravelly and springy places along cold subalpine streams.

No. 5684c. July 25, Mt. Ellen, Henry Mts., Utah, 10-000° alt., in similar situations.

This has the wide leaves of the variety *glaucum* as well as the large flowers, it has the viscid pubescence of the variety *subalpinum;* sepals linear and attenuate, 12″ long, about 2″ wide at base, three times as long as the petals, nearly glabrous; spur shorter than the sepals, about 10″ long, ascending; racemes compound below; flowers deep blue. This approaches nearest to *D. elatum* of any American form, but the petals of that species are very dark and the sepals wider; in this species the upper petals are white and veiny and the lower light-blue and hairy.

This grows at timber-line along brooks, has a very strong odor of musk, and grows in large tufts about a yard high.

No. 5759 is a transition form toward the type. Gathered August 7, Fish Lake, Utah, 11,000° alt., in hollows near snow banks.

Delphinium pauciflorum var. *depauperatum* (Nutt.) Gray. My specimens No. 5391, June 5, 1895, Marysvale, Utah, in Bullion Cañon, 9000° alt., in gravel, have the roots of this and the habit and flowers of *D. bicolor*, so do many from Colorado. The following numbers are much alike also:

5409a. June 7, 1895, Marysvale, Utah, in gravel, 6000° alt.

5441a. June 15, 1895, Ireland's Ranch, Salina Cañon, Utah, 8,000° alt., in gravel.

5176a. May 4, 1895, Silver Reef, Utah, in gravel. 6000° alt.

It is well nigh hopeless to try to bring order out of the chaos in this genus for half the recognized species run together.

Aquilegia flavescens Wats. King's Rep. 5, 10.

No. 5911h. August 25, Tate Mine near Marysvale, Utah, 9300° alt., along Bullion Creek among willows.

No. 5338b. May 31, Marysvale, Utah, in marshy places, 6000° alt.

An examination of the type specimen as well as the original description shows that Mr. Watson took for the type of his *A. flavescens* what he has since regarded as only a form of *A. cærulea* and which certainly intergrades with that species. The plants which he has referred to this species, which grow at a lower altitude, are quite different and characterized by very short and hooked spurs, very small flowers, hardly more than 6" long, always yellow; upper leaves reduced to minute bracts; stems very long, 1–3° long, slender, at least three times as long as the short and small leaves. It is probable that the labels of the two species have been changed in the National Herbarium, but must have been changed by Mr. Watson himself before his descriptions were drawn up, for the plant which he calls "subalpine" form is seldom found higher than 7500 ft. altitude, but is common as low down as 6000 ft. altitude along streams, while the plant which he has labeled as growing at 5000 ft. altitude is seldom, if ever, found at a lower elevation than 8500 ft., nor is any Aquilegia known to exist in the Wasatch Mountains at 5000 ft. altitude, for that altitude is reached before you touch the base of the mountains.

This low altitude species, above described, may take the name of *flavescens*, since it was considered belonging to the species, rather than to make a new name. It is always found in cold springs or streams at low elevations. This is my number 1348 and to it I also refer Watson's No. 36.

AQUILEGIA CÆRULEA var. CALCAREA.

No. 5312a. May 28, at the head of the cañon above Cannonville, Utah, 7000° alt., in very compact and barren clay, among trees of *Pinus ponderosa*.

No. 5312am. Same locality as above.

This variety has the habit of the typical form, but the leaves are reduced to three clusters of three leaflets each, and these again cleft, parted or lobed; leaflets only 6‴ long, and as wide, glaucous, minutely notched at the rounded apex, faintly nerved, thick, overlapping each other, and so the whole leaf seeming to be composed of three nearly round clusters of leaflets, the whole not over an inch in diameter; petioles very glandular-hairy, as well as the stems and peduncles; flowers about half the size of typical *cærulea;* sepals blue-purple, oval, acutish, 6‴ long by 4‴ wide, abruptly contracted at the insertion; petals nearly square but rounded at apex, 4‴ long, pinkish, about equalling the stamens; anthers oval, ¼‴ long; stem leaves absent, but there is one 3-lobed bract at the base of the lowest peduncle, 6‴ long; spurs about twice the sepals, with abortive gland at the tip; leaf-sheaths very thick and fibrous on the top of the root.

This appears to be a well-defined new species, but in

view of the great variability in this genus and in view of the peculiar soil in which it grows, I prefer to consider it as a mere variety. Found growing among the spruces at 8000 feet altitude, on the most barren clay soil.

Platystemon Californicus Benth. This is *P. crinitus* Greene, but the characters given by Greene do not hold out. No. 5121. Jones, Diamond Valley, Utah, April 28, 1895, 3500° alt., in sand.

Cardamine cordifolia Gray.

No. 5341. June 1, 1894, Marysvale, Utah, 7000° alt., in cold spring.

This is a form with stems and often the leaves short-shaggy, with white hairs.

Cardamine cordifolia var. *incana* Gray.

No. 5341a. June 1, Marysvale, Utah, 7000° alt., in gravel, in cold springs.

Very pubescent above, with spreading white hairs.

Arabis hirsuta var. *glabrata* T. & G., Fl. 1, 80.

No. 5683. July 25, 1894, Mt. Ellen Park, Henry Mountains, Utah, 9000° alt., in gravel.

No. 5743b. August 4, 1894, Fish Lake, Utah, 9000° alt., in meadows.

No. 5601b. July 6, 1894, Soldier Summit, Utah, 7300° alt., in gravel.

No. 5537c. June 29, 1894, Thistle, Utah, in gravel, 5300° alt.

No. 6054e. September 17, 1894, Nagle's Ranch, Arizona, Buckskin Mountains, 7800° alt., on the edge of streamlets from springs.

No. 5731f. Fish Lake, Utah, 9000° alt., in gravel.

This seems to be a well-marked variety, differing from the typical *hirsuta* of Europe in the leaves being either entire or sparsely dentate, while the European plant seems to be characterized by very long and often hooked teeth, making the leaf appear almost laciniate. I have seen no true forms of the typical species in the United States, except a few found along the eastern base of the Rocky Mountains, in Colorado and Wyoming.

ARABIS ARCUATA var. PERENNANS (Pringle, Watson, P. A. A. 22, 467).

It is impossible to separate this from typical *A. arcuata*, though extremes seem sufficiently distinct. Coville's No. 1747 is surely a reduced form of this; root leaves oblanceolate, entire, 6–8″ long, rosulate; stem leaves oblong to oblanceolate, sessile, or nearly so, acutish, reduced above; racemes 3–5′ long, rather closely flowered; pods a little arcuate, and below horizontal on recurved pedicels, 1–1½′ long, 1″ wide, barely acute; pedicels slender, 2″ long; seeds obscurely in two rows, narrowly winged above and not at all at very base, oval, ⅓″ long; plants cæspitose and many stemmed. Watson's specimen from the Uintas, at 12,000° alt., is taller and more lax, with longer root leaves, otherwise about the same. Rose's specimen from the Yellowstone Park is the same. My No. 5330, from Marysvale, Utah, May 31, 1895, 6000° alt., in gravel, has longer pedicels and wingless seeds, but I fail to find any valid separating character.

DRABA NEMOROSA var. STENOLOBA (Ledeb. Fl. Ross. 1, 154).

There is no character assigned by Watson for this species which holds; generally the pods are linear-oblong and acute at each end, usually curved most on the outer side, and about as long as the pedicel, glabrous, but all these vary, some pods are half the pedicel or even a third the pedicel, others are longer; some are oblong-ovate, and some are blunt at apex, all ours are annuals. *Draba hirta*, at least specimens so named, from Spitzbergen, is perennial with twisted pods.

ERYSIMUM LINIFOLIUM (Nutt. T. & G., Fl. 1, 91).

Sisymbrium linifolium Nutt. l. c.

No. 5485. June 23, Castle Gate, Utah, 6000° alt., in gravel, along the river in open places in the juniper belt.

No. 5486b. Same locality as above.

No. 5309. May 28, Tropic, Utah, 6000° alt., in fields, in gravel.

It has always seemed to me that this species is wrongly referred to *Sisymbrium* and more properly belongs to *Erysimum*. It is near to *E. cheiranthoides*, but verging toward *E. asperum*.

ERYSIMUM ASPERUM var. PARVIFLORUM (Nutt. T. & G., Fl. 1, 95).

This is certainly only a variety of *E. asperum* as given in Bot. King by Watson, though Watson's specimen is intermediate between this and the type.

Erysimum asperum var. *pumilum* Watson, Bot. King's Exp. is nothing but typical *E. asperum* as it is found on the Great Plateau; his specimen is biennial and not perennial.

Erysimum pumilum var. *perenne* Watson, Coville, Death Valley Rep. is not surely perennial, but seems to be an ordinary form of the type with orange flowers nearest to the var. *Arkansanum*.

THELYPODIUM WRIGHTII var. TENELLUM.

No. 5559. Provo, Utah, in Slate Cañon, on rocks, July 2, 1895, 6000° alt.

No. 5308ah. Marysvale, Utah, in gravel, 6000° alt., June 4, 1895.

Many stemmed from a perennial root, stem slender, intricately branching above, 2–3° high, retrorsely hispid below; stem leaves all entire or faintly sinuate toothed, thin and delicate, 1½–3′ long, lanceolate, obtuse, cuneate at base; pods 2′ long, almost filiform, knotty, with long beak 1–2″ long, with a pseudostipe 1–2″ long; filiform, pedicels 4″ long, reflexed, rarely ascending; racemes 6–12′ long; root leaves and rarely, lower stem leaves lyrate.

STREPTANTHUS HOWELLII (Watson, P. A. A. 21, 445). *Thelypodium Howellii* Watson l. c.

This seems to be a well marked *Streptanthus* nearest to *S. cordatus*. Leaves sagittate, broadly linear, acute, 1′ long, sessile, oblanceolate, sinuate toothed, root leaves with almost no petiole; stems ascending from the crown of a perennial root, 1–1½° high: racemes long and narrow, 1° long or less; flowers purple or purlish; sepals saccate, triangular-oblong, 3″ long, tips barely spreading; petals with narrow blade, twisted and coiled, twice the sepals, purple, claw enlarged below; anthers coiled, sagittate, 1″ long, exserted 1½″: immature pods with long beak, 1″ long, narrow; pods sessile, 1′ long, nearly erect; pedicels 2–2½″ long, greatly enlarged at apex; plants pubescent below and glabrous above. Has the habit of an Arabis. Harney Valley, Or., June 8, 1885, Howell.

CAULANTHUS CRASSICAULIS var. MAJOR.

No. 5685. Jones, Bromide Pass, Henry Mts., Utah, in gravel, on mountain sides, 9,000° alt.

Short lived perennial, 2½–3° high, erect, stems with barely a trace of inflation, simple but tufted, glabrous throughout; leaves 3–5′ long, clustered at the root, variously sinuate-lyrate-pinnatifid, with acutish often curved lobes below, terminal lobe half the whole leaf, lanceolate to oblong, 1–2½′ long, petioles shorter than the blade; stem leaves with still shorter petioles, linear to lanceolate, mostly entire, 2–3′ long, racemes naked, 1–1½° long or less; flowers about their own length apart and less than half the pods apart in fruit; pedicels ½" thick. 1½" long. very stout. ascending. apex 1" wide; pods sessile. 1" wide. 3–4" long. tipped by the capitate sessile stigma. erect: flowers yellow and tipped with purple. narrowly urceolate: sepals 3" long. oblong-lanceolate. obtuse. scarcely saccate. tips recurved: petals linear. nearly double the sepals, recurved and twisted with purple center and white hyaline margins. blade scarcely dilated; filaments equaling the sepals. winged. widening below; anthers yellow. 1½–2" long. linear. not coiled. falling away from the tips of the filaments. barely acute: immature seeds narrowly oblong to nearly linear.

Smelowskia ovalis. Mt. Adams. Washington. above snow line. August 12. 1892.

Type in the National Herbarium. I have forgotten the collector. With the habit of *S. calycina* but leaves coarser and thicker. floccose hoary throughout except the pods: pods about 1" long and nearly as broad. a trifle narrowed at the apiculate apex. erect on pedicels about 3" long: septum obovate. 1" long: flowers small 1–1½" long. white. petals spreading: style about ½ the length of *S. calycina*. This is very well marked by the short pod.

Physaria didymocarpa var. Newberryi (Gray. Bot. Ives. 6).

P. Newberryi l. c.

No. 5276g. May 5. Silver Reef. Utah. 4500" alt.. in red sand. in dry and hot places.

No. [illegible]. June [illegible]. near Orangeville. Utah. in clay [illegible]

No. [illegible]. May [illegible]. [illegible] alt.. in red sand.

No. [illegible]. May [illegible]. [illegible] alt. in clay.

No. 5224a. May 15, Rockville, Utah, in red sand, 4000° alt.

No. 5163d. May 4, Silver Reef, Utah, in gravel, 4500° alt.

The only difference which I can see between this variety and the type is in the shorter style, which is generally shorter than the septum of the mature pod; the plant grows in a different zone, the type species being found among the junipers of the Great Basin region, and the variety occurring chiefly in the regions below the juniper belt as far down as the upper Larrea belt.

LEPIDIUM SCOPULORUM. *L. heterophyllum* (Watson, Am. Nat. 9, 268) Jones, Zoe, 4, 267.

The latter name is preoccupied. Dr. B. L. Robinson has called my attention to this and I have suggested this name in its place.

Lepidium integrifolium Nutt. *L. Utahense* Jones. Having an opportunity to compare my material with authentic specimens I find no appreciable difference.

Lepidium Oreganum Howell seems to be the same as *L. dictyotum*. His specimen examined is from Rogue River, Oregon, April.

Viola pinetorum Greene is only a coarsely toothed form of *V. aurea*, it has a long thick root, 6′ long. Some specimens of undoubted *V. aurea* without the other characters of *V. pinetorum* have coarsely toothed leaves.

Biscutella Californica is deliciously fragrant.

GREGGIA CAMPORUM var. LINEARIFOLIA (Watson, P. A. A. 18, 191). *G. linearifolia* Watson l. c.

CLEOME INTEGRIFOLIA var. ANGUSTA. No. 6057a.

Pods very narrowly-linear and cylindrical. This is the usual form indigenous to the West. Plants introduced from Mexico are like the typical form.

ARENARIA CONGESTA var. ACULEATA (Watson. Bot. King 40).

ARENARIA CONGESTA var. MACRADENIA (Watson. P. A. A. 17. 367. in part Robinson P. A. A. 29. 296).

The character of the sepals even in Parish's type specimen in the National Herbarium fails. for the sepals are oblong-ovate. barely acute. with many nerves. and are not more acute than Watson's type of *A. aculeata* in the National Herbarium. Watson's specimen has slender stems and knotty joints. while the more southern specimens in my collection do or do not have the joints conspicuous: the sharpness of the leaves varies greatly. There is no crucial character separating these three recognized species.

ARENARIA NUTTALLII var. GRACILIPES.

No. 5951. August 29. Brigham Peak. Utah. near Marysvale. 11.500′ alt.. in gravel. on exposed slopes above timber line.

No. 5770c. August 7. Fish Lake. Utah. 10.800′ alt.. on exposed ridges.

Sepals ovate to lanceolate. acute. sometimes slightly pungent. with narrow hyaline margin. midvein rather prominent. 1½ long: leaves channeled as in the type. fasciculate. usually arched outward. blunt. or sometimes abruptly tipped with a short awn. 2–4 long: stems many. very slender from the erect tap-root. sometimes 18′ long and filiform. making a loose mat on the ground: flowers widely spreading or reflexed: petals shorter than the sepals: leaves scarcely connate: pubescence variable but glandular-hairy through the season: flowering stems usually 2–5′ long. rarely longer. spreading: bracts like the sepals and leaves. narrowly linear. thick. I also refer to this variety the following specimens in the National Herbarium: Cascade Mountains. Oregon. Howell:

Yellowstone Park, Tweedy; Mt. Adams, Washington, Howell; Yellowstone Park, Rose.

This grows in volcanic gravel above timber line, forming loose mats 7–15′ in diameter, and grows along with *Stellaria longipes* and *Phlox Douglasii*.

The variety *gracilis* Robinson, occupying the same zone as this variety, has slender-tipped, pungent leaves and bracts, and subulate-lanceolate sepals, with a very prominent midvein, which runs off into a sharp and rather long awn, the sepals being 2–2½″ long. This is Bolander's No. 4976. Coville and Funston's No. 1546 has little broader sepals and condensed habit, but the sepals, leaves and bracts are all pungent. Palmer's No. 195 seems to be about the same. All these are high altitude forms, growing in a very cold alpine zone, while the true *Nuttallii* grows in low, warm and arid altitudes, in a wholly different zone.

Arenaria Kingii (Wats. King's Rep. 39, t. 6).

Stellaria Kingii Wats. l. c.

No. 5515. Manti, Utah, June 27, 6000° alt., in gravel, on dry slopes.

Having at last collected this species, it is manifest that it is a true *Arenaria* instead of a *Stellaria*, in spite of the bifid petals; in fact, it cannot be distinguished from the allied species of *Arenaria*, except by the petals. There is no *Stellaria* with which it has anything in common. It grows on gravely hills.

Acer glabrum var. tripartitum (Nutt. T. & G., Fl. 1, 247).

This would seem worthy of recognition as a variety. It is the usual Rocky Mountain form, with leaves 1–1½′ long, with three leaflets, or at least nearly parted, short racemes, 1–2′ long; wing of fruit 6″ long, and peduncles scarcely ever shorter than the leaflets. The typical form occurs only northward from California to Idaho and Wyoming, with large simply lobed leaves, longer than the petiole; racemes 2–4′ long: fruit wings about 1′ long. The variety I have from as far north as Helena, Mont., Kelsey, where it is a curiosity. Also 5396, from Marysvale, Utah, in Bullion Cañon, 10,000° alt., along streams, June 5, 1895, grows in clumps, with reddish bark, seldom over 15° high. No. 5663g, Marvine laccolite, Henry Mountains, Utah, 6000° alt., in gravel, in box cañons.

LINUM KINGII var. PINETORUM.

No. 5306. May 28, 1894, cañon above Tropic, Utah, 6500° alt., in gravelly clay.

No. 6015g. September 6, 1894, Panguitch Lake, Utah, 8400° alt., in gravelly soil.

Low, 3–7′ high; much branched from the base, with very many erect, simple stems, or the outer ones decumbent at base; all the lower part of the stems densely clad with imbricated, 5-ranked, obtuse, glaucous, thick, oblong (to narrowly oblong above) leaves, 1–2″ long, or rarely longer: the upper leaves more distant and longer, but always longer than the internodes, rarely acute, 3–4″ long; flowers racemose or racemosely-clustered, rarely corymbose; sepals broad, barely acute or obtuse, rather broadly scarious and ciliate-dentate, 1-nerved, or with 1–2 very faint additional ones: petals yellow, obovate and rounded at apex, widely spreading, 3″ long. In other respects this agrees with the type of *L. Kingii*. It appears to be remarkably distinct and grows in a wholly different zone, at 8000 ft. alt., among the pines, in very porous volcanic soil. In view of the abnormal conditions under which it grows, it is probable that it is only a good variety, though the plant is remarkably abundant in the pine forests.

CEANOTHUS FENDLERI var. VIRIDIS.

No. 6039n. September 12, 1894, Elk Ranch, Utah, 7000° alt., in gravel.

No. 6042d. September 13, 1894, on grade below Carmel, Utah, 6500° alt., in gravel.

No. 5822h. August 10, 1894, Fish Lake, Utah, 10,000° alt., in gravel.

No. 5405f. June 6, 1894, Marysvale, at Jugtown, in gravel, 7000° alt.

No. 5312ao. May 29, 1894, head of cañon, above Tropic, 7000° alt., in clay.

No. 5308. May 28, at 6500° alt., same locality.

No. 5208b. May 11, 1894, Cedar City, Utah, 6000° alt., in gravel.

Whole plant glabrous throughout, or only minutely and sparsely pubescent along the veins of the leaves.

CEANOTHUS GREGGII var. LANUGINOSA.

This has more oblong leaves which are white-woolly below, and rather gray above. This is Pringle's No. 708, collected March 30, 1886, in the Santa Eulalia Mountains, Mexico. Also Palmer's plant from Coahuila, Mexico, both specimens in the National Herbarium.

PTELEA TRIFOLIATA var. ANGUSTIFOLIA (Benth. & Pl. Hartweg, 9).

No. 6048. September 15, 1894, Nagle's Ranch, Buckskin Mountains, Arizona, 7600° alt., in gravel.

After having examined a large suite of specimens from many localities, I find it is utterly impossible to keep up these two species, as there is no assigned character which holds, and I can discover no other valid one.

Trifolium Haydeni Porter Hayden's Rep. 1871. I think this is erroneously referred to *T. Kingii* by Coulter. The proper stems are only 1–2′ long; root leaflets round, to broadly obovate, 3–4" long, apiculate, very nervose, rather coarsely denticulate-serrate; stem leaves oval to elliptical and a trifle longer; peduncles 3–4′ long, terminal and single; whole plant glabrous and shining; flowers white, 6" long, 2" wide at apex, radiating in all directions, in a head, rachis not produced; calyx teeth as long as, or one-third longer, than the tube, subulate; tube ¾" long; densely cæspitose and lower stipules imbricated; petioles 1–2′ long. Cook City, Montana, Kelsey, No. 345, Rose, N. W. Wyoming.

Trifolium gymnocarpum I cannot separate from *T. Plummeræ.*

Trifolium Harneyensis Howell I cannot separate from *T. eriocephalum.*

Lupinus Sileri Watson seems to me a good species. Specimens in the National Herbarium are Newberry in McComb's Exp., Ward, Utah; Capt. Bishop same. This is the same as *L. capitatus* Greene, Pitt. 1, 171. Other specimens are Knowlton, San Francisco Mountains, Arizona; Rusby, Cosnino, Arizona, which is in the same region.

LUPINUS MICENSIS.

No. 50640. April 14, Mica Spring, Nevada, 4000° alt., in granitic gravel.

No. 5149h. May 3, Silver Reef, Utah, 3500° alt., on slopes in red stand.

No. 5163g. May 4, Silver Reef, Utah, 3500° alt., on slopes.

No. 5045f. April 13, Mica Spring, Nevada, 4000° alt., in granitic gravel.

No. 5095b. April 21, Pagumpa, Arizona, 4000° alt., in gravel.

No. 5072b. April 16, Mica Mine, Arizona, 4000° alt., in gravel.

No. 5176j. May 5, Silver Reef, Utah, 3500° alt., in gravel.

Allied to *L. brevicaulis;* annual, 2–6′ high, branching from the base and lateral stems decumbent; whole plant even to the calyx and pods sparsely long- and silky-villous, or in some cases quite densely so, the hairs always spreading and soft; petioles usually 3 to 4 times the leaflets; leaflets spatulate, 8″ long or less, and about 2½″ wide, rounded, often apiculate, about 8, not reduced above; whole plant very leafy; flowers 3½″ long, in short, spike-like racemes, reddish-purple, subtended by short, triangular bracts; calyx lobes lanceolate, 1½–2″ long and the calyx cleft nearly to the base, banner oval and shorter than the keel; keel 1½″ wide; pods narrowly-oblong, about 8″ long, 2½″ wide, deeply cross-wrinkled between the seeds; seeds 3–4, nearly square, about 1½″ wide. This differs from *L. brevicaulis* in the narrow pods, much larger flowers, in racemes instead of heads, and the caulescent stems. It seems to be intermediate between *L. brevicaulis* and *L. Arizonicus.* Having gathered it in very many localities and finding that its characters remain constant I do not hesitate to separate it as a good species.

This grows in the Larrea belt in red sand and on gravelly slopes.

Petalostemon flavescens Watson seems to be a white flowered form of *P. Searlsiæ.* I can see no other valid difference.

Psoralea castorea Watson.

No. 5024j. April 5, 1894, Beaverdam, Arizona, 1800° alt., in drifting sand.

This plant is found only within the Larrea belt on drifting sand dunes, growing singly, from a deep seated, erect, nearly spherical root which is fleshy and with only scattered woody fibers within; the root is 2′ or more in diameter.

Psoralea mephitica Wats.

No. 5082a. April 20, 1894, Pagumpa, Arizona, 4000° alt., in gravelly clay.

No. 5098b. April 23, 1894, top of grade above Pagumpa, Arizona, 5,000° alt., in gravelly clay.

No. 5095. April 21, 1894, Pagumpa, Arizona, 4000° alt., in gravelly clay.

This is the plant referred to by me in my Contributions in Zoe as *P. castorea*. An examination of duplicate type specimens of the latter species shows that Watson was mistaken in both his description of the latter species and in the locality at which it was collected. Watson mistook the bracts for the calyx lobes, and gave as the locality where the plants were collected "Beaver City, Utah," adding (doubtless on the authority of Dr. Palmer) the statement that both species grew together. This is erroneous. Dr. Palmer never collected any plants at Beaver City, Utah, during the year in which he collected these species, but he did collect in that year in the Beaverdam Mountains on the northeastern corner of Arizona at a place called Beaverdam. On the drifting sand dunes at Beaverdam Dr. Palmer collected *P. castorea* probably at the end of his day's journey, that being the first place in which water can be secured west of the Beaverdam Mountains, and therefore must have been his camping place for the night. During the earlier part of the same day along the road he must have collected *P. mephitica* among the junipers high up on the mountains in a wholly different zone from that of the *Larrea* belt in which the former was collected. The difference in elevation at which the two species grow is over 3000 ft. *P. mephitica* never grows in loose sand but always on rocky or gravelly places and is remarkable for its long, tuberous-thickened, branched, woods, and interlaced roots, forming broad patches among the junipers.

HOSACKIA RIGIDA var. NUMMULARIA.

No. 5224k. May 15, 1894, Rockville, Utah, 4000° alt., in red sand.

No. 5125d. April 28, 1894, Diamond valley, Utah, 4000° alt., in red sand.

No. 5098d. April 23, 1894, 10 miles south of Black Rock Spring, Arizona, 4500° alt., in gravel, among junipers.

No. 5128. April 30, 1894, Santa Clara valley, Utah, 3000° alt., in gravel, along the river bed.

Herbaceous throughout, prostrate, lower leaflets round, all the leaves except the very uppermost short-petioled, whole plant ashy, root woody, stems 1–2½° long, forming loose mats and decumbent or procumbent.

This abounds in sandy or gravelly places, mostly along the streams. To this I also refer Wright's No. 1357, Palmer's specimen from Fort Huachuca, Arizona, both being pubescent; also a specimen of the Mexican Boundary Survey, a specimen from Parish collected at Lowell, Arizona, Rusby's specimen from the San Francisco Mountains, Arizona, collected in April, and a specimen from western Texas by Neally, all in the National Herbarium.

Phaca L. Among the genera recently revived for American plants there is none which deserves so little attention as this; but this genus is revived, apparently, in order to avoid the odium attaching to relegating so large and common a genus as *Astragalus* to synonymy because of priority by position. It is curious to see the twins *Phaca bisulcata* and *Astragalus scobinatulus* in separate genera; also *Astragalus Shortianus* and *Phaca pectinata*, while *Astragalus racemosus*, *oroboides* and *gracilis*, with strictly one cell, are relegated to *Astragalus*, and *A. aboriginum*, *alpinus*, *Robbinsii*, with rudiments of a septum, are also retained. It would be interesting to see what kind of a generic character the authors of the "Check List" would make for *Phaca*.

ASTRAGALUS ANGUSTUS.

Astragalus pictus Gray, var. *angustus* Jones, Zoe, 4, 37.

Mr. Sheldon, in making a new name for the type species, refers my var. *angustus* to it under the name of var. *Jonesii* without previously consulting me as to the desirability my accepting the high honor of having a variety dedicated to me which in the very nature of the case could never be more than a synonym. Mr. Sheldon makes the name to supplant *angustus*, because, as he assumes, it was used previously by Boissier, but Boissier never made the name *Astragalus angustus*, as far as I know. In his second attempt at correcting names, Mr. Sheldon, departing from his uniform rule of giving specific names to varieties and mere forms, reduces (without having seen my type) my var. to a synonym of the var. *foliolosus* Gray, raising that form to a species, stating that its synonyms are "*A. pictus* var. *angustatus* Jones, not *A. angustatus* Boissier, *A. ceramicus* var. *Jonesii* Sheldon." Now, I never published any *Astragalus angustatus*, and it appears that Boissier never published an *Astragalus angustus*. Hence the resulting batch of synonyms is wholly useless. Curiously enough, *Astragalus pictus* var. *angustus* differs from the type and from the vars. *foliolosus* and *filifolius* more than they do from each other, and under the methods so common in closet botanizing would be called a good species.

The same method of making useless synonyms was employed in renaming *Astragalus strigosus* Coulter & Fisher (*A. griscopubens* Sheldon). Mr. Sheldon being wholly unaware that that name is a synonym for *Astragalus serotinus* Gray.

The synonym of this species and its varieties, on the basis recommended by the recent circular of representative American botanists is as follows:

ASTRALAGUS ANGUSTUS.

Astragalus pictus var. *angustus* Jones, Zoe, 4, 37, 1893.

Astragalus ceramicus var. *Jonesii* Sheldon, Bull. Minn. Geol. and Nat. Hist. Surv., 9, 19, 1894.

A. pictus var. *foliolosus* Sheldon, I. c. 9, 138, 1894.

Not *A. pictus* var. *foliolosus* Gray.

ASTRAGALUS ANGUSTUS var. PICTUS (Gray, Pl. Fend. 37).

Phaca picta Gray, Pl. Fend. 37, 1849.

Astragalus pictus Gray, P. A. A. 6, 214, 1866.

Tragacantha picta O. K. Rev. Gen. Pl. 2, 947, 1891.

Not *Astragalus pictus* Steud. Nom. Ed. 2 I, 163, 1840.

Astragalus ceramicus Sheldon, l. c. 9, 19, 1894.

Astragalus pictus var. *foliolosus* Gray, P. A. A. 6, 215, 1866.

An examination of the type of this variety in the National Herbarium shows that it differs from the typical species in no respect worthy of varietal rank.

Astragalus foliolosus Sheldon, l. c. 9, 138, 1894.

Not *Astragalus foliolosus* Bunge, Gen. Astrag. Geront. 2, 125, 1886; a recognized species.

ASTRAGALUS ANGUSTUS var. LONGIFOLIUS (Pursh, Fl. A. Sept. 2, 741, 1814.

Psoralea longifolia Pursh, l. c.

Orobus longifolius Nutt. T. & G. Fl. 1, 346, 1838.

Not *Astragalus longifolius* Lam. Ency. Meth. 1, 322, 1783.

Astragalus filifolius Gray, Pac. R. R. Rep. 12, 42, 1860.

Astragalus pictus var. *filifolius* Gray, P. A. A. 6, 214, 1866.

Not *Astragalus filifolius* Clos, C. Gay, Fl. Chil. 2, 111, 1846.

Astragalus ceramicus var. *imperfectus* Sheldon, l. c. 9, 19, 1894. All the above belong to *A. angustus*.

Astragalus subcinereus Gray. An examination of a duplicate type specimen of *Astragalus Wootoni* Sheldon shows that it is the same, the pods varying from deeply mottled to uncolored.

Astragalus acerbus Sheldon seems to be identical with *A. Dodgianus* Jones, and the latter is not surely separable from *A. Wingatensis* Watson.

Astragalus recurvus Greene proves to be the same as *A. lanceareus* Gray, and not *A. obscurus*, as I had supposed, the original type specimen being very poor and in fruit only. My specimen "with crimped edges to the pod" barely in fruit is this species.

Astragalus Pattersoni Gray. An examination of the type of *Astragalus diphysus* var. *albiflorus* Gray, Bot. Ives, shows that it is identical with *A. Pattersoni*. Now, here is a chance to immortalize oneself by making a brand new name for this nauseous and poisonous weed, on the once-a-synonym-always-a-synonym system.

ASTRAGALUS PATTERSONI var. PROCERUS (Gray, P. A. A. 13, 369, 1878).

Astragalus procerus Gray, l. c.

Astragalus prælongus Sheldon, l. c. 9, 19, 1894.

Since this is only a large flowered form with unusually broad pods, but which vary into the typical species, and since there is no reason for not using Gray's original name, which as a variety is unused, I place it as above.

Astragalus cerussatus Sheldon. This species I think should stand, as it is not the *Astragalus triflorus* Gray, Pl. Wright 2, 45, 1853, though it may be proved to vary into it, and therefore may have to be reduced to a variety of it. It is the *Astragalus triflorus* of Watson in King's Rep., at least in part, of Coulter's Manual, and of most western collectors, and it may prove to be only an extreme form of *A. Wetherilli* Jones, in which case the latter name will prevail, but at present it seems sufficiently distinct. It seems to be confined to the mountainous regions of Colorado and adjacent States and Territories. It blooms throughout the season, often in bloom quite late.

Astragalus triflorus Gray, l. c. This seems to be close to *A. Candolleanus* (H. B. K.) Sheldon, and being the first name in the genus, should stand. It is not at all certain that it is distinct from *Phaca triflora* DC. Ast. 62, t. 1, 1802. It seems to be a very variable species. The typical form is annual, with the habit of *A. Geyeri* Gray, and seems to be an early bloomer. Specimens belonging here are C. Wright, New Mexico, the smallest specimen on the sheet from the Mexican Boundary Survey; one sheet of Dr. Mearns, Carrigallilo Mountains, New Mexico, April 18, 1892. A taller form (which may be *A. cerussatus* Sheldon) from El Paso, Texas, G. R. Vasey, is apparently perennial, and has purple flowers.

The various forms of this species which may deserve varietal rank are A. TRIFLORUS var. CANDOLLEANUS (H. B. K. Nov. Gen. et Sp. 6; 495. 1823), with oblique pods and many leaflets (usually 8 to 15 pairs).

A. TRIFLORUS var. INSULARIS(Kellogg, Bull. Cal. Acad. Sci. 1; 6. 1884), with few leaflets and scarcely oblique small pods. It is quite possible that this will still hold as a good species.

Astragalus Pondii Greene is close to the two above varieties and may deserve specific rank. It was poorly described, but authentic specimens show differences. It is the second specimen on the sheet in Cal. Acad. Sci. along with *A. insularis*, described by me in Contributions 4, p. 28, last form described.

Astragalus triflorus Sheldon (not DC.), Death Valley Rep., is *A. Tejouensis* or near it.

ASTRAGALUS HOLOSERICEUS. Between King City and Jolon, Monterey County, California. Miss Eastwood, June 7, 1893.

This is probably a variety of *A. cerussatus* or *A. triflorus*, but until the limits of those species are clearly made out this must stand, for there are no transitions yet known. Perennial seemingly, short-shaggy throughout but the stems and mature pods less so, hairs white, fixed by the base, spreading, tangled; stems a little flexuous, 1–2° high, with nodes 1–2′ apart, barely striate, ascending, unbranched; lower leaves small, 2–3′ long, with 5–7 pairs of elliptical and acute leaflets 3″ long, and petioles equaling the rachis; stipules subulate with short filiform tip, 3″ long, reflexed, thick, adnate, not connate; upper leaves 4–6′ long, with short petiole 1′ long or less, leaflets about 10 pairs, linear-lanceolate, very acute, 6–8″ long; flowers nearly white, 3″ long, on stout pedicels 1½″ long which are twice the ovate and minute bract, in a short close raceme 1–2′ long; peduncles and rachis about 4–6′ long, rather stout; calyx tube campanulate, 1½″ long and nearly as wide, with subulate to triangular teeth nearly as long and curved; banner short and rounded, blade 2″ long, arched in short curve to 90° just beyond calyx tube, erect part less than 1″ high, sides reflexed a little; wings narrowly oblong, arched to 45° or less, a trifle longer than keel and but little shorter than the banner, narrower than the keel; keel a little arched below, rather abruptly bent at apex to 90° and vertical portion with straight edge, the very tip a little recurved so as to make a short boss, erect part as long as base and 1½″ high, a trifle longer than the calyx teeth; pods oval 1′ long, 9″ wide, ventral suture less arched than the dorsal and a trifle sulcate at times, otherwise cross-section round, chartaceous, much inflated, jointed at base but sessile or with stipe shorter than broad, 1-celled, reflexed or spreading; flowers ascending.

Astragalus debilis (Nutt.) Gray. Mr. Sheldon seems to have renamed this species under the name of *A. Bodini*, without noticing their similarity. Having seen a duplicate type specimen I discover that Mr. Sheldon's description is very inaccurate. He says pod "flat"; "species nearest to *Astragalus tenellus* Pursh, but the habit is more nearly that of *Astragalus flexuosus* Douglas." In fact the specimens on which Mr. Sheldon bases his species are so near to *A. leptaleus* that it is very difficult to separate them. The species has no relation to the *Homalobi*. The pods are triangular or nearly round in cross-section. Below I append a description from my field notes of the same species. Ward also collected it long ago in the same locality as mine and distributed it as *A. oroboides*.

ASTRAGALUS DEBILIS (Nutt.) Gray.

No. 5649. Loa, Utah, 7000° alt., in clayey meadows, July 18, 1894.

No. 5709b. August 1, 1894, same locality.

Prostrate, in open mats, often 3° in diameter, from an erect and woody root; stems very slender; stipules large, foliaceous, triangular, mostly reflexed, 4″ long, 2″ wide; leaflets oval to lanceolate, acute; peduncles slender, 2–4′ long, surpassing the leaves, capitately flowered and spicate in fruit, but spike short; pedicel short, stout and black, ½″ long; bracts green, lanceolate, 1½″ long; flowers 3–4″ long, light purple, numerous, spreading, the pedicel inclined to be twisted in fruit; calyx tube 1½″ long, almost cylindrical, but a little compressed below and obcompressed above on the deeper cleft upper side, hyaline, nigrescent with short, black and appressed hairs, lower side straight, upper side a trifle convex, neither oblique nor obliquely attached; calyx teeth equal, about 1″ long, subulate; sinuses except the rounded upper one acute; banner oval to oblong-oval, ascending abruptly at calyx tips to 45°, sides reflexed at point opposite the keel, ¼″ wide, not at all at apex, which is emarginate and rounded, erect part of banner 2″ long, sulcus reduced to a groove above but semicircular below keel tip and ⅓″ wide, white spot comes to within ⅓″ of the sides and ½″ of the tip, obovate to fan-shaped, and lacerate above by the intruding purple veins, which are very fine, and unite below in twos, but do not form a ring, white spot 1″ wide and 1⅓″ long, all of the banner is veined with dark purple; blade of the wings 1½″ long, sometimes 1″ wide, obliquely oblong, with narrowed tip, but obtuse, left hand one spreading and concave to the keel, and turned nearly horizontal, with the concave side down, right hand one incurved over the keel tip and ½″ longer, light purple and with purple veins near the lower side, keel much inflated near the calyx tips, but flat beyond and ½″ longer, incurved to 100°, very obtuse, dark purple; pods triquetrous, 1″ thick, oblong, 4″ long, acute at each end, on a stipe ¼″ long, which is as thick as long or nearly so, spreading, minutely pubescent, double the calyx tube, which it splits as it matures.

Astragalus Tolucanus Robinson & Seaton. Pringle, No. 4238, on dry ridges under pines, Nevada de Toluca, Mexico, 12,000′ alt., September 6, 1892. With the inflorescence of *A. agrestis* and the pods of *A. aboriginum* nearly. Perennial, with many delicate, decumbent stems from an erect root, 6′ long or less; leaves 2–3′ long, delicate, thin, with very short petiole; leaflets 10–12 pairs, 2–3″ long, 1–1½″ wide, elliptical, obtuse, glabrous above, puberulent below; root leaves smaller; lower stipules small, upper enlarging, all connate opposite the petioles, lanceolate, acute, uppermost 3″ long, somewhat hyaline; nodes 1′ apart; peduncles at least equaling the leaves, fully as thick as the stems, erect or ascending, sulcate; flowers capitate, about 20, 6″ long, spreading, purple, on pedicels, 1″ long; bracts large, oval to obovate, green, nigrescent, 2″ long; calyx tube campanulate, 1½″ long, 1″ wide, green, with broad triangular green lobes nearly as long; banner obovate to oblong, blade 3″ long, ascending 45° in a gentle curve from calyx tips, sides reflexed ⅓″ wide; wings 1″ wide, narrowly oblong, ascending, concealing keel, ¾″ shorter than banner, fully 1″ longer than keel; keel a little downwardly arched, bent at tip abruptly to 90° and vertical edge straight, 1½″ high, acutish, purple tipped, veined; pod membranous, 1-celled, 6″ long, 2″ wide, elliptical, apiculate, round in cross-section apparently, not sulcate, stipitate on stipe as long as calyx tube, pendent, smooth, seed bearing throughout. Manifestly closely allied to *A. aboriginum*. Described from duplicate type in National Herbarium.

Astragalus serpens.

No. 5639i. July 17, 1894, Loa Pass, Utah, 7500° alt., in gravel, among sagebrush.

Perennial, from a thick, erect, woody, much-branched root, whose stems for several inches in length endure over winter and are covered with dead leaf-petioles and even leaves and peduncles; stems prostrate at base, ascending at tip, 2–10′ long, in rather dense tufts, much branched below, leaves reduced above and also below, all petioled, petioles nearly equaling the rachis, the whole being seldom over 1½′ long; leaflets about 6 pairs, elliptical and folded, 2–3″ long; whole plant finely pubescent with appressed hairs fixed at the base; nodes very short, 6″ long or less; peduncles 1½′ long or less; flowers 2–5 on stout pedicels; pedicels 1″ long, much longer than the ovate bract, reflexed or spreading, 3″ long; calyx tube broadly campanulate, 1″ long, with triangular teeth half as long; banner nearly round, blade 1½″ long and wide, with sides a little reflexed, just equaling the wings and keel, arched abruptly to 110° at calyx tips, white spot large and purple-veined, flowers greenish-purple; wings obovate-oval, very oblique, 1½″ wide, nearly 2″ long, just the shape of the keel; keel blunt, a trifle arched below and tip incurved 90°, with a minute boss at tip, keel about 1½″ high; pods 1-celled, oval-ovate to half-oval, 1½″–3″ long, cross-section triangular to roundish, pod barely sulcate ventrally, thin-papery, acutish, rounded at base and very shortly stipitate (stipe half the calyx tube), purple-spotted; seeds oblong, on stalk ½″ long; ventral suture seed-bearing only in the middle. The stipules are triangular, small and seemingly free.

ASTRAGALUS SILERANUS var. CARIACUS.

No. 6036. September 12, 1894, Elk Ranch, Utah, 7000° alt., in gravel.

No. 6033f. September 11, 1894, 4 miles below Ranch Utah, 7400° alt., in gravel.

This plant differs from the type in having pods which are 1½′ long or less, variously acuminate-pointed and often much contracted at the base; the leaflets are elliptical-oblong, 4″ long or very much less on the same plant. with a petiole often 1½″ long; the stipules are rather rigid, green, reflexed, triangular or with a triangular base and the upper ones with a long subulate apex: the plants are often 3 long and either flat on the ground or (as in the type) ascending among the bushes: the pods are from narrowly-oblong to oval and often with an upcurved apex and a downwardly-curved base in the forms with longer pods.

ASTRAGALUS MEGACARPUS var. CAULESCENS.

No. 5639f. July 17, 1894, Loa Pass, Utah, 8000° alt., in gravel under sagebrush.

This has large leaves, 6' long, stems a foot high, with large, green triangular stipules 2–3'' long and stipe as long as the calyx.

The variety *Parryi* Gray has elliptical leaflets and pods 1½' long on a stipe ½'' long. What appears to be nearly the same, from Peach Springs, Arizona, by Lemmon, has a stipe as long as the calyx, and teeth, and stems from 4–6' long; the pod is as described above, but elliptical-oval. It is quite probable that both varieties will prove to be only forms not deserving varietal rank.

ASTRAGALUS STRIATIFLORUS.

No. 6080k. September 25, 1894, above Springdale, Utah, 4000° alt., in red sand.

Perennial from an erect root; stems prostrate, 2–6' long, herbaceous; nodes 6'' long or less; stipules hyaline, light chestnut colored, connate and cup-like, 2'' high, a little pointed; whole plant densely appressed-hairy with rather long slender hairs fixed by the base, and stems short-shaggy with white hairs; leaves on slender petioles which are a trifle longer than the rachis, leaves 2' long; leaflets contiguous, 4–6 pairs, orbicular, folded, 2'' long; peduncles slender, bearing few subcapitate flowers at the apex, 2½–5' long; pedicels ½'' long, shorter than the ovate bracts; calyx tube campanulate, 1–1½'' long, 1'' wide; triangular teeth a little shorter; flowers about 4'' long, banner greenish white, purple below and purple veined, oval, the blade 2'' long, its sides reflexed a little, arched abruptly at tip of calyx tube to 90°; wings oblong, equaling the keel, ½'' wide, arched; keel ½'' wide at base and with the apex produced gradually into a long, narrow, nearly erect sharp tip; ovary linear; fruit not seen, but certainly belonging to the *Inflati* and probably near to *A. serpens.* Ordinarily I should not think of describing an *Astragalus* without the pods, but this is so distinct in its characters that I venture to publish it.

This grows in sandy soil on gravelly slopes.

Astragalus Tejonensis. Allied to *A. oocarpus.*

Pringle, Mojave desert, May 13, 1882, on hills bordering the desert. Tehachapi, Cal., June, 1884, hills near Tejon Pass, Cal., May 13, 1882. This seems like a hybrid between *Parishii* and *allochrous*, but cannot be, as *allochrous* does not grow there; the only possible parents are *Parishii* and *Douglasii.* Green throughout, but on close inspection there is the same ashy pubescence as of the allied species, which is composed of flattish, narrow, short hairs, fixed by the base and closely appressed; leaflets and leaves as in *A. Parishii*, but leaves 3–5′ long and ascending, and leaflets 10 pairs, 1′ long or less, elliptical to oblong lanceolate, rounded at apex, 2–2½″ wide; proper petiole present in all; peduncles and rachis 5–8′ long, finely sulcate as well as the stems; flowers racemose, usually ascending, often distant, inflorescence 2–3′ long in flower and 4–6′ long in fruit; pods ascending usually, half ovate-oval, ventral suture straight, 1–1½′ long, ¾ to 1′ wide, nearly round, sessile, variously reticulated, nearly glabrous when ripe; flowers like *A. Parishii* but banner not elongated; keel tip but little incurved and broader: calyx lobes subulate and half the tube; decumbent, many stemmed. *A. Parishii* seems to have longer nodes, narrower leaflets, and is more open and erect. This is instantly recognized from allied species by its small yellow flowers, long calyx lobes, broad leaflets and green appearance. Specimens from Palmer from Edgewood, Cal., July, 1892, have pods 1′ long, emarginate leaflets, but otherwise the same. Type in National Herbarium.

To this species I refer with much doubt Palmer's specimen from Lagoon Head, Lower California, March 6–15, 1889. Stems erect; petioles absent; leaves and the stiff long peduncles ascending; whole plant nearly glabrous; stipules minute, adnate, not connate; spikes 2–4′ long; peduncles floriferous on the upper third; flowers loosely spicate-racemose, almost sessile, the pedicels being half the short ovate bracts; calyx broadly campanulate, tube 1″ long and wide, triangular teeth half as long; flowers 4″ long, purple; keel large, blade 3″ long, erect portion as long as the horizontal part and bent to it at an angle of 90°; banner ascending sharply to 45° and remote from calyx lobes, sides reflexed; wings 3″ long, lanceolate, arched 30°, a trifle longer than the keel which is fully 1″ shorter than the banner; pods obliquely ovate-oblong, 9″ long, 6″ wide, sessile, rounded at base, papery, sulcate ventrally, shortly acute, circular in cross-section, nearly smooth, ascending. Should this prove to be a new species, as it is most likely to be, it may be called *Astragalus piscinus.*

ASTRAGALUS DOUGLASII var. GLABERRIMUS. Los Huevelos, Lower California, 1889, Brandegee; also at San Fernando. Plants 1–1½° high, rather bushy, branched at base, perennial, whole plant glabrous; leaflets about 7 pairs, linear-lanceolate and very sharp, 6–8″ long and 1″ wide, distant, all but the very uppermost petioled; racemes very lax, 4–5′ long including the rachis, floriferous on the upper two-thirds; flowers 6–10; pods half-oval, shortly-flat-triangular beaked, 1–1½′ long, round in cross-section, deflexed; otherwise as in *A. Douglasii.* Type in the California Academy Herbarium.

Astragalus Haydenianus Gray. From a large amount

of material now on hand, it appears that *A. grallator* Watson is only an abnormal form, due either to insects or some fungoid agency. The normal young pods are vetch-like, and with age become greatly obcompressed. The corrugations are due in their intensity or faintness to moisture and shade, or the opposite. I find that these two causes, which are purely accidental, produce all the forms hitherto separated as species and varieties, and so have no distributional significance. The synonyms are: *A. Haydenianus* vars. *major* and *Nevadensis* Jones, *A. scobinatulus* Sheldon, *A. demissus* Greene, *A. Jepsoni* Sheldon, and *A. grallator* Watson.

Astragalus nitidus Douglas, Herb. Hort. Soc. Hooker, Fl. 1, 149. A careful examination of all the figures and descriptions of *A. adsurgens* Pall. and *A. Laxmanni* Jacq., together with specimens of Maximowics from Japan, lead me to feel quite certain that our plant is distinct from both of them, and is the same as *A. striatus* Nutt. T. & G., Fl. 1, 330.

Astragalus agrestis Douglas, Hook., Fl. 1, 148. This has been erroneously referred to *A. hypoglottis* L. Our species has oblong to linear green bracts, which are very conspicuous, obtuse at the base of the head of flowers, and acute toward the top of the head, often with a hyaline margin, about equaling the calyx tube; calyx cylindric, teeth linear-subulate, 1″ long; banner obovate, 6″ long, slightly ascending (15°), sides reflexed more or less; wings linear, nearly as long as the banner, which is 2″ longer than the keel; banner arched from a point beyond the calyx tips; pods very deeply sulcate, often almost to ventral suture, and septum narrow, usually white-woolly, oval to oblong, splitting the calyx, shortly stipitate; leaves narrowly elliptical, never acute, usually emarginate when mature: stipules long-sheathing opposite the petioles, green at least at tip, blunt, or rarely acute, often 6″ long, and resembling the bracts; pedicels about ¾″ long. Throughout the Great Plateau, in subalpine meadows, or even as low as 6000° alt., in meadows. Apparently from Bolivia, Rusby, but specimens more like those of Europe. *Astragalus virgultulus* Sheldon is the same as *A. agrestis* apparently.

Astragalus hypoglottis L. as figured in Pall. Astrag. has subulate pointed bracts; campanulate calyx, teeth subulate and shorter than the short tube; banner short and oval; wings oblong and barely longer than the keel; leaflets lanceolate and always acute; stipules much smaller, bracts ovate to linear-lanceolate; calyx lobes less than half the tube; pods with longer stipe and broader, ovate, very blunt at both ends; plants sparsely hairy with long slender hairs (ours are usually nearly glabrous); pods simply hairy; leaflets about 8 pairs; peduncles longer than the leaves.

ASTRAGALUS CANADENSIS L. var. CAROLINIANUS (L).

This seems to be a very good geographical variety, abounding from North Carolina to the Ohio River, and rarely beyond. This has long open spikes of small, not greenish nor thick flowers: oblong-oval, apiculate, 2-celled pods, 4″ long.

Astragalus simplicifolius (Nutt. T. & G., Fl. 1, 350, 1838) Gray. It is manifest that this is a reduced form of what has heretofore been called *Astragalus cæspitosus.*

ASTRAGALUS SIMPLICIFOLIUS var. CÆSPITOSUS (Nutt. T. & G., Fl. 1, 352).

To this must be referred the very common and normal form of the species. The synonymy is *A. cæspitosus* (Nutt.) Gray, P. A. A. 6, 230; *Tragacantha cæspitosa* OK. Rev. Gen. Pl. 2, 943: *Homalobus canescens* Nutt. T. & G., Fl. 1, 352; *Homalobus brachycarpus* Nutt. l. c.;

A. spatulatus Sheldon, l. c. 9, 19; *A. lingulatus* Sheldon, l. c. 9, 118. The last seems to be identical with *Homalobus canescens* Nutt. This species is quite variable.

ASTRAGALUS HYALINUS. Nearest to *A. triphyllus* Pursh. Stems loosely matted, usually erect, 1 to 2′ high, at the ends of the much branched thick root; stipules very conspicuous, large, much imbricated, at least 9″ long, smooth except at the very base, where there is a tuft of long, straight, white hairs; leaflets narrowly elliptical to oblanceolate, about 6″ long and 2″ wide, obtuse or barely acute, 3, densely silky all over, with hairs attached by the middle, on a varying petiole; flowers, one or two in a place, sessile at the base of the leaf, apparently white with a dark keel tip; calyx very white-villous, cylindrical, 6″ long, 1½″ wide, little exceeding the stipules, teeth subulate, 1″ long; corolla very pubescent outside, banner oblanceolate, emarginate, about 1′ long and 1½″ wide, proper blade 3″ long; wings narrowly linear and as long as the banner; keel 3″ shorter than the wings, very narrow, apex but little arched; fruit immature but manifestly sessile, very white-silky, ovate or triangular, round in cross-section, 3″ long. At first I took this to be an abnormal form of *A. triphyllus*, due to a fungus, but careful examination failed to show any fungus growth, while some specimens showed normal vigorous pods. This differs from *A. triphyllus* in the pubescent corolla, broader leaflets, stems, and the very conspicuous stipules. *A. triphyllus* is densely congested, like *Krynitzkia aretioides*. Upper Lawrence Fork, Kimball County, Nebraska, No. 80; Cliffs, Banner County, Nebraska, August, 1890; Hills, Kiowa valley, Scott's Bluff County, Nebraska; all collected by Rydberg. Type in University of Nebraska and duplicate types in National Herbarium.

Astragalus Californicus (Gray) Greene seems to me to be a good species, the distinguishing characters being the more numerous leaflets, the long, tapering, mottled pods, and the shorter calyx. More abundant material might, however, prove these characters to be invalid, but so far they are good.

Astragalus Tweedyi Canby seems to be very distinct, but close to *A. collinus*. Calyx ascending; pods 1-celled, erect, on a stout stipe, which is ½ longer than the calyx, with the shape of *A. arrectus*, nearly cylindrical, shortly acute at both ends, nearly straight, 6 to 8″ long, 1½″ wide, about 1″ thick; cartilaginous, and like all the rest of its group, filled with pulp; leaflets fully linear, blunt, 6–8 pairs, 1″ wide, 1′ long, rather distant, almost no proper petiole; stipules very small; plants erect, 2° high; peduncles strict, 6–12′ long, sulcate; whole plant finely pubescent, with slender hairs fixed by the base.

These notes taken from a duplicate type.

ASTRAGALIS HUMISTRATUS var. TENERRIMUS.

No. 6052f. September 17, 1894, Buckskin Mountains, Arizona, 9000° alt., in gravel, under conifers.

No. 6064. September 20, 1894, road to Nagle's Ranch, Buckskin Mountains, Arizona, 9000° alt., in gravel.

No. 6056bm. Same locality and date.

This variety is characterized by having very many slender stems lying perfectly flat on the ground and covered with soil towards the base, the whole forming a mat from 1½–3° in diameter; the leaflets are mostly in 4–6 pairs, ovate to obovate and rounded, 1½″ or less long and nearly glabrous; the flowers are very few, in a loose, short raceme, nearly white, with filiform calyx lobes longer than the tube, and the pods of *A. Sonoræ* Gray (which is only a form of *humistratus*).

ASTRAGALUS TEGETARIUS var. ROTUNDUS.

No. 5649b. July 18, 1894, Loa, Utah, 7000° alt., in clay.

No. 6002. September 6, 1894, Panguitch Lake, Utah, 8400° alt., in gravel.

This has the habit of *Astragalus tegetarius*, but the leaves are stiffer and more pungent, much after the fashion of *A. Kentrophyta*, but lies strictly flat on the ground; the pods are generally not in the least flattened, oval-ovate and usually straight, though the immature pods are occasionally flattish and sometimes a little curved.

Astragalus tegetarius var. *implexus* Canby does not seem to be worthy of varietal rank as it is only a condensed form.

This species is very close to *A. Kentrophyta*, but the habitat is very different, most of the differences in appearance can be explained by the different habitat.

Astragalus Kentrophyta var. *ungulatus*. This is the white hoary plant with lanceolate, curved and acuminate-tipped pods with round cross-section, represented by Watson's specimen in the National Herbarium from Monitor valley, Nevada, and by mine from Sprucemont, Nevada.

ASTRAGALUS CYMBOIDES.

No. 5658q. July 21, 1894, Cottrell's Ranch, Henry Mountains, Utah, 6000° alt,, in gravel.

No. 5464j. June 19, 1894, Huntington, Utah, in clay, 5000° alt.

No. 5445f. June 16, 1894, near Emery, Utah, 7000° alt., in clay.

Perennial, flat and matted on the ground; peduncle shorter than the leaves in flower; stipules triangular, rather large, adnate, not connate, rather longer than the very short nodes, densely appressed-strigose and thus giving the stems a shaggy appearance; leaflets 4–5 pairs, elliptical, obtuse, usually about 3″ long, finely appressed-pubescent, with hairs fixed by the middle; petioles longer than the rachis, 2½–5′ long; pods prostrate, oblong-ovate, cross-section almost round, 3″ wide and 8″ long, apiculate, straight, very fleshy, inner wall woody, the outer composed of a pulp over ½″ thick which at the ventral suture is 1″ thick, the ripe pods when dry have a thickened and raised ventral and dorsal suture, the latter the smaller, the general appearance of the pod being that of *A. Missouriensis*, the middle of the pod is swelled longitudinally so much so that the pod seems to be 4-sided, but the two sides are always rounded and not acute as they are at the sutures, pods reddish; seeds many, filling the hairy cavity, long stalked; calyx laterally flattened and much deeper cleft above; teeth unequal; banner white, bent rather abruptly at a point ½″ beyond the calyx tips to 45° and arched above, oblong-oval, sides reflexed at the keel 1″ deep, and banner usually fiddle-shaped by the sides not being reflexed above and below, notched ½″ deep; sulcus decidedly narrowed below and nearly V-shaped, 1″ deep, ½″ wide; above it is 1″ wide and U-shaped and vanishes about 1″ below the tip of the banner; banner water-lined; wings linear-oblong, arched a trifle, notched below the apex, as wide at the blunt apex as below, close-pressed and convex to the keel to a point ½″ below the tip of the keel, then spreading and tips horizontal and incurved and often touching each other, the concave side is downwards, pink-purple and streaked with darker color, ½″ wide; keel blunt and rounded at the apex to 95°, purple-tipped, 1″ high and nearly as much shorter than the wings; calyx tube ½″ thick, 1″ wide and 3″ long, the upper side a little convex, lower straight, oblique at base and a little thickened there; pedicels very short and as long as the bracts; the stems are usually 2–6′ long.

This grows in clayey and rather alkaline soil in desert places, and always seems to have white flowers; the pods frequently have the papery surface split away from the woody inner wall, especially at the sutures after the fashion of *A. cicadæ*.

ASTRAGALUS ZIONIS.

No. 5261w. May 17, 1894, Springdale, Utah, 4000° alt., in red sand.

No. 5249h. May 16, same locality.

No. 5224d. May 15, 1894, Rockville, Utah, in red sand, 4000° alt.

No. 5061b. March 30, 1894, Bellevue, Utah, in red sand at 3600° alt.

No. 5239. May 16, 1894, Springdale, Utah, 4000° alt., in red sand.

No. 5249g. Same date and locality.

This is a tufted perennial with the habit of *A. amphioxys*, but more slender, while the spreading, rather longer pubescence of very delicate hairs is fixed by the base and not by the middle as in that species; stipules very broad, 1–2″ long, adnate to the petiole but free from each other, hyaline below; stems densely tufted from a deep, perennial, erect root, wholly herbaceous, with nodes 3′ long or less, ascending; leaves 5–12′ long, with petiole about ⅓ the length and slender; leaflets about 10 pairs, ovate to lanceolate, 6″ long, very acute, not contiguous; peduncles about as long as the leaves, and the rachis ¼ as long as the peduncle; pods ascending, arcuate, abruptly long-acute, with flat subulate style, linear-oblong, 2″ wide and about 1¼′ long, a trifle sulcate and rather triangular in cross-section, at least when dry, but when fresh much rounded, ventral suture not raised but pod much flattened on each side of it, narrow below, sessile, with a complete joint at base, short-shaggy, mottled, pubescence very fine and soft, rarely the pods are obcompressed when much arcuate so as to be linear in cross-section; this is a common character, however, in pods which are nearly round, or even in pods which are much compressed laterally, they being greatly obcompressed when much arcuate; banner brilliant pink purple or darker, oval-ovate, sides reflexed 1½″ wide at a point opposite the keel tip and narrowing to nothing both ways; sulcus in the banner nearly cylindrical, 1½″ wide, and forming about ⅔ of a circle, 1″ deep and very broad and large, gradually shallowing upwards to the apex of the banner which is deeply notched; the white spot being ½″ above the keel and goes far below it, filling the sulcus and is M-shaped, purple-veined below and inclined to be stippled above, it reaches within 2″ of the tip of the banner; wings linear, 1″ wide, ascending, a little obtuse, tips horizontal and connivent over the keel, purple throughout, 1″ longer than the keel; keel straight, bent sharply to 90° and acutish, purple; calyx tube 3″ long and about 1½″ wide, ½″ thick. The stems are spreading or prostrate, 2–12′ long, and are often much tufted, growing in sandy or gravelly places from 4200° to 4600° altitude, near rocks on gravelly slopes, and occasionally hanging in festoons from crevices of the rocks. When growing this can only be separated from *A. amphioxys* by the pubescence and by the rather diamond-shaped leaflets, though in the dried plants the pod is very different.

ASTRAGALUS ARIETINUS.

No. 5554o. June 30, 1894, Fairview, Utah, in gravel, 6500° alt.

No. 5208a. May 11, 1894, Cedar City, Utah, in gravel, 6000° alt.

This is the plant referred to by Watson in King's Report, p. 71, as a form of his *A. iodanthus*, but is reasonably distinct from that species, as an examination of his type shows. It is characterized by the large, hyaline, veiny, lower stipules, which are round to reniform or even obovate, 2–3'' long and nearly as wide, connate below; leaflets about 9 pairs: wings very long and straight, with white tips which are horizontal: calyx long: pods fleshy, the pulp often 1½'' thick, the innermost wall of the pod being almost woody, pods from nearly straight to coiled in a circle, cross-section from nearly round to almost didymous, according to the sulcation at the two sutures. The flowers are always purple, and the plants grow in tufts with decumbent stems and are very variable in the pubescence and pods, but seldom approach the type of *iodanthus*. It is No. 270 of Watson from Utah, and No. 269 from the West Humboldt Mountains, Nevada. It also occurs at Mammoth Hot Springs, Wyoming, on dry, rocky hills. It is very abundant throughout the Great Basin, but does not seem to occur outside of it, except at the north and northeast. For a fuller description of this plant, see Zoe, vol. iii, p. 294, under *A. iodanthus*.

Astragalus arietinus var. stipularis. Miss Eastwood, along McElmo Creek, S.W. Colorado, June, 1892.

Proper stems an inch or less long, densely covered with large round to oval hyaline stipules, 2–2½'' long, rarely broadly ovate and acute; perennial and cæspitose, strigose with very short hairs fixed by the base: leaflets 4–6 pairs, elliptical to obovate, obtuse, 4'' long or less, rather thick, proper petiole twice the rachis, slender, leaves 3' long or less: peduncles subscapose, 2–3' long, stout, capitately few-flowered: bracts ovate and hyaline, 1'' long: calyx tube cylindrical, a little oblique, 3'' long, 1–1½'' wide, oblique and a little narrowed below, reddish, teeth about ⅓ the tube, triangular; flowers purple and like those of the type species; pods linear-oblong, 1′ long, 2½″ wide, either triquetrous or obcompressed, so that the cross-section is linear, sulcate deeply in the triquetrous forms except at base and apex, coriaceous, acute at apex and a little narrowed at base, nearly straight to arcuate to ⅓ circle, 1-celled, ventral suture raised, thin and sharp externally, pods green or mottled; perennial from an erect root; growing in sand. This is liable to prove a new species, and in that case may take the name *A. stipularis.*

Astragalus dorycnioides Douglas. Mr. Sheldon, l. c. 9, 145, says: "The difficulty which many botanists seem to have had in determining the limits of *Astragalus inflexus* Douglas and *Astragalus Purshii* Douglas has probably arisen from the nonconsideration of this species, which is intermediate between the two." Now since there is no room for any intermediate species between these two which almost shade together it becomes interesting to know the character of *A. dorycnioides* Douglas; the following is the description in G. Don, Gen. Syst. Gard. and Bot. 2, 258: "Plant erect, densely clothed with hoary silky villi; leaflets obovate-linear, obtuse; calyx smooth; racemes longer than the flowers; perennial. Native of North America near the Columbia River. Flowers purple. An elegant plant. Stipules distinct and free from the petioles; flowers in dense heads; banner linear and elongated. Pods straight." This plant is undoubtedly *A. succumbens*, but the specimens to which Mr. Sheldon has applied this name in the National Herbarium are nearly typical *A. Purshii*, and have not a single character in common with Douglas' description of *A. dorycnioides.*

It seems like stretching a point to extreme tenuity to make a new name for *Astragalus Thompsonæ* Watson on the ground that there is an *Astragalus Thompsonianus* which is neither the same word in its spelling nor its origin and is at best only a synonym for *Astragalus nivalis.* On the same basis doubtless it was wise to load up *Astragalus nudus* Watson with another synonym since there is already another *Astragalus Watsonianus* (OK) which means the same as *A. Serenoi* (OK), *i.e.*, is named after the same person and so means exactly the same. Mr. Sheldon has also complimented Mr. Watson with *A. Watsoni* for *A. Hendersoni.*

ASTRAGALUS INTERMEDIUS, Arizona, Palmer. Type in National Herbarium.

Plants with the habit and general appearance of *A. amphioxys*, but pubescence with hairs fixed by the base. but pods much like *A. Bigelovii.* Perennial, cæspitose or tufted; stems very short, 1′ long, or less, densely covered with large imbricated, hyaline, puberulent, ovate stipules, 3–4″ long, or less: leaves 2–4′ long, silvery silky with short appressed white hairs; petioles half the leaf; leaflets 10–15 pairs, contiguous, oval, 3″ long, obtuse; peduncles scapiform, about 6′ long, rather loosely spicately flowered on the upper third, erect, not slender, sparsely pubescent: bracts ovate, hairy, hyaline, 2″ long, or less, much longer than the very short pedicels: flowers horizontal or ascending; calyx tube cylindrical, 3″ long, 1½″ wide, a little wider below, obliquely attached, very pubescent with short white or dark hairs; teeth subulate, 1″ long: banner oval, gently arched to 45° from calyx tips, blade about 4″ long, sides reflexed ½″ wide, claw rather long: wings oblanceolate, gently arched, a little longer than keel and 1″ shorter than the banner: keel straight, apex arched to 90° in a gentle arc to the blunt tip: flowers purple: pods oblong, 9″ long, straight, shortly acuminate with an oblique tip, truncate to emarginate at base, sessile or very nearly so, 2½″ wide, 1½″ thick, obcompressed, very slightly sulcate ventrally, and broadly so dorsally, but not deeply, 2-celled except at the apex, velvety-pubescent with very short hairs, about one-third longer than the calyx and teeth, barely splitting the calyx tube.

Astragalus asclepiadoides Jones. Mr. Sheldon places this species under a new section, "*Asclepiadodes*" (l. c. 9, 159), in his provisional list, completely ignoring the fact that I had previously (Zoe 2, 238) made a section for it and called it "*Pachyphyllus*" (*Pachyphylla* by typographical error).

Astragalus ursinus Gray.

No. 5095ah. April 23, 1894, top of grade, four miles above Pagumpa, Arizona, 5000° alt., in gravel, among junipers.

This plant appears to be identical with *A. arietinus* when growing, but the 2-celled pod would place it elsewhere; when dry the greatly compressed, almost vetch-like pod is strikingly different. Stems often single, never many in a place, prostrate or ascending, always ascending in flower; flowers seem to be identical with those of *arietinus*, but paler; pods linear-oblong, cross section obovate-emarginate, wall fleshy and soft, the pulp ½″ thick and the pods 1½″ thick, ventral edge grooved, dorsal edge neither grooved nor ridged externally, but dorsal septum produced nearly to the ventral suture, the septum thin and white, and manifestly double; pod reddish, 1½–2′ long, arcuate to ⅓ of a circle, contracted but obtuse at both ends; seed cavity small and completely filled by the seeds, which are on stalks 1″ long; pods on reflexed pedicels, but arcuate so that the apex is ascending; stipules very large, blunt and sheathing below, mostly hyaline, almost reniform below, ovate above; bracts ovate, 1½″ long; pedicels almost none; peduncles 4′ long;

proper stems rather short; leaves 3–4 on the stem, about 7′ long, about 6–8 pairs of oval, obtuse, almost contiguous leaflets, 8″ long, 3″ wide, glabrous above, hoary below, as well as the stems; peduncles and petioles with white, short, narrow, appressed hairs, fixed by the base. This is a perennial, growing in gravelly soil among the junipers, on the mesa above the Grand Wash.

This seems to differ well from *A. ursinus*, and in case it should prove distinct, may bear the name *A. ensiformis*.

ASTRAGALIS NEWBERRYI var.' CASTOREUS.

No. 5006. April 4, 1894, Copper Mine, 18 miles west of St. George, Utah, in gravel, 5000° alt.

Leaflets 2–3 pairs; calyx usually black-hairy, 8″ long, and teeth 1½″ long; corolla dark purple like the type, 5″ long; pod 1½′ long, arched into three-fourths of a circle and long-acuminate from the base, completely obcompressed at the base and not at all so at the apex, but rather compressed, shortly shaggy all over, with rather sparse hairs. In other respects this appears like the type, but the pod is very different. Because of the variability of the pod in this species I cannot consider this as more than a good variety.

This may prove to be too near to *A. cibarius* Sheldon.

ASTRAGALUS REMULCUS. Rusby, No. 576, Bangharte's Ranch, Arizona, May, 1883. Probably also G. R. Vasey, Kingman, Arizona, June, 1881. With the habit of *A. Shortianus*, and somewhat related to it.· Perennial, rather stout, stems spreading over the ground, with erect or ascending peduncles: proper stems 6′ long, but may be much longer: puberulent: stipules triangular, green, adnate, not connate, reflexed, 2″ long: leaves 6′ or less long: the petiole one-third the rachis, and stout and sulcate: leaflets glabrous above, strigose below, with fine wavy short hairs, fixed by the base, oval, 4″ long, thick, about 15 pairs, nearly contiguous, smaller above; peduncles 6′ long or less, stout, sulcate, capitately 14–20 flowered; not much elongating in fruit; bracts lanceolate hyaline, 1–1½″ long, twice the very short and stout pedicels; flowers spreading, purple, 8″ long, large; calyx tube a little obliquely cylindrical, 4″ long, 1½″ wide, somewhat compressed, obliquely attached below and a little narrower, light colored; teeth deltoid to triangular, and a little longer than wide; banner elliptical, blade 4″ long, sides reflexed 1″ wide in the middle, ascending beyond calyx tips to 45–60° in gentle curve; wings narrowly oblong, obtuse, arched 45°, 1″ wide, a trifle longer than the keel and 1″ shorter than banner; keel straight, apex gently bent to nearly 90°, rounded and obtuse, 3″ longer than calyx; pods oval-ovate, 6″ long, 4″ wide, sessile, glabrous, fleshy, much obcompressed, not sulcate, 1-celled, base rounded, apex upcurved, compressed, triangular-acute, 2″ long; fruiting peduncles apparently decumbent.

Astragalus Shockleyi. Perennial, apparently tall and coarse, stems coarsely sulcate, branches zigzag with nodes 2–4′ long; stipules very short and wide, very broadly deltoid, adnate, ½″ high or less, with a short apiculation, 1–2″ wide, with green tip; leaves with thick terete rachis, nearly as thick as the stems, 4–6′ long, stiff and ascending, scarcely sulcate, leaflets usually 2 pairs, distant, linear, 6–12″ long, thick, often deciduous; plants racemosely branched above, and almost glabrous throughout; the scant leaflets give the plants the appearance of being all stems and peduncles; pods obliquely oblong, about 6″ long and 3″ wide, round (when fresh) in cross-section, obliquely apiculate and obliquely inserted at base; dorsal suture nearly straight in the middle, but abruptly bent at each end; ventral suture a little concave and thin, a trifle intruded, not at all sulcate; dorsal thicker, both sutures raised externally, dorsal suture intruded at least half way into the pod, except at the tip, where it is not at all intruded; walls cartilaginous, wrinkled both ways externally, pod sessile, but a trifle narrowed at the insertion; pedicels stout, 2″ long, ascending, and pods erect on pedicel. Very close to *A. nudus* Watson and *A. pachypus* Greene, differing from the former in the nearly 2–celled and smaller pod, from the latter in the short, sessile, round pod and few leaflets. Fish Lake valley, Nevada. Shockley, July 20, 1886.

Astragalus pruniformis. This belongs to the *A. pachypus* group. Stems slender, ascending from a decumbent base, 1–1½′ high, very coarsely sulcate, nodes 1–2½′ apart; leaves 3–4′ long; leaflets 6″ long, oblong, with a cuneate base, emarginate, 2″ wide, thin, pilose with fine, loosely appressed hairs, fixed by an enlarged base; peduncle stout, 6′ long, floriferous on the upper third; stipules subulate, green, 1″ long, reflexed in fruit; calyx campanulate, tube about 1½″ long, teeth subulate, and as long: pedicels stout, ½″ long, equaling the triangular-subulate bracts; stipe slender, 3″ long, ascending; pod in the fleshy state probably round in cross-section, and almost exact oval, 4″ long by 3″ wide, with much external pulp, when dry pods are lenticular in cross-section, parallel (transversely) ridged and reticulated, ventral suture ⅓″ thick throughout, dorsal also raised but thin, pod strongly apiculate, the point being a little above the middle of the end, 2-celled to the very apex, with the septum double and the parts separate; flowers not seen.

Butte County, Oregon, July, 1893. Mrs. Austin. Type in the National Herbarium.

ASTRAGALUS BERNARDINUS. Moronyo King Mine, east side of San Bernardino Mountains, California, 5000° alt., June 16, 1894, Parish. Type in National Herbarium.

This is closely allied to *A. tricarinatus* Gray, having the same rushlike round, rather tortuous stems, very similar pods, and peculiar pubescence of short, flat, closely appressed hairs fixed by the base, but it differs in being only 6′ high; stems flexuous and ascending; nodes 1′ apart or less; stipules triangular, green, adnate, free, evident; leaves 2–3′ long, with rachis like stems but smaller, green, tapering upward, arcuate, leaflets only ashy and equally so on both sides, flat, thickish, about 4 pairs, smaller toward tip of rachis, with very weak petiolules, distant, narrowly oblong, 6″ long by 1½″ wide, obtuse but not emarginate, equally rounded at both ends; peduncles proper only 1′ or less long, with rachis 3–4′ longer and racemosely flowered; flowers 6–10, reflexed; pods linear and narrowing below, apiculate, 1′ long, 1½″ wide at the end, ½″ wide at base, smooth, triquetrous, the back being a little narrower than the two sides and shallow sulcate, the sides a little concave, ventral angle acute, the others obtuse, completely 2-celled, cross-section Y-shaped, very finely netted, jointed at base, on a stipe nearly as long as calyx tube; calyx tube campanulate, 1″ long; teeth half as long, triangular; flowers not seen. Perennial, with many stems from a thick erect root.

ASTRAGALUS REVENTOIDES. Tweedy, No. 7, dry sagebrush areas, Grasshopper Creek, Beaver Head County, Montana, 5000° alt., August, 1888. This would pass for a broad leaved form of *A. reventus* were not the pod 2-celled or nearly so. Perennial, erect, stems short, 6′ long or less; leaves silvery on both sides with short fine appressed hairs fixed near the middle, 6′ long or less, stiff, rachis stout, equaling the petiole; leaflets oval to elliptical, emarginate, a little narrower below, 6″ long, 8 pairs or less; peduncles subscapiform, 1° long, erect, not stout, slightly sulcate, racemosely flowered on the upper third in fruit, pedicels slender, erect, 1″ long; calyx campanulate 1½″ long, with very short teeth; pods sessile, oblong-oval, 4-5″ long, 3″ thick, coriaceous, corrugated, somewhat fleshy when green, erect, obliquely apiculate, truncate at base, ventral suture straight, thick and rather prominent externally, dorsal narrow, not sulcate, nearly 2-celled by intrusion of dorsal suture, except at apex, pod nearly round in cross-section.

ASTRAGALUS RUSBYI var. LONGISSIMUS. This has the proper stems erect and only a few inches long, from a woody base; peduncles angled 1–1½° long, sparsely racemose on the upper two-thirds; flowers very small, about 1½″ long; yellowish keel; wings and banner very short and blunt; calyx teeth minute; puberulent throughout even to the pods; leaflets 14–18 pairs, oblong-oblanceolate, emarginate; with no proper petiole. Type specimens in the National Herbarium. Dr. Palmer, Chihuahua, Mexico, 1885; Pringle same, by streams in the Sierra Madre Mountains, September 23, No. 1219. This has the pods more or less sulcate dorsally and broadly, septum a little intruded in the middle of the pod.

ASTRAGALUS RACEMOSUS var. BREVISETUS. Ramos, Zacatecas, Mexico, May 5, 1892, Jones. Calyx gibbous above, broadly oblong, 1½″ long and 1″ wide, spurred, setæ at base about 1″ long; pedicels 1″ long in flower, stout, longer than the ovate-subulate bracts; leaflets about 20 pairs; plants hoary with minute hairs. To this I refer Carleton's specimen, No. 221 from Cimarron Valley, Indian Territory, with few leaflets, and pedicels 2 long. Also Rydberg's from Curtis Creek, Fremont County, Nebraska, June 22. This is nearer to the type species however.

ASTRAGALUS RACEMOSUS var. LONGISETUS. This has very long hyaline, setaceous bracts as long as the calyx tube; often robust and with leaflets 3″ wide by 8″ long, elliptical, and large reflexed stipules. The specimens which form the types of this variety are in the National Herbarium. Wolf, No. 216, Apex, Colorado, June; Fort Collins, Colorado, on prairies; Sheldon, Pueblo, Colorado; Bodin and Eastwood, Denver, Colorado; Idaho, Hayden's Survey; Moose Jaw Creek, N. W. Q., on open prairies, Macoun.

Astragalus iodanthus Watson, Bot. King's Exp. 5, 70. This plant, which Watson confounded with a much better marked species *A. arietinus*, so far as the type is concerned belongs to western Nevada and adjacent California. It has slender stems; nearly glabrous throughout; flowers white, narrow, smaller; wings and banner usually equal; calyx lobes filiform-subulate nearly equaling the short tube; pods lanceolate-acuminate, completely obcompressed till the opposite sides touch throughout except at the tip; walls thin, merely coriaceous, pod mottled, minutely puberulent, 3–4″ wide, 1–1½′ long, less than 1″ thick. In some of its forms this approaches *A. lentiginosus*. Specimens in the National Herbarium besides the type are Sierras, California, Lemmon; Virginia City, Nevada, Curran, July, 1884.

Astragalus lotiflorus var. *brachypus* Gray. This plant seems to be a good species, but those most competent to pass an opinion on it, the field collectors of the plains, say that both this form and the type form are found on the same plant, the length of the peduncle being only accidental, this will reduce the variety therefore.

ASTRAGALUS LEIBERGI. No. 354, Sandberg and Leiberg, Egbert Spring, Douglas County, Washington, alt. 3500°, July 1, 1893. Type in National Herbarium.

Related to *A. arrectus*. Perennial; proper stems woody below, a few inches high, then branched in tufts, and branched 1′ or less long, covered with hyaline, imbricated, glabrous stipules 3″ long, adnate, not connate, subulate triangular; leaves filiform, 6′ long, petiole and rachis equal, tapering to a needle-like but not pungent point, erect; leaflets about 4 pairs, nearly filiform, 9″ or less long, deciduous, reduced toward tip of rachis, distant; peduncles subscapiform, 1–1½° high, erect, straight, thick, striate, floriferous on the upper fourth; flowers racemose, ascending, usually ochroleucous; pedicels stout, ascending, 1″ long, equaling the subulate bract; calyx campanulate, 1½″ long, about double the subulate teeth, hyaline; pods linear-oblong, 1′ long, 2″ wide, 1″ high, shortly-acute at both ends, a little arcuate, ventral suture thick and prominent externally, a little concave, dorsal suture convex, thin, broadly sulcate, intruded as a thin hyaline partition half way to the ventral suture, or at times almost touching the ventral suture, partition absent at the tip of the pod, walls of pod coriaceous, fleshy when green; stipe nearly double the calyx tube, thick especially above, pods erect.

To this I would also refer a more caulescent specimen from Spipen River, Washington, Wilkes' Exp., National Herbarium.

Astragalus arrectus var. scaphoides. Dry sage-brush areas, 5500° alt., Beaver Head County, Montana, on hills west of Clark's Cañon, July, 1888. Very coarse and stiff, apparently 2° high or more, ascending, stems 3″ thick; peduncles 1° long, coarsely grooved; leaflets about 10 pairs, elliptical, 1′ long, ashy below, glabrous above, leaves 6′ long, proper petiole short; pods almost exactly those of *A. asclepiadoides*, but 2-celled except at apex, obcompressed, 1′ long, coriaceous, on a stipe 9 long, which is arcuate so that pod is erect, pods 3 thick and 1½″ high; calyx tube 4″ long, with teeth 1″ long or less, triangular; flowers white; peduncles floriferous on the upper third. Type in National Herbarium.

Astragalus arrectus Gray. Mr. Sheldon has redescribed this species under the name of *A. eremiticus*, l. c. 9, 161, and Coulter and Fisher have done the same under the name of *A. atro-pubescens* in Bot. Gaz. I have seen the types of both species. Mr. Sheldon's name *eremiticus* I think can be maintained as var. EREMITICUS (Sheldon, l. c.) for the southern plant growing among the junipers in the Beaverdam Mts., Ariz., and vicinity (my numbers 5071, 5003, 5082b, 5098m, 5004e), which is a part of his type, but my specimens from Sprucemont, Nevada, included in his type are true *A. arrectus*.

ASTRAGALUS ATRATUS var. MENSANUS. No. 792, Coville and Funston. Along the Darwin road about two miles southward from Mill Cañon Divide, Darwin Mesa, Inyo County, California, May 20, 1891. Pods narrowly oblong, 1′ long, 2″ wide, obliquely apiculate, narrowed below into a very short stipe, puberulent, decidedly flattened, ventral suture very thick, dorsal thin and not at all sulcate; keel not produced; lower leaves elliptical, upper linear; whole plant ashy. Type in National Herbarium.

ASTRAGALUS CONJUNCTUS var. OXYTROPIDOIDES. No. 798, Howell, near the Dalles, Oregon, May 8, 1885. Leaflets 8–10 pairs, linear, rachis not greatly prolonged, leaves 6′ long or less; flowers white, 15–20; calyx teeth over half the tube, 1½″ long, subulate, hairs silky and white, flowers long and narrow; banner elongated, 2″ longer than the wings, blade at least 6″ long, oblong, ascending 30° remotely from keel; wings oblanceolate, nearly straight, 1″ longer than the keel; keel about 2″ longer than calyx tips, straight or nearly so, apex bent in gentle curve to 90°, obtuse; pods 1′ long, 2″ wide, sulcate dorsally, narrowly oblong, shortly acuminate, very much resembling those of *Oxytropis Lamberti;* peduncles 1–1½′ long, sulcate, rather slender.

ASTRAGALUS FRANCISQUITENSIS. Brandegee, San Francisquito, Lower California, October 18, 1891. Perennial, with many delicate stems, ascending, with the habit of *A. leptaleus*, but seeming near to *A. Nuttallianus*, nearly glabrous throughout; nodes 2–3′ long; stipules subulate, not united, adnate, 2″ long; leaflets 8–11 pairs, obovate-oblong to elliptical, emarginate, delicate, finely petiolulate; proper petiole 6–9″ long; leaves 4′ long; peduncles filiform, 6′ long, subcapitately 10–15 flowered; pedicels slender, ½″ long, about equaled by the triangular bracts; calyx campanulate-cylindric, tube 1½″ long by ¾″ wide, subulate teeth about half the tube, tube hyaline, nigrescent; banner blade 2½″ long, oblong, ascending 45°, the curve beginning beyond the calyx tips, sides reflexed, ⅓″ wide; wings linear, straight, ½″ wide, just about as long as banner, not at all arched (thus resembling those of *A. arietinus*), 1½″ longer than keel; keel short, very obtuse and rounded; pods linear-oblong, 6″ long by 1–1½″ wide, obtuse, sessile, completely 2-celled, glabrous, not sulcate, chartaceous.

Astragalus metanus. Hanson's Ranch, Lower California, near the border, Brandegee, April 18, 1885. Has the habit of *A. Inyoensis.* Prostrate from a perennial root, 2–3′ long; stems rather stout and sulcate, minutely pubescent throughout with fine appressed hairs, fixed by the base, nodes about 3′ long, rather flexuous; stipules very small, green, reflexed, 1″ long, adnate, not connate; leaves 4–7′ long; proper petiole 1′ long on the lower leaves, but none on the upper ones; leaflets linear-lanceolate, barely acute, not contiguous, 1⅓′ long; peduncles about 1° long, floriferous on the upper third; flowers racemose, nearly sessile, with ovate bracts shorter than the pedicels; calyx broadly campanulate, tube 1″ long; teeth triangular and about one-third the tube; flowers purple; banner 3″ long, bent abruptly and remotely from the calyx tips to 45–90°, oval, sides reflexed; wings arched 60°, and exposing the base of keel, oblong, ¾″ shorter than the banner; keel abruptly bent to 110°, and apex produced to a point, 1⅓″ high, base of blade a trifle arched, about ½″ shorter than the wings, light colored; flowers horizontal, about 3″ apart; pods lenticular, much the shape of the *Kentrophyta* group when flat beaked, the upper ½ or ⅓ perfectly flat and triangular-acute, the ventral suture nearly straight, cross-section oval near the base of the pod, chartaceous, narrowed at very base, but sessile, lunate, 1-celled, seed-bearing in the middle, 4″ long, 2½″ wide.

The type is in the California Academy of Sciences.

Astragalus Julianus. San Julio, Lower California, Brandegee, April 19, 1889. This appears to be near to *A. sabulonum.* Annual, erect; inflorescence corymbiform by the shortening of the upper nodes; stems mostly shaggy with partly spreading hairs, except the rather shortly and sparsely pubescent pods; nodes 6″ to 1′ apart; leaves 4′ long, of 7–8 pairs of narrowly elliptical acutish leaflets, which are 6″ long and 1 to 1½″ wide; proper petiole 6–9″ long; peduncles stout, as long as the leaves, the rachis as long in fruit; flowers in a close raceme, lax in fruit, nearly sessile, reflexed and fruit also; calyx campanulate, tube 1″ long, subulate teeth about the same; pods obliquely-oval, swelled to circular in the middle, contracted at base, with a flat, triangular, acuminate beak, 2″ long, which is nearly central; ventral suture thick, not sulcate, seed-bearing in the middle and seed stalks long, pods 6" long by 3" wide, papery; flowers nearly white, but purple tipped, apparently.

ASTRAGALUS SEROTINUS Gray, Pac. R. R. Rep. 12, 18 and 51, 1860, var. CAMPESTRIS (Nutt. T. & G. Fl. 1, 351, 1838.

Astragalus campestris Gray, P. A. A. 6, 229.

Astragalus convallarius Greene.

No. 6032b. September 11, head of the Sevier River, Utah, 8000° alt., in gravel, under *Pinus ponderosa.*

No. 6015n. September 7, Panguitch Lake, Utah, in volcanic gravel, under *Pinus ponderosa.*

No. 5958q. August 29, Brigham Peak, Utah, 11,700° alt., on open slopes, at and above timber line, in gravel.

No. 6002n. September 6, Panguitch Lake, Utah, in gravel, 8400° alt., under pines.

No. 5695h. July 27, Mt. Ellen, Henry Mountains, Utah, 10,500° alt., on open slopes, above timber line and below it, in volcanic soil; never seems to grow in any but well drained soil.

It is manifest, from an abundance of material from many localities, that this is only a variety of *A. serotinus*, since the calyx teeth and keel of the flower vary much, and show all sorts of intergrades.

As has been indicated above, *Astragalus strigosus* Coulter and Fisher, with its synonym, *A. griseopubcns* Sheldon, is the same as *A. serotinus.*

ASTRAGALUS HOOKERIANUS (T. & G.) Gray, var. WHITNEYI (Gray P. A. A. 6, 526). An examination of a large amount of material from the California Academy of Sciences shows that this is only a variety of *A. Hookerianus*, as the characters do not hold.

Astragalus Virgineus Sheldon, Death Valley Rep., seems to be the same as *A. sabulonum* Gray.

Astragalus eremicus Sheldon, Death Valley Rep., was based on a very imperfect and small specimen of *A. Coulteri* var. *Fremonti.*

ASTRAGALUS COULTERI var. FREMONTI (T. & G. Pac. R. R. Rep. 4, 80). From a large amount of material I find transitions between this and *A. Coulteri*, but find that many specimens referred to this are really only forms of *A. lentiginosus*, and not truly referable to the type of *A. Fremonti.*

Astragalus No. 1961, Coville, Death Valley Rep. 277, is *Sophora stenophylla.*

Astragalus amphioxys Sheldon (not Gray) in Coville, Death Valley Rep. 85, Nos. 496 and 543 is *A. leucolobus* Jones, which is not the same as *A. lectulus* Watson.

Astragalus Purshii Sheldon (not Hooker), Death Valley Rep. 87, No. 119.6, from the Tejon Mountains, California, is also *A. leucolobus* Jones. Pringle's, from the same region, May 13, 1882, is the same. *A. leucolobus* is well marked by the roundish, long-villous pubescent leaflets; densely cæspitose habit; large lanceolate, hyaline bracts, 2″ long; usually black-hairy calyx (the pods are those of *A. Purshii)*; hairy stipules; purple flowers, with banner having a striate purple-veined white spot; peduncles 2 to 6′ long, erect or spreading, always capitately 6 to 12 flowered, and flowers ascending.

Astragalus cibarius Sheldon. This well marked species was fully described by me in my Contributions, No. 3, 291, and I provisionally referred it to *A. glareosus*, but indicated its variance from that species, preferring to do that rather than to make a possible synonym. Mr. Sheldon, however, with very little material to work on and without exercising the usual courtesy in such cases, publishes it without even mentioning the fact that it had been previously and fully described by me. It is possible, however, that its hasty publication prevented a proper examination of the literature on the species.

Astragalus Chamæleuce Gray. This name belongs to the *Phaca pygmæa* of Nuttall, but so far I can find no one who knows what that species is. The only character that seems assured is the habitat which is "Rocky Mountains, on the hills of Ham's Fork of the Colorado of the West," which means Ham's Fork, southwest Wyoming, on the northern slopes of the Uinta Mountains. So far as the description goes this might be some form of *A. amphioxys*, *A. cicadæ*, or *A. glareosus*. The forms collected by me have all been referred by others to *A. glareosus*, but an examination of authentic material of the latter species shows that my plants are not *A. glareosus*. They are identical with Watson's *A. Chamæleuce* Bot. King's Exp. 74. The stems are woody, flat on the ground, as if rooting; pods very variable, always pubescent, but sometimes sparsely so, walls fleshy and pulpy, 1″ thick, ventral suture thick and barely if at all intruded, a trifle sulcate ventrally, usually flattish on the dorsal suture, cross-section oval, the pod being obcompressed, usually 3″ thick and 4″ wide, vertical longitudinal section oblong and arcuate, 9″ or more long, horizontal longitudinal section oval-ovate to ovate, pods seldom over an inch long, jointed to a very short stipe, but often very deeply sulcate ventrally and even dorsally at base, the inner wall is thick and woody, but the outer is soft pulpy, cavity is full of hairs, showing that it is filled with juice during part of its growth, the cavity is double the width of the seeds when they are green; calyx ½″ thick, 1″ wide and 4″ long, hyaline, thicker at base, cleft deeper above and lower teeth the longer. For further notes see my Contributions 3. 294. Should this prove to be distinct from *A. Chamæleuce* as is quite likely it may bear the name of *A. Uintensis*.

ASTRAGALUS CHAMÆLEUCE var. PANGUICENSIS.

This puzzling plant has the habit of the type and the pods of some forms of *A. amphioxys;* pubescence almost shaggy, very dense, of very fine and slender hairs fixed by the base; nodes usually longer than the triangular, hairy, green stipules, which are 2–3″ long; stems prostrate, 2–6′ long; root woody and with many short branches; leaves 2–3′ long, slender petiole longer than the rachis; leaflets contiguous, 6–8 pairs, elliptical, 2–3″ long, obtuse; pods linear-lanceolate, 1′ long by 2–3″ wide, short-acuminate, very much obcompressed till cross-section is linear, doubly sulcate throughout ventrally, ventral suture prominent externally, not at all sulcate dorsally, 1-celled, coriaceous and fleshy when young, finely and closely appressed-pubescent; flowers light purple, 9″ long; banner 4″ longer than the calyx and 1–2″ longer than the keel; calyx nearly cylindrical, but larger at the mouth, tube 4–5″ long, 1″ wide, subulate teeth 1″ long, calyx pubescent like the leaves, but less densely so, green, nearly sessile; bracts lanceolate hyaline, 2–3″ long; flowers in heads 6–8, on short peduncles, ascending; corolla like that of *A. Chamæleuce.* No. 6002m and 6023f, September 8, 1894, Panguitch Lake, Utah, 8400′ alt., in gravel.

ASTRAGALUS MUSINIENSIS.

With the habit of *A. Newberryi* Gray, but allied to *A. cicadæ* Jones. Densely cæspitose in small tufts, crown thick with coarse petioles and large, glabrous, triangular, nervose stipules, 4″ long, which, with the petioles, form heads at the crown; petioles stout, 1′ long, crowned by 1–3 lanceolate to elliptical-lanceolate leaflets, 1′ or less long; pubescence minute, dense and appressed throughout; pods ovate, 9″ long, 3–4″ wide, with an incurved and acute apex, very shortly-strigose pubescent, 1-celled, sulcate ventrally at base, cross-section nearly round, outer coat loose; flowers not seen. This would pass for some form of *A. Newberryi* var. *eriocarpus*, but for the loose outer coat of the pod. It may be a reduced form of *A. cicadæ* or even of *A. Newberryi* var. *eriocarpus*, but the leaflets are of different shape, though in the latter species they are often reduced to three. No. 5454a. June 18. 1894, two miles south of Ferron, Utah, on clay slopes, at about 6000° alt.

ASTRAGALUS CICADÆ var. LACCOLITICUS.

Plants tufted, perennial, proper stems very short and nodes not longer than the triangular, green, very strigose stipules, which are 3″ long, adnate, not connate; leaves 3–4′ long, the petiole as long or longer than the rachis; leaflets 5–7 pairs, acute at both ends, nearly oval but apparently diamond-shaped, contiguous, 4–5″ long, 3″ wide; pubescence short, stiff, dense, appressed, fixed by the middle or near it; peduncles shorter or barely longer than the leaves, erect in flower and decumbent in fruit, capitately few-flowered; flowers like *amphioxys*, purple, 1′ long; calyx nearly half the whole, tube 4″ long, subulate teeth 1″ long, calyx tube cleft deeper above by a broad sinus, finely pubescent; flowers and fruit ascending; pods ovate to elliptical, acute, 1′ long, 5–6″ wide, rounded at base, puberulent when young, outer coat separating, but reticulations not evident. No. 5658q. July 21, 1894, at Cottrell's Ranch, Henry Mountains, Utah, 6000° alt., in volcanic gravel.

Astragalus lentiginosus Dougl. G. Don, Gen. Syst. Gard. and Bot. 2, 257, 1832; Hook, Fl. 1, 151. This is the most annoying group of forms in the genus, comprising all perennial forms of the group with 2-celled pods. The pods vary from membranous to cartilaginous and are connected throughout, the flowers vary from white to purple, and the shape of the pods from lanceolate to oval-ovate at least, always acute and often acuminate. I have tried in vain to separate the forms into valid species. As a rule the sutures meet near the upper side of the pod by the great intrusion of the dorsal suture, but sometimes they do not quite meet, especially when old, pods often didymous by being deeper sulcate than usual.

ASTRAGALUS LENTIGINOSUS var. DIPHYSUS (Gray, Pl. Fend. 34, 1849. This includes all the purple and large flowered forms of eastern Nevada and Utah to Colorado and New Mexico and south to Arizona having membranous to chartaceous pods.

ASTRAGALUS LENTIGINOSUS var. McDOUGALI (Sheldon, I. c. 9, 169). This seems to include *A. Rothrockii* Sheldon, l. c. 9, 174. This includes all forms with small and coriaceous pods.

A. LENTIGINOSUS var. CUSPIDOCARPUS (Sheldon, l. c. 9, 147). This includes all forms with acuminate, long and coriaceous pods. It is described by Mr. Sheldon as 1-celled, but authentic specimens from Leiberg's collection named by Mr. Sheldon are 2-celled. The range of this variety is northward, while the var. *McDougali* is mostly southward.

ASTRAGALUS LENTIGINOSUS var. CHARTACEUS.

No. 5627m. July 13, 1894, Ephraim, Utah, in clay, 6000° alt.

This plant is very nearly like the type in pubescence, the leaves being rather thick and almost wholly glabrous, from almost obcordate to obovate emarginate; the pods vary on the same plants in the width of the septum from less than ½–2″ wide, but always 2-celled by being deeply sulcate ventrally, the walls are chartaceous, very delicately veined externally with parallel veins which throw off many veinlets which connect with each other, the pods are perfectly smooth with a triangular, incurved, short tip, the cross-section is nearly didymous or very often with the re-entering angle on the dorsal side absent, pods 1–1½′ long and about 8″ wide, 3–5 in a very loose, subcapitate raceme which is shorter than the leaves; the flowers are like those of the type. This plant grows on flat clay land, in Sanpete valley, at about 6000 ft. altitude. It grows in clumps with nearly prostrate stems 4–8′ long. This may prove to be a mere form not worthy of varietal rank.

ASTRAGALUS LENTIGINOSUS var. NIGRICALYCIS. Bakersfield, Kern County, California, Miss Eastwood, March 24, 1893. Densely short-woolly-pubescent throughout, or velvety, except the less pubescent older leaves and mature pods; calyx densely black-hairy; leaflets obovate to oblanceolate, emarginate to truncate, 5–9 pairs; petioles and peduncles various; stipules triangular, small, reflexed; flowers white, 6″ long, in heads in flower and in short spikes in fruit; calyx campanulate to short-cylindric, tube 1½″–2″ long, 1″ wide, scarcely gibbous at base, teeth half the tube or less, subulate to triangular, calyx sessile or nearly so; bracts ovate, 1″ long; banner elongated, oblong, 3–4″ long, ascending 45–60° in gentle curve remote from calyx tips, sides reflexed ½″ wide most in the middle; wings narrowly oblong, nearly straight, a trifle longer than the keel; keel straight, apex gently arched to 90° in a uniform curve from apex to end of curve, shortly acute, 2–3″ shorter than banner; pods 1′ long, oval to ovate, always more or less lunate, chartaceous, mottled, tip short-triangular to short-acuminate, always incurved, 2-celled, sulcate ventrally; pods 3–7, flowers 10–15; perennial with ascending stems, 1–2 high. Also collected at Alcalde, California, May, 1881, Brandegee. Type in the California Academy of Sciences Herbarium, as well as my own.

Astragalus lentiginosus var. diaphanus (Dougl. Hook. Fl. 1, 151). This precedes *lentiginosus* on the page, but I follow the usual rule of keeping up the long used name. This variety includes all papery pods which are linear, compressed and falcate, and may include *A. palans* Jones with its variety *araneosus* (Sheldon, I. c. 9, 170), though *A. palans* seldom has papery pods.

Astragalus lentiginosus var. latus Jones, Zoe 4, 272. *A. salinus* Howell.

This includes the prostrate forms with oval pods.

Astragalus palans Jones var. araneosus (Sheldon, l. c. 9, 171).

No. 5215e. May 14, 1894, Rockville, Utah, 3500° alt., in red sand.

No. 5163h. May 4, 1894, Silver Reef, Utah, in gravel, 3500° alt.

No. 5218. May 15, 1894, Rockville, Utah, in red sand, 3500° alt.

This is a variety of the type approaching *lentiginosus* var. *diphysus*, and appears to be the same as *A. araneosus* Sheldon, the type of which is in my collection.

Astragalus amphioxys Gray has for a synonym *A. crescenticarpus* Sheldon, being exactly typical; also *A. vespertinus* Sheldon, l. c. 9, 150, in which the pod is obcompressed. The flattening either into compressed or obcompressed is purely accidental in this species, and varies with the development and bending of the pod.

All the forms of *Astragalus amphioxys* are at once distinguishable from any forms of *A. Shortianus* by the pubescence, which is in the latter species fixed by the base, while in the former it is fixed near the middle, and therefore giving the plants a silvery appearance.

Astragalus Newberryi Gray. Having now gone over again and again the ground where this species was found, and having gathered very many specimens of all forms, I am certain that this is identical specifically with *A. eriocarpus* Watson, and must therefore take the name. Unfortunately Watson's type is an abnormal form of his species, having grown in the shade, and therefore having abnormally long peduncles, but having collected it abundantly near his typical localities, and finding both long and short peduncles, according to the exposure, I have no hesitation in placing it here as a variety of *A. Newberryi.*

ASTRAGALUS NEWBERRYI Gray var. ERIOCARPUS (Watson, Bot. King 5, 71). This includes *A. candelarius* and var. *exiguus* Sheldon.

Astragalus Purshii var. *tinctus* Jones, Contributions 5, 269. *Astragalus lanocarpus* Sheldon is the same.

ASTRAGALUS SEATONI n. sp. Near to *Hartwegi*, but with slender stems. Collected by H. E. Seaton, on Mt. Orizaba, Mexico, 10,000° alt., August 6, 1891. Also by Bourgeau, at Cacubaya, near Mexico City, 1866. The latter plant is stouter, strigose and with fewer leaflets; pods papery; root thick. Seaton's plants have very slender ascending or spreading stems; leaflets linear or nearly so, 6–8″ long, 10–12 pairs; peduncles long and slender, 4–8′ long; spikes linear, loose, 1–2′ long; bracts scarcely visible; calyx almost globular, the tube about ½″ long and teeth one-third as long, triangular and black-hairy; flowers yellowish or white, sessile; banner round, not over 1″ long or less, sides reflexed, a little longer than the narrow-oblong wings, which are arcuate, and about one-fourth longer than the semicircular oblique keel: pods oblong-oval, 2″ long, 1″ wide, and cross-section nearly round, deeply sulcate and septum produced, but pods not wholly 2-celled, apiculate, very short stipitate, apparently jointed to the stipe: pods and flowers reflexed and nigrescent, with short stiff hair; plants otherwise nearly glabrous.

ASTRAGALUS HORNII var. MINUTIFLORUS. Flowers about 2½″ long; calyx teeth blunt and short; tube 1″ long; pods nearly oval, with an acuminate beak, 2½″ wide, 4″ long, including the beak, 3–6 in a close head; a trifle sulcate ventrally, nearly smooth, and whole plant nearly smooth; peduncles slender, 2–2½′ long, shorter than the leaves; this has the habit of *A. lentiginosus* var. *diaphanus*.

San Jorge, Lower California, Brandegee, March 17, 1889, on saline flats.

OXYTROPIS ACUTIROSTRIS (Watson, P. A. A. 20, 360, 1885).

Astragalus acutirostris Watson, l. c.

Astragalus streptopus Greene.

An examination of flowering specimens of Greene's species and of the duplicate type specimens of *A. acutirostris* make it certain that this is a true *Oxytropis* and not an *Astragalus*. A very valuable character separating this genus from *Astragalus*, which seems to have been at least partially overlooked, lies in the wings, which are always enlarged and lobed at the tip, generally very much enlarged and crumpled, and little like any American *Astragali*, though *A. calycosus* has lobed wings.

Should it be necessary to reduce this genus to *Spiesia*, the name must be *S. acutirostris* (Watson).

OXYTROPIS NOTHOXYS (Gray, P. A. A. 6, 232, 1866).

Astragalus nothoxys (Gray, l. c.)

Spiesia nothoxys (Gray).

This species is manifestly an *Oxytropis*, and has been confounded by most Arizona collectors with *A. Arizonicus*, having been distributed widely as that species.

HEDYSARUM BOREALE var. LEUCANTHUM (Greene, Pitt. September 20, 1892, page 294). *Hedysarum flavescens*

Coult. and Fisher, Bot. Gaz. 18, 300. Greene's name seems to be the older, and though the description is so meager that no one can tell what it belongs to, yet from the locality it is quite liable to belong to *H. boreale*. At any rate, *H. flavescens* is manifestly a form of *H. boreale*. Prof. Kelsey thinks this is more than an albino form, and so it seems to be. N. L. Britton, in the "Check List," considers *H. boreale* and *Mackenzii* as forms of the same species, but I certainly would not so consider them. *H. boreale* is separable from the other by the areolæ of the pods, being nearly as wide as long, by the calyx teeth very short and triangular, and by the short banner; in *H. Mackenzii*, the calyx teeth are long, the areolæ linear or nearly so, and the banner as long as the keel; there is some variation in the characters, but I have never seen specimens which could not at once be separated by the areolæ of the pods. The sporadic appearance of *H. boreale* in the southern part of its range is in favor of its identity with *H. Mackenzii*, but until its characters fail I see no way but to keep it up. The eastern forms of *H. Mackenzii* have smaller flowers and long filiform calyx teeth, so far as I have seen them.

Lathyrus Utahensis.

No. 5441l. June 15, 1894, Ireland's Ranch, Utah, at the head of Salina Cañon, in gravel, at 8000° alt.

This is the plant referred to, *L. paluster* var. *myrtifolius*, by Watson in King's Report and is Watson's No. 296. Whatever may be done with the apparently interminable forms of this genus in the Sierras and northward this species seems to be sufficiently distinct from them all to warrant recognition. Leaflets oval, 2′ long, obtuse at both ends and apiculate above, 4–6 pairs: stipules large and leafy, reniform and variously cut, 6″ high or less and nearly 2′ wide: peduncles 4–8′ long, usually about as long as the leaves, rarely more, often less, 4–6 flowered; flowers 9″ long, dirty purple or white; calyx lobes subulate-triangular, the lower about equaling the tube which is 1½″ long; fruit half-elliptical, 2′ long; whole plant smooth, plants straggling upwards from 3–6° high, among oak brush. This plant which is very common throughout Utah at middle elevations varies but little. I refer here also Miss Eastwood's specimen, No. 1, from Durango, Colorado, and Ward's specimen from Utah collected in 1875. *L. pauciflorus* may be a starved form of this species, but from the imperfect material of that species it is impossible to tell what its limits are.

POTENTILLA UTAHENSIS var. CAMPESTRIS.

Ivesia Utahensis var. *campestris.*

No. 1624. Whitney Meadows, Sierra Nevada Mountains, California, 8500° alt., August 19. Coville.

Young leaves silky, long-villous: narrow leaflets 2½″ long or less; plants less glandular, but otherwise the same, though a little more slender and leaflets not compacted.

AMELANCHIER ALNIFOLIA var. UTAHENSIS (Koehne, Die Gattungen der Pomaceen. Berlin, Ostern, 32, t. 2, 1890).

No. 5204d. May 10, 1894, Cedar City, Utah, in gravel, 6000° alt.

No. 5224n. May 15, 1894, Rockville, Utah, in red sand, at 3500 alt.

No. 5286k. May 22, 1894, Kanab, Utah, 5300° alt., on sandstone rocks.

No. 5149l. May 3, 1894, Silver Reef, Utah, on slopes, 3500° alt.

A large number of specimens from different localities and elevations show all sorts of variations in this most variable species, the variations being due to aridity and exposure. I therefore place this variety where I did the original specimen collected by me in 1880, No. 1716, at Leeds, S. Utah, upon which the species of Koehne was founded. This would seem to be *A. pallida* var. *arguta* Greene.

All the numbers given above have acute leaflets, the following have obtuse and rounded leaflets and are nearer the type of *A. Utahensis*.

No. 5006a. April 4, 1894, Copper Mine, 18 miles west of St. George, Utah, 5200° alt., on rocky slopes.

No. 5289k. May 23, 1894, Johnson, Utah, 5000° alt., in rocky places.

No. 5312n. May 28, 1894, cañon above Tropic, Utah, 6500° alt., on slopes.

No. 5291e. May 25, 1894, 15 miles south of Pahria, Ariz., in gravel, 5000° alt.

No. 5663k. July 23, 1894, Marvine Laccolite, Henry Mountains, Utah, 6000° alt., in gravel.

No. 5095h. April 21, 1894, Pagumpa, Arizona. 4000° alt., in gravel.

No. 5082d. April 20, same locality.

PURSHIA TRIDENTATA var. GLANDULOSA (Curran, Bul. Cal. Acad. 1, 153). I agree with Mrs. Brandegee (Mrs. Curran) that this is only a variety of *P. tridentata*. Having examined all of Mr. Coville's specimens in the National Herbarium I failed to find constant characters.

POTENTILLA SABULOSA. IVESIA SABULOSA.

No. 6032. Sept. 11, 1894, head of the Sevier River among pines, 8000° alt., in compact gravelly clay.

This plant ranks next to *P. Kingii*, but very much resembles *Utahensis*. Stems erect, 6–9′ long, rather slender, growing in patches much after the fashion of *Arenaria Kingii*, and *Fendleri*: petals linear to oblanceolate, about as long as the calyx lobes: accessory calyx lobes minute.

obtuse, thickish; main calyx lobes ovate to triangular and acute; seeds 1–2, orbicular, nearly equaling the calyx, smooth; receptacle very setose; whole plant glabrous (but not glaucous) except the sparingly hairy upper stems and calyx tube; leaves 1½–3′ long, linear, with broad, thick, densely imbricated sheaths below and thus forming close mats.

This plant abounds in pine forests at the head of the Sevier River, at about 8000 ft. alt., growing in very poor, somewhat gravelly, clay soil.

RIBES LACUSTRE var. LENTUM.

No. 5695o. July 27, 1894, Bromide Pass, Henry Moúntains, Utah, 10,000° alt., in gravel.

No. 5397k. June 5, 1894, Head of Bullion Creek, Utah, in gravel, at 10,000° alt.

No. 5684t. July 25, 1894, Mt. Ellen Park, Henry Mountains, Utah, 10,000° alt., in gravel.

Densely covered throughout with a yellowish, viscous pubescence, as well as soft hairs. This is the common form throughout the mountain region of Utah and Nevada, the pubescence being so viscous as to stain the sheets yellow in which the plants are collected. It is possible that the glutinous pubescence was overlooked in Gray's type of var. *molle*, in that case this will be identical with it.

ŒNOTHERA TRILOBA var. ECRISTATA.

No. 6015t. September 6, 1894, Panguitch Lake, Utah, on muddy shores, 8400° alt.

No. 5893r. August 23, 1894, head of Bullion Creek, Utah, 11,500° alt., in subalpine parks, in gravel.

No. 5174. May 22, 1894, Kanab, Utah, 5300° alt., in meadows.

No. 5638d. July 16, 1894, Burrville Sink, Utah, on muddy shores, 7500° alt.

No. 5397n. June 5, 1894, near Tate Mine, on Bullion Creek, Utah, 10.000° alt., in gravel, along stream.

No. 5312q. May 28, 1894, cañon above Tropic, Utah. in gravel, 6500° alt.

No. 6056ah. September 18, 1894, De Motte Park. Buckskin Mountains, Arizona, 9000° alt.. in gravel.

No. 5957b. August 29, 1894, Brigham Peak, near Marysvale, Utah, 10,500° alt., in alpine meadows.

The western forms of this species have the pod destitute of a crest or lobe at the top. I have seen but one specimen west of the Plains (from California) that is truly typical, the rest belong to this variety. This species, like *brachycarpa*, has the wings reduced to nothing at the very base of the pod.

Œnothera brachycarpa Gray is said by Watson to be the same as *Œ. marginata* var. *purpurea* Watson Bot. King's Exp., but an examination of the type in the National Herbarium fails to establish the fact. The specimen is so immature and so poor that no one can say that it is not *Œ. cæspitosa*, and which it is far more liable to be. This would clear up the synonymy very much, as it is highly improbable that the species to which it has been referred would exist in subalpine situations. while its known home is the arid deserts. 4000° to 5000° above the sea.

Œ. SCAPOIDEA var. PARRYI (Watson, Am. Nat. 9, 270). *Œ. Parryi* Wat. l. c. This is manifestly only a variety with pedicels an inch or less long and a pod 3″ long.

ŒNOTHERA CARDIOPHYLLA var. PETIOLARIS.

Perennial, erect, stems usually simple, 1–2° high: rather finely pubescent and glandular: leaves cordate to nearly round or even ovate, about 1′ long: thick, veiny, coarsely dentate: petiole 1–3′ long: flowers in rather short terminal racemes, reddish, 6″ wide: stigma capitate, 1″ wide:

calyx tube narrowly obconic, 4″ long; calyx tips free; buds oval, 2″ long; petals round to rhombic, 3″ long; pods nearly sessile, straight or arcuate, 12–18″ long, 1½″ wide, blunt, scarcely narrowed at tip, bluntly 4-angled; seeds very small. The plants are inclined to be floccose-woolly.

Rosario, Lower California, Orcutt, April 30, 1886; Mojave Desert, California, May, 1884, Lemmon; Surprise Cañon, Panamint Mountains, California, Coville & Funston, Nos. 624 and 725; also No. 208, Funeral Mountains, California, 1891, same collectors. Types in National Herbarium.

ŒNOTHERA TENUISSIMA

No. 6083. September 26, 1894, Rockville, Utah, 4000° alt., in clay washes.

This plant belongs to the section *Chylismia* of Watson. Plants annual, very much branched throughout, forming a large, rounded, erect, bushy clump, 3° high or less, with very slender branches and long axillary and terminal racemes, about 8′ long; these are floriferous throughout; each flower is bracteate with a minutely pediceled, green, ovate to triangular bract below each pedicel; the bracts become minute above and gradually enlarged into small, acute, shortly-petioled leaves below. The main leaves are lanceolate, sparsely and very slightly and irregularly dentate, acute at both ends, on a varying petiole, with prominent primary and secondary veins below, which are often purplish; leaves minutely pubescent below, and sometimes with very delicate, sparse, white hairs; plants otherwise wholly glabrous, the blade of the leaf is usually about 10″ long, and the petiole as long or much shorter; pedicels usually 5–8″ long and filiform; pods clavate-oblanceolate, about half the pedicels, inclined to be 4-angled, obtuse at apex, a little less than 1″ wide; the

the body elliptical, wings 2′ wide or more, fruit deeply emarginate above and below, oil tubes 2–3 in the intervals and 4 on the commissure; leaves ovate in outline, the blade 1½–2½′ long, glaucous, thick, segments in the typical form nearly contiguous, very small, ovate to nearly square; old leaf-sheaths very persistent at the crown, the green sheaths large, hyaline, nerved and usually colored; root long, erect and fleshy. In other respects this plant closely resembles *C. montanus*, having the same very thin wings without any enlargement except next the seed and all nearly equally developed.

This is No. 1685 of my Utah collection, and abounds on the clayey and gravelly plains, valleys and lower hillsides throughout Utah and Nevada.

A common form of this species is the variety MONOCEPHALUS Jones in Herbarium, having the umbellets densely congested into a single close head, especially in fruit; the leaves are much reduced and the segments contiguous or imbricated. This form is *C. montanus* var. *globosus* Wats. in King's report as to the fruit; it is also his specimen from the Trinity Mountains, Nevada. The type of this form is my specimen gathered at Terminus, Utah, in 1890. To this I also refer a specimen from Candelaria, Nevada; also Palmer's specimen from Pocatello, Idaho, probably Marsh's specimen from Ft. Wingate, New Mexico, and possibly a specimen of Whipple's Expedition ticketed "California." All the above specimens are in the National Herbarium, there being a duplicate of my type in the Herbarium.

CYMOPTERUS UTAHENSIS var. EASTWOODÆ.

Much more robust than the type with stems sometimes 16′ high and occasionally somewhat caulescent; rays 1–2′ long; fruit narrower and often much longer proportionally, usually with thin wings 1–1¼″ wide, generally truncate or barely emarginate at both ends; oil tubes 2 to 4 in the intervals and 6 to 8 on the commissure; wings somewhat corky-thickened next the seed; leaves lanceolate and the segments rather distant and larger; involucres and involucels as in the type species. The type of this variety is Miss Eastwood's specimen from Durango, Colorado, collected June 3, 1890. This variety was previously collected by me on May 2, 1890, at Cisco, Utah.

Mr. Rose had studied this form and given it a name in the National Herbarium, but now considers it to belong to this species.

Cymopterus montanus Nutt.

What I take to be the typical form of this species, whose home is on the plains of Colorado, Wyoming and New Mexico, has the flowers sessile, or nearly so, among the leaves, and even in fruit the peduncles are usually very short, not as long as the leaves and decumbent; involucre a narrow, irregular, hyaline border; involucels always conspicuous, with very wide hyaline border and triangular and green center, but variable in size; flowers white; fruit never wider than oval, usually 3″ long; leaves never sheathing above the ground, prostrate on the ground. Specimens of this species in the National Herbarium are from Fort Lyon, Arkansas, Palmer; Denver, Colorado, Eastwood; Pueblo, Colorado, Hicks; No. 210 from Colorado, Hall and Harbour; Fort Russel, Wyoming, Ruby; and Laramie, Wyoming, Nelson.

C. MONTANUS var. PEDUNCULATUS.

This is a peculiar form or quite possibly a new species with long, erect peduncles often double the leaves, with rays 1′ long; involucres reduced to a mere rudiment; involucels linear to oblong and small; fruit small, about 2½ long, oblong, emarginate above and below; wings nearly 1 wide, rather wider below, and gradually thickened towards the seed; oil tubes 3-4 in the intervals and about 6 on the commissure; involucels divided to the base and pinnate-nerved; root abruptly tuberous-enlarged. The types of this variety are Woolson's specimen from Dallas, Texas; Reverchon's, same locality, with narrower wings.

I refer here with much hesitation a specimen from Fort Belknap, collected by Sutton Hayes, for this may belong to *Utahensis* var. *Eastwoodæ*, described above.

Cymopterus purpurascens (Gray, Bot. Ives, 15) Jones.

No. 5002. March 30, 1894, at the foot of the grade above Bellevue, Utah, in gravel, 3700° alt.

No. 5140c. May 1, 1894, Washington, Utah, in alkaline clay, on flat, 3000° alt.,

No. 5196m. May 8, 1894, Le Verkin, Utah, in gravel, 3700° alt.

To this I also refer a specimen from the Uinkaret Mountains, South Utah, collected by Mrs. Thompson in 1892; Rusby's from Mangus Spring, New Mexico; Lemmon's from Peach Spring, Arizona; Bishop's from south Utah; McDougal's from the San Francisco Mountains, Arizona; Dr. Mearns' from the Carrigallilo Mountains, New Mexico, collected April 17; and the central specimen on the sheet marked *C. montanus* from the Mexican Boundary Survey; all these being in the National Herbarium.

This species is very marked by its long peduncles, at least longer than the leaves; by its purple and greenish, not hyaline (except on the very margin), many nerved, barely lobed, large, rounded involucels, which in flower are cup-like and inclose the brilliant-purple flowers, simulating one of the *Compositæ;* involucre similar and either of ovate and acute or wider and obtuse lobes which are either very large or sometimes reduced even to a rudiment, but always with more or less of a hyaline margin; fruit about 6″ long, nearly round to oval and deeply emarginate above; wings 2–3″ wide, with a narrow, corky margin next the seed; otherwise very thin; oil tubes 3–4 in the intervals and 8 on the commissure, not evident externally; leaf sheaths enlarged at the base and stems covered below with long, hyaline leafless sheaths; leaves from fully to barely bipinnate, with obovate, often lobed divisions, always glaucous and thick, 3–4′ long, mostly ovate in outline, with petiole equaling or exceeding the blade.

CYMOPTERUS GLOMERATUS var. PARRYI (C. & R. Umbell. 50) Jones.

Coloptera Parryi C. & R. Umbell. p. 50.

Cymopterus Parryi (C. & R.) Jones, Zoe, 4, p. 49.

I do not adopt the obsolete name *C. acaule* (Pursh).

An examination of the material referred to this species in the National Herbarium shows that the specific character does not agree with the generic character given by Coulter & Rose under *Coloptera*, the flowers being white instead of yellow. The other two species described by them, as I have already indicated in Zoe, 4, p. 49, have a minute hyaline involucre, while this species has no trace of any, and therefore must be compared with *Cymopterus glomeratus* and not with *C. Fendleri*. On comparison with a large suite of specimens of *C. glomeratus* there is absolutely no character of leaf, habit, or inflorescence to separate this species from that. The only character, and that a variable one, lies in the wing of the fruit, which in some specimens is quite thick on the outer edge and with only a rudimentary thin prolongation beyond, but in other specimens the prolongation is more pronounced. In some specimens of *C. glomeratus* in the National Herbarium from the plains, the corky portion of the wing is quite narrow and without the slightest evidence of a thin prolongation beyond, but the specimens are referred to *C. glomeratus* because of the less thickened wing. I have shown in Zoe that *C. glomeratus* shares with *C. Fendleri* the character of a thickened central portion of the wing, that portion next the seed being fully as thin as the portion outside of the corky part. In view of these facts I see no escape from making this a variety of *C. glomeratus*, whose geographical range seems to be that portion of the Great Plateau west of the plains, including western Wyoming, Idaho and Montana.

Peucedanum triternatum var. *alatum* Coulter and Rose is a true *Pseudocymopterus*, with raised winged ribs, and certainly connects *Peucedanum* with *Cymopterus*, as the latter genus is now received. If *Pseudocymopterus* is to be retained its limits certainly must be changed.

MENTZELIA MULTIFLORA var. INTEGRA.

No. 6082c. September 26, 1894, Rockville, Utah, 4000° alt., in red sand.

No. 5419e. June 14, 1894, Salina Cañon, Utah, four miles up, at 5300° alt.

No. 5263. May 19, 1894, on grade south of Rockville, Utah, in clay, 4500° alt.

Leaves oblanceolate, simply and bluntly toothed or lobed; flowers long pediceled and bractless; seeds nearly orbicular, large, broadly winged. To this I refer Palmer's specimen from southern Utah, No. 171, referred in the National Herbarium to *M. chrysantha*.

This grows on very barren clay soil.

Symphoricarpus rotundifolius Gray.

No. 5261m. May 17, 1894, Springdale, Utah, 4000° alt., in red sand.

No. 5447d. Salina Pass, June 16, 1894, 8000° alt., in gravel.

No. 5312ay. May 28, 1894, cañon above Tropic, Utah, in clay, 6000° alt.

No. 5441t. June 16, 1894, Salina Cañon, in gravel, 8000° alt.

No. 5639. July 17, 1894, Loa Pass, Utah, 8000° alt., in gravel.

No. 6015w. September 7, 1894, Panguitch Lake, Utah, 8400° alt., in gravel.

No. 5289n. May 23, 1894, Johnson, Utah, 5000° alt., in gravel.

No. 5137. April 30, 1894, Santa Clara valley, Utah, 5000° alt., in gravel.

SYMPHORICARPUS ROTUNDIFOLIUS var. OREOPHILUS (Gray).

Symphoricarpus oreophilus Gray, Jour. Lin. Soc. xiv.

The characters given by Gray do not hold out, but there is shading from one into the other at all points.

BIGELOVIA DOUGLASII var. SPATHULATA.

No. 5758m. August 6, 1894, Fish Lake, Utah, in gravel, 9000° alt.

This shrub has the habit of *B. Douglasii*, but all the lower leaves are spatulate to oblanceolate, while the upper leaves are linear, acute, and generally a little twisted, thus approaching *B. Vaseyi* on the one hand, and *B. Douglasii* on the other; the flowers are a little glutinous, and the leaves sparsely ciliate on the margin; the stem is scabrous, otherwise nearly glabrous; the heads are sometimes slightly floccose; scales usually in threes, never more, obtuse or barely acute, not at all keeled, the outer ones green towards the apex and oblong, and the inner ones linear and almost wholly hyaline; corolla lobes reflexed and triangular; style branches much exserted, and hispid not quite half their length; akenes pubescent throughout.

To this I also refer the two doubtful specimens:

No. 5095aj. April 23, 1894, four miles above Pagumpa, Arizona, 5000° alt., in gravel.

No. 5264c. May 19, 1894, near Smithsonian Butte, Utah, 5000° alt., in gravel.

BIGELOVIA TURBINATA.

No. 6066c. September 24, 1894, Canaan Ranch, Utah, 5000° alt., in gravel.

This species seems to be next to *B. juncea*. Bracts 5–6 in each row, the lowest minute and often loose, all obtuse or only apiculate, and all with a darker center, as if keeled, oblong to linear, innermost 4″ long and 2″ wide, shorter than the flowers; pappus white; corolla oblanceolate-cylindrical, with minute, ovate, appressed lobes; style appendages filiform; anther tips nearly linear; plants glabrous and a little glutinous even to the flowers; leaves sparse, long, canaliculate, uppermost reduced to mere rudiments. This has the habit of the allied species, being about 4° high, in a rounded, bushy tuft or shrub, and grows on clay soil on the borders of an old sink.

BIGELOVIA HOWARDI var. ATTENUATA.

No. 5847a. August 21, 1894, Marysvale, Utah, 6500° alt., in gravel.

No. 6106k. October 7, 1894, divide north of Beaver, Utah, 7000° alt.

No. 6052k. September 17, 1894, Buckskin Mountains, Arizona, 9000° alt., in gravel.

No. 5912. August 27, 1894, Marysvale, Utah, 7000° alt., in clay.

Usually with green stems, rarely whitened; heads viscous; leaves linear to filiform; all the bracts long-attenuate, not coriaceous, passing into green and similar involucral leaves; flowers rather inclined to occur in heads or short corymbs, fully equaling the leaves, light-yellow; bracts as long or longer than the pappus. I take as the type of this variety my Nos. 5912 and 5847a, and also refer to it No. 6052k and possibly No. 6106k. I also refer Newberry's specimen gathered in McComb's Expedition. Other specimens which belong near here, but with white or whitish stems, are:

No. 6002aa. September 6, 1894. Panguitch Lake, Utah, 8400° alt., in gravel.

No. 6039b. September 12, 1894, Elk Ranch, Utah, in gravel, 7000° alt.

No. 5989i. September 4, 1894, Circle Valley Cañon, in gravel, 7000° alt.

BIGELOVIA MENZIESII var. SCOPULORUM.

No. 5204v. May 10, 1894, near Cedar City, Utah, 6000° alt., in gravel (altitude probably too high).

No. 6074. September 23, 1894, above Springdale, Utah, 4000° alt., on sandy and gravelly slopes.

This plant is near to *B. Menziesii*, but is characterized by very long nearly linear, 3-nerved entire leaves, from 2–3′ long and 2–4″ wide, acuminate at each end; the plants are woody below, very sparingly branched and the branches very long and almost rush-like, erect, and with gradually reduced leaves towards the inflorescence which is corymbose, of few clusters of 3–5 heads; heads yellow, 5″ long, 10 or more flowered; bracts gradually reduced to minute scales at the base, all being rounded and without distinct, green tips, the inner bracts narrowly oblong and only a little shorter than the flowers, all glabrous; fruit linear, hispid; pappus of very delicate white and strongly barbellate, unequal hairs, nearly as long as the flowers; calyx lobes linear, and as long as the anthers; style branches very hispid and long, and stigmatic portion short.

This plant grows on rocks and on rocky hillsides in the cañon of the upper Virgen river above Springdale, Utah.

BIGELOVIA SQUARROSA (H. & A.) *Aplopappus squarrosus* H. & A. Bot. Beech. 146. This seems to me much better placed in *Bigelovia* and is close to *B. Menziesii*.

BIGELOVIA MACRONEMA (Gray).

Aplopappus macronema Gray, P. A. A. 16, 80.

Macronema discoidea Nutt., Tr. Phil. Soc. n. s. 7, 322.

No. 5940r. August 28, 1894, head of Bullion Creek, near Marysvale, Utah, in gravel at 11,500° alt.

No. 5895. August 24, 1894, Falls of Bullion Creek, near Marysvale, Utah, 9500° alt., in gravel.

It has always seemed to me that this species is a better *Bigelovia* than *Aplopappus;* in fact it is so close to *B. Bolanderi* that it is very difficult to separate the two species. I leave it for other botanists to determine whether the other species in the section *Macronema* of *Aplopappus* of Gray shall be transferred to *Bigelovia*. In my judgment, however, the dividing line is better left here. In discussing the species of *Macronema*, E. L. Greene curiously enough says that this group is most nearly allied to *Chrysopsis*. He fails to see that *Macronema* passes directly into *Bigelovia* and that it is well nigh impossible to separate specifically *Bigelovia Bolanderi* and *Aplopappus macronema*, while there is a direct transition to true *Bigelovia* through these species and *B. Parryi*, *Nevadensis*, *Howardi*, etc. I caunot follow Mr. Greene in his work on *Bigelovia*, *Aplopappus*, and allied genera.

BIGELOVIA LEIOSPERMA var. ABBREVIATA.

No. 6105. October 7, 1894, mouth of Clear Creek Cañon, Utah, 6000° alt., in clay.

Very minutely pubescent throughout, not at all floccose; leaves mostly very much reduced and almost scale-like, linear, 4″ or less long, apiculate, barely flattened; scales nearly all linear and with only a trace of green at the tip, not at all keeled; flowers mostly double the rather short pappus, the upper half of the tube being inflated and fusiform and with rudimentary closed teeth; nerves very prominent, glabrous; style branches subulate-filiform and hispid for more than half their length. This is a small shrub about 1° high, branched at the base and forming a rounded tuft of rather slender branches covered with the scale-like leaves, the flowers are cream-colored or light yellow and are clustered at the end of the branches, rather corymbosely, in heads of five or more.

This grows on very barren clay soil, at the foot of the mountains where almost no other plant will grow.

CHRYSOPSIS CÆSPITOSA.

No. 5249u. May 16, 1894, Springdale, Utah, in red sand, 4000° alt.

Cæspitose from a woody much branched root, forming a dense mat; leaves spatulate 6″ long, clustered on very short (13–20″ long) ascending stems, very villous but not hispid, apiculate; heads 3″ high, sessile: scales ashy-green, linear, somewhat unequal, inner with purple and hyaline margins and acute; flowers yellow, rays dark and short: outer pappus setulose. This is a very striking plant, but may be an extreme form of *C. villosa*.

This plant grows in very sandy soil covering sandstone rocks, and seemed to be very scarce.

ASTER GLAUCUS var. WASATCHENSIS.

No. 5861. August 22, 1894, Tate Mine, near Marysvale, Utah, 9000° alt., in gravel, on mountain slopes.

Plants glandular above: scales broad and with green, foliaceous, spreading tips: plants densely tufted, branched above, 1–2° high.

ASTER THERMALIS.

No. 5410. June 7, 1894, Monroe, Utah, 5500° alt., in warm springs on lime tufa.

Perennial, with rather fleshy-fibrous roots, growing in

tufts near alkaline hot springs; stems not in the least woody, rather flexuous and rush-like, erect or ascending, 9′–1½° high, glabrous below, glandular above and very heavily scented; leaves grass-like, 2–7′ long, 1″ wide, fleshy, sessile, a little clasping, the uppermost leaves bract-like; scales in about 3 irregular series, green, appressed, acute, linear; rays ½″ wide, 6″ long, purple, very conspicuous; disk-flowers light yellow; akenes linear, ribbed, villous; pappus very frail and rather scanty; flowers racemose, few, long-peduncled, 6″ wide and disk 3″ high.

This plant is never found except in water flowing from alkaline hot springs which are heavily charged with lime, bicarbonate of soda and chloride of sodium. The soil in which it grows is always very poor and thin, covering lime tufa. It seems to be the only aster which will grow in such places. It is never found in high elevations nor does it occur outside of the plateau region of Utah.

ASTER TORTIFOLIUS var. FUNEREUS. No. 358, Coville, Death Valley Rep. Furnace Cr. Can. Funeral Mts., Cal., Jan. 30, 1891, and No. 863, near Keeler, Cal., May 16. An intergrading form is his No. 184 from Brown's Peak, Cal., Jan. 16. This variety has closely imbricated scales, in many series, gradually reduced below, not attenuate, closely appressed.

Aster xylorhiza T. & G. This differs in no real respect from *A. Wrightii* and *A. venustus* except that it is lower and has longer leaves. *A. venustus* Jones answers to *A. Parryi* of the same section.

ERIGERON CINEREUS var. ARIDUS.

No. 5149v. May 3, 1894, Washington, Utah, 3500° alt., on sandstone rocks.

This plant seems to be near *E. divergens* in general appearance, but is shrubby at the base; heads 8″ wide, including the rays which are very many and narrow, 3″ long, purple; bracts green but sparsely hispid; lower leaves pinnatifid with distant linear lobes; leaves oblanceolate-linear, on long petioles; upper leaves sessile, linear, 20″ long; flowers corymbose above, heads long-peduncled, with gradually reduced, bract-like leaves on the peduncles; plants erect, 9–18′ high, growing in clumps in the crevices of hot sandstone rocks in the most exposed situations. I also refer to this a plant collected by Vasey in 1881 in California, and by Dr. Smart on the Verde River, Arizona, in 1867, both being referred to *E. divergens* and both being barely woody at base.

ERIGERON CÆSPITOSUS var. NAUSEOSUS.

No. 5386. June 2, Marysvale, Utah, 8300° alt., in gravel at the base of the mountains. Grows in tufts in exposed and hot places.

Very leafy, root leaves 3′ long or less, spatulate to obovate, 12″ or less, upper ones obovate, 1′ long by 6″ wide, sessile, almost clasping; whole plant scabrous and the leaves rough ciliate; heads mostly single, on the long, ascending, very leafy stems; proper peduncles about 1′ long; plants glutinous above and with a very nauseating odor; scales abruptly attenuate at tip, rough, not hairy.

This abounds on dry, rocky slopes and cliffs at the foot of the mountains.

ERIGERON CÆSPITOSUS var. LACCOLITICUS.

No. 5661. July 23, Marvine Laccolite, Henry Mts., Utah, at 6000° alt., in volcanic gravel. Grows in tufts.

Stems many from a much branched, woody root, covered below with imbricated leaf-petioles, stems about 1° high, mostly erect or ascending, several flowered; flowers small, 3″ wide exclusive of the rays; root-leaves 4′ or less long, oblanceolate, on a long and winged petiole, rounded, upper leaves all sessile, oblong or narrower, 1′ long or less by 2–2½″ wide; heads short-peduncled; whole plant ashy and very rank smelling.

This grows in very dry places at the base of the mountains. I also refer to this variety Ward's No. 560 from the same region.

Townsendia montana Jones was again collected in 1894 and seems to hold its characters well.

TOWNSENDIA FLORIFER var. COMMUNIS.

No. 5322f. May 30, 1894, Kingston, Utah, on basaltic knolls, 5300° alt.

No. 5315b. May 28, 1894, ten miles south of Coyote, Utah, on lava, 6500° alt.

No. 5323. May 31, 1894, Marysvale, Utah, 6000° alt, in gravel.

This is the cæspitose form, with simple stems and large flowers, 12″ wide or less, with acuminate rays. To this variety I also refer a specimen from Howell, collected on the Columbia River, opposite Umatilla, April 29, 1882, this specimen being in the National Herbarium.

The typical form of *Townsendia florifer* has ascending and branching stems, with 2–4 long-peduncled, small heads, about 6″ wide and high, with rays 3–4″ long; scales ashy, strigose outside, and imbricated; pappus equal; not more than a winter annual, about 5½′ high; flowers purple; leaves linear-spatulate, 3′ long or less.

APLOPAPPUS LINEARIFOLIUS var. INTERIOR (Coville, Biol. Soc. Wash. 7, 65, 1892).

Aplopappus interior Coville, l. c.

No. 5149u. May 3, 1894, Silver Reef, Utah, on rocks at 3500° alt.

No. 5060. April 14, 1894, Mica Spring, Nevada, in granitic gravel, 4000° alt.

No. 5045n. Same locality and date.

No. 5297l. May 26, 1894, Pahria Cañon, Utah, 5300^c alt., in sand.

After looking over all the specimens obtainable on these two species, I can only consider *A. interior* Coville as a geographical variety of *A. linearifolius*, as transitions seem to occur at all points. Mrs. Brandegee has already called attention to this as a variety.

GYMNOLOMIA MULTIFLORA var. ANNUA. All the forms of this species growing in our northern mountains are perennial, as described by Nuttall; those growing on the arid plains of the south are annual, and with quite a different habit; the heads are decidedly conical, the plants are more delicate and slender, often only a few inches high. The variety ranges from southeastern Utah to Mexico and westward, but I have never seen it in the Great Basin.

TETRADYMIA SPINOSA var. LONGISPINA.

No. 5110. April 26, St. George, Utah, in red sand, on the slopes of Triassic rocks, at 3000^c alt., growing in tangled clumps about three feet high. Soil a little alkaline.

Shrub 3–6^c high, with rounded top and long, slender, intricately interlaced branches, closely and permanently woolly, branches rather tortuous; primary leaves 6–12″ long, linear-spatulate, tomentose, straight or nearly so, horizontal to somewhat deflexed, never hooked, becoming long slender spines, 1′ long, like those of *Opuntia rutila*, but not so long: secondary leaves fleshy, smooth, linear, but a trifle widened above, 6″ long: scales 3″ long, oblong: heads 4″ long, on a peduncle 1″ long or less: flowers about 6–10 in a head: young branches never elongated. This differs very markedly from the type in the long spines and in its habitat, but there are transitions in the National Herbarium.

This grows only in the Larrea belt, and abounds on drifting sand dunes in the hottest places.

CHÆNACTIS CARPHOCLINIA var. ATTENUATA (Gray, P. A. A. 10, 73).

Chænactis attenuata Gray, l. c.

No. 5036au. April 13, ten miles below Mica Spring, Nevada, in gravel, at 2000° alt.

No. 5045r. April 13, three miles below Mica Spring, Nevada, in gravel, 4000° alt.

These specimens have very many awns on the receptacle. It would seem that the presence or absence of awns is accidental rather than specific. This fact also noted by Mrs. Brandegee, in Zoe.

F. V. Coville, in the Botany of the Death Valley Exp., p. 134, says: "This plant has not been reported since its original discovery." It was collected in 1884, by me, at Yucca, Arizona, and reported in my lists published in the same year. See "Flora of Arizona, First Fascicle."

Chænactis carphoclinia Gray.

No. 5036z. April 13, near Hole in the Rock, 8 miles above Stone's Ferry, Nevada, at 1500° alt., in gravel.

There is a great deal of doubt in my mind of the validity of the characters in this group. In this particular number there are no awns on the receptacle, and the bracts have a prominent midrib, but in other respects it corresponds with this species.

CHÆNACTIS ALPINA (Gray, Syn. Flora, 341).

Chænactis Douglasii var. *alpina* Gray, l. c.

This species is very well defined, as to the high alpine specimens, shows little variation, but it has been confused with perennial forms of *C. Douglasii*. It is characterized by being cæspitose, often forming large mats above timber line, stems not filiform; leaves about 2′ long, with many pinnate, narrowly oblong segments, which are rounded and often lobed at apex, petiole equaling the blade, younger parts floccose; stems subscapiform, 2–4′ long, erect or ascending, rough, scarcely at all glutinous, little enlarged above; heads large, nearly 1′ high, with broadly linear green but rough scales; pappus oblong to oblong-cuneate, about half the corolla tube, lacerate, glabrous. This is my No. 1232. Watson's specimen from the Uinta Mountains, Utah, is much the same. The plants are perennials with very many long branches which are prostrate. This is close to *C. Nevadensis.*

CHÆNACTIS DOUGLASII var. MONTANA. This includes most of the forms usually referred to the var. *alpina* of Gray. It includes all the forms so referred in the National Herbarium, except Watson's specimen mentioned above, and Patterson's, from Gray's Peak, Colorado. All are perennial; peduncles not scapiform, but plants much reduced in the extreme forms; pubescence less floccose, otherwise much as in the type. This is the usual form of the higher mountains, from 7000° to 9000° alt., and ranges from the Rocky Mountains of Colorado to the Sierras.

RIDDELLIA TAGETINA var. PUMILA.

No. 5474. June 21, 1894. Grand Junction, Colorado, 4400° alt., in gravel, in open places.

Depressed from a thick, branched, woody base; stems flexuous, 4–8′ long; leaves all entire, obovate to oblanceolate, 1′ or less wide and 2–3′ long, villous or rarely woolly; heads never clustered, about 10″ long, very villous; peduncles 2½–4′ long; rays 6″ long, broadly oblong; pappus oblong, truncate to erose, ½ the smooth angled, sulcate akene.

This grows on gravelly and high banks of the Grand River, in very dry situations. It was also collected by Miss Eastwood in the same locality before I found it.

Encelia nudicaulis Gray.

Helianthella nudicaulis Gray.

No. 5069k. April 17, near Horse Spring, Arizona, 2500° alt., in compact clay, on a bare and rather alkaline knoll, in poor soil.

No. 5376. June 2, Marysvale, Utah, on loose talus, in very dry and hot places, volcanic soil, 6500° alt.

No. 5095ak. April 23, top of grade, four miles above Pagumpa, Arizona, on barren clay slope, in juniper belt.

An examination of the type specimen of this species shows that it is not the same as *E. argophylla*. All the specimens described by me in Zoe, iii, 304, belong to this species, apparently, or else there are two species included in this one. It is very probable that this will prove to be an exceedingly variable species, having a wide range. The northern plants which I have described in Zoe have a very different habitat, since they grow on dry and rocky ridges, in very exposed situations, where the soil is free from alkali. On the other hand, all the specimens given above under this species flourish in an alkaline, clayey soil, where no other plants will grow. The allied species, *E. argophylla*, I have never seen anywhere except in very salty soil, along the salt deposits of eastern Nevada, in the Larrea belt. It appears that it is never found growing in any other situations.

An examination of nearly all the species of *Encelia* shows that the character of wingless akenes separating this from *Verbesina* is not good, as several species, such as *E. eriophylla*, *E. viscida*, etc., have corky wings. In *E. nutans* Eastwood, which is probably the same as, or a variety of, *Verbesina scaposa* Jones, though rayless, the outer disk akenes are triquetrous and winged, and doubtless represent ray flowers; the species was, therefore, properly placed in that genus with winged akenes, though the habit is that of *Encelia*.

I can see no resemblance in this group of plants to *Helianthella* proper, but there is a strong likeness to *Encelia;* in addition, the former genus is confined to the cold mountains at high altitudes, while the latter is confined to the dry and very hot regions at low elevations, or when the elevations are above 5000° it is due to isolated peaks or ridges in hot regions. To put this plant and its congeners into *Helianthella* does violence to their generic relationship as well as habitat.

Encelia argophylla (Eaton) Gray. *Tithonia argophylla* Eaton, Bot. King Exp.

No. 5032q, April 10, 1894, at salt mine, in hard alkaline clay, 10 miles above Stone's Ferry, Nevada, 1200° alt.

This appears to be exactly like the type of the species. The scales seldom exceed the disk, ovate acuminate; akenes very villous, cuneate-oblong, wings narrow, awns stout and nearly as long as the corolla tube; rays about 1′ long, 1–2″ wide, hairy externally like all the other species of this group; ray akenes abortive, triquetrous with or without pappus (this character also belongs to the group); stems very thick and tufted, branched and very short, woody, densely covered with the very thick leaves; peduncles scape-like from the crown of the leaves, 2° long.

ENCELIA GRANDIFLORA. *Helianthella argophylla* Coville, Death Valley Rep., p. 132, not Gray, Proc. A. A. xix, 9.

This differs from the above in the glabrous or barely pubescent not villous akene, which is deltoid, 3″ long by 2″ wide, with wing conspicuously developed and minute awns; bracts lanceolate, long-acuminate, longer than the disk, often 6″ longer; rays 4″ wide and 2′ long, heads with the rays 5′ wide. This is a much larger plant than

E. argophylla, and differs in seemingly good characters.

LAPHAMIA PALMERI var. TENELLA.

No. 5249aa. May 16, 1894, Springdale, Utah, 4000° alt., in crevices of sandstone rocks.

Stems and petioles filiform; leaves all alternate, doubly and coarsely dentate, thin; whole plant sparsely floccose-villous, even to the head; otherwise much as the type. This plant grows on rocks in delicate tufts, about 6′ long, the stems are weakly, and ascending or pendent.

LAPHAMIA CONGESTA.

About 6′ high, in dense tufts from woody base, stems branched above and branches 2–3′ long and crowned by the solitary head whose proper peduncle is 1′ or less long; leaves all alternate, lanceolate to broadly ovate, all with a cuneate base, petiole equaling or half shorter than blade, 2–6″ long exclusive of the petiole, not veiny, entire or in the forms with larger leaves coarsely few-dentate, acutish; stems terete, slender; whole plant very rough-scabrous; heads campanulate, about 3″ high, 15–20 flowered, cream-colored, rayless; scales narrowly oblong to nearly linear, about 12, acute; akènes very scabrous throughout, with one slender awn, nearly equaling the akene. Whether this and *L. tenella*, which is also rayless, vary into *L. Palmeri* can only be conjectured. If the leaf characters hold they are certainly good species.

No. 6063. September 21, 1894, in clefts of rocks on the mesa below the Buckskin Mountains, Arizona, 7000° alt.

LAPHAMIA GRACILIS.

No. 6050c. September 15, 1894, below Nagle's Ranch, on edge of Buckskin Mountains, Arizona, in crevices of limestone rocks, at 7000° alt., in very dry places.

Slender from a shrubby base, hanging in festoons from overhanging rocks, 6–12′ long, weak, ashy; leaves pedately lobed and lobes oblong, rarely again toothed, blades 6″ or less long, all on a slender petiole about half their length, and obtuse or barely acute, alternate, thin; plants racemosely branched above with solitary heads on slender peduncles, 2′ or less long; flowers light yellow, few, in small heads, 2″ long and 2″ wide, flowers and scales pubescent; akenes with thick lateral nerves and usually a slender awn at each angle, narrowly oblong, scabrous pubescent.

Senecio Bigelovii Gray. I can find no crucial character separating this from *S. Rushyi* Greene.

CNICUS CLAVATUS.

No. 5715. August 2, Fish Lake, Utah, 10,000° alt., in gravel.

Plants tufted from a deep perennial root, erect, 2–2½° high, glabrous throughout; leaves lanceolate, root leaves about 1° long, stem leaves 6–8′ long, all deeply pinnatifid with lanceolate, very shortly spinose lobes, spines 1″ long or less; leaves percurrent by a narrow wing (3′ high) below the petiole; heads on peduncles 2′ long, involucrate with several leafy bracts; scales imbricated, coriaceous except at tips of the innermost, all but the uppermost spinose with short spines, and these also with lateral spines; bracts imbricated, the outer somewhat shorter, the upper all with dark tips; flowers white, corolla lobes ⅓ the tube; at least some of the pappus awns clavate thickened at the apex in each flower, short, barely exceeding the corolla tube; heads 2′ high, nearly hemisperical to campanulate. This grows at high elevations close to snow, in the same situations in which *C. Eatoni* is found, but it is conspicuously different from that species.

CNICUS CALCAREUS.

No. 5695bh. July 27, Bromide Pass, Henry Mountains, Utah, 10,000° alt., in gravel.

No. 5696. July 30, Cainville, Utah, 4000° alt., in clay.

Glabrous throughout except some floccose wool on the lower bracts, involucral scales scabrous; stems tufted from a thick perennial root, with the habit of *C. Eatoni*, 2° high, erect, stout, somewhat branched above; leaves 8–12′ long and 10–15″ wide, running down on the stem nearly the length of the node, pinnatifid into triangular or oblong lobes, which are short and stout-spinose at tip and minutely so on the margin; heads sessile or nearly so, about 1½′ long, turbinate-cylindrical, purple flowered; scales regularly imbricated and outer gradually shorter, thin, inner subulate-acuminate, very acute, but not prickly, outermost scales ovate, tapering into a short erect awn; anther tips apiculate; corolla about equaling the lobe; all the scales have a dark line in the middle, are not closely appressed nor rigid; the plants are very leafy throughout, with heads 1–3 in a cluster, which is sessile or nearly so.

This grows on alkaline clay soil, above Cainville, along the Fremont River.

If the published characters for other species of this genus hold, then all these here described are good species, but I fear that many recognized species will eventually prove to be forms of polymorphous species, into which some of these may fall.

Cnicus nidulus.

No. 5290a. May 25, Pahria, Utah, in red sand along the river bed, 5000° alt.

Erect from a thick woody, perennial root, 2° high, floccose-woolly throughout, except the nearly glabrous, seemingly viscid scales; leaves lanceolate, pinnatifid with triangular lobes, which are very stout-spiny; spines yellow, 6″ long; heads ovate, 10″ long; scales coriaceous, regularly imbricated and close-pressed, without green tips, outer ovate to lanceolate, the body short, but tipped with a very stout and long, yellow, roundish spine, 1′ long, or equaling the head, the inner scales with shorter spines and the innermost series of delicate, linear, subulate, thin, red-tipped scales, shorter than the outer spines; corollas not developed. This species is very striking and seemingly different from any other. It grows in red, alkaline sand, along the bottoms of the Pahria river, at Pahria, Arizona.

Crepis occidentalis Nutt.

No. 5568h. July 2, Provo, Utah, 6500° alt., in gravel.

No. 5455e. June 18, two miles north of Ferron, Utah, on clay, 5500° alt.

No. 5432. June 15, Ireland's Ranch, in Salina Cañon, Utah, 8000° alt., in gravel.

This differs from the type in having only 8 scales to the involucre, and tends to break down the distinction between this and the allied species.

Nemacladus ramosissimus Nutt.

No. 5077ae. April 19, at spring, 15 miles above Pierce's Spring, Arizona, 1700° alt., in sand.

No. 5045u. April 15, Mica Spring, Nevada, 4000° alt., in gravel.

This is *N. capillaris* Greene, but I can find no valid characters on which to separate the two species.

PRIMULA INCANA.

No. 5312av. May 29, Beaver Coöp ranch, at the head of the South Fork of the East Fork of the Sevier river, 7000° alt., in cold bogs. A very early bloomer.

Plants erect and single from fleshy or flesby-thickened roots, no tap root, 6′ high, simple, scapose; root-leaves rosulate, oval to elliptical-oblong, obtuse, rounded, minutely-denticulate, 1–2′ long, without petioles, green above, whole plant otherwise white-farinose except the corolla; scapes rather stout for the plant; capitate heads of flowers subtended by obtuse, narrowly-oblong bracts, 2–3″ long; flowers sessile or nearly so, 8 or more, in a dense head; calyx narrowly-oblong, the lobes rounded and obtuse; corolla purple, the tube barely surpassing the calyx; lobes about 1″ long, throat yellow. This species seems to be nearest to *P. farinosa*, and may be only a form of it, but it seems distinct.

FRAXINUS ANOMALA var. TRIPHYLLA.

No. 5082w. April 20, Pagumpa, Arizona, in the Grand Wash, 4000° alt., among rocks, at the lower edge of the juniper belt.

Leaflets 3 and stalked.

GENTIANA TORTUOSA.

No. 6008. September 7, 1894, Panguitch Lake, Utah, 8400° alt., in meadows.

Prostrate from an annual root, tortuous stems rather slender and entangled, 2–6′ long; leaves linear-lanceolate and acuminate, 2′ long, little reduced above; plants floriferous from base to apex; peduncles slender, 1¼′ or less long, tortuous, often deflexed in fruit; flowers several to each node, white, 3–4″ long, oblong-campanulate; lobes elliptical, acutish, nearly as long as the tube; fringe reduced to scattered setæ on the base of the lobes; anthers nearly round, extrorse; capsules short-stipitate, as long as the calyx, oblong; seeds oval, ½″ long, smooth and coat very close, yellowish; calyx tube almost none, lobes linear, 2–3″ long, rather unequal, nearly equaling the corolla or 1½″ shorter.

This delicate little plant is near to *G. Amarella* and grows in similar situations in gravelly meadows.

Apocynum androsæmifolium var. *pumilum* Gray.

No. 5684ak. July 25, Mt. Ellen Park, Henry Mts., Utah, 9400° alt., in volcanic gravel, on open slopes.

No. 5560. July 2, Provo, Utah, in gravel, 6500° alt., among oak brush on mountain sides.

Plants erect; branches widely spreading, 2° long; leaves elliptical-ovate, 2′ long, 1′ wide; flowers many, cymose-panicled at the ends of the branches, purple, campanulate, 2½″ long; calyx lobes triangular, acute; corolla lobes triangular, 1″ long. This is intermediate between *A. cannabinum* and *A. androsæmifolium*. It is the only form of this species which I have seen in the West. It grows in more open places along roadsides in the mountains, but never grows in meadows or in moist places frequented by *A. cannabinum*, nor can it be a hybrid with that species. This seems to be *A. floribundum* Greene, Erythea, i, 151.

ASCLEPIAS LABRIFORMIS.

No. 5650. July 19, 1894, Capitol Wash, near the Henry Mts., Utah, 5000° alt., in sandy gulch.

Erect, glabrous except the floccose-woolly pedicels; leaves thick and leathery and all but the lowermost alternate, all short-petioled, the petiole being 3″ long, leaves lanceolate and tapering from the base to the shortly-acute apex, 1–3″ wide, 3–5′ long, overtopping the short-peduncled umbels (peduncle 6″ long), umbel with many filiform bracts; pedicels about 9″ long, slender, about 20; calyx green, lobes ovate to elliptical, 1″ long; corolla lobes oval, white, 2″ long; column ½″ long; hoods oblong, truncate, greenish white, with a green midrib, just equalling the anther tube; horns much exserted, touching or overlapping in the center; anther wings very broad at base and truncate, then quickly narrowing above; pods on recurved pedicels, mostly erect, glabrous, ovate-oblong, taper-pointed, 15–20″ long.

Phacelia glechomæfolia Gray, *P. perityloides* Coville in Death Valley Rep. Mr. Coville's plants seem to be identical with Gray's.

PECTOCARYA LINEARIS var. PENICILLATA (H. & A. Bot. Beechy, 371).

Pectocarya penicillata H. & A., l. c.

The constant recurrence of intermediate forms makes it no longer possible to keep up this as a distinct species. Among other intermediate forms is one from Wilcox from Idaho in the National Herbarium.

Pectocarya setosa Gray. The nutlets are described as equally divergent, but they are not; they are geminate. In *P. pusilla* they are equally divergent in my specimens.

KRYNITZKIA ECHINOIDES.

No. 5297p. May 26, 1894, Pahria Cañon, Utah, 5300° alt., in red sand.

No. 5312ac. May 28, 1894, Cannonville, Utah, 6000° alt., on clay slopes.

This is a cæspitose perennial, 6′ high, with erect stem, and is very fulvous, except the lowest leaves; pubescence of the leaves close, short, dense, appressed, upper stem and calyx very setose with spreading hairs; corolla white or cream-colored, usually 1–2″ longer than the nearly filiform calyx lobes, which are 3–4″ long, corolla lobes rounded, rotate, short, about 1½″ long; nutlets sharp-angled, wingless, muricate, corrugated, papillose, and short-setose on the backs and sides. This is very close to *K. fulvocanescens*, to which some specimens have been referred. The synonymy of this species is very much confused, the original *Eritrichium fulvocanescens* of Gray in Herb. based on Fendler's specimen from New Mexico is *K. echinoides*, though the specimen is only in flower. Watson, in King's Report, took up the name of Gray for a Utah plant, and erroneously referred Fendler's plant to it, but the specimen on which Watson's description and figure were founded is now in the National Herbarium, and is clearly a low altitude variety of *K. sericea* (Gray) *i. e.*

var. FULVOCANESCENS (Wats.), being scarcely at all fulvous, and the fruit with warty (never acute) corrugations and intermediate papillæ, and the short edge of *K. glomerata*, the nutlets are not carinate, but are somewhat convex on the back and abruptly narrowed towards the rounded apex. The nutlets do not fall off in any species of this group and are pediceled below. My No. 5163ac is almost exactly this variety.

KRYNITZKIA LEUCOPHÆA var. ALATA.

No. 5289t. May 23, 1894, Johnson, Utah, 5000° alt., on sandstone cliffs.

No. 5261j. May 17, 1894, Springdale, Utah, 4000° alt., in red sand, or gravel.

No. 5455c. June 18, 1894, two miles north of Ferron, Utah, in clay, 5700° alt.

No. 5144. May 3, 1894, Silver Reef, Utah, 4500° alt., on sandstone cliffs.

Nutlets bordered by a thick, entire, narrow, raised wing.

To this I refer No. 632, Fendler and M. M. Palmer's specimen from Fort Defiance, New Mexico.

Echinospermum floribundum Lehm. Dr. Gray upholds Greene's *E. ursinum* from Northern Arizona and adjacent Utah, but I fail to find any character which is permanent, even in Greene's duplicate type in the National Herbarium the characters do not hold.

Polemonium cærulcum L.

No. 5441ac. June 15, 1894, Ireland's Ranch, Salina Cañon, Utah, 8000° alt., in gravel, along shaded streams, in springy places. This is *P. filicinum* Greene. I have never been able to see any valid character separating this species from *P. foliosissimum* Gray. If the flowers are light colored then it is the one species, if not it is the other, the leaf and floral characters vary.

Phlox longifolia var. *brevifolia* Gray.

No. 5098j. April 23, 1895, ten miles south of Black Rock Spring, Arizona, in gravel, in the juniper belt, 4500° alt. No. 5286v. May 23, 1895, Kanab, Utah, 5300° alt., in red sand. These forms have linear-oblong, obtusish leaves, 1′ or less; the calyx has long and hyaline interspaces greatly enlarged, and folded outward so as to make the calyx seem ovate at base. The replication of the calyx is a character of no permanent value.

PHLOX LONGIFOLIA var. GLADIFORMIS.

No. 5208c. May 11, 1894, Cedar City, Utah, 6500° alt., in gravel, on slopes.

This has the habit of *P. Douglasii*, but is less compact; leaves densely clustered around the sessile flowers, loosely imbricated below, or the nodes 3″ long, leaves about 1½′ long, all subulate-lanceolate, pungently acute, thick and stiff, midrib and margins prominent, the latter a little involute; whole plant, even to the flower, glandular and sparsely floccose-hairy, but green; calyx lobes a little over half the tube; corolla tube nearly double the calyx, lobes oval, entire. The plants are loosely cæspitose, and would seem to be hybrids between *longifolia* and *Douglasii*, were it not for the fact that they are very abundant in the locality where found. It grows on north slopes of gravelly hills at the mouth of the cañon of Cedar Creek.

There seems to be a complete transition from *P. longifolia* to *P. Douglasii* through the above variety and the var. *brevifolia*.

Phlox austromontana Coville seems to deviate in no constant character from *P. Douglasii*, the replication of the calyx relied upon as a crucial character, proves of no value, as it varies from nothing to a wide fold. I cannot refer this to *P. speciosa*, as Gray has done.

GILIA CONGESTA var. PANICULATA. .

No. 5464m. June 19, 1894, Huntington, Utah, in clay soil, 6000° alt.

Densely floccose pubescent, with many small heads in a rounded panicle and very much branched above, 6–12′ high.

GILIA MCVICKERÆ.

No. 5378. June 2, 1894, Marysvale, Utah, in gravel. on arid slopes, 7000° alt.

No. 5972g. August 31, 1894, Marysvale, Utah, in loose gravel, 6500° alt.

No. 5989m. September 4, 1894, Circle Valley Cañon. 7000° alt., in gravel.

This belongs to the section *Giliandra*, biennial, erect, 2–2½° high, branched above into a very wide, open. corymbose panicle, which is often 2° across, and sometimes practically flat-topped; all the upper leaves are reduced to arcuate bracts, the root leave and lower stem leaves are pinnatifid with oblong, entire, rounded, acutish lobes, leaves 2–3′ long; glabrous throughout, except the glandular, campanulate, very short pediceled calyx; calyx tube about 1″ long, the blunt lobes minute; corolla about 6″ long, sky blue, tube 3–5″ long, rather ampliate above and campanulate, with oval lobes, which are 1½″ long and surpass the widened upper tube, and are rotate-spreading: filaments long-exserted, double the tube, blue, crowned with oval, minute anthers, filaments spreading. This plant has the habit of very robust forms of *G. latiflora* and *G. inconspicua*. It is a very beautiful plant, and was first discovered some ten years ago by Miss Kelley, now Mrs. McVicker, of Salt Lake City, at Panguitch Lake, Utah. Her specimen was without the root leaves. I sent a portion of it to Dr. Gray with the manuscript name. He also regarded it as a new species, but thought it might prove to be a form of *G. pinnatifida*. In view of the imperfection of the original specimen, I thought best to leave it unpublished till better material could be collected. Abundant material is now at hand, and I take pleasure in dedicating this species to its discoverer. This grows on very hot and dry talus and even on cliffs.

GILIA FLORIBUNDA var. ARIDA.

No. 5701a. July 31, 1894, Capitol Wash, Utah, near the Henry Mountains, 5000° alt., on sandstone rocks, in very arid places.

This plant has the habit of *G. Watsoni*, but is nearest to *G. floribunda;* whole plant greenish, but rough, with short, stiff, not dense pubescence throughout; tufted in dense bunches and branched below, stems 2–4′ long, terminated by a dense head of sessile flowers, 5–10 in the head; stems white; leaves pedately 3–5 parted, filiform, rigid, straight, acerose, involute, 1-nerved, 5–6″ long, somewhat longer than the internodes, not reduced above, and heads with similar bracts; nodes enlarged; calyx cylindrical, tube 3″ long and lobe 1″ long, very hyaline between the angles; corolla white, 1–2″ longer than the calyx, narrowly oblong lobes, 2–2½″ long; seeds 2 in each cell, linear. The heads are seemingly a little glandular and the flowers are vespertine.

This plant grows on the hottest red sandstone rocks in little crevices and pockets, and is quite fragrant.

GILIA LEPTOMERIA var. TRIDENTATA.

No. 5445n. June 16, 1894, near Emery, Utah, in clay, 7000° alt.

Corolla cuneate and tridentate, rotate-spreading, and flowers smaller and shorter than in the type. This is figured by Watson in King's Report along with the type, but when growing it is conspicuously different from it, though often found growing with it.

Gilia scopulorum Jones.

No. 5032l. April 11, 1894, lava ridges, ten miles above Stone's Ferry, Nevada, 1000′ alt.

No. 5024ao. April 5, 1894, on divide 12 miles east of Beaverdam, in Nevada, 6000′ alt., in gravel.

No. 5029m. Bunkerville, Nevada, April 6, 1894, 1550′ alt., in gravel.

No. 5029k. Same date and locality.

No. 5110ac. April 26, 1894, St. George, Utah, 3000′ alt., in red sand.

No. 5036ah. April 13, 1894, near Hole in the Rock, above Stone's Ferry, Nevada, 1550′ alt., in gravel.

No. 5045x. April 13, 1894, Mica Spring, Nevada, 4000′ alt., in granitic gravel.

This well marked species, which Dr. Gray has referred to *G. inconspicua*, shows no intergrades in any of my collections for the last fifteen years, and there are no intergrades in the National Herbarium.

LYCIUM TORREYI var. FILIFORME.

No. 5015. April 5, 1894, Beaverdam, Arizona, 1700′ alt., in sand.

Leaves spatulate-linear, small, 9″ or less long; flowers on filiform pedicels 9″ or less long.

PENTSTEMON ACUMINATUS var. CONGESTUS.

No. 5262. May 19, 1894, near Canaan Ranch, Utah, 5000′ alt., in sand.

No. 5467c. June 20, 1894, Price, Utah, in clay soil, 5500′ alt.

No. 5441ai. June 15, 1894, Ireland's Ranch, in Salina Cañon, 8000′ alt., in gravel.

Strict, erect, 2′ high: spikes simple, long, dense, linear: leaves large, oblong, mostly rounded and obtuse: sepals ovate to oval, acutish, scarious-margined: flowers 8′ long, blue: sterile filament densely long-hairy: corolla hairy within; anthers glabrous. This seems to differ from *P. Parryi* chiefly in the blue flowers.

P. acuminatus Douglas is a plant of the northwest ranging through the Great Basin, while the plant of the plains (*P. Fendleri* Gray) is certainly as distinct as most of the reputed species.

P. confusus Jones connects *acuminatus* and *Parryi* and may run into one or the other. *P. Wrightii* Gray, which is the same as *P. Utahensis* Eastwood, seems sufficiently distinct, but these with *P. puniceus* form a very closely related group.

PENTSTEMON EATONI var. UNDOSUS.

No. 5110ah. April 26, 1894, in red sand at St. George, Utah, 2700° alt.

No. 5289u. May 23, 1894, Johnson, Utah, 5000° alt., in red sand, among junipers.

Scabrous or short-pubescent, except the flowers and uppermost stems; lower stem leaves 3–5′ long, narrowly oblong, 1½′ wide, wavy on the margin.

This was also collected by Capt. Bishop in the same region in 1872. It grows among the junipers in gravelly soil throughout northern Arizona along the Colorado River and into Utah as far as the rim of the Great Basin.

PENTSTEMON CONFERTUS var. ABERRANS.

No. 5601i. July 6, 1894, Soldier Summit, Utah, 7300° alt., in gravel.

No. 5740. August 4, 1894, Fish Lake, Utah, 9000° alt., in meadows, in gravel.

Flowers about the size of *P. humilis* var. *breviflorus* or a trifle larger and plants with the habit of *P. humilis;* 1° high or less, not glandular, otherwise almost exactly as *P. confertus.* This variety abounds throughout the mountains of central Utah, being more common than any other form. It occurs most frequently in subalpine meadows and shady woods in rich soil.

BUDDLEIA MARRUBIIFOLIA var. UTAHENSIS (Coville, Proc. Biol. Soc. Wash., 7, 69). This seems like a narrow leaved form, as the wider primary leaves are absent in Coville's specimens.

Eriogonum nivale Canby seems to differ in no constant respect from *E. ovalifolium*.

Eriogonum Rusbyi Greene is *E. Jonesii* Watson.

ABRONIA MICRANTHA var. PEDUNCULATA.

No. 5101. April 26, 1894, St. George, Utah, 2700° alt., in red sand.

No. 5482m. June 22, 1894, Green River, Utah, 4500° alt., in clay.

No. 5183t. May 7, 1894, Le Verkin, Utah, in gravel, 3400° alt.

Peduncle double the petiole; fruit often 3-nerved, emarginate at both ends, often very large, 9″ long; flowers small; fruit often red.

This seems like a hybrid with *A. cycloptera*, but the latter is not known in the region. The bracts of both these species are ovate, small, 1–2″ long, abruptly acute or at least acuminate at apex.

ATRIPLEX SUBDECUMBENS.

No. 5745. August 6, 1894, Fish Lake, Utah, 9000° alt., in gravelly meadows.

This species is nearest *A. microcarpa* and *A. argentea*: plants slender, subdecumbent, annual, 2–8′ long, much branched at the base, mealy throughout, leafy throughout; leaves ovate to lanceolate, with a cuneate base, barely petioled, 6–9″ long, rather thin, entire; plants fructiferous throughout in small, sessile clusters: flowers minute, greenish; fruiting bracts 1″ long, rather cuneate-orbicular to obovate, apex crimped and obscurely dentate, or sometimes toothed, green, sides smooth, back nervose to muricate, seldom if ever with green points, bracts united below.

This grows in gravelly, dry patches in meadows, at 9500 feet altitude.

ATRIPLEX GRACILIFLORA.

No. 5697. July 30, 1894, Blue Valley, near Henry Mts., Utah, 4000° alt., in clay.

No. 5656e. July 20, 1894, Cainville, Utah, 4500° alt., in clay.

Annual, racemosely branched throughout and these branches again branched, forming a large tuft 1–2° in diameter, plants about 1° high, inner stems erect and outer ones ascending or spreading, sparingly mealy throughout, the younger parts more so, stems round; leaves all cordate-ovate, 10″ long or less, on a petiole 1″ long, entire, fleshy when fresh and rather thick even when dry; pistillate flowers few, scattered singly among the upper leaves; staminate flowers in a slender, dichotomous, bractless panicle, yellowish, in heads of 5 or more at the ends of short pedicel-like branchlets 1–2″ long, the main branches of the panicle 2–4′ long; flowers minute; fruit on a stalk 3″ long, bracts united except at the top and produced down the stalk to within ½″ of the base in a broad wing, also extended on all sides of the fruit into a green wing which is barely sinuous above, this wing is 2″ wide on the sides and 1″ wide at the apex of the fruit, fruit 5–8″ wide, orbicular to reniform, not warty nor appendaged, the body of the fruit is ovate to elliptical, 1″ wide and 2″ long, closely invested by the bracts which are separate only at the apex. This remarkable plant must rank near to the shrubby *A. canescens* in the fruit character, though so unlike it in all other respects.

This grows in alkaline soil on the flats bordering the Fremont River, where the soil is a very compact clay during the growing season.

ATRIPLEX CORNUTA.

No. 5481. June 22, 1894, Green River, Utah, 4500' alt., in clay.

This has much the habit of *A. graciliflora*, annual, branched chiefly below, erect, 6–12' high, very mealy and whitish, stems round; leaves ovate to deltoid with all three angles sharp when deltoid, often cordate, 1' long or less, thick, petioles 1" long; pistillate flowers in clusters of 1–3 in the lower axils, in fruit all on pedicels 2–3" long and often pendent; staminate flowers in small, sessile clusters in simple, terminal, leafy racemes, reddish-white; fruit forming a ball 2–3" in diameter by the innumerable horns or corrugated and lobed processes which cover the bracts completely and equal their bract lobes.

This frequents the same alkaline clayey soil as *A. graciliflora*.

ERIOGONUM AUREUM.

No. 6091. September 28, 1894, St. George, Utah, 2700' alt., in sand.

This is an intricately branched shrub, 1–3° high, widely branched and stout, with rounded top; the stems are rather short, and quite leafy to the base of the short peduncle; leaves elliptical, shortly contracted into a petiole 2–6" long, the whole leaf being 12–20" long, densely woolly below and much greener above, entire; peduncles 1–2½' long, then trichotomously and repeatedly branched, not angled, branchlets short; bracts all subulate and 1–2" long, but the upper ones minute; involucres and uppermost divisions appearing glabrous, but really minutely and sparsely woolly; involucres oblong, 1" long, divided about ⅓ the way into rounded, obtuse, erect lobes; flowers golden, 1" long, outer lobes oval, obtuse, inner lobes oblong, with a green midrib, lobes widely spreading; flowers with a nipple-like projection at the base and involucres rounded, not angled; the pedicels are exserted.

This is very close to *E. corymbosum* and *E. Thomsonæ.*

To this species I would refer two specimens collected by Rusby at Holbrook, Arizona, as var. *glutinosum*, being glutinous above; having oblong, crenate leaves, 2′ long; involucre somewhat angled; flowers smaller.

To this species I would refer as var. *ambiguum*, No. 1688, Coville, collected on the east slope of the Sierras, Inyo County, California, having narrow and entire leaves, peduncles 4–6′ long, small clusters of flowers and only secondary and very short rays, the involucres much angled and flowers very short, ½″ long.

ERIOGONUM RENIFORME var. COMOSUM.

No. 5036ao. April 12, 1894, near Hole in the Rock, 10 miles above Stone's Ferry, Nevada, in gravel, 1500˘ alt.

No. 5036bj. April 12, 1894, 10 miles below Mica Spring, Nevada, in gravel, 2000° alt.

Annual, 3–6′ high; leaves radical, 10″ long, and with a slender petiole 1′ long, the blade round or somewhat oblate, truncate to cordate at base, and fully developed ones emarginate at the apex; the younger leaves are very white, with long, comose, tangled hairs, the upper side less so or even green; inflorescence trichotomously branched, mostly at the very base of the stem; small bracts at least woolly within and with hyaline margins; involucres erect on pedicels 1′ long or less, over 1″ high, hemispherical, glaucous, with very short hyaline teeth; flowers very small and exserted, ½″ long, pubescent with rough, flat, short, scale-like hairs; flowers reddish or greenish yellow, outer lobes elliptical, inner lanceolate and barely acute; stems and pedicels glabrous and flowers many in each involucre.

This is readily separable from *E. Thurberi* by the leaves and flowers and the more open and slender habit; it is easily separable from *E. reniforme* by the pubescent flowers.

ERIOGONUM LONGILOBUM.

No. 5590j. July 5, 1894, near Price, Utah, in clay, 6000° alt.

Densely cæspitose in large perennial mats, very woolly-hairy throughout, except the glabrous tips of the flowers; leaves broadly oblanceolate, all on a petiole nearly as long as the blade, 1–2′ long, thick, margins inclined to be revolute or at least the leaf convex, obtuse; peduncle scapose, 1–2′ long, with 3–5 sessile involucres in a dense head; involucres parted nearly to the base, lobes 2″ long, leaf-like, triangular, erect, equaling the flowers; flowers abruptly contracted and with a minute prolongation above the joints; pedicels 1″ long, but the flowers seem to be sessile; lobes of the flowers obovate-oblong, rounded, barely erose, yellow, with darker or green midrid, often tinged with red, barely 1½″ long; heads not bracteate; akenes very woolly. This appears nearest to *E. villiflorum*, but somewhat resembles *E. ovalifolium* in habit and leaves.

RUMEX SUBALPINA.

No. 5957. August 29, 1894, Brigham Peak, near Marysvale, Utah, 10,500° alt., in gravel, in the bed of a subalpine stream.

No. 5893ai. August 23, 1894, near the head of Bullion Creek, above Marysvale, Utah, 11,000° alt., in gravelly soil, along the bed of a subalpine stream.

Erect in large clumps, 3–5° high, smooth throughout, stems 1′ thick or less, coarsely sulcate; root leaves 1° or less long, 2–5′ wide, on petioles nearly as long, oblong-lanceolate, smooth, entire, truncate at base or abruptly contracted, hardly acute at apex, petiole stout and marginless, stem leaves similar but narrower, uppermost often linear, none of the leaves at all sheathing, panicle narrow, 1–2° long, branches appressed, densely flowered; flowers on filiform pedicels 4–6″ long, pendent; fruit wings without tubercles, triangular to deltoid, 2–3″ long, coarsely reticulated, sometimes rhombic, but in that case always with a contracted slightly produced apex, always fimbriate toothed below, and in the broader-winged forms very conspicuously so nearly to the apex.

This grows in cold, gravelly, springy places along subalpine streams, at from 10,000 to 11,000 feet altitude, along with *Oxyria digyna*, *Polygonum bistorta* and *Aquilegia cærulea.*

CROTON LONGIPES.

No. 5213. May 12, 1894, two miles east of Leeds, Utah, in sand, at 3500° alt.

No. 5024au. April 5, 1894, west side of Copper Mine, in Beaverdam Mountains, Nevada, 3000° alt., in gravel.

No. 5149am. May 3, 1894, Silver Reef, Utah, in sand, at 3500° alt.

Shrubby at base, erect or ascending, 1–2° high, slender, freely branched, stems white-stellate, leaves oblong-elliptical, apiculate, obtuse, rounded or short-cuneate at base, 1′ long, on petioles nearly as long, densely or sparsely stellate below, often glabrous above; flowers in very short, umbellate racemes, on long, slender pedicels, 2–3″ long; staminate flowers many, 1″ wide, with triangular lobes; fertile flowers with pedicels subtended by linear bracts; bracts as long or longer than the pedicels; calyx of the pistillate flowers similar to that of the staminate ones, but nearly 3″ wide; fruit nearly round, 1″ long, densely white-stellate. This can neither be placed with *C. Californicus*, *C. gracilis* nor *C. Neo-Mexicanus*, though it is related to them all. It is nearest *C. corymbulosus*, but the flowers are smaller, the calyx lobes very different, the styles very short, the leaves different, staminate flowers long-pediceled, pistillate less so and erect, not reflexed. In *C. corymbulosus* the calyx lobes are oblong to obovate and shortly acute, fruit reflexed, flowers large, styles long, leaves acute and generally ovate, and nearly equally white-lepidote.

This abounds in sandy places, especially on drifting sand dunes in the valley of the Virgen and southward.

COMANDRA UMBELLATA var. PALLIDA (A. DC. Prod. 14, 636).

Comandra pallida A. DC., l. c.

There is no constant character separating this from the type that I can discover.

TRIGLOCHIN MARITIMUM var. DEBILE.

No. 5289. May 23, 1894, Johnson, Utah, 5000° alt., in alkaline clay.

Flowers simply racemose on the slender, weak stems, which are 6–12′ high; roots very thick; leaves all radical and short. This plant is much smaller than *T. palustre*, but in the character of the flowers and fruit certainly belongs to *T. maritimum*. Watson's *T. palustre* of King's Report is a taller form of the same.

This grows on clayey alkaline flats at Johnson, Utah, where no other plant will grow.

CALAMAGROSTIS SCOPULORUM. Densely tufted, about 2° high, erect, stems slender; leaves about a foot long, coarse, prominently striate-nerved, 3″ wide, flat, tapering to a slender point, very light colored as if glaucous, glabrous throughout except the nerves upwardly are very scabrous; ligule scabrous, 2″ long, entire and truncate to lacerate: inflorescence spicate, broadly linear 4–6′ long and about equaling the uppermost narrow leaf, 6″ to 1′ broad, occasionally a little lax, usually strict and dense; spikelets nearly white, appressed: rays about five, the inner ones 1′ or more long and the outer very short, pubescent with ascending short hairs, the nodes in the lower part of the spike about 9″ apart; spikelets about 2″ long, lower glumes spreading in flower, equal, the lower 1-nerved and nerve very prominent and green, ovate-lanceolate, simply acute, glabrous and hyaline except the very scabrous nerve, usually; upper glume the same but with two additional nerves extending only to the middle, occasionally these are wanting; palet barely shorter than the others, 2-toothed, lanceolate, very faintly 2-nerved, glabrous, simply acute, like the others very thin and hyaline; floral glume ovate-lanceolate, 4-toothed at apex, faintly 4-nerved, scabrous throughout, equaling the palet; awn attached below the middle, straight, shorter than the glume; hairs ⅔ the palet, sparse. This is manifestly allied to *C. sylvatica*.

No. 6075. September 25, 1894, Springdale, Utah, growing in clumps at the base of sandstone cliffs along the Virgen river, 4000° alt.

Poa festucoides. Tufted like *Festuca ovina*, perennial, erect 2–3° high, slender, minutely upwardly scabrous throughout except the spikelets; root leaves clustered and with enlarged short sheaths, the blade 3–5′ long, involute-filiform and tapering to a point, stem leaves similar but a little longer, sheaths half the internodes which are a foot or less long; ligules usually 2″ long and lacerate at top; panicle on a peduncle as long as the upper leaf, 4–6′ long and rachis as much more; rays single about 1½′ apart, filiform, widely spreading, proper ray 1′ long and half its prolongation as a rachis; raylets single, racemose, 3–4, 1′ or less long, branched above the lower ⅓, bearing 2–3 narrowly elliptical spikelets which are 4–5″ long, 1–1½″ wide, with about 5 florets, spikelets widely spreading and pendulous, all pediceled; lower glume 1-

nerved narrowly elliptical, acute, somewhat keeled, hyaline, 2″ long, nearly glabrous; upper glume 2½″ long, elliptical, barely acute, rounded, nearly smooth; flowering glume with two evident marginal nerves, two very faint additional inner ones, and three others which are visible only at the top, the midnerve being obsolete except at the very tip, oblong-lanceolate, with a short-acuminate tip, scabrous above and hyaline, but on the sides hyaline border very narrow, 3″ long, rounded; palet concave in center, with two green nerves, acute, nearly linear, equaling the flowering glume, hyaline; rachis of the spikelet with joints about 1″ apart, pubescent; spikelets somewhat flattened.

No. 5671. July 25, 1894, Mt. Ellen, Henry Mts., Utah, 10,000° alt., on open slopes, forming a very conspicuous and important part of the vegetation.

STIPE PINETORUM. Close to *S. Sibirica* Watson, Bot. Cal. Allied to *S. comata*. Densely tufted like *Bouteloua oligostachya* and *Aristida purpurea*, perennial, about 1° high, erect; root leaves 2–3′ long, filiform-involute, sheaths 6–12″ long thickened, brown; ligule obsolete; small wiry stems with about 3 leaves whose sheaths are longer than the internodes, 3–4′ long, with a very short filiform blade 1–2′ long; panicle linear, 6′ long or less, barely exserted; rays 2–3, appressed, the lateral ones almost none, the central one 3–6″ long and 2–3 flowered; lower glume 3-nerved, subulate, tapering into a thread-like tip, 4″ long, smooth; upper glume 1-nerved and narrower and as long, smooth; flowering glume 2½″ long, linear, pungent at base and sparsely long hairy, hairs denser above forming a tuft 2″ long: awn slightly twice-geniculate, 6–9″ long, glabrous.

No. 6023p. September 8, 1894, Panguitch Lake, Utah, 8400 alt., growing in open places among the pine forests.

STIPA ARIDA. With the habit and general appearance of *S. viridula*, also leaves and ligules; sheaths as long or longer than the internodes; rays in fives below, very short or almost none, appressed, with 3–4 spikelets, filiform, the whole, exclusive of the awns, being 1′ or less long; outer glume lanceolate-acuminate, 4″ long, very thin and hyaline, 3-nerved; inner glume 3″ long, narrower and acuminate, 3-nerved; flowering glume 2″ long, fusiform, very narrow, glabrous above and shortly-pubescent below; awn 2–3′ long, thread-like, flexuous but hardly geniculate, smooth; panicle 6′ long, hardly exserted, dense, wand-like.

No. 5377. Marysvale, Utah, on very dry talus slopes in shingle, 6000° alt., June 4, 1894.

ELYMUS SALINA. With the habit of *Sporobolus airoides*, but culms very different; forming coarse and very close tufts, 1–2° in diameter; stems much thickened below the fibrous leaf-sheaths, perennial, erect, 1–2° high; lower leaf-sheaths loose and somewhat enlarged, very coarsely nerved, nearly smooth, about half as long as the nodes; ligule abortive; leaves pubescent on the under side, especially so at the throat, thick, involute, acute, linear, 4–6′ long (the root leaves), stem leaves 2–4′ long, nodes about 3; inflorescence a simple, loose spike like *Agropyrum glaucum* which it much resembles; spikelets single at each joint, barely contiguous, placed flatwise to the rachis, 6–10; empty glumes 2, subulate, pungent, 2″ long, generally falcate, 1-nerved; spikelet pediceled, pedicel about ½″ long and very thick and stout, often very short; spikelets about 6″ long, 2″ wide and 1″ thick, 7–9-flowered: flowering glume ovate-lanceolate, shortly acuminate, about 4″ long, thick, indistinctly 5-nerved, smooth and rounded, with narrow, hyaline, slightly lacerate margin, palet 2-nerved, concave in the middle and with broad hyaline margin folded around the stamen, lanceolate, acuminate, bidentate, equaling the flowering glume, joints of the rachis of spikelet about 2″ apart. This remarkable plant has most of the technical characters of Hystrix, but is manifestly allied to *Elymus condensatus*, and may prove to be only a form of it.

No. 5447. Top of Salina Pass, Utah, 8200° alt., in clay, in rather alkaline soil.

EPHEDRA NEVADENSIS var. VIRIDIS (Coville, Death Valley Rep., 220).

Ephedra viridis Coville, l. c.

No. 5213d. May 12, ten miles below Kanarra, 4500° alt., on rocks.

No. 5590e. Head of Soldier Cañon, 6700° alt., in clay, July 5.

No. 5001k. At foot of grade above Bellevue, Utah, 3700° alt., among rocks, March 30.

No. 5163at. Silver Reef, Utah, in gravel, 4500° alt., May 4.

No. 5124. Diamond Valley, Utah, 4500° alt., on rocks, April 28.

No. 5110ao. April 26, St. George, Utah, 3000° alt., in red sand.

No. 5338ah. Marysvale, Utah, 6000° alt., in gravel, May 31.

No. 5297v. Pahria Cañon, Utah, 5300° alt., in red sand, May 26.

No. 5289y. Johnson, Utah, 5000° alt., in red sand, May 23.

No. 5476v. June 21, Grand Junction, Col., 4500° alt., in gravel.

No. 5663bt. Marvine Laccolite, Utah, 6000° alt., in gravel, July 23.

This plant, which I have collected in various places in

Utah and Nevada since 1880, has never seemed to me anything more than the normal form of the species, while the form on which the type *Nevadensis* was based is the less frequent one found in the valleys in less gravelly or clay soil. The variety abounds on sandy or gravelly hills, especially in rocky places and among rocks; it is even found on cliffs. It abounds throughout Utah as far north as the lower end of Salt Lake Valley and westward, and as far eastward as western Colorado.

Cylindrosporium glycyrrhizæ Hark.

No. 5572. July 3, Provo, Utah, in Slate Cañon, 8000° alt.

On *Vicia Americana.*

Spores somewhat smaller than as described, but not otherwise different.

PLEOSPORA UTAHENSIS E. & E.

No. 5902. August 24, Falls of Bullion Creek, Utah, 9500° alt.

On dead stems of *Eupatorium occidentale.* Perithecia scattered, erumpent-superficial, depressed-globose, sparingly fringed around the base with short, coarse, brown hyphæ, finally collapsing above, 150–250 m. diam., with a papilliform ostiolum. Asci oblong, rounded above, 75–90x20–23 m., paraphysate, 8-spored, with only a short rudimentary stipe. Sporidia crowded, biseriate, oblong-elliptical, at first yellow, uniseptate and constricted, then 3-septate, and finally about 7-septate, muriform and dark brown, 20–23x14–16 m.

This differs from *P. Richtophensis* E. & E. in its smaller, more distinctly rumpent perithecia and comparatively broader sporidia, and from *P. alpestris* E. & E. in its 8-spored asci and smaller sporidia.

Puccinia Pentstemonis Peck.

I.

No. 5739. August 4, Fish Lake, Utah, 9000° alt., in meadows.

On *Pentstemon confertus* var. *cæruleo-purpureus.*

No. 5015bg. September 7, Panguitch Lake, Utah, 8400° alt.

Same host.

II.

No. 6015bg.

Same host.

III.

No. 6002ax. September 6, Panguitch Lake, Utah. 8400° alt.

Same host.

No. 6015bg.

Same host, *æcidia* found on No. 5739 and 6015bg are the same as *Æcidium Palmeri* And., and while the teleutospores on No. 6002ax and 6015bg differ slightly from the description given by Peck, they undoubtedly belong to his species. No. 6015bg has the three forms well developed on the same leaves, and as we find no description of the uredo-spores, we make the following:

Puccinia Pentstemonis Peck. Amphigenous. I. Spots and spermagonia wanting; pseudosporidia usually single, but sometimes in small clusters, three or four times as long as broad, and from a much thickened base, which is often bright purple, upper part bright yellow, becoming white at maturity and splitting nearly or quite to the base; spores subglobose or oval, contents granular, epispore thick, slightly echinulate, 18–20x20–22 m. II, III. Sori small, round, scattered, nearly black; uredo-spores in the sori with the teleutospores, globose or oval, smooth, 15–18x18–26 m. Teleutospores broadly oval, somewhat constricted; epispore smooth, slightly but distinctly thickened at the apex, 27–30x20–22 m; pedicel hyaline, fragile.

UROMYCES LYCHNIDIS T. & E.

No. 5851. August 22, near Tate Mine, above Marysvale, Utah, 9000° alt.

On *Lychnis Drummondii.*

I, II, III on the same leaves. I. Spots indefinite; pseudosporidia hypophyllous, irregularly clustered on discolored areas, short, rupturing irregularly; æcidial spores irregularly rounded, epispores smooth, contents bright yellow, 15–18 m.

II, III, from the same sori; amphigenous, mostly epiphyllous, scattered, irregularly rounded, black or dark brown, bordered by the ruptured epidermis; uredal spores globose, light yellow, minutely echinulate, 12–14 m; teleutal spores ovate, dark brown, epispores not thickened at the apex, longitudinally wrinkled, about 15x20 m; pedicel about equaling the spore, hyaline, very fragile.

Puccinia aberrans Peck.

I.

No. 5064bi. April 14, Mica Spring, Nevada, 4000° alt.

On *Arabis arcuata* var. *perennans.*

No. 5163av. May 4, Silver Reef, Utah, 4000° alt:

On same host.

No. 5165. May 5, Silver Reef, Utah, 4500° alt.

On *Arabis Holbœllii.*

II.

No. 5165.

On same host.

III.

No. 5165.

Same host.

No. 5338ai. May 31, Marysvale, Utah, 6000° alt.

Same host.

No. 5354. June 1, Marysvale, Utah, 7000[c] alt.

Same host.

The three forms of this species all occur on No. 5165. The æcidial form is the same as *A. monoicum* Peck. and *A. drabæ* T. & E., and as the specific name *aberrans* has priority in publication, it must stand. An examination of the several forms on diffent hosts from nine localities show the following characters:

Puccinia aberrans Peck. I. Hypophyllous; spots none; pseudosporidia often covering the entire lower surface of the leaf, long cylindrical, lacerate at the top, bright yellow; æcidial spores subglobose, smooth, yellow, epispore thick, smooth, 20–28 m. II. In the same sori with the teleutospores; globose, epispore thin, smooth, 28–33 m. III. Usually hypophyllous, but with occasional sori on the upper surface; spots none; sori small, round, sometimes confluent, reddish-brown; teleutospores light colored, oblong, or oblong-clavate, obtuse or round-pointed, much thickened at the apex, constricted, smooth, lower cell very thin-walled, 40–50x20–25 m. Pedicel hyaline, about as long as the spore, quite fragile.

This occurs on *Arabis*, *Draba* and *Smelowskia*.

Puccinia globosipes Peck.

No. 5193. May 8, Le Verkin, Utah, 3000° alt.

II. III.

On *Lycium Andersoni*.

In these specimens the sori occur on both the stems and leaves, and both II and III are found in the same sori. The uredo-spores do not appear to have been collected previously, and are described as follows: II. In the same sori with III. Spores oval or obovate, epispore rather thick, sharply echinulate toward the apex and smooth at the base, 15–20x30–38 m.

Uredo Castilleiæ T. & E.

No. 5651. July 19, Capitol Wash, Utah, 5000° alt.

On *Castilleia affinis.*

Amphigenous; in small, round or oblong sori long, covered by the epidermis; spores oval or globose, reddish brown; epispore finely and distinctly aculeolate, rather thick, contents granular, 20–24x15–18 m.

Synchytrium fulgens Schrœt.

No. 6012. September 12, Panguitch Lake, Utah, 8400° alt.

On *Epilobium adenocaulon.*

This is referred here somewhat doubtfully, as the sporangia are much smaller than in the type, but it does not seem to differ otherwise.

Synchytrium caricis T. & E.

Amphigenous, in elliptical or oblong clusters forming distinct, reddish brown spots scattered over the entire leaf; sporangia numerous, globose or oval and often angular, light yellow, 12–15x20–25 m.

No. 5867a. August 23, Tate Mine, at the head of Bullion Creek, Utah, 1150° alt., on *Carex Pyrenaica.*

Erysiphe cichoricearum DC.

No. 5988. September 4, Circle Valley Cañon, Utah, 7000° alt.

On *Bigelovia graveolens.*

This form has been so determined by Burrill (N. A. *Pyrenomycetes*, p. 13), but, as he points out, it differs widely from the type in the greater number of sporidia. In these specimens the appendages are slender, delicate and hyaline; the reticulations of the perithecium are very small, the cell wall thin and delicate; asci numerous, 15–20 or more; sporidia uniformly 4–6, much smaller than in the type. The name *E. sepulta* has been provisionally proposed for this form by Ellis and Everhart,

Bot. Gaz. xiv, p. 286, but without an adequate description. The characters given above would suggest a relationship to *E. communis* rather than to *E. cichoriacearum*, but no forms of that species have so far been reported on *Compositæ*.

CYLINDROSPORIUM ACERINUM T. & E.

No. 5917. August 27, Marysvale, Utah, 8000° alt.

On *Acer glabrum*.

Spots yellowish, small, irregular, not bordered; acervuli epiphyllous, large, black, scattered; spores strongly curved, hyaline, granular, 35–40x1½–2 m.

LOPHIDIUM INCISUM E. & E.

No. 5754. August 9, Fish Lake, Utah, 9000° alt., in gravel.

On *Symphoricarpus oreophilus*. Perithecia erumpent-superficial, sometimes subseriate or subconnate, depressed globose, black, ½ to ¾ mm. diam., at first rounded at the apex, without any appearance of an ostiolum, then gashed or cleft across the top and the broad compressed ostiolum rising from the bottom of the cleft, or sometimes (in the early stage of growth) 3–4 radiate-sulcate-cleft at the summit. Asci cylindrical, 120 to 150x12–14 m.; short stipitate, 6–8 spored, with abundant, filiform paraphyses. Sporidia uniseriate, oblong-elliptical, 5–6 septate and muriform, 23–27x12–14 m. This has the habit of *Cuburbitaria*.

All localities given in this article are in Utah, unless otherwise stated.

ERRATA.—Page 613, line 14, for *Astragalus foliosus* read *A. foliolosus* (Gray) Sheldon: erase "Sheldon, not Gray."

Page 622, line 21, for *Erysimum pumilum* var. *perenne* read *E. aspeum* var. *perenne*.

Page 632, next to last line, for "woods" read "woody."

EXPLORATIONS IN THE CAPE REGION OF BAJA CALIFORNIA IN 1894, WITH REFERENCES TO FORMER EXPEDITIONS OF THE CALIFORNIA ACADEMY OF SCIENCES.

BY GUSTAV EISEN.

[With Plates lxxii–lxxv]

INTRODUCTORY.

Until quite recently the peninsula of Baja California, Mexico, and especially its Cape Region, has been to natural science in general a terra incognita. This refers especially to all lower forms of animal life, although in the higher classes very little work had been done, and none that could be in any way called exhaustive. Of some groups a few species had been collected and described, but the real scientific points regarding the connections of its fauna and flora with those of other regions in the immediate vicinity were almost entirely unknown.

There remained then and remains yet an immense amount of facts to be recorded, collections to be made of groups of animals and plants, of the nature of which science had no knowledge. The zoological features of Baja California, and the very many and great questions of general interest connected with them remains yet to be established.

In a thorough exploration of this so unknown field the California Academy of Sciences is the pioneer. The many and various papers by specialists upon the zoology of the Cape Region of Baja California, which now appear and soon will appear in the Proceedings and Memoirs of this Academy are evidences that these statements are no mere words. Through this work now being done by the Academy, the latter is gaining a most enviable reputation among the scientific bodies of this country, and is establishing itself on a footing equal with the best.

An opportunity to study and explore a country so little and so erroneously known as our Baja California does not frequently offer itself. Such an opportunity is in our days a very rare one, so rare indeed that it is safe to say that this is one of the very few left, and scientifically almost any other country was within a few years better known than this peninsula at our very doors. Indeed when the Academy began these explorations, there remained no other country within our reach which was less known, more misjudged, less understood, and about which more conjectures were made and less real facts known.

We could not possibly have chosen a better field, none richer, none less worked, none more interesting, none so inexhaustive, even if it had been in the Academy's power to visit more distant lands. The Academy could have had no other country entirely for itself.

The great advantage of a thorough exploration of the Cape Region is evident when we remember that the value of every scientific collection consists in the completeness of series of the species from a certain limited territory, and not in the possession of scattered species or incomplete series from widely separate localities, no matter how attractive and otherwise interesting these specimens might be.

To the Academy of Sciences then belongs the honor to have grasped this opportunity at a time when the scientific bodies and academies of the world are vieing with each other to be the first ones in new fields of explorations and scientific researches. Our terra incognita of Baja California and Mexico has been a most fruitful one. Large collections have been made and while their value may not always be money value, it is safe to say that even in this respect no other outlay in this direction could possibly have brought more satisfactory results.

Nevertheless, a few more words of explanation may be necessary. The question has often been asked why have not the energies of the Academy been more or exclusively devoted to home work. The question is appropriate, but the answer is not a difficult one. California and its sister States are now being rapidly explored, and their fauna and flora are becoming correspondingly well known. With two universities of high standing, with several smaller and sectarian institutions of learning, with a number of private scientific societies, and with a very large number of private collectors and students of natural history, it is thought that the State is well provided for, and that within a very few years all the various groups of animals and plants will be in a zoological sense very well known, if not pretty well exhausted. With Mexico these conditions are entirely different. In the sparsely settled territory of our sister republic naturalists are yet few, and explorations most incomplete. While thus our own country is having its fauna and flora rapidly described, it has become more and more important to know their relations with the large Mexican region. From this region our United States has received a large part of its fauna and flora, tropical and semitropical species, which in course of their immigrations to the north, have been more or less modified, according to the requirements of their new surroundings. On the other hand, northern species have penetrated into Mexico, and undergone more or less changes there. In other words, the interchange of faunas and floras which has been carried on since the ice age between the two countries is one of extreme interest, and the many problems connected with it are well worthy of our greatest energies and studies.

Mexico of to-day, under a most enlightened government, is making tremendous strides forward in civilization and national prosperity. Immigrants are pouring in from all sides, but with them come unhappily foreign weeds and plants, and foreign animals, which are bound to, in a few years, considerably change the aspect of the fauna and flora of the country. In the vicinity of the cities near the coast, or along the highways, we find everywhere foreign plants and animals well established, which are driving the native ones further back. If we, therefore, wish to learn the original aspect of Mexican animal and plant life, before man begins to interfere too much, we must commence our studies at once. In a few years it will be too late, as much which is the most interesting now will then have changed or been exterminated in the same manner as in this and many other countries. But there are besides other reasons why explorations in Mexico have been considered desirable, one of these is that in that country animal and vegetable life is a hundred times richer and more luxuriant than with us, and the collections acquired are thus correspondingly large. Another point of very great importance is, that explorations in Mexico are comparatively very cheap, and with the very limited means at our disposal we have been able to bring together collections the size and value of which could not have been duplicated in our own country in ten times the time and with ten times the cost. It has then simply been a question of the largest collections with the least possible cost.

Our last three expeditions have been greatly assisted by the courtesy of the Mexican Government, which, through its Ministro de Hacienda, Hon. José Y. Limantour, has caused special facilities to be extended to us in all places visited, without which, it is safe to say, our success would have been correspondingly less, and been difficult to achieve. The expedition of 1894 was also very materially assisted by the kindness of General Romano, Governor of the Territory of Tepic.

The object of this paper is to furnish a short account of the last expedition sent out by the California Academy of Sciences to the Baja California Cape Region, and Tepic on the main land of Mexico; to furnish a more detailed map of the said Cape Region, as well as a general map ot the peninsula of Baja California.

The need of a special map of the Cape Region has been apparent for some time. The localities visited by the members of the various expeditions were only indifferently marked down on any previously existing map, and these mostly incorrectly located. The rivers and creeks were nowhere even hinted at and the mountain regions were everywhere found only indicated.

The Academy is constantly asked to furnish copies of the map compiled by T. S. Brandegee to travelers, miners, merchants, scientists, etc., intending to visit Baja California. Of late it has not been able to fill these requests, because all the maps at its disposal were bound in the volumes of the Proceedings of the Academy.

A short reference will also be made to previous expeditions by the Academy to these localities, to their results, both as regards collections, new species and types, as well as the papers published by various specialists describing the collections thus made.

In the summer of 1894, the California Academy of Sciences appropriated money for an expedition to Mexico, the main object being to collect various forms of lower animals for scientific study. This expedition consisted of Mr. Frank H. Vaslit and the writer. The expedition started from San Francisco in August; spent one month in exploration and collecting in the mountains of the Cape Region of Baja California, then crossed over to

Mazatlan, from there to San Blas, thence to the City of Tepic, inland. After a month or more exploration in this vicinity, the members of the expedition returned via Mazatlan, Guaymas, etc., to San Francisco in December.

This, however, was not the first expedition of its kind, the Academy having previously sent out five different expeditions, the writer being a member of two of them, having thus had occasion to visit the Cape Region of Baja California for scientific purposes three times.

The expeditions have received many and constant courtesies from the Pacific Coast Steamship Company and its agents in Mexican ports, and also from the officers of their steamers, particularly from Captain John von Helms and Purser W. A. Childs.

TEMPERATURE AND CLIMATE.

As might be expected from a small peninsula situated just within the boundaries of the tropics and surrounded by water, the temperature is a moderate one. All along the low lands up to 800 feet or thereabouts, we meet at no time of the year with frosts. In the San José Valley the frost free belt extends as far north as La Palma, but as the valley steadily but slowly rises towards the upper end, we find that at Caduaño and Miraflores during the cold winter months, January and February, light frosts may now and then occur. In the vicinity of these places, situated about 1000 feet above the ocean, frosts have been known to kill back the native vegetation, especially the yellow flowering shrubs known as "palo de arco" (*Tecoma stans*). But from San José to La Palma no such frosts have ever been noticed. All along the coast, however, on one side to Todos Santos, on the other to La Paz, frosts at any time are unheard of. Similarly, as we ascend in the mountains, light frosts may be expected. Both at El Taste and at Sierra Laguna ice is frequently formed during the night, even in the month of March. But from that month to December, even at this altitude, of from 4000 to 6000 feet, no frosts occur. As during the cold months rain seldom falls, snow is never formed, in fact, snow has never been noticed in any part even on the highest peaks. The frost evidently only settles in low and damp places. Even during an occasional rain storm, in January, it is never sufficiently cold to precipitate snow on the highest peaks, some of which undoubtedly reach 8000 feet, the highest one measured by us being over 7000 feet. As a consequence of this favorable temperature, tender, tropical, horticultural plants, such as pineapples, coffee, etc., thrive anywhere along the low coasts and up the San José Valley as far as La Palma or higher. The nights are always tempered by breezes from the surrounding ocean, and even the days are never excessively hot, while in winter the temperature on the lower levels is such that blankets are needed during the nights, and light overcoats during morning and evening. From October to June the days, even the warmest, are very pleasant and most enjoyable. The warm months are June to September, though by the middle of September the heat is rapidly decreasing. Even during the warm first half of September, the warmest part of the year, the temperature seldom rises above 90. During September we thus found generally about 88 Fah. in the shade in the house, during the hottest part of the day, while at night the temperature in the house averaged 82 Fah., cooling off towards morning. The change is thus slow and gradual. It must be remembered that this was considered as exceptionally warm, no such hot weather having occurred for years.

Surrounded as the Cape Region is by the ocean, it enjoys an exceptionally clear atmosphere. The sky is always, except of course during the short rain storms, marvelously clear and brilliant. We observed at no time that hazy, yellow dust which is so common in inland countries generally where summer rains are absent. Even from the lower levels, whenever the ocean could be seen, the horizon always presented itself as a sharp, well-defined line, and one of the most beautiful sights is the advent of a distant rain storm with thunder clouds, rising above the horizon. Even the most distant clouds are then seen as sharply as the nearest ones, the whole offering a most beautiful perspective panorama. This is also the character of the whole Pacific Coast of Mexico and Central America.

In the various sierras of the Cape Region this clearness and brilliancy of the air is increased, and we could not help but think that if ever a very superior place for an observatory is desired, the sierra of the Cape Region, especially that of El Taste, is pre-eminently one that I think cannot be surpassed in any part of the world. The cause is, as I have said, the nearness of the ocean, which on almost every side surrounds this mountainous country. This prevents the dust from distant plains from reaching here, while high temperature prevents condensation of the moisture in the shape of fogs. Fogs are entirely unknown in the Cape Region.

SANITARY CONDITIONS.

The Cape Region of Baja California is one of the very few places where the various conditions of temperature, moisture and other climatic conditions are almost perfect. The humidity of the air is never great, and still never so low as to become irritating to the lungs. The long distance to the mainland, and the directions of the prevailing winds are such that the dust and smoke which they might carry along are precipitated long before they reach the Cape Region. The pureness of the air is noticeable at once and gives the visitor or invalid an indescribable feeling of pleasure and relief.

Malarious fevers are almost entirely unknown. The water in the creeks and springs is exceptionally pure and good tasting. The even temperature, the slow changes and small variations in the heat of day and night, winter and summer, are especially favorable. If we add that the three dreaded Mexican diseases, smallpox, yellow fever and cholera, have never visited San José del Cabo, it will be seen that we have good reasons to consider the southern part of the Cape Region as one especially favored, a real nature's sanitarium.

RAINFALL.

The rainfall in the Cape Region is much more abundant and certain than in any other part of Baja California. Located within and on the border of the tropics, the region receives its rain at the time and from the same place of the balance of tropical Mexico, that is during the summer months and from the south, the contrary in many respects being the case with that part of Baja California lying to the north. But the rainfall in the Cape Region commences later than on the opposite mainland and is considerably less. The rain commences in July or August and lasts until October or November. By the end of that month, or before, the summer rain is generally over; there has seldom been December rain. But in January there may be another period of rain from the north, the tail end, so to say, of our Alaska cyclones, which may in a couple of showers precipitate several inches of rain and greatly to the benefit of the country. The grass starts then anew and many shrubs and plants burst out in leaves and flowers. But this winter rain is generally scant and of short duration. As regards the summer rains, September is the wettest month. There is then sometimes rain for two or three days in succession in the lowland, while in the high mountains the thunder may be heard daily for weeks and the rain be seen precipitated in showers when the mountains are viewed from the plains below. In the various parts of the Cape Region the rainfall is most unequal. In a general way it may be said to increase in quantity from the north southwards and from the lowlands towards the sierras or mountains. Thus in the vicinity of La Paz the rain is less than two inches a year. Here there may be no rainfall for three years except a slight drizzling, not enough to start the grass. But even a few leagues southward the rainfall has so increased that there is yearly pasture for the stock. This character extends along the gulf coast southward, the rain always being much scarcer along the shore than a few miles inland, but it gradually increases towards the Bay of San José del Cabo.

The valley of San José del Cabo is about forty miles long by two and three wide, and is by far the most abundantly watered on the whole peninsula. The upper end of the valley is at Miraflores, which is about 1000 feet higher than the ocean or San José.

At this place the rain is every year abundant and certain, and while no rain-gauge has ever been used it is safe to say that the summer rainfall amounts to about twenty inches, while at San José it probably does not reach twelve inches. At Miraflores the rain commences much earlier, sometimes several weeks earlier than at San José. But even within the very narrow valley of San José the rainfall is remarkably local. Thus the rain may be seen to reach from the north Santa Anita, Santa Catarina and other places within ten miles of San José several weeks before it reaches San José. This rain from the north seldom reaches San José. This latter place receives its local rains from the southeast, where the clouds are seen gathering over the ocean shortly before the rain begins.

As regards the rainfall of the mountains, two points are of interest. The western crest of the high mountains receives more rain and is more moist than the eastern crest, but owing to the steeper slopes and quicker drainage and greater evaporation from north winds, the creeks on the western slope are of much less duration and much smaller in volume than those on the eastern side, while all unite in the San José Creek or River. Another point is that the central part of the sierra receives more rainfall than the southern and northern parts.

As a consequence of this the western slope of the high sierra is more moist than the eastern slope during the rainy season, but when the rains are over it dries up much quicker and is thus able to sustain much less large vegetation. The greater fertility of the valley of San José is due to the course of the San José River which runs in a general direction from north to south, parallel with the sierra, thus collecting the combined water from all the creeks from San José to Miraflores, while on the western side there is no such central stream to collect the waters from the mountains, each creek emptying independently in the Pacific Ocean at floodtime, while in the dry season the waters sink before they reach the coast.

I have already stated that Miraflores, at the upper end of the San José Valley, is situated at the northern divide of the valley. The Miraflores Creek is the most northern one but one which empties into San José River, as all creeks from the sierra north of this place empty into the Gulf or into the Pacific. But strange enough, this divide of streams, although not much over 1200 feet, is also a divide of rainfall. A league or so—five, six miles

—north of Miraflores, which, as we know, possesses the greatest rainfall of any place on the lowlands of Baja California, or in the vicinity of Santiago, the rainfall suddenly dwindles down to five or six inches. While thus the pasture is most luxuriant at Miraflores, it is most scant in the vicinity of Santiago. At Agua Caliente the greater vegetation is due to the fact that it is situated higher up in the mountains than Santiago.

As to the actual rainfall in the mountains nothing, of course, is known with certainty, but it is safe to assume that from El Taste to Sierra de La Laguna it is not less than 20 inches, while in the vicinity of Santa Genoveva it probably reaches 25 inches or more.

Towards the south the rainfall gradually becomes less, and after leaving the region of El Chinche and San Nicholas and San Felipe, it is decidedly scant, and along the coast at Cape San Lucas, and from there on to La Palmilla it is probably not over five inches yearly. In the accompanying map I have endeavored to show the distribution of rainfall in the Cape Region.

While in the northern part of Baja California the rainfall is precipitated during the cold season, or in the winter months, it falls in the Cape Region during the warm season or summer months, or at a time when it can be immediately utilized by the then growing vegetation. In the north the winter rainfall is mostly lost to vegetation, as it is only the part that is stored up as snow and moisture in the soil until spring that can be utilized for vegetation in general, the exception, of course, being some grasses and herbs which sprout as soon as the rain commences, though they generally flower later in the spring.

In the Cape Region this is different. The very first shower causes a marvelous change in the country. The hills and slopes and much of the valley or mesa lands are covered by a dense mass or jungle of shrubs and low trees, which in the mountains are much higher. During the dry season they become mostly bare and dormant, but with the first shower they cover themselves with leaves and flowers, and the whole country assumes an appearance of marvelous beauty and verdure. There is thus little moisture actually wasted, none during a storage period for future use, as in the northern part. But, on the other hand, when the rain ceases, the vegetation dries up very quickly, and, except in the high mountains, no indications are left of the beautiful foliage and flowers of a few months previous.

The slow, drizzling rains, common in all temperate countries, including the larger part of Baja California north of the Cape Region, give place in the latter to tropical showers, which suddenly gather and in a few hours may precipitate several inches of rain, after which the sky clears and the sun comes out warm.

RIVERS, CREEKS, ETC.

As might be expected, from the increased rainfall in the Cape Region, we also meet here with more creeks and springs than in any other part of the peninsula. In the sierra water is found in almost every gulch, even during the dry season, while in the rainy one every creeklet is running full. But strange to say, there are in this sierra few real springs. The watering places during the dry season are invariably in the otherwise dry beds of creeks and gulches, except, of course, in places where the creeks run continually. Real springs, such as we are accustomed to find in countries where the winter's rainfall is stored up in the soil for summer use, springs gushing out of the soil or from under rocks on the hillside, etc., are almost entirely absent, or at least very scarce. I cannot remember having seen more than three in my travels in the Cape Region. One of these is at Agua Caliente, north of Miraflores, where there is a fine hot sulphur spring near the bottom of the creek; the other is the famous spring at San Bartolo, between Santiago and La Paz. This spring is one of the marvels of the peninsula, and the finest spring I have seen in any country. It courses out through one or two holes from under an alluvial and glacial mesa, in probably about five cubic feet of water per second. It is of exceptional purity and coolness, always retaining its low temperature and its even and undiminished flow, winter and summer. The spring empties in an otherwise dry gulch and cañon known as San Bartolo Cañon. In the dry season the cañon or creek bed is dry immediately above the spring, and also some four miles below it. But for four miles or more the flow from the spring is sufficient to cause the appearance of a small stream in the bottom of the cañon, besides giving sufficient surplus to constantly irrigate several hundred acres of land, terraced on the steep slopes of the cañon. The cañon itself is very narrow and precipitous, its sides are terraced in places, and everywhere are seen fields of sugar cane, bananas, oranges, etc., and other tropical fruit trees and plants, making this one of the most charming spots imaginable, in great contrast with the surrounding hills which are quite barren, as compared to those around the San José Valley.

Another spring is found at La Palma, nearer San José, and a few others are scattered about the country, here and there.

The San José River is the largest water course in the Cape Region. It is some thirty miles long, or, if we count in its main tributary at Miraflores, it may be said to be about forty-five miles long. It receives during the rainy season a number of tributaries from the sierra on the west side. But in the dry season most of these do not reach the main river in the way of surface water, but there is always a sub- or underflow, which keeps up the water in the main stream. The water in this is everywhere taken out for irrigation, but the underflow comes again repeatedly to the surface, so that at the outlet of the "estero" at San José there was 500 cubic feet of water running to waste into the ocean in the month of March, while three or four miles up the valley the river had only the appearance of a large ditch, entirely under control for irrigation. But in the rainy season the tributaries to the San José River come down like torrents from the Sierra and after an unusual "aguacero" even the San José River with its shallow bed, half a mile wide, may be impassable for three or more days, sometimes even for a week. The water for irrigation seldom fails, and only once in twelve years has it become alarmingly scarce so that crops were a partial failure. This refers also to the annual rainfall with the same force.

On the east the San José River receives no tributaries, on the west however there is a number smaller and several more respectable ones. The latter are counted from San José northward: Santa Rosa; San Lazaro (at Santa Anita); San Miguel; San Ignacio (at La Palma); Caduaño. Miraflores and San Bernardo.

The San José River irrigates a great many thousand acres from one end of the valley to the other, but its waters are badly managed and much wasted, and could, if properly cared for, irrigate thirty times more land than at present.

The only other permanent river in the Cape Region which always reaches the sea is the Todos Santos River. This river heads in the Sierra Laguna and from there runs straight down to the Pacific. It is used for irrigating several thousand acres of bottom land at Todos Santos, the third important town in the Cape Region. Other creeks of importance on the Pacific side are San Jacinto and Palmar to the south and Carissal to the north of Todos Santos, but while they may be raging torrents, dangerous and impassable in the rainy season, they dry up during the dry season and flowing water is found in their beds only high up in the sierra, while near their mouths sundry lands may be irrigated from seepage water and underflow. The country around El Chinche, Calaveras, San Felipe and San Nicolas, all south of El Taste, form the watershed of a large cañon heading for Cabo San Lucas. But I understand that the waters seldom reach the Pacific Ocean, at least around San Lucas there is no stream, though much of the land appears subirrigated.

As regards the quality of the water in the San José River I may remark on its exceeding purity. It is remarkably brilliant and pure, free from sediment and quite crystaline in appearance. It is good tasting and very healthful and one of the best waters I have tasted, though much inferior to that of San Bartolo.

The ridge north of Miraflores divides the waters of the San José Valley from those of Santiago. The Santiago Creek, which heads between San Bernardo and Sierra Laguna, somewhere around Chuparosa, runs directly into the Gulf, when it runs at all. But generally the waters of the streams stop in a lake at Santiago, the largest and probably the only lake in the Cape Region. It is about one-half mile long and one-eighth mile wide, and it never dries up. The waters of this river are also used to irrigate considerable land, but in a crude and most unsatisfactory way.

Besides these creeks which show constant water, there

are hundreds of others, which consist merely of dry cañons, with flat dry river beds, giving undisputed evidence of a former period of great rainfall, which probably has been constantly diminishing since the great ice age.

THE SIERRA.

The most interesting part of the Cape Region is the great sierra, which towers above everything, and which imparts its character to the whole country, whether it is seen from aboard the vessels, far out at sea, or from the high mountain crests of the sierra itself. The sierra may be said to begin slowly rising in the vicinity of Cabo San Lucas and ending immediately north of Sierra Laguna, but the true and high sierra proper begins with Mt. Troyer and El Taste in the south, and ends with Mt. Limantour in the north. This sierra, which possesses no general name, consists of a granite mass or upheaval, very precipitous on the western side, and little less so on the eastern side. Thus the highest points are situated nearer the western than the eastern side. The eastern side of this sierra ends at or borders on the San José Valley, while on the western side it reaches the Pacific, with a slight rise near the shore line. There is no main crest or backbone running north or south. On the contrary, the sierra is composed of a number of ridges running parallel east and west, and separated by passes 3000 to 4000 feet high, while the high peaks of the Sierra reach 6000, 7000 and possibly 8000 feet. This feature of the sierra makes it impossible to travel with pack-animals any great distance north and south. Thus, if we are once landed at El Taste in the south, and wish to reach Santa Genoveva and Sierra Laguna in the center and north, it is absolutely necessary to first descend to the plains, and then to ascend the mountains again at another point. The many ridges sloping down to the east are as a rule very precipitous and very narrow, with crests so narrow that if one should fall from these he might from the same point tumble down to either side.

The sierra is most imperfectly known. Few, if any, of the educated people of the region have ever visited the higher mountains, which to them are a terra incognita. Only those peaks which can readily be seen from the valley have been named, and of them only those which appear as very prominent landmarks. The interior peaks are entirely unknown, and many first explored and ascended by us, have not previously appeared on any map, nor were they designated by names by any of the inhabitants in the vicinity.

With the right possessed by every original explorer, we have named some of those peaks, as will be stated further on in a more detailed description. Each one of the many ridges which slope down towards the east, north and south are known as or designated as a "Sierra." As these names of the various sierras have not previously been mentioned in print, nor appear on any maps, I will here enumerate them.

South of El Taste and Cerro la Calavera we have three distinct groups of sierras: El Chinche, San Felipe and San Nicolas, the position of which may be seen from the map. If we again start from San José towards the north, the various and principal ridges or sierras are as follows:

Cerro la Calavera, with Mt. Troyer at the upper or western end.

La Ballena, with Mt. Molera and El Taste or Candelario at the upper or western end.

San Lazaro, with the highest peak on the eastern side.

El Coyote, Los Angeles, H[illegible], Cajoncito, La Communidad, San Miguel, San Ignacio, Cerro Blanco and

San Pablo, with the very high peak of Santa Genoveva at the western end. This is the highest peak of the Sierras, probably reaching near 8000 feet.

San Bernardo, Sierra de la Laguna, which again is a cluster of very high peaks, comprising La Aguja or El Picacho, Porfirio Diaz, Mt. Limantour, San Rafael, etc., while of smaller ridges there are El Molero, Sirvuelar, La Torra, San Leonicio, all situated between San Bernardo and Sierra la Laguna proper.

Of this very large number of sierras we have only explored with any accuracy the most southern and the most northern, viz.: El Taste and Sierra Laguna. These parts resemble each other in the one respect that they contain each a level flat or meadow surrounded by higher peaks and wooded hills. El Taste and Sierra Laguna signify in each case not a high peak isolated from others, but merely a collective name for a group of mountains around these two meadows. El Taste is undoubtedly one of the few Indian names that has survived in the Cape Region. In Mayo, the Sonora dialect, it means a flat, level piece of ground, where the Indians run their horse races. The level meadow in El Taste is also known as La Carrerita, meaning the same thing in Spanish. It consists of one larger and several smaller flats, none being over thirty acres in extent. To the west and several miles distant is a peak, which on the Coast Survey maps is marked as the El Candelero, but which name is not known by any inhabitants of the district. On that account we found it best to retain for this, the most visible peak from the west, the name El Taste. My aneroid barometer showed the peak to be 5500 feet high. It is the last high peak to the west, very precipitous towards the Pacific and quite a landmark from the coast. On the east side of La Carrerita the sierra rises gradually into a very steep narrow ridge crowned by three rounded peaks, the middle one of which we estimated at 6200 feet, we having not been able to reach the very highest point. We named this mountain Mt. Molera, after Mr. E. J. Molera of our Academy. This mountain can only be seen from the San José Valley half way between Santa Catarina and Santa Anita. The view is from other points covered by the much lower La Ballena, etc.

On the south side of El Taste, Mt. Molera and La Ballena there runs a respectable creek towards the east and into the San José river. It heads up on El Taste and separates from this peak another of prominence but somewhat lower. This peak, which is the most southern one of all the high peaks in the Cape Region Sierra, we named Mt. Troyer, after Mr. Carlos Troyer of our Academy. The peak is somewhat lower than El Taste, about 5200 feet. It is situated about due west from La Calavera.

On the southern slope, on the ridge between El Taste and Mt. Molera, is situated a ranch, Santo Domingo del Taste, 3200 feet, and further down on the creek is a fine camping place known as Corral de Piedras, 2000 feet. From this place San José may be reached in one day's ride.

On the northwestern slope of El Taste there is another camp El Saltillo, 3200 feet, one of the best and most interesting for the naturalist. Further down, below the sierra, is the Rancho San Jacinto.

Sierra Laguna.—By this name is understood in a general way the most northern part of the high sierra north of San Francisquito, the latter never being included in the Sierra Laguna, though it is not far distant. The name of Laguna is derived from a lake or lagoon which formerly existed there, but which some fifteen or more years ago broke through and emptied into the creek, which carried off its surplus waters to the Gulf. To-day the lagoon is dry, and forms an oblong somewhat irregular flat, on account of its dryness not worthy of the name of meadow. Through its center courses a tiny spring or two in the deeper channel which at the rainy season, or after heavy storms, undoubtedly has the appearance of a brook, but which at the dry season becomes almost entirely dry, most of the water which comes from the upper end sinking before it reaches the center of the flat. The flat contains probably about a hundred acres. About one mile or so northwest of the flat is situated one of the most prominent landmarks in the Cape Region, the Picacho or La Aguja. This is a high needle-like peak, almost perpendicular on all sides, but especially so on the west side, where it falls about a thousand feet down in the cañons and slopes below. The top is bare and narrow, like a sugar loaf. A few hundred feet back of it, to the east, is another but less perpendicular cone, partly covered with trees and vegetation, and about fifty feet less high. My aneroid indicated the height to be 6200 feet. But two or three miles further east, on the southeast side of the former lagoon, is situated the highest mountain in this particular sierra. It being unnamed and not especially designated by any of the inhabitants on either the Todos Santos or the Miraflores side, we named it Mt. Porfirio Diaz. Its wanting previously a name may be accounted for because its top is partly covered by vegetation, and it is thus not as prominently visible as the Picacho, but it is almost a thousand feet higher, or about 7050 feet. Right opposite, on the north side of the Laguna, the sierra crest is a succession of bold and rough tops, the highest one reaches about 6000 feet. Unhappily the aneroid was left behind at the ascent, but I estimated the height at fully 6000 feet. We named this Mt. Limantour.

The ascent and descent to and from Sierra Laguna is one of the most arduous in the Cape Region, especially when pack-mules are to be brought along.

Geological Features.—Although no special attention was given the geological structure of the Cape Region, a few observations taken on the road may not be without interest. The main sierra from El Chinche to Sierra Laguna, and beyond it to Triumfo, consists of a granite upheaval, which almost everywhere shows signs of the glacial period. This is especially evident on the east side. Here, all along from San José to Rodeo, we meet with enormous morains, which all run more or less parallel from west to east. Especially are the morains prominent between Miraflores and San Bartolo. At the lower end of the cañon of San Bartolo are large steep morains, known as the Quebradas de San Bartolo, consisting of enormous boulders heaped on the top of each other, several hundred feet high. The coast mountains east of San José River and from there on to La Paz in the north, appear to consist chiefly of volcanic stratified red rocks. East of San José, however, comprising a district from near San José River and running east, we meet with hard, crystalline nonfossiliferous lime formation. This formation gives to the country an entirely different aspect; the mountains, instead of being rounded, have the form of table-mountains, crowned by very sharp, needle-like, pyramids. How far this region extends northward I am unable to say.

I think there is every evidence that the whole of the Cape Region is in a state of upheaval, and probably has been so ever since the end of the glacial period. At Magdalena settlements I found successive sea beaches several hundred feet high, with the same shells as now living in the ocean below, in very good state of preservation.

At the end of the glacial period the sierra of the Cape Region probably consisted of a low island, with morains ten to fifteen miles long, sloping down to the sea. These morains did not cross the narrow valley of San José, nor did I find any remains of such morains east of this valley, though I may remark that my explorations in this region have been very imperfect.

RELATIONS AND ORIGIN OF THE FAUNA.

From our geological and other observations, it is evident that at the end of the great ice period the Cape Region of Baja California existed as an isolated island, separated by a broad sound, perhaps several hundred miles wide, from the main part of the peninsula, while from the main land of Mexico there probably existed the same distance as at present. This island must have had little or no animal life, there being an entire absence of lowlands on which a milder climate would have made it possible for animal life to subsist and retain itself, while the higher mountains were wrapped in snow and ice. This rocky island must have been several thousand feet high, with no bare ground exposed to the warmth of the sun. As the ice melted away and the soil was exposed, the land gradually rose, bringing with it a lowland surrounding the mountains. The first immigrants of animal life must have been temperate forms, which again, as the climate became warmer, ascended to the mountains, while the later and more tropical forms remained in the lower lands.

To begin with, only such animals could have immigrated to the Cape Region as were able by some means or other to cross the ocean from the mainland. But as later on the Cape Region became connected with the peninsula by a low stretch of land, immigration of animals became much easier and permitted an inroad of northern species in much larger numbers. Through these various and continuous immigrations from surrounding countries, it might be inferred that the fauna and flora of the Cape Region must contain a mixture of temperate and tropical forms. This is also the case. It is yet too early in our rather imperfect knowledge of the entire number of species inhabiting the Cape Region and surrounding countries, to summarize too liberally. but through the collections acquired by this Academy we now know that a large percentage are temperate forms. a small percentage tropical. while a great number of species are endemic, that is, have so modified themselves from their ancestors in the Cape Region, that they are classified as new species. As not all of our collections have been worked up. a fur her discussion of this subject must be left for a future time. I will here state that while we have collected thousands of different species in this our new field. it is evident that as many more yet exist there. these being scarce forms, which probably are more dependent on the peculiarities of the seasons and localities for their existtence and numbers. Believing that the Cape Region of Baja California is one of the most interesting isolated points in the world as regards its fauna and flora, it is my intention to treat of this subject—of the faunal relations of this region—more exhaustively at a future time.

NATIVE WILD FRUITS AND ECONOMIC PLANTS.

A few words may be said about the native wild fruits of the Cape Region. These fruits are not many. but several of them are valuable. one or two are very fine. The head of the list must be given to the red "pitahaya agria" (*Cereus* [illegible]), or red-fruited cactus. It is one of the very finest fruits I know. In shape it is round and of the size of an orange, in color it is red; its pulp is red with many very small black seeds and is very juicy, high flavored, slightly acid, but also very sweet.

This cactus grows everywhere on the low mesa land to an altitude of about 1200 feet, and the fruit can be had for the picking or bought for next to nothing. This fruit is, like all cactus fruit, covered with spines, but they may be scraped off easily and are not troublesome. A most beautiful preserve is made from the pitahaya with sugar. In taste the pitahaya reminds one of a very fine watermelon. There is another kind known as "pitahaya dulce." It is smaller, sweeter, but not quite as fine. The pitahaya agria ripens in August and September, the pitahaya dulce shortly before.

A fruit universally distributed over the Cape Region is the so-called "ciruela" *(Crytocarpa procera)*. This is is a shrub or small tree which for several months in the year during the rainy season produces enormous masses of small yellow oblong plums, with a single round seed. The fruit is juicy, yellow, and very refreshing, though the flavor is not always fine and rather odd. The seeds contain a very fine kernel, in my opinion the finest nut I have tasted. Miss Alice Eastwood has called my attention to the fact that this plant belongs to the same family as the Pistacia, and that this will account for the exceedingly high flavor of the nut. I have no doubt but that the kernel may be introduced into commerce and be a valuable substitute for the real *Pistacia vera*. There is a great difference between the various trees of this kind. Some bear very large, well flavored and handsome fruit, others again have small fruit with a decidedly turpentine-like flavor. This tree should be cultivated and improved.

The ciruela is very common from the sand dunes along the shore to the mountains, where at an elevation of several thousand feet larger trees may be found, some with palatable fruit.

Another fruit is the "guayparin," a species of persimmon, of a brown color when ripe. It is not astringent and is very highly flavored and really more palatable than the Japan persimmons. It should be useful for crossing with the Japan persimmons.

Among other wild useful plants two may be mentioned. One is the "palo blanco" *(Lysiloma candida)*, the bark of which is peeled, dried and exported in large quantities to England and United States for tanning purposes. It is said to be one of the very best tan-barks known, giving the leather a fine russet color and making it very soft. This tree grows everywhere in the mountains, in almost inexhaustible quantities, It is a slender tree reaching a height of thirty to fifty feet.

The herb known as "damiana," used for the production of damiana bitters, is common in many places. Its botanical name is *Turnera diffusa*.

Aspects of Vegetation.—For a detailed botanical account of the botany of this region, the various papers published by T. S. Brandegee will be found most exhaustive. Here I will only point to a few features of the general landscape as they present themselves to the traveler. With very few exceptions the whole of the Cape Region is densely covered with shrubs and low trees, among which in the lower elevations are mixed numerous cacti, some tall and rigid, others of spreading habits and forms.

During the rainy season all this vegetation is intensely green, the foliage is fine and feathery and the hills and mountains present at a distance a mass of green, which may nearest be compared to the lace-like appearance of the maiden-hair fern. The trees as a rule are small, slender and low, of an average of twenty to thirty feet. Some few varieties are taller. The finest and densest vegetation begins at 1000 feet and extends to 6000 feet. We may distinguish several distinct regions, which, however, are not in every locality found at the same altitude. The lower one of this region, if we except the sandy low land along the shores, are the mesas and the lower hills. Here the trees are low with fine feathery foliage, numerous species of acacias and allied genera. Next region is the region of the figs. At about 3000 to 4000 feet wild figs *(Ficus Palmeri)* become numerous and form a feature of the landscape. At about 4000 feet oaks became very prominent, and we may call this region the region of oaks.

Above the oak region we enter, in the most elevated sierra, the region of the pines, especially at Sierra Laguna. The pines are spreading, without central or standard trunks, branching low down like our digger pines *(Pinus Sabiniana)*, but otherwise in shape recalling the oaks.

All the trees in the Cape Region show a spreading form of their crowns, the effort evidently being to protect their stems, roots and the ground from the heat of the sun and its drying out effects.

There are two other trees which give a great prominence to the landscape. I mean the two palms found here, the species of which have not yet been very critically examined into. The lower cañons in the sierra as well as in the lowlands, where cultivation is carried on, are here and there covered with groves of the tall and most beautiful fan palm *(Washingtonia Sonoræ?)*. This palm is not seen above a few thousand feet, and is probably not indigenous. Higher up, at an elevation of 4000 to 6000 feet, or in places lower yet, we find everywhere in the gulches and along the streams the blue fan palm *(Erythca armata?)* with stems a hundred or more feet high, with a diameter of frequently only six inches at the height of four feet from the ground. This palm is very graceful, it rises high above the other vegetation, and is very prominent, especially as it grows in small groves on the hillsides where there is mosture.

The richness in flowering plants in the Cape Region is remarkable. The whole country is frequently ablaze with the yellow flowers of the "palo de arco" (*Tecoma stans*), or with the lovely large morning-glory (*Ipomea aurea*). The blue morning-glories are also exceedingly numerous in places, every bush and tree being woven together with a dense net-work of these green vines, covered by innumerable blue and violet flowers, in many sizes, from those larger than a dollar to those smaller than a pea.

TIME FOR COLLECTING.

The best time for collecting and studying animal life in the Cape Region varies greatly with the seasons. But as a general rule it may be stated that very soon after the first heavy shower of rain is the proper time to begin, as at that time all animal life starts anew. After the first rains nine-tenths of all chrysalises and larvæ hatch at once, and the whole country is then teeming with animal life. The bushes and trees cover themselves with leaves and flowers, giving ample food for caterpillars and insects of all kinds. As this begins in the sierra, the sierra is the proper place in which to begin explorations and collections. A week after the first rain the fauna is at its height. Myriads of butterflies are then seen filling the air in daytime, while during the evening hours micro- [illegible] or hiding places to feed on the tender leaves of the new vegetation. But, as there is a great irregularity, both in the quantity of the rainfall as well as in the time when it commences, so is there considerable variation in the commencement of this exuberance of animal life. It may begin in July, but it may not begin until September. I have seen so many butterflies filling the air in one of the high valleys of the Sierra El Taste that the air seemed thick with them, and this continued for several weeks. When at such a time a cloud over the sun caused a temporary shadow, this immense and innumerable host of butterflies suddenly vanished, having taken refuge on the under side of the leaves of trees, bushes and herbs. When in an hour the sun again shone out in all its warmth and brightness, the butterflies all at once left their hiding places and again filled the air. One day, when I climbed the El Taste peak, I beheld just such a wondrous sight. The whole valley between El Taste and Mt. Troyer was filled to a height of 3000 feet, or from the bottom of the valley to the top of the peaks, and several miles in width, with butterflies in almost every color of the rainbow, but principally white and yellow. As I sat there gazing at this marvelous spectacle, suddenly a cloud overshadowed the sun. In fifteen minutes the butterflies were gone, the air was clear; but when in an hour the sun came out bright and warm, the butterflies also came out in almost as large quantity as before. Such days and nights must be used without rest by the naturalist, as they may vanish very soon. Thus our principal collecting of beetles and other insects was done in a week's time. The second shower of rain spoiled everything. After a rainstorm of four or five hours duration, we were astonished to find that there were scarcely any insects left of any kind. Butterflies, beetles, moths, wasps, almost every kind had vanished as by magic. One evening before, we had caught a thousand insects by the aid of the lamp in a few hours. The first evening after the rainstorm had passed away we caught about ten in twice the time, and it did not improve with time. The swarms of insects did not return, and a few more showers made it worse yet. Finally the dry season set in, and then there was hardly anything more to be found anywhère except under rocks.

One of the features of the fauna of the Cape Region is the immense quantity of land shells found in some places. In certain evidently favored localities the ground is literally covered with the dead and white shells of land mollusks. We ride along for hours through cañons, where the ground is thus strewn. Then as we turn into another cañon, we find no shells at all, not even after close search under rocks and trunks of trees. Then again as we pass on, we may enter a locality where we again find an abundance and are able to collect thousands in the space of a very short time, there being actually no limit to the quantity. Another peculiarity as regards the land shells is that every sierra and every cañon almost possesses peculiar forms not found anywhere else. All the shells are white or nearly so, a few are pinkish white. In places they glimmer on the ground as close and as prominent as white pebbles on a beach.

A great deal has been said about the number of rattlesnakes found in Baja California, but most stories about them are greatly exaggerated. We may find several during a day, or we may not see any for a week. It is very rarely that any one is bitten by a snake, and scarcer yet that the bite is fatal. A universal, and, as I am told, a sure cure for the rattlesnake bite is the remedy used by the natives. They take a piece of the pitahaya cactus, roast is over the fire and apply it to the wound. It is said hat it extracts the snake poison in a very short time, and efore any serious trouble has set it. The most feared nimal in Baja California is the skunk. The popular beleve is that its bite causes hydrophobia. No native will leep out of doors without covering himself, head and all, n his serape or blanket in order to prevent the skunks rom attacking him. To my knowledge, however, no ase of hydrophobia has occurred for years.

There are no other poisonous animals in the Cape Region, the scorpions being no more harmful than wasps. The big tarantulas are feared, and probably with good eason, though I have heard of no cases where people or animals were bitten.

Below will be found a short record of the various expeditions sent out by the California Academy of Sciences to Baja California and other parts of Mexico:

1. Expedition in March, 1888. W. E. Bryant. Magdalena Island, San Jorge to Comondu and across the peninsula to La Giganta and Loreto. Back by La Giganta, San Gabriel, San Juan. Back through Comondu.

2. Expedition spring of 1889. W. E. Bryant and Chas. D. Haines. Magdalena Island, Santa Margarita Island, San Jorge, Comondu, from there overland to San Gregorio, San Ignacio, Calmalli, Santa Borgia, El Rosario, San Quintin.

3. Expedition September and October, 1890. W. E. Bryant. San José del Cabo, Agua Caliente, Sierra, Triumfo, La Paz.

4. Expedition March to May, 1892. W. E. Bryant, Gustav Eisen. San José del Cabo, Miraflores, Agua Caliente, Santiago, Gulf shore, Sierra Laguna, San Francisquito, La Paz, Espiritu Santa Island, Guaymas, Sonora, Hermosillo, Durasnillas, San Miguel.

5. Expedition September and October, 1893. Gustav

Eisen. San José del Cabo, Sierra El Taste, across to Pescadero and Todos Santos, Cabo San Lucas, and back to San José, Miraflores, San Francisquito, Sierra Laguna, Todos Santos.

6. Expedition September, October, November, 1894. Gustav Eisen, Frank H. Vaslit. San José del Cabo, Miraflores, Santa Anita, La Palma, Sierra San Lazaro, El Taste, Piedra Corral. Overland from San José to La Paz by Santiago, San Bartolo, Triumfo, La Paz. Mazatlan, by steamer to San Blas. Overland to Tepic, by land to Mazatlan, via Santiago Ixtquintla, Squinapa, El Rosario, etc.

Expeditions two, three, four and five were made in company with Mr. T. S. Brandegee, who traveled exclusively at his own expense. While his specialty was plants he also collected a few insects, which were donated to the Academy. Part of expedition No. 5 was also in company with Mrs. T. S. Brandegee.

The last two expeditions collected no birds and but few mammals. Specialties were reptiles, mollusca, insects, etc. During the last expedition, of 1894, it is safe to say that of some of these groups of animals about four times as many specimens and species were collected as during all the other expeditions together.

DISTRIBUTIONS OF COLLECTIONS.

Until the expedition of 1893 the collections brought home had remained undescribed, except as regards birds and mammals. These collections consisted of reptiles, beetles, butterflies, wasps, flies and dragon-flies, collected by the members of the expeditions of 1888, 1889 and [illegible]. By reference to the various papers already published and by noticing the routes of these expeditions it may be seen where these collections were made and what they were.

But beginning with the expedition of 1891, the collections of lower animals became so large that it was decided by the President of the Academy, H. W. Harkness, and the governing Council, to distribute collections to various specialists for descriptions in the Proceedings of the Academy. This was done in 1892, 1893 and 1894. The conditions upon which collections were thus distributed for scientific work were very liberal and regarded principally three points:

1. The MS. of the respective authors regarding the collections sent them should be published by the California Academy of Sciences.

2. All types * to be deposited in the collections of the California Academy of Sciences.

3. The respective investigators to retain a set of duplicates, the first set always going to the Academy.

Under the above conditions the collections were distributed and worked up as follows. References, unless otherwise specified, are to Proceedings California Academy of Sciences, second series.

PROTOZOA.—A large material for the study of parasitic protozoa was collected.

EISEN, GUSTAV. On the Various Stages of Development of Spermatobium, with Notes on other Parasitic Sporozoa. Vol. v, May 18, 1895.

OLIGOCHÆTA OR LAND ANNELIDS.—A very large collection of this class of worms in alcohol is being worked up by the writer. Very great interest is attached to this

* As this paper is intended also for non-specialists it may not be unnecessary to state that with "types" are understood one or more specimens of any species upon which the first description of the species was based. The great scientific value of types is well understood by naturalists and every scientific collection is chiefly valued according to the "types" it contains. The types will always retain their scientific value, no matter how common will be found the species to which they belong. A perusal of the papers describing the fauna of Baia California will show the great number of types thus possessed by our Academy.

group as no species were formerly known from the region. A large number of new forms have been found. The oligochætological fauna of the Cape Region is essentially a neotropic one, only one or two upper California species being found.

EISEN, GUSTAV. Anatomical Studies on New Species of Ocnerodrilus. Vol. iii, July 9, 1893.

EISEN, GUSTAV. On the Anatomical Structure of Two Species of Kerria. Vol. iii, April 15, 1893.

EISEN, GUSTAV. Pacific Coast Oligochæta, No. 1. Memoirs of the California Academy of Sciences. 1895. 18 plates.

EISEN, GUSTAV. Pacific Coast Oligochæta, No. 2. MS. now in press.

LAND AND FRESH WATER MOLLUSCA.—The collections were worked up by Dr. J. G. Cooper, who has kindly furnished the following summary. In all, the various expeditions brought home sixty-three species from the peninsula. Of these seventeen were new to science. Of these latter seven were taken by the three first expeditions, the other ten by the last ones.

Of all the species collected, twenty-nine are found north of the U. S. boundary line, on the peninsula only twenty-eight, east of the Gulf also six.

Around Mazatlan and Tepic seventeen species were found.

COOPER, J. G. On Land and Fresh Water Shells of Lower California. Vol. iii, April 23, 1891.

COOPER, J. G. On Land and Fresh Water Mollusca of Lower California, No. 2. Vol. iii, October 6, 1892.

COOPER, J. G. On Land and Fresh Water Mollusca of Lower California, No. 3. Vol. iii, May 5, 1893.

COOPER, J. G. On Land and Fresh Water Mollusca of Lower California. No. 4. Vol. iv, April 28, 1894.

COOPER, J. G. On Land and Fresh Water Shells of Lower California, No. 5. Vol. [illegible] June 8, 1895.

COOPER, J. G. On West Mexican Land and Fresh Water Mollusca. Vol. v, June 8, 1895.

COOPER, J. G. [illegible] of the [illegible] Land and Fresh Water Mollusca of Lower California. Zoe, v [illegible]

MARINE MOLLUSCA.—Described by Dr. J. G. Cooper. The collection was principally made by Mr. W. E. Bryant, during the expedition of 1892.

COOPER, J. G. Catalogue of Marine Shells, collected chiefly on the Eastern Shore of Lower California, for the California Academy of Sciences, during 1891-2. Vol. v, May 21, 1895.

SPIDERS.—During the last two expeditions a specialty was made of spiders. These were turned over to that most prominent arachnologist, Prof. George Marx, of Washington, D. C., but who, through his untimely death, was prevented from finishing the work. In private letters to the writer he stated that there were about three hundred species, mostly new, from the Cape Region alone. These collections, together with those collected during the expedition of 1894, in the Cape Region and at Tepic, were, after the death of Prof. Marx, forwarded to Prof. Nathan Banks, whose eminent services have been secured in working up the collections. His manuscript is not yet ready.

PHALANGIDEA OR HARVESTMEN.—All collections made have been sent to Prof. Clarence M. Weed, who has undertaken their description.

SCORPIONS.—A very large collection from Baja California, Cape Region and Tepic, preserved in alcohol, has not yet been distributed.

ACARINA OR MITES.—Are being worked up by Dr. Otto Stoll. Manuscript not yet received.

MYRIAPODA OR CENTIPEDES.—A very full collection from the Cape Region. The working up of this collection has been undertaken by Prof. R. I. Pocock. The manuscript has not yet reached us.

CRUSTACEA.—A collection in alcohol of fresh water and land living species have been sent to Prof. Walter Faxon. His manuscript not yet received.

Other species have been described by Professor Samuel J. Holmes, as below:

HOLMES, SAMUEL J. Notes on West American Crustacea. Vol. iv, May 30, 1894.

This paper contains description of a few Baja California Crustacea.

ONISCIDÆ OR SOW-BUGS.—A very full and carefully made collection in alcohol from the Cape Region, Tepic, etc., has not yet been distributed.

INSECTS.—The following statement of the various collections of insects brought together by Mr. Frank H. Vaslit and myself during the last expeditions, as well as of those collected by previous expeditions, has been compiled by Prof. Charles Fuchs:

THYSANURA AND COLLEMBOLA

SCHÖTT, HARALD. In MS.

Thirty species, sixteen n. sp. types, one hundred specimens, collected in 1893 and 1894.

ODONATA

CALVERT, PHILIP P. The Odonata of Baja California, Mexico. Vol. iv, February 4, 1895.

Fifty species, six n. sp. types, two thousand six hundred specimens.

Of the fifty species, nine are widely distributed over temperate America, eighteen are neotropical, eighteen nearctic, three restricted to the peninsula.

The collections were mostly made during the last two expeditions and were preserved in [illegible]. The very the illustrations are ready, but the manuscripts have not yet reached us.

HETEROPTEROUS HEMIPTERA.

UHLER, P. R. Observations upon the Heteropterous Hemiptera of Lower California, with Descriptions of New Species. Vol. iv, June 20, 1894.

One hundred and fifty-four species, thirty-two n. sp. types, eight thousand specimens.

Seventy-three species, fourteen n. sp. types, collected by expeditions 1888–1889.

Fifty-nine species, seven n. sp. types, collected by expeditions 1892–1893.

Twenty-two species, eleven n. sp. types, previously in the Academy collection, collected in California.

The collection made by the expedition of 1894 has not yet been reported on by Prof. Uhler.

NEUROPTERA.

BANKS, NATHAN. Some Mexican Neuroptera. Vol. v, August 20, 1895.

Sixteen species, eight n. sp. types, ninety specimens. Collected by the expedition of 1894, mostly in the Cape Region of Baja California and Tepic, Mexico.

DIPTERA.

TOWNSEND, C. H. TYLER. Notes on the Diptera of Baja California, including some Species from Adjacent Regions. Vol. iv, April 8, 1895.

Sixty-one species, sixteen n. sp. types, one thousand specimens.

Eight species, one n. sp. type, collected by expeditions 1888–1889.

Thirty-five species, thirteen n. sp. types, collected by expeditions 1892–1893.

Eighteen species, two n. sp. types, previously in the Academy collection, collected in California.

Collections by the expedition of 1894 are now being worked up by Prof. Townsend.

Coleoptera.

Horn, George H. Coleoptera of Baja California. Vol. iv, August 8, 1894.

Six hundred and eighty species, eighty-four n. sp. types, thirty-four n. sp. typical varieties.

The collections upon which this paper is based were made by the expeditions of 1888 to 1893.

Horn, George H. Coleoptera of Baja California. Supplement I. Vol. v, July 10, 1895.

Fifty-two species, twenty-three n. sp. types and typical varieties.

The collections upon which this paper is based were made exclusively in 1894.

The species collected up to October, 1895, are represented by one hundred and seven n. sp. types, sixty-seven n. sp. typical varieties, seven hundred fifty-five species and about ten thousand specimens.

The collection of beetles by the different expeditions is by far the largest aggregate of material from Baja California submitted for scientific study. In addition to this the Academy possesses a collection of beetles from the Pacific mainland of Mexico, State of Sinaloa to Tepic, collected by the expeditions of 1893-1894, amounting to seven hundred and fifty-five species, three thousand specimens.

Hymenoptera.

Fox, William J. Report on some Mexican Hymenoptera, principally from Lower California. Vol. v, September [illegible], 1895.

Seventy-eight species, thirteen n. sp. types.

Fourteen species [illegible] n. sp. types, collected by expeditions [illegible]

[illegible]

Nine species, including one n. sp. type, collected by expeditions 1888–1889.

Ninety species, including thirteen n. sp. types, collected by expeditions 1892–1893.

FOX, WILLIAM J. Third Report on some Mexican Hymenoptera, principally from Lower California. Vol. v, July 20, 1895.

The collections upon which this paper is based were made exclusively in 1894, amounting to sixty-seven species including five n. sp. types.

The Hymenoptera, including Parasitic Hymenoptera and Formicidæ, are represented in all by seventy n. sp. types, three hundred and thirty-two species, four thousand five hundred specimens.

PARASITIC HYMENOPTERA.

ASHMEAD, WILLIAM H. Some Parasitic Hymenoptera from Lower California. Vol. iv. April 25, 1894.

Forty-four specimens, twenty-one species, nine n. sp. types.

Seven species, including one n. sp. type, collected by expeditions 1888–1889.

Fourteen species, including eight n. sp. types, collected by expedition 1893.

ASHMEAD, WILLIAM H. Some Parasitic Hymenoptera from Baja California and Tepic, Mexico. Vol. v, September 7, 1895.

Thirty-eight species, twenty-two n. sp. types, collected by the expedition 1894.

This very extensive series comprises some of the largest families of the order, the members of which in their larval state, excepting the gall-feeding Cynipidæ, are parasitic upon or within the bodies of other insects, using the words of Westwood, "are of vast importance in the economy of nature by preventing the too great increase of different species of insects, especially of the caterpillars and moths, of which they destroy a great number."

BLASTOPHAGINÆ, or Fig Insects.

A large collection of these minute wasps, with the various species of wild figs inhabited by them, was made in 1893 and 1894. The collection was delivered to Prof. C. V. Riley, at his urgent request, but since his untimely death Prof. W. H. Ashmead has kindly undertaken the description of this interesting and difficult group.

FORMICIDÆ, or Ants.

PERGANDE, THEO. On a Collection of Formicidæ from Lower California and Sonora, Mexico. Proc. [illegible] September [illegible], 18[illegible].

Sixteen species, four n. sp. types.

Eleven species, including three n. sp. types, collected by the expeditions of 1888–1890.

Five species, including one n. sp. type, collected by the expeditions of 1892–1893.

PERGANDE, THEO. Formicidæ of Lower California, Mexico. Proc. [illegible] May [illegible], 1894.

Seven species, including three n. sp. types, collected by the expedition 1893.

These papers refer only to collections made previous to 1894.

The very large collections made by us in 1894, both in the Cape Region and at Tepic, are now worked up by Prof. Pergande, who writes that the manuscript will be ready before very long, the very large size of the collections having made delay necessary.

VAN DENBURGH, JOHN. Description of Three New Lizards from California and Lower California, with note on Phrynosoma Blainvillii. Vol. iv, July 12, 1894.

VAN DENBURGH, JOHN. Notes on Crotolus Mitchellii and Crotolus Pyrrhus. Vol. iv, September 25, 1894.

VAN DENBURGH, JOHN. Phrynosoma Solaris, with a Note on its Distribution. Vol. iv, September 25, 1894.

FISHES.—Only fresh water fishes were collected, which have been handed to Prof. C. H. Gilbert for determination and description. In this connection it may not be out of place to call attention to the following very important paper published by our Academy and bearing upon the fishes of the region:

JORDAN, DAVID STARR. The Fishes of Sinaloa; 31 plates. Vol. v, August 15, 1895.

BIRDS.—Collected during the first four expeditions only. Worked up by Mr. W. E. Bryant, who published:

BRYANT, WALTER E. Description of a Subspecies of Song Sparrow from Lower California. Vol. i, September 29, 1888.

BRYANT, WALTER E. Description of the Nests and Eggs of Some Lower California Birds, with a Description of the Young Plumage of Geothlypis beldingi. Vol. ii, June 20, 1889.

BRYANT, WALTER E. A Catalogue of the Birds of Lower California, Mexico. Vol. ii, December 17, 1889.

ANTHONY, ALFRED W. New Birds from Lower California, Mexico. Vol. ii, October 11, 1889.

Mr. Anthony's collections were made entirely independent of the Academy explorations.

MAMMALS.

BRYANT, WALTER E. Preliminary Description of a New Species of the Genus Lepus, from Mexico. Vol. iii, April 23, 1891.

MERRIAM, C. HART. Description of a New Kangaroo Rat from Lower California (Dipodomys merriami melanurus subsp. nov.) collected by Walter E. Bryant. Vol. iii, June 5, 1893.

MERRIAM C. HART. Description of Four New Pocket Mice from Lower California, collected by Walter E. Bryant. Vol. iv, September 25, 1894.

STOWELL, JOHN M. Description of a new Jack-Rabbit from San Pedro Martir Mountain, Lower California. Vol. v, May 28, 1895.

BOTANY.—Although all the botanical collections were made by Mr. T. S. Brandegee in a private capacity and therefore not strictly coming under the explorations of the Academy, still I have thought it well to include a list of the botanical papers, especially as they are published by the Academy in its Proceedings. A nearly full collection of Mr. Brandegee's plants was donated to the Herbarium of the Academy by him.

BRANDEGEE, T. S. A Collection of Plants from Baja California, 1889. Vol. ii, Nov. 12, 1889.

BRANDEGEE, T. S. Flora of the Cape Region of Baja California. Vol. iii, July 14, 1891.

BRANDEGEE, T. S. Additions to the Flora of the Cape Region of Baja California. Vol. iii, November 10, 1892.

HARKNESS, H. W. Fungi collected by T. S. Brandegee in Lower California in 1889. Vol. ii, Dec. 20, 1888.

VASEY, GEO. Grasses from Lower California. Vol. ii, Dec. 17, 1889.

COGNIAUX, A. Cucurbitacearum novum Genus et Species. Vol. iii, July 7, 1890.

GEOLOGY.

LINDGREN, WALDEMAR. Petrographical Notes on Baja California, Mexico. Vol. ii, June 29, 1889.

LINDGREN, WALDEMAR. Notes on the Geology and Petrography of Baja California, Mexico. Vol. iii, April 16, 1890.

VARIOUS DISTANCES IN THE CAPE REGION.

	Leagues.
Cabo San Lucas to San José del Cabo	9
San José to Santa Catarina	2
Santa Catarina to Santa Anita	1
Santa Anita to La Palma	3
La Palma to Caduaño	2
Caduaño to Miraflores	1
Miraflores to Agua Caliente	2
Miraflores to Santiago	3
Santiago to Ensenada de la Palma	5
Ensenada to San Bartolo	3
San Bartolo to Rodeo	2
Rodeo to San Antonio	5
San Antonio to Triunfo	2
Triunfo to San Pedro	6
San Pedro to La Paz	8

EXPLANATION OF PLATES.

PLATE LXXII. *General Map of the Peninsula of Baja California.* Compiled from the U. S. Hydrographic Office charts, and from the Brandegee map elsewhere alluded to, as well as from various other sources.

PLATE LXXIII. *Map of the Cape Region of Baja California.* The outlines have been compiled from the U. S. Hydrographic Office charts. The nomenclature has been furnished by Capt. John von Helms and by Capt. E. Labastida of La Paz. The interior topography from notes and memoranda of the explorations of the writer and Mr. Frank H. Vaslit.

PLATE LXXIV. *Map of the Region of Sierra El Taste.* Compiled from notes and sketches of Frank H. Vaslit and the writer, made during the expedition of 1894.

Map of the Region of Sierra Laguna. Compiled from notes, etc., by the writer, made during the years of 1892 and 1893.

PLATE LXXV. *Map of the Rainfall and General Geological features of the Cape Region.* Compiled by Frank H. Vaslit and the writer.

NOTES ON THE HABITS AND DISTRIBUTION OF AUTODAX IËCANUS.

BY JOHN VAN DENBURGH,

Curator of the Department of Herpetology.

This black salamander was originally described, by Prof. E. D. Cope, from a half grown specimen collected at Baird, Shasta County, California. Two others, one of which was adult, were afterwards secured at the same place by Messrs. C. H. Townsend and Livingston Stone.

A considerable number of specimens collected by myself at Los Gatos, Santa Clara County, and by the members of the Zoölogical Club of the Leland Stanford Junior University, at Steven's Creek, in Santa Clara County, and Glenwood and Boulder, in Santa Cruz County, California, greatly increase the known range of this species. These specimens were found under boards, decaying logs, stones in the vicinity of running water, and in the drain from a spring.

Autodax iëcanus, doubtless, is a nocturnal forager. I have seen it upon the surface of the ground only twice, and in each instance night was so nearly at hand that objects near my feet could just be distinguished. If liberated during the day, or if confronted with a light at night, when it is much more active, this species will proceed, almost invariably, toward the nearest spot of darkness or shadow. It usually walks along quite slowly, moving but one foot at a time, but is capable of motion surprisingly rapid for a salamander. When moving rapidly, it aids the action of its legs by a sinuous movement of its whole body and tail.

The tail of this Autodax is prehensile. Several individuals, when held with their heads down, coiled their tails around my finger, and, when the original hold was released, sustained themselves for some time by this means alone. One even raised itself high enough to secure a foothold.

The animal's tail is also of use to it in another way. When caught, *Autodax iëcanus* will often remain motionless, but if touched will either run a short distance with great speed, or, quickly raising its tail and striking it forcibly against the surface on which it rests, and accompanying this act with a quick motion of its hind limbs, will jump from four to six inches, rising as high as two or three.

Several specimens, which were kept alive for some time, climbed up the vertical sides of the glass bottle in which they were confined, and sometimes even passed the incurved portion of its neck. They were aided in doing this by the extremely viscid mucus with which they are covered.

One, caught on the evening of August first, struggled violently trying to escape from the hand in which I held it. A few minutes later, I noticed that it was very quiet, and, after carefully examining it, decided that the heat of my hand had caused its death. In order to test this more thoroughly the animal was thrown into water, but still it showed no signs of life, and floated, as placed, either on its back or belly. It was then carried home and laid upon my table. Ten minutes later it began to walk slowly away.

A large *Autodax iëcanus* and fifteen eggs were sent me from Los Gatos, July 23, 1895. The eggs were evidently those of a batrachian, doubtless of this species. Each egg was about 6 mm. in diameter, almost spherical, and inclosed in a thin, tough, gelatinous sheath. Each of these sheaths was drawn out, at one place, into a slender peduncle, which was attached to a basal mass of the same gelatinous substance. In this way, each egg was at the

end of an individual stalk, and all were fastened to a common base. This base had evidently been anchored to a stone or lump of earth. The eggs were in the early stages of segmentation. The following note accompanied them: "The salamander and eggs were found under the platform in front of a barn, in dry earth next the foundation wall, and about fifteen inches or more below the surface. The ground had been filled in, and was full of spaces. There was some dry rotten wood near the eggs. One or two smaller salamanders were near. About twice as many eggs were found as sent. There was no water within ten or fifteen feet." The salamander sent with these eggs was a female, and had a very large number of minute eggs in its ovaries.

On July 30, 1895, I killed a very large Autodax which had been sent me, from Los Gatos, several days before. It contained twenty-five eggs exactly like those described above, except that they lacked the gelatinous covering. These eggs appeared to be still in the ovaries. There were twelve on the right side of the animal, and thirteen on the left. Besides these enlarged ones, there were many minute ova.

DESCRIPTION OF A NEW SPECIES OF RANZANIA FROM THE HAWAIIAN ISLANDS.

BY OLIVER P. JENKINS.

[With Frontispiece.]

Ranzania makua n. sp.

D. 17, A. 18, C. 19, P. 13. Depth in length to base of caudal 2⅛. Head in length 2⅝, eye in head 6. Eye in snout, 2⅓. Body much compressed, the ventral margin presenting a sharp, evenly curved keel. In a lateral view it is deepest just behind the pectoral fin, narrowing but slightly to the abruptly truncated posterior extremity, but anterior from this point both dorsal and ventral margins curve gradually to the truncated snout. The eye placed considerably above the axis of the body, and a little nearer the snout than the base of pectoral, and very close to the margin of the dorsal outline. Teeth formed into a beak like that of a turtle, completely hidden by projecting folds of skin, which form a truncated opening to mouth.

Gill opening just in front of the upper base of the pectoral, covered by a two-lobed valve.

Body covered by an armor of small plates, more or less regularly hexagonal. This structure is more or less concealed in the fresh specimen.

Pectoral fin in height slightly less than ⅔ length of head. It is well above the axis of the body. Dorsal in height nearly equaling length of head. Height of anal somewhat less. Dorsal and anal each separated from the caudal by a notch.

The form of the rays of the fins is interesting. During most of their extent they form flat horny rods, but at the outer extremity each divides into a great number of branches, spreading out like a fan, the edges of which reach those of the contiguous rays. Fig. 1 gives the detail of the three lower rays of the pectoral, Fig. 2 of the extremity of two and parts of two others of the dorsal, and Fig. 3 three rays of the caudal fin.

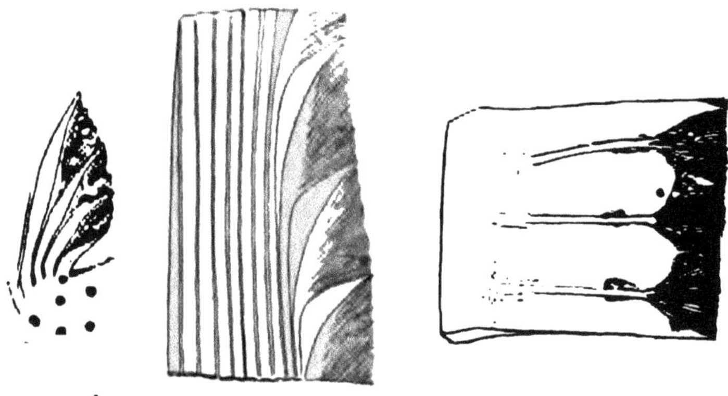

Fig. 1. Fig. 2. Fig. 3.

Coloration very brilliant in fresh state. Sides bright silvery. Upper part of body dark; the sides of the body are decorated by bright silvery bands, which have the following disposition: The anterior ones more distinct and definite. The anterior ones are convex anteriorly, and nearly parallel. The first three silver bands have distinct black bands as borders to the bright silver bands. First band silver portion 5 mm. broad; the most anterior point of its curve is 2 c. m. from end of snout. Second band begins at upper anterior margin of eye, bends forward to within 4 c. m. of end of snout, then curves gently backward and downward, becoming indistinct near ventral margin of the body. Third band begins at lower anterior margin of eye, bends but slightly forward then downward and backward, becoming indistinct near ventral margin of body. The black band about posterior margin of third silver band on part of its course gives way to a row of black spots. The fourth, fifth, sixth and seventh silver bands are branched above and in some cases below. Along their middle courses they are not bordered by the black bands, but possess numerous distinct black spots, the black bands being retained as margins to their lower portions and in some places along their upper portions. Fourth lies behind the eye. It begins near dorsal line, runs slightly forward, just reaching the posterior margin of the eye, then downwards and backwards, branching below into two bands; fifth band forms two branches on one side, three on other side of body between eye and opercular opening; sixth extends from base of pectoral; seventh indistinct, arising from just behind base of pectoral. All these bands just described pursuing nearly parallel courses downwards, curving backwards and becoming indistinct near ventral margin of body. The remaining silver bands are irregular, forming an indistinct network, and system of silver spots, with but few black dots.

This species is in general form much like *Ranzania truncata* Retzius, but differs from it: (1) in having a smaller eye; (2) in having the eye well above the level of the mouth; (3) in having both eye and pectoral fins placed above the axis of the body; (4) in possessing higher vertical fins, and (5) in the coloration. The following is a table of measurements taken from the fresh specimen:

	c. m.
Total length	51.
Length of body to base of caudal fin	47.5
Head	18.7
Depth	25.
Breadth of body just above pectoral fin	7.
Snout	7.
Eye	3.
Vertical diameter of opercular opening	3.
Vertical diameter of mouth opening	3.5
Posterior margin of eye to origin of pectoral	9.5

Height of dorsal fin	17.5
Height of anal fin	15.5
Length of caudal	3.5
Base of caudal	19.
Length of pectoral	11.5
Base of pectoral	3.

The proposed specific name *makua* is, according to Mr. C. B. Wilson, of Honolulu, a native name of the fish, and signifies "the source from which the Bonito and Albicore sprung in after ages." The specimen of which the above is a description is now in the Museum of the Leland Stanford Jr. University. The University is under obligations to Mr. Chas. B. Wilson, of Honolulu, for this valuable contribution. The fish was caught January 25, 1892, by Mr. Hiel Kapu, at the mouth of Pearl Harbor, and was frozen in ice and sent to the University by Mr. Wilson. It arrived in an excellent state, which allowed me to make a study of it while it was still fresh. It was immediately drawn in colors by Miss Anna L. Brown. In answer to a letter of inquiry, Mr. Wilson gave me the following interesting account of the fish and its capture:

"It was taken in shallow water three or four feet deep. It is a deep-sea fish by habit. It was seen by a party of fishermen in a canoe going from shore to a deep-sea fishing ground, when they were not more than a hundred feet from the beach near the entrance to Pearl Harbor, Oahu. The man who first saw it, drew the attention of the leader of the fishing party to the appearance of this strange object close to the canoe, at the surface of the water. The leader told him to 'hit it with his paddle,' but the man refused, saying he was afraid that it was an 'Akua' (Spirit or Deity). The leader himself then hit it with his paddle on the side of the head, when it immediately shot off in a semi-circular path, through the breakers, in front of the canoe, and ultimately landed (in its terror) on the beach about 600 yards away, where the fishermen saw it and captured it still alive, a few minutes afterwards. It frequents the deep ocean alone, and is believed by the natives to be be the 'MAKUA' source from which sprang in after ages the Bonito and Albicore. In its habits it is like them, and is only seen in shallow water when in flight from its natural enemies, the shark, sword-fish, etc., and as they when fleeing from their enemies rise as near the surface of the water as they can and seek the shelter and protection of some floating object such as a log, a ship's hull, or the like, so doubtless in this case the Apahu was seeking the refuge of the canoe's side when it was first seen. I have discovered by inquiry from the native Hawaiians, that all three species of fish when driven into shallow water seem to be dazed and lose control of themselves, and ultimately are forced on the beach by the action of the waves. One other specimen of the 'APAHU,' the second besides the one I sent to the University, was cast up on the beach at Waikiki, near the residence of Edmund Hart. The finders cooked it and ate it. They said it was very fine eating. These are the only two specimens seen here in ages. It is a very rapid swimmer quite as swift or swifter probably than the dolphin. When it was struck, it disappeared like a flash of lightning and the fishermen did not expect to see it again. It was on account of this property of remarkable speed which it possessed that I recommend in my original communication to Prof. Jordan, that the attention of prominent yacht builders be called to its lines in the hope that they might find something of use therefrom."

Since the above was written Rev. E. B. Tuthill has sent to Dr. Jordan for examination a number of drawings of fishes of Honolulu. Among them is a drawing of a specimen of this species. The date of its capture is not given, and it is possible that the drawing was taken from the specimen referred to by Mr. Wilson as the second one known on the Islands besides the one here described. According to Mr. Tuthill, the species is very rare and not known to the fishermen.

THE FISHES OF PUGET SOUND.*

BY DAVID STARR JORDAN AND EDWIN CHAPIN STARKS.

(With Plates lxxvi-civ.)

The present paper contains an enumeration of the fishes known to inhabit the waters of Puget Sound, a large estuary or fjord entering the northwestern part of the State of Washington. The paper is based primarily on a collection made by the junior author in July, 1895, under the auspices of the Hopkins Laboratory of the Leland Stanford Junior University, he being the guest of the "Young Naturalists' Society of Seattle."

This society undertook at this time a dredging expedition for the special purpose of collecting invertebrates. Through the interest of Mr. Timothy Hopkins, the junior author was enabled to take part in this work.

Nearly two weeks in July were devoted to dredging. A small steamer was chartered for this purpose. A camp was established at Point Orchard on Admiralty Inlet, and collecting and dredging were carried on within a radius of twenty miles from that point.

Besides the fishes that were brought up in the dredge, collections were made of "rock-pool" fishes at low tide, and seines were worked along the beaches.

After the return of the dredging expedition, the fresh waters about Seattle were seined, with the help of different members of the Naturalists' Society. This fresh water collection is described by Mr. Alvin Seale, in an appendix to the present paper.

Besides the work done about Seattle, a week was spent by the junior author at Neah Bay, near Cape Flattery, in the Straits of Juan de Fuca. Here a collection of the rock-pool fishes was made. A rich field for this work

*Contributions to Biology from the Hopkins Seaside Laboratory, No. 3.

was found on Waadda Island, a small rocky islet, lying about half a mile from the shore, near Neah Bay. In this same locality large collections were made in 1880 by Professors Jordan and Gilbert.

Previous to this expedition a small but very valuable collection of fishes had been presented to the Leland Stanford Jr. University by the Young Naturalists' Society. The new forms in this collection are described in the present paper by Jordan and Williams. In the present list are also included the species enumerated by previous writers as occurring in Puget Sound and the Straits of Juan de Fuca. In the list published in 1880 by Jordan and Gilbert ninety species are mentioned as found in these waters. From this list we have drawn freely in our present records of the habits of species. In a later list by Dr. Carl H. Eigenmann (1892), 106 species are recorded. In the present list 141 species are recorded from these waters.

The junior author wishes to express here his obligations for the many favors extended to him by the different members of the Young Naturalists' Society, who did all that was in their power to make his part of the expedition a success. He is under particular obligations to Mr. Charles L. Denny and Mr. Edward S. Meany, who helped him in many ways, both in the dredging trip and on his trip to Neah Bay. He is indebted also to Mr. Henry H. Hindshaw for entertainment in Seattle and help of various kinds. Valuable aid was also given by Mr. Adam Hubbert, Miss Adella M. Parker, Miss Maud Parker, Mr. Trevor Kincaid, Mr. J. W. Busby, Mr. Al- [illegible] Miss R[illegible] Mrs. [illegible] E. [illegible], Mrs. H. [illegible] C. V. Piper, [illegible] is due to the kindly interest of Messrs. Goodall & Perkins, representing the Pacific Coast Steamship Company.

The following species are here described as new to science, the types of all of them being deposited in the Museum of the Leland Stanford Junior University as gifts from the Hopkins Laboratory or from the Young Naturalists' Society of Seattle. The new genera are indicated in full-face type.

Ruscarius *meanyi*. No. 3127.
Oligocottus embryum. No. 3128.
Gilbertina *sigalutes*. No. 3129.
Averruncus *emmelane*. No. 3135.
Xystes *axinophrys*. No. 3130.
Lethotremus *vinolentus*. No. 3131.
Neoliparis florœ. No. 3019, 3133.
Liparis dennyi. No. 3703.
Bryostemma *nugator*. No. 3134.
Xiphistes *ulvœ*. No. 3132.

Besides these species, the following additional new species are described from other localities:

Zalarges *nimbarius*. No. 3125. Open sea.
Hexagrammus otakii. Tokio, Japan.
Podothecus veternus. Robin Island, Alaska.
Podothecus accipiter. Robin Island, Alaska.

The following additional generic names are here used for the first time:

Astrolytes, Pallasina, Stelgis, Quietula, Ronquilus and **Xererpes.**

The fish fauna of Puget Sound marks a transition from the California fauna characterized by the abundance of *Scorpœnidœ*, *Embiotocidœ*, etc., to that of Alaska, in which *Cottidœ*, *Agonidœ* and the Arctic types of *Blennies* are dominant. Here both classes occur, though less abundant than in their respective regions. The present collection is chiefly from depths greater than those reached by Jordan and Gilbert, who collected largely in Puget

Sound in 1880. The extensive collections made by the Albatross in the north have been mostly from much greater depths.

The plates of the present paper are all drawn by Miss Anna L. Brown, artist of the Hopkins Seaside Laboratory.

Family PETROMYZONIDÆ.

1. **Entosphenus tridentatus** (Gairdner).

Common; ascending the fresh waters in spring to spawn, reaching a length of over 2 feet. It is not used as food. Not taken by us.

2. **Lampetra cibaria** (Girard).

Not rare; ascending streams, reaching a length of 8 inches; not used as food. Not taken by us.

Family HEXANCHIDÆ.

3. **Notorhynchus maculatus** Ayres.

Recorded (as *Notorhynchus borealis*) from Nisqually, Washington, by Dr. Gill. Not taken by us.

4. **Hexanchus corinus** Jordan & Gilbert.

Originally described from Neah Bay and from the Bay of Monterey. Not seen by us.

Family GALEIDÆ.

5. **Prionace glauca** (Linnæus). BLUE SHARK.

Recorded by Jordan & Gilbert; rare. Not seen by us.

Family DALATIIDÆ.

6. **Somniosus microcephalus** (Bloch). GROUND SHARK.

Not uncommon. A very sluggish shark. Recorded by Jordan & Gilbert from Victoria. A stuffed specimen from Seattle in the Museum of the Young Naturalists' Society.

Family SQUALIDÆ.

7. **Squalus sucklii** (Girard). DOG-FISH.

Exceedingly abundant. Taken in great numbers with set lines. It is valued for the oil extracted from its liver.

Family RAJIDÆ.

8. **Raja rhina** Jordan & Gilbert.

Not uncommon; reaches a length of 32 inches. Not taken by us.

9. **Raja binoculata** (Girard). COMMON SKATE; RAY.

Common on sandy shores. Reaches a length of 6 feet and a weight of over 60 pounds. One small specimen obtained, very prettily marked with large ocellated spot on the base of pectorals, which fades in the adult. Several of the egg cases of this species were dredged from deep water, where they lie apparently unprotected on the sandy bottom.

Family CHIMÆRIDÆ.

10. **Hydrolagus colliæi** (Lay & Bennett). RAT-FISH.

Numerous specimens taken on sandy beaches at night with a seine, where they were attracted by a camp-fire. It reaches a length of 2½ feet.

Family ACIPENSERIDÆ.

11. **Acipenser transmontanus** Richardson. WHITE STURGEON.

Common; running up the rivers in the spring. It reaches a length of 15 feet and a weight of 300 to 400 pounds. Used largely as food, although its flesh is coarse. Not taken by us.

12. **Acipenser medirostris** Ayres. GREEN STURGEON.

Not common. Reaches a large size, but is not used as food. Not taken by us.

Family NEMICHTHYIDÆ.

13. **Nemichthys avocetta** Jordan & Gilbert.

The type of this species was taken near Port Gamble in 1880, by Prof. O. B. Johnson of the University of Washington. It was presented to the U. S. National Museum by President A. J. Anderson. Mr. Ashdown H. Green of Victoria, B. C., reports a second specimen as recently taken near Victoria and preserved in the museum of that town.

Family CLUPEIDÆ.

14. **Clupea pallasi** Cuvier & Valenciennes. HERRING.

Exceedingly abundant. Smoked and salted in large numbers. Mr. J. P. Hammond* states that from 18 to 25 years ago it was not an uncommon occurrence for a "gang" of fishermen to catch from 200 to 300 barrels of herring in a night on Puget Sound. Now the largest night's work is 20 barrels.

15. **Clupanodon cæruleus** (Girard). SARDINE.

This sardine occurs in large numbers in the warmer part of the season.

[**Alosa sapidissima** (Wilson). SHAD.

This species was introduced into the Pacific about 1878, and was first noticed in Puget Sound in 1884. They are slowly increasing in number, although the catch is as yet unimportant. Specimens of 6½ pounds in weight have been taken in the Sound. Not seen by us.]

Family ENGRAULIDÆ.

16. **Engraulis mordax** Girard. ANCHOVY.

Abundant; occurring in immense schools. Chiefly used for bait. Not taken by us.

---

*American Angler, December 18, 1886

Family SALMONIDÆ.

17. **Oncorhynchus tschawytscha** (Walbaum). QUINNAT SALMON; CHINNOOK SALMON; TYEE* SALMON.

The first salmon to appear each season, abundant from August to October. It commonly weighs about 17 (11 to 20) pounds, but specimens weighing 70 pounds are on record. The most important fish on the Pacific Coast. In Puget Sound it is not very abundant, and being obtained late in the season, its flesh is somewhat lean and dry, ranking with the silver salmon, with which it is usually canned. In the Columbia River this species is canned early in the season, and its quality then is much superior to that of any salmon canned in Puget Sound.

18. **Oncorhynchus kisutch** (Walbaum). SILVER SALMON; SKOWITZ.

Abundant from August to November. It reaches a length of 30 inches and a weight of 4 to 8 pounds. It is largely canned at Seattle under the name of Red Salmon. Its flesh is very red, but dry and not richly flavored, being much inferior to the Quinnat or "Tyee."

19. **Oncorhynchus keta** (Walbaum). DOG SALMON; LE KAY.

Abundant; reaches a weight of 20 pounds. It is only eaten by the Indians, as it runs late in the fall when its flesh is very dry and poor. One small specimen taken.

20. **Oncorhynchus gorbuscha** (Walbaum). HUMPBACK SALMON; HADDO.

The smallest of the salmon, reaching a weight of 7 pounds. It is very abundant on alternate years in the Sound (1893, 1895, etc.), being wholly unknown in even

* Tyee, the common Chinnook name for this species on Puget Sound, is said to mean *king* or *chief*.

years. It is dark in color, with pale flesh and is regarded as the poorest of the salmon, although its inferiority to the silver salmon is in appearance rather than in taste. It is, however, canned in large numbers, and is of economic importance.

21. **Oncorhynchus nerka** (Walbaum). SUKKEGH; BLUEBACK SALMON.

Abundant, reaching a weight of from 4 to 8 pounds. Often landlocked in the lakes. In value intermediate between the "Tyee" and the "Skowitz" or Silver Salmon. The male in the fall is known as "red-fish."

22. **Salmo mykiss** Walbaum. CUT-THROAT TROUT.

Found in abundance in salt water in Puget Sound. It often reaches 8 or 10 pounds, but specimens weighing much more have been taken.

23. **Salmo gairdneri** Richardson. STEELHEAD.

Common near the head of Puget Sound. Considerable quantities are taken for the market. It sometimes reaches 14 to 18 pounds in weight. It is now canned regularly with the silver salmon.

24. **Salvelinus malma** (Walbaum). DOLLY VARDEN TROUT.

Abundant. In Puget Sound it is taken from salt water in large numbers. An excellent food fish, reaching in salt water a weight of 11 pounds or more. Locally known as bull trout or salmon trout.

Family ARGENTINIDÆ.

25. **Hypomesus pretiosus** (Girard). SURF SMELT.

Very common on sandy beaches in Puget Sound. It reaches a length of a foot and becomes very fat. It is a food fish of great value. Several specimens obtained. A beautiful, symmetrical little fish.

26. **Thaleichthys pacificus** (Richardson). EULACHON; CANDLE-FISH.

Abundant in the spring; not taken by us. A fine food fish. Reaches a length of about 10 inches. A fisherman at Olympia says that this species buries itself in the sand of the beach, in the same fashion as the species of *Ammodytes*.

27. **Osmerus thaleichthys** Ayres. SMELT.

Common, but not of great importance as a food fish. Length about 6 inches. Not taken by us.

Family MYCTOPHIDÆ.

28. **Tarletonbeania crenularis** (Jordan & Gilbert).

A specimen taken off Vancouver Island in 1880 by Dr. Tarleton H. Bean, who gave it the manuscript name of *Myctophum procellarum*. Not taken by us.

29. **Myctophum californiense** Eigenmann & Eigenmann.

Recorded from Vancouver Island by Dr. Günther under the name of "*Scopelus boops*;" more common southward in deep water. It is perhaps not distinct from *Myctophum humboldti*.

Family CHAULIODONTIDÆ.

30. **Zalarges nimbarius** Jordan & Williams, n. gen. and sp. Plate lxxvi.

Head 4 in length to base of caudal; depth 5; D. 9, A. 15. Scales probably present in life, but no traces left except a few impressions. Muscular bands about 42.

Body moderately elongate, subfusiform, formed somewhat as in a stickleback, the tail tapering and slender, the belly broad and not carinate, the sides moderately compressed. Anterior profile of head rising evenly, not convex; a slight depression before eye. Mouth large, low, oblique, the lower jaw somewhat projecting. Premaxillaries short; maxillaries long, expanded, the lower edge curved, overlapping the dentary bones. Maxillaries extending beyond eye, to angle of preopercle, as in *Stolephorus*, their length $1\frac{2}{5}$ in head, their tip acutish. Eye very large, 3 in head; snout 4. Bones of lower jaw thin, broadly expanded, meeting across the throat at the articular joint, leaving a club-shaped naked area under the chin. Entire edge of maxillary armed with a single series of slender sharp teeth, somewhat unequal, some of them forming slender canines, which are however but little longer than the other teeth and not fang-like. Teeth in lower jaw similar, those of both jaws largely directed forward. No teeth on vomer or tongue; a row of small slender teeth on each palatine bone. No pseudobranchiæ. Gill-rakers rather long and slender, about 5+17 in number, the longest about half eye. Branchiostegals short, 8 in number. Opercle short and thin; scarcely striated; subopercle and interopercle developed.

Photophores large and conspicuous, forming convex pearly bodies on a dark background. Two series in a straight line along lower part of sides, making four series in all. The two lower series run from chin to the caudal fin, 47 in each series, 10+13+9+8+7=47. The two upper rows begin under chin at front of isthmus and extend to front of anal fin, 24 in each row, 10+13+11; 8 photophores along branchiostegal membranes, one for each ray, all overlapped but not hidden by the broad transparent rami of the lower jaw: one photophore on preopercle, one on subopercle, one on preorbital, and one at lower posterior margin of eye: 2 under tip of chin.

Dorsal fin low, inserted on posterior half of body, somewhat behind ventrals, at a point midway between preopercle and base of caudal, its last rays extending over the anterior third or fourth of anal; adipose fin not evident, perhaps obliterated. Caudal apparently lunate, $1\frac{3}{5}$ in head. Anal low, its base $1\frac{1}{3}$ in head. Ventrals $2\frac{1}{4}$ in head, inserted midway between front of eye and base of caudal. Pectorals inserted very low, narrow and pointed, $1\frac{3}{5}$ in head.

Back brownish, the sides burnished silvery; silvery area on cheeks Y-shaped, the Y placed obliquely. Fins with some dark dots, these forming obscure bars across caudal; dark specks on back of caudal peduncle, and across base of caudal; some dark dots elsewhere on body.

Type two specimens, each $2\frac{3}{8}$ inches long, and in good condition, numbered 3125 on the register of Leland Stanford Jr. Museum. They were cast up in a storm and thrown by the waves on the deck of a vessel coming in from Australia. The exact locality in the open Pacific is not known. The types were presented by the Young Naturalists' Society to the Museum of Stanford University.

The new genus *Zalarges* seems to belong to the *Chauliodontidæ*, near the Atlantic genus *Yarrella* Goode & Bean. It may be thus defined: Body subfusiform, moderately compressed, probably covered in life with thin caducous scales. Head subacute, the membrane bones normal, thin; mouth large, with expanded maxillary and mandibular bones; lower jaw projecting. Teeth very slender, unequal, uniseral, none on tongue or vomer; no fangs. Eye large. Gill openings very wide; gill-rakers long and slender; branchiostegals 8; no pseudobranchiæ. Photophores conspicuous, in two rows on each side of belly, the upper row ceasing at front of anal; some detached photophores on head. Dorsal short, on posterior half of body, slightly overlapping the short anal. Ventrals inserted before dorsal. Pectorals narrow and low. Coloration silvery. (*Ζάλη*, surges; *ἀργής*, silvery.).

Family ALEPISAURIDÆ.

31. **Alepisaurus borealis** Gill.

Very rare; in deep water. A head from Puget Sound is in the Museum of the California Academy of Sciences.

Family PARALEPIDÆ.

32. **Arctozenus coruscans** (Jordan & Gilbert).

The sole specimen known was taken at Port Townsend in 1880, by Jordan & Gilbert. It is in the U. S. National Museum.

Family AMMODYTIDÆ.

33. **Ammodytes personatus** Girard. SAND LANCE.

Found in immense schools along sandy beaches in Puget Sound. It burrows in the sand between tide marks. It reaches a length of 5 or 6 inches. Two specimens taken.

Family AULORHYNCHIDÆ.

34. **Aulorhynchus flavidus** Gill.

Abundant in sheltered bays. It reaches a length of 5 or 6 inches.

Family GASTEROSTEIDÆ.

35. **Gasterosteus microcephalus** Girard.

Abundant. Specimens obtained in brackish water near Ballard, Seattle. Length 2 inches.

36. **Gasterosteus catraphractus** Pallas.

Specimens obtained in abundance, from 3 to 3½ inches in length. It lives on sandy beaches and spawns in the latter part of July and in August.

Family SYNGNATHIDÆ.

37. **Siphostoma californiense** (Storer). PIPE FISH.

Not very common. It reaches a length of 18 inches.

Family STROMATEIDÆ.

38. **Rhombus simillimus** (Ayres). PÁMPANO.

Rare in Puget Sound. Not taken by us.

Family BRAMIDÆ.

39. **Brama raii** Bloch. POMFRET.

A specimen taken at Port Townsend by Mr. James G. Swan, and reported by him as being not uncommon off Vancouver Island. It reaches a length of about 20 inches. Recently numerous specimens have been taken off San Francisco and Monterey. Not taken by us.

Family EMBIOTOCIDÆ.

40. **Damalichthys argyrosomus** (Girard). WHITE PERCH.

Very abundant; many specimens obtained. It reaches a weight of 2 pounds, and is a common food fish, though not of high quality.

41. **Tæniotoca lateralis** (Agassiz). STRIPED PERCH.

Very common; a brilliantly colored fish. A number of specimens taken. It reaches a weight of 2 pounds, and is an important food fish, finding a ready sale, although the flesh is somewhat poor.

42. **Embiotoca jacksoni** Agassiz. BLUE PERCH; SURF FISH.

Somewhat scarce. It reaches a weight of 1½ pounds. Its flesh is poor. A few specimens obtained.

43. **Brachyistius frenatus** Gill.

Not very abundant; not used as food. Weight ¼ pound.

44. **Cymatogaster aggregatus** Gibbons. SHINER.

The most abundant species of the group. It is small in size and is only used for bait. Several specimens taken.

Family SCORPÆNIDÆ.

45. **Sebastodes melanops** (Girard). "BLACK BASS."

Abundant in Puget Sound and a food fish of value.

46. **Sebastodes mystinus** (Jordan & Gilbert). PRIEST FISH.

Scarce, but more common farther south. No specimens taken by us.

47. **Sebastodes pinniger** (Gill). RED ROCK COD.

Abundant in rather deep water. Not taken by us.

48. **Sebastodes ruberrimus** Cramer. RED ROCK FISH; TAMBOR.

Taken with hook and line in some abundance in Puget Sound.

49. **Sebastodes caurinus** (Richardson).

Very common; brought into the market in abundance. This species has not been found south of Puget Sound, being replaced southward by the very closely allied *Sebastodes vexillaris*. Several specimens obtained by us in the seine.

50. **Sebastodes auriculatus dalli** (Eigenmann & Beeson).

Common; a shallow water species. Many specimens taken with a seine. The specimens of *Sebastodes auriculatus* from Puget Sound are very dark in color, and about half of them lack the coronal spines which are especially characteristic of *Sebastodes auriculatus* on the coast of California. The name *dalli* seems to have been given to a specimen of this type taken at San Francisco. Pending investigation we may adopt the subspecific name *dalli* for the Puget Sound form of this species.

51. **Sebastodes maliger** (Jordan & Gilbert).

Found by Jordan & Gilbert to be a common species in the Straits of Juan de Fuca. Reaches a weight of 6 pounds. Not taken by us.

52. **Sebastodes nebulosus** (Ayres). ROCK COD.

Rather common. No specimens taken by us.

53. **Sebastodes nigrocinctus** (Ayres).

This peculiarly marked rock fish was found by Jordan & Gilbert to be common in the entrance to the Straits of Juan de Fuca, in deep water. No specimens obtained in Puget Sound.

Family HEXAGRAMMIDÆ.

54. **Hexagrammus decagrammus** (Pallas).

Said to be quite common, but less abundant than *Hexagrammus asper*. Not taken by us.

55. **Hexagrammus ordinatus** (Cope).

Taken at Port Angeles, on the south shore of the Straits of Juan de Fuca, by the Albatross. Not seen by us.

56. **Hexagrammus asper** Steller. *Hexagrammus superciliosus* (Pallas).

Not abundant and not taken by us.

57. **Hexagrammus hexagrammus** (Pallas). STARLING.

Abundant everywhere in Puget Sound. It lives about rocky places, and is taken in large numbers with hook and line and nets. It reaches a length of about 16 inches and a weight of 2 or 3 pounds. It is a food fish of fair quality, but inferior to the species of Sebastodes. This is the southern limit of this species. Several specimens were obtained by us.

We may here record a new species of Hexagrammus from Tokio, Japan, hitherto confounded with the American species *Hexagrammus hexagrammus.*

Hexagrammus otakii Jordan & Starks, n sp. Plate lxxvii.

Head 4 in length to base of caudal; depth 4⅓: dorsal XIX–23; anal 21; scales 21–106–34; eye 4⅘ in head; snout 3⅓; highest dorsal spine 2⅘; highest dorsal ray 2⅓; highest anal ray 3; pectoral 1¼; ventral 1⅔; caudal 1⅘.

Body elongate, not greatly compressed, the head small and pointed. Mouth not large, the maxillary reaching slightly past the vertical from anterior rim of orbit; jaws subequal; teeth conical and sharp, the outer row enlarged, smaller on vomer, none on palatines; interorbital space broadly convex; a wide, short, multifid dermal flap over posterior edge of each eye.

Head scaled above to slightly in front of eyes, opercle and cheek entirely and densely covered with small scales; snout, preorbital, suborbital, lower jaw and interopercle naked. Scales everywhere, except on cheeks and opercles, strongly ctenoid.

Lateral lines 5 on each side; the upper one from nape parallel with dorsal, stopping under the beginning of posterior fourth of soft dorsal, sometimes uniting with the second lateral line at this point, not joining its fellow of the opposite side in front of dorsal; the second running parallel with it, continued to base of caudal, situated below the first in distance equal to half eye; the third, parallel with curve of back, running from the upper end of the gill-opening to the base of the caudal; the fourth short, beginning slightly in front of ventral, past the outer edge of its base, not reaching to its tips; the fifth parallel to anal, in distance above it equal to space between upper lateral lines, barely reaching base of caudal posteriorly, anteriorly joining its fellow of the opposite side between vent and base of anal, continuing simple forward. Of four specimens examined, in two it stops at the base of ventrals, in the others it ends midway between that point and isthmus.

First and second dorsal and anal subequal in length of base; spinous dorsal very slightly higher than soft dorsal, its origin slightly behind upper end of gill-opening, the notch between it and soft dorsal shallow; origin of anal midway between front of eye and base of caudal, its rays shorter than those of soft dorsal; pectoral short and wide, the rays toward the upper edge the longest, the tips of which reach to end of ventrals; origin of ventrals behind that of pectorals a space equal in distance to length of snout; caudal short, shallowly lunate.

Color light brown above, white or yellowish below, variously marked with irregular dark brown mottlings and spots arranged chain-like; top of head and snout dark; dorsals dark and mottled; pectorals crossed with irregular bars; ventrals dusky, not black at tips; anal dusky and mottled, the end of each ray white.

Four specimens, collected by Mr. Keinosuke Otaki, a graduate of the Department of Zoology in Stanford University, now a member of the Fish Commission of Japan. They are from the markets at Tokio, Japan, the largest about 9 inches in length.

This is the species recorded from Tokio by Dr. Steindachner (Beitr. Kenntniss Fische Japans, iv, 66) as *Hexagrammus asper*. It is not likely that the latter American species occurs in Japan.

58. **Ophiodon elongatus** Girard. CULTUS COD; BLUE COD.

Abundant. An important food fish, reaching a weight of 60 pounds.

59. Oxylebius pictus Gill. Plate lxxviii.

Not uncommon, living among the rocks near shore. Not taken by us.

60. Zaniolepis latipinnis Girard.

Rare in Puget Sound. It reaches a length of a foot. Two specimens obtained by Prof. O. B. Johnson are in the Museum of the Young Naturalists' Society.

61. Anoplopoma fimbria (Pallas). Black Cod; Beshowe.

Common in Puget Sound, where it is valued as a food fish. It reaches a length of 40 inches.

Family COTTIDÆ.

62. Jordania zonope Starks. Plate lxxix.

Jordania zonope Starks, Proc. Ac. Nat. Sci., Phila., 1895, p. [illegible]. The three type specimens of this singular fish were collected in channel rocks near Point Orchard. The largest specimen (No. [illegible] L. S. Jr. Univ.) is 4 inches long. This species has [illegible]=40 vertebræ, a number considerably in excess of that found in the related genera *Icelus* and *Artedius*.

The following is the original description of *Jordania zonope*:

Genus Jordania Starks.

Allied but not closely to *Triglops* and *Chitonotus*.

and sharply serrate. Gill-membranes united, free from isthmus; a slit behind last gill. Spinous dorsal with very long base of about 17 spines, longer than the soft dorsal; anal long; ventrals 1, 5, inserted behind base of pectorals.

JORDANIA ZONOPE Starks.

Head $3\frac{1}{2}$ in length of body; depth $5\frac{1}{2}$; dorsal XVII–15; anal 22; lateral line 50; orbit $3\frac{2}{3}$ in head; maxillary $3\frac{1}{3}$; longest dorsal spine $1\frac{5}{8}$; longest dorsal ray $2\frac{1}{8}$; longest anal ray $2\frac{1}{3}$; length of ventrals $1\frac{3}{4}$; pectorals $\frac{1}{4}$ longer than head; caudal $1\frac{4}{5}$.

Body rather elongate, compressed posteriorly, not much anteriorly, the back not elevated; dorsal and ventral outlines almost straight from head to caudal peduncle.

Head not large, profile from front of dorsal to eyes nearly horizontal and straight, then abruptly turning steeply downward to end of snout, lower profile gently curved from chin to ventral fins.

Mouth small, the maxillary not reaching the vertical from front of orbit; jaws about equal or the lower slightly projecting; teeth in villiform bands on jaws, vomer, and palatines; eyes large, set high in head, a little shorter than snout; interorbital space deeply concave, half as wide as eye; a slip of skin, half as long as the diameter of the eye, over the anterior edge of each eye, and a longer one over the posterior edge; a few minute fleshy slips on nape; nasal spines long and sharp, somewhat curved back; spine on preopercle simple, hooked up, a minute spine above it, and a blunt spine below; posterior end of interopercle prominent, forming a blunt spine; opercle produced posteriorly in a flap, which lies in a shallow groove in the shoulder girdle; no opercular spine; gill-membranes united, but not joined to the isthmus; a distinct slit behind fourth gill arch; branchiostegals 5.

Top of head to middle of eyes, opercles and upper part of preopercles closely covered with small rough scales: head otherwise naked; body above lateral line completely covered with ctenoid scales, not very regular in size, arranged in about 67 series; lower half of body covered to within a short distance of anal with about 50 oblique plate-like folds of skin, the posterior edges of which are finely and sharply serrate, the pores of lateral line are situated in the upper end of these folds; pectoral base, belly and a narrow space along base of anal, naked; fins, with the exception of pectoral, which has a few rough scales on the rays, naked.

Dorsal spines slender, the first one inserted in advance of pectoral base, directly over the upper end of gill-opening, the fin somewhat round in outline, the spines not varying greatly in length, with the exception of two or three on each side; soft dorsal a little lower than spinous, the rays subequal, its base is a little shorter than the base of first dorsal, and slightly longer than the length of head: ventral fins long, their tips reaching past front of anal fin, their length equal to the distance from snout to edge of preopercle, the pubic bone very prominent; pectoral fins long and curved upward, the middle rays the longest, reaching far past tips of ventrals and front of anal to the space between dorsals; the ends of lower rays free, the width of the fin at its base is contained three times in the length of the head; caudal rounded.

Color in spirits blackish, with traces of 4 or 5 darker cross-bars on back, sides below lateral line mottled, faint dark spots along lateral line, more conspicuous posteriorly; a dark bar half as wide as eye, running from eye downward across cheek to anterior end of interopercle; bordered on each side by a light streak, a similar bordered bar running across top of head, slightly turning around posterior margin of orbit, downward along margin of preopercle, and ending on posterior end of interopercle; snout abruptly black, lips dark; fins all dark and slightly mottled, tips of ventral, anal, and caudal rays a little lighter; caudal and pectoral dark at base; slips on top of head black; belly very finely dusted with minute dark points.

This species is not uncommon in Puget Sound; the types are three specimens taken in channel rocks at Point Orchard, near Seattle, by Miss Maud Parker and Mr. Adam Hubbert, members of the Young Naturalists' Society of Seattle. The largest of these is 4 inches in length. The types are in the Museum of the Leland Stanford Junior University, numbered 3124. Unfortunately the life colors of this brilliant species were not taken. There is in life much red on the lateral plates and elsewhere on the body and fins. This disappears at once in alcohol.

63. **Radulinus asprellus** Gilbert. Plate lxxxi.

Not common; two specimens dredged near Seattle, the larger about 4 inches in length.

64. **Chitonotus pugettensis** (Steindachner).

. Not common; two specimens obtained with a seine. It reaches a length of 9 inches.

65. **Ruscarius meanyi** Jordan & Starks, n. gen. and sp. Plate lxxx.

Head 2½ in length; depth 3½; dorsal X–14; anal 12; lateral line 6–32; orbit 4 in head; maxillary 2; snout 4; highest dorsal spine 3; highest dorsal ray 3; pectoral 1½; ventrals 2⅔; caudal 2⅓.

Body robust, deepest and broadest at shoulders, tapering quickly backwards into a slender caudal peduncle; back somewhat elevated; ventral outline nearly straight from chin to caudal fin; dorsal outline gently and evenly curved from snout to caudal peduncle.

Mouth terminal and nearly horizontal, maxillary reaching past pupil nearly to posterior edge of orbit; jaws subequal; teeth in narrow villiform bands on jaws, vomer and palatines; process of premaxillary prominent, extending between and above nasal spines; preopercular process well developed, long, near its tip a very small second spine is developed, making the process bifurcate, 3 or 4 short spines below on edge of preopercle; opercle ending in a flap; top of head with dermal flaps, one over anterior margin of eye, and a group of 2 or 3 over posterior margin; a few shorter ones on nape; mucous pores around mandible, large; opercle, upper part of preopercle, top of head to eyes, and the orbital ring covered with sharply ctenoid scales, upper part of eyeball with small rough scales, balance of head naked.

Lateral line with a row of rough plates; upper half of body completely covered with scales, their anterior edge imbedded, coarsely ctenoid on their posterior edge; lower half of body naked.

Dorsal spines slender, those in the middle highest, the fin without a notch, the longest spines reaching to front of soft dorsal where fin is depressed, well separated from soft dorsal; first dorsal ray inserted over first anal ray, the fin longer and higher than anal; pectorals somewhat pointed posteriorly, reaching just past the space between dorsals; ventrals inserted behind the base of pectorals a distance equal to the length [illegible] their tips reaching [illegible] shades above; a dark bar from eye to side of snout, one from eye downwards past end of maxillary, another behind it across posterior edge of preopercle; some dark markings on maxillary; lower lip dark; pectorals light, with dark wavy lines across them; dorsal fins dark and mottled; anal and ventrals varying from white to black; caudal with a dark bar at base, light with irregular dark cross markings.

Two specimens dredged, about 1½ inches in length. They are in the Leland Stanford Jr. University Museum, No. 3127.

This species is the type of the new genus *Ruscarius*, allied to *Chitonotus*, but distinguished by the continuous dorsal, scaly back, and weak armature of the preopercle. It is named for Mr. Edmond Stephen Meany, Secretary of the University of Washington, in recognition of his work in the Young Naturalists' Society.

66. **Astrolytes fenestralis** (Jordan & Gilbert).

Common; several specimens obtained with a seine. It is not found in rock pools. It reaches a length of 5 inches. Vertebræ 8 + 25 = 33. This species is the type of a distinct genus, *Astrolytes*, distinguished from *Artedius* by the scaly, rough, uneven cranium, and more strongly armed preopercle.

67. **Artedius lateralis** Girard.

Two specimens obtained with a seine; probably not abundant.

Color in alcohol very dark; the head black; the body dark olive green, with faint pale markings on sides above lateral line; below with numerous clear-cut white spots, irregular in size, none of them half as large as pupil; belly dusky or white; fins all jet black; first dorsal with 2 or 3 faint light bars across the spines running backward and downward; soft dorsal with 7 or 8 series of spots on the membrane, not involving the rays, running obliquely backward and downward; other fins plain black.

68. **Hemilepidotus hemilepidotus** (Tilesius).

Very abundant in shallow water among weeds, and in rocky places. It reaches a length of 15 inches; rarely used for food. Several specimens obtained with hook and line.

69. **Acanthocottus polyacanthocephalus** (Pallas).

Abundant. One of the largest cottoids, reaching a length of 2 feet. Specimens collected with the seine.

70. **Enophrys bison** (Girard).

Abundant. An exceedingly ugly-looking fish, reaching a length of 12 inches. It is not used for food. Several specimens obtained with the seine on sandy beaches.

71. **Leptocottus armatus** Girard.

The most common large cottoid in Puget Sound. It reaches the length of a foot, and is seldom used for food. Specimens obtained in abundance.

72. **Scorpænichthys marmoratus** (Girard).

Not uncommon; said to reach a weight of 20 to 25 pounds in Puget Sound. It is not valued as a food fish.

73. **Blennicottus globiceps** (Girard).

Rather common, in pools left in the sand by the tide. Several specimens taken near Neah Bay. The largest was 6½ inches long, this being the largest of this species on record. These specimens (subspecies *bryosus*) have many more cirri on the head than southern specimens.

74. **Oligocottus embryum** Jordan & Starks, n. sp. Plate lxxxii.

Head 4 in length; depth 4¼: dorsal IX–15; anal 10;

orbit 4 in head; snout 4; maxillary $2\frac{5}{8}$; highest dorsal spine $2\frac{1}{2}$; dorsal ray $1\frac{3}{4}$; anal ray $1\frac{3}{4}$; length of caudal fin $1\frac{2}{3}$; ventrals $1\frac{1}{8}$; pectorals $2\frac{1}{2}$ in body.

Body elongate, compressed, back slightly elevated, deepest under spinous dorsal; caudal peduncle moderately slender. Skin perfectly smooth.

Head small, tapering rapidly forward to the rather sharp snout as viewed from above; profile of head, straight below, acutely and evenly rounded above; mouth terminal, horizontal; maxillary reaching to the vertical from the middle of pupil; lower jaw included; teeth on jaws, vomer and palatines, in narrow villiform bands; process of premaxillary prominent, extending slightly above nasal spines, giving the appearance of three spines above snout; eye set high in head, the orbit as long as snout; preopercular spine short, blunt and triangular, entirely covered with the skin; edge of preopercle below, entire; opercle ending in a short flap; top of head with two rows of "mossy" cirri, running from the superior orbital margin, curving over head and continuing on lateral line; they disappear on its anterior third.

Dorsal spines rather stout, the fin lower than soft dorsal, rounded in outline; soft dorsal well separated from spinous, the front of fin the highest; pectorals long, the eighth ray the longest, rendering the fin pointed behind; it reaches to the base of about the seventh dorsal ray. The pectoral rays below the eighth are swollen, and posteriorly free from the membrane; anal about as high as soft dorsal, the rays somewhat swollen and more or less free; ventrals long, reaching about to front of anal, their insertion behind base of pectoral, a distance equal to the snout and eye; caudal fin slightly rounded.

Color varying from light green to a rich maroon; traces of 5 or 6 dark cross-bars on back, lower parts dusky with small light spots; belly white; a dark bar from eye to side of snout, one from eye to edge of preopercle behind end of maxillary, and another from eye to below preopercular spine; lips black; lower rays of pectorals crossed with black and white bars, which fade out above; ventrals light with some dusky mottlings; dorsal dark above, light at base, no markings; anal with black and white bars running across the rays, caudal [illegible] mottled.

Two specimens collected in the tide pools left in the sand on a beach a couple of miles east of Neah Bay, the largest 2½ inches in length. They are now in the Leland Stanford Jr. University collection, No. [illegible].

A third specimen has been collected at Point Lobos, Monterey County, California, on Carmelo Bay, by Mr. John O. Snyder. This specimen is considerably brighter in color and the markings are more distinct.

75. Oligocottus maculosus Girard.

Very abundant. Specimens taken in large numbers in a muddy lagoon near Point Orchard. It is one of the smallest of the marine Cottidae, not over 3 inches in length. A number of specimens were also taken at Neah Bay in tide pools. These differ from the others in being lighter in color and in having many more cirri on the top of the head.

76. Dasycottus setiger Bean. Plate [illegible].

One small specimen brought up in the dredge, 2½ inches in length. [illegible]

77. Nautichthys oculofasciatus [illegible]

[illegible]

79. **Ascelichthys rhodorus** Jordan & Gilbert. Plate lxxxiv.

Plentiful at Waadda Island, Neah Bay. It is found under rocks between tide marks. Not a very active fish. This is the type locality where it was first taken by Jordan & Gilbert in 1880. It reaches a length of 3 inches. It occurs also on the rocky coast about Cape Mendocino in California.

Family PSYCHROLUTIDÆ.

80. **Psychrolutes paradoxus** Günther.

The original type from the Gulf of Georgia. Not obtained by recent collectors. Dr. Boulenger informs us that twelve dorsal rays are present in the original type, three of them entirely hidden by the skin.

81. **Psychrolutes zebra** Bean. Plate lxxxv.

Probably rare. One small specimen obtained, about an inch and a half in length.

82. **Gilbertina sigalutes** Jordan & Starks, n. gen. and sp. Plate lxxxvi.

Head 3 in length of body; depth 4; dorsal VIII, 18; anal 14; ventrals I, 3; pectoral 15; eye 6 in head; interorbital $2\frac{1}{2}$; maxillary $2\frac{1}{5}$; ventrals 2; pectorals 1; caudal $2\frac{1}{4}$; base of dorsal $1\frac{3}{5}$ in length of body; base of anal 3.

Body rather slender, robust anteriorly, compressed posteriorly, the greatest breadth and depth at shoulders. Head large, the nape slightly produced; mouth large and broadly rounded, oblique, the jaws about equal; maxillary extending to posterior margin of eye, its end buried under the skin of the cheek; eyes placed high, the interorbital space very wide and slightly convex, its width about $2\frac{1}{2}$ times that of the eye; the posterior end of mandible very prominent: bones of head cavernous, largely made up of cartilage: anterior end of preorbital forming a blunt spine over mouth: process of premaxillary prominent: a couple of blunt projections behind each eye: upper part of shoulder girdle projecting, forming a blunt spine on nape above gill-slit: a row of large pores around suborbital ring, and along under part of mandible: no opercular spines.

Head and body covered with a very loose, naked, movable skin; dorsal fin continuous: no notch between spines and soft rays; the spines very slender, the first one inserted over end of opercular flap: the last rays reach to the base of caudal fin: anal lower than dorsal, its origin midway between base of caudal fin and posterior margin of eye, ending at about the same point that dorsal does, but not reaching so far: pectorals long and slender, reaching past front of anal and over half way between their bases and base of caudal fin: they are adnate to the body for the anterior third or fourth of their length: ventrals long, not quite reaching to vent, adnate to the body for half their length: caudal fin rounded.

Color light olivaceous; body and head with innumerable dark points giving the fish a dusky appearance: a large dark blotch across body at the posterior end of the dorsal and anal: a similar spot under pectoral: head uniform dusky, lighter below: belly white, middle of pectoral dark: dorsals dark: lower fins white.

A single small specimen dredged, $1\frac{1}{2}$ inches in length. It is numbered [illegible] on the register of the Leland Stanford Jr. University Museum.

[illegible] marked genus. [illegible] long dorsal [illegible] It is named [illegible] contributed more than any one else to the knowledge of the fishes of the Northern Pacific.

Family RHAMPHOCOTTIDÆ.

83. **Rhamphocottus richardsoni** Günther. Plate lxxxvii.

Head 2; depth 2; dorsal VII, 13; anal 6; pectoral 14; orbit 6 in head; maxillary 4; snout 3; highest dorsal spine 6⅓; dorsal ray 4; anal ray 4; pectoral 2⅙; ventral 2; caudal 3.

Body short, compressed, the back elevated, its greatest depth just in front of spinous dorsal.

Head large, as long as the rest of the body; snout long and narrow; mouth U-shaped, its gape longer than wide, lips thick, their surface broken up into papillæ; maxillary reaching the nasal spine; lower jaw included; teeth in villiform bands on jaws and vomer, none on palatines; eye placed high, its diameter contained twice in the snout, once and a half in the interorbital; a branched dermal flap, as long as pupil, at tip of the snout; head with two large bony ridges above, continuous with the orbital rim and ending in strong blunt spines at occiput, head deeply concave between these ridges; nasal spine sharp and recurved; a pair of strong spines over the eyes; a sharp spine just above opercle, a blunt one on opercle below flap, and a long sharp one at angle of preopercle; a low bony ridge leads to each of these spines; a long sharp spine on clavicle just behind gill-opening; a blunt bony knob at posterior end of mandible; gill-openings extending upward from upper pectoral ray, their length equal to the length of the snout.

The entire head and body covered with multifid spines, those on head much smaller than the ones on sides; a few simple spines along cephalic ridges; the first dorsal spine covered with spinules, and each dorsal ray has a row on its side; a few spines on the base of the pectoral and anal rays.

Dorsal spines very weak, fitting in a groove in the back; soft dorsal higher than spinous, the tips of the rays reaching the base of caudal fin; anal short, few rayed, reaching slightly beyond soft dorsal; pectorals pointed, their lower rays entirely free, reaching about to the base of the third anal ray; ventrals reaching to ends of pectorals, their origin behind the lower part of pectoral base a distance equal to the length of snout; caudal rounded behind.

Body creamy yellow, with conspicuous irregular dark stripes, edged with black, running obliquely across the body; similar stripes radiating from the eye in all directions, one to end of snout, a triangular one downwards, one running backwards and downwards to middle of preopercle, then turning upwards and running nearly to occipital spine, two or three short ones above; each of these involving the membrane of eye; 2 or 3 black-bordered dark spots on edge of opercle; a light yellow streak surrounded by black across caudal peduncle, behind which all is bright cherry-red to the end of caudal fin; two similar spots on base of pectoral; top of head crossed with wavy black-edged dark bars; tip of lower jaw black; a line of black spots running along under parts of mandible; fins all bright red, each ray of dorsal with a sharp black spot at its base; a few spots on dorsal spines; anal, pectorals and ventrals [illegible]

[illegible]

The posterior end of the prominent ridge, which runs backward from the superior orbital rim on each side, is formed by the epiotic process. It ends in the form of a long "occipital spine;" almost directly under it is the short parotic process.

The post-temporal is short, wide and flat; its upper end is attached to the inner side of the epiotic spine, and for the whole length of its anterior edge, to the skull between the epiotic and parotic processes. From its lower inner surface it sends a wide thin bone, which is firmly fastened to the base of the skull. It bears a backward projecting spine on its lower end, inside of which the supra-clavicle is attached.

Actinosts large, wide and thin, without an opening between them. Subopercle absent; preopercle large, sending a spine backwards; opercle triangular on its lower inner angle, the interopercle is developed and strongly coossified with it; it sends a slender process forward under the preopercle; a projection downward from the posterior end of the articular; suborbital wide, thin and concavo-convex, its convex surface outwards. Skull without basal chamber; vertebræ 10 + 14.

Family AGONIDÆ.

84. **Aspidophoroides inermis** Günther.

The type from Vancouver Island recorded by Günther.

85. **Bothragonus swani** (Steindachner).

Known only from the type taken near Port Townsend.

86. **Pallasina barbata** (Steindachner).

Taken at Port Angeles by the Albatross.

This species is the type of the genus *Pallasina* Cramer, distinguished from *Brachyopsis* by the long, Syngnathus-like body, and by the presence of a long barbel at the chin. The genus *Siphagonus*, to which Dr. Steindachner refers it, is based on *Agonus segaliensis*, which seems to be a true *Brachyopsis*.

87. **Podothecus acipenserinus** (Pallas).

Very abundant on sand beaches, where it is taken with seines. It reaches a foot in length. Many specimens taken.

Two additional new species of *Podothecus*, presented by the Alaska Commercial Company, collected by Capt. J. G. Blair at Robin Island, in the Gulf of Patience, Saghalien, may be here recorded:

Podothecus accipiter Jordan & Starks, n. sp. Plate lxxxviii.

Head $3\frac{3}{8}$ in length; depth $6\frac{1}{2}$; dorsal VIII–9; anal 10; pectoral 15; lateral plates 36; eye $4\frac{1}{2}$ in head; snout $2\frac{1}{8}$; second dorsal spine $1\frac{5}{8}$; second dorsal ray $1\frac{1}{2}$; third anal ray $1\frac{3}{5}$; caudal $1\frac{4}{5}$; upper ray of pectoral $1\frac{1}{7}$; ventrals $2\frac{1}{3}$.

Body elongate, not compressed; head triangular as viewed from above; the mouth wide, entirely inferior, ∩-shaped, the lower jaw shutting behind the upper by a distance equal to half eye; maxillary not reaching quite to anterior orbital rim; distance of anterior edge of upper lip from tip of rostral spines a little more than half eye; teeth in upper jaw almost obsolete; villiform band of teeth in lower jaw, wide in front becoming narrow at sides; vomer and palatines toothless; a patch of thick barbels below snout in front of mouth, the longest equal to vertical diameter of eye, a similar patch at end of maxillary, about equal in length to the shortest on snout; two short barbels on each side of lower lip between symphysis and angle of mouth. A pair of short, sharp, rostral spines, pointing directly forwards; at their base and much wider apart is a pair of spines which point upwards, backwards and slightly outwards; running backwards from these are the ridges that bound the wide groove in which the maxillary process fits; these approach each other behind and end in sharp spines which point backwards and upwards; these spines are midway between middle of eye and the spines behind rostral spines; no median or movable spine at tip of snout; a pair of large spines above posterior third of eye and a pair of larger ones at occiput, these are continuous with the dorsal ridges; a curved ridge running from superior orbital rim and ending in a small spine just above opercle; a small ridge on opercle; preopercle with a large spine; a couple of spines below eye at lower edge of suborbitals, running from them to tip of snout is a ridge along lower edge of preorbitals; it is somewhat irregular but without spines; interorbital space wide and deeply concave, a pair of ridges on each side, converging forwards; supraorbital rim prominent; anterior nostril ending in a short, wide, conical papilla, with a small opening at the apex; no noticeable depression at occiput.

Dorsal ridges converging from the occiput to behind the soft dorsal; they unite on the second plate behind the base of last dorsal ray, this is continued as a single ridge on about 8 plates where it becomes obsolete; the upper lateral ridge follows the course of the lateral line to about the middle of spinous dorsal, where it slants sharply upward and is continued to tail above lateral line; lateral line midway between upper and lower lateral ridges posteriorly; a single spine above base of pectoral indicating an obsolete ridge between the lateral ridges; lower lateral ridge becoming obsolete under pectoral on 2 or 3 plates behind its base; abdominal ridges widest apart behind base of ventrals, uniting directly behind anal base and running simple backwards, becoming obsolete on caudal peduncle; all the ridges with sharp recurved spines, with the exception of abdominal ridges behind part of anal; where the dorsal and anal ridges disappear the caudal peduncle assumes a quadrangular shape, the corners being formed by the spines of the lateral ridges; no row of spines around base of caudal or pectoral.

Fins all very high, origin of dorsal between the fourth and fifth dorsal plates, the fin to base of last spine covering 6 plates, the membrane covering 2½ more; the second and third spines the longest, a membrane connecting the last spine to the body for its whole length; when fin is depressed the ends of the last spines reach to the front of second dorsal; the second dorsal to end of last ray covers 8 plates, the membrane covers one more; the second and third rays are the longest, the last ray is connected to the body for about a third of its length; base of anal covering 8½ plates; the rays are very long and not differing much in length, the last ray not connected to body by a membrane; the fin begins in front of soft dorsal but is about coterminous with it, its rays when depressed reaching past ends of soft dorsal, 6 plates past base of its last ray; pectorals barely reaching to tip of last dorsal spine, the fin pointed above, first and second rays the longest, the lower rays produced beyond the membrane, making a notch in posterior outline of fin; origin of ventrals directly below base of pectoral, their tips reaching 6 plates beyond their base: caudal long and truncated; vent directly behind base of ventrals.

Color light brown above, white below: back with many narrow brown bars placed at irregular distances apart: head with many blended brown spots, one under eye, one on front margin of eye, one or two on top of head, one behind eye, one on preorbital, a similar spot on base of pectoral rays; pectoral dusky. First dorsal with 3 rows of spots across the rays, a very narrow brown border to fin; second dorsal with similar spots, not arranged in rows; anal light above, uniform brown below; ventrals white; caudal fin dark at base with 3 or four dark spots towards middle of fin.

One specimen collected at Robin Island, by Capt. J. G. Blair. It is 8 inches in length.

Podothecus veternus Jordan & Starks, n. sp. Plate lxxxix.

Head $3\frac{2}{3}$ in length; depth $7\frac{1}{3}$; dorsal IX–8; anal 8; pectoral 15; lateral plates 36; orbit $4\frac{1}{5}$ in head; snout $2\frac{1}{5}$; upper rays of pectoral $1\frac{1}{2}$; highest dorsal spine $2\frac{1}{3}$; highest dorsal ray $2\frac{3}{5}$; highest anal ray $2\frac{3}{5}$; caudal $2\frac{1}{2}$.

Body elongate, about as wide as deep anteriorly, much wider than deep posteriorly; mouth inferior, the lower jaw shutting far behind the upper; teeth on jaws, vomer, and palatines obsolete; a few short barbels beneath snout in front of mouth, and at angle of mouth; their length about equal to pupil; bones of lower jaw extensively cavernous.

A pair of short blunt rostral spines pointing directly forwards; at their base and wider apart is a pair of sharp spines curving outwards, backwards and upwards; at the posterior end of the rather wide rostral groove are a pair of small spines pointing upwards and backwards; from their base a pair of diverging ridges run through the interorbital to above posterior margin of orbit. No median or movable spine at tip of snout. A strong spine over eye, and a longer one at occiput; a low sharp ridge on side of head, running from ocular spine and ending in a low spine at upper end of gill-opening; a very low ridge on opercle not ending in a spine; preopercle with a strong spine with a wide keel-like base; a hooked spine below eye on suborbital, from which a ridge runs along lower edge of preorbital to end of snout, below posterior end of rostral groove; on this ridge is a triangular spine pointing backwards; between this and the suborbital spine is an acute outward pointing spine not much widened at its base; interorbital concave, its width equal to the length of the eye, 2 in snout; supraorbital rim prominent. The dorsal ridge of body is continuous with occipital and supraorbital spines, it joins its fellow of the opposite side posteriorly, directly behind the second dorsal, and is continued simple on caudal peduncle; the spines are large and strongly hooked back anteriorly, becoming nearly obsolete posteriorly, only traceable on caudal peduncle by the center of each plate on the median line being slightly produced; spines on lateral ridges with stronger spines near middle of body than anteriorly or posteriorly; two or three blunt spines above base of pectoral, indicating an obsolete ridge between lateral ridges; lateral line at end of pectoral fin running along the upper lateral ridge a short distance, and becoming obsolete anteriorly; spines of abdominal ridge low and blunt, nearly obsolete posteriorly; the ridge joins its fellow of the opposite side directly behind base of anal fin and continues as a single low ridge on caudal peduncle; a small plate before base of each ventral; a median row of three running forward to gill membrane, three on each side of these, a row around base of pectorals. Origin of dorsal behind the fourth dorsal plate; including the membrane behind, it covers 9 plates; one plate between dorsals; the second dorsal covers 8 plates, behind which are 14 plates; the last ray of first and second dorsal and anal, are connected to the body by a membrane: upper ray of pectoral the longest, reaching to below the ninth or tenth spine of dorsal ridge, the lower rays slightly produced beyond the membrane.

Color in spirits, reddish-brown above, light below; narrow, irregular, transverse dark streaks across back and sides; a longitudinal dark bar along each side of base of both dorsals; a dark streak forward from eye; margin of spinous dorsal blackish; soft dorsal with a small spot behind, a dark spot on pectoral rays near their base and some dark bars behind it across rays; anal and ventrals colorless; caudal dusky.

A single specimen collected by Capt. J. G. Blair at Robin Island, about 8 inches in length.

Related to *P. acipenserinus* and *P. gilberti.* It differs from the former in having fewer and shorter barbels, teeth on jaws obsolete, keel and preopercle larger; dorsal ridges without spines posteriorly, and the spines on the preorbital ridge different in shape; from the latter in having the body different in shape, not everywhere deeper than wide, but the reverse posteriorly; anal much shorter and lower; no teeth on jaws, and the spines on preorbital ridge better developed and different in shape.

Allied to *Podothecus* is the genus *Stelgis* Cramer, of which *Podothecus vulsus* is the type. It is distinguished from *Podothecus* mainly by the comparative lack of barbels and by details of armature. We present a figure of the species drawn from the original type of *Stelgis vulsus*, the only specimen yet known. (Plate xc.)

88. Averruncus emmelane Jordan & Starks, n. gen. and sp. Plate xci.

Head from tips of rostral spines 4 in length of body; depth 7½; dorsal VIII or IX–8; anal 11; pectoral 14; lateral line 35; orbit 4 in head; snout to tips of rostral spines 3⅛; maxillary 3¾; interorbital 6¼; pectoral $1\frac{2}{5}$; second dorsal spine $2\frac{3}{5}$; third dorsal ray 2⅓; longest anal ray $2\frac{3}{5}$; caudal fin 2.

Body elongate, subcylindrical, the caudal peduncle long

and slender, very slightly depressed. about three plates in front of caudal fin it widens slightly and is compressed; belly somewhat prominent. breaking the otherwise straight vertical outline from chin to caudal fin: dorsal outline straight from occiput to caudal fin.

Head as viewed from above almost regularly triangular. the prominent preopercular spines and the snout forming the angles: its dorsal profile irregular. much broken by spines.

Mouth inferior. broadly U-shaped. maxillary reaching just past the vertical from front of orbit: lips thick, covered with short. fine papillæ: upper jaw protractile; teeth small, in villiform bands. on the jaws. vomer and palatines; the distance from the anterior edge of premaxillary to end of the rostral spines is less than half the length of snout; two cirri as long as pupil under rostral spines, anterior lower edge of preorbitals with cirri. a group of 3 cirri on end of maxillary. and a group of 4 or 5 on posterior end of mandible; one on the middle of each branchiostegal ray. these forming a line from isthmus to opercle an area on chin from the mouth to the hyal bones "woolly" with short cirri; 2 or 3 cirri on lower edge of opercle and interopercle. A pair of parallel rostral spines pointing forward, their tips covered with skin: behind them is a deep oval pit, on the anterior outer edge of which are a pair of spines that point upward and outward and are slightly hooked backwards: at the posterior end of the pit are two spines. smaller than those above. and slightly curved backward: no median nor movable spine at tip of snout: a group of four short spines around anterior edge of eye. and one large triangular spine over posterior edge: the interorbital space is deeply concave. with a low sharp ridge on each side of the median line: preopercle very rough with irregular spines and tubercles:

middle of suborbital stay with a strong hooked spine; below the stay, on the naked area, are 2 or 3 plates with spines on their centers; angle of preopercle with a large sharp spine; along the lower edge of preopercle are 3 or 4 blunt spines; a ridge of 4 spines running back from each eye, corresponding with the dorsal keels of body; below this on each side is a ridge, somewhat irregular but not broken into spines, terminating in a spine that points between the dorsal and upper lateral keel of body; a small ridge on upper edge of opercle which does not end in a spine; a few small spines around posterior edge of opercle; a few minute spines along median line of top of head, the upper part of the eye covered with minute prickles. At the occiput is a deep pit, broader and deeper than long, divided by a low ridge through its middle.

Body with four ridges on each side, formed by the body plates, each plate ending in a strong recurved spine, except those of the abdominal ridge, which are smooth; a row of minute spines along median dorsal line from first dorsal to occiput; small spines following the lateral line; no trace of keels or spines in front of ventrals. The abdominal ridges are widest apart on the belly, they unite on the tenth plate in front of the caudal fin. The dorsal ridges unite on the ninth scale in front of the caudal fin, but the spines continue double to the tail; a row of sharp, small spines around the base of the pectoral and caudal fins.

Dorsal spines slender, the fin highest in front, the second spine the longest, its tip reaching to the base of the next to the last spine when the fin is depressed; third dorsal ray the highest, its tip reaching nearly to the last ray when depressed; the last ray is very short and adnate to the body for the whole of its length. Lower rays of pectoral fins produced, extending beyond the membrane, the longest extending beyond the upper ray of the fin; anal longer and lower than soft dorsal, ending at the same corresponding place; last ray reaching to the fifteenth plate before caudal fin. Ventrals differing in length in the different sexes, reaching from slightly beyond vent to nearly half its length beyond; inserted slightly behind pectorals; caudal fin rounded behind; vent anterior, situated on the tenth plate in front of anal.

Color dark brown, belly white; sides crossed with irregular white bars, giving the fish a mottled appearance, besides dark dashes as though the fish had been bathed in ink ([illegible]); snout black; a black streak along lower edge of preopercle; a black spot on iris above; dorsals light, mottled with black; anal white with dark mottlings, a dark bar across the posterior rays, the tips of all the rays white; ventrals black, abruptly white at tips; pectoral and caudal dark with a white border, a light spot in the center of fins, and many white spots on the rays; a black spot at base of pectoral.

Two specimens collected with a seine near Point Orchard, the largest 7 inches in length. They are in the Museum of the Leland Stanford Jr. University, No. 3135.

This species is the type of a distinct genus, *Averruncus*, allied to *Podothecus*, but with teeth on the vomer and palatines. The lack of the median movable rostral spine separates it from *Odontopyxis*, in which genus the dorsal fins are very small.

89. **Xystes axinophrys** [illegible] n. gen and [illegible]

[illegible]

dorsal ray 2; highest anal ray 2¾; length of caudal fin 1⅔.

Body elongate, subcylindrical, deepest and broadest at shoulders; belly prominent; dorsal outline straight from first dorsal spine to caudal fin, curved up anteriorly to occiput. Head very irregular, much broken by large spines; mouth inferior, rather broad, maxillary reaching to the vertical from front of orbit; lips thin, not broken up in papillæ; upper jaw protractile; teeth small, in villiform bands on jaws, vomer and palatines; the anterior edge of premaxillary is directly under the base of rostral spines; a few very small blunt papilliæ behind chin; a barbel at end of maxillary, not half so long as diameter of pupil.

A pair of sharp rostral spines pointing forward and upward; behind these is a pair of curved spines pointing upward, outward and backward; no median spine or movable spine at tip of snout; between these and behind the rostral spines is an almost circular pit, which is entirely occupied by the upper end of the premaxillary process; interorbital wide and concave, a slight median ridge running from the rostral pit to a point above pupil, on each side of which is an outward curved ridge ending in a minute spine; over each eye is the largest spine on the head or body, the large triangular orbital spine, its base occupying nearly the whole space above eye; it is sharp, compressed and strongly hooked back; on the anterior part of its base is a small, sharp, preorbital spine, pointing upward; a series of minute spines running medially along the top of the head and body from a point between the orbital spines to the first dorsal spine; on each side of these are two large blunt spines, with the traces of a smaller one between them, they are continuous with the dorsal keels of body; farther down and continuous with the upper lateral keel of body is a ridge broken up into 4 irregular spines, larger than the body spines; 4 triangular spines on edge of preopercle, the upper one the largest; a very irregular ridge running from upper preopercular spine, under eye, to snout; a ridge on upper part of opercle.

Body with 4 ridges on each side, formed by the scales, each of which ends in a spine; traces of a ridge between lateral ridges; the spines on abdominal ridges as sharp as those on rest of body; a Y-shaped ridge of spines in front of ventrals, the forks toward the ventrals and the base ending at gill-membrane; a raised area between ventral fins, running from their base to opposite their tips, which is entirely covered with small prickles; the anus is in the posterior end of this; the dorsal and abdominal ridges coalesce with their fellows of the opposite side, but they come together so gradually that it is impossible to tell exactly where they unite, as the spines continue distinct to the caudal fin. Small spines covering the outer part of the base of the pectoral; a ring of spines around caudal base; a few minute spines on eye above pupil; occiput abruptly lower than body, but scarcely forming a pit, as the body is about level behind it.

Spinous dorsal highest in front, the second spine reaching to base of last spine when fin is depressed; the dorsal rays subequal in length, the last not shortened and not adnate to body; last ray reaching to the tenth plate before caudal fin; pectoral fin posteriorly rounded in outline, the lower rays not produced; it reaches to the second plate before anal fin; ventrals small, reaching just past vent; anal longer and lower than soft dorsal; dorsal and anal ending at the same corresponding place; caudal fin rounded behind.

Color, in spirits, gray, with 7 or 8 dark cross-bars; head

uniform gray with the exception of a dark spot at occiput; belly dusky; dorsals somewhat mottled; anal black, with a white spot near its middle; pectorals white, with a large black spot on base of rays; ventral black, abruptly white at tips; caudal black, edged with white.

One specimen brought up in the dredge, 1½ inches in length. It is in the Leland Stanford Jr. University Museum, number 3130.

This species seems to represent a new subgeneric or generic type, allied to *Averruncus*, distinguished by the supraocular spine and by the subequal rays of both dorsals, the last of each not adnate to the body.

90. **Xenochirus triacanthus** Gilbert. Plate xciii.

Rare; brought up in the dredge in company with *Odontopyxis trispinosus*. One specimen obtained, 3½ inches in length. In this young example, here figured, the lower rays of the pectoral are not produced.

91. **Odontopyxis trispinosus** Lockington.

Abundant in deep water; the most common species brought up by the dredge. Length 4 inches. In this genus and in *Xenochirus* there is a movable spine or long plate on median line at tip of snout.

Family CYCLOPTERIDÆ.

92. **Lethotremus vinolentus** Jordan and Starks, n. sp. Plate xciv.

Head 2¼ in length; depth 2¼; dorsal V–7; anal 6; eye 3 in head; snout nearly 4; maxillary 2½; interorbital 2⅓; ventral disk 1⅓; height of spinous dorsal 2⅓; length of pectoral 2¾.

Body short and thick, broadest at head, deepest in front of first dorsal spine, abruptly compressed at vent; back somewhat elevated.

Mouth terminal, oblique, the jaws about equal; snout very blunt; maxillary reaching slightly past the vertical from front of eye; teeth in narrow villiform bands; teeth on vomer (the specimen is so small, we cannot be sure of the palatine teeth); eye large, set high in the head, its diameter greater than the length of the snout; interorbital wide and flat, the diameter contained 1½ times in the width; gill-opening oblique, about as wide as eye and on a level with eye; disk ¼ longer than broad, its length about equal to distance from gill-opening to anterior edge of eye.

Skin thick; head and body nearly naked, a few spines scattered over it; spinous dorsal with 3 or 4 small spines, a minute simple spine in front of eye and 2 or 3 above it; 4 multifid spines following the curve of back, under spinous dorsal, and 1 under the front of second dorsal, 2 similar spines on each side of nape, just above opercles; 2 on edge of opercle and 3 on edge of preopercle; an irregular row of 6 running from above base of pectoral to front of anal fin, and a couple of small ones behind gill-opening; body otherwise naked. All the above spines, with the exception of those noted as simple, are long sharp spines in groups of from 3 to 6 with a common base, generally the length of each spine exceeds the length of the base. No lateral line.

Spinous dorsal reaching to the first ray of soft dorsal when fin is depressed, higher than soft dorsal; anal and soft dorsal similar; caudal small, truncate or slightly rounded: pectorals very short, reaching to the posterior edge of ventral disk.

Color bright wine-red, slightly lighter below, without markings, sides dusted over with very small dark points: spinous dorsal dusky: other fins colorless. Colors disappear in alcohol.

One specimen brought up from deep water in the dredge, ½ inch in length. Numbered 3131 on the register of the Leland Stanford Jr. University Museum.

This species seems to belong to the genus *Lethotremus* Gilbert, MS. From *L. muticus*, type of the genus, it is distinguished by its few-rayed fins and by its scanty prickles.

93. **Eumicrotremus orbis** (Günther).

One specimen of this interesting fish taken, 2 inches in length.

Family LIPARIDIDÆ.

94. **Neoliparis greeni** Jordan and Starks, n. sp. Plate xcvi.

Head $3\frac{4}{8}$; depth 4; depth at disk 5; dorsal VI–34; anal 30; pectoral 35; caudal 15; eye small, about 10 in head; snout 2¾; longest pectoral ray 1¾; disk 2½; longest dorsal ray 2⅙; longest anal ray 2⅙; caudal 1¾.

Body elongate, posteriorly compressed; profile undulate, over snout blunt and rounded, depressed over eyes, well rounded from eyes over occipital region. Skin thin and exceedingly loose, nearly to the end of the dorsal and anal rays.

Jaws equal; maxillary extending to posterior margin of eye; teeth small, nearly simple, depressible and blunt, slightly hooked back, arranged in oblique series, those in the front running nearly straight in, but towards the sides of the jaw they grow more and more oblique till they are nearly parallel with the jaw at the sides; superior pharyngeal teeth conical and sharp, slightly longer than the teeth in the jaws, arranged in a single round patch on each side; inferior pharyngeals separate, with small teeth. (Teeth probably tricuspid in young specimens.)

Posterior nostrils in a short wide tube; cheeks well

rounded; gill-rakers short and thick, no longer on the outer side of the first arch than on the other arches, fourth arch not free; gill-slit short, its length contained about 3 times in head, its lower edge extending in front of pectoral to about the third ray; opercles with a blunt spine which is covered by the skin.

Dorsals two, connected by a low membrane; first dorsal about twice as high as anterior part of second dorsal; the first rays of pectorals inserted under eye and in front of disk; the anterior rays short, graduated to the sixth ray, which is about 4 times longer than the first, the next few rays again short and gradually lengthening posteriorly; posterior rays $\frac{1}{3}$ longer than anterior, fin broadly rounded behind; ventral disk nearly round, its posterior edge reaching the vertical from gill-slit; its distance from tip of lower jaw $1\frac{1}{3}$ times its length; caudal truncate or slightly rounded; vent under ends of pectorals.

Color, in alcohol, uniform brown, breast and lower parts of head creamy, fins slightly darker. When fresh the sides were blotched with pinkish.

The type of this species is a specimen 10 inches long, in the Leland Stanford Jr. University Museum, number 3019. It was collected in the Harbor of Victoria by Mr. Ashdown H. Green, of Victoria, President of the Natural History Society of that town.

95. **Neoliparis floræ** Jordan & Starks n. sp. Plate xcvi.

Head $3\frac{3}{5}$ in length of body; depth at ventral disk $5\frac{4}{5}$; depth under middle of soft dorsal $4\frac{1}{2}$; dorsal VI–27; anal 21 to 23; caudal 15; pectoral 30; eye 7 in head; interorbital space $2\frac{3}{4}$; maxillary $2\frac{1}{2}$; pectoral $1\frac{3}{5}$; ventral disk $2\frac{1}{4}$.

A small specimen collected at Waadda Island, Neah

Bay. No. 3133, Leland Stanford Jr. University Museum.

Body moderately elongate, much compressed posteriorly, about as wide as deep anteriorly, its greatest depth under middle of soft dorsal where the back is elevated. Flesh very firm, the body retaining its form, the skin loose but not flaccid.

Head small, the nape not produced; mouth moderate, the maxillary extending to below the anterior margin of orbit; jaws subequal; teeth tricuspid, arranged in series which are nearly transverse on middle of jaws, becoming more and more oblique towards the sides, the outermost series nearly parallel with the sides of jaws; nostrils ending in a short wide tube; gill-opening short, extending downward to about the fifth pectoral ray, its length about half interorbital space; opercle ending in a flap, which extends over middle of gill-opening; ventral disk slightly longer than wide, its distance from tip of lower jaw once and a half its length; vent equidistant from posterior edge of ventral disk and front of anal; skin thick and not very loose.

Origin of spinous dorsal a little in front of the vertical from vent, its distance from snout 3 in length of body; anterior part of dorsal separated by a notch; origin of anal about equidistant from snout and base of caudal fin; some of the lower rays of pectoral produced forming a notch in the lower posterior margin of fin, the fourth to the tenth of the upper rays the longest, forming a rounded point behind, extending slightly past the vertical from snout; dorsal and anal scarcely connected with the caudal; caudal long and slender, rounded behind.

Color a uniform dark olive green, under parts white, a light streak medially along back from dorsal to occiput, a light spot over opercle; pectoral light at base, dusky behind; other fins colored like the body; lips white.

In the Museum of the California Academy is the only specimen we have ever seen of the rare *Neoliparis mucosus* (Ayres), likewise obtained at San Francisco by Mr. H. D. Dunn. We here present a description and figure of this specimen (No. 360):

Neoliparis mucosus (Ayres). Plate xcv.

Head 4 in length; depth 4⅓; dorsal VI–26; anal 26; pectoral 29; caudal 12; eye 7 in head; snout 3; ventral disk 1½; pectoral 1⅜; longest dorsal ray 2; highest anal ray 2; caudal 1¼.

Body not greatly elongate, rather robust, compressed posteriorly, holding its width well past middle of body; head short and thick, broader than body, ⅓ longer than broad, its length 1⅜ times its depth; mouth small, truncate, its cleft almost entirely anterior, scarcely extending laterally; end of maxillary buried under the skin, barely reaching to eye; nostrils not ending in tubes; lower jaw slightly the shorter; teeth sharp, tricuspid, the middle cusp much the highest and longest, arranged in nine oblique series in both jaws, becoming more and more oblique toward the sides; interorbital space moderately wide, about 3½ in head, a little convex; gill-slit not extending below upper edge of pectoral, its length about 1½ times eye and 3 in ventral disk.

Pectoral broadly rounded when spread, its notch comparatively very shallow, its tip reaches past vent but not to notch in dorsal; ventral disk large, slightly longer than broad, its posterior margin almost midway between its anterior and front of anal, its anterior margin half its length from chin; dorsal with a shallow notch; origin of fin over posterior margin of ventral disk, its longest rays in its posterior half; origin of anal a little nearer snout than base of caudal, the last four or five rays rapidly shortened, making the fin truncate behind; dorsal and anal scarcely joined to caudal; caudal long and slender, rounded behind.

Color olive brown, light below; indistinctly mottled; dorsal and anal darker at their margins; pectorals uniform dark brown; caudal light, with indistinct cross-lines; lips dark.

Here described from the only specimen known to us, five inches in length; from near San Francisco. It is now in the collection of the California Academy of Sciences (No. 360). Collected by H. D. Dunn, off San Francisco.

96. **Neoliparis callyodon** (Pallas).

Obtained by the Albatross at Port Angeles.

This is the species figured by Mr. Garman (monograph of the *Discoboli*) as *Liparis mucosus*. His description seems, in part at least, to have been drawn from *Neoliparis floræ*. The latter has larger gill-openings than either *Neoliparis mucosus* or *N. callyodon*.

Neoliparis callyodon is extremely abundant about the Aleutian Islands. The coloration, form of mouth, small gill-opening and the number of fin-rays all point out this as the original *callyodon* of Pallas.

The following is an analysis of the species of *Neoliparis*, as far as known:

a. **Gill-opening very narrow, almost entirely above base of pectoral, the lower edge not below third pectoral ray.**

b. **Anterior nostrils with distinct tubes.**

c. **Dorsal rays about 30; anal rays about 24; form robust; ventral disk 2¼ in head; color brownish, clouded or banded. North Atlantic on both coasts, south to Cape Cod.** ***montagui.****

cc. **Dorsal rays 34 to 36; anal rays 25 to 28; lower jaw included; form rather elongate, the head depressed; ventral disk 2⅓ in head; color pale, irregularly dotted with darker, sometimes plain brownish. Alaska, south to Puget Sound.** ***callyodon.***

bb. **Anterior nostrils with a raised rim, and without distinct tubes; head short, blunt, 4 in length; ventral disk very large, 1½ in head; snout blunt; mouth very short, its cleft almost entirely anterior, the maxillary scarcely reaching eye. Dorsal rays 32; anal 26. Color plain rosy or brownish, not spotted. Off San Francisco.** ***mucosus.***

aa. **Gill-opening rather large, its base opposite 4 or 5 upper rays of pectoral; body deep posteriorly; nostrils with raised rim, but without distinct tubes; ventral disk moderate, 2¼ to 2½ in head; head about 3⅓ in body, depressed above; cleft of mouth broader, partly lateral, nearly 3 in head; color plain brownish or reddish.**

d. **Dorsal rays VI–27; anal 21 to 23; pectoral 30; flesh firm. Puget Sound to Monterey.** ***floræ.***

dd. **Dorsal rays VI–34; anal 30; pectoral 35; flesh lax. Puget Sound.** ***greeni.***

97. Liparis cyclopus Günther. Plate xcvii.

Head 4⅙; depth 4½; dorsal 34; anal 29; pectoral 30; caudal 12.

Body much depressed and rather broad anteriorly, deep and much compressed posteriorly; head a third longer than broad and a third broader than deep. Flesh much more firm and the skin less lax than in most species of *Liparis*. Opercles with a rather strong spine concealed by the skin; mouth rather large, terminal; jaws subequal; teeth small, tricuspid, in broad bands; eye small, 6 in head; snout 3, flattish and broad above; interorbital space 4⅔ in head; ventral disk oval, 2⅓ in head, its anterior edge half the length of the eye behind postorbital margin; gill-opening moderate, 3½ in head, extending

* *Liparis montagui* Donovan.

downward to the third or fourth ray of pectoral. Dorsal fin low, continuous, not joined to caudal, beginning slightly before anal, on a vertical with vent; vent midway between edge of ventral disk and front of anal. Pectoral fin emarginate, the upper lobe $1\frac{3}{8}$ in head, the lower 2, the shortest intervening rays 3. Anal long and low, barely joined to caudal. Caudal $1\frac{1}{2}$ in head.

Color olivaceous, darker above; body and pectoral fin finely speckled with olive brown; fins dotted; bases of the fins paler than their tips; belly white.

Two specimens $4\frac{1}{2}$ inches long, in excellent condition, taken in Elliot Bay, near Seattle, were received from the Young Naturalists' Society. They are numbered 3126 in the register of the Leland Stanford Jr. University Museum. This species, not been previously recognized since its original description, is recorded by Dr. Gilbert from Unalaska.

98. **Liparis dennyi** Jordan and Starks, **n. sp.** Plate xcviii.

Head $3\frac{3}{8}$ in length of body; depth $4\frac{1}{2}$; dorsal 39; anal 30; pectoral 36; caudal 12; eye 8 in head; maxillary $2\frac{1}{8}$; snout $2\frac{3}{4}$; gill-opening $2\frac{2}{8}$; upper pectoral lobe $\frac{1}{3}$; lower lobe $1\frac{1}{2}$; intervening rays $2\frac{1}{4}$; ventral disk $2\frac{1}{3}$; highest dorsal rays $2\frac{2}{8}$; highest anal rays $2\frac{2}{8}$; caudal rays $1\frac{3}{4}$.

Body moderately elongate, much compressed posteriorly, slightly so anteriorly; head moderate, the cheeks and nape prominent. Mouth wide, with little lateral cleft; maxillary extending to below the anterior margin of eye, its end covered with the skin of the head; the lower jaw slightly the longer; the teeth tricuspid, those on the inner part of jaw largest, arranged in about 14 series in each jaw; series nearly transverse on middle of jaw, becoming more and more oblique towards the sides, where they are nearly parallel with the sides of the jaws; interorbital wide. slightly concave: nostrils ending in very short, wide tubes. the posterior over the anterior margin of eye, the anterior in front of it a distance equal to the diameter of eye: opercle ending in a short. wide spine covered with skin; it is situated slightly above the middle of gill-opening: gill-opening running from about the eleventh pectoral ray to a level with the eye.

Origin of dorsal slightly behind base of pectoral, its distance from the snout $3\frac{1}{5}$ in length of body. its anterior rays short. gradually lengthening posteriorly, the rays from the anterior third to near the end about equal, the last ray abruptly shortened. forming a slight notch where the fin joins the caudal: posterior two-thirds of caudal free above; anal similar to dorsal. about the same height, its origin nearer snout than base of caudal, about under the base of the tenth dorsal ray. posteriorly it is longer than the dorsal. joining the caudal at about half its length; ventral disk nearly round. its distance from tip of lower jaw $1\frac{1}{3}$ in its diameter. 1 in distance from vent, 2 from first anal ray; vent midway between front of anal and edge of disk; upper lobe of pectoral broadly rounded, reaching to two-thirds of the distance between vent and front of anal: lower lobe long. reaching nearly to vent; caudal long and slender. rounded behind. Skin very thin and loose on body and head. covering the anterior parts of dorsal and anal. attached at about the middle of rays posteriorly and covering the base of caudal rays.

Color light brown. lighter below. thickly covered with minute brown points. which form spots and mottlings on sides: upper part of head dark. lips spotted with brown: dorsal and anal dark brown. slightly mottled with lighter: pectoral light. with irregular brown spots and bars running across it. Caudal dark brown. mottled at base. two light bars crossing it towards its end. leaving a narrow posterior margin of brown.

The type specimen, 8 inches in length, was collected in Admiralty Inlet by the Young Naturalists' Society and presented by them to the Leland Stanford Jr. University. The species is named for Mr. Charles L. Denny, of Seattle, in recognition of his active and intelligent interest in the natural history of Washington.

99. **Liparis fucensis** Gilbert.

Taken in the Straits of Juan de Fuca by the Albatross. Locally abundant. This seems to be the species described and figured by Mr. Garman (Monograph of the *Discoboli*), under the erroneous name of *Liparis calliodon*. It will be described by Dr. Gilbert in the current number of the Proceedings of the United States National Museum.

100. **Liparis pulchellus** Ayres.

Rather rare. Three or four small specimens brought up in the dredge.

The following analysis will serve to distinguish the North American species of *Liparis*:

a. *Liparis*. Vertebræ in moderate number, about 39; dorsal rays about 35; anal rays 27 to 30.

- *b*. Gill-openings very narrow, entirely above base of pectoral; pectoral rays from 34 to 37; head a little shorter than broad, and a little longer than deep; dorsal and anal slightly joined to caudal; caudal narrow, its rays 12. North Atlantic, south to Cape Cod. *liparis*.
- *bb*. Gill-openings broad, the lower part considerably below base of upper ray of pectoral.
  - *c*. Pectoral rays 30; head low, flattish, a third longer than broad, a third broader than deep; jaws subequal; dorsal free from caudal, which is slightly joined to anal; caudal narrow, of 12 rays. Puget Sound to Unalaska. *cyclopus*.
  - *cc*. Pectoral rays 41 to 43; head short, not quite as wide as long; caudal 15 to 20; the dorsal and anal slightly joined to its base. Puget Sound. *fucensis*.

aa. *Careliparis* Garman. Vertebræ about 46; dorsal rays 40 to 44; anal rays 35 or 36; dorsal and anal largely joined to caudal.

- *d*. Pectoral rays 35 or 36.
  - *e*. Gill-opening small, its lower edge not below first ray of pectoral; nostrils small, the tubes short or absent.

f. Fins plain, not distinctly mottled or barred; body robust, its color plain brownish or with dark spots. Coasts of Greenland. *tunicatus.*

ff. Fins more or less mottled or barred, body moderately elongate; lower rays of pectoral rather short, not half head, not reaching beyond ventral disk; body mottled, usually with concentric rings. Aleutian Islands to Eastern Siberia. *agassizii.*

ee. Gill-opening rather large, extending downward to about fourth ray of pectoral; nostrils with short tubes; lower lobe of pectoral long, reaching much beyond disk, nearly to vent; color brown, the body and fins mottled and clouded. Puget Sound. *dennyi.*

dd. Pectoral rays 42; gill-opening large, its lower edge below upper part of pectoral; body robust; surface covered with round yellowish spots. Aleutian Islands. *cyclostigma.*

aaa. *Actinochir* Gill. Vertebræ about 52; dorsal rays 45 to 48; anal rays 38 to 40; pectoral rays 34 to 37; dorsal and anal largely joined to caudal; gill-opening large, about one-third its length before pectoral; anterior nostril tubular.

g. Head broad, flattened above; body rather elongate; skin usually with wavy, concentric longitudinal streaks, sometimes spotted. Unalaska to Monterey. *pulchellus.*

gg. Head high, boldly rounded, with prominent nape; color olivaceous, clouded and dotted, but without wavy streaks. Coasts of Greenland. *major.*

Family BATHYMASTERIDÆ.

101. **Ronquilus jordani** (Gilbert). Plate xcix.

Not common; occurring in deep water. Reaches a length of 8 inches. A fine specimen presented by the Young Naturalists' Society.

The genus *Ronquilus* is distinguished from *Bathymaster* by its scaly cheeks, enlarged scales on lateral line, and especially by its increased number of simple rays or spines in the dorsal.

Family GOBIIDÆ.

102. Gobius nicholsi Bean.

Not rare about Vancouver Island. Not taken by us.

103. Lepidogobius lepidus (Girard).

Three specimens dredged, the largest 4 inches in length.

104. Gillichthys mirabilis Cooper. MUD FISH.

Not common this far north. Found in the mud in lagoons. No specimens taken by us.

105. Quietula y-cauda (Jenkins & Evermann).

This little goby was taken in Saanich Arm, Vancouver Island, by Jordan & Gilbert. One of the two specimens taken from the stomach of *Hexagrammus hexagrammus* and recorded as *Gobiosoma ios* belongs to it. The other is the type of the latter species. This species is the type of the genus *Quietula* Jordan & Evermann, distinguished from *Gillichthys* by the presence of dermal flaps on the shoulder girdle.

106. Clevelandia ios (Jordan & Gilbert). Plate c.

The original type of this species was obtained from the stomach of *Hexagrammus hexagrammus*, in Saanich Arm, by Jordan & Gilbert, in 1880. It was not in good condition and the description is defective. Two specimens were dredged near Seattle by us. A description is here appended:

Head $3\frac{1}{2}$ in length of body; depth 6; D. V–16; A. 14; eye $6\frac{1}{2}$ in head; maxillary $1\frac{4}{5}$; pectoral $1\frac{3}{5}$; ventrals $1\frac{7}{8}$; caudal $1\frac{1}{3}$; base of soft dorsal 3 in length of body; base of anal $3\frac{1}{2}$.

Body long and slender, compressed, the back not elevated; caudal peduncle moderately wide. Head long, profile steep to within a short distance of the front of the eye, thence horizontal. Mouth very large, not very oblique, the maxillary projecting to opposite the middle of the cheek; jaws subequal; teeth in narrow villiform bands, eye small, longer than wide, set high in head; interorbital space narrow, about as wide as eye. Body covered with very small cycloid scales, impossible to count. Spinous dorsal well separated from soft dorsal, the spines slender; soft dorsal the higher, its origin a little nearer base of caudal fin than tip of snout; anal about equal to soft dorsal in height, its origin a little behind first dorsal ray, nearly coterminous with soft dorsal; ventrals inserted slightly behind pectorals, reaching midway between their base and front of anal; caudal short, its end rounded.

Color light olivaceous, the cheeks and sides with many dark points which form mottlings; snout dark; a dark spot on upper part of opercle; top of head black; dorsals pale, with three or four dark lines running across the rays; some dark spots on base of anal; pectorals crossed with dark wavy lines; caudal with about five irregular cross-bars.

Two specimens obtained, each 2 inches in length.

Family BATRACHIDÆ.

107. **Porichthys notatus** Girard.

Very common in shallow water. It attaches its eggs to the rocks just above low-tide mark, and watches them till they hatch and the young are quite well matured. The young fasten themselves to the rocks by means of a ventral disk, which soon disappears. It makes a peculiar grunting noise when disturbed. It reaches a length of over a foot. Several specimens taken.

Apparently the type of *Porichthys margaritatus* Richardson was the tropical species since described as *Porichthys nautopædium*. The name *margaritatus* should not be used for the northern form.

Family GOBIESOCIDÆ.

108. **Caularchus mæandricus** (Girard).

Very abundant under the rocks between tide marks. It feeds on small shells and crustacea. A large number of specimens obtained at Neah Bay and in the vicinity of Seattle; the largest 4½ inches in length; said to reach a length of 6 inches.

Specimens from Neah Bay varied from light olive to bright cherry-red.

This species has 13 + 19 = 32 vertebræ. The species referred to *Gobiesox* have, so far as known, 10 + 16 = 26. This increased number, associated with its northern distribution, may define the genus *Caularchus*.

Family XIPHIDIONTIDÆ.

109. **Bryostemma polyactocephalum** (Pallas).

This species has been referred to the genus *Chirolophis* (*Blenniops*). It, however, differs from the latter in the entire absence of the true or median lateral line, and may be made the type of a distinct genus, for which we suggest the name of *Bryostemma*. In *Bryostemma*, as in *Chirolophis*, there is a short series of large pores above the pectoral.

The following is a description of our specimen from Seattle:

Head 6½; depth 6; D. LXI; A. 61. P. 14. V. 1, 3. Fifteen pores above pectoral.

Body elongate, much compressed, covered with small, smooth, imbedded scales. Head very short, blunt in profile; mouth short, terminal; lower jaw heavy, projecting, its lip with two small cirri; teeth subequal, small, bluntish, close set, in one row in each jaw; eyes 4 in head, near together; snout 4; supraorbital cirri, 2½ in head; interorbital space with two large superciliary cirri; top longest nearly as long as eye; about 15 minute cirri along dorsal edge of lateral pores, one on each pore. Rows of pores running around eye, under preopercle, and along entire length of the short lateral series; about 15 in lateral series, which is 2 in length of head; gill-rakers not developed; gill-membranes not joined to the isthmus. Dorsal fin beginning over pectoral and running to caudal; anterior rays fringed with fleshy cirri; first ray, including cirri, 2 in length of head; anal beginning close behind vent and running to caudal, lower than dorsal; vent about ⅓ distance from tip of snout to tip of caudal; distance from base of ventral to vent 4¾ in length of body; pectoral fin but little shorter than head, its breadth at base not half its length.

Color, in spirits, pale brownish, with about 13 dark blotches along dorsal and anal fins; more distinct on dorsal; a black spot on fourth to sixth dorsal spines very distinct; a very faint spot on anterior part of anal; a few dark markings about head and nape. Cirri mostly pale.

One fine specimen, 6½ inches long, from Point Orchard, near Seattle. Collected by Prof. O. B. Johnson.

This species seems to belong to *Bryostemma polyactocephalum*. As figured by Mr. Nelson, the latter species seems to differ in the absence of the lateral pores and in the much shorter and broader pectoral fin; the proportions of the body before the vent are also different.

A number of young specimens collected by the Albatross in Alaska seem to belong to the same species. These are more elongate and less compressed, the body much mottled and vaguely barred, the ventral fins chequred in fine pattern; head sand color; a black blotch on fourth to sixth dorsal spine; anterior dorsal spines little elevated and with few fringes; sides of head with few cirri, except in one specimen in which the cheeks are covered with cirri densely matted. Evidently the species is very variable.

110. **Bryostemma nugator** Jordan and Williams, n. sp. Plate ci.

Head 5½; depth 5½; dorsal LIV; anal 41; ventral I, 3; pores of lateral series 25.

Body elongate, formed as in *Pholis*, less compressed than in *Bryostemma polyactocephalum*, covered with small, smooth, imbedded scales. Head short, very obtuse, almost truncate; top of head from nostrils to near front of dorsal covered with fleshy cirri, much smaller than in *Bryostemma polyactocephalum;* only two or three small ones extending on first dorsal spine; supraorbital cirrus short, 4 to 5 in head; two small cirri placed at the sides of snout, with a larger median one behind them, forming a triangle; jaws equal; mouth horizontal, the angle extending to below pupil; eyes small, 4 in head; snout very short, almost vertically truncate, ⅔ of eye; teeth of both jaws subequal, short, bluntish and close set. Lateral series short, 7½ in length of body, concurrent with the dorsal outline of body. A line of pores begins in front of eye on a level with pupil, runs under eye and to a level with pupil again, then back to and along the entire length of the short lateral series. Gill-rakers not developed; gill-membranes free from isthmus. Vent ⅓ distance from tip of snout to tip of caudal; distance from origin of ventral to anus 4½ in length of body. Pectoral fin 5½ in body, as long as head. Dorsal fin beginning in front of the pectoral, highest along the posterior half; the longest spine, 2⅔ in head, the fin higher than anal; dorsal slightly joined to caudal; anal separated from caudal; caudal rounded, 1⅗ in head; first dorsal spine 4½ in head, its surface with 2 or 3 small cirri.

Color of one specimen, probably male, dark brown, with 13 pale cross-bars along back, extending on dorsal fin; along sides these become obsolete; on belly they become increased in number and broadened below; dorsal fin with 13 large, very distinct, black ocelli, with yellowish rings, one between each pair of the pale blotches; anal with about 7 small blackish spots at base on posterior part, the fin otherwise nearly plain; caudal faintly barred with light and dark; pectorals pale, with two dark pale-edged oblique bars before them; sides of head with irregular dark vertical bars, one of them forming an inverted λ below eye, this and others extending across lower jaw; cirri mostly black.

The other specimen, probably the female, has the body nearly plain brown, the dorsal with but 4 ocelli, the anterior nine being replaced by dark bars on the fin; anal with dark oblique cross-bars; pectorals barred with black. Markings on head more sharply defined, coloration otherwise similar. This second specimen is 4¾ inches in length; the other, 4.

These two specimens were taken near Seattle and presented by the Young Naturalists' Society. They are numbered 3134 on the register of the Leland Stanford Jr. University Museum.

Three additional specimens of *Bryostemma nugator* were taken by Mr. Starks in rock pools on Channel Rocks. The life colors of these were as follows:

Color, dark red above, orange-brown below, belly cream color; sides below with cream-colored cross-bars, wider than eye, running from the axis of body downward and fading into the general color below: a λ-shaped mark downwards from eye, across branchiostegals to isthmus, a similar mark behind eyes, across edge of preopercle this last sometimes broken up and chain-like; top of head dark; snout light; 2 oblique dark bars at base of pectoral; dorsal with 12 or 13 sharp dark brown spots as large as eye, edged with bright red, these arranged regularly along the whole length of fin; pectorals and caudal bright red, with wavy irregular brown lines running across the rays; anal red, with dark brown bars as wide as the interspaces running obliquely downwards and forwards; ventrals light brown.

111. **Pholis ornatus** (Girard).

This blenny is extremely abundant in Puget Sound, where many specimens were taken. It is found under rocks between tide marks, reaching a length of a foot. No specimens were found at Neah Bay. The species varies much in color, being typically yellowish-green with dark markings, but varying to brown or cherry red with the markings faint or obsolete. One specimen is notably different in color: Body purplish red, lighter on the belly; two conspicuous black-bordered white spots on front of dorsal; a light streak bordered with black from eye to nape; pectorals one-fourth shorter than in the others. Dr. Gill tells us that the generic name *Pholio Scopoli* is equivalent to the later *Murænoides*.

112. **Apodichthys flavidus** Girard.

Common in shallow water among the kelp. It varies from bright green to red, orange or violet. Two specimens belonging to the green form (var. *virescens*) were taken by us in Puget Sound; the larger 10 inches in length, the smaller 3 inches. These differ in color from the typical examples. The large one is a bright grass-green, mottled with light gray; a series of blended white spots, as large as eye, along the axis of body from the pectoral fin to the middle of caudal peduncle; belly with many similar spots smaller in size and somewhat sharper in outline; a row of conspicuous black spots, irregular in size, shape and position, along back at the base of dorsal spines; a black line as wide as pupil from nape to eye, a similar line from eye to posterior end of maxillary; a faint light streak across cheek posteriorly; cheek and base of pectoral dusted with fine dark points.

The small one is bright green without distinct markings on body; a silvery bar, running posteriorly from tip of snout through eye, across cheek, to the middle of opercle; no bar downward from eye to maxillary, or from eye to nape as in the large one.

113. **Xererpes fucorum** (Jordan & Gilbert).

Recorded by Jordan & Gilbert as rather rare on Waadda Island. No specimens obtained by us. The new genus *Xererpes* Jordan & Gilbert is distinguished from *Apodichthys* by the small anal spine, which is not grooved in front.

114. **Anoplarchus atropurpureus** (Kittlitz).

Taken at Neah Bay and in the vicinity of Seattle. Abundant under rocks, above low tide mark, in company with *Xiphidion xiphistes* and *Pholis*. It reaches a length of 8 inches. These specimens are scaled on the posterior half of the body only, which is probably true of the genus as a whole.

115. **Xiphistes chirus** (Jordan & Gilbert).

The most common blenny in Puget Sound, where we obtained specimens in abundance. At Neah Bay *Xiphidion mucosum* and *rupestre* were found. We found neither of these in the vicinity of Seattle. Variable in color, running from dull brown to bright red. This species is the type of a distinct genus, *Xiphistes*, distinguished from *Xiphidion* by the well-developed pectoral.

116. **Xiphistes ulvæ** Jordan & Starks, n. gen. and n. sp. Plate cii.

Head 8 in length of body; depth 10; dorsal LXXIV; anal III, 48; eye 5 in head; maxillary 2¾; pectoral 3½.

Body eel-shaped as in the related species *X. chirus*; head short; mouth small, oblique, maxillary extending to below posterior margin of eye; jaws subequal, with canine teeth; 4 enlarged canines in front of lower jaw; teeth in upper jaw gradually enlarged from behind forwards; eye moderate, equal to length of snout; interorbital space prominent, sharply convex, narrower than width of eye; nape not constricted.

Five mucous canals radiating downwards and backwards from eye, not reaching to edge of preopercle; the branches running upwards from upper lateral line ending on the membrane of dorsal, the lower lateral line not connected with the abdominal line. Lateral line otherwise as in *Xiphistes chirus*. Origin of dorsal at a distance behind nape equal to distance from nape to middle of eye; the fin posteriorly barely connecting with caudal, anal with 3 spines, its origin about a head's length nearer snout than base of caudal, connected with caudal posteriorly; pectorals equal in length to snout and half eye, slightly shorter than caudal; caudal rounded, fan-shaped.

Color olive-green above, very bright green below; sides along middle and lateral line posteriorly, with conspicuous white spots, half as large as pupil, each with a black spot before and behind it; a black streak from tip of snout, through eye, to nape, a streak starting from eye behind quickly fading out; dorsal darker than body, unmarked; the anterior third of anal green without markings, behind this, faint cross-bars of brown appear, these grow broader and darker posteriorly; caudal olive green with a light bar across base; pectorals green without markings.

One specimen obtained on Waadda Island, Neah Bay. It was found high on the rocks, among algæ just below high water mark. Length 5 inches.

This species is very closely related to *Xiphistes chirus*. It differs from it chiefly in having 3 anal spines, in the branches of the upper lateral line running higher, and in coloration. It is numbered 3132 on the register of the Leland Stanford Jr. University Museum.

117. **Xiphidion rupestre** (Jordan & Gilbert). Plate ciii.

Equally abundant with *Xiphidion mucosum* under rocks about Neah Bay. It does not reach such a large size as the latter.

118. **Xiphidion mucosum** Girard.

Abundant at Neah Bay, where it was found under rocks between tide marks, in company with *X. rupestre*. Reaches a length of 18 inches.

Family STICHÆIDÆ.

119. **Lumpenus anguillaris** (Pallas).

Taken in abundance with seines along sandy beaches in Puget Sound. It reaches a length of 20 inches.

Family CRYPTACANTHODIDÆ.

120. **Delolepis virgatus** Bean.

A stuffed skin from near Seattle is in the collection of the Young Naturalists' Society, collected by Prof. O. B. Johnson.

Family ANARRHICHADIDÆ.

121. **Anarrhichthys ocellatus** (Ayres). WOLF FISH.

Rare in Puget Sound: more common southward. It reaches a length of 8 feet, and is sometimes eaten. It feeds on crustacea and mussels, which it pulls off from the rocks and crushes between its powerful jaws.

Family ZOARCIDÆ.

122. **Lycodopsis paucidens** (Lockington).

Abundant in Puget Sound. Length about 10 inches. Not taken by us. The large-mouthed specimens, called by Jordan & Gilbert *paucidens*, are the male and the small-mouthed ones, called *pacificus*, the female of the same species.

Family SCYTALINIDÆ.

123. **Scytalina cerdale** Jordan & Gilbert. Plate civ.

Specimens were found in abundance in the loose gravel under boulders at Waadda Island, Neah Bay. It has not been taken since 1881, when Drs. Jordan & Gilbert took the two type specimens (one of which was afterwards destroyed by fire) in this locality.

The skeleton does not differ essentially from that of *Lycodopsis paucidens*, with which it has been compared. The skull is not at all depressed, the wide depressed form of the head of the fish is due to the fleshy cheeks. The frontals take up the greater part of the top of the skull, the parietals are separated by the supraoccipital, which extends forward to the frontals. Opercles all present. Lower jaw large and strong, Post-temporal scarcely so firmly attached as in *Lycodes;* the clavicle long and slender.

Family GADIDÆ.

124. **Microgadus proximus** (Girard). TOMCOD.

A few specimens obtained. Very abundant. Taken in large numbers by the fishermen. It is a food fish of some value, and meets with a ready sale. It reaches a length of a foot.

125. **Gadus macrocephalus** Tilesius. ALASKA COD.

Not uncommon in certain localities near Cape Flattery. This is probably its southern limit.

126. Pollachius fucencis (Jordan & Gilbert).

Occasionally taken in deep water in Puget Sound. Not obtained by us.

Family MERLUCCIIDÆ.

127. Merluccius productus Ayres. HAKE.

Abundant. It does not rank high as a food fish, as its flesh is soft and watery. It reaches a length of over 2 feet.

Family TRACHYPTERIDÆ.

128. Trachypterus rex-salmonorum Jordan & Gilbert. KING OF THE SALMON.

Very rare. Two specimens recorded from Neah Bay, where it is regarded by the Indians as a sacred fish, the King of the Salmon.

Family PLEURONECTIDÆ.

129. Hippoglossus hippoglossus (Linnæus). HALIBUT.

One of the most valuable fish taken in the region. It is found most abundant off Cape Flattery. Several fishing schooners are engaged in the halibut fishery. It reaches a weight of over 200 pounds, and a length of 5 or 6 feet.

130. Eopsetta jordani (Lockington).

Not abundant in Puget Sound. It reaches a length of 18 inches and a weight of 3 to 5 pounds. A fine food fish. Not seen by us.

131. Hippoglossoides elassodon Jordan & Gilbert.

Common. The types of this species were first obtained at Seattle and Tacoma, where it was taken with hook and line from the wharves. Length about a foot.

132. **Lyopsetta exilis** (Jordan & Gilbert).

A small flounder; not very abundant. It does not reach a length of over 9 inches. It is of no value as a food fish. A specimen in the collection of the Young Naturalists' Society has the right pectoral black, but it is not otherwise peculiar.

133. **Psettichthys melanostictus** Girard.

Abundant. It is one of the best of the flounders for food. It reaches a length of 20 inches.

134. **Citharichthys sordidus** (Girard).

Very common in deep water in the sound; weight 1½ pounds.

135. **Isopsetta isolepis** (Lockington).

Common in rather deep water. It reaches a length of 15 inches. Not taken by us.

136. **Isopsetta ischyra** (Jordan & Gilbert).

Not common. The type from Elliot Bay, near Seattle, where it was taken with a seine. Length 18 inches.

137. **Parophrys vetulus** Girard.

Very abundant. Many specimens collected with a seine in shallow water. It is a good market fish, and reaches a length of 14 inches. The young are spotted with blackish.

138. **Lepidopsetta bilineata** (Ayres).

Very common. Specimens secured in abundance on sandy beaches. About 18 inches in length. Puget Sound specimens are rougher than those found farther south.

139. **Platichthys stellatus** (Pallas). DIAMOND FLOUNDER.

The commonest flounder in the sound. It is not held in as high esteem as a food fish as some of the other

8. **Gasterosteus microcephalus** Girard.

One adult 58 mm. in length, and three young 34 mm. in length, were taken in Green Lake. The adult had 7 well developed plates, the young had but 5. It was of interest to note that these specimens seemed to indicate that the young have the plates well developed first on the sides below and between the dorsal spines, and that the dorsal portions of the plates were the first to develop. The young were much lighter in color than the adult. Eleven specimens, apparently all adults, were taken in Lake Union. These were apparently similar to those from Green Lake, except they had 9 or 10 plates.

9. **Cottus asper** Richardson.

Twenty-eight specimens, 2 to 5 inches in length, taken July 26, in Lake Washington. A common species.

LIST OF PLATES.

MEXICAN FORMICIDÆ.

BY THEO. PERGANDE.

The present paper is based upon a collection of Formicidæ, made by Messrs. Eisen and Vaslit during September, 1894, in the Cape Region of Lower California and during October and November on the mainland of Mexico; the majority having been taken in the territory of Tepic, and on an overland trip from the city of Tepic to Mazatlan, all on the Pacific Coast of Mexico.

In addition to a few species which have been previously recorded by me, the collection contains quite a number of new or otherwise interesting forms, some of which have not heretofore been observed to occur so far north, while a few extend their territory into Texas and even as far east as Missouri.

Subfamily CAMPONOTINI.

1. BRACHYMYRMEX ADMOTUS Mayr.

Brachymyrmex admotus Mayr., Verh. zool. bot. Ges. Wien, xxxvii, 1887, p. 523.

Worker: Length, 1.6 mm. Head somewhat longer than broad, its sides nearly parallel and slightly arcuate; the occiput emarginate. Clypeus broader than long, strongly convex, arcuate in front. Scape, reaching beyond the occiput. Ocelli very minute. Mandibles narrow and furnished with four acute teeth; their surface polished, indistinctly striated and sparsely punctured. Thorax not much longer than the head, stout, of nearly equal width; the prothorax about one-fourth narrower than the head: constriction between the meso- and metanotum rather deep, the declivity of the metanotum flat and longer than the basal section. Scale small, narrow, rounded at apex. Entire surface of body polished, the thorax and abdomen with extremely fine and transverse striæ. Pubescence yellowish, the erect hairs rather coarse and sparse, most dense on the abdomen; antennæ and legs without erect hairs. Appressed pubescence most dense on the head, antennæ and legs.

Color of the head, antennæ and thorax brownish-yellow, the abdomen somewhat darker brown, with the sutures paler. Clypeus, mandibles and legs paler yellow, the femora often more or less distinctly dusky.

Five specimens. Tepic, Mexico.

2. PRENOLEPIS LONGICORNIS (Latr.) Roger.

Formica longicornis Latreille, Hist. nat. Fourmis, 1802, p. 113.
Prenolepis longicornis Roger, Verz. d. Formiciden, 1863, p. 10.
For synonyms and references, see Dalla Torre, Cat. Hymenopterorum, vii, Formicidæ, 1893, p. 179.

Worker: Length, 3 mm. Head about twice as long as broad, rounded behind, its sides nearly parallel. Clypeus strongly convex and with a rather sharp median carina. Mandibles narrow and with about four or five acute teeth. Antennæ very long and slender, the scape about three times the length of the head. Thorax elongated, slender; dorsal surface of the pro-mesonotum slightly arcuate, the suture dividing them nearly obsolete; metathorax slightly elevated posteriorly, gently and uniformly arcuated and furnished each side, above the coxæ, with a small conical tubercle.

Scale stout, narrow, slightly broadest at apex. Abdomen ovoid, pointed posteriorly. Legs long and slender. Pubescence whitish; erect hairs coarse and rather sparse; a few semi-erect hairs on femora and tibiæ; appressed pubescence observed only on legs and antennæ. Head and body polished and faintly shagreened.

Color black, with a bluish reflection on the head and thorax; scape and legs black or brownish; palpi whitish; tarsi and posterior margin of the abdominal segments yellowish or brownish.

Numerous specimens. Tepic, Mexico.

A cosmopolitan species: common in the tropical regions of Asia, Africa, Australia and America, and in many of the hothouses of Europe and this country.

3. PRENOLEPIS ANTHRACINA Rog. var. NODIFERA Mayr.

Prenolepis nodifera Mayr, Sitzber. Acad. Wien, lxi, 1870, p. 388.
Prenolepis nodifera Mayr, Verh. zool. bot. Ges., xx, 1870, p. 948.
Prenolepis nodifera Forel, Mittheil. Munch. ent. Ver., v, 7, 1881, p. 2.
Prenolepis nodifera Forel, Bull. soc. Vaud. sc. nat. (2) xx, P. 91, 1884, p. 348.

Worker: Length, 2.4–2.8 mm. Head about twice as long as broad, the sides parallel and nearly straight, rounded behind the eyes, the occiput slightly emarginate. Clypeus of the usual form, rather strongly convex and carinated along the middle. Mandibles furnished with six acute teeth. Antennæ rather stout, the scape about one-fourth longer than the head. Thorax of the usual shape in this genus, the meso-metanotal constriction deep; the metanotum convex, elevated, nodiform, with an acute angle each side at base of the declivity. Scale narrow, wedge-shaped, somewhat broadest and slightly rounded at the apex. Abdomen of the usual shape. Legs rather stout.

Erect hairs blackish, rather coarse and quite dense, though less dense on the scape and legs. Appressed pubescence whitish, most dense on the legs and antennæ.

Color black, polished, the thorax and scale sometimes dark brown. Mandibles, base of scape, trochanters, tarsi and sometimes the apex of femora and tibiæ yellowish or brownish-yellow, posterior margin of abdominal segments, if extended, whitish.

Female: The characters of the female, which I judge as belonging to this species, are as follows:

Length, 4 mm. Head about as broad as long; eyes larger, the antennæ more slender, the clypeus shorter, broader and more distinctly truncate in front than in the worker. Thorax broader than the head and but slightly convex above. Scutellum broader than long; the metanotum short, convex, with two more or less distinct foveæ near its anterior margin. Scale broader than in the worker, its upper edge arcuate. Abdomen elongate ovate. Legs stouter than in the worker. Erect hairs rather short and fine, the appressed pubescence very dense on the head and abdomen.

Color of the head, antennæ, mandibles, some parts of the thorax and legs brownish-yellow; the rest dark brownish.

Wings wanting.

Male: Length, 2.6 mm. Head slightly shorter and the eyes larger and more projecting than in the worker. Mandibles narrow and with but one tooth at the apex. Metanotum not elevated or nodiform, gently sloping towards the scale, with two, posteriorly uniting, median carinæ. Scale broader than in the worker, its apex arcuate. Genital claspers long, slender, curved slightly inward, and furnished with numerous rather long and slender hairs. Wings brownish, the stigma and veins darker.

Erect hairs finer than in the worker, especially those of the abdomen. Scape without erect hairs. Appressed pubescence dense, particularly so on the head and thorax.

Coloration as in the worker.

Many workers, twenty males and two females.

Tepic, Mexico; San José del Cabo and Sierra San Lazaro, Cape Region, Lower California.

4. CAMPONOTUS ESURIENS (Smith) Mayr.

Formica esuriens Smith, Cat. Hym. Ins. Brit. Mus., vi, 1858, p. 54.
Camponotus vulpinus Mayr, Verh. zool. bot. Ges., Wien, xii, 1862, pp. 658 and 770.
Camponotus esuriens Mayr, Verh. zool. bot. Ges., Wien, xiii, 1863, p. 398.
Formica esuriens Norton, Am. Nat., ii, 1868, p. 59.
Formica (Camponotus) esuriens Norton, Proc. Essex Inst., vi, 1868, Comm., p. 1.
Camponotus atriceps st. *esuriens* Forel, Bull. soc. Vaud. sc. nat. (2), xvi, P. 81, 1879, p. 76.
Camponotus esuriens McCook, Proc. Acad. Nat. Sc., Philad., 1879, p. 140.
Camponotus atriceps st. *esuriens* Forel, Bull. soc. Vaud. sc. nat. (2), xx, P. 91, 1884, p. 340.

Numerous workers. Tepic, Mexico.

5. CAMPONOTUS FRONTALIS n. sp.

Large worker: Length, 7.8 mm. Head quadrangular, its sides parallel and gently curved anteriorly, the occiput slightly emarginate. Clypeus about twice as long as broad, its sides slightly diverging anteriorly, the anterior margin nearly straight. Frontal area small, obsolete. Eyes rather small, oval and but slightly projecting. Mandibles robust, furnished apparently with six teeth. Scale wedge-shaped, stoutest at base, broadest above, slightly convex in front, nearly straight behind, the apex rounded. Head and thorax opaque and densely and finely granulated; sparsely punctured; punctures of the prothorax somewhat coarser and more numerous and the surface slightly reticulate. Clypeus rugoso-granulate. Mandibles sub-opaque, finely and densely striated and sparsely punctured. Abdomen slightly polished, densely and finely granulated, the punctures rather numerous and coarse.

Erect hairs white and glistening, that of the head short and not readily observed and intermixed with few minute, appressed yellowish hairs. Pubescence of the thorax and first abdominal segment rather dense, long and fine; pubescence of the antennæ and legs shorter and semi-erect; that of the scape intermixed with a few longer, erect hairs.

Color black, the abdomen with a slight bluish reflection. Head, in front of insertion of antennæ, including the clypeus and anterior part of face between the frontal carinæ, cherry-brown, the brown color extending obliquely to the lower external angle of the base of the mandibles; scape, first joint of the flagellum and joints two to four of the tarsi somewhat paler brown; posterior margin of the segments of the abdomen somewhat yellowish in a certain light.

Small worker: Length, 5.4 mm. Head slightly longer than broad, broadest just behind the eyes and slightly narrower towards the mouth; the occiput rounded; clypeus broadest in front and with a distinct median carina; the frontal area more distinct, the eyes larger and more convex and the antennæ longer and more slender than in the large worker. Head and thorax opaque, the sculpturing as in the large worker, except that the clypeus is not rugose; the abdomen is more distinctly polished and the bluish reflection more pronounced, the surface faintly but densely wrinkled. Pubescence similar to that of the large worker, though longer and denser on the head, coarsest in front of the antennæ and sides of the head.

Color entirely black, excepting the flagellum and tarsal joints, which are darker or lighter brown.

Two large, fifteen small workers. Tepic, Mexico.

This species resembles very much *Camp. novogranadense*, differing from it, however, in the larger and broader head and darker and differently distributed brown color and denser pubescence of the large worker, and the entirely black head of the small worker. It seems also to be related to *Camp. abscisus* and *andrei*, but differs from both in the bi-colored head.

6. CAMPONOTUS PUNCTULATUS Mayr var. RUFICORNIS n. var.

Small worker: Length, 5 mm. Head longer than broad, slightly narrowed anteriorly, its sides straight, rounded behind the eyes, the occiput but faintly emarginate. Clypeus triangular and with a rather acute, median carina; its anterior margin angulated. Frontal area minute, the frontal carinæ nearly parallel; mandibles with four or five teeth. Antennæ long and slender, the scape reaching considerably beyond the occiput. Prothorax about as broad as the head behind; suture between the meso- and metanotum obsolete, both segments descending in an almost straight line to the base of the scale. Scale very stout, of equal thickness and nearly quadrangular, the anterior face somewhat shorter than the posterior one, the upper edge thick and slightly rounded. Abdomen elongate oval. Legs rather long and quite slender. Head, thorax and abdomen delicately shagreened; mandibles smooth and sparsely punctured.

Erect hairs whitish, rather scarce on the head and thorax, more dense and slightly coarser on the abdomen; hairs of antennæ and legs short and appressed, those along the inner edge of posterior tibiæ longer and semierect. Appressed pubescence quite dense, yellowish and glistening.

Color black. Face in front of eyes, antennæ and anterior angle of prothorax reddish-yellow, clypeus, anterior [illegible] mandibles yellow-

[illegible]

[illegible] of the [illegible]scence.

7. Camponotus tepicanus n. sp.

Large worker: Length, 5 mm. Head slightly longer than broad and slightly narrowest anteriorly, the sides nearly straight, the occiput but slightly emarginate. Clypeus somewhat longer than broad, slightly broadest anteriorly, convex above and without a median carina, its anterior margin arcuate. Frontal area obsolete. Mandibles very robust and provided with five or six rather blunt teeth. Antennæ short, the scape reaching barely beyond the occiput. Thorax uniformly arcuate above, broadest in front and gradually diminishing in width towards the scale. The prothorax is about two-thirds the width of the head and slightly convex above; sides of metathorax nearly parallel; sutures between the segments distinct; scale stout and of nearly equal thickness, broadest above, the apex slightly arcuate and bluntly rounded. Legs stout.

Head and thorax semi-opaque, the occiput polished, densely and finely granulated and sparsely punctured, the punctures most dense on the vertex and thorax; an impressed longitudinal line between the frontal carinæ. Clypeus somewhat polished, the granulation extremely fine, almost obsolete, the surface sparsely punctured. Mandibles nearly smooth, rather indistinctly and finely striated and sparsely punctured. Abdomen densely granulate-punctate.

Erect hairs reddish and quite dense and fine on the thorax and scale, somewhat coarser and more sparse on the abdomen, except on the posterior margin of the segments; very sparse on the head. Posterior tibiæ without erect hairs: very few on the scape. Appressed pubescence minute, most dense on the thorax and abdomen.

Head, scape, first joint of the flagellum, thorax and legs reddish-yellow. A large, somewhat squarish, posteriorly broadening spot on the vertex, remaining joints of the flagellum, teeth of mandibles, two, not well defined spots on the pronotum, the dorsum of the meso- and metanotum, the scale and abdomen black; the posterior margin of the abdominal segments somewhat yellowish.

Small worker: Length, 4 mm. In coloration and sculpturing almost identical with the large worker. The scape is somewhat longer; the coloration of the head, thorax, legs, etc., paler yellow, while the black color of the head extends over the entire vertex, the occiput and the space between the frontal carinæ.

Three specimens. Tepic.

This may be a race of *Camp. marginatus*, though it is more hairy than any of the races of that species known to me.

Subfamily Dolichoderini.

8. Dolichoderus granulatus n. sp.

Worker: Length, 3.6 mm. Head about as broad as long, broadest just behind the eyes and narrowing slightly towards the mouth, the sides nearly straight. Clypeus broader than long, convex and with a median carina, obtusely triangular in front. Frontal area very small, obsolete. Frontal carinæ strongly diverging and but slightly curved posteriorly. Scape reaching beyond the occiput. Mandibles robust and armed with four or five acute teeth. Thorax stout, compact, narrowing posteriorly, the prothorax nearly as broad as the head, somewhat convex above, its lateral margin forming a blunt carina; the dorsal surface of the pro- and mesonotum evenly arcuated, the suture between them distinct; constriction between the meso- and metanotum deep, the metanotum somewhat elevated, nodiform, the declivity steep, not excavated; the lateral margins of the dorsal or basal section form quite distinct carinæ. Scale stout, broad, convex in front and behind, broadest above, the edge sharp and semicircular. Abdomen short, broadly oval, about as long as the thorax and rather broader than the head. Legs stout. Head densely and quite coarsely granulated. Thorax densely rugoso-granulate, the sculpturing slightly stronger than that of the head and somewhat concentric towards the sides of the pronotum; declivity of the metonotum as well as both sides of the scale transversely striated. Abdomen with dense and fine transverse wrinkles. Head and thorax semi-opaque; mandibles, legs and abdomen polished.

Erect hairs white, most dense and rather coarse on the abdomen, longer and finer on the thorax and base of the abdomen; quite profuse on the scape and legs. Appressed pubescence very scarce, most dense on the flagellum.

Color black; mandibles, and sometimes the entire antennæ reddish-yellow, the flagellum generally brownish towards the end. Eyes brown. Legs paler or darker brown, the tarsi and trochanters generally brownish-yellow.

Twelve specimens. Tepic.

9. AZTEKA CŒRULEIPENNIS Emery var. FASCIATA n. var.

Large worker: Length, 3.4–3.6 mm. Head somewhat longer than broad, slightly narrowest anteriorly, the sides gently arcuate, posterior angles rounded, the occiput quite deeply emarginate. Clypeus very broad, its lateral lobes extending to the sides of the head, the anterior margin but slightly arcuate. Mandibles robust and armed with eight or nine acute teeth. Scape about as long as the head and reaching somewhat beyond the occiput. Prothorax about one-fourth narrower than the head, convex above and at the sides, rounded in front. The mesonotum is somewhat elevated above the pro- and metanotum, convex above, compressed at the sides, and much narrower than the prothorax; the metanotum broadens posteriorly; the declivity is quite abrupt and shorter than the basal section. The scale is wedge-shaped, broadest above, the apex acute and arcuate. Abdomen small, broadly oval. Legs stout.

Body polished; mandibles finely and densely striated and sparsely punctured; head, thorax and abdomen finely and densely punctured.

Erect hairs quite dense, including legs and antennæ. Appressed pubescence yellowish, minute and rather dense.

Color of the head, antennæ, mandibles and sometimes the entire legs, reddish-yellow; sometimes the entire face or only a spot between the frontal carinæ, the upper margin of the vertex and the thorax brownish; the legs, except tarsi, brown or blackish: the abdomen paler than the head, with pale brownish bands. Teeth of mandibles and the eyes black.

Small worker: Length, 2.4 mm. The head is proportionally smaller and the occiput less strongly emarginated than in the large worker. It is almost uniformly brown, except the face each side of the frontal carinæ, the mandibles and scape, which are reddish, and the tarsi yellowish, while the whole surface is more highly polished.

Numerous specimens. Santiago Ixtquintla.

The discovery of the sexes may prove this to be a distinct species.

10. AZTECA [illegible] n. sp.

Large worker. Length, 4.8–5 mm. Head longer than broad, the sides parallel and slightly arcuated, the occiput deeply emarginated; vertex with a short, impressed, median line. Clypeus of the usual shape and slightly arcuated in front. Mandibles robust and armed with seven acute teeth. Antennæ rather short, the scape stout, not reaching to the occiput; the two basal joints and the last joint of the flagellum longest, joints three and four slightly longer than wide, the remaining joints about as long as wide and increasing gradually in width towards the end of the antennæ. Prothorax about one-fourth narrower than the head, widest in front of the middle and very convex; pro-mesonotal suture rather deep; the mesonotum distinctly elevated and convex; metanotum not elevated, convex and with a transverse impression each side near the anterior margin. Scale stout, nearly as broad above as below, the apex bluntly rounded. Abdomen broadly ovate and about as long as the thorax. Legs stout.

Polished. Head and thorax extremely fine and densely punctured, the punctures slightly stronger on the clypeus. Mandibles finely but rather indistinctly striated and finely and densely punctured, intermixed with few coarser punctures. Abdomen finely shagreened.

Erect hairs yellowish, sparse and almost wanting on the abdomen, most numerous on the head in front of the antennæ and mandibles, the thorax and around the end of the abdomen; antennæ and legs without erect hairs, except a few at the apex of the scape. Appressed pubescence whitish and dense.

Color lighter or darker brown, the thorax, legs and abdomen sometimes quite pale yellowish. Mandibles reddish-brown, the teeth and eyes black. Clypeus, the scape along its front edge, declivity of the metanotum, the base and sutures of the abdomen and the legs more or less distinctly yellow or brownish-yellow.

Small worker: Length, about 3.4 mm. Resembles in every respect the large worker, except that the head is proportionately shorter and its sides more strongly arcuated, while the coloration is more uniformly brownish-yellow.

Twenty-two specimens. Santiago Ixtquintla.

This species resembles very much *Azteka angusticeps*, differing from it, however, in the larger size, the sparser erect hairs and the absence of such hairs on the antennæ and legs.

11. TAPINOMA MELANOCEPHALUM (Fab.) Mayr.

Formica melanocephala Fabricius, Entom. Syst., ii, 1793, p. 353.

Lasius melanocephalus Fabricius, Syst. Piez., 1804, p. 417.

Myrmica melanocephala Lepeletier, Hist. nat. Ins. Hym., i, 1836, p. 185.

Formica nana Jerdon, Madras Journ. of Litt. & Sc., xvii, 1851, p. 125.

Myrmica pellucida Smith, Journ. Proc. Linn. Soc. Zool., ii, 1857, p. 71.

Myrmica (Monomorium) pellucida Smith, Cat. Hym. Ins. Brit. Mus., vi, 1858, p. 124.

Formica familiaris Smith, Journ. Proc. Linn. Soc. Zool., iv, 1860, Suppl. p. 96.

Tapinoma melanocephalum Mayr., Verh. zool. bot. Ges. Wien, xii, 1862. p. 651.

Micromyrma melanocephala Roger, Berlin entom. Zeitschr., vi, 1862, p. 258.

One specimen. San Blas.

This species is quite common on all of the West Indian islands and has also been found in a hothouse of the botanical garden at Washington, D. C.

12. DORYMYRMEX PYRAMICUS Rog. var. FLAVUS McCook.

Dorymyrmex flavus McCook, Comstock, Rep. Cotton-worm, 1879, p. 188.

Dorymyrmex pyramicus var. *flavus* Mayr, Verh. zool. bot. Ges. Wien, xxxvi, 1886, p. 433.

Three specimens. San José del Cabo and Sierra San Lazaro, Cape Region, Lower California.

13. DORYMYRMEX PYRAMICUS Rog. var. NIGRA n. var.

Worker: Length, 2.8–3.4 mm. Black; the mandibles reddish-yellow, the under side of the head and the tarsi brownish. This form is somewhat more robust, though otherwise like *flavus*.

Twenty-five specimens. Tepic.

Subfamily PONERINI.

14. ECTATOMMA RUIDUM Roger.

Ponera (Ectatomma) ruida Roger, Berl. entom. Zeitsch., iv, 1860, p. 306.

Ectatomma ruidum Mayr, Verh. zool. bot. Ges., Wien, xii, 1862, p. 732.

Ectatomma scabrosa Smith, Trans. Ent. Soc., London (3) i, 1, 1862, p. 31.

Ectatomma scabrosa Roger, Berl. entom. Zeitsch., vi, 1862, p. 292.

Ectatomma ruidum Mayr, Verh. zool. bot. Ges., Wien, xxxvii, 1887, p. 539.

Eleven specimens. Tepic.

This species is quite common in Brazil, Central America and the West Indies.

15. ECTATOMMA (GNAMPTOGENYS) RIMULOSUM Roger var. SPLENDIDUM n. var.

Worker: Length, 4–4.6 mm. Head longer than broad, its sides parallel, the posterior angles rounded, the occiput slightly emarginate. Eyes convex and projecting, placed beyond the middle. Antennæ rather short, the scape barely reaching the occiput, the flagellum clavate, joints two to eight shortest and about as long as broad. Clypeus triangular and about as broad as long, the anterior margin straight, the upper surface faintly concave. Mandibles elongated, narrow, leaving a large triangular opening between them when closed, the terminal half curved downward; the cutting edge smooth or faintly denticulate.

Thorax compressed at the sides, uniformly and gently curved and slightly convex above. Suture between the pro- and mesonotum obsolete, both segments forming one piece which is slightly broadest in front; pronotum somewhat angulated posteriorly and prolonged in front into a short neck. Meso-metanotal suture slight and most readily observed if viewed from the front. Metanotum of equal width, the declivity convex and bordered on each side by a curved carina, which at its upper end forms a minute and slightly projecting angle. Node very stout, longer than broad, truncate in front and behind, broadest and highest posteriorly, where it is as broad or somewhat broader than the metathorax; it is longitudinally arcuate, convex above and at the sides; on the anterior end of its under side is a large, flattened and backward curved tooth, the basal portion of which extends carina-like to near the end of the segment.

Abdomen longer than the thorax, anteriorly slightly broader than the scale and strongly constricted between the first and second segment. First segment with a tooth below the insertion of the scale.

Entire insect polished. Head, thorax, scale and the two basal segments of the abdomen closely and longitudinally striated above and at the sides, the striæ somewhat finer on the second segment of the abdomen, the remaining segments smooth and highly polished. Mandibles sparsely punctured. Erect hairs of the body, legs and scape fine and quite profuse.

Color red or yellowish-red, the abdomen somewhat paler: the head and thorax with a red reflection and the scale and two basal segments of the abdomen with a beautiful golden reflection. Cutting edge of mandibles black or dark brown; eyes black; scape and legs yellow.

Many specimens. Tepic.

This variety differs from *rimulosum* besides the somewhat smaller size, in the larger scale and golden reflection. It comes apparently also near *Ectatomma (Gnamptogenys) continuum*, differing from it, however, in the larger size and other minor characters.

16. PACHYCONDYLA VILLOSA (Fab.) Mayr.

Formica villosa Fabricius, Syst. Piez., 1804, p. 409.
Ponera villosa Illiger, Mag. f. Insektenk. vi, 1807, p. 194.
Ponera bicolor Guérin, Iconogr. règn. anim., vii, Insect., 1845, p. 242.
Ponera pilosa Smith, Cat. Hym. Brit. Mus., vi, 1858, p. 95.
Ponera pedunculata Smith, Cat. Hym. Brit. Mus., vi, 1858, p. 96.
Pachycondyla villosa Mayr, Verh. zool. bot. Ges., Wien, xii, 1862, p. 720.
Formica villosa Roger, Berl. ent. Zeitsch., v, 1862, p. 287.

Two specimens. Tepic.

This species was also discovered by Mr. E. A. Schwarz at San Diego, Texas.

17. PACHYCONDYLA HARPAX (Fab.) Mayr.

Formica harpax Fabricius, Syst. Piez., 1804, p. 401.
Pachycondyla Montezumia Smith, Cat. Hym. Brit. Mus., vi, 1858, p. 108.
Ponera amplinoda Buckley, Proc. Ent. Soc., Philad., vi, 1866, p. 171.
Pachycondyla orizabana Norton, Am. Nat., ii, 1868, p. 64.
Pachycondyla harpax Mayr, Sitzb. Akad. Wiss., Wien, lxi, 1871, p. 397.

Two specimens. Tepic.

Specimens of this species, agreeing exactly with Buckley's description of *Ponera amplinoda*, were also discovered by Mr. E. A. Schwarz at Beeville, Texas, living under dried cow dung.

18. ODONTOMACHUS HÆMATODA (L.) Latr.

For synonyms and references, see Dalla Torre, Catalogus Hymenopterorum, vol. vii, Formicidæ, p. 50.

Two specimens. Tepic.

Subfamily Dorylini.

19. Eciton (Acamatus) schmitti Emery.

Eciton Schmitti Emery, Bull. Soc. Entom. Ital., v, 26, 1894, p. 183.
Eciton Schmitti Emery, Zool. Jahrb., viii, 1894, p. 258.

Many specimens. Tepic.

20. Eciton (Acamatus) melanocephalum Emery.

Eciton melanocephalum Emery, Zool. Jahrb., viii, 1894, p. 260.

Eight specimens. Tepic.

21. Eciton (Acamatus) californicum Mayr subsp. opacithorax Emery var.

Eciton californicum Mayr, subsp. *opacithorax* Emery, Zool. Jahrb., viii, 1894, p. 260.

Numerous specimens. San José del Cabo.

It differs from the typical form in the slightly coarser sculpturing of the thorax.

22. Eciton nitens Mayr.

Eciton nitens Mayr, Annu. Soc. natural., Modena, iii, 1868, p. 168.
Eciton nitens Mayr, Sitzb. Akad. Wiss., Wien, lxi, 1870, p. 398.

Two specimens. San Lazaro, Lower California.

These specimens, notwithstanding their rather small size, measuring only 2.4 mm. in length, agree apparently in every essential point with the description of the above species. I deem it therefore advisable, until a larger series has been obtained, to regard them as but a small form of that species.

Subfamily Myrmicini.

23. Pseudomyrma gracilis (Fab.) Mayr.

Formica gracilis Fabricius, Syst. Piez., 1804, p. 405.
Leptalea gracilis [illegible] Arch. f. Naturg., v, P. 2, 1839, p. 309.
P[illegible] [illegible] Insect., 1845, p. 427.
P[illegible] [illegible] London (2) iii, 4, 1855, p. 157.
E[illegible] [illegible] 1862, p. 289.

Pseudomyrma bicolor Norton, Am. Natur., ii, 1868, p. 65.
Pseudomyrma gracilis Mayr, Sitzber. Akad. Wiss., Wien, lxi, 1870, p. 407.
Pseudomyrma bicolor Rothney, Trans. Ent. Soc., London, 1889, p. 352.
Pseudomyrma gracilis Emery, Bull. Soc. Ent. Ital., xxii, 1890, p. 59.

Ten workers. Tepic.

This is an extremely variable species. In some of its forms the entire head, the thorax, both nodes, anterior and median legs, are reddish-yellow, the antennæ brown or blackish and the posterior legs and abdomen black. In another form, only the thorax, with the exception of a round spot on the mesonotum, more or less of the anterior and median legs and the first node are reddish, the rest black. A third form is black, with the anterior margin of the head, mandibles, base and apex of the scape, prothorex, margin of the mesonotum, apical third of anterior femora, the tibiæ and tarsi, petiole and apex of first node and base of abdomen reddish-yellow, while in a fourth form the anterior margin of the head, the mandibles, the pro- and metathorax, the anterior femora, except at base, and the anterior tibiæ and tarsi are only reddish-yellow and all the rest black.

24. PSEUDOMYRMA PALLIDA Smith.

Pseudomyrma pallida Smith, Trans. Ent. Soc., London (2) iii, 4, 1855, p. 159.
Pseudomyrma pallida Smith, Cat. Hym. Ins. Brit. Mus., vi, 1858, p. 155.

Ten specimens. Tepic.

25. MONOMORIUM MINUTUM Mayr, race EBENINUM Forel.

Monomorium minutum Mayr, race *ebeninum* Forel, Hist. Phys. Nat. Polit., Madagascar, Formicides, xx, 1891, p. 165.

Numerous specimens. Santiago Ixtquintla, Tepic.

Very similar in appearance to *Monomorium* race *car-*

bonarium Smith, from which, however, it may be readily distinguished by the stronger meso-metanotal constriction and the more elevated posterior portion of the metanotum, the declivity of which forms nearly a sharp angle with the dorsal face of the segment, and in the more slender and longer pedicel of the first node.

This race is quite common on the West Indian islands and the Central American republics.

26. CREMASTOGASTER SUMICHRASTI Mayr, var.

Cremastogaster sumichrasti Mayr, Verh. zool. bot. Ges., Wien, xx, 1870, pp. 990 and 993.

Many specimens. Ixtquintla, Tepic.

Very similar in appearance and coloration to *Cr. missouriensis* Emery, but differing from it in the shorter terminal joint of the antennæ, the not emarginate posterior edge of the first node, the more transverse second node and the much longer and finer hairs of the head, thorax and abdomen.

27. CREMASTOGASTER OPACA Mayr.

Cremastogaster opaca Mayr, Verh. zool. bot. Ges., Wien, xx, 1870, pp. 989 and 992.

Twenty-one specimens. Tepic.

28. CREMASTOGASTER SCULPTURATA n. sp.

Worker: Length, 2.2–2.4 mm. Head polished, finely and sparsely striated above and below; granulated at inner side of eyes. Antennal foveæ finely reticulated. Clypeus with a flattened, smooth and polished median area which is bordered each side by a slightly elevated and posteriorly diverging carina: its lateral pieces sparsely striated. Mandibles polished, faintly striated and sparsely punctured. Club of antennæ two-jointed.

Pronotum with a few coarse and somewhat irregular carinæ. Mesonotum slightly elevated in front, sparsely and coarsely sculptured. Basal section of the metanotum with a few, posteriorly diverging carinæ; the declivity small, triangular and smooth; metanotal spines rather long, slender, directed backward and upward. Sides of thorax coarsely granulated.

First node elongate-quadrate, flat above, rounded in front, truncate behind and with the posterior angles quite acute; surface smooth and polished. Second node small, rounded and with two rather deep longitudinal lines above, the space between them granulated. Abdomen highly polished and rather long and pointed; faintly transversely wrinkled.

Erect hairs of the head, thorax and abdomen rather sparse, whitish, long and slender: hairs of legs and antennæ shorter and sub-erect. Appressed pubescence apparently none.

Color black. Mandibles yellowish; antennæ and legs dark brownish, the coxæ and tarsi somewhat paler.

Four specimens. Tepic.

This species resembles somewhat *Crem. crinosa* Mayr, but differs from it in the sculpturing of the head, thorax and nodes: it comes also near to *Crem. carinata* and *curvispinosa* Mayr, from which it, on account of the peculiar formation of the clypeus and the differently sculptured thorax, may be readily separated.

29. CREMASTOGASTER LINEOLATA (Say) Mayr, race CERASI Fitch, var.

Myrmica cerasi Fitch, Trans. N. Y. State Agr. Soc., xiv, 1854, p. 835.

Myrmica cerasi Fitch, First and Second Rep. Ins. N. Y., 1856, p. 130.

Cremastogaster lineolata Say, var. *cerasi* Emery, Zool. Jahrb., VIII, 1894, p. 282.

Numerous specimens. Sierra San Lazaro and San José del Cabo, Cape Region, Lower California.

Differs from the typical form in the slightly rougher

pro- and mesothorax and the somewhat coarser striæ of the metathorax.

Cremastogaster sanguinea Roger is evidently but a variety of *cerasi*.

30. PHEIDOLE PUNCTATISSIMA Mayr.

Pheidole punctatissima Mayr, Sitzb. Akad. Wiss., Wien, lxi, 1870, p. 400.
Pheidole punctatissima Mayr, Verh. zool. bot. Ges., Wien, xx, 1870, pp. 980 and 983.
Pheidole punctatissima Mayr, Verh. zool. bot. Ges., Wien, xxxvii, 1887, pp. 583 and 598.

Five soldiers and six workers. San Blas, Tepic.

31. PHEIDOLE TEPICANA n. sp.

Soldier: Length, 3 mm. Head about twice as long as broad, the sides parallel, faintly narrower posteriorly and with the angles rounded, the occiput deeply emarginate and with a deep median channel extending nearly to the clypeus. Frontal area minute, oval, deeply impressed. Clypeus very short, transverse, sinuate in front and behind; the anterior margin deeply emarginate at the middle. Antennæ short, the scape about one-half the length of the head. Mandibles robust, the cutting edge with two or three teeth at the apex.

A deep constriction between the meso- and metanotum. Lateral angles of the pronotum obtusely rounded; metanotum with two short, erect, stout spines. Nodes as in the worker.

Vertex finely and transversely striated and more or less strongly rugose at the posterior angles, the striæ interspersed with coarse punctures: face and clypeus longitudinally striated, the striæ becoming more rugulose between the eyes and frontal carinæ. Mandibles polished, sparsely punctured and with a row of coarser, elongated punctures near the cutting edge.

Thorax polished, its sides and dorsal face of the metanotum densely granulated, the declivity and channel between the spines finely reticulated or granulated. Sides of first node very finely granulate, the second node without sculpturing. Erect hairs quite dense and rather coarse, especially so on the head and abdomen.

Color dark brown or black. The anterior third of the head, extending between the frontal carinæ to about the middle of the head, and sometimes almost the entire sides of the head, the clypeus, mandibles, flagellum, tibiæ and tarsi, more or less distinctly reddish-yellow, the scape and femora brownish. Sometimes the entire head, thorax and nodes are reddish; the anterior half or more of the abdomen yellowish-brown, and legs and antennæ yellowish; or the posterior angles of the head and the thorax brownish and the legs and antennæ reddish.

Worker: Length, 2.2–2.4 mm. Head longer than broad, the sides arcuated, the occiput rounded, the vertex with a faintly impressed median line. Clypeus convex and with a slender median carina, the anterior margin arquate. Mandibles of the usual shape, armed with a number of acute teeth. Pronotum convex above and at the sides, prolonged into a neck and without lateral angles; there is a rather deep depression across the anterior end of the mesonotum and a deep constriction between the meso- and metanotum; upper face of metanotum quite flat, the thorus short, stout, acute and about one-fourth the length of the basal section. Nodes of the usual shape, the second node rather small and nearly globular. Legs and antennæ rather long and slender.

Head faintly and indistinctly shagreened, granulated between the eyes and frontal carinæ. Sculpture of the clypeus indistinct; mandibles finely striated and sparsely punctured. Thorax densely granulated, the pronotum almost smooth.

Erect hairs fine and sparse, slightly more dense on the scape and abdomen.

Polished, black or dark brown, mandibles and tarsi yellowish.

Ten soldiers, twenty-five workers. Type.

The worker resembles very much that of *Pheidole commutata* Mayr, but differs from it in the less strongly constricted suture between the meso- and metanotum, the longer metanotum and its spines and smaller second node; the soldier has a much longer head and entirely different sculpturing.

[illegible]. PHEIDOLE [illegible] n. sp.

Soldier: Length, [illegible] mm. Head very large; much longer and broader than the thorax; its sides parallel, the posterior angles rounded, the occiput deeply emarginate and with a deep furrow extending to the clypeus. Clypeus short, convex, its posterior angle extending far between the antennæ, the anterior margin deeply emarginate each side, arcuate in front and deeply emarginate at the middle. Antennæ very short, the scape not reaching to the middle of the head. Mandibles very robust and with one or two blunt teeth at the apex. Thorax rather deeply constricted between the meso- and metanotum, the latter with two short and stout, acute and erect spines; the declivity concave; prothoracic angles obtusely rounded. Upper edge of first node straight or slightly concave; second node transverse, broader than long, with the lateral angles bluntly rounded.

Head [illegible] -striate, the [illegible] and more or [illegible] between [illegible] striate. M[illegible] striated and sparse[illegible] and mesonotum transversely and finely striated above, sides of prothorax, anteriorly, rugoso-striate; metanotum and sides of the thorax densely granulated, the declivity polished and with fine, transverse reticulations. First node, smooth above, its sides finely granulated; second node finely and densely granulated and with a few longitudinal, impressed lines. Abdomen polished and sparsely punctured. Erect hairs whitish, rather dense, short and fine, though somewhat stouter on the abdomen.

Head red, with the posterior angles and median furrow sometimes brown or black, or with a blackish stripe each side of the occiput. Scape black or dark brown, the flagellum yellowish-brown, darker towards the end. Thorax and nodes black or brownish; the abdomen black. Femora dusky, or with a brownish tinge, the tibiæ generally paler and the tarsi more or less distinctly yellowish.

It resembles very much the soldier of *Ph. tepicana*, but differs from it in the larger size, the rougher head, deeper emargination of the clypeus and more transverse second node.

Seven specimens. Tepic.

33. PHEIDOLE CARBONARIA n. sp.

Soldier: Length, 2.2 mm. Head longer than broad, slightly broadest anteriorly, the sides nearly straight, the posterior angles rounded, the occiput deeply emarginate; a deep median furrow which reaches nearly to the frontal area. Frontal area elongate triangular and deeply impressed. Clypeus sinuate in front and behind and deeply emarginate at the middle. Antennæ short, the scape reaching slightly beyond the middle. Prothorax about one-half the width of the head and shaped like that of the worker, the sutures between it and the mesonotum but faintly indicated, the transverse depression of the mesonotum very slight, meso-metanotal constriction deep, the metanotum very similar to that of the worker, the spines short and stout. Second node nearly globular.

Polished; the anterior third or more of the head quite densely and longitudinally striated, the rest sparsely punctured, the occiput with faint transverse wrinkles. Median section of the clypeus finely striated, the lateral section smooth. Mandibles polished and with but few punctures. Prothorax smooth or very faintly sculptured in front; the mesothorax smooth above and densely granulated at the sides; metathorax granulated above and at the sides. Nodes and abdomen smooth. Erect hairs white and rather sparse.

Black or dark brown; the anterior margin of the head, clypeus, mandibles, antennae and legs yellowish or brownish-yellow.

Worker: Length, [illegible] mm. Head somewhat longer than broad, the posterior angles broadly rounded, the occiput distinctly emarginate; the face with a delicate median furrow. Frontal area distinct, triangular. Clypeus with a faint median carina and delicate oblique striae each side of it, the anterior margin not emarginate. Antennae rather short, the scape reaching but slightly beyond the occiput, the club longer than the remaining joints of the flagellum. Prothorax somewhat narrower than the head, rounded above and at the sides. Pro- and mesonotum without an apparent dividing suture; the mesonotum tapering posteriorly and without a transverse depression; meso-metanotal constriction quite deep; metanotum flat [illegible] sparsely punctured. Prothorax without sculpturing, except faintly so on the neck, the mesothorax smooth above and densely granulated at the sides; metathorax granulated above and at the sides. Nodes and abdomen smooth. Erect hairs very sparse.

Black. Antennæ, clypeus and legs brown; teeth of mandibles black or brown; mandibles and tarsi yellowish.

The worker resembles very much in general appearance the genus Monomorium, but differs from it in the two-jointed palpi.

Four soldiers and seven workers. Tepic.

34. Pheidole floridana Emery, var. deplanata n. var.

The soldier of this variety differs from the typical form in the somewhat smaller size, the smoother and anteriorly less distinctly emarginate clypeus, more strongly granulated thorax, the flattened dorsal surface of the metanotum and the shorter and stouter thorus.

The worker is smaller and more strongly granulated. The head, metathorax and end of body dark brown, pro- and mesothorax reddish-brown; antennæ, legs, nodes and basal segments of the abdomen brownish-yellow.

Five soldiers and two workers. Tepic.

35. Pheidole vaslitii n. sp.

Soldier: Length, 3.8–4.4 mm. Head somewhat longer than broad, deeply emarginate behind and with a deep median furrow, extending nearly to the clypeus. Clypeus convex, the anterior margin arcuate and notched at the middle. Antennæ rather stout, the scape about two-thirds the length of the head. Mandibles robust and with but two stout teeth at the apex. Pronotum slightly angulated at the sides; the mesonotum with a deep and broad transverse depression, its posterior section with a sma blunt tubercle each side anteriorly: meso-metanotal co striction deep and narrow, the metanotum with a de channel along the middle, the lateral margins of the cha nel quite sharp: spines stout, acute and about one-four the length of the basal section. First node as usual: t second node transversely oval and with two longitudin impressed lines.

Head reticulate-striate and densely and finely gran lated; area between the frontal carinæ not granulate the striæ diverging posteriorly: median area of the cl peus polished, and with coarse striæ along the anteri margin. Mandibles polished, sparsely and quite coarse punctured. Pronotum transversely striated and indi tinctly granulated, and with hair-bearing points. M sonotum densely but rather indistinctly granulated, t anterior section with a few hair-bearing points; the enti metathorax as well as the sides of the other two segmen densely granulated. Abdomen polished and very fi and densely shagreened and with hair-bearing point Erect hairs rather long, fine and quite dense, those of t scape and legs somewhat shorter, appressed pubescen most dense on the head and abdomen.

Color yellowish-red or brownish-red, the head and a domen frequently darker in the larger specimens.

Worker: Length, 2.6 mm. Head longer than broa the sides gently arcuate. Clypeus convex, the anteri margin but slightly arcuated and slightly notched at t middle, with a slender median carina and a shorter o [illegible] Antennæ [illegible] the scape reachi [illegible] of the flagell [illegible] usual shape, w [illegible] number of mini [illegible] four projecting late angles; the mesonotum with a rather deep median depression, the posterior half slightly carinated each side and minutely angulated in front; meso-metanotal constriction quite deep; the basal section of the metanotum is longer than the declivity and somewhat broadest posteriorly, the upper surface is quite flat and slightly concave between the thorus, which are rather small, acute and nearly erect. Upper face of the first node nearly straight, the second node nearly globular.

Surface of the head, except between the frontal carinæ, quite densely and finely granulated, and with a few fine striæ in front and behind, and at inner and outer side of eyes; space between the frontal carinæ smooth; clypeus sparsely and finely granulated; mandibles faintly striated and quite coarsely punctured. Thorax densely granulated, the pronotum in front with two or three transverse striæ. Nodes densely granulated. Abdomen smooth, very fine and obscurely reticulated. Erect hairs rather sparse, long and fine, somewhat shorter on legs and antennæ; appressed pubescence sparse, most dense on the head.

Polished; thorax yellow or reddish-yellow; the head and nodes either of the same color or lighter or darker brownish, the abdomen brown or black. Antennæ and femora brownish; the clypeus, tibiæ and tarsi generally somewhat paler; mandibles yellowish, the teeth black or brown.

Nine soldiers and thirteen workers. Sierra San Lazaro.

This species resembles very much *Ph. breviconus* and *cubaensis* Mayr, but differs from both in sculpturing and other characters.

36. PHEIDOLE SUSANNÆ Forel, race LONGIPES n. race.

Soldier: Length, 4.4–4.6 mm. Head longer than

broad, the sides gently arcuated, the occiput deeply emarginate; the face with a rather deep median channel. Frontal carinæ nearly half the length of the head. Frontal area triangular. Clypeus with a sharp median carina; the anterior margin slightly sinuate each side and with a shallow and broad emargination at the middle. Mandibles with two large teeth at the apex. Antennæ rather slender, the scape reaching somewhat beyond the occiput. Pronotum without lateral tubercles; mesonotal depression not as strong and the metanotal spines longer and stouter than in the worker; second node nearly globular.

Head sub-opaque and densely granulated, the anterior half distantly striated and faintly reticulated, the posterior half with elongated, shallow depressions; space between the frontal carinæ quite smooth and faintly shagreened; the frontal area polished; clypeus faintly rugose about the middle and somewhat more coarsely so each side. Mandibles polished, rather densely and quite coarsely striated, sparsely, and very coarsely punctured near the cutting edge. Thorax finely and densely granulated, the granulation slightly coarser in the longitudinal depression of the metanotum; second node distinctly shagreened. Abdomen densely granulated, more coarsely so on the first segment and with numerous, more or less elongated shallow depressions and hair-bearing points. Erect hairs yellowish, coarsest and densest on the abdomen, very sparse on the thorax.

Color reddish-yellow; antennæ, mandibles and abdomen slightly brownish, anterior margin of the clypeus, the cutting edge and teeth of the mandibles brown. Eyes black.

Worker: Length, 3 mm. Head much longer than broad, much prolonged and tapering beyond the eyes, with a deep constriction at the end, forming a distinct neck. Clypeus truncate behind, slightly arcuated anteriorly and slightly emarginate at the middle. Mandibles with two large and acute teeth at the apex, the rest finely denticulate. Antennæ very long and slender, the scape about twice as long as the head. Thorax much elongated, the prothorax elongated pyriform, broadest behind and with a neck-like constriction anteriorly; it is convex above and at the sides; transverse mesonotal depression quite strong, extending down to the lateral margin, the posterior half sloping gently and uniformly to the metanotum; meso-metanotal constriction somewhat stronger than that of the mesonotum. Metanotum elongated and gently curved towards the base of the first node, the basal section much longer than the declivity; declivity slightly concave along the middle; the thorus very minute. Second node subglobose.

Head finely and densely shagreened, the space between the frontal carinæ smooth; clypeus faintly granulated. Mandibles sparsely punctured. Pro- and mesonotum finely and densely shagreened, the metanotum granulated. Nodes apparently smooth. Abdomen smooth, the basal half or more finely shagreened. Erect hairs sparse and slender, shorter and suberect on the scape and legs; appressed pubescence very minute and sparse.

Color yellow; the mandibles and legs slightly paler, the abdomen faintly brownish. Eyes black. The whole surface polished.

Sixteen soldiers and numerous workers. Cape Region, Sierra San Lazaro.

This form differs from *Ph. susannæ* in the somewhat larger size, longer antennæ, shallower transverse depression and not abrupt posterior half of the mesonotum, the less deep meso-metanotal constriction and the longer metathorax.

It seems also to be related to *Pheidole (Atta) testacea* Smith.

37. PHEIDOLE SUBDENTATA n. sp.

Worker: Length, 2.6 mm. Head nearly twice as long as broad, the sides parallel and slightly arcuated, rounded behind; clypeus convex, rounded behind, arcuate in front, the lateral angles reaching to the mandibles, without a median carina; frontal area but feebly indicated; mandibles of the usual shape, armed with two long teeth at the apex and six or seven minute teeth behind them. Antennæ rather long and slender, the scape almost twice as long as the head. All joints of the flagellum longer than thick, the club being about one-half the length. Pronotum about one-fourth narrower than the head, rounded above and at the sides, prolonged into a neck anteriorly; mesonotal depression shallow; meso-metanotal constriction deep; metanotum somewhat elevated posteriorly and slightly curved longitudinally, convex above and with a slight longitudinal furrow at base, the basal section much longer than the declivity and furnished with two minute teeth. Nodes of the usual shape.

Head smooth, sparsely striated in front of eyes and between base of antennæ; antennal grooves slightly granulated. Clypeus smooth, with few and indistinct striæ. Mandibles smooth, faintly and sparsely punctured. Pronotum and anterior half of mesonotum above, faintly shagreened and with a few hair-bearing points, posterior half of the mesothorax and the metathorax densely granulated; nodes and abdomen smooth, the petiole of the first node granulated laterally; the first segment of the abdomen sometimes with a few elongated, shallow, obsolete depressions. Erect hairs rather sparse, pale and slender.

Polished yellowish-brown, the head and abdomen

generally somewhat darker, the clypeus, mandibles, antennæ and legs somewhat paler.

Many specimens. Tepic.

38. PHEIDOLE OBTUSOSPINOSA n. sp.

Soldier: Length, 6.5 mm. This species resembles very much the soldiers of *Ph. vaslitii*, though it is much larger, the scape is shorter, and only about one-half the length of the head, which is also more densely and more sharply reticulate-striate; the pro- and metanotum more distinctly transversely reticulate-striate; the metanotum, besides being granulated, is also finely and transversely striated between the spines and on the declivity; the spines are much stouter, obtuse at tip and more or less distinctly curved inwards; the second node is much broader, transversely oval and with numerous, rather deep, longitudinal striæ; the sculpturing of the abdomen is very similar but much coarser, and the hairs on every part of the body denser.

Color ferruginous, the abdomen varying from light brown to nearly black. Anterior margin of head black. Mandibles brownish-red.

Many specimens. Tepic.

PHEIDOLE Westwood, subg. CERATOPHEIDOLE, n. subg.

This remarkable species, of which but two specimens were obtained, agrees in almost all of its characters with those of soldiers of the typical Pheidole, from which it principally differs in the unusually long and slender antennæ, *the scape of which reaches to the occiput*, whereas the club of the flagellum, instead of being three-jointed as in Pheidole proper, *is composed of four long and subequal joints.* The number of joints in the palpi could not be ascertained. The discovery of the sexes and workers, if such exist, may exhibit additional characters, which might entitle this form to a generic rank; for t present, however, it may be considered as being but subgenus of Pheidole.

39. PHEIDOLE (CERATOPHEIDOLE) GRANULATA n. s

Soldier? Length, 5.6 mm. Head nearly twice as lo as broad and somewhat broadest in front, the sides ge tly arcuate, posterior angles rounded, the occiput deep and somewhat angularly emarginate; a deep median fu row extending to the frontal carinæ. Frontal carin about one-third the length of the head, converging post riorly. Frontal area distinct, elongate triangular. Cl peus transverse, extending posteriorly between the ba of the antennæ, the posterior margin arcuate, posteri margin of the lateral pieces deeply sinuate, the anteri margin arcuate and angularly emarginate at the middl Eyes in front of the middle. Mandibles very robust ar provided with two stout teeth at the apex. Antenn twelve-jointed, slender, the scape reaching to the occipu the seven basal joints of the flagellum subequal in lengt each about four times as long as thick, joints eight to t also subequal in length, each of them about one-thi longer than any of the preceding joints, the last joi slightly the longest. Prothorax not quite one-half broad as the head, broadest posteriorly, prolonged into neck anteriorly, convex above and at the sides; meson tum slightly elevated anteriorly and with a rather bro and shallow, transverse, median depression; meso-met notal constriction deep and rather broad; metanotum fl tened above, somewhat concave between the spines, t basal section about twice the length of the declivit spines stout, directed back- and upward and about on fourth the length of the basal section of the segme Nodes as in Pheidole, the second node broader than lo and obtusely angulated.

Head, densely and finely granulated and longitudinally reticulate-striate, the striæ becoming obsolete towards the occiput. Frontal area polished and with a median carina. Median section of the clypeus indistinctly granulated, with a median carina and a few anteriorly diverging striæ, the anterior margin coarsely punctured, giving it a serrated appearance. Mandibles polished, coarsely and rather sparsely striated and punctured. Entire thorax densely granulated, the neck of the pronotum with a few transverse striæ; both nodes and the abdomen also densely granulated, and the first abdominal segment with a few shallow and somewhat elongated depressions. Erect hairs yellowish, rather stiff and quite dense, especially so on the head and abdomen, intermixed quite evenly with shorter, suberect, stiff hairs. Legs and antennæ with appressed pubescence, intermixed with a few erect hairs on the scape and along inner edge of femora.

Color reddish-yellow, the legs paler; mandibles reddish, with the external margin and cutting edge, the anterior margin of the clypeus and the eyes black.

Two specimens. Tepic.

This form appears to be the connecting link between Pheidole and Messor.

40. APHÆNOGASTER MUTICA n. sp.

Worker: Length, 5 mm. Head longer than broad, slightly narrowest in front, rounded behind. Frontal area oval and with a few longitudinal striæ. Clypeus triangular, arcuate in front and with a broad and somewhat angular emargination at the middle. Mandibles armed with three large teeth at the apex and a number of smaller ones behind them.

Prothorax about one-fourth narrower than the head, very convex and with a short neck; the meso- and metathorax narrower and nearly parallel; anterior half of the mesonotum oval, convex and somewhat elevated ab the pronotum, its posterior half almost in a plane with metanotum; constriction between the meso-metanot quite deep and narrow; basal section of the metanot rather more than twice the length of the declivity, its terior half rounded above, the posterior half with a pressed, elongated triangular area which merges pos riorly with the declivity. Spines wanting, their positi indicated by an angle at the upper edge of the declivi First node stout, erect, convex in front, above and hind; second node pyriform if viewed from above, e vated and broadest behind, convex above and round behind.

Head, densely and finely granulated and finely retic late-striate; striæ between the frontal carinæ divergi posteriorly; clypeus with a median carina and irregula striated; mandibles densely and quite coarsely striat and sparsely punctured, the punctures becoming coars towards the cutting edge. Prothorax highly polished a faintly shagreened; elevated section of the mesonot highly polished and without sculpturing; the posteri section finely and irregularly striated and granulated, t lateral pieces densely granulated and with faint obliq striæ; metathorax delicately shagreened, with faint a sparse transverse striæ at base above and a few coars longitudinal striæ at sides above the coxæ. Nodes p ished and but faintly shagreened. Abdomen polishe the basal half or more of the first segment finely sh greened.

Erect hairs yellowish, sparse, more dense and coars on the abdomen; pubescence of antennæ and legs sh and suberect, with a few longer hairs along inner edge femora.

Color of the head, antennæ, thorax and nodes reddis the clypeus, mandibles and legs more yellowish; anteri margin of the clypeus, the teeth and external edge of the mandibles and the abdomen black.

Eight specimens. San José del Cabo.

41. ISCHNOMYRMEX MEXICANUM n. sp.

Worker: Length, 7 mm. Head elongated and more than twice as long behind than in front of the eyes, tapering to a neck and terminating in a sharp and elevated collar. Frontal area triangular and depressed. Clypeus transversely triangular, its posterior angle rounded, the lateral pieces deeply sinuate, the anterior margin arcuate and broadly emarginate at the middle. Mandibles armed with three stout teeth at the apex and a series of smaller teeth behind them. Prothorax about one-half as broad as the head in the region of the eyes, prolonged anteriorly into a neck, the mesonotum with a shallow, transverse depression beyond the middle; metathorax much elongated, its basal section about four times the length of the declivity, it is nearly straight above and with a shallow, longitudinal channel along the middle, the spines are short, stout and toothlike. Nodes rather stout. Legs very long and slender, the first tarsal joint of the posterior legs at least as long as the tibiæ.

Head, densely granulated, its anterior half somewhat irregularly rugoso-striate, with the striæ between the frontal carinæ somewhat finer; there is also a deep, elongated, median depression just beyond the carinæ; frontal area with a median and several posteriorly converging carinæ; clypeus indistinctly granulated; mandibles quite densely striated and sparsely punctured. Prothorax polished, faintly shagreened and very sparsely punctured; the mesonotum granulated along the middle, the anterior lateral half obliquely striated and the posterior half densely granulated; surface of the metathorax quite coarse and transversely rugoso-striate, the striæ at the sides, posteriorly, fine and longitudinal. Nodes faintly shagreened. Abdomen smooth, the first segment with a few narrow and elongated depressions. Erect hairs yellowish, short and stiff, most dense on the abdomen, those of the antennæ and legs shorter and finer.

Polished, brown, the mandibles and tarsi reddish-yellow.

Two specimens. Tepic.

This species appears to be related to *Ischnomyrmex (Myrmica) longipes* Smith, the description of which is, however, so obscure as to leave considerable doubt about it.

42. POGONOMYRMEX BARBATUS (Smith) Mayr.

Myrmica barbata Smith, Cat. Hym. Ins. Brit. Mus., vi, 1858, p. 1[illegible].
For additional references, see Dalla Torre, Cat. Hym. vol. vii, Formicidæ, 1893, p. 118.

Twenty-three specimens. Miraflores and Sierra San Lazaro, Cape Region, Lower California.

43. POGONOMYRMEX CALIFORNICUS (Buckley) Emery.

Myrmica californica Buckley, Proc. Ent. Soc., Philad., 1867, p. [illegible].
Pogonomyrmex badius Mayr, Verh. zool. bot. Ges., Wien, xx, 1870, p. 971; xxxvi, 1886, p. 450; xxxvii, 1887, p. 610.
Pogonomyrmex badius Pergande, Proc. Cal. Acad. Sci. (2), iv, 1[illegible], p. 33.
Pogonomyrmex californicus Emery, Zool. Jahrb., viii, 1894, p. 311.

Numerous specimens. San José del Cabo.

A somewhat larger, more robust and more strongly sculptured variety of this species, has also been found by Dr. Gustav Eisen at Tucson, Arizona.

44. XIPHOMYRMEX SPINOSUM n. sp.

Worker: Length, 3.2–3.4 mm. Head, thorax and nodes coarsely and longitudinally rugose, those of the head somewhat finer and the spaces between them distinctly reticulated, the clypeus quite coarsely striated;

striation of mandibles somewhat finer and denser. Antennæ stout, the scape not quite reaching to the occiput. Meso-metanotal constriction faint, the thorns stout, acute and straight, directed back- and upward and diverging towards the end; they are somewhat longer than the distance between them at base; declivity deeply concave, and with an acute, rather long and broad tooth each side of the insertion of the petiole of the first node. Nodes very stout, the first one almost cubical, convex above and at the sides, narrowest in front, with the anterior and posterior face perpendicular. Second node transversely oval, convex. Abdomen polished, the first segment finely and densely granulated and sparsely punctured. Erect hairs yellowish, shortest and stoutest on the abdomen, legs and antennæ.

Fourteen specimens. Sierra San Lazaro, Cape Region, Lower California.

45. Cyphomyrmex flavidus n. sp.

Worker: Length, 2.2–2.8 mm. Head, as usual in this genus; the antennal grooves deep and extending to the apex of the lateral angles of the occiput, the frontal carinæ with a deep and rounded emargination opposite the eyes; the occiput obtusely and deeply emarginate, the vertex with a depressed and flattened triangular area at the posterior angles and a circular area each side of the triangular frontal area. Clypeus slightly arcuate; the scape of the antennæ reaching slightly beyond the apex of the posterior angles; joints two to seven of the flagellum about as long as broad.

Pronotum convex above, carinated laterally and provided each side, posteriorly, with a prominent, stout and obtusely rounded tubercle, and with an acute angle anteriorly at insertion of the coxæ. Mesononotum oval and with a longitudinal furrow, the margins bordering it, a rounded, tubercular elevation. Meso-metanotal co striction deep, the metanotum concave along the middl the spines are reduced to short and flattened, bluntly t angular teeth, a small tubercle in front of each tooth a a tubercle above the coxæ. First node, not counting t pedicel, broader than long, more or less distinctly arc ate in front, angulated posteriorly, with the upper su face quite flat; second node transverse, narrowest i front, and with a deep, posteriorly broadening chann along the middle. First abdominal segment with a d pressed median line, reaching to or beyond the middle.

Face and clypeus sparsely but distinctly granulate rest of body opaque and sparsely covered with minut glistening, yellowish and appressed scale-like hairs, whic are most dense on the abdomen, legs and antennæ.

Color yellow or reddish-yellow; generally the clype and a more or less well-defined, broader or narrow stripe along the middle of the face, of a lighter or dark brown; teeth of mandibles and eyes black.

Seven specimens. Santiago Ixtquintla, Tepic.

This species appears to be related to *Cyp. kirbyi* a *morschi*, but differs apparently from both in some of t characters mentioned.

46. ATTA LÆVIGATA (Smith) Mayr.

(*Ecodoma lævigata* Smith, Cat. Hym. Ins. Brit. Mus., vi, 1858, 182.

Atta sexdens var. *lævigata* Mayr, Reise d. Novara, Zool. ii, 7, F micidæ, 1865, p. 80.

Numerous specimens. Tepic.

47. ATTA (ACROMYRMEX) SAUSSUREI Forel.

Atta (Acromyrmex) tardigrada Buckley, st. *saussurei* Forel, B Soc. Vaud. sc. nat. (2), xx, P. 91, 1884, p. 361.

Many specimens. Tepic.

BIOLOGICAL STUDIES ON FIGS, CAPRIFIGS AND CAPRIFICATION.

BY GUSTAV EISEN, PH. D.

INTRODUCTORY.

Caprification of figs is a practical process based on scientific principles, which latter are as interesting and have been as misunderstood as those connected with the practical part of the process. Since time immemorial caprification has been practiced in certain countries, and practical results have been claimed for it. As regards the practical value of caprification, there are two distinct and opposite views held by different investigators. Some claim that caprification is necessary and valuable, others hold that it is useless. As regards the scientific principles involved, there are also various views put forward, as will be explained further on, some of which are radically opposite to the others. The chief reason why this question has not been solved long ago has been two-fold. First, many of the scientific investigators have not been practical horticulturists, while others have not been aware that they experimented on figs which really did not require caprification, and which would not be benefited by it.

Every investigator began and ended his researches with the erroneous idea that all cultivated figs were alike, and he drew his conclusions accordingly. This alone explains the indifferent results achieved so far.

The many points involved in these interesting questions are both practical and scientific, and the two groups are so interwoven that the one cannot possibly be understood without a full knowledge of the other.

I am anxious that this may be understood in the beginning, as in the following pages practical details will be found hand in hand with scientific studies. The practical cultivator who knows but little of scientific phraseology, would not understand the terms unavoidably used below, except they were properly explained. Similarly, the scientific investigator, whose interest in this subject lies principally in the process of caprification and in its supposed value or uselessness, would not properly understand the practical details connected with the horticultural crops of the figs, except that they be explained in a way that may seem too elementary to the horticultural student or practical botanist.

My studies and experiments concern principally the following varieties of figs:

Caprifig (Ficus carica silvestris).—This is the wild fig, in which the Blastophaga breeds and goes through its various transformation. This fig tree species contains three crops of figs, of which only the male flowers attain maturity without caprification. The female flowers 'require pollination in order to produce seed.

Smyrna Fig (Ficus carica smirniaca.—The fig trees of this class possess no male flowers, no mule flowers, no gall flowers, only female flowers. The latter require always pollination or caprification, in order to come to any kind of maturity. Only cultivated varieties.

Common Fig (Ficus carica hortensis).—The common fig of our orchards. This fig tree contains no male flowers in any crop. The figs require no caprification or pollination in order to mature fruit, and it is yet undecided to what extent this class can be benefited by these processes. Only cultivated varieties.

San Pedro Fig (Ficus carica intermedia).—This class contain no male flowers. The first crop contains mule flowers, is not susceptible of caprification and not benefited by it. The second crop of this class requires caprification and pollination in order to attain maturity, as its flowers are all perfect female flowers. Only cultivated varieties.

CROPS OF THE FIG.

General Remarks.—The caprifig as well as the edible fig, bears several distinct crops every year. So distinct are these crops, and so important does the distinction between them appear to those nations which depend upon fig culture as an article of food and commerce that the various crops have been given separate and characteristic names.

In order to understand these names, a detailed description of the various fig crops is necessary. We must bear in mind that while the fig and the caprifig crops in a general way resemble each other, they still disagree in some important points. This may also be said to be the case with the principal types of the edible fig. In a general way, it may be stated that we have three distinct crops, appearing each one at a separate time, ranging from spring, summer and fall, according to the season in the respective countries. But each one of these crops is characterized in a distinct way, and without a full knowledge of them, a perfect understanding of caprification is impossible.

The Various Crops of the Fig.—While the edible fig tree as a rule possesses three distinct crops, we do not always find all these crops following each other on the same tree. This may be and is often the case, but fig trees and fig varieties exist in which one or more crops are wanting. The first, second or third crops may be respectively suppressed or one of these crops may be present, while the other two are suppressed.

Shortly after the fig tree begins to leaf out in the spring, small button figs are seen pushing out from the wood of last year, below the young leaves of the present season.

The place were these figs develop is the place where d ing last season existed a leaf, and which fell off last f These figs grow rapidly and mature generally in the of May in all southern countries, or in June in north ones. This is the first crop of figs, also known as ea figs, first figs or summer figs. This crop of figs has yet matured, or, in some varieties, has hardly matur when other young figs are seen to push out from the l joints of the present year. In course of a month or t these figs ripen and constitute then the second or m crop. With most figs this crop ripens in August, later earlier according to variety. This crop is also known second figs, autumn figs or late figs.

A third or later crop is found in some varieties formi in August and ripening in November. This may be cal the third crop. But this third crop is not greatly disti from the second crop; both develop from the leaf joi of the same season. In reality this third crop of edil figs can only be considered as the last of the second cr The first crop is, however, entirely distinct from the s ond crop, as it is produced on the old wood. Sometin the last figs of the third crop do not fall in the autun but winter over and ripen. early next spring just as first crop, and are thus hardly distinguishable from it. the caprifig the three crops are more distinct than in a cultivated varieties of the edible fig.

The Crops of the Caprifig.—In the caprifig the th crops are more distinct than in the edible fig, but, as that fig, they are not always all present in the same tr Thus caprifig trees exist which develop only one cr while others possess two or three crops. The variat in crops may be confined to individual caprifigs trees the same variety, or it may characterize some special riety, in which all the trees are exactly alike.

The time at which these respective crops of the caprifig come to maturity does not exactly correspond with the time of the ripening of the edible figs. The first crop of both figs appears and ripens at about the same time, but the second crop of the caprifig ripens before the second crop of the edible fig. The first and second crop figs of the edible varieties are continuous in their appearance at least, and continue so until late in the fall. Figs of almost every size may always be found on the edible fig tree during its period of vegetation. But in the caprifig the various crops are more distinct and separate, there being often short time between the maturing of the second crop and the appearance of the third crop on the same tree.

In Smyrna the various crops of the caprifig are confined to distinct trees, which again have received distinct names, though both kinds are undoubtedly only distinct sexes of the same variety of caprifig. The tree which bears the first crop, *boghadhes*, are known as *orginos boghadhes*, while those trees which bear the second crop or *ashmadhes* are known as *orginos ashmadhes*.

The first crop or the *orginos boghadhes* never contain any male flowers and pollen. This tree may, however, have a later crop which bears male flowers.

The *orginos ashmadhes* again which produce the figs used for caprification, which crop is the second crop or the ashmadhes, do as a rule never possess any first crop. It will therefore be seen that in order to possess a complete succession of crops of the caprifig, we must either cultivate varieties which bear three crops on the same tree, or if we grow the Smyrna *orginos* we must have both the boghadhes and the ashmadhes. The former breed the first crop of blastophagas, the ashmadhes again breed the second crop of blastophagas from eggs laid by the wasps hatching from the boghadhes.

As the boghadhes or first crop and the ashmadhes or second crop in Smyrna are produced on different trees, it will be seen that either we must have both of these trees in the same orchard, or we must caprificate the trees bearing only the second crop. The latter plan is adopted in Smyrna, where only orginos ashmadhes are cultivated. There two or three strings with caprifigs are hung on the edible fig, while in order to produce crops of figs and fig wasps on the caprifigs eight or ten strings with figs are required, the conditions and sizes of trees being equal.

Names of the Crops.—In order to avoid misunderstandings, the various crops are given distinct names in all foreign countries where fig culture is prominent. The crops of the caprifig, which not always correspond with the crops of the edible figs, are named differently. The following table will give a clearer idea of these names. As the English language has no suitable names for the various crops of the caprifig and the fig, I propose that we for the early first crop of edible figs adopt the Spanish name "brebas," and that we simply call the second crop edible figs, "figs," or autumn figs. For the caprifig I believe we can do no better than adopt the nomenclature of the German specialists who now use the Neapolitan names: *mamme*, *profichi* and *mammoni*, respectively for the first, second and third crops. There can thus be no misunderstanding as to what is meant. These names are rapidly becoming international and would admirably serve their purpose. In the following pages of this paper I shall as much as possible avail myself of those names. As our fig industry develops the words "brebas" and "profichi" (48) will become household words just as for instance the word "dehesas" has been adopted by both raisin-growers, raisin-packers and by the public generally.

NAMES OF VARIOUS CROPS OF EDIBLE FIGS (FICUS CARICA).

	April–August. First Crop.	June–August. Second Crop	November. to Late.
France.....	Figues-fleurs or Florones.	Figues d'autome; Figues ordinaire; Figues automnales.	
Italy.......	Fiori. Fichi primattici, Fioroni.	Pedagnuoli; Forniti.	Cimaruoli.
Spain......	Brebas.	Higos.	
Portugal...	Figos lampos.	Figos vendimos.	
Morocco....	Bukor.	Karmus.	
Algiers	Boccôre.	Kermez or Kermouse.	
Venice.....	Bolos.	....	
Dalmatia...	....	....	
Greece.....	Prodromoi ornos.	Fornites.	
California..	Brebas.	Figs, autumn figs.	
Latin	Grossi.	Forniti.	
Niçoise.....	Figa flore.	Oustinchi.	

		(Sept.-May.) First Crop.	(May-July.) Second Crop.	(Aug.-[illegible] Third Cr[illegible]
France	Caprifiguier.			
Italy	Caprifico.	Mamme.	Profichi.	Mammoni
Spain.........	Caprahigo.			
Portugal......	Figo de toca.			
Arabic Spain..	Obaakar.			
Greece	....	Oratitires.		
Morocco......	Tokkar.			
Asia Minor....	Ilek, or Orginos.	Boghadhes.	Ashmadhes.	
Malta	Tokar.	Tokar ta-nosa.	Tokar-tayeb.	Tokar leou
Ancient Greek.	....	Olynthoi.		
California.....	Caprifig.	Mamme.	Profichi.	Mammoni.

Characteristics of the Various Crops of the Caprifig. In the foregoing it has already been pointed out that t various crops of the caprifig differ from each other several respects. Here it is only necessary to generali The mamme form in the fall, remain on the trees ov winter and come to maturity early next spring. Th crop contains only a few or no male flowers, many g flowers, but no true female flowers, as seeds have nev been found in this crop. The time of maturity is June

The profichi appear in May or earlier. and mature June or July, according to climatic conditions. Th contain an abundance of male flowers, no female flowe and a large number of gall flowers. The mammoni (third crop) produce only a few male flowers, numero gall flowers and a few female flowers, capable of producing seed after having been pollinated by the male flowers of the profichi or previous crop.

The various crops of the caprifig do not always succeed each other continuously. There is frequently a lapse of time between the falling of the profichi and the appearance of the mammoni. No account has here been taken of the female caprifig tree, as yet almost unknown.

CHARACTERISTICS OF THE CROPS OF THE CAPRIFIG.

	Male Flowers.	Female Flowers.	Gall Flowers.
Mamme, or 1st crop.	Wanting, or very few.	Wanting.	Many.
Profichi, or 2d crop.	Very many.	Wanting.	Many.
Mammoni, or 3d crop.	Few.	Very few.	Many.

Characteristics of the Crops of the Edible Fig.—As to the edible figs the different crops are different in size, quality, flavor, sweetness and sometimes in color. The first crop, the "fiori," fig fleurs, ficos lampas, brebas, etc., are large figs, not very sweet, but pulpy and luscious for eating fresh, and they are highly prized on that account. The different names given to these large figs indicate the value in which they are held. The difference is considered so important that for instance in Spain and Mexico the common people will insist that the "brebas" are not figs. In California, however, no great distinction is made as to the three crops. When fig culture becomes as important here as it now is in Europe and Asia, names may be required for the first crop of edible figs. We have already proposed for the first crop the name "brebas," now used in all Spanish-speaking countries.

A large number of figs do not produce any first crop

or brebas, some give very few, and others again, like San Pedro, produce only brebas, the second crop o rarely maturing any figs, which even then never prove any great value.

The second crop, known in France as "figues-or naires," in Spain as "higos," in Portugal as "vendimos and in English-speaking countries only as "figs," ne here no special reference. It is this crop alone which used for drying in Smyrna or in other foreign fig-growi countries, as well as with us in California. These fi are sweet or very sweet, and, compared with the breb much smaller in size. In Italy a difference is made as the first or lowest figs of the second crop, which are call pedagnuoli or low figs, while the later or upper figs the same branches are considered less valuable and a known as cimaruoli or top figs. In the edible figs t third crop can hardly be said to exist as a separate cro as the last figs are only a continuation of the second cro The fig tree continues often to bear until frost sets in, until the tree becomes otherwise dormant. Some varieties, like the Natalino, ripen their last figs in midwi ter, if properly protected.

POLLINATION.

The process of fecundation or pollination is necessa in order that the ovary may be fertilized and produ seed. The pollen grains, when ripe, appear to the u aided eye as a fine dust. But under the microscope ea grain may be seen to be beautifully and characteristical sculptured. These pollen grains are brought on t stigma either by wind, transmitted by insects, or fall gravity. As soon as the pollen grains are on the surfa of the stigma—provided the latter is in proper conditio neither too old or too young, that is receptive—they gin at once to grow, sending out one or more pollen tub like long roots, which penetrate through the style, and following its canal, finally through the funnel-shaped opening in the ovule, reach the inner nucellus. The fertilization has then taken place, and immediately afterwards changes take place in the ovule and nucellus, which in short time lead to the production of a fertile seed. As a rule, we find that in the same flower the pollen grains and the stigma are not fully developed at one and the same time. It is therefore evident that the pollen in a flower cannot be useful for fertilizing the stigma in the same flower. This is nature's remedy against self-fertilization, requiring that the pollen be brought from some other flower or from some other tree of the same kind. In the majority of flowers the pollen can only be transported from one flower to another by means of insects, and often the flowers are so peculiarly constructed that only a certain kind of insect can reach the pollen, or rather, can reach the honey glands at the base of the anthers, as without the presence of these glands the insects would have no occasion to visit the flowers, which in such a case would remain sterile.

Nearly every flower we see in the field, and certainly every bright colored flower, requires the visit of some insect, in order that its stigma may be fertilized by the pollen which adhered to the insect when it left the last flower visited. Thus the insects and the flowers stand in close intimacy. The honey glands of the flowers furnish food for the insects, which are attracted to the flowers by their size, color, scent, or by the odor of the honey. The insects pay for their visit and for their meal by unknowingly carrying the pollen from one flower to the other. The insects are fed, the flowers pollinated. Only in very few instances do the insects live and breed in the flowers. One such instance is the fig, in which the that the edible fig is in some way descended from caprifig (29). The caprifig is the wild fig of the Medit ranean region, though its original native home must searched for in the mountain regions of southern Arab From its original habitat the caprifig tree was spread cultivation, or at least by transplantation, to other distri and finding suitable conditions, soon established itself a wild tree in the forests and mountains of the respect counties suitable to its multiplication through seedlings. is now generally known to botanists that the capri carries figs which contain three distict kinds of flowe male, female and gall flowers, all in the same fruit, will be described later on. But, besides, it is also kno (20) that there exists also a caprifig tree which only be fruit in which all the flowers are female or pistilla though trees of this kind are comparatively very ra Cuttings taken from either one of these trees would o produce its kind, but seedlings might produce both kin though probably the majority of the offspring would like the parent tree.

Through cultivation and selection by man several ty of the caprifig tree have been originated, though they not at present well understood or described. The Itali botanist Pontedera, and after him Gallesio, were the fi ones to mention this fact, and although other botani

Fico della natura, the original wild caprifig, with only one crop a year, this crop developing during the summer and ripening in the fall (22).

Fico mostro, all caprifigs which bear no fruit or which drop all their figs while these are yet young; also trees in which the male flowers only arrive at development.

Fico mula, with female flowers, which do not develop fertile seed, and which, as he expresses himself, become pomologically but not botanically ripe.

Fico semi-mula, with no male and with only female flowers, which, when pollinated, become botanically ripe, and consequently also pomologically ripe. This fig is undoubtedly the female tree of the caprifig.

From the descriptions of the other kinds, we may at least conclude that there exist numerous races or variations among the caprifigs. To what extent these variations of the caprifig will prove constant can only be determined by further investigations.

Among the caprifigs imported to California from various places, we can distinguish several varieties, though on account of the age of the young trees, it is yet too early to properly describe them. One variety possesses large, almost entire or shallow lobed leaves, others have the leaves more lobed.

The herbarium of the Academy of Sciences in San Francisco possesses specimens of caprifigs from France, with as deeply lobed leaves as any variety I have seen.

It is evident the varieties of caprifigs are many, distinct as to habits, number of crops, shape and quality of fruit, some even being edible, shape and size of leaves, etc. The importance of the different varieties of caprifigs cannot be overestimated, as it will certainly be found that a variety which will be suitable in one place, will be a failure in another. Home raised seedlings should therefore be resorted to.

Those caprifigs already imported to California produ an abundance of male flowers in the profichi. A majori or at least a great quantity, of these profichi come pomological and botanical maturity without pollinati and caprification. They produce male flowers with pe fect pollen, but as far as I have seen, no female flowe with fertile seed. The caprifig at Niles produced a nur ber of soft, yellow and large figs in the end of July, a containing pollen. These figs were much larger tha any dry caprifigs imported from Italy and Smyrna, whi may possibly be explained by the latter having be picked in a somewhat earlier stage of development. so, the pollen in the imported figs must have perfect itself after the picking of the fruit, a very doubtf theory (24).

The fact that the caprifigs at Niles do not produce ai fertile seeds, although they have both perfect male ai female flowers, depends upon the fact that, as in tl edible fig, the male flowers shed their pollen first lo after the female flowers have past their state of receptivit As this Niles fig only produces one crop a year, it is ev dent that it is impossible for the female flowers to ha been fertilized from the pollen of a previous crop; thi however, being the only way in which seed in any fig ca be produced.

The Fig.—The fruit which we call a fig is really not o single fruit, but a large number of fruits (or flower placed on a common receptacle. The fig itself is this r ceptacle, and in its interior are seen the small fruits, the flowers if the fig is unripe.

If we cut open a fig lengthwise, we see first exterior a fleshy homogenous mass, the receptacle proper enclc ing a central hollow, which connects with the outsi through a narrow passage at the eye. Lining this centr hollow on the inner surface of the receptacle are seen an almost innumerable quantity of small apparently similar flowers, which are fleshy, of unequal size and a little deformed, and which apparently only slightly resemble flowers with which we are generally acquainted. These are, however, the true flowers of the fig. They fill the whole interior surface of the receptacle, except close to and at the "eye" where they are replaced by "scales" or small leaflets, which latter interlock and form a thatched obstruction in the throat of the fig. This is generally the appearance of the fruit of the common or edible fig tree.

The wild or the caprifig is slightly differently constructed, a difference, however, which is of the utmost importance and interest.

In the caprifig we find, besides the scales at the eye and in the throat, not less than three different and distinct flowers covering the interior of the receptacle: male, female and gall flowers. The male flowers occupy the place nearest below the scales of the throat, while the lower part of the receptacle is filled with mostly gall flowers and with a few female flowers. The proportion of these flowers is different in the different crops of the figs. The hibernating "mamme" or first crop have a few male flowers and many gall flowers, but no female flowers. The second crop or "profichi" has many male flowers and many gall flowers, but no female flowers. The third crop or the "mammoni" has no male flowers, a few female flowers and many gall flowers. There are, however, exceptions to this rule, but this proportion is the most common one and is generally constant. There is also a purely female plant of the wild caprifig which possesses only female flowers, but this plant is as yet almost unknown. It has already been mentioned that this form was first described by Pontedera, but has not been describ by later botanists. Its existence, however, is entirely accordance with what is the rule in other fig species (25 The different crops of the fig will be more minutely d scribed presently.

If we consider the fig pomologically it will be seen tha as it is principally the receptacle that is eaten, the variou flowers found in the fig greatly detract from the value the fig, as they are never as juicy as the receptacle par Especially is this the case with the male flowers, whic are never edible; and whenever they occur they must b cut away before eating.

The Male Flowers.—The male or staminate flowers the caprifig are as just stated situated immediately belo the throat of the fig, variously occupying from one-half c two-thirds of the space in the receptacle of the secon crop, are rarely found in the third crop and are com paratively few in the first crop.

The flowers, though small and sometimes somewh irregular, are still perfect. (1). They possesses fou petals, generally shorter than the anthers, and short than those of the female flowers. Inside these petals a seen four stamens carrying larger pollen producing c pollen bearing anthers.

In the second crop these stamens attain their full de velopment in the months of June or July according locality, or about two months after the time that the femal flowers have reached their perfection in the same fig. is evident, therefore, that in usual cases, the pollen fro the anthers cannot fertilize or pollinate the female flowe in the same fig. Their function is to pollinate the fema flowers of the succeeding crop. Thus the pollen fro the second crop or "profichi," pollinates the "man moni" or third crop, the female flowers of which are their prime and receptive at a time when the pollen of the profichi is ripe. The pollen in the profichi is very abundant, of a pale yellow color, resembling a flowery yellow powder, which may easily be shaken out and collected without injury to its vital qualities.

The above refers only to the caprifig, or, if we wish to be more distinct yet, to the male tree of the caprifig. The edible fig, as cultivated in our orchards, does not possess any male flowers (26) except in extremely rare cases, as will be mentioned below.

The anthers in the male flowers are not always properly developed. This is especially the case in seedlings raised from Smyrna fig seeds, which originated from a pollination with the caprifig. Such seedlings do not all possess male flowers, those that do are more or less similar to the caprifig flowers, the anthers frequently being as well developed as in the real wild fig (27).

Female Flowers.—In the common caprifig female flowers have been found with certainty only in the third crop or mammoni. In this crop alone have fertile seeds been found, but always in very small quantities: hardly more than one fertile seed in every fig (47). In the edible figs perfect female flowers capable of producing developed embryos are more common. Generally it has been supposed that all flowers found in the edible figs were female flowers capable of producing fertile seeds. But this is undoubtedly not the case. All flowers of the edible figs in a general way resemble the female flowers, but, as I will shortly demonstrate, they are not all alike, and they differ in the various crops and in different varieties.

In the second crop of the genuine Smyrna figs nearly all flowers are perfectly developed female flowers, which only require pollination in order to bring fertile seed. This appears also to be the case in San Pedro and other figs, which regularly drop their second crop figs. As far as microscopical structure is concerned their flowers are entirely similar to those in the genuine Smyrna figs. That common edible figs possess at least some female flowers is clearly demonstrated by the finding of fertile seed in many such figs in localities where caprifigs are grown spontaneously. But the small quantity of seeds found indicate that the quantity of real female flowers is always small.

In places were caprifigs are not growing wild, that is where they are not growing spontaneously from seed, it is very difficult to decide whether a flower is a true female flower or not, and the only possible way to ascertain it is to pollinate it and await the results of fertilization. A wild caprifig always indicates that pollination is taking place through the agency of wasps, as even the caprifig will not propagate itself spontaneously and become wild without their agency, as the pollen cannot be transferred through the wind either to the female flowers of the capri or the edible fig.

As regards the structure of the female flowers some slight variation is noticeable. The petals are generally four in number, but sometimes three or five. According to Solms the number is quite variable within the above limits, but according to my own observations the number four is the most constant. In size the petals vary some, one pair often being a little longer than the other, and all four are always longer than the petals of the male flowers. All are more or less fleshy and sometimes they are furnished with short hairs at the margin. In the center between these petals projects a single pistil, at the base enlarged, forming the ovary. The central part is elongated two or three times longer than the ovary. This part is the style. The upper part of the style is bent and funnel-shaped, often, or perhaps generally, divided, one projection of the stigma being longer than the other. With a higher magnifying power the margin and upper surface of the stigma is seen to 'consist of a layer of minute glands, of a warty appearance, while from the center of the stigmatic funnel extends downwards a narrow canal or lumen, which passes through the whole length of the style and down through one side of the ovary, here bending upward and touching the very embryo. When the female flowers are receptive, that is when they are in condition to receive the pollen from the male flowers, these glands become greatly swollen and somewhat glossy, of a green or light green color, which after the receptive stage is passed changes to a bright brown. The inner surface of figs in such a stage are seen to be spotted brown when cut open. The stigma attains its recepitvity long before the male flowers are ripe in the same fig receptacle. This difference in the maturity of the flowers makes it impossible for the female flowers to be fertilized or pollinated by the male flowers of the same fig. Thus the female flowers of the mammoni can only be pollinated by the male flowers of the preceding crop—the profichi.

The crops of the edible figs do not exactly correspond with those of the caprifig. Thus when the male flowers of the profichi are ripe, and at a time when the other flowers in this fig have passed their prime months before the female flowers of the second crop Smyrna figs have just attained the state of receptivity. They can therefore be pollinated by the male flowers of the profichi of the caprifigs. The time for this pollination is June or July according to climatic conditions in various countries. This rule as to the difference in time of ripening of the male and female flowers in the caprifig holds also good in resemble the female flower.

The petals in the gall flowers are smaller and mo unequal in size. The chief difference, however, betwe these flowers and the female flowers is found partly the stylus of the pistil, which is not as elongated as in the male flowers, and partly in the stigma, which is very mu smaller and entirely wanting the glands at its upper m gin. The gall flowers cannot be pollinated, or if they a neither does the pollen develop pollen tubes nor does t embryo or egg in the lower parts or ovary become ferti

While it is true that the gall flowers do not produ seed, still it is a fact that they develop to a certain exte if punctured by the wasp, or more correctly after the e of the Blastophaga wasp has been properly deposi They then develop into galls, that is the lower part of t stigma swells up, the integuments of the embryo-s harden, forming a glossy and brittle covering as a prot tion for the larvæ of the wasp.

Those gall flowers which are not thus wounded by Blastophaga egg, do not develop any further, but at o wither and shrink up. Gall flowers are found in all w fig species, though in some species their nature is not parent until the egg of the Blastophaga is laid. In the edible fig no gall flowers have been found with certainty, at least the Blastophaga wasp, for whose special benefit these gall flowers seem to have originated, has never been found breeding in the edible figs. It has been supposed that the cause of this was to be sought in the sugary juices of the edible fig, which killed the eggs or embryo of the wasps, but I am satisfied that this is not exactly true. Many varieties of wild fig species produce very sweet fruits, edible and quite palatable, and still these figs serve as home for Blastophagas. The cause for the inability of the wasp to breed in common figs must be sought for elsewhere, and, as I will presently point out, is due to the fact that the edible figs contain flowers modified to such an extent that they are unsuitable as breeding places for the wasps.

The gall flowers are characterized by a much shorter style, by an undeveloped stigma, devoid of receptive glands, and by an imperfect embryo which never develops more than to a certain limited degree. The discovery of the distinction between gall flowers and female flowers is due to Solms-Laubach (25).

Until his researches were made known it was supposed that the female flowers turned into galls when stung by the wasps. He again proved that the distinction existed independent of the wasps, which however only select the peculiar gall flowers as the only ones suitable to receive their eggs.

Mule Flowers.—Under this name I arrange the majority of the flowers of that class of edible figs, varieties which mature their figs regularly without the presence of the caprifig and its pollen. These flowers are, as far as I know, not found in the caprifig, nor in any other wild fig species. They are undoubtedly a product of culture

and must be considered either as more highly develo
gall flowers, which, bereft of the Blastophaga influer
have partially regained their original structure, but wh
just on that account have lost the capability of produc
galls; or they may be considered as degenerated fem
flowers which have lost their fecundity by inertion—
other words, by not being pollinated for ages, so to sa
in the same way as many other cultivated flowers h
degenerated. I am inclined to consider the latter as
more probable, though at present no direct proof can
given. That the great majority of the flowers in
edible figs (except the Smyrna race) is different from
true female flowers, both in structure and nature, is
doubted, whether we assign as a cause one or the ot
of the above theories. These mule flowers never re
any botanical maturity, and are really something half-w
between the true female flower and the true gall flowe

The mule flowers are characterized by an imperf
stigma, by a style in length intermediate between that
the gall and the female flower, by imperfect embryo,
by the property of becoming fleshy, sweet and edi
without pollination. I have so far not found any in
Smyrna figs, comparatively few in the second crop of
San Pedro class, but almost exclusively occurring in
first crop of this class. The stigma of the mule flow
has no developed glands and is not receptive.

Male Flowers in Edible Figs.—It has frequently b
stated that male flowers are not found in edible figs.
this must be considered [illegible]. However, there

to botanists, and, strange enough, it was first described from specimens found in the edible fig. The male flower of the fig was first described by the prominent botanist La Hire in the year 1714, from figs grown in Paris either under glass or in the open ground (36). Unfortunately La Hire does not give particulars as to the variety from which the flowers were taken, and it is not even certain that La Hire got his flowers from the edible fig. Another variety of edible fig which regularly produces seed is the "Croisic," cultivated in the vicinity of the ocean bathing place Croisic, on the coast of Brittany, in the Department of Loire inferieure. This fig has been mentioned by Solms-Laubach (37), and described as being green when ripe, with white or pale pulp, very juicy and sweet, but with poor aroma. The male flowers occupy the same place and distribution as in the profichi of the caprifig. The place they occupy on the receptacle ripens less perfectly than the balance of the fig, and remains always somewhat hard and dry, generally to such an extent that it becomes necessary to remove that part of the fig before eating.

Another somewhat similar edible fig was observed by the same author as cultivated at Cherbourg in France, also on the Atlantic coast. The male flowers in this fig were, however, degenerated or improperly developed (36). The finder of these figs believes them to be only highly developed caprifigs which have become edible. He is even tempted to trace their introduction to France to the time when the Phœnician traders extended their ocean voyages to the northern coast of France, a time when supposedly the edible figs were yet in a semi-wild or undeveloped condition.

Another fig with numerous male flowers was found by Mr. B. M. Lelong at Los Gatos, October 20, 1891. The fig, judging from the photograph, is of medium to la size and edible. Mr. Lelong describes the pollen as v abundant and that the fig possessed numerous fertile se (38), which he says must have been produced by pollen of the male flowers above.

The Cordelia Fig.—The only certain instance of m flowers having been found in an edible fig in Califor is the one I am about to mention below.

In July, 1893, I found a box of figs in the market San Francisco, marked as having come from Cordelia Solano County, containing very large yellow figs, a s larger than our largest Adriatic. Upon opening th figs I found every one with a fully developed zone male flowers, fully ripe and with an abundant, perfec developed pollen. In other respects the figs resembl very much the Italian Gentile, which is now also growi in California in various localities. These figs belong to a distinct variety and were propagated as table fi though the dry zone of male flowers greatly detract from the quality of the fig. The fig was juicy and ve sweet. It is not impossible that this fig is identical w the Croisic fig described by Solms-Loubach, and that has been brought here by setlers from Croisic in Franc

Finally, it may be stated that both myself and Mr. W. Maslin, of California, have raised seedlings of Smyr figs. Some of those raised by the latter came to *part* maturity at least, and contained male flowers in great or lesser abundance. Such figs, however, must be co sidered as improved caprifigs—improved by being rais [illegible] and Croisic f [illegible] male caprif [illegible] they also ha [illegible] greater deg [illegible] expected t these figs will develop perfect seeds without the aid of the Blastophaga, as it is probable that they, as other figs will, bring their male and female flowers to perfection at widely different times, in other words, that when their female flowers will be receptive, their male flowers will not yet have developed their pollen.

It must be clearly understood that edible figs possessing male flowers are inferior to those which do not possess any, and the presence of male flowers is without any importance, from a horticultural point of view. The Blastophaga cannot live in those figs, because they do not possess perfect gall flowers; the pollen cannot be utilized for pollination or caprification, because there is no practical way of getting it out of the fig and onto the flowers of the next crop, and finally such figs are inferior for eating, as the male zone is dry and not eatable.

In the caprifig we have three kinds of flowers. Male flowers, which, on account of their time of ripening of the pollen, can only pollinate female flowers of the succeeding crop. Female flowers which produce seed, but which, on account of the early time at which they are receptive, can only be pollinated from the pollen of the preceding crop. Gall flowers, which resemble the female flowers, but which are at no time receptive, and which serve no other purpose than breeding places for the Blastophaga wasp. The caprifig possesses also a purely female plant with only female flowers.

The edible figs consist of two or three distinct types.

The Smyrna type *(Ficus carica smirniaca)*, with only female flowers, capable of producing seed by pollination.

The Common type *(Ficus carica hortensis)*, with principally male flowers, neither capable of producing seed nor able to serve as galls or home for the Blastophaga wasps.

The San Pedro type (*Ficus carica intermedia*), wi mule flowers in the first crop and female flowers in t second crop.

The Cordelia type (*Ficus carica relicta*), which is very rare one, which carry some male flowers, and whi must be considered as semi-capri, or reverted edible fig

Various Kinds of Maturity.—In the fig as well as other fruits, we can distinguish between two kinds of m turity. Gallesio was the first one to make the distinctio and I here adopt it, somewhat modified, as being particular use in demonstrating the nature of the fig. W find that some or most edible fig varieties set and matu their figs without pollination, but that as a consequen such figs contain no perfect flowers with fertile embryo This state of maturity may be called pomological maturit as it does not necessarily require the botanical perfe tion of the flowers. Pomological maturity is attained b the great majority of edible figs, and is undoubtedly a inheritance from the caprifig, which becomes similar pomologically mature. This pomological maturity is n necessarily accompanied by any botanical maturity, a for instance, is proven by our California figs, which nev contain any fertile seed (35).

The other kind of maturity may be called botanic maturity, as it requires the female flowers to be deve oped with perfect embryos, in order that the fruit ma set and become also pomologically mature. If the fru is edible or cultivated as a fruit, the pomological maturit will always be effected by the botanical maturity. Th Smyrna figs can only attain pomological maturity by fir [illegible] all other figs b [illegible] necessarily or generall [illegible] again, like the Sa [illegible] first crop, but th second crop, which possesses perfect female flowers, does never become pomologically ripe, and can only be botanically ripe by pollination.

The pomological maturity always indicates and implies a long continued cultivation of the fruit by man, and can be applied only to cultivated fruits. Among other fruits, besides the fig, which attain pomological maturity without botanical maturity at the same time, we may mention some varieties of dates, one variety of pomegrenate, the seedless orange, many apples and pears, the common edible banana, the pepino Solanum of Central and South America, seedless grapes, and a number of other fruits and vegetables in which the seeds are abortive, and have become so, partly through the continued asexual propagations of the plant, partly from other causes. Botanical maturity is attained by all fruits which produce perfect seed, and if the fruit is edible, it is also pomologically mature.

But it must be remembered that the fruits here enumerated as attaining pomological maturity only, are all such as have been developed from pollinated flowers. As far as is known, no other fruit than the fig develops without previous pollination. The development of the common edible fig receptacle must, therefore, be considered somewhat in the same light as the maturity and development reached by a tuber, or by the stems of the sugar cane, etc. Pomological maturity merely indicates that the fruit becomes edible, while botanical maturity means that the fruit has developed fertile seeds.

Seeds in Smyrna Figs.—We have already several times referred to the fact that all edible figs may be divided in two distinct classes or types, one which, when ripe, does not necessarily contain fertile seed, and one which, when ripe, always contains fertile seed, as otherwise it would

not be ripe or mature. There are also other differen
The Smyrna figs belong to the latter class, and the
ways contain ripe and fertile seeds.

But, as the cultivated Smyrna fig never contains
male, and as caprification with the wild fig is always
sorted to in order to cause the figs to mature, it is evi
that the seeds thus produced must, when growing,
us hybrid plants, plants which more or less partak
both parents, the wild as well as the Smyrna fig.

Artificial pollination of figs is no new or remark
discovery. Gasparrini relates how (40) he repeat
introduced the pollen of the caprifig into the edible
especially of the Lardaro variety. But his pollina
produced no decided results. No increase in the num
of fertile seeds was noticed, either because the flow
of the Lardaro variety were principally mule flowers,
which the pollen could have no effect, or because the fen
flowers had all been previously pollinated. From
Gasparrini draws the illogical conclusion repeatedly quo
by later writers, that the caprifig is of a different spe
from the edible fig, that its pollen cannot influence
fecundate the female flowers of the edible fig, and
consequently the practice of caprification is illusion
and of no value whatever. Gasparrini did not know
the class of figs which I have designated as the Smy
type, and which, unlike any other class, produces p
cipally receptive female flowers, which do not proc
seed without the aid of pollen from the caprifig.
Gasparrini had opportunity to extend his interesting
in detail [illegible] investigations [illegible] this class of figs, the
[illegible] have
[illegible]

[illegible] in California is inten
[illegible] this point of

production of fertile seeds. Imported in 1880 (for details see the historical part), and quite extensively propagated and planted in the most dissimilar parts of California, those figs failed to bear one single ripe fruit during a period of (10) ten years. The fruit would form in abundance, the flowers would develop and become apparently receptive, as shown by the glands of the stigma, and by the length of the style, but the fruit would invariably fall, when apparently one-third or one-half grown. It was this fact, together with my observation that imported Smyrna figs always possessed numerous fertile seeds, while such were never found in our (other) edible figs, that made me a strong advocate of caprification, and which satisfied me that pollination was necessary and not illusionary, as almost every one else (41) believed, principally on the testimony of Gasparrini and Olivier. It would indeed have been strange that Smyrna figs should not ripen their fruit in California, if the maturing only depended on climatic conditions or differences in soil. Those figs, consisting of three distinct varieties, were planted in the most dissimilar localities and in greatly different soils, and exposed to varied climatic conditions, found in the northern, central and southern parts of California, in the interior valleys, in the foothills, as well as on the coast. All the old world fig districts together would hardly show more variations in climatic and other conditions, than did those various localities in which the Smyrna figs were tried in this State. Still not one tree properly matured a single fruit. A few of the first crop became half ripened, that is, became yellow, soft, but insipid and not sweet; and besides never attained a proper size, or a size at all approaching that of the imported dried figs. I had no opportunity of trying direct pollination (from want of caprifig pollen) until 1891, in the last days of July. The experiment, I afterward learned, was tried the year before in Fresno, and successfully produced a few ripe Smyrna figs (40). On the 26th of July I requested Mr. E. W. Maslin to accompany me to the Shinn orchard, situated near Niles, not far from the San Francisco Bay (42). The only caprifig tree there possessed a few very ripe fruit, large and pulpy, in which the pollen was fully developed and very abundant. We shook the pollen out in the palm of Mr. Maslin's hand, and from there transferred it to the Smyrna figs, of which there were various sizes. Not then knowing in what size the flowers would be receptive, I pollinated various sizes, in all about thirty figs, which were properly marked by strings. Figs on all the three different varieties were pollinated. As means to introduce the pollen we used a goose quill, the end of which was pared off obliquely.

This open part of the quill was filled with pollen, then pushed through the scales closing the eye of the Smyrna fig and the pollen shaken down, in probably about one thousand times larger quantity than was actually needed to fertilize the fig. I found that figs of a certain size, about three-quarters of an inch in diameter, allowed the quill to readily penetrate, the scales on them giving away quite readily at the slightest push of the quill. It was not necessary to cut the scales or to open the eye artificially and violently, simply the pushing the quill in would allow the pollen to drop down in the receptacle. Many of the figs thus pollinated came to perfection as large, ripe and luscious figs, in every way perfectly developed with numerous perfect seeds. But out of the *many* thousand similar Smyrna figs on the same as well as on immediately adjoining trees, *not one single fig* that was not pollinated by Mr. Maslin and myself came to maturity; all fell from the trees just as they had been doing during ten years previously. I consider this experiment absolutely conclusive. It shows: *That* the true type of Smyrna figs does not set and mature fruit in California if left to themselves, but that the figs invariably fall off. *That* they contain perfectly developed female flowers, receptive in the end of July. Of these facts I had already satisfied myself long before through microscopical study of the fig flowers.

That these flowers if pollinated will cause the fruit to mature, while they themselves (the flowers) develop fertile seed. Our experiment also proved that the pollen from the caprifig tree is not and cannot be transferred by the wind, or by other insects than the Blastophaga, to the edible figs. As this caprifig tree had born ripe pollen for years (it being ten years old or more) still not one of the surrounding Smyrna figs had been pollinated and had come to maturity before our experiment was made. Still they grew so near to the caprifig tree that their branches closely interlocked, almost forming one single tree. The principal value of this experiment depends upon this very fact and upon the age of the trees, which were old enough to have matured fruit, if they could have done so without pollination.

Caprification is only one step further, it is the pollination by the aid of a wasp, semiartificially introduced in the fig by hanging the caprifigs in their immediate vicinity. An account of this experiment is found in the "Annual Report of the State Board of Horticulture of the State of California, 1891, page 230 to 231." The account is substantially correct, but the part relating to the Blastophaga contains some errors which will be noted in a different place. The figs so pollinated were exhibited by Mr. James Shinn at the Horticultural Convention in Marysville, also by E. W. Maslin at the rooms of the

State Board of Trade and have also been photograph by B. M. Lelong in the State Reports so often referr to (for 1891).

Experiments in pollinating Smyrna figs have since be carried on by me yearly with the same results—bo horticultural and botanical maturity. The proper tir for pollination in California changes frequently from ye to year according to seasons. I haye seen a differen in the time when the figs were ready for the pollen, var ing between several days to several weeks or a mont Near the coast this difference is much greater than in tl interior valleys, where one year with the other the vari tion in time seldom extends to more than two weeks.

Our experiment further proved without a doubt that tl figs in question, consisting of several hundred trees, ir ported by G. P. Rixford in 1880 and 1882, through tl aid of Consul E. J. Smithers in Smyrna, were genui Smyrna figs, On account of the persistent dropping the figs the idea originated by the late Dr. Stillman b came prevalent that these figs were not what they pr tended to be, but simply wild figs sent us by the jealo Smyrna growers, who were afraid that our fig productic would come in opposition to their own products. On th account most of the "Bulletin" figs were rooted o (43).

I grieve to tell that I once shared this idea and did M Rixford a great injustice in publishing it in the Rur Press, the retraction in the following number, upon fin ing out my mistake, hardly undoing the wrong. Howev Mr. Rixford has lived to see himself and those concern in the introduction righted, and his efforts and success being the first one to bring the genuine Smyrna fig California cannot be too greatly appreciated. Only t future will demonstrate the true significance of this ir portation and of Mr. Rixford's work.

Pollination of San Pedro and Gentile Figs.—Since the MS. of this paper was completed, or almost so, I have been able to successfully experiment with pollination of the second crop of San Pedro and Gentile figs. My experiments were made in Kern County, California, on two fig trees, one a San Pedro and the other a Gentile, planted by me some years previously.

As I have stated, these two fig varieties mature only a first crop under any circumstances, at least no case has come under my observation where a fig of the second crop came to perfect maturity, the crop generally dropping as a whole when one-fifth grown. The time for experiment was July 11. The pollen had been secured from a caprifig tree of the Bulletin variety grown about 200 miles away and transported in a glass-stoppered bottle.

The San Pedro tree contained about 420 figs. Thirty-six of these were pollinated with a goose-quill by injecting caprifig pollen through the eye in liberal quantities, many times more than would have been brought there by inquilines. A few weeks afterwards the majority of the pollinated figs were turning soft, while the non-pollinated figs remained hard, many falling off. By the 16th of August I again visited this tree. Ten mature figs had been taken off. Eight remained on the tree fully ripe and very sweet, but somewhat smaller than the first crop, and six figs were partly mature. All of these figs were among those caprificated or rather pollinated by me and marked. Of all the other, nearly 400 figs, on this tree, which had not been pollinated, not a single one showed any sign of maturity, and later on all these dropped off.

With the Gentile tree the case was quite similar. This tree contained 86 second crop figs. Of these 19 were pollinated and 10 of these came to full maturity, two to partial maturity. The balance all dropped off bei even beginning to soften. Microscopic examination the ripe fruit of these pollinated second crop San Pe and Gentile showed that a large number, more than o third (about), of the flowers possessed fertile seed. T effect of pollen had thus been to produce a botanical well as a horticultural maturity, the latter having be effected by the former. My experiments in pollinati have not been repeated, but I think enough has be demonstrated to show that the regular and periodic dr ping of figs is—in varieties where the whole crop fails come to maturity—invariably caused by the want of p lination of the female flowers. In other words, wh perfect female flowers are found in the fig, they m have pollination in order to produce a horticultural ma rity of receptacle. Where again the flowers in the rece tacle are imperfect as far as their generative capacity concerned, then a horticultural maturity will ensue, pol nation and caprification being useless, and impotent produce any effect. Horticultural maturity will be effect without pollination or caprification.

Pollination of First Crop San Pedro.—May 16th pollinated nineteen figs of first crop San Pedro, the fi being of very much the same size as those ready for p lination of the second crop. In the middle of June th teen of these figs had matured, the others had fallen o Examination showed that no fertile seeds had been forme all the ovaries being shrunk and abortive. This was a

[illegible]

[illegible]

less. As regards the figs on this tree which I did not pollinate, I may state that nearly all arrived at perfect horticultural maturity.

Seeds in the Common Edible Figs.—Under this class I arrange, as has already been stated, all fig varieties with only mule or with principally mule flowers, which set and mature their figs without the aid of pollen. If the seeds of such figs are examined under the microscope, it will at once be seen that they are only seeds in appearance, but not in reality. They are mere glossy hulls of a yellow or brown color, but with no kernels and embryo capable of development. Even without the aid of a microscope this may be ascertained by crushing the seeds with the point of a knife. The shell will then be seen to collapse, the interior being absolutely empty without any kernel. Although I have examined many thousands of figs grown in California during the past ten years or more, I have failed in finding a single seed properly developed. I at first attributed this alone to the former total absence of caprifigs in this State. I now believe it to be due in equal degree to the absence of or at least scarcity of female receptive flowers in our figs, generally speaking. This same observation as regards the absence of seeds in common figs has been repeatedly made in Europe. In France, Solms-Laubach found no figs which contained developed embryos (44). Gasparrini, however, found repeatedly seed in several of the Italian figs. However, he says that in the early figs, probably meaning first crop figs, he never found any fertile seed or seeds with embryo. But in the "pedagnuoli" of the white figs and of the Dottato he found frequently fertile seed, even in places where caprification did not take place. As has been previously stated, he pollinated the flowers of the Lardaro variety, but did not succeed in producing any more seed than what would otherwise have been the case. To account for this **l** assumes that the seed had formed through what **by na** uralists is called "parthenogenesis," or self-**developmen** However, Gasparrini's experiments upon this subject **a** defective and not at all conclusive. It is much **more prol** able that those varieties did contain some perfect **fema** flowers, which had in some way been pollinated. **Pa** thenogenesis is too rare an occurrence to be **accepte** without thorough experiments (45). The fact that **Ca** ifornia figs, which formerly at least could **not possib** have been pollinated, never exhibited fertile seeds, **speal** strongly against the parthenogenesis theory of **Gaspa** rini and for the belief that even for ordinary **figs bot** pollen and female flowers are required for the **productio** of seed. Solms-Laubach found fertile seed in **many N** apolitan figs and frankly admits that their **presence ca** only be explained by the influence of pollen. **Figs** various edible varieties, which were sent from **Brazil b** Fr. Müller to Professor Solms-Laubach (44) were **inv** riably found void of embryos. In that country **no capr** figs existed, just as in California. Until **further the pa** thenogenesis theory must be disregarded for the ***Fic*** *carica* tribe. In northern Italy G. Arcangeli **foun** "some" fertile seeds in *Fico biancolino*, which **he cal** a semi-wild fig, the majority of its seeds, **however, bein** merely shells (46). In the other figs **growing in th** vicinity of Pisa, such as the *Fico piombinese* and the ***Fi*** *verdino*, no fertile seed were ever found.

The conclusion which I draw from the above and oth investigations in regard to the perfect and fertile see [illegible] produce fruit wit [illegible] with only mule flov [illegible] embryos or wi [illegible] so-called seeds a mere shells without kernel. When this class of figs are found to contain some seeds with developed embryo, it is to be explained by the presence among the mule flowers of perfect female flowers, which again have been brought to development only by the introduction of pollen, either by the Blastophaga wasps or by some other means.

Parthenogenesis or seed-development without pollen has been proven to exist in at least one tropical fig, *Ficus Roxburghii;* here, however, only by excitement caused by the sting of a Blastophaga, without pollination. That this is not the process of seed production in the Smyrna tribe of our edible fig is absolutely and conclusively proven by the experiments of Mr. E. W. Maslin and myself. We produced seedlings from Smyrna fig seeds which all when arriving at maturity proved to be hybrids between the edible Smyrna and the wild caprifig. This hybridization could of course not possibly have taken place except by the introduction of pollen to the female flowers. If their seeds had developed by parthenogenesis the seedlings would not have been hybrids but would have been varieties of the Smyrna fig. The seeds were of course taken from imported Smyrna figs, which had been caprificated in Aïdin. The very fact that hybrid figs were produced on the seedlings showed conclusively that the mother figs had been caprificated with wild pollen, and that the process had been effective.

FLOWERS IN OTHER FIG SPECIES.

In connection with what has been said above, it may be of interest to shortly consider the structure of the fruits of other fig species. Those who wish to more especially study the flowers of the fig, we refer to the special works enumerated in the "Literature." Here we can only mention this subject in a passing way. In a great number of fig species there are two distinct individual tre The male tree, with figs which contain male flowers gall flowers, the meaning of which will by this time fully understood. Other trees of the same species ca only figs which contain nothing but female flowers. So species of this class have the male and gall flowers p miscuouly distributed over the surface of the receptac in others again the male flowers occupy an upper zo around the eye of the fig, while the gall flowers are co fined to the lower and opposite end of the fig.

Other species of figs again (such as *Ficus elastic* produce figs which possess both male and female flow promiscuously placed on the same receptacle. This rangement of the figs is probably the most ancient one all. In this fig the differentiation between gall flow and seed flowers begins first after the Blastophaga h laid its egg and depends apparently upon chance only those which have not been pierced but only pollinated the wasp, develop seed, while those in which the Blas phaga egg has been laid develop into gall flowers.

If we again consider only the peculiar bottle-like ceptacle of the fig, we find plant genera related to t fig which have an open flat or slightly convex receptac such as is the case with *Dorstenia*. Others again, li the mulberry, have a very convex receptacle, on the o side of which are found the individual fruits (28) inste of inside, as in the fig.

EVOLUTION OF THE FIG.

The theory of evolution now generally accepted nearly all naturalists can readily be applied to the While we cannot absolutely prove the various stages development of our edible fig from more ancient and le perfect form, we can, nevertheless follow these devel ments by studying various figs and nearly allied pl genera. Such a study, the details of which are outside of the scope of this treatise, will clearly show us how, from mere simple forms, through adaptation to surrounding influences, our more complex figs have gradually developed from more simply organized forms, or if we will, from ancestors differently constructed. In order to simplify our study, we might profitably divide it in two parts, and first consider the development of the "fig" or the fig receptacle, which we call the (fig) fruit, and secondly the development of the group of plants which we call "fig trees," more particularly the edible fig tree, as being the one which concerns us the most. Not only will this study show us that our figs have developed from less highly constructed ancestors, but also that in some instances, as regards the flowers, a certain retrogression has taken place, in which some flowers, through want of use of certain organs, have degenerated from more perfect ones.

EVOLUTION OF THE FIG FLOWERS AND THE FIG RECEPTACLE.

In order to reach its present form, both the fig flower and the fig receptacle must have, in course of time, undergone many changes, nature having always in view to prevent self-fertilization and produce as perfect seeds as possible. This change and gradual development must have taken place very much as follows:

The first form of receptacle was convex, as it is yet in the mulberry. This surface exposed the flowers too much to the adverse influences of wind, insects, etc., and the receptacle became more flattened out, as it is yet in *Dorstenia*, a plant related to the fig. But the change kept going on, and the receptacle became more and more concave, thus exposing it less and less to outside influences.

Finally, fig forms appeared with a flask-like recepta only open at the top or eye, and in order to exclude majority of depredating insects, this opening beca covered over with scales, as in our present figs, wilc cultivated.

If we again consider the flowers alone, we find tha the lowest forms of figs, which must also have most sembled the oldest forms, the female and male flow were promiscuously scattered over the surface of the ceptacle. But in order to further prevent self-fertilizati the male flowers matured later than the female flowe and this again necessitated first the introduction of poll from other figs, later on from figs of other crops of t same tree, and later on yet from figs on different tre The differentiation as to time of maturity of the male a female flowers is probably anterior to the closing of t eye of the receptacle by scales. A further developme and differentiation took place as regards the respecti location of the flowers. The male flowers were gradua made to occupy the upper part of the receptacle arou the eye, while the female flowers were assigned the low or bottom part of the receptacle opposite the eye.

A further step in development was a differentiation the female flowers under the influence of the wasps whi had come to inhabit the flowers. Some flowers prolong their styles in order to make it impossible for the Blas phagas to injure them by the deposition of eggs. Oth female flowers again shortened their styles in order facilitate the deposition of the Blastophaga eggs. T stigmas of these flower became useless, gradually creased in size and changed their shape, at the same ti losing their receptive glands. These latter flowers a the gall flowers, as we see them at the present time [illegible] varieties. That the gall flower is a degenerat female flower is certain, as in some varieties it yet depends upon chance which flowers are to be gall flowers and which are to remain female flowers. Those pierced by the wasps develop into galls, those which are not pierced remain female flowers. In our present caprifig the female flowers, even if pierced by the ovipositer of the Blastophaga, will never become gall flowers.

This was the state of development of the wild fig when man appeared to take an active part in the development of the fig and in the production of new varieties, suitable as food or luxuries. How this evolution by cultural selection must have taken place will be presently considered. Here it may only be remarked that the mule flowers may have originated in two different ways. Either they may be explained as a degeneration of female flowers, which have lost their power of producing seed, by not being regularly pollinated, or they may have originated from gall flowers which, from want of wasps, gradually lost the power of producing galls, or which lost that power with increased sweetness and edibility of the receptacle. The latter two theories combined seem to me the most plausible.

ORIGIN OF THE EDIBLE FIG.

Like all other fruits cultivated by man, the fig tree, as we find it to-day in our orchards, improved and bearing edible, luscious fruits, must have descended from wild ancestors, less edible and less valuable for the use of man. In most all other fruits it is easy enough to point out the wild ancestors, as we yet find the original cherries, plums, peaches, apples, pears, etc., growing wild in the forests and fields of our respective continents. But with the fig it is somewhat different. Our edible figs differ considerably from the wild fig, the caprifig of course being the only fig tree from which we may possibly suppose a descent through ages of cultivation. But as has be shown in these pages and as has been well known f nearly two centuries, the caprifig differs in having m: and gall flowers, while our edible figs possess principal flowers of a different kind. In order to explain the d velopment through cultivation by man of our edible fig several different theories have been put forward by pror inent investigators, each theory in its turn to be throw down if not fully disproved by more recent research.

We will here shortly consider each one of these th ories separately, as they are of great interest both in p mological as well as scientific respects.

First Theory.—The oldest theory perhaps was the brought out by the well-known Italian investigator horticulturist Gasparrini. He held that the fig and caprifig are specifically distinct, or in other words he not believe that the caprifig is the male tree and the fig the female tree of the same species. He would deri the edible fig from some unknown ancestor not yet foun perhaps from some species which in course of ages ha entirely disappeared in its wild natural state.

Gasparrini based his opinion principally upon his fa ure to produce or rather to increase the number of ferti seed in the edible fig, either by pollination or by capri cation with the pollen of the caprifig. His experimen were numerous and fairly carefully performed, and as f as they go quite valuable. But they prove an entire different thing from what Gasparrini claimed, and it impossible to logically draw the conclusions from his e periments which he unhesitatingly did.

The force and value of his arguments become less ir portant, convincing and conclusive, when we can sho that his many experiments, upon which alone he based h theory, were made on fig varieties which possess few no receptive female flowers, but which we now know are characterized by their small number of female flowers, by their large number of mule flowers, and by the total absence of gall flowers. To Gasparrini the large class of Smyrna figs, which possess only fully developed female flowers, was entirely unknown. At his time as well as in our day this class of figs do not grow in Italy, and he had no idea of their existence anywhere else. A theory, therefore, which does not take in consideration all classes of edible figs cannot be considered as absolutely plausible and convincing. We now know through my own experiments in pollination, and through the production of hybrid figs from Smyrna fig seeds by Mr. Maslin and myself, that the pollen of the caprifig really is capable of producing fertile seed when applied on the stigma of the female flowers of the Smyrna fig. This does not, it is true, prove with absolute certainty that the caprifig and the edible fig are of the same botanical species, but it does disprove Gasparrini's conclusion that the pollen of the caprifig is incapable of fertilizing the female flowers of the edible fig, when these flowers are properly developed. The presence of fertile seeds in many figs was explained by Gasparrini through what is known botanically as parthenogenesis or unsexual development (30). That parthenogenesis is a possibility cannot be denied, as it is proven to exist in at least one tropical fig, and probably exists in several, but it is a rare occurrence. And even if taken in consideration it must be remembered that it is now proven that in the species in which it does exist it is caused by the sting of a Blastophaga wasp, which stimulates a growth in the nucellus, which might be called internal or seed budding (98). The existence of parthenogenesis or self-budding without the impulse of outside influences has not been shown to exist in the edible fig, and until it is shown we must leave it without serious consi-eration.

Second Theory—Professor Solms-Laubach the mo prominent of all late investigators of the edible fig, b lieves that the caprifig is the wild species from which th edible fig has been originated by cultural selection b man. When he speaks of caprifig he refers alone t the caprifig tree, which produces male and gall flowe in its various crops, but does not take in consideratio any purely female tree of the caprifig. To Solms-Laubac the existence of a purely female tree of the edible fi was unknown at the time he put forward his theory Short as his theory is and without going into details an without efforts to explain everything, it must be con sidered extremely plausible. But later on Professo Solms-Laubach gave up this theory or changed it to som degree, adopting the one here described as the thir theory.

Third Theory.—This theory as regards the origin of th edible figs was developed by the eminent naturalist, Prof Fritz Müller (31). He considers the caprifig to be th original wild male fig tree and the edible fig the femal tree of the same species, both sexes having existed sepa rately and originately as wild trees, before their cultiva tion was begun by man. This cultivation must then hav been entirely confined to the female tree, and any improve ment in the fruit must have been brought about throug bud variation (99). This third theory was already held by Linneus (32), and was the one which Gasparrini especiall endeavored to disprove. Prof. Fritz Müller founded hi theory on the fact that the caprifig tree is to a remarkabl degree barren, producing fertile seeds always few in num bers, and only in his third crops, the "mammoni," whil the edible fig tree is supposed to show a greater fertility through an increased production of fertile seeds. This theory was at once adopted by Professor Solms-Laubach after he had in Java studied a number of wild fig species. He had there discovered that most of these fig species possessed sexually separate individual trees. In other words some fig trees produced fruits with mixed flowers, both male and gall flowers, while other trees of the same fig species produced only fruit with female flowers (33).

According to this theory the caprifig produces two sexually distinct trees. The male tree with male and gall flowers and a few female flowers, and the female tree with only female flowers.

The existence of such a tree was not shown by Prof. Müller, but it had already been mentioned and described by the Italian investigators, Pontedera and Gallesio, the former describing it as "Erinosyce," the latter as "fico semi-mula."

In his description Gallesio adds that this fig has no male flowers, but only female flowers, which when fecundated produce seed or become botanically ripe, while the pomological ripeness also takes place as an effect of the foregoing fertilization (20).

Gallesio's description has been doubted, though I think that in this, as well as in his other classification of the various forms of the caprifig, he is entirely correct.

This theory is strengthened by the fact, already referred to, that seedlings from Smyrna figs, fertilized by the pollen of the caprifig, do to some extent show a distinction of sexes on different trees. But an objection of some consequence to this theory is borne out by the fact that not all edible figs produce seeds, that some contain no fully developed female flowers, and that some again contain male flowers, the latter, however, rarely. Only in the Smyrna tribe of figs do we find fertile seed in very great quantity, though it is true that also some Italian figs the edible kind produce mature seed.

This has led me to propose a new theory of the orig of the edible fig, based on the occurrence of differe flowers in different varieties of figs, which proves to n mind that not all of our figs are, strictly speaking, d scended in the same manner from the original ancesto which however in every instance is the wild or caprifig.

Fourth Theory.—According to my own views the edib figs are of several different kinds, which in their extrem or types are well characterized. I distinguish at lea four different types.

First Type.—The common edible figs, which produc ripe fruits (receptacles) without caprification and pollin tion. This type becomes pomologically mature, but do not become botanically mature, or at least the latter rarel Its flowers are mainly mule flowers and a few fema flowers, but no perfect gall flowers and no male flower This class includes nearly all our common edible figs Europe and California, and all those propagated in ho houses. This class of figs bears several crops, but the is no great or important difference either in the receptac or in the flowers of the receptive crops. Some of the f varieties belonging to this type produce a few fertile see when pollinated or caprificated.

Second Type.—This type or group comprises the Smyrn figs and is characterized by its flowers which are only f male ones, perfectly developed. They produce in abu dance when pollinated or caprificated. They have r mule flowers, no gall flowers and no male flowers. Th fruit becomes botanically ripe and as a consequence the botanical maturity the receptacle becomes also pom logically ripe. This is a purely female type, all the flov ers being perfectly developed female flowers. This typ of figs are at present confined to the Smyrna district, being there the fig or the only fig cultivated for commercial purposes. In the other Mediterranean districts this type of figs is entirely unknown. Introduced to California it never produced fruit until pollinated.

Third Type.—This is the San Pedro type, with different flowers in the different crops. The first crop or fiori contains only mule flowers. This crop becomes consequently pomologically ripe without pollination or caprification, but even if pollinated it will never become botanically ripe or produce seeds, as the flowers are all with abortive embryos or ovaries.

The second crop contains only fully developed female flowers which require pollination in order to set fruit or become botanically mature, the receptacle never becoming horticulturally mature as long as the flowers are not pollinated.

To this class belongs a limited number of figs, which are especially valuable on account of their brebas or first crop. Among varieties belonging to this crop are the San Pedro (yellow), the Gentile, the Bitontoni, the Portuguese, and a few others.

Fourth Type.—To this class belong very few edible figs which are characterized by having more or less perfectly developed male flowers in a zone around the eye. The other flowers are principally mule flowers. This class becomes pomologically mature as well as botanically mature, the latter referring to the male flowers. If the gall flowers and female flowers are developed properly is not known. To this crop belongs the *Croisic* fig, the *Cordelia* fig, and a few others.

The origin of these various types I derive from the various crops of the caprifig, through artificial or horticultural selection. The first type may either have descended

from the male tree of the caprifig, through elimination
the male flowers. The gall flowers in not being used i
galls would naturally endeavor to regain their female na
ure, while the female flowers by now and then being po
linated would more or less retain their female or see
producing nature. This theory was first suggested b
me in letter to Prof. Solms-Laubach, who however think
that the origin of this class may be equally well explaine
by supposing their descent from the female caprifig,
which case the female flowers through non-use have b
come degenerated. But the fact that this class contai
both mule flowers and perfect female flowers speaks,
think, in favor of my theory that the mule flowers are
reality only degenerated gall flowers, or perhaps mo
correctly gall flowers which through non-use are regai
ing their female nature. If descended from the fema
tree of the caprifig there is nothing to explain why som
of the flowers are capable of producing seed while th
majority are not.

The second type of Smyrna figs must have descende
directly from the female tree of the caprifig, their flowe
having retained their female nature through constant ca
fication.

The third class is more difficult to explain. Howeve
I think it may have descended from a cross between a
improved Smyrna and a caprifig.

The fourth class is nothing else than a direct descen

efforts of man to cultivate and propagate only the best or what proves most suited to his purposes has caused him to gradually discard, first all inferior trees, later all inferior varieties, all which either did not suit his taste or which in other respects did not prove as profitable as others. This progress in selecting varieties has been continued to our own day with nearly all kinds of fruit, progressing more or less rapidly according to the intelligence and civilization of the cultivators. As the fig is one of the oldest of fruits mentioned in the history of the human race, the selection and improvement of varieties must have taken place at an early date; in fact, at the dawn of higher civilization. No barbarous people could evolve the luscious edible fig from the insignificant and, for eating, worthless caprifig, even if we suppose that some chance seedling of the female type with superior fruits had been found. The likelihood that caprification was invented simultaneously with the cultivation of the first edible fig makes it more probable that the civilization of the people in question was considerable. The origin of the edible fig of the Smyrna kind must be traced to some one of those ancient nations whose history and remains are the most obscure and the least unravelled of any.

It is more than probable that the Smyrna race was first originated and that later the other class of edible figs was evolved. Or it may be possible that both originated simultaneously, or nearly so, in separate countries. The truth and facts of this we will never know, and our assertions can only have the value of more or less probable conjectures.

The first figs of either class must have been very inferior to those now considered our best. The class which descended from seeds of the male caprifig must, to begin with, have possessed some male flowers in at least one of its crops. The first effort in selection must have bee eliminate these male flowers, as both they and the pa the receptacle on which they grow are hard, dry and erwise not palatable. Thus in the Croisic fig (26) male flowers, together with their part of the recepta is always removed before eating, and this necessary cess must have first stimulated efforts to produce a without the objectionable parts. While this selection improvement of edible figs was being carried on by ancient cultivators the wild fig was not entirely left to self. It was found necessary in some instances to pagate even the wild fig in order to procure the figs caprification. What would be more natural than to s pose that those figs were especially propagated which duced greater abundance of pollen? This selection i small way would in time give rise to several types e among the wild figs, similar to those perhaps as have b described by Pontedera, Gallesio and others.

After the first objectionable features of male flow were eliminated other improvements followed as to fla taste and sweetness, etc.

GENERAL REMARKS ON CAPRIFICATION.

Caprification is a horticultural process, which cons in suspending the profichi or summer figs of the capr on the branches of the edible fig. The object of cap cation is to produce seed in the edible figs and to ca these latter to set and mature. Only such profichi as c tain fig wasps—*Blastophaga psenes*—are of any value caprification. Shortly after the profichi have been s pended the female Blastophagas hatch out of their g and in their efforts to leave the fig become covered w the ripe pollen of the caprifig. Once outside of the c rifig the Blastophagas search for other caprifigs in or to lay their eggs in them. But not finding any caprifigs, they enter the edible figs in mistake. The effect of this visit is the pollination of the Smyrna fig flowers with the caprifig pollen brought along by the wasps. The pollination again causes the edible figs of a certain class to mature seed and to set its fruit. In order that pollination may be properly accomplished, it is necessary that the figs practiced on should have female flowers in a proper state of development with receptive stigmas, and that the pollen of the caprifigs should be properly developed and in a good condition. Not all edible figs are equally susceptible of caprification. The time for caprification is in June and July, according to locality. Caprification is nothing else than an artificial pollination accomplished partly by man, who suspends the caprifigs; partly by the wasps, which carry the pollen from the caprifig to the female flowers of the edible fig.

HISTORICAL NOTES ON CAPRIFICATION.

There are very good reasons for supposing that caprification is as old as the cultivation of the fig by man. That it originated in some of the oldest agricultural countries is much more probable than that the practice is of comparatively modern origin, for instance invented by the Greeks during the time intervening between the Homeric songs and the era of Alexander. For this belief speaks the fact that the caprifig is not a native of Greece nor of any other Mediterranean country, but one of southern Arabia, and possibly also of other countries in the vicinity of the Red Sea and the Persian Gulf. The fig was introduced to Greece, as has already been shown (49), and whether we presume that the first introduced fig race required caprification or not, it follows that this caprification was not and could not have been invented in Greece nor in any other country where the caprifig was not originally wild, and wild at the time the first figs requiring caprification were grown under cultivation. If the self-setting fig race had been the one first introduced to Greece then the Greeks would never have thought of caprification, or if some uncommon genius had done so, he would have been obliged to go to distant countries in order to see, find and bring home the caprifig of which he could otherwise have had no possible knowledge. The discovery of caprification in Greece, as has been held by the majority of investigators, except Solms, would be as improbable and as impossible as the discovery of the placer mining of gold in a country where native gold only occurs in solid veins of ore. Caprification must have originated in a country where the caprifig was wild. But particulars about the discovery will never be forthcoming, the records having been forever lost. Even in the oldest books of the Semitic races no mention is made of any process which can with any certainty be explained as referring to caprification. As is stated elsewhere, in the Book of Amos (50), we read of botes schiquaim, which means "one who operates on the wild fig." But if this operation refers to caprification, or to the oiling of the fig, or to the yet common and necessary practice of cutting the "Sycomore figs" with a knife in order to give an opportunity to their inquilines escape, will always remain an uncertainty, with some probability that the last explanation is the correct one. A circumstance which makes it probable that caprification was in very ancient times practiced in Asia is the fact that Syria is yet the country which grows principally or almost exclusively figs requiring caprification in order to set and mature. In nearly all other countries other, though inferior, varieties have been or are being substituted, varieties which mature without pollination and caprification (51).

For the oldest written record of caprification we must go to the oldest Greek writers. Aristotle, the teacher of Alexander, and one of the best-informed scholars of ancient times, describes caprification in very much the same way as it is practiced to this day. Aristotle explains the effect of caprification through the bite of the wasp, which causes the air to enter the fig, etc. He, as well as all writers, for a period of 2000 years or until the time of Linnæus, were unable to give a true explanation of the effects of caprification (52).

The most minute description of caprification as practiced and understood by the ancients is given by Theophrast (53). Not only does he correctly describe the process of caprification, but he informs us of certain facts of great interest. One of these is that there are two races of figs, one which requires caprification in order to set fruit, and one which sets fruit without caprification. Theophrast was the first one to point out this, and he must have learned it through observation of the various fig varieties grown at his time. Another statement made by this writer is to the effect that caprificated figs had a lesser commercial value than figs not thus caprificated (54). Whatever may have been the case at his time, it is not so now. If Theophrast's statement is correct it can be explained by the Smyrna tribe not thriving in Greece, or by their unimproved state at that time.

Theophrast also mentions how ignorant cultivators instead of using caprifigs suspended other substances in the trees, such as galls from elm trees, the peasant believing that the wasps emerging from these elm galls would have the same effect as fig wasps. Of course if the fig tree in which they were suspended belonged to a race which did not require caprification, the effect of either varieties of wasps (or of any other foreign substance) would be the same or none. Theophrast's explanation of the effects of capr cation is similar to that given by Aristotle. He reje the theory that the wasps close the eye of the fig a through the prevention of the entrance of the air caus maturity. On the contrary, he maintains that the was enlarge the eye of the fig. cause its juices to flow, su up the superfluous "humors" of the fig. and that t warm and fermentation producing air effects the matu ing of the figs. The differences between the two rac of figs of which one requires caprification and the oth not is explained by this author through the influence soil and climate, as well as by a different nature of tl fig, which enables it to ripen its fruit without the aid the wasp. The circumstance that in Italy no caprificati was practiced at his time, he explains by the suppos drier soil and climate of that country, which absorbs tl superfluous juices of the fig. The humid climate Greece, he contends, makes it necessary to employ tl aid of the wasps in order to relieve the figs of their supe fluous moisture.

Pliny, the great Roman naturalist and compilato follows Theophrast closely (55). He classes the caprifi as the wild fig, wanting in the juices necessary for tl food of the wasps. The latter not finding the necessa food flies to the edible fig and through nibbling enlarg the mouth of the fig, and allows the fertilizing air to ente which again transforms the milky juices of the fig to swe honey. Pliny believed that caprification was only pra ticed in the Archipelago (from which it was later intr duced to Italy). At the time of Pliny caprification w. unknown in Italy. The account given by the great La naturalist is evidently only a compilation from othe authors, especially from Theophrast (55). He appears not to hav made any personal investigations or examinations.

Through all the mediæval ages, or for over fifteen hundred years after Pliny, horticulture and natural science made little progress, and the opinions of the ancient writers were adopted as regards almost all points of human knowledge. So also their theories about caprification. For fifteen hundred years after Pliny this process was practiced by the cultivators of the soil in the same way as in the time of ancient Greece; no one was found to inquire in its nature and value, much less to solve the enigma of this, the most interesting of all horticultural usages of all times.

In 1583, Cæsalpinus discovered the sexual organs of plants and was able to point out their functions, but his discovery bore no fruit as regards a better understanding of caprification, and all writers after him for nearly two hundred years followed the teachings of Theophrast, Pliny and Plutarch.

In the early part of the eighteenth century two botanists occupied themselves with a closer study of the fig. One of them was Giulio Pontedera (56), who was the first one to describe the flowers of the caprifig and their structure, though he did not recognize their sexal nature. He also studied the fig wasps and caprification, but little suspected the true nature and influence of the wasp. Pontedera ascribes the effects of caprification to the bitings of the wasps, which cause the air and light to enter the fig. This is the more remarkable when we consider how very minute are the wounds caused by even a large quantity of wasps. As seldom more than very few wasps enter one fig, it will be seen that the extra air that can penetrate on account of the wasp bites is very small indeed, if any at all.

Another investigator, one of the most prominent botanists of the early part of the eighteenth century, was

Tournefort. He traveled in the Levant and in Greece and made special study of caprification as practiced there. Being well acquainted with fig culture in Provence, in France, he was well qualified for his time to take up the study of caprification. Tournefort had studied Theophrast and tried to explain his statement about the lesser value of the caprificated figs, through the necessity of drying such caprificated figs in ovens which again cause their aroma to disappear. As Solms-Laubach points out Tournefort confounded the wasps with moths which infest dried figs, just as so frequently happens in our day. Tournefort describes the three crops of the caprifig, mentions the two races of the edible figs, of which one requires caprification, while the other will set fruit without it. The effects of caprification he explains in the same way as every one before him, by the biting of the wasps, which causes the superfluous juices to escape. Finally, he mentions that a fig which in Provence without caprification produces 25 pounds of figs, in the Island of Zea gives 200 pounds (57). A very unsatisfactory statement, when we consider the distance of the two localities and the uncertainty that the two trees were actually of the same variety; not to speak of climate, soil, age, cultivation, etc.

It was reserved for Linnæus to discover the true nature of caprification (32). While previous to his time, the nature of the sexes in flowers had been described and generally accepted, still no one had thought of the possibility of an insect transmitting the pollen from one flower to another and thus cause fecundation. As Pliny of old had forshadowed the theory of evolution so did Linnæus a century before its rediscovery indicate how, at least, in one instance, flowers were dependent on insects for their pollination. Linnæus points out how, in order that the female flowers of the fig may be properly fecundated, it becomes absolutely necessary for the pollen of the anthers to be distributed through the cavity of the fig. And this could not be accomplished, if nature had not supplied the fig with a wasp, which could carry the pollen from the male flowers to the female tree. And this wasp, he says, is the "psen" of the ancients, or the fig insect. The opinion of Linnæus was published in 1749. But Linnæus was not aware of the fact that some figs ripened their fruit without fecundation; want of material for investigation caused him to think that the fig was absolutely diœcious, in other words that it possessed sexes distinctly separate only on different trees.

John Hill again, who published his great work A History of Plants, in London, 1751. refers only shortly to the fig and its caprification. He condemns Tournefort's theory of puncture and irritation, and states that pollination is the real effect of caprification; but he does not refer to Linnæus, though it is probable that he must have heard of the latter's views upon the subject.

Later in the century both Milne and Cavolini, independently from each other, discovered that a difference must be made between the maturing of the seed and the maturing of the receptacle, and that the former maturity, at least, must require pollination, even if the latter (or pomological maturity) could be accomplished without it. Milne clearly defines this by saying (59): "The question supposes that the fig trees in this country bring fruit to maturity without assistance of caprification, and the fact cannot be denied. The same thing, we have seen, obtains in Spain, Provence and Malta; but the fruit, or more properly, the fruit vessel, is in all cases to be distinguished from the seed contained within it. If the male be wanting, the seed will not vegetate when sown; but the fruit may, nevertheless, swell and come to an appearance of perfection; and so it is observed to do in the instance in question, and in many others, especially when the fruit is formed of one of the parts less connected with seed, as the calix, receptacle, etc."

Filippo Cavolini published his work on caprification in 1782, or twelve years later than Milne, whose opinion he had, however, not read. Cavolini believes the caprifig to be the male tree, and the fig the female of the same species. He further notes the difference between the fig receptacle and the seed (60), and how the former can come to maturity on account of its greater attachment to the stem of the tree, while the seed, which is only attached to the pericarp by its vessels, requires pollination in order to mature. This pollination causes the juices in the fig to flow more freely, bringing both the seed and receptacle to maturity. That some figs mature their receptacles and others not depends on a defective structure, by which the juices from the stem of the tree are more or less obstructed in their flow into the fig's receptacle. As this obstruction is lesser or greater, the fig requires more or less pollination, in order to cause more or less sap to flow, while the seed, in order to attain maturity, always requires pollination. That the same variety of fig can mature in one locality without caprification, while in a different district it must be caprificated in order to mature its receptacle, depends upon differences in locality and soil. Cavolini's ideas are clearly expressed and to the point.

At the very end of the century a French botanist, Olivier, traveled in the Ottoman Empire, Egypt, Persia and in Greece, making a particular study of the fig. His descriptive work of his travels was published in Paris (year 9). Olivier came to the conclusion that caprification was a useless and ignorant proceeding, which should be abandoned. "This operation," he says, "of which some authors, both ancient and modern, have spoken with admiration, appeared to me to be nothing else than a tribute which man pays to ignorance and prejudice. Caprification is unknown in many parts of the Levant, in Italy, in France and in Spain, and begins to be abandoned in the Archipelago, where it used to be practiced, and which, nevertheless, still produce excellent figs for eating. If the operation was necessary, whether fecundation be effected by the fertilizing pollen dispersed in the air introducing itself into the mouth of the fig, or whether nature makes use of a little fly to transmit it from one fig to another, as is commonly believed, it is evident that the first fig in flower could not fecundate at the same time those that have already attained a certain size, and those which are only just appearing, in order to ripen two months later." The knowledge which Olivier possessed of caprification was in reality most superficial and defective, and some of his statements are even false and misleading, and not worthy of quotation, except for the fact that disbelievers in caprification have pointed to him as an eminent botanist, who had conclusively proved the delusiveness of the process in question. Olivier did not even know that it was the caprifig which was used for caprification, but stated that it was the common "figues fleurs," the brebas, or first crop edible figs, which were hung on the trees. This also appears again in the last lines of his statement quoted above, beginning: "First fig or flower," etc. His statement that caprification was unknown in Italy and Spain is also (61) incorrect.

In 1820, Giorgio Gallesio, a prominent Italian horticulturist, published his treatise on the fig. How far Gallesio's statements were based on investigations in nature, are not known. Later writers on figs have endeavored to sho that his theories were founded principally on book lear ing, and not on observation (62). I am not of that opinio as his statements show a frankness and fairness entire indicative of truthfulness. Gallesio holds that there a two races of figs, one which requires caprification order to mature its fruit, and one which matures witho the aid of this operation. The different requiremen between the two fig races depend upon a difference i construction of the figs, and each race retains its chara teristics, regardless ot the influences of soil and climat The difference in construction lies in the ovary of the fi Some figs have ovaries without ovules, and those fi which cannot be fertilized, can also not feel the action the pollen from the caprifig. These are the *mule* fig The other class of figs, with perfect ovules, are sensitiv of the pollen, and under its influence develop perfe seeds. These he calls semi-mules. The fecundatio causes the juices to flow to the fig and effects its maturit The caprifig alone containing the pollen is, therefor necessary, and the only way to apply it is through capr fication.

Gallesio also describes a caprifig with only fema flowers—"the fico semi-mula" (63). His statement th the original wild caprifig bore only one crop of figs. shown by Solms-Laubach (64) to be erroneous, or at lea very improbable.

In the middle of our century the Italian botanis Guglielmo Gasparrini, published a series of four diffe ent treatises upon figs and caprification, extending in tin [illegible] against caprification, believing himself warranted to do so on account of the result of the experiments made by himself. Gasparrini's experiments have been by many considered conclusive and almost final, and his views have been adopted almost unchanged by later writers on the subject of caprification. While conceding that Gasparrini's experiments were scientific and fairly carefully made, and highly interesting and demonstrative, I hold that the main conclusion which he drew was singularly illogical, though it may have been warranted by the insufficiency of his experiments. Gasparrini's almost only, but fatal, error was, that he experimented only on a few Italian figs, not suspecting even that there might be other figs differently constructed. From his observation he concluded that because "a few were so," therefore, "all must be so." Gasparrini's experiments are too elaborate to be here noticed in detail. Those who wish to further study the subject are referred to his respective works, one of which is partly translated in the California State Horticultural Reports for 1891.

Gasparrini formulates his conclusions in nineteen different paragraphs, answering as many different questions, but principally concerning three different points:

1. Does the caprifig fecundate the domestic fig, and cause them to set (65)?

2. Does the caprifig fecundate the female flowers of the fig and produce seed (66)?

3. Does the caprifig hasten the maturity of the fig?

We will shortly consider some of the more important points, as answered by Gasparrini, his answers being given here:

1. Does caprification hasten the maturity of late figs? Answer, no. The experiments made by Gasparrini are good, and I consider his conclusion correct (67).

2. Does caprification cause late figs to set in grea number? No. The conclusion is illogical. The periments only show that caprification had no effect the varieties experimented with—the Colombro, Lard and Sarnese.

3. Does the caprifig, by the assistance of its insec fecundate the female flowers of the late fig? No. I experiments were partly defective, partly insufficient, a the conclusion drawn is illogical and incorrect. T varieties experimented on were the last ones mention besides the Dottato. Gasparrini found seeds in figs caprificated, but as they had not been absolutely isolat they may have been pollinated, nevertheless. He expla the production of seed without pollination by "parthe genesis." The fact, as has been already shown, th seeds never form in edible figs growing in California a Brazil, where the caprifig is not generally distributed (6 is sufficient evidence for rejecting parthenogenesis, and adhering to the pollination theory (69).

4. Does the fly cause the setting and afterwards earlier maturity of the fig by the puncture it makes in No. In this I fully agree, the experiments being app ently conclusive (70). The general conclusions to whi Gasparrini came he summarises as follows:

1. That to understand well the effects of caprificati it is in the first instance necessary to know the nature the fig and of the caprifig, and what connection they ha with each other. We have seen that the caprifig is the male of the fig, as has been hitherto believed, bu species so different from it that it may well be taken type of a new distinct genus.

2. The structure of domestic figs, as well as of th to which the caprifig is applied, is perfectly similar in far as concerns the organs of the flower, the structure

the seed and of the receptacle; so that it does not appear how the insect of the caprifig can be necessary to some varieties only.

3. We have seen by experiments that the insect neither hastens the maturity nor causes the fruit to set, whether of early or late figs, nor yet is it necessary for fecundation.

4. That the circumstance of the caprifig losing many of the fruits in which the fly has not been bred, does not serve to prove the necessity of caprification, but rather to refute the doctrine completely, as the fly does not breed in the domestic fig; and besides, we have seen that when the caprifig bears a large crop of fruits many of them fall unripe, even though the insect has been in it, and the grub be found in the ovaries.

5. In respect to the dropping of some figs, the causes must be sought for chiefly in constitution and mode of vegetation of those varieties, also in the soil, climate and adverse conditions of the season.

6. That this caprification is useless for the setting and ripening of fruit, and therefore this custom, which entails expense and deteriorates the flavor of the fig, ought to be abolished from our agriculture.

To the above conclusions of Gasparrini I will offer the following remarks. My own experiments (to be detailed further on) in pollinating Smyrna figs with the pollen of the caprifig, show conclusively that the caprifig and the edible fig are closely related, though this fact does not necessarily imply that they belong to the same species, of which, however, there is no doubt (71).

No. 2 of above can only refer to those varieties on which Gasparrini experimented and knew. My experiments, just referred to, show that not all figs are constructed exactly alike, and that accordingly the fig insect is necessary to some varieties, while it may not be so others.

Point No. 3. The insect may not hasten the matur of the fruit, but still cause it to set. My pollination e periments show that pollination does cause the of a certain class to set, and as the caprification is tl only *practical* way by which this pollination can be pr duced, this caprification must be necessary for this cla of figs, though not for any other class differently co structed.

Point 4. The fact that the caprifig loses much of i fruit does not prove caprification an error. It only shov that pollinated fruits are susceptible to influences, climat and others, which affect the setting of all figs (and a fruits), whether they be Smyrna figs or not.

Point 5. The causes enumerated as effecting the dro ping of the fruit, do not exclude other causes from havi the same effect. Caprification, if effective under favo able circumstances, cannot prevent caprificated figs fall, if climate and soil are unfavorable. If not capri cated, they would, under the above conditions, have falle anyhow.

Point 6. In order to be correct, this point should on refer to some or certain figs, or to all figs similar to tho experimented upon.

Before leaving Gasparrini and his work, I will short state what his experiments, so interesting, and laborious performed, have really proven.

They have shown us that caprification does not hast the maturity of the fig. Further, they have proven, th it is not the sting itself of the wasp which influences tl setting of the fruit [illegible] fig. They also sh that many figs, which [illegible] caprificated by the growers, require no [illegible], and that, as far as the figs are concerned, caprification should be abandoned. This point is confirmed by the circumstance that these same figs mature in countries where no caprifigs are grown.

I cannot see how any other conclusions of importance can be drawn from the experiments of Gasparrini. It must always be regretted that he never thought of the possibility of their being any other race of figs than that one which he happened to have under his eye. How different, for instance, would his conclusions have been if he had had the *true* Smyrna figs to experiment on.

In our own times no one has given as much study to the fig question as Professor H. Count Solms-Laubach. His researches were published in 1882, and contain a perfect mine of knowledge, partly compiled, partly his own investigations. While scientifically investigating his subject and studying the figs and the fig insects, both in Italy, Java and France, it appears that he had no opportunity to make direct experiments in caprification, but founded his opinions principally on the experiments of Gasparrini. He sifts the knowledge of others with rare ability and patience, adds numerous and interesting observations of his own. His researches are of the utmost importance. As a botanist, he rejects, as insufficiently proven, Gasparrini's theory of parthenogenesis, and showing that Brazilian figs produce no fertile seed, concludes that caprification is necessary for that purpose.

During his investigations in Java he discovered that most figs growing there consisted of female trees as well as of male trees, and he found that the male tree possessed a flower especially adapted to foster the Blastophaga, a degenerated or differentiated female flower, which he calls the "gall flower." This gall flower has probably lost its power to produce seed. Returning home and investigating the caprifig, he found that even this fig contained this gall flower, almost exclusive of any r female flower. He further shows how different spec of figs are inhabited by different species of Blastopha; He also unconditionally adheres to the theory of caprifig and the fig being of the same species. Later he adopts the theory of Fr. Müller, that the edible fig the female plant and the caprifig the male plant. But is entirely unaware of the existence of a race of fi differently constructed than the common edible figs whi he had investigated, and he shared the opinion of Ga parrini, that all figs were affected by caprification in t same manner, though he recognized the absolute nece sity of pollination and caprification in order that fert seeds may be produced. But if caprification is not need any more, it was once a necessity, ages ago, when the was first brought into cultivation by man, and before t present race of figs, which requires no caprification order to set and mature, had originated. The class of fi which the Italians considered as requiring caprificati had been shown to set fruit without this operation; t class that once at a time required caprification must, ther fore, have been lost, and superseded by a better, mo modern class, evolved from the former. He com to the following conclusion: "Caprification was onc ages ago, a necessity; it is now no more useful, but on a horticultural operation, transmitted from generation generation, down to our time, and in its original forr Its scientific importance as means for judging the mod but can be proven to be yet extant and to constitute our *best* figs, the conclusion arrived at by Solms must fall. Prof. Solms-Laubach has since acknowledged the correctness of this.

THE FIG WASPS OR BLASTOPHAGAS.

All known wild fig trees, and there are over one hundred distinct species described (72) by botanists, are inhabited by very minute wasps known as "inquilines" or parasites, and scientifically described as *Blastophagæ*. These Blastophagas not only visit the figs, but they live, breed and develop in them, nay more, if deprived of their fig-hosts, these Blastophagas could not live, breed and develop anywhere else. The organization of these little wasps is such that while they may possible feed themselves for a short time (though it is doubtful if they do feed at all), they could not possible deposit their eggs elsewhere than in the "gall flower" of their respective fig, not even in the female flowers of the same fig. Even if they did succeed in doing so, their eggs would not develop elsewhere and the brood would soon perish. The species would thus become extinct. But this is not all. The relation between the fig and the Blastophogas is so intimate that in order to foster those little wasps nature provides the fig with flowers especially constructed to their use, the "gall flowers" elsewhere described. But if the Blastophaga is dependent upon the fig for its existence, the fig is hardly the less so upon the Blastophaga. Without the Blastophaga no fertile seed would be produced with any regularity (if at all) and the fig species would be in danger of perishing. The influence of the Blastophaga is somewhat different in different fig species. It has been shown that in one species at least, not only the female flowers are dependent for their pollination upon the Blastaphaga, but that the male flowers actually do not develop their fertilizing pollen without having first been stimulated by the mechanical action of the Blastophaga. This discovery was made by Dr. Cunningham, a result from his experiments on *Ficus Roxburghii*, in India.

Figs are visited by many insects, but a distinction must be made between "visitors" and "regular boarders." The former visit the figs in order to feed, either on the fruit and its juices, or upon other insects. The boarder or "inquilines" again breed in the fig, and cannot breed anywhere else. The former will injure the fig in various degrees, while the latter are absolutely necessary for the material maintainance and multiplication of the fig tree species they inhabit.

LIFE-HISTORY OF THE CAPRIFIG WASP, BLASTOPHAGA PSENES.

If we during the month of June or July (73) cut open a "profico" or scond crop caprifig just when it is full grown, we may notice that it contains a large number of gall flowers or galls around which crawl numerous little insects, some of which resemble minute black wasps, while others of the same size are wingless and very differently shaped, as well as being of a yellow or brown color. A closer inspection will reveal to us that a number of the galls are perforated by a single ,round hole through which may either be seen the hollow of the gall, or the wasp itself, not having yet escaped. A closer inspection may even show us how the light colored insects, which are the male wasps, are enlarging the holes in order to enable the females or winged wasps to escape. These respective insects are male and female of the same species known to naturalists variously as *Blastophaga psenes*, *Blastophaga grossorum*, or *Cynips psenes*.

The male insects are the first ones to hatch and escape, with their powerful mandibles or jaws they easily cut through their galls and then set to work to liberate the females. Before the latter escape they are fecundated while yet in the gall, by the males, Each gall contains only one single wasp. The male wasps never leave the fig. They are so constructed that they could not very well live outside, and even inside the fig they soon perish, their life-work having been accomplished when liberating and fecundating the females. The females even do not tarry long in the fig and soon find their way out through the eye of the fig, which has opened sufficiently to let them pass through without injury to their wings (74). In case the fig has been injured and compressed in such a way as to close the eye, the wasps will remain as prisoners until otherwise let out, for instance, by cutting the fig.

With care and aided by a magnifying glass, we may further follow the female Blastophagas as they escape from their old habitation. Their first work is to look for figs suitable to lay their eggs in, the only object of the wasps now being to propagate their species, it being doubtful if they feed at all. As soon as outside of the old caprifig the female Blastophaga halts on the outside of the fig and endeavors to free herself of a whitish powder with which she appears to be literally covered. This powder is the pollen from the anthers of the male flowers of the caprifig in which she hatched and with which she came in contact when she escaped from the fig. This process of cleaning she performs in very much the same way as does a house-fly, streaking herself with her front legs, bending at the same time the head, body and wings. She never succeeds in getting entirely clean, as a large portion of the pollen will adhere in spite of all her efforts. But when she considers herself sufficiently clean she flies away and lights on a less than half-grown caprifig of the same or some other tree. The mammoni or third cro the caprifig has by this time advanced so far in devel ment that its interior flowers are just of the proper and age to suit the wasps (75). If there are no such at hand the wasps will soon perish. Having lit on a m moni the Blastophaga finds the fig-eye closed by sca (76). But these scales are not impossible to penetr In order to enter the fig the Blastophaga saws out a t little piece of the outside edge of a top scale, which ope to her an entrance between two scales. Next she pus herself under the scale and then zigzags herself throu until she reaches the interior hollow of the fig. But efforts to get through between the scales have been t mendous, and in so doing she almost invariably loses wings. They are always lost in the very beginning of work, and can be seen remaining, wedged in between outside scales just like feathers stuck under the band a hat. In order to ascertain the presence of the Blas phaga in a green fig it is not always necessary to cut fig open, as the presence of the wings of the wasp sti ing between the scales is a sure sign that the wasp succeeded in getting in. And even if the wings ha fallen off the little wound caused by the gnawing of wasp can be told by the minute drop of sap that has ooz out and hardened. It is this drop of sap which was, markably enough, for ages considered as being the r cause of the setting of the figs. If no wings and no g is seen on the scale it may be safely assumed that Blastophaga has entered the fig in question.

As soon as the now wingless Blastophaga has enter [illegible] gall flowers there to [illegible] inserts one in ea [illegible] the same way a [illegible] This partic lar place lies between the nucleus of the fig ovary and the integument surrounding it. If left anywhere else observation shows that the egg will not develop. In order to accomplish this the wasp first alights on the stigma of the fig flower. Then she extends her ovipositor and runs it down through the canal which from the center of the stigma leads through the whole length of the style to the funnel or entrance to the ovary of the flower. This is penetrated by the ovipositor and the egg is laid and securely wedged in between the nucleus of the ovary and the integument surrounding it.

As soon as the egg is deposited, the ovipositor of the wasp is withdrawn. The lower part of the canal is filled by a filiform appendage of the egg, while the upper part of the canal fills with a brown exudation from the wounded cells. As soon as one egg has been laid, the wasp immediately departs to another flower there to repeat the process. The egg depositing power of a wasp is simply enormous, and one wasp is capable of laying an egg in each of the many gall flowers of a fig. After the eggs have been all deposited the Blastophaga endeavors to regain the outside of the fig through the same way she entered. But in this she rarely succeeds. Being by the egg-laying process completely exhausted, she generally succumbs before she regains her liberty and her dead body may be found in the opened fig. The work of the Blastophaga has not alone been that of depositing eggs. Involuntarily she has rubbed against some of the female flowers of the fig, and the pollen which adhered to her body when she entered has been deposited on the stigmas of these flowers. The effect of this pollination is the development of seeds in the female flowers. This would not otherwise have taken place without the aid of the wasp, because the pollen from another fig could not very well have penetrated through the closely thatched scales the eye: and the pollen from a male flower in the san fig would only be ripe from a month to six weeks afte wards, at a time when the stigmas of the female flowe will have attained their full development and receptivit After the egg has been deposited the gall flower does no at once cease to develop. The embryo and kernel of th seed keep on growing for a month. After that the egg o the Blastophaga begins to develop, and when it passe into the larva stage it begins to feed on the embryo of th fig, which thus soon perishes. The integument of th ovary again grows and assumes the form of a large, hard brownish and glass-like gall. In two months the youn female Blastophaga wasps have attained their full devel ment and after copulation with the wingless males ar ready to leave the caprifigs. And this they do in th same way as they left the previous crop, the profichi The males die within the figs in all the crops. They hav performed their function and are of no more use. It ma here be incidentally stated that even if the wasp's egg i not deposited in a gall-flower, the latter will after a cer tain time cease to develop. It will never produce seed.

At this time the winter figs or the "mamme" destine to become the first crop of the following year are of th proper size and development required by the Blastopha gas, which enter them in the same way as described abov in order to deposit eggs.

Next spring these develop and ripen and the young Blastophagas leave them in April. They immediatel the caprifig through the year. Each crop of fig has, as we have seen, had its own crop or brood of wasps, from which follows that if a crop of figs should entirely fail the crop or brood of the wasps would also perish.

To facilitate the understanding of the life history of the wasp the following diagrammatic table has been prepared:

A. Second Crop or Profichi of the Caprifig. April to June.

1. In April these figs are less than quarter grown.
2. The Blastophaga females enter the figs and here deposit their eggs in the gall flowers.
3. In June, or two months later, these Blastophaga eggs have become fully developed, and the perfect wasps emerge to seek other figs.
4. In emerging the wasps are covered with pollen.

B. Third Crop or Mammoni of the Caprifig. June to October.

1. In June the third crop is quarter grown.
2. The Blastophagas emerging from the previous crop penetrate into these third crop figs and deposit their eggs in the gall flowers.
3. In doing so they also pollinate the female flowers.
4. In October the Blastophaga eggs are fully developed and the perfect wasps emerge hunting for the young figs of the mamme.
5. A few seed fully developed are found in this the third crop of the caprifig, none being found in the two other crops.

C. The First Crop or the Mamme of the Caprifig. October, through winter, to April.

1. In October the mamme are quarter grown.
2. The Blastophaga, hatching from the preceding or third crop, enter the mamme and there deposit their eggs in the gall flowers.
3. The mamme, with the gall flowers and the eggs of the Blastophaga, hibernate on the tree without further development.
4. With the advent of spring the mamme and the Blastophaga eggs resume development.
5. In April the Blastophaga eggs have developed into full grown wasps, which emerge from the figs seeking the young figs of the next crop, the profichi, in order to deposit their eggs in them (78).

PROCESS OF CAPRIFICATION.

The process of caprification consists in bringing the caprifigs, of the proper age and crop, in close proximity to the edible figs, in order that the wasps, as soon as they leave the caprifigs, may be lured into the edible figs. Practically this is accomplished in different ways, more or less proper and economical. In Smyrna, Syria, Greece, Italy and Africa the caprifigs are pulled at the proper time in June, the profichi being the only crop used for this purpose. The caprifigs are then almost full grown and the male flowers ready to shed their pollen. The caprifigs are at once strung on split reeds or rough straws in quantities of three to five on each straw. These straws are thrown over and suspended among the branches of the edible fig tree. Another method, much inferior to the former, consists simply to cut branches from the caprifig trees and suspend them in the edible fig trees. This injures the caprifig trees and does not enable the cultivator to regulate the number of caprifigs according to the quantity actually needed. This method is only in use in certain parts of Portugal and Spain. A third way to accomplish caprification is to plant a few caprifig trees in among the edible fig trees, and to simply depend upon the wasps themselves finding their way in sufficient number to the edible figs. This method is the least proper of any for many reasons, one of which is that the caprifig thrives equally well or even better in poor rocky soil, and it would be more economical to give the good soil over to the edible fig trees. Another reason is, that the Blastophaga would preferably go to the young caprifig instead of to the young edible fig, if the former were to be found close by, as they of course would, if the caprifig tree was planted in among the edible fig trees. If, again, the caprifig trees were such that they bore no young caprifigs at the time [illegible] wasps hatch out, then, [illegible] their propagation [illegible] would be impossible [illegible] to bring new Blastophagas from other caprifig trees. The growing of caprifig trees in among the edible figs would, therefore, not only be useless, but would be entirely improper. This leads us to the necessity of having separate plantations for the caprifig.

Separate Plantations of Caprifigs.—In most countries where caprification is practiced it is a general saying that it is necessary to go to the hills for the caprifigs (79). It is not only more economical, as has just been pointed out, to grow the caprifigs separately, but it is in many instances necessary. The soil and moisture in the orchard is not always suitable to the caprifig. It must be remembered that the caprifig is yet in its comparatively wild state, while the edible fig is a horticultural product or creation which only can be expected to attain its proper qualities under the most favorable conditions. Practically this is true. It frequently happens that when the edible figs are receptive, or their female flowers ready for pollination through caprification, that at this very time the caprifigs growing in the vicinity of the edible figs are not properly developed, while other caprifigs grown in different soil and at a different elevation are just of the proper size and condition to furnish both Blastophagas and pollen. But another even more serious objection to having a few caprifigs growing in rich soil is that their crops are not always following each other in immediate succession. The Blastophaga, in order to properly propagate her brood, requires young caprifigs ready to receive her eggs as soon as she hatches out. Frequently the tree from which she hatches does not possess these figs, while other caprifig trees do. It is therefore necessary, in order to procure a constant supply of Blastophagas, to have a large number of caprifig trees growing together. In a grove of such trees there will always be some that bear figs of proper

size in which the wasps may breed. Such plantations ‹
caprifigs should be made separately and in various loca
ities, in order that the supply of figs may never fai
Quite frequently it also happens that the caprifig crop i
a certain locality fails, while in a different one, where th
climatic conditions have been dissimilar, the crop of cap
rifigs may be abundant. This shows the necessity to hav
caprifig plantations in various localities, especially in
country like California, where the caprifig trees are nc
wild and where frequent importations of caprifigs wit
inquilines are difficult if not impossible. When the cap
rifig crop fails in Smyrna fresh caprifigs are imported b
the vessel load from the Grecian islands, and in all coun
tries where caprification is considered necessary in orde
to procure a crop of figs a regular and profitable trade i
carried on in caprifigs, which often bring much more tha
the edible figs (80).

Quantities of Caprifigs Required.—The quantity c
caprifigs needed to caprificate a fig tree varies with th
size of the tree. In Smyrna some thirty figs are require
to caprificate a large full-grown edible fig. If too fe
are hung on the tree an insufficient quantity of wasps ar
had and the edible figs are not all pollinated. If too man
are hung too many wasps will enter the edible figs, injur
their flowers, and, according to some observers, caus
the figs to prematurely drop and decay.

For smaller trees a less quantity of caprifigs are re
quired than for larger trees, but even the largest i
Smyrna are not given more than thirty figs to the tree
The caprifigs are hung on the limbs of the fig trees befor
sunrise and when the wind is not blowing. The proce:
of hanging up the profichi must be repeated several time:
as only those edible figs are pollinated which are of prop
size and receptivity. For the younger figs new profich

must be supplied later on, if their maturity is required, and the quantity of caprifigs necessary at any time must be regulated according to the number of figs to be pollinated.

Proper Time for Caprification.—Caprification should always take place when the female flowers of the fig are receptive. This can be easily ascertained by cutting open the fig. With a magnifying glass the stigmas of the female flowers should be seen to be bright and light greenish with a peculiar fresh luster, as if they had been lightly varnished or moistened. Further the stigmas and styles should be erect, if bent and brownish they are too old to receive the pollen and past their receptive state. This occurs in Smyrna and in the Mediterranean districts in the middle of June. The second crop, the only one used, is then in proper state of development to be caprificated. In California the time for caprification will vary with the locality. In the Bay district, around San Francisco, the Smyrna figs are receptive in the end of July, in the interior earlier. The edible figs when ready for the caprifigs are about one-third grown, hard and green. The caprifig again should be cut when the pollen is properly developed and just before it has burst from the anthers. The caprifigs are then almost full grown, though in opening them the female and gall flowers will not be found developed. The size of the fig varies with variety—as there are many varieties in caprifigs—but the average size would be an inch and a half long by three-quarter inch wide. Some caprifigs are much smaller. Thus the caprifigs received from Smyrna and taken there from the fig trees in which they had been suspended, were almost twice as large as those brought from Palermo (81). The various races of caprifig vary in size and softness, some remaining always hard, while others grow larger and become soft and pulpy.

As regards the Smyrna figs at the time of caprification it may be stated that at the moment when their female flowers are receptive, the scales at the eye loosen or rather becomes flexible, allowing the wasps to enter. This softness of the scales may be easily ascertained as by pushing a quill or a stick against the fig eye, it easily penetrates between the scales, without injury to them and without causing any juice to exude, if the fig is of proper size. While on the contrary if the fig is too young the scales will be found to be hard and fixed, cannot be pushed back, and the least wounding will cause an abundance of milky juice to exude. If again the fig is too old, the scales will be equally hard and fixed and the fig will be yellow and will early drop.

If the fig is cut open, the stigmas should be fresh and moist, the styles erect and greenish, not brown.

What Takes Place in Caprification.—We have already followed the life history of the Blastophaga in the caprifig. Its history in the edible fig is somewhat different. The wasps cannot breed in the edible figs, they can only visit them. Shortly after the caprifigs have been suspended, or sometimes even before, the Blastophagas begin to hatch. It even appears that the pulling of the caprifigs hastens the maturity and escape of the wasp. As soon as these have hatched they crawl out of the caprifigs in search of young caprifig mammoni, in them to lay their eggs, as has been already described. But as the caprifigs are not near, no such mammoni are to be found. In place of them the wasps only encounter edible figs, and not being aware of the deception practiced, they enter these edible figs for the purpose of breeding. The flowers of the edible figs are, however, so constructed that the intentions of the wasps are completely frustrated. Instead of the necessary gall flowers, which are especially adapted to the ovipositor organs of the wasps, only female flowers with long styles are found and which are otherwise so modified that the wasps find it impossible to properly lay their eggs. All their frantic efforts to penetrate the canal of the style and to reach the fig ovary and its nucellus are in vain. The Blastophaga cannot breed in any edible fig. Still, her visit has a very great effect on the edible female fig flowers, provided these are of the proper age and development. The pollen from the caprifig, with which the wasps were liberally dusted, adheres to the female stigmas, the effect being pollination and fecundation of the flowers. The Blastophaga herself dies and her dead body may be seen upon opening a fig which has not advanced too far in maturity.

It is here assumed, as is really the case, that the wasp cannot properly place its egg in the female flower, but even if she could do or would accidentally do so, the egg would not properly develop, as it is only the gall flower which is suitable to the growth of the larva of the wasp. But even if by chance such development would take place the young wasp would quickly perish by being enveloped in the sugary liquid of the mature fig. A certainty is, however, that I have never found any gall in the mature Smyrna figs, which shows that no such development takes place.

What Does Not Take Place in Caprification.—Since the most remote time, so many opinions have been expressed as regards the consequences of caprification, that it may be proper to here point out what does not take place. The old opinion that the gnawing of the wasp relieves the fig of its superfluous juices, and thus causes it to mature, is too absurd to be given much thought. The gnawing done by the wasps is so infinitely small that the fig, through the combined efforts of twenty wasps, would not lose one single ordinary drop of sap. Fig wounded by a needle in such a way that many drops (juice escape do not show any tendency to set better, as have repeatedly demonstrated. The gnawing of a fe wasps can, therefore, not have any effect on the recep tacle of the fig. Nor would caprification affect figs whic regularly set their fruit without the process. Thus, c all the figs which we have tried in California, some fift or more varieties, only some seven or eight kinds do nc set their fruit, all others do. To caprificate the regula and common kind of edible figs would, therefore, be useless waste of time and work. They would probabl produce some fertile seeds, but it is doubtful if thei quantity would be sufficient to greatly improve the fig. I has been said that the Blastophaga produces a gall in th edible figs, and that this gall formation would cause th figs to set and mature, in the same way as a worm-eate pear or apple ripens sooner than the uninjured fruit. Bu we have already seen that no such gall is produced i edible figs, and experiments in Italy have almost conclus ively demonstrated that the entrance of the wasps doe not hasten the maturity of the fig fruit (81).

There remains only one point more. It has lately bee shown that in one East Indian fig the wasp causes th female flowers to set seed without pollination, supposedl only by piercing them, and thus causing an irritation. have good reason to believe that this is not the case a regards the Smyrna fig. The wasps first received by Mı Shinn, at Niles, were let loose among the Smyrna figs i his grove at Niles. I examined the wasps as they hatche out from the figs and found no pollen on any of then the male flowers having dried up on the passage fro Smyrna without producing pollen. Some twelve hou after the wasps were let loose I cut open numerou

Smyrna figs and found that the wasps had already entered them and were moving over the stigmas of the flowers. Still, not a single fig that had not been pollinated by hand developed that season, although probably a thousand wasps had been let loose in one place among the fig trees. If irritation alone would cause maturity, we would hardly have failed to receive at least a couple of mature figs through the visit of the wasps. The quality of the seedlings grown by myself and Mr. Maslin have, on the contrary, shown that pollination actually takes place.

We may, therefore, with a fair degree of certainty, establish the following facts:

1. The visit of the wasps to the female flowers of the Smyrna figs is powerless to produce fertility or maturity, except accompanied by pollination.

2. The gnawing of the wasps on the scales of the eye, or the mere irritation of the flowers does not produce a flow of sap sufficient to stimulate the fig in any unusual way.

The Effects of Caprification.—Caprification can, therefore, only be effective and profitable in varieties which contain a majority of developed female flowers. If such figs are not caprificated they will drop off, shortly after the receptivity of the female flowers is past. On such figs the immediate effect of caprification is: first, the setting and the coming to full maturity of the fig receptacle (the fig); second, the development and maturity of the female flowers and their ovaries and seeds. Another important effect of caprification is the dropping at full maturity of caprificated figs, or rather of figs in which caprification has been successful. All Smyrna figs drop of themselves when ripe, while all other fig varieties in which caprification is not an absolute necessity, must be cut or pulled from the tree at harvest time, as they will fall only when past their prime. The advantage of havin figs requiring caprification is, therefore, evident in a districts where such figs will grow.

The expense of caprification is much smaller an requires less labor than the pulling or cutting off of th figs when ripe, provided, of course, that the figs woul set without being caprificated, which they will not do.

Besides the pomological or horticultural maturity of th receptacle, the caprification causes the botanical maturit of the female flowers, which, as we will see, is of grea importance to the cultivator.

The Importance of Seeds in Dried Figs.—The greate value of caprificated varieties over those which do not re quire the process is to be sought in the development o fertile seed. The seeds in our common figs consist onl of empty glossy shells with no trace of kernel. All suc seeds have no taste and can in no way contribute to th flavor of the dried fig. Not so, however, Smyrna fig which have been caprificated. They all contain seed o large size with a full, oily kernel, which when crushed i found to be in the highest degree aromatic and "nutty." Such seed when present in sufficient quantity greatly con tribute to the quality of the fig, giving them an intensel aromatic flavor. It is only during the process of dryin that the aromatic taste of the seed is permeated througl the pulp of the fig in very much the same manner as al monds and other nuts communicate their flavors to pud dings, preserves, or canned fruits generally. Smyrn figs when dried are therefore more highly flavored tha any other figs. To the fresh fig the seeds do not com [illegible]

Which Figs should be Caprificated.—The shortest answer to this question is: all figs which drop off if not caprificated. It has not yet been fully ascertained which these figs are. It is only certain that the great majority of figs will mature their receptacles without caprification. In California we have, however, since some ten years had growing several varieties imported from Smyrna and of these none perfected fruit until they were artificially pollinated. This class then requires pollination and caprification, and must be caprificated if fruit is to be expected. We have also had other figs in this State which have never matured fruit, though twenty years old. One of these varieties is growing on the place of R. B. Blowers of Woodland. The trees must be some twenty or more in number and at this date must be about twenty-five years old. Only one tree of this kind once produced a ripe fig. It is evident that this variety requires caprification, both in order to set fruit and to mature its seed. If the variety came from Smyrna is not known, but it undoubtedly belongs to that class of figs.

Another class of figs require caprification for their second crop. Among such varieties San Pedro is the most prominent one (86). But there are other varieties like the San Pedro, such as the Portuguese of Italy, the Gentile, etc., all of which set their first crop but drop their second crop. Microscopic examination shows that the second crop of these figs possess fully developed female flowers, while the first crop which matures have only flowers with abortive ovaries. In another place in this paper I have related my experiments in caprificating the second crop of San Pedro and Gentile, and the success achieved, undoubtedly proving that caprification is necessary for a certain crop while it is not necessary for another crop.

Where Caprification is Practiced.—Nowhere is capri cation practiced more thoroughly, more constantly a more successfully than in the home of the fig, Syria a Asia Minor. In the vicinity of Smyrna, the foremost region of the world, the figs of which are acknowledg superior to any grown elsewhere, caprification is a nece sity. The fig crop without it would fail, at least the cro from all varieties which produce the Smyrna figs of com merce. The fact that some figs may be produced wit out caprification even there must be attributed to the san cause which produces some fertile seed in the Italian fi without direct fertilization by caprification. The re cause of the setting of figs in either case is the presen of caprifigs in the vicinity, from which the wasps car the pollen irregularly and sparingly, but sufficiently produce a few figs and a few seed. The importance caprifigs in Syria and Smyrna is so great that they oft command a higher price than the edible figs, and in cas of failure of the caprifig crop sailing vessels are sent distant ports to the Grecian islands to bring whole cargo of the fruit. This bringing of cargoes of caprifigs great expense by intelligent growers must point to t value of caprification there, and is in glaring contra with the occasional practice of some ignorant cultivato in Greece and Italy, who, failing to procure caprifig suspended galls of elm trees among their figs (87). the culture of figs followed the immigration of the Phœn cians and later on that of the Arabs, so do we to this da find caprification practiced in all countries formerly occ pied by those nations. [illegible] along the north coast [illegible]

[illegible]

To this day caprifigs are highly valued and bring a high price in Tripoli, Tunis, Algiers and Morocco, and parts of the Iberian Peninsula, especially when the crop is scarce. Leclerc (61) tells us that in Algiers the profichi of the caprifig bring two sous per dozen (not quite a half cent).

In Greece caprification has been in vogue since very ancient times, as has been mentioned elsewhere. From that country it spread to southern Italy first after the time of Pliny, and has there been practiced ever since, principally in the territory of the old kingdom of Naples or in southern Italy generally.

To the general rule that caprification is practiced in Greece and Grecian colonies, one exception is mentioned by Solms-Laubach. In Marseille and vicinity caprification is unknown. It is also not practiced in central and northern Italy, or in the territories occupied anciently by the old Umbrians, Etrurians and Latins, nor is it practiced anywhere in southern France and the Riviera. Solms declares two causes for this to be possible. Either in ancient times caprification was practiced even there, and later on abandoned, or it was never introduced, fig culture having been only lately brought to these regions, and at a time when caprification was no more necessary, varieties in the meantime having appeared which would ripen their receptacles without it. But as from the descriptions of Pliny and Cato it becomes evident that caprification was not known in Italy in their time, it is almost certain that in countries where caprification is not now practiced, it has never been introduced. This is the case in all fig districts of America, as neither in the Southern States, in California nor in Brazil, the Argentine or in Peru and Chile, has caprification ever been even advocated until within the last ten years, or after the late introduction of the real Smyrna figs from Smyrna. This has also been the case in Australia and New Zealand. It is probable that to countries, within easy reach of Syria, the first varieties introduced were those requiring caprification; later only the self-ripening kinds followed, or were originated on the spot. The kinds which require caprification are much more exacting of climate, soils and conditions generally than the self-ripening kinds, as we know that Smyrna figs if transplanted to less favored localities loose their superior qualities, even if caprified. As the self-ripening kinds became more common and distributed the Smyrna varieties were allowed to gradually die out, but the caprification had taken such hold and had become so deeply rooted, that it continued to be practiced on varieties which did not require it.

The circumstance again that caprification was not introduced in the more northern provinces, such as north Italy, south France, and north of Spain, must be sought in the unsuitability of those places for those varieties which required caprification. It is more than probable, that in the above countries fig culture never assumed any degree of development until the advent of figs which did not require caprification in order to bear. The variety of caprifig which carries its mamme over winter is more susceptible to frost than other figs. At least it will be impossible for the fig wasps to survive in countries where the caprifig crops are interrupted by heavy winter or spring frosts. This would also make caprification impossible, except the caprifigs were yearly imported from more favored districts, a proceeding that would not prove practical and remunerativeness.

On other Insects [illegible] the Blastophaga.—It is well known that figs are visited by numerous insects other than the regular Blastophaga, and the question arises to what degree could they be depended upon to carry the pollen from the caprifig to the edible fig, or could they do so at all. Some entomogists not acquainted with the practical side of the question have claimed that the Blastophaga was not required, but that any insects would do the work. In order to carry the pollen from one fig to another it is by no means necessary that the insect should breed in the fig. All it has to do is to crawl in to the caprifig *at the proper time*, and then to crawl out and in into the edible fig, and the pollination is accomplished. But in order that any practical result to the crop at large may come from this visit, several circumstances are imperatively necessary and must coincide. The insects must make these visits at the proper time. They must be of proper size to be able to enter the closed eye of the fig. The insects must be present in sufficiently large number to pollinate the fig crop, not single figs. As to the first point it will be seen that no other insect can be found, which will have any business in the caprifig at the time when required, and even if the pollen of the caprifig would serve as its food, it would have no cause to afterwards visit the edible fig, which at the period when such visit is required does not produce any food, it being green and hard, with no trace of sugar. Insects only visit flowers in search of food or to lay their eggs. No other insect than the Blastophaga has been found to do the latter properly and at the time when required. The second point is readily understood. The eye of the edible fig is closed, and only an insect with a peculiarly developed instict would know how to push its way between the closed scales. At this stage of the development of the fig no insects have been found which visit the figs, except the Blastophagæ and some parasitical wasps which prey on her brood, and which would not enter the fig except they knew the Blastophagas were a ready there. The most important point, however, is th quantity of insects required at a given time. Only a insect which will actually breed in the caprifig can be de pended on, and it must breed in countless numbers. few visitors would have no practical influence on the fi crops. They may fertilize or pollinate a few flowers but they would be of no practical value to the grower an would not produce a crop. Taking it all in all no insec has been known, and no one is likely to ever be know that can be substituted for the *Blastophaga psenes.*

Different Species of Blastophaga in Different Figs.— As far as is known, different species of figs are, as a rule inhabited by distinct and characteristic species of inqui lines. Thus one, *Blastophaga psenes*, has only been foun in one or two nearly related fig species, and no othe Blastophaga species has been found in our caprifigs Parasitical wasps are always found together with th Blastophagæ, preying on and developing in them, just a the Blastophaga preys on and develops in the embryo o the fig. Even when different fig species grow clos together, do the wasps keep to their respective fig hosts accidentally the wasps may visit other figs, but they d not breed in them. It appears as almost certain that ever fig species is inhabited by Blastophagæ. Thus, in th botanical garden of Java (88) a row of fig trees, consist ing of five different species of figs, was found to b inhabited by as many different species of Blastophagæ each variety in its own fig host, to which it was strictl confined (90). The cause of this localization of specie must be sought in the organization of the wasps an their ovipository organs, which only enables the insec to deposit its eggs in a certain kind of flower, whicl again has been changed so as to accommodate the pecu liarities of the wasp, her size and capabilities. Under such circumstances, there is but little hope that, for instance, the wasp inhabiting the Lower California and Sonora fig species can be made to inhabit and breed in our caprifigs (89). Even the sycamore fig is inhabited by its species of inquilines, but which have never been found in the caprifigs (91). It may, therefore, be assumed with great certainty, that only closely allied fig species are inhabited by the same species of Blastophagæ. But in many species of figs we find more than one species of Blastophaga. Some figs even are inhabited not only by different species, but also by different genera of true Blastophaga, while the latter again are preyed on by parasitical wasps often equal to them in size.

SUMMARY.

Caprification, then, is an horticultural process, based on scientific principles. It has been used since very ancient times, and is yet in vogue in many countries. It is an absolute necessity in places where Smyrna figs are grown, or in places where it is of importance to pollinate such figs as possess receptive female flowers. Caprification causes such figs to set and mature, when otherwise they would fall off immature. This horticultural maturity is caused by and preceded by the botanical maturity of the female flowers. Again, caprification is not required for that great class of figs which sets and ripens fruit without it, except, indeed, it should be found practical, profitable and possible, to produce sufficient seed, in such varieties of this class as possess receptive female flowers in sufficient number.

LITERATURE.

For access to some of the books enumerated below, my thanks are due t Mr. Adolph Sutro and his librarian, Mr. George Moss, who not only place the Sutro Library at my disposal, but also imported rare books for m special benefit.

Several references to books which I had no opportunity to inspect, hav been made after Solms-Laubach.

As a great number of the following works are books of travel and con tain only few pages or a few lines devoted to our subject, I have adde the pages on which are found these special references.

My very sincere thanks are due to my esteemed friend, the well know entomologist, Dr. E. A. Schwartz, in Washington, D. C., for first callin my attention to Dr. Cunningham's experiments on *Ficus Roxburghii* an for copying extracts of his memoir at a time when the original was not a my disposal. To my knowledge, there is no one in this country who ha so carefully studied the subject of caprification as Dr. Schwartz.

ANNALI del Minist. d'Agricolt. Ind. e commercio, Quinquennio, 1870 1874, Vol. I, Roma, 1876, page 696. Contains reports from various fi districts in Italy

ARCANGELI, G. Sulla caprificazione e sopra uno caso de sviluppo ano male neifiori del Ficus Stipulata Thunberg. Societa Toscana d Scienze Naturali, Nov. 2, 1882.

ARISTOTLE. Historia animal. lib. V, Cap. XVI, 3, Caprification.

BARY, ERWIN DE. Tagebuch der Reise von Tripoli nach Ghât und Aïr Zeitschr. d. Gesellschaft f. Erdkunde, Berlin, 1880. Bd. 15, ch. 3 page 2301.

BASINER, T. F. Naturw. Reise durch die Kirgisen Steppe nach Khiva in von Baer u. Helmersen, Beitr. zur Kentn. der russich. Reichs, Vol IV, Petersburg, 1848, page 237.

BEHR, H. H. The Smyrna Fig Insect. Pacific Rural Press, San Francisco Calif., Feb. 20, 1892.

BERNARD, MRS. BOYLE. Our Common Fruits. London, 1866, page 232.

BERNARD, M. Mémoire sur l'histoire naturelle du Figuier, in Mémoire pour Servir a l'Hist. Nature. de la Provence, Vol. I, Paris, 1782.

BENSON, MARTIN. Guide to Fig Culture and Catalogue of Rare Tropica Fruits and Plants. Cutler, Dade County, Florida.

BELLEW, H. W. Journal of a Political Mission to Afghanistan. London 1862, page 9

BOISGELIN, L. DE. Malte, ancienne et moderne. Paris, 1809, Vol III page 277.

BRAUN, A. Die Pflanzenreste d. Egypt. Museums Berlin. Berlin, 1877 page 14.

BRAUN, ALEXANDER. Ueber Parthenogenesis bei Pflanzen, Abh. der K Akad. d. Wissenschaften zu Berlin 1856. Berlin, 1857

Brandis, D. Forest Flora of Northwest and Central India. London, 1874, p. 419.

Brotero, Felix de Avellar. Compendio de Botanica. Paris et Lisboa, 1788, Vol. II, page 159.

Bussato, Marco, da Ravenna. Giardino d'agricoltura ed. 5. Venezia, 1781 (1592).

Buch, Leopold von. Physikalische Beschreibung d. Canarischen Inseln. Berlin, 1825, p. 120.

Cato. Scri. rei rustici, Ed. Schneider, Vol. I, page 19, Cap. 8, 1.

Cavollini, Filippo. Memoria per servire alla storia compiuta del Fico e della proficazione. Opuscoli scelli sulle scienze e sulli arti, Vol. V. Milano, 1782.

Cella, Paolo della. Viaggio da Tripoli da Barberia alle frontiere occidentali dell Egipto. Genova, 1819, page 30, 120.

Celsius, Olaus. Hierobotanicon, Upsala, 1747, Vol., page 370.

Chabas. Etudes sur l'antiquité historique d'apres les sources Egyptiennes. Paris, 1872, p. 105.

Cox, William. A view of the Cultivation of Fruit Trees. Philadelphia, 1817.

Coquerel. Description des Parasites d'un Figuier de l'île de Bourbon. Guérin, Rev. et Voy. Zool. VII, 1855, page 365, etc.

Cronan, P. L. et H. M. Florule d. Finistere. Paris, 1867, page 210.

Davy, John. Notes and observations on Ionian Islands and Malta. London, 1842. He experimented with caprification and states that, when caprification by his order was discontinued, the crop failed, only few figs ripening and these of inferior quality. He says that the statement of De Candolle (Physiologie Vegetal Tome II, page 580) that caprificated figs are inferior, is entirely erroneous. De Candolle having made the statement from hearsay, and not from personal experience.

Denotovich, Anthony C. Caprification of the Smyrna fig. The Pacific Rural Press, May 18, 1895. Mr. Denotovich tells of the orginos boghadhes and the ashmadhes. He mentions that also the male or caprifig there are *caprificated*. From Mr. Denotovich I learn that it was from his father's place in Aidin that the Bulletin cuttings were secured.

Delpino, Federico. Note critiche sull. opera la distribuzione dei sessi nelle piante, etc., del Prof. F. Hildebrand. Atti Soc. Ital. Sc. Nat., Vol. X, Milano, 1867, pages 272-303. Also Vol. 16-17, Milano, 1873-74, page 239.

Du Breuil, M. A. Culture des Arbres et Arbrisseaux a fruits d. table. Paris, 1876, page 602.

Duhamel du Monceau, Henri, Louis. Traité des Arbes Fruitiers. Paris, 1768, 2 Vol. 4; 1762 Ed. is inferior and of no value.

DUHAMEL DU MONCEAU. Traité des Arbres et Arbustes que l'on cult. e France. Paris, 1809.

DUVEYRIER, H. Exploration du Sahara, Vol. Les Touâreg du Nord. Paris 1864, page 193.

EISEN, GUSTAV. The Fig of Commerce, its Culture and Curing; and a de scriptive catalogue of all its known varieties. Los Angeles, Califor nia, 1887.

EISEN, GUSTAV. Caprification of the Fig. Citrograph, Redlands, Cali fornia.

EISEN, GUSTAV. The Fig and its Culture and Curing, with Special Ref erence to California. Fresno, Calif., 1885.

ENGLER. Versuch e. Entwickelungsgesch. d. Pflanzenwelt. Leipzig, 1879 page 57.

FIEDLER, DR. CARL GUSTAV. Uebersicht d. Gewachse d. Königreiche Griechenland. Dresden, 1840, pages 606 to 613. This work, whic has some pretensions as regards importance and advice, contains ab solutely nothing of original research. The author has the bad an inexcusable fault of translating all names of varieties into German and speaks of Zuckerfeige, Brustfeige, Südfeige, etc.

FIGARI, BEY. Studii scientifici sull Egitto, Lucca, 1865, Vol. II, pag 217 and 80.

FORSKÅL, PETRUS. Flora ægypto-arabica, ed. Karsten Niebuhr Haunie 1775.

FORSYTH, F. D. Report of a Mission to Yarkand. Calcutta, 1875, page 79

FRANK, B. Die Pflanzen Krankheiten. Encyclopadie d. Naturw. I Abth 13th Lief., pages 552–568.

GALLESIO, G. Pomona Italiana. Pisa, 1817, 3 vol.

GALLO, AGOSTINO. Venti gionate d'agricoltura. Bergamo, 1757, pag 112. Fent. Ed., 1588.

GARIDEL, PIERRE. Histoire des plantes que croissent aux environs d'Aix etc. Aix, 1715, fol. 100 planches.

GARIDEL, PIERRE. Histoire des Plantes de la Provence, 1715.

GARDENERS' CHRONICLE, London, contains a large number of articles o figs, caprification, etc , but as their enumeration would make this lis too bulky it is here omitted. None of the articles is of any specia value as regards caprification.

GASPARRINI, GUGLIELMO Ricerche sulla natura del Caprifico e del fico sulla caprificazione Rendiconte dell Acad d Napoli, Vol IV, 1845 pages 321 412, tab. 8.

GASPARRINI, GUGLIELMO Nuove Ricerche sopra [illegible] punti di anatomia [illegible] caprifico From [illegible] Vol VII, 1848, pages 394 417 tab. 3.

GASPARRINI, GUGLIELMO [illegible] fichi del Con [illegible] Vol IX, Napoli [illegible]

GASPARRINI, GUGLIELMO. Nuove asservazione su taluni agenti artifiziali che accelerano la maturazione nel fico. Atti della reale Acad. d. sc. fisiche e mathem. Vol. II, Napoli, 1865.

GAUDIN, CH. ET C. STROZZI. Contributions a la flore fossile italienne, mén 4; Neue Denksch. d. allg. Schweizerischen Ges. f. d. ges. Naturwiss. Vol. XVII. Zürich, 1860, page 10.

GENY, PH. Les Figuiers spontanés et cultives dans les Alpes Maritimes. Nice, 1867. Only in MS. colored plates but no text. The present whereabouts of the MS.—which is of great value—is unknown.

GIACINTO, CARLO. Agricultero de Malta. Not seen by me.

GLAS, GEORGE. History of the discovery and conquest of the Canary Islands. London, 1764, page 81.

GUSSONE, J. Enumeratio plant. vasc. in insula Inarime provenientium. Ficus auct. G. Gasparini, Napoli, 1854.

HALEVY, JOSEPH. Voyage au Nedjran, Bullet. de la soc de George d. Paris. Ser. VI, Vol 6, 1873, p. 271.

HANOTEAU ET LETOURNEUX. La Kabylie et les coutumes Kabyles I. Paris, 1878, page 434.

HASSELQUIST, FRIED. Reise nach Palestina, 1749–1752, ed. c. Linnæus. Rostock, 1762, page 221.

HELDREICH, THEODOR VON. Die Nutzpflanzen Griechenlands, mit besondere Berücksichtigung der neugriechischen und pelasgischen Vulgarnamen, 1862, pages 20 and 21.

HEGARDT, CORNEL. Ficus in C. Linné Amœnitates Acad., Vol I, page 41. Holmiæ et Lipsiæ, 1749.

HEHN, VICTOR. Culturpflanzen und Hausthiere in ihrem Uebergang aus Asien nach Griechenland und Italien, 2 aufl. Berlin, 1877, p. VII.

HERRERA, GABRIEL, ALONSO DE. Agricultura general corregida segun el testo original de la prima ed. publ., 1513, por el mismo autor, Real sociedad economica Madrilense. Madrid, 1818, Vol. II, page 245.

HILL, JOHN. A history of Plants, London, 1751, page 134. In this very fine work, the author criticises Tournefort's explanation of the effects of the Blastophagæ, and clearly states that it is easily seen that the real result of the Blastophaga visit to the female fig is a pollination of the flowers, by means of the pollen from the male flowers.

HÖST, G. Nachrichten von Marokos und Fes. Copenhagen, 1781.

HUGHES. The country of Belochistan, etc. London, 1877, page 19.

HUC, M. Souvenirs l'un voyage dans la Tartarie el la Chine, 1846. Paris, 1850, Vol. II.

IBN EL AWWÂM. Livre de l'agriculture, traduit par J. J. Clem. Mullet. Paris, 1864, Vol I, page 336.

JOURNAL OF THE HORTICULTURAL SOCIETY OF LONDON, Vol. III, 1846. Translation of Gasparrini's Memoir of the Caprification of the Fig.

KNOOP JOHANN, HERMAN. Beschrijvingen, afbeeldingen von de beste soorten von Appelen en Peeren. Amsterdam, 1790, pages 32–35.

KNORR. Thesaurus rei herbarie hortensisque. Norimbergæ, 1770.

KLAPROTH. Description du Tubet—du chinois en russe—par l. Père Hyacinth Bitchouriu. Paris, 1831, page 139.

LAGARDE DE. Ueber die Semitischen Namen des Feigenbaums und der Feige. Göttinger-Nachrichten, Jahrg. 1881, page 388.

LA HIRE. Observation sur les figues. Hist. d. l'acad. roy. d. sc. Mem. d. Math. et Physique, 1712. Paris, 1714. The male flower of the fig is first described here.

LECLERC. De la caprification au fecondation artificielle de figuiers. Comptes rendues d. l'acad. des sc., Vol. 47, 1858, page 330.

LELONG, B. M. The Caprification or the Setting of the Fruit. Sacramento, Calif., 1891.

LELONG, B. M. The fig. Ann. Report of the State Board of Hortcult. of the State of California, 1889. Sacramento, 1890.

MAYR, DR. GUSTAV. Feigeninsecten, beschrieben von. Verhandlungen der Kais-Königl. Zoolog-Botanische Gesellschaft in Wien. Jahrg. 1885, XXXV, Bd. Wien, 1886, pages 147-250. Tafl. XI-XII. The standard work on fig insects.

MAYER, PAUL. Zur Naturgeschichte der Feigeninsecten.

MANCA DELL'ARCA, L'ANDREA. Agric. de Sardegna.

MASLIN, E. W. Caprification of the Fig. Placer County Republican, Dec. 29, 1886.

MEYER, E. H. F. Geschichte d. Botanik, Bd. III, page 278. Königsberg, 1856.

MEYERDORFF. Voyage d'Orembourg a Bokhara. Paris, 1826, page 203.

MILNE COLIN, A. Botanical Dictionary or Elements of Systematic Botany. London, 1770. Caprification.

MIQUEL, F. A. G. Prodromus Monographiæ Ficuum. Hooker's London Journal of Botany, Vol. VII.

MOVERS, F. C. Die Phonizier t. II. 2. Berlin, 1850, page 528.

MÜLLER, FRITZ. Caprificus und Feigenbaum. Kosmos, VI Jahrg. Bd. xii, 1882, page 342, etc.

MÜLLER, FRITZ. Zur Naturgeschichte d. Feigeninsecten. Mitheil. a. d. Zoolog. Station, Neapel, III B. 1882.

MÜLLER, FRITZ. Referate of Solms-Laubach. D. Herkumft etc. Kosmos vi, Jahrg. Bd xi, 1882, page 306.

MUELLER, BARON FERD. VON. Select Extra Tropical Plants, 7 ed. Melbourne, 1888.

MÜLLER, H. Die Befruchtung der Blumen durch Insecten, etc. Leipzig, 1873, page 90.

NIEBUHR, CARSTEN. Reisebeschreibungen nach Arabien. Copenhagen, 1774, Vol. I, page 420.

NATURE. London, Oct. 16, 1890.

OLIVIER. Voyage dans l'Empire Ottoman, l'Egypte et la Perse. Paris an 9. Vol. I, page 313.

ORTEGA, CASIMIRO GOMEZ DE. Continuacion de la Flora Española que escribia Don Joseph Quer., Vol. iv, page 103. Madrid, 1784.

PACHO, J. R. Voyage dans la Marmarique et l. Cyrenaique. Paris, 1827, page 32.

PALGRAVE, W. G. Narrative of a year's journey through Central and Eastern Arabia. London, 1865, Vol. I, pages 59, 85, 327, 342.

PASSA, JAUBERT DE. Voyage en Espagne. Paris, 1823, Vol. 2, page 226.

PASQUALE, GUISEPPE, ANTONIO. Relazione sulla Statofisico-economica agrario d. prima Calabria ult. Napoli, 1863, page 307.

PETERMANN, MITTHEILUNGEN. Vol. 18, page 171, 1872; 1861, page 255. Erg. No. 64, page 77; Erg. 10, Heft 47. Gotha, 1876, page 20.

PETZHOLD, A. Der Caucasus, Leipzig, 1867. Bd. 2, page 238.

PHILLIPS, HENRY. The Companion for the Orchard. London, 1831.

PLANCHON. Etude d'estuffs d. Montpellier. Paris, 1864, p. 44.

PLINIUS. Lib. xv., Cap. 19.

POLLINI, CIRO. Viaggio al Lago d. Garda e al Monte Baldo. Verona, 1816, page 31.

POLYTECHN. INSTITUTE Wien. Jahrbucher des K. K. Polyt. Inst. by J. J. Prechtl. Bd. 9, Wien, 1826, pages 131–134. Dalmatian figs.

PONTEDERA, GIULIO. Anthologia, sive de floris natura. Patavia, 1720, lib. iii.

PORTÆ, VILLÆ, JOH. BASTISTAE. Napolitani libri xii. Francofurti, 1602, page 307.

RILEY, C. V. Some Entomological Problems bearing on California Pomology, *Caprification*. Read at the meeting of the American Pomological Society, Los Angeles, January 30, 1895.

RILEY, C. V. Fertilization of the Fig and Caprification. Read at the 1891 meeting of the American Asso. Advancement of Science. Both of these papers present an extensive review of the caprification, the best so far compiled in this country.

RICHARD, A. Tentamen floræ Abyssiniæ, vol. ii, page 265. Paris, 1851.

RICHHOFEN, FR. VON. China, Bd. 1, page 559. Berlin, 1877.

RISSO, A. Histoire Naturelle d. principales productiones d. l'Europe meridionale. Paris, 1826. Vol. ii, pages 130–171.

RIVILLE, GODEHEU DE. Memoire s. la Caprification. Mem. d. Math. e. d. Physique pres. p. div. Savants a l'Academ. Paris. Vol. ii, 1735, page 369.

RITTER, CARL. Die Erdkunde von Asien, Bd. vii, Abth. 2, page 534. Berlin, 1844. Ed. ii.

RITTER, CARL. Erdkunde i. B. i. Africa. Berlin, 1822, pages 907–998.

REED, H. The Fig Industries in Florida. Proceedings American Pomological Society, 1889.

REYNIER, L. De l'Economie publique et rurale des Grecs. Genéve et Paris, 1825, page 456.

REYNIER, L. D. l'Economie publique et Rural d. Arabes et d. Juifs Genéve and Paris, 1820, page 770.

ROHLFS, GERHARD. Mein Erster Aufenhalt in Morocco. Bremen, 1873 pages 271, 368.

ROHLFS, GERHARD, Reise nach Kufra. Mith. d. Afric. Gesellsch. i. Deutsch land. Vol. ii, heft 1, pages 23, 27. Berlin, 1880.

ROSSI, DR. FERDINANDO. Produzione d. Fichi Secchi in Italia. Napoli 1881.

ROZIER, FRANC. Cours complet d'agriculture theorique practique, etc. ou dictionnaire universel d'agriculture par une société d'agriculteurs Paris, 1781-1805, 12 vol. 4to and supplement. A standard work of considerable value in identifying old varieties of figs.

RUSSEL, A. Naturgeschichte von Aleppo. German trans. von J. F. Gmelin. Göttingen, 1797, vol. i, page 108.

THE SMYRNA FIG HARVEST, in Harpers' Monthly Magazine, 1890.

SAUVAIGO, DR. E. Notes sur les Figuiers introduits et cultivés dans les environs de Nice. Bulletin-Journal d. l. Société Centr. d'Agricult d. Nice e. d'Alpes-Maritimes. Nice, 1889. J. Ventre & Co. A most valuable work.

SAPORTA, GASTON. Sur l'Existence Constatée du Figuier dans Environs de Paris a l'Epoque quartem. Bull. soc. geol. d. fr. ser. iii, Vol. 2 1873-74, page 442.

SAPORTA, GASTON DE. La Monde des Plantes avant l'apparition de l'homme. Paris, 1879, page 317.

SAVASTANO, DOTT. L. Il Marciume del Fico. Annuario d. R. Sc. Sup d'Agricultura in Portici, Vol. iii.

SAUNDERS, S. Description of three n. gen. and sp. of fig insects. Trans Ent. Soc. London, 1883. Also in the same place, Cynips Caricæ of Hasselquist, etc., page 383.

SAUNDERS, S. S. On the habits and affinities of Apocrypta and Sycophaga Transact. Entom. Society, 1878, page 313.

SAINT LAURENT, JOAM. DE. Della Caprificazione. Mem. del So. Columbaria Fiorientina, Vol. ii, page 243, seq. Livorno, 173, 2.

SEMNOLA, VICENZO. Della Caprificazione. Rendiconte della Academia d Napoli, Vol. iv, 1845.

SEMLER, HEINRICH. In San Francisco, Brazil. Die Tropische Agrikultur, Ein Handbuch fur Pflanzer und Kaufleute Wismar, 1887, 2d Band pages 106 to 142 Semler's account of the fig is greatly inferior to any of the other parts of this otherwise great work It contains many errors evidently the result of compilation.

S[illegible] A. H. [illegible] Persien. Zeitschrift d. Gesellsch. [illegible] Berlin, Bd. 14 [illegible] 122, 1879.

S[illegible] [illegible] to several parts of Barbary [illegible] 1757.

SIMMONS, P. L. Tropical Agriculture. London, 1889, pages 478–480.

SOUSA, FIGUEIREDO A. DE. Manoal d'Arboricultura tractado epratico da cultura dos arboles fructiferas, page 296.

SOLMS-LAUBACH. H. GRAF ZU. Die Geschlechter differenzierung, bei d. Feigenbaumen. Botanische Zeitung, 1885. No. 33–36, Leipzig.

SOLMS-LAUBACH. H. GRAF ZU. Die Herkunft, Domestication und Verbreitung d. gewönlichen Feigenbaums (Ficus Carica L.) Göttingen, 1882. Abhandlungen d. K. Gesch. d. Wissenschaften, B. xxviii.

SPRENGLE, KURT. Theophrastus Naturgeschich. d. Gew. Altona, 1822. Bd. ii, page 80.

TOURNEFORT, PITTON DE. Relation d'un Voyage du Levant, Vol. 1, page 130. Amsterdam, 1718.

TOURNEFORT, PITTON DE. Observation sur les Maladies des Plantes. Hist. d. Academie, etc., 1705. Paris, 1706.

TOURNEFORT, JOSEPH, PITTON DE. Elements de Botanique. Paris, 1694, 3 Vol., 451 planches.

TANARA, VINCENZO. L'economia del cittadino in villa. Venezia, 1644, page 378; ed. 1661, page 376.

TARELLO. Ricordo d'Agricoltura. Venezia, 1572.

TATTI, GIOVANNI. Agricoltura. Venezia, appr. Zansovino, 1561.

THOMSON, DAVID. Handbook of Fruit Culture under Glass.

THEOPHRASTUS. C. pl. iii, 6, 6 ed.; v. 2, 8 ed. Wim.

TOUQUEVILLE. Voyage en Moreé. Paris, 1805, Vol. 1, page 449.

UNGER, F. Die Pflanzen des Alten Aegyptens. Sitzungsber. d. K. Akad. d. Wissensch. zu Wien. Math-Natw. Cl. Bd. xxxviii, 1859 (pages 83, and 110).

U. S. PATENT OFFICE REPORTS. Washington, D. C., 1858, page 384; 1859, page 129; 1862, page 501; 1870, page 205.

Commercial Relations of the United States. Reports from the Consuls. No. 15, 1882.

United States Consular Report. Fruit culture in the several countries. No. 41½. June, 1884.

Same. No. 44. Aug., 1884.

Reports from the Consuls. No. 88. January, 1888.

U. S. DEPARTMENT OF AGRICULTURE, Division of Pomology, Bulletin No. 1, 1887. Washington, D. C., pages 89 to 95. By Prof. H. E. Van Deman. Gives principally an account of the fig in the Southern States of the U. S. A.

U. S. DEPT. AGRICULTURE, Wash., D. C. Special Rept. No. 4. Cultivation of The Fig and the Method Of Preparing The Fruit For Commerce.

WEBB et BERTHELOT. Histoire naturelle des iles Canaries, vol. i, part i. Paris, 1842.

WELLSTED, J. R. Reisen in Arabien. Halle, 1842. Bel. ii, p. 103).

WELLSTED, J. R. Travels to the City of the Caliphs, along the Shores the Persian Gulf. Vol. ii. Lond., 1840.

WENJUKOW. Die Russich-Asiat. Grenzlande. Leipzig, 1874(page 464).

WESTWOOD. Trans. Ent. Soc. London, iv, 1847, page 260. Pl. x. Al same 1883. Pl. x. Also same 1882, page 47.

WICKHAM, WILLIAM. Memoranda respecting the culture of the fig tre in the open air in England. February, 1818.

WICKSON, E. J. California Fruits and how to grow them. San Fra cisco, Dewey & Co., 1889. Fig, pages 402–413. A most excelle expose of the fig as cultivated in California. The illustration repr senting the Smyrna fig as grown in Placer Co., California, represent really thé White Adriatic and not the Smyrna, as has been prove afterwards, the error being caused by a misrepresentation of th grower.

WOHLTMANN, F. Dr. Handbuch der Tropischen Agrikultur, etc. 1 B die Natürlichen Faktoren der Trop. Agricult. Leipzig, 1892.

VARRO. Script. rei rust. ed Scheider. Vol. i, page 268, lib. ii, cap. xi, 5.

VENUTO, ANTONIO. L'Agricoltura. Napoli, 1516, cap. 9, Del Fico.

NOTES.

For full titles of works mentioned below, and for a fuller reference t the researches and publications of the various authors, see the list of LIT ERATURE above.

1. As the flowers of the fig species are generally and well known t botanists, I have considered best to hold my description of the fig flowe in a more popular form, so as to be more easily understood by non-bota nists.

20.. Pontedera, p. 175. This female tree he calls *Erinosyce*. Gallesi also mentions such tree under the name of *Fico Semi-mula*, but it is un certain if he himself has seen it. A somewhat similar form of the caprifi is described by Solms-Laubach, p. 35, as having grown wild in a garde at Chiaja, near Naples. As all, or at least nearly all other fig species whic have been particularly described possess such an exclusively female form it is more than likely that Pontedera's description is correct. Müller an Solms-Laubach assume that the edible fig is the female tree and the caprifi the male tree, which I can only understand to mean that the edible fig i descended from the female tree.

21. Gallesio, p. 46. Solms-Laubach doubts the correctness of thes descriptions and calls them most artificial, p. 33. But after his discover of purely female trees of the Java fig varieties, he may have somewha modified his opinion.

23. According to Solms-Laubach there is absolutely no foundation f this description, p. 33.

24. This caprifig tree grows in Shinn's orchard at Niles; was importe

by G. P. Rixford. It is the only large caprifig tree known to me in California.

25. The female tree of the fig was first described by Solms-Laubach in species from Java. See his Die Geschlechtesdifferenzirung der Figenbaumen.

26. As will be seen in a different place, so far only very few exceptions have been noted, among them the Cordelia fig in Solano county, Cal., and the Croisic fig growing at the mouth of the Loire river in France, and the fig found by B. M. Lelong at Los Gatos, in California. See Solms-Laubach, I, p. 14.

27. French authors generally describe the caprifig male flowers as having only three petals, which is an error, undoubtedly originated by describing the figure in "Du Breuil," where the figure of the male flower is erroneously drawn.

28. Solms-Laubach was the first to thoroughly study the arrangement and structure of the fig flowers. He was the discoverer of the gall-flowers and the distinction between them and the female flowers. See his last cited work (25).

29. Both varities are known as *Ficus carica* Linneus, and to belong to the same botanical species.

30. For Gasparrini's theory, see his work vol. I, p. 378. Parthenogenesis is an extremely rare occurrence and cannot be accepted without positive proof, which Gasparrini fails to give. He also moderates the force of his arguments by saying that it may be possible for the pollen to have entered in some way.

31. Fritz Muller, Cosmos, 1882, p. 342, seq.

32. Cornel Hegardt, Ficus in Linné, p. 41. This most important and interesting reference made by Linneus to the sexes of the fig reads as follows: "Quod si jam fructus feminæ fœcundetur e cavitate fructificationis caprifici sc. maris adscendens farina antherarum penetret, tandemque per totam cavitatem disseminetur necesse est. Hæc omnia naturæ viribus impassibilia apparerent, nisi supremus genitos Ficui huic propriam assignasset cupidinem. Cupidoficus nobis dicitur quem antiqui psenem seu insectum ficarium vocarunt."

33. Solms-Laubach. Die Geschlechterdiff. etc., p. 1. Prof. Solms-Laubach adopts this view without any discussion and refers to it as being as easily seen as soon as pointed out as the "egg of Columbus."

34. Solms-Laubach relates, I, p. 17, that also in Naples it is well known that seedlings produce partly caprifigs, partly a number of varieties which are poor for eating purposes. This practical demonstration of the relationship of the two figs is of importance as demonstrating that they are different sexes only of the same species. My California experiments have resulted similarly.

35. The fact that one tree has been found which produces male flowers

and according to him also seed, does not in the least detract from this theory. The Lelong fig, the Cordelia fig, as well as the Croisic fig, must be considered as improved caprifigs which have not yet lost their male flowers.

36. LaHire, page 287. But Colin Milne was the first one to closely point out that the cultivated figs contained no male flowers, only female flowers (1770.) See Milne's dictionary, article, "caprification."

37. Solms-Laubach, I, p. 14.

38. B. M. Lelong in Report of State Board of Horticulture, 1891, p. 234. Mr. Lelong says that the flowers "were so grouped that the pollen from one was freely conveyed to the other. Thus fertilized the female blossoms had developed into hundreds of perfect, seeds with well defined kernels." Here then is no proterogynic dichogamy, a notable exception if true.

39. Solms-Labach, I, p. 14, also Gasparrini, II, p. 400, tab. 2.

40. Gasparrini l. c. under point No. 8. He says he impregnated artificially thirty flower heads on a Ladaro fig by introducing into the aperture the pollen of the caprifig. In California this experiment was first tried in 1890 by Mr. G. Roeding.

41. As is customary with unpopular theories, the first remarks on caprification in California were simply sneered at and at the best considered illusionary, and heated discussions were entered into.

42. For further account of this visit, see the introduction of the Blastophaga in California.

43. So called because they were distributed to the subscribers of the San Francisco Bulletin, an evening daily paper, which, during Mr. G. P. Rixford's management, did much for horticulture in California. A full history of this introduction will be found in a work on Fig Culture, which I soon hope to publish.

44. Solms-Laubach, I, p. 64 and 65. No seeds found in figs from Angoulême, St. Savinier, La Mothe, as well as in the "wild" figs from these districts. Nor did he discover seeds in figs sent from Brazil, by Prof. F. Müller, same No. I, p. 39. The seedling referred to in Prof. Müller's letter probably originated from imported Smyrna figs, similar seedlings being quite common in, for instance, Washington, D. C.

45. Solms-Laubach, p. 34, is inclined to reject the parthenogenesis theory entirely, until proven by conclusive experiments. Even Gasparrini himself half doubts it, saying that pollen may possibly have entered in some unaccountable way.

46. Arcangeli, p. 2.

47. Solms-Laubach, I, p. 11, found only twenty fertile seeds in forty caprifigs (mammoni). Gasparrini, I, p. 328.

48. Profichi is pronounced as proféekée.

49. See historical part.

50. Amos, cap. 7, v. 14; see further Solms-Laubach, I, p. 75; Lagarde, p. 370, 395, 283.

51. Investigations are required. The North African States may possess many of the Smyrna tribe.

52. Aristote, Hist. anim. lib. v, cap. xvi, 3.

53. Theophrast, II, c. 9, 5, etc.

54. This statement of Theophrast appears to me to indicate that none of the finer Smyrna kinds were thriving in Greece. This is yet the case in our day, neither in Greece, Italy, France or Spain do the caprification requiring kinds of Smyrna succeed. Indeed, they are said to give but inferior fruit as soon as planted outside of a few localities near Smyrna. None of the cultivators or nurserymen of Europe seem even to be aware of there being a type of Smyrna figs different from the kind grown by them. Leclerc, p. 332, is the first writer to point out that caprificated figs are superior to those not caprificated. But this can only refer to dried figs, nor is it likely to refer to figs of the same variety.

55. Pliny, Nat. Hist., L. 15, c. 21.

56. Pontedera denied the nature of the sexes of the plants generally. From his point of view, it was, of course, impossible to recognize any reciprocity between the flowers of the fig and the Blastophaga.

57. Tournefort, I, p. 130.

58. Solms-Laubach, I, p. 26.

59. Milne, p. 13.

60. Cavolini, p. 240; cap. 28, p. 238.

61. Gasparrini, who especially points to Olivier as a great French naturalist, appears, in his great enthusiasm over a similarity in views, to have overlooked the fact that both Tournefort and Olivier were grossly ignorant of what they wrote about. Any one who, in writing of caprification, believes that the "figues fleurs" are hung in the trees, cannot possibly possess an opinion of caprification worthy of being quoted and adopted.

62. Gasparrini, I, p. 365; Semmola, 7c., p. 422; Solms-Laubach, I, p. 34.

63. Gallesio, p. 47.

64. Solms-Laubach, I, p. 31.

65. See paragraph on pomological maturity, p. .

66. See paragraph on Botanical maturity, p. .

67. The answers quoted here are Gasparrini's; the comments those of the author.

68. The caprifig has not been introduced to Brazil, and only recently to California, where as yet comparatively few trees are found, all without the wasps.

69. In this Solms-Laubach unconditionally, almost, agrees. See I, p. 36 to 40.

the caprifig. A second inquiline (the Apocrypta paradoxa Coquerel) is foun associated with the former in this fig, and it appears that also in othe Sycomore fig species are found several species of inquilines living togethe Wild fig species which I found in Mexico, were also inhabited by differen Blastophagæ.

92. This statement I take from B. M. Lelong's report. My own ex perience is that trees grown from cuttings sucker as much as any others.

93. E. W. Maslin, of Placer county, has grown a large number o seedlings from Smyrna figs, but none of them has proved valuable or ha even properly matured its fruit. W. M. Williams has told me of a seedlin fig originated in Los Angeles county, and Prof. E. J. Wickson in hi California fruits, etc., refers to a "seedling fig grown by Major Readin in 1858, which bore figs of uncommon size." In the Mediterranea countries figs are frequently originated from seed accidentally, but fe varieties prove of any value. Solms-Laubach refers to figs growing wild i France, which must have come from seeds, I, pp. 64, 65, and which di not show fertile seeds. He attributes their origin to the aid of birds which must have brought the seed with them from southern district where the caprifig exists, probably on their migration flights from Afric and south Italy towards the north. The caprifig seeds itself regularly i all countries where the Blastophaga is found, but not in other places which indicates that even the caprifig must be caprificated.

94. These notes are principally from Prof. Van Deman's account in th U. S. Depart. of Agriculture, Divis. of Pomology, Bulletin No. 1, 1887, p. 90, and from.

95. P. J. Berkman, in the U. S. Dept. of Agriculture, Special Report No. 4, p. 8.

96. Denotowitch, Anthony C., now of Fresno, California, late of Aidin Smyrna, has given me several points of interest in regard to Smyrna figs He has imported to Fresno several varieties of figs from Smyrna, an during the months of May and June, 1895, he received regular shipment of caprifigs with Blastophagæ every fifteen days. These were placed in a fig orchard some six miles east of Fresno, but no results have been ye recorded, nor could any well be expected for several months to come.

97. As has already been stated, the first introduction of Blastophag psenes was made by Mr. Shinn, at Niles, but want of sufficient caprifig for their propagation made the venture a failure There was at the time only one caprifig tree on the place. This tree had just finished shedding a crop of ripe caprifigs of which only [illegible] yet remained on the tree. But there was [illegible] crop, and the [illegible] their eggs. The insects would [illegible] perish, as they [illegible] not possibly survive until [illegible]. It is absolutely nec essary to [illegible] immediately upon their

hatching have new caprifigs of proper size in which to lay their eggs. If these new figs are not found on the same tree from which the wasps hatch they must be present on other trees in the vicinity. The failure of the first importation of Blastophaga to California can therefore not be wondered at. With only one caprifig tree and that one bearing only one crop a year, this importation could not possibly have been a success.

98. The most interesting and astonishing discovery of parthenogenesis in the wild fig was first made by D. Cunningham on the various forms of flowers of Ficus Roxburghii. His experiments and observations on the flowers of this fig species, both before and after the access of the Blastophagas, show conclusively that parthenogenesis takes place in this species and that the Blastophaga is necessary not only to produce seed in the female flowers, but also to perfect the male flowers. For a full account of this see his work: D. D. Cunningham, on the Phenomena of Fertilization in Ficus Roxburghii, Wall. Annals of the Royal Botanic Garden, Calcutta, Vol. I.

99. Seedlings would invariably have a tendency to bring male flowers in the fig.

CONTENTS.

ADDITIONAL NOTES ON THE HERPETOLOGY OF LOWER CALIFORNIA.

BY JOHN VAN DENBURGH,

[Curator of the Department of Herpetology.]

The collection upon which these notes are based was made by J. M. Stowell and S. C. C. Lunt while on a zoological expedition to San Pedro Martir Mountain in the northern part of Lower California. The specimens which they secured now belong to the Leland Stanford Junior University, and I am greatly indebted to Dr. Chas. H. Gilbert of that institution for the privilege of examining them. Several species and one genus are included which have not been recorded previously from the peninsula. These are *Crotaphytus wislizenii*, *Sceloporus orcutti*, *Lampropeltis boylii*, *Bascanion piceum*, *Thamnophis hammondii*, *Crotalus ruber*, and *Rana draytonii*.

A few specimens in the Academy's collection also are recorded, adding one more species, *Lepidochelys olivacea*, to the known fauna of this territory.

STOWELL AND LUNT COLLECTION.

CROTAPHYTUS WISLIZENII B. & G.

One specimen (No. 1087) of this species was secured at San Tomas, July 15, 1893.

UTA STANSBURIANA B. & G.

The collection contains specimens (Nos. 1437–1444) taken at San Telmo, June 17; San Rafael Valley, June 19; and in the foothills of San Pedro Martir Mt., June 20–21, 1893.

SCELOPORUS ZOSTEROMUS Cope.

This species was found at San Telmo, and in the foothills of San Pedro Martir Mt.

Sceloporus orcutti Stejn.

This very distinct form was found in the San Rafael Valley, July 15; at Wasson's Ranch (in San Rafael Valley 68 miles southeast of Ensenada), July 14; between Ensenada and San Rafael Valley, June 8; and in the foothills of San Pedro Martir Mt., June 20–21, 1893.

Sceloporus bi-seriatus Hallow.

This lizard was secured between Ensenada and San Rafael Valley, June 8, 1893; at San Telmo, July 15; and on San Pedro Martir Mt., July 6, 1893.

Sceloporus graciosus B. & G.

Several brightly colored males and females were collected on San Pedro Martir Mt. The number of scales on the back ranges from fifty-nine to sixty-six, fourteen to seventeen being equal to the length of the shielded part of the head. The femoral pores vary from fifteen to nineteen.

Phrynosoma blainvillii Gray.

Blainville's horned-toad was found at Ensenada, June 7 to 9; San Telmo, June 17; and at Wasson's Ranch in San Rafael Valley, July 15, 1893.

Gerrhonotus scincicauda (Skilt.)

Five typical specimens were taken on San Pedro Martir Mt., June 20 to July 5, 1893.

Cnemidophorus stejnegeri Van D.

The numerous examples of this species collected by Stowell and Lunt have been recorded elsewhere. (Proc. Cal. Acad. ser. 2, vol. iv, pt. i, p. 301, 1894.)

Verticaria hyperythra beldingi (Stejn.)

The thirty-one Verticarias secured by Messrs. Stowell and Lunt have been included in the table given on page

131 of this volume. They were collected at San Telmo, June 17; near Ensenada, June 12; and between San Vincente and Salado, June 15, 1893.

LICHANURA ROSEOFUSCA Cope.

A young specimen of this snake (L. S. Jr. U. No. 1125, vicinity of Ensenada, L. C., June 8, 1893) is interesting because it shows the instability of the chief character upon which *L. orcutti* is based. On one side of the head there are two true loreals, as in *L. orcutti*, while on the other there are three, as in *L. roseofusca*. The scale rows are forty-one in number.

LAMPROPELTIS BOYLII (B. & G.)

The single specimen (L. S. Jr. U. No. 1724) is typical of this species. It was secured in the foothills of San Pedro Martir Mt., July 11, 1893.

SALVADORA GRAHAMIÆ B. & G.

One typical specimen (L. S. Jr. U. No. 1723) was collected on San Pedro Martir Mt., July 6, 1893. With it is the following note: "Sage brush—partly buried in sand."

BASCANION PICEUM Cope.

That *B. piceum* is based on anything more than melanistic individuals of *B. flagellum frenatum*, I have great doubt. The evidence at hand, however, is not quite sufficient to prove their identity, and it seems better to recognize them as distinct forms than to run the risk of premature 'lumping.' The difference seems to be purely one of coloration, for although the type of *B. piceum* (from Camp Grant, Arizona) has nineteen rows of scales, all the (5) specimens that have been found since have seventeen.

Of the three adult specimens secured by Messrs. Stowell and Lunt, one is as dark as the type of *B. piceum*, while the others are somewhat lighter, especially on the tail. A young specimen (No. 1632) is not distinguishable from individuals of *B. f. frenatum* of the same size.

L. S. Jr. U. Nos. 1132, 1133, 1632, 1720, vicinity of Ensenada, L. C., June 11, 1893.

THAMNOPHIS HAMMONDII (Kenn.)

The two young garter-snakes (L. S. Jr. U. Nos. 1721 and 1722) collected on San Pedro Martir Mt., July 3, 1893, evidently belong to this species, although each has a well-developed dorsal line extending over its whole length. The supralabial plates are 7–7 and 7–8; the scale rows, 21.

CROTALUS RUBER (Cope).

A head (L. S. Jr. U. No. 1718) from Ensenada must be referred to this form if it be admitted that *ruber* is separable from *atrox*. The difference is purely one of color, but at present seems to be quite constant, however additional specimens may affect our views. The mere fact that the difference is slight seems no reason for using a trinomial.

CROTALUS LUCIFER B. & G.

A head (L. S. Jr. U. No. 1719) from San Pedro Martir Mt. seems to belong to this species, although it is so black as almost to conceal the characteristic markings.

Dr. Streets's "*Crotalus adamanteus atrox*" from Los Coronados Islands, referred to on page 156, really belongs to this species, as Dr. Stejneger has already shown.*

* U. S. N. M. Report, 1893, p. 445, 1895.

RANA DRAYTONII B. & G.

Except that the dorso-lateral ridge is probably less developed, the frogs which Mr. Stowell collected on San Pedro Martir Mt. seem to agree in structural characters with specimens of *Rana draytonii* from San Francisco, Oakland and Monterey, California. The dark dorsal blotches, however, are indistinct, small, and much less numerous than in typical *R. draytonii*, and the light spots on the posterior surfaces of the thighs are more regular and distinct. Two specimens from Colton, San Bernardino Co., Cal., agree in color with the Lower Californian frogs, and one of the examples from Monterey is not very dissimilar.

ACADEMY OF SCIENCES COLLECTION.

LEPIDOCHELYS OLIVACEA (Esch.)

One young specimen (No. 2248) was collected at San José del Cabo, by Gustav Eisen, Jan. 25, 1893.

TANTILLA PLANICEPS (Blain.)

A specimen collected by F. Billa at San José del Cabo, Nov., 1895, has but one postocular plate on each side of the head, as in one of the specimens already recorded. The entire lower surface is red.

THAMNOPHIS HAMMONDII (Kenn.)

A single garter-snake, typical of this species, was brought from Comondu by Mr. W. E. Bryant.

PROCEEDINGS.

January 21, 1895.—STATED MEETING.

The PRESIDENT in the chair.

George Otis Mitchell and R. H. Freund were proposed for membership.

A vote of thanks was extended to Mr. Owen A. Wells, member of Congress from Wisconsin, for his interest and earnest labors in behalf of forest preservation.

John Van Denbergh read a paper on Poisonous Reptiles of California.

February 4, 1895.—STATED MEETING.

The PRESIDENT in the chair.

Donations to the Museum were reported from Charles M. Tyler and George F. Breninger.

Additions to the Library:

From correspondents	184
By purchase	29
By donation	4

Leverett M. Loomis read a paper on Birds of the Ocean off Monterey in Midwinter.

Miss Alice Eastwood exhibited foliage and cones of the newly-discovered redwood from the head of Redwood Cañon and made some remarks concerning the trees.

A history of the donation of the fund of $5,000 for the purchase of books, by the terms of the will of Amariah Pierce, was given by Mr. S. W. Holliday, and on his motion the thanks of the Academy were voted to President Harkness for obtaining the donation for the Academy from his old friend.

President Harkness gave a sketch of the life of Mr. Pierce and how he came to be a life member of the Academy, also the circumstances attending the making of the donation.

February 18, 1895.—STATED MEETING.

The PRESIDENT in the chair.

Samuel J. Holmes read a paper on the Crustacea of the Pacific Coast.

March 4, 1895.—STATED MEETING.

The PRESIDENT in the chair.

R. H. Freund, George O. Mitchell, Frederick A. Woodworth, S. J. Holmes and John Hornung were elected resident members.

Donations to the Museum were reported from Chas. Holm, R. E. Wood, Capt. J. N. Knowles, A. Krause, Wm. G. Barrett, Mr. Ferem and H. W. Fairbanks.

Additions to the Herbarium:

354 plants collected in Mexico, presented by Agricultural Department, Washington, D.C.

70 Japanese plants and 186 Atlantic species from Dr. Wm. M. Canby, by exchange.

50 plants from Behring Sea, donated by the curator.

300 Canadian plants, presented by the Geological Survey of Canada.

180 Florida plants, donated by the curator.

116 unnamed plants from the Agricultural College, Las Cruces, N. M.

25 Japanese and North American Characeæ from Dr. T. F. Allen.

1479 Pringle's Mexican plants, by purchase.

Additions to the Library:

From correspondents 121
By purchase .... [illegible]
By donation 23

Prof. H. P. Johnson read a paper on the Structure and Life-history of the Infusoria as Illustrated by the Genus Stentor.

March 18, 1895.—STATED MEETING.

The PRESIDENT in the chair.

H. W. Fairbanks read a paper on An Ascent of Mount Whitney in May, illustrated with stereopticon views.

April 1, 1895.—STATED MEETING.

The PRESIDENT in the chair.

Donations to the Museum were reported from Wm. F. Nolte, J. Z. Davis, L. Belding and Edward McCue.

Additions to the Library:

From correspondents 116
By purchase 47
By donation 9

Prof. R. H. Freund read a paper on Blood, its Macroscopic Characters and their Importance, illustrated under the microscope.

April 15, 1895.—STATED MEETING.

The PRESIDENT in the chair.

The committee on public reservations presented the following resolutions, which were adopted:

WHEREAS, It is rumored that the Government intends to construct a road by filling in part of Mountain Lake in the Presidio Government Reservation at San Francisco. This road, which would cross the lake at a place where it is thirty feet deep, would be many hundreds of feet long and would sadly disfigure this beautiful lake, and undoubtedly lead to its final destruction by subsequent filling in and by greatly increased growth of tules and other weeds.

Mountain Lake, which contains fourteen acres, more or less, is one of the prettiest landmarks on the peninsula and its shores could with little expense be converted into a beautiful park. A portion of the southerly part of the lake is in the public park of the City and County of San Francisco, known as Mountain Lake Park, which, in the course of a few years, will doubtless be beautifully improved by the city and county. Its water is pure and good. It seems incredible that while we are spending $500,000 in creating a lake in the Golden Gate Park of San Francisco any one should conceive the idea of destroying that most beautiful Mountain Lake, situated close by, only for the purpose of securing a straight road from the Marine Hospital to a public street. A road around the lake would cost many times less, would be more durable and more beautiful, and would serve every purpose, as going around the small lake would take but a few minutes more time than crossing it on the newly proposed road. But in order to reach the nearest street it is not necessary to cross the lake at all, as a short cut already existing through a small sand-bank, if somewhat widened, together with filling in a few feet, would give immediate access to one of the city highways. To destroy the beautiful Mountain Lake would be, we think, a great wrong, besides there can be really no necessity for it. It would be far better to dredge out the shallow weedy portion of it and stock it with fresh-water fish, plant trees around its margin and make it "a thing of beauty" which would be "a joy forever," than to destroy it by making a highway across it.

To put a grade eighty feet wide on top crossing that lake would require a filling of at least one hundred and sixty feet wide on the bottom; besides, its weight would squeeze the mud from the bottom and in fact ruin the lake as such. It would be as well to fill it in at once.

We think it would be as sane to undertake to destroy the Farallones or any other small island in the Pacific, so that a ship might pass directly over it, as to grade across this beautiful lake for a direct road to the city when no one could be to any extent accommodated by such change.

In our judgment it would be a wrong, a great wrong, one that would be regretted for all future time.

We say improve Mountain Lake, beautify it, and make its ever-flowing springs useful by stocking it with fish. Therefore be it

Resolved, That the California Academy of Sciences strongly opposes the proposed filling in of any part of the Mountain Lake, and urgently requests the authorities of the War and Interior Departments of our Government at Washington to preserve the lake intact in order that it may some time in the future be beautified and utilized.

Resolved, That copies of this resolution be forwarded to the Secretary of the Interior and the Secretary of War, also to the Senators and Representatives in Congress, as well as to the Mayor and Supervisors of this city.

W. S. CHAPMAN,
GUSTAV EISEN,
Committee.

May 6, 1895.—STATED MEETING.

The PRESIDENT in the chair.

Donations to the Museum were reported from Wm. F. Nolte, Rev. F. H. Wales, Olaf Olsen, J. B. Walker, Mrs. Fannie V. Hubbard, M. Braverman and Dr. David S. Jordan.

Additions to the Library:

From correspondents 146
By purchase 124
By donation .... [illegible]

Mr. D. S. Richardson gave an illustrated lecture entitled "Mexico, an Hour below the Border."

May 20, 1895.—STATED MEETING.

The PRESIDENT in the chair.

Donations to the Herbarium were reported from Dr. Wm. M. Canby and Frank W. Hubby.

Accessions to the Department of Conchology:

Donation from D. Thaanum, Sandwich Islands, 21 species.

Exchange from Buffalo Society of Natural Science, 37 species.

Exchange from Naturalists Society of Seattle, 22 species.

From State Mining Bureau, 110 species of fossils, collected by W. L. Watts in Ventura and Los Angeles counties.

Dr. Gustav Eisen lectured on the Expedition to Tepic, Mexico, in 1894, with stereopticon illustration.

June 3, 1895.—STATED MEETING.

The PRESIDENT in the chair.

Donations to the Museum were reported from H. Müller and Lillian Thompson.

Additions to the Library:

From correspondents 119
By purchase 52
By donation 14

President Harkness exhibited specimens and described an interesting fungus found on Madroño in Mill Valley.

July 1, 1895.—STATED MEETING.

The PRESIDENT in the chair.

Donations to the Museum were reported from Oscar Kunath, H. E. Weeden, J. Z. Davis, Alex. H. Lyons, W. W. Price, Lester L. Edner, A. C. Wright, J. M. Hyde and A. Krause.

Additions to the Library:

From correspondents 117
By purchase 62
By donation 2

The President called attention to the summary dismissal of Prof. Geo. Davidson from his position at the head of the Coast and Geodetic Survey on this coast, and, on motion, W. S. Chapman, Gustav Eisen and G. P. Rixford were appointed a committee to present suitable resolutions on the subject.

Edward S. Jones described the preparation of calcium carbide and gave an exhibition of the acetylene gas derived from it.

Prof. Gustav Eisen gave a lecture on Lower California and Sonora, illustrated with stereopticon views.

August 5, 1895.—STATED MEETING.

The PRESIDENT in the chair.

Donations to the Museum were reported from John L. Howard and A. C. Wright.

Additions to the Library:

From correspondents 168
By purchase 68
By donation 8

George Otis Mitchell read a paper on Our Modern Conception of Matter and Force.

Announcement was made of a recent discovery on Alcatraz Island of a fossil, the existence of which disproves the idea that the San Francisco sandstones are pre-cretaceous. The cast of the fossil was found by Captain A. W. Vogdes and it may be referred to the genus Venericardia, a genus which ranges from the cretaceous to the tertiary formation. This discovery, with that by Major Elliott of an Inoceramus, named by Gabb after its discoverer, clearly indicates that the San Francisco sandstones probably belong to the cretaceous period.

Louis A. Robertson read a poem on evolution.

The committee appointed to draft appropriate resolutions on the dismissal of Prof. Davidson from the Coast Survey presented its report, which was received and placed on file.

September 2, 1895.—Stated Meeting.

The President in the chair.

Donations to the Museum were reported from L. Belding, Mrs. C. W. Geiser, Henry Helfrich, John M. Curtis and C. E. Hayes.

Additions to the Library:

From correspondents 117
By purchase .... [illegible]
By donation .... [illegible]

Leverett M. Loomis read a paper entitled "Aerial Voyagers."

Amendments to Article III of the Constitution, proposed by the Trustees, were read and adopted.

September 16, 1895.—Stated Meeting.

The President in the chair.

Donations to the Museum were reported from Henry Hemphill.

Additions to the Herbarium:

216 specimens of Hawaiian plants, presented by the Botanical Club.

37 specimens from Ojai Valley, Cal., presented by F. W. Hubby.

75 species from W. G. Wright, in exchange.

72 specimens from the herbarium of the University of Minnesota, in exchange.

Prof. William E. Ritter read a paper on the Zoological Station of Naples and what it has done for the promotion of biological science.

President Harkness reported that the Council had decided to report back to the Academy the proposed amendments to Article III of the Constitution without change.

The proposed amendments were then read, and, after the words "or in approved interest-bearing corporation bonds" were stricken out, were adopted, by sections.

As adopted the proposed amendments read:

"Section 5 of Article III of the Constitution of the California Academy of Sciences is hereby amended so as to read as follows:

"Section 5. Whenever the Trustees shall have in their hands funds that in their opinion are not needed for the immediate use of the Academy, they shall have the power to loan the same in the name of the Corporation upon such terms as they may deem advisable.

"No loan, however, shall be made except the same shall be secured by mortgage of unincumbered real estate in the State of California; or by a pledge of bonds of the United States, of the State of California, of a County or City and County of this State, whose value shall, in the opinion of said Trustees, be ample security for the amount of the loan and the interest thereon.

"The funds of the Academy shall not be loaned to any of its Trustees, nor shall any loan be made except upon the vote of not less than five of the Trustees, entered upon the record of their proceedings, and specifying the amount, terms and security, and the person to whom the loan is made. If any loan is made contrary to the provisions of this section, the Trustees making the same shall be individually and severally liable to the Corporation for the amount so loaned.

"Section 6 of Article III of said Constitution is hereby amended so as to read as follows:

"Section 6. The Trustees shall have power, if in their judgment it is advisable, to invest any of the funds of the Academy not needed for immediate use, in bonds of the United States, of the State of California, of any County or City and County of this State. Such investments, however, shall be made only by the unanimous vote of all the Trustees, entered upon the record of their proceedings and specifying the amount and character of the investment."

Mr. E. J. Molera read certain proposed amendments to Articles II, III, IV and VI of the Constitution, consideration of which was postponed until the next meeting.

October 7, 1895.—STATED MEETING.

The PRESIDENT in the chair.

H. P. Johnson, John C. Merriam, W. A. Setchell, Vernon L. Kellogg, David C. Booth and A. Van Der Naillen were proposed for membership.

Donations to the Museum were reported from Mrs. Sophia Casey, G. P. Rixford, C. S. Capp, Henry Helfrich and Henry Hemphill.

Additions to the Library:

From correspondents 113
By purchase 49
By donation 4

Dr. David Starr Jordan delivered a lecture on The Value of Faunal Studies.

The amendments to the Constitution proposed by Mr. Molera were presented and on vote the matter was indefinitely postponed.

October 21, 1895.—Stated Meeting.

The President in the chair.

Otto von Geldern, John Hornung and Hermann Kower were proposed for membership.

Additions to the Herbarium:

358 specimens collected by Dr. Edw. Palmer at Acapulco, Mexico, donated by California Botanical Club.

195 specimens collected in the Hawaiian Islands by A. A. Heller, donated by the curator.

17 specimens from Catalina Island, donated by Mrs. W. J. Trask.

Dr. H. H. Behr made "Some Remarks on Extinct Animals and the causes which led to their Extinction."

The following papers were read by title:

Notes on a Specimen of *Alepisaurus æsculapius* Bean, from the Coast of San Luis Obispo Co., Cal. By Flora Hartley.

Description of a New Jack Rabbit from San Pedro Martir Mountain, Lower California. By John M. Stowell.

A Supplement to the Bibliography of the Paleozoic Crustacea. By Anthony W. Vogdes.

A Review of the Herpetology of Lower California. Part 1—Reptiles. By John Van Denburgh.

On Land and Fresh Water Shell of Lower California. No. 5. On West Mexican Land and Fresh Water Mollusca. By J. G. Cooper.

On Heteromorphic Organs of *Sequoia sempervirens*. By Alice Eastwood.

California Water Birds, No. 1. By Leverett M. Loomis.

Coleoptera of Baja California. Supplement. By Geo. H. Horn.

[illegible] Mexican Hymenoptera. By Wm. J. Fox.

The Fishes of Sinaloa. By David Starr Jordan.

Some Mexican Neuroptera. By Nathan Banks.

The Species of the Genus Xantusia. By John Van Denburgh.

The Neocene Stratigraphy of the Santa Cruz Mountains. By George H. Ashley.

Changes in Fauna and Flora of California—On the Power of Adaptation in Insects. By H. H. Behr.

A List of Lichens collected by Mr. Robert Reuleaux in the Western Part of North America. By Dr. Stizenberger.

Notes on the Habits and Distribution of *Autodax iecanus*. By John Van Denburgh.

The Californian Phryganidian (*Phryganidia californica* Pack.) By Vernon L. Kellogg and F. J. Jack.

Cranial Characters of the Genus Sebastodes. By Frank Cramer.

A Review of the Herpetology of Lower California. Part II.—Batrachians. By John Van Denburgh.

Description of a New Species of Gobiesox from Monterey Bay, Cal. By Seth Eugene Meek and Charles J. Pierson.

Some Parasitic Hymenoptera from Baja California and Tepic, Mexico. By Wm. H. Ashmead.

Contributions to Western Botany. No. 7. By Marcus E. Jones.

Explorations in the Cape Region of Baja California in 1894. By Gustav Eisen.

Description of a New Species of Ranzania from the Hawaiian Islands. By O. P. Jenkins.

November 4, 1895.—STATED MEETING.

The PRESIDENT in the chair.

H. P. Johnson, John C. Merriam, W. A. Setchell, Vernon L. Kellogg, David C. Booth and A. Van Der Naillen were elected to resident membership.

Additions to the Library:

From correspondents	174
By purchase	473
By donation	4

Dr. Gustav Eisen delivered a lecture, illustrated with lantern slides, entitled "A Glimpse at the Ancient and Modern Civilizations in Guatemala."

November 18, 1895.—STATED MEETING.

The PRESIDENT in the chair.

A paper by Captain A. W. Vogdes, entitled "Typical Military Roads, with Illustrations of the French Road System," was read by the acting Secretary.

December 2, 1895.—STATED MEETING.

The PRESIDENT in the chair.

Donations to the Museum were reported from J. F. Bekeart, Samuel B. Doten and G. P. Rixford.

Additions to the Library:

From correspondents .... [illegible]
By purchase .... [illegible]
By donation .... [illegible]

Also Harmonica Macrocosmica, an astronomical atlas, published in Amsterdam in 1708, presented by Carlos Troyer.

Dr. O. P. Jenkins delivered a lecture on "Glaciers, Past and Present," illustrated with stereopticon views.

December 16, 1895.—STATED MEETING.

The PRESIDENT in the chair.

Additions to the Herbarium:

81 specimens from Atlantic States from Wm. M. Canby, in exchange.

41 Hawaiian plants, donated by Miss H. A. Spaulding.

30 specimens from the Summit, donated by G. W. Dunn.

22 specimens from Lake, Shasta and Kern counties, presented by members of the Botanical Club.

81 plants from Calaveras County, donated by Dr. F. E. Blaisdell.

Mr. Leverett M. Loomis reported the donation of several hundred carefully prepared bird specimens and a representative series of eggs of California birds from Dr. T. S. Palmer of the Division of Ornithology and Mammalogy of the U. S. Department of Agriculture.

A vote of thanks was tendered to Dr. T. S. Palmer for this highly valuable addition to the Academy's scientific collection.

The nominating committee reported the following ticket:

For *President*, David Starr Jordan.

First Vice-President, William E. Ritter.

Second Vice-President, J. G. Cooper.

Corresponding Secretary, Theodore H. Hittell.

Recording Secretary, Gulian P. Rixford.

Treasurer, Lucius H. Foote.

Librarian, Carlos Troyer.

Director of Museum, J. Z. Davis.

Trustees, W. C. Burnett, W. S. Chapman, Charles F. Crocker, H. W. Harkness, W. S. Keyes, George C. Perkins, Granville W. Stewart.

January 6, 1896.—ANNUAL MEETING.

The PRESIDENT in the chair.

George C. Edwards and Frank H. Vaslit were elected life members, on recommendation of the Council.

Otto von Geldern and Hermann Kower were elected to resident membership.

The reports of the Officers and Board of Trustees were read and ordered on file.

As the result of the annual election, the following were declared officers for the ensuing year:

DAVID STARR JORDAN, *President.*
WILLIAM E. RITTER, *First Vice-President.*
H. H. BEHR, *Second Vice-President.*
J. O'B. GUNN, *Corresponding Secretary.*
G. P. RIXFORD, *Recording Secretary.*
L. H. FOOTE, *Treasurer.*
CHARLES A. KEELER, *Librarian.*
J. Z. DAVIS, *Director of Museum.*

Trustees:

W. C. BURNETT, W. S. CHAPMAN, CHARLES F. CROCKER,
W. S. KEYES, E. J. MOLERA, GEORGE C. PERKINS,
G. W. STEWART.

REPORT OF THE LIBRARIAN FOR 1895.

The additions to the Library for the year 1895 have been as follows:

From correspondents	1646
By purchase	1394
By donation	109
Total	3149

Valuable donations of books and pamphlets have been received from T. F. Allen, G. H. Barber, Frank Campbell, T. C. Chamberlain, Geo. K. Cherrie, Frank M. Comstock, E. D. Cope, M. A. Cornu, Alice Eastwood, Dr. Gustav Eisen, Thos. Hanbury, S. W. Holladay, Robert L. Jack, Charles Janet, Vernon L. Kellogg, Dr. Otto Kuntze, Dr. F. Kurtz, D. T. MacDougal, George H. MacKay, James M. Macoun, A. M. McClatchie, George Lane Mullins, A. S. Packard, R. A. Philippi, H. A. Pilsbry, William E. Ritter, Benjamin L. Robinson, H. C. Russell, Federico Sacco, Erwin F. Smith, C. A. Townsend, S. M. Tracy and F. S. Earle, Carlos Troyer, H. W. Turner, Frank H. Vaslit, A. W. Vogdes, Henry B. Ward, Herbert J. Webber and Don E. S. Zeballos.

Early in the year the bequest made by Mr. Amariah Pierce of $5,000 for the purchase of books, became available. In February a list of desirable publications was carefully prepared, which was printed in convenient form and a copy sent to a large number of the principal dealers in scientific books, with a request that each would indicate on the list the works which he could furnish, with the lowest price for cash, and return the marked lists. The returns were then tabulated for the purpose of comparison and orders were sent out in accordance with the most favorable bids. The orders for such books as could be procured in the United States went to a local house and a large order was sent to Leipzig. The remainder was placed in the hands of our London bookseller, who undertook to procure the books from the different dealers designated on the list furnished to him, collate them and ship to the Academy in convenient quantities.

The additions to the Library thus far from this fund number 643 complete volumes, besides numerous parts of volumes not yet completed. In accordance with the provisions of Mr. Pierce's will each volume bears upon the title page:

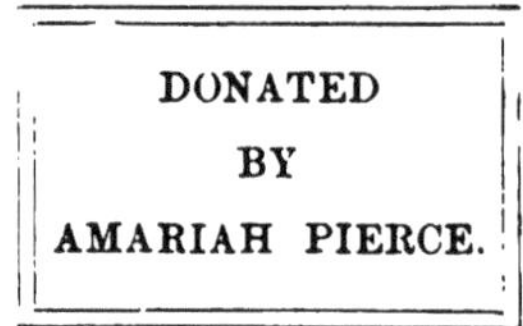
DONATED
BY
AMARIAH PIERCE.

The cost of the books already received has been about $3,500. The final shipment on the orders placed is now on the way from London. It includes a complete set of the publications of the Museum d'Histoire Naturelle of Paris and also of the Société Géologique de France, besides many other rare and valuable works, the value of the consignment being about $1,200.

The following is a list of the publications, purchased with this fund, received up to January 1, 1896:

Anatomische Hefte. Weisbaden. Arbeiten, Bande i–v; vi, Heft 1; Ergebnisse, Bande i, ii.

Archives de Biologie. Paris. Vols. i–xiii.

Archives de Zoologie Experimentale et Générale. Paris. Series 1, vols. i–x; series 2, vols. i–x and supp. vols. to iii, v; series 3, vols. i, ii.

Archives Italiennes de Biologie. Turin. Tomes i–xxiii; xxiv, Nos. 1, 2.

Archiv fur Mikroskopische Anatomie und Entwicklungsgeschichte. Bonn. Bande i–xlv; Namen und Sachregister zu Bande i–xl.

Bibliotheca Zoologica. Stuttgart. Hefte i–xviii; xix, Lief. 1.

Biologisches Centralblatt. Leipzig. Vols. i–xv.

Hooker's Botanical Miscellany, vols. i–iii; Journal of Botany, vols. i–iv; London Journal of Botany, vols. i–vii; Hooker's Journal of Botany and Kew Gardens Miscellany, vols. i–ix.

Isis von Oken. Leipzig. 1817–1848.

Jahresbericht über die Fortschritte in der Lehre von den Pathogenen Mikroorganismen. Braunschweig. Jahrgang i–viii; ix, abt. 1; Register Jahrgang i–v.

Jenaische Zeitschrift für Naturwissenschaft. Jena. Vols. i-xxix.

Journal of Physiology. London. Vols. i–xvii.

Morphologische Arbeiten. Jena. Vols. i-iv; v, Nos. 1, 2.

Morphologisches Jahrbuch. Leipsic. Vols. i–xxii.

Proceedings of the Zoological Society of London, 1830–1894. Index 1830–1890.

Transactions of the American Philosophical Society. Philadelphia. New series, vols. i–xvi.

Zeitschrift fur Wissenchaftliche Mikroskopie und für Mikroskopische Technik. Braunschweig. Bande i–xi.

Zoologische Beitrage. Breslau. Bande i–iii.

Agassiz, Louis. Iconographie des Coquilles Tertiaires. Nenchatel, 1845.

Apgar, Austin C. Pocket Key of Birds. 1893.

Arago, Jacques. Voyage Autour du Monde. 5 vols. Paris, 1840.

Baillon, H. Histoire des Plantes. Vols. i–xiii. Paris, 1867–1895.

Baird, Brewer & Ridgway. North American Birds. 3 vols. Boston, 1874.

Bajon. Mémoires pour servir a l'Histoire de Cayenne, et de la Guiane Françoise. Paris, 1778.

Baker, J. G. Handbook of the Fern-Allies. London, 1887.

Baker, J. G. The Amaryllideæ. London, 1888.

Baker, J. G. Handbook of the Irideæ London, 1892.

Balfour, F. M. A Treatise on Comparative Embryology. Vols. i, ii. London, 1880–81.

Baudrillart. Dictionnaire des Pêches. With atlas. Paris, 1827.

Beddard, F. E. Animal Coloration. London, 1892.

Beechey's Voyage in H. M. S. Blossom. Zoology. London, 1839.

Behrens, Wilhelm. Tabellen zum Gebrauch bei Mikroskopischen Arbeiten.

Bloxam and Huntington. Metals; Their Properties and Treatment. London, 1885.

Boas, J. E. V. Lehrbuch der Zoologie. Jena, 1894.

Boveri, Theodor. Zellen-Studien. Hefte i–iii. Jena, 1887–90.

Brand and Coxe's Dictionary of Science, Literature and Art. Vols. i–iii. London, 1865.

British Museum. Catalogue of Birds. Vols. i–xiv.

Catalogue of Lizards. 1845.

Catalogue of Lizards, 2d ed. Vols. i–iii, 1885–87.

Catalogue of Shield Reptiles. Part i, 1855. Appendix, 1872. Part ii, 1872.

Hand-List of the Specimens of Shield Reptiles. 1873.

Catalogue of Batrachia Gradientia. 1850; 1882.

Catalogue of Batrachia Salientia. 1858.

British Museum. Catalogue of Colubrine Snakes. 1858.
Catalogue of Gigantic Land Tortoises. 1877.
Catalogue of Tortoises, Crocodiles and Amphisbænians. 1844.
British Ornithologists' Union List of British Birds. London, 1883.
Buller, Walter L. A History of the Birds of New Zealand. Vols. i-ii. London, 1888.
Bush, Richard L. Reindeer, Dogs and Snow-Shoes. New York, 1871.
Cavanilles, A. J. Icones et Descriptiones Plantarum quæ aut Sponte in Hispania Crescunt. Vols. i-vi. Matriti, 1791-1801..
Chapman, Frank M. Birds of Eastern North America. New York, 1895.
Chastellux, Marquis de. Voyages dans l'Amerique. 2 vols. Paris, 1786.
Cheney, Simeon Pease. Wood Notes Wild. Boston, 1892.
Claus, Carl. Grundzüge der Zoologie. Baude i, ii. 4th edition. Marburg, 1880-82.
Claus, Carl. Lehrbuch der Zoologie. 5th edition. Marburg, 1891.
Comstock, John H. A Manual for the Study of Insects. Ithaca, 1895.
Daudin, F. M. Histoire Naturelle des Reptiles. Vols. i-viii. Paris, 1803.
De Candolle, Alphonse et Casimir. Monographiæ Phanerogamarum. Vols. i-viii. Paris, 1878-1894.
Delessert, E. Voyages dans les deux Océans Atlantique et Pacifique. Paris, 1848.
Dixon, Charles. The Migration of Birds. London, 1892.
D'Orbigny, Alcide. Voyage dans les deux Ameriques. Paris, 1853.
Doubleday, Edward. The Genera of Diurnal Lepidoptera. Vols. i, ii. London, 1846-1852.
Dumeril. Prodrome de la Classification des Reptiles Ophidiaus. Paris, 1853.
Duhamel du Monceau. Traité Général des Pêches. 2 vols. Paris, 1769.
Elliot, D. G. Monograph of the Tetraoninæ or Grouse Family. 5 parts. New York, 1864-65.
Engler and Prantl. Pflanzenfamilien. Lief. 1-125. Leipzig.
Festschrift zum Siebenzigsten Geburtstage Rudolf Leuckarts. Leipzig, 1892.
Fitzinger, L. I. Neue Classification der Reptilien. Wien, 1826.
Fitzinger, L. I. Systema Reptilium. Fasc. 1. Vindobonæ, 1843.
Flügel, Felix. English-German and German-English Dictionary. 3 vols. London, 1894.
Foster, M. Text Book of Physiology. 5 vols. New York, 1893.
Gatke, Heinrich. Heligoland as an Ornithological Observatory. Translated by Rudolph Rosenstock. Edinburgh, 1895.
Gibson, W. Hamilton. Our Edible Toadstools and Mushrooms. New York, 1895.
Goebel, K. Outlines of Classification and Special Morphology of Plants. Oxford, 1887.

Goss, N. S. History of the Birds of Kansas. Topeka, 1891.

Gray, Jane Loring. Letters of Asa Gray. Vols. i, ii. Cambridge, 1894.

Gray, Maria Emma. Figures of Molluscous Animals selected from Various Authors. Vols. i–v. London, 1854–1874.

Grisebach, A. H. K. Flora of the British West Indian Islands. London, 1864.

Haeckel, Ernst. Systematische Phylogenie der Protisten und Pflanzen. Theil i. Berlin, 1894.

Harvey and Sowder. Flora Capensis. Vols. i–iii. Dublin, 1859–65.

Hertwig, Oscar. Die Zelle und die Gewebe. Jena, 1893.

Hertwig, Oscar. Lehrbuch der Entwicklungsgeschichte des Menschen und der Wirbelthiere. Jena, 1893.

Hewiston, William C. Illustrations of New Species of Exotic Butterflies. Vols. i–v. London, 1856–76.

Hooker, W. J. Synopsis Filicum. 2d edition. London, 1883.

Hooker, W. J. Niger Flora and Enumeration of the Plants of Western Tropical Africa. London, 1849.

Hooker's Icones Plantarum. Series 1, vols. i–iv; 2, vols. i–vi; 3, vols. i–x; 4, vols. i–iv. London, 1837–94.

Hornaday, William T. Taxidermy and Zoological Collecting. New York, 1893.

Huxley and Martin. A Course of Elementary Instruction in Practical Biology. London, 1888.

Jan, G. Elenco Sistematico degli Ofidi. Milano, 1863.

Jonge, J. K. J. de. The Barents Relics. Translation by S. R. van Campen. London, 1877.

Kingsley, John S. The Riverside Natural History. Vols. i–vi. Cambridge, 1888.

Koch, Ludwig. Die Arachniden-Familie der Drassiden. Hefte i–vii. Nürnberg, 1866.

Krefft, Gerard. The Snakes of Australia. Sydney, 1869.

Kuntze, Otto. Revisio Generum Plantarum. Parts i, ii, iii[1]. Leipzig, 1891–93.

Kutzing. F. T. Species Algarum. Leipzig, 1849.

Lacépède, Comte de. Histoire Naturelle des Serpens. Tomes i, ii. Paris, 1788–89.

Latreille. Familles Naturelles du Règne Animal. Paris, 1825.

Latreille et Sonnini. Histoire Naturelle des Reptiles. Vols. i–iv. Nouv. ed. Paris, 1830.

Lindley, J. Tradescantia undata.

Mabille et Vuillot. Novitates Lepidopterologicæ. Fasc. 1–12. Paris, 1890–95.

Marshall and Hurst. A Junior Course of Practical Zoology. Fourth edition. London, 1895.

McIlwraith, Thomas. The Birds of Ontario. Toronto, 1894.

Merrem, Blasius. Versuch eines System der Amphibien. Marburg, 1820.

Minot, H. D. The Land Birds and Game Birds of New England. Cambridge, 1895.

Molino, P. Fr. Alonso de. Vocabulario de la Lengua Mexicana. Leipzig, 1880.

Muller, Herman. The Fertilisation of Flowers. London, 1883.

Museum Richterianum. Leipzig, 1743.

Museum Senckenbergianum. Vols. i-iii. Frankfurt-am-Main, 1833-45.

Newton, Alfred. A Dictionary of Birds. Parts i-iii. London, 1893.

Orton, James. Comparative Zoology. New York, 1894.

Palæontologische Mittheilungen aus dem Museum des Königl. Bayer. Staats. 2 vols. and atlas. Stuttgart, 1864-84.

Pappe, L. Silva Capensis. Cape Town, 1854.

Parker, T. Jeffery. Lessons in Elementary Biology. London, 1893.

Payer, J. L'Expédition du Tegetthoff. Translation in French by Jules Gourdault. Paris, 1878.

Pfeiffer, L. Die Protozoen als Krankheitserreger. Zweite Auflage. Jena, 1891. Nachträge. Jena, 1895.

Rothschild, Walter. The Avifauna of Laysan and the Neighboring Islands; with a Complete History to Date of the Birds of the Hawaiian Possessions. Parts i, ii. London, 1893.

Sachs, Julius von. Lectures on the Physiology of Plants. Oxford, 1887.

Sachs, Julius von. History of Botany. Oxford, 1890.

Salviani, Hippolyto. Aquatilium Animalium Historia. Roma, 1554.

Sargent, Charles Sprague. Forest Flora of Japan. Cambridge, 1894.

Sargent, Charles Sprague. Scientific Papers of Asa Gray. Vols. i, ii. Cambridge, 1889.

Saunders, Howard. An Illustrated Manual of British Birds. London, 1889.

Schäfer, E. A. The Essentials of Histology. Philadelphia, 1894.

Schlegel, H. Essai sur la Physionomie des Serpens. 2 vols. and atlas. La Haye, 1837.

Schwalbe, G. Lehrbuch der Anatomie der Sinnesorgane. Erlangen, 1887.

Sclater, P. L. Argentine Ornithology. Vols. i, ii. London, 1888-89.

Seebohm, Henry. The Birds of the Japanese Empire. London, 1890.

Shaw, George. General Zoology or Systematic Natural History. Vols. i-xiv. London, 1802-24.

Shufeldt, R. W. The Myology of the Raven. London, 1890.

Standard Dictionary of the English Language. Vols. i, ii. New York, 1895.

Staudinger and Schatz. Exotische Schmetterlinge. I Theil, Zweite Auflage, Lief. 1-10; ii Theil, Lief. 1-6.

Stearns, W. A. New England Bird Life. Vols. i, ii. Revised and Edited by Dr. Elliott Coues. Boston, 1893.

Stone, Witmer. The Birds of Eastern Pennsylvania and New Jersey. Philadelphia, 1894.

Strasburger, Eduard. Das Botanische Practicum. Revised edition. Jena, 1887.

Strasburger, E. Handbook of Practical Botany. Edited by W. Hillhouse. London, 1893.

Sundevall's Tentamen. London, 1889.

Swainson, Wm. Exotic Conchology. Second edition. London, 1841.

Swainson's Zoological Illustrations. Series 1, vols. i, ii; series 2, vols. i–iii. London, 1820–29.

Tchihatcheff, Pierre de. Voyage Scientifique dans l'Altai Oriental. With Atlas. Paris, 1845.

Trumbull, Gurdon. Names and Portraits of Birds. New York, 1888.

Tradescantia, 6 plates; Commelinaceæ, 4 plates.

Tuckerman, Edward. Synopsis of the North American Lichens. Part i; Boston, 1882. Part ii; New Bedford, 1888.

Villefosse, Héron de. De la Richesse Minérale. 2 vols. and atlas. Paris, 1819.

Voyage of H. M. S. Sulpher. Zoology, 2 vols.; London, 1843–45. Botany, London, 1844–46.

Waterhouse, F. H. Index Generum Avium. London, 1889.

Wilson, Alexander. American Ornithology. 3 vols. and atlas. Philadelphia, 1828.

Wilson, Scott and Evans. Aves Hawiiensis. Parts i–v. London, 1890–94.

Wood, W. Index Testaceologicus. London, 1856.

Zimmerman, A. Das Mikroskop. Leipzig, 1895.

INDEX.

ı **full face**, new species and varieties in SMALL CAPITALS, synonyms in *italics*.

INDEX.

New genera in **full face**, new species and varieties in SMALL CAPITALS, synonyms in *italics*.

ERRATA.

Page 34, after title, add By J. G. Cooper.
Page 41, line 17, *lamarck* read Lamarck.
Page 46, the species numbered 160 should follow No. 79 on page 42.
Page 77, in list, for Gyalopum read Gyalopium.
Pages 77, 81, 142, for Rhinochilus read Rhinocheilus.
Page 156, for Crotalis read Crotalus.
Page 164, line 11, for *H. californiensis* Young read young *H. Californiensis*.
Page 525, for *Xautusia* read *Xantusia*.
Page 528, sixth line from bottom, for ovoviparous read ovoviviparous.
Page 555, line 8, add from Tepic.
Page 627, after var. tripartitum (Nutt.) add Pax.
Page 638, read *Tejonensis* for *Tejouensis*.
Page 639, read *debilis* for DEBILIS.
Page 645, read A. PISCINUS for *piscinus*.
Page 650, read var. UNGULATUS for var. *ungulatus*.
Page 655, read A. STIPULARIS for *stipularis*.
Page 658, read A. ENSIFORMIS for *ensiformis*.
Page 670, read A. UINTENSIS for *Uintensis*.
Page 666, read A. METANUS for *A. metanus*.
Page 667, read A. JULIANUS for *A. Julianus*.
Page 693, line 25, read cannot for caunot.
Page 693, lines 4, 5, 23, read *Macronema* for *macronema*.
Page 703, line 23, read var. *tenella* for *L. tenella*.
Page 716, *Eriogonum* should be placed on page 718.
Page 719, read GLUTINOSUM for *glutinosum*.
Page 719, read AMBIGUUM for *ambiguum*.
Page 727, read *Glycyrrhizæ* for *glycyrrhizæ*.
Page 730, line 5, *drabæ* should be *Drabæ*.
Page 731, line 22, *cichoricearum* should be *cichoriacearum*.
Page 732, next to last line, *aspeum* should be *asperum*.
Page 773, line 4, for Crotolus read Crotalus.
Page 913, line 3, for flowery read floury.
Page 921, third line from bottom, for male read mule.

PLATE III

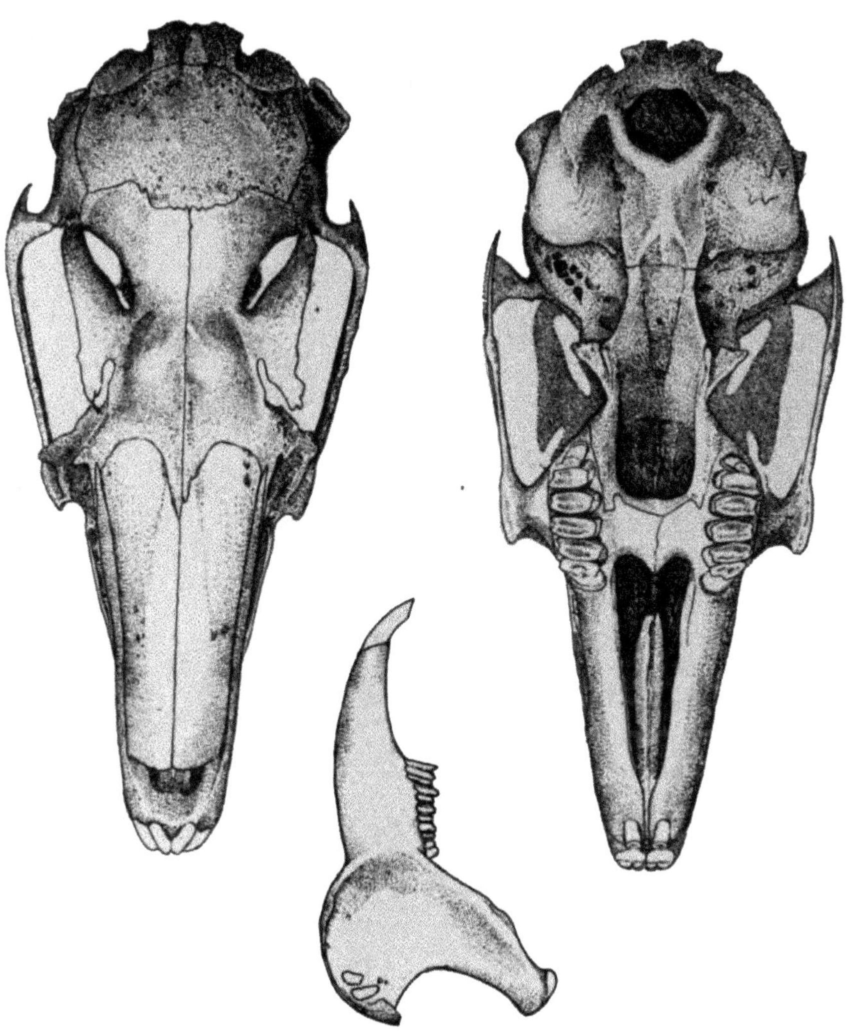

PLATE I

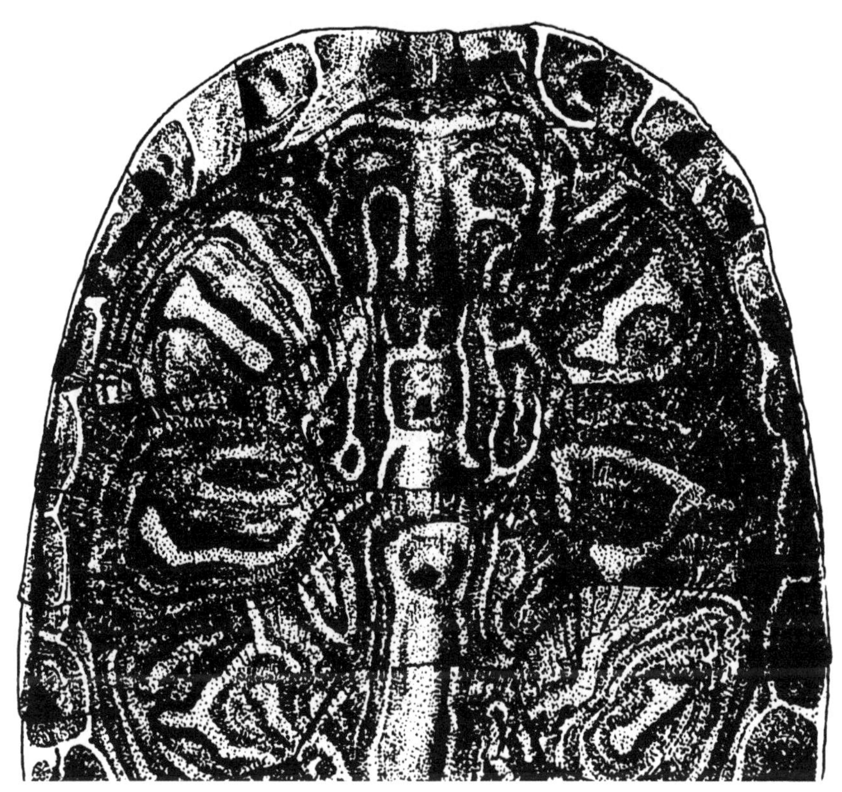

Lightning Source UK Ltd.
Milton Keynes UK
UKHW020432210219
337573UK00007B/1610/P